Anonymus

Descriptive Index of current Engineering Literature

Volume I.

Anonymus

Descriptive Index of current Engineering Literature
Volume I.

ISBN/EAN: 9783742800169

Manufactured in Europe, USA, Canada, Australia, Japa

Cover: Foto ©berggeist007 / pixelio.de

Manufactured and distributed by brebook publishing software
(www.brebook.com)

Anonymus

Descriptive Index of current Engineering Literature

Descriptive Index

OF CURRENT

ENGINEERING LITERATURE.

VOLUME I.

1884 - 1891.
(INCLUSIVE.)

CHICAGO:
Published by the Board of Managers of the
ASSOCIATION OF ENGINEERING SOCIETIES.
JOHN W. WESTON, Secretary,
51 Lakeside Building.

PREFACE.

The material composing this volume has appeared in the monthly numbers of the JOURNAL OF THE ASSOCIATION OF ENGINEERING SOCIETIES since October 1884, under the title of "*Index Notes.*" The notes published in the twelve numbers of one year have been republished, with cross-references, arranged in one alphabetical list, in an appendix to each volume. Seven of these annual summaries are now brought together, re-arranged alphabetically, with numerous cross-references, and with a more systematic arrangement of subject matter. It is published by the Board of Managers of the JOURNAL and placed on sale at a price barely sufficient to pay the cost of arrangement and publication.

Although there are over eleven (11) thousand notes and cross-references in this list, no claim is made for its completeness, even for the period covered by it. It is thought to contain, however, nearly all periodical, society, and fragmentary matter of permanent value not only for the period 1884 to 1891, inclusive, but a great deal which appeared earlier. Thus the entire proceedings of the *American Society of Civil Engineers*, of the *American Society of Mechanical Engineers*, and of the *Association of Engineering Societies*, have been indexed so far as they seemed deserving. Also the complete files of *Van Nostrand's Engineering Magazine*, and of *The Engineering News.* The *Proceedings of the Institution of Civil*

Engineers are here indexed only for the period covered for the current literature, viz 1884—1891 inclusive. A complete index of these proceedings is issued every few years by the Institution.

The object kept constantly in view in the preparation of these Index Notes was to put, in as small a space as possible, such short descriptions of the scope and general character of the current engineering literature of a periodical or fragmentary character as would enable one in search of valuable information on a particular subject to decide whether or not it would be worth his while to obtain or consult the original article, paper, or report. No abstracts, or results, or summaries have been made, and no conclusions given. In other words these notes only serve to indicate where valuable data can be obtained. It is mostly limited also to the fields of civil and mechanical engineering. That is to say, chemical and metallurgical subjects in mining engineering have not been indexed, while other subjects in this general field more closely related to the work of civil and mechanical engineers have been taken. Articles of a casual or passing interest, but of no permanent value have not often been indexed, and such matter as was thought to lead to erroneous conclusions or as were based on false assumptions or theories, have always been excluded.

Since its inception the preparation of these notes has been under the direct control of the undersigned, nearly all the notes for the first two years having been prepared by him. Since that time the work has been performed in succession by Mr. C. V. Mersereau (C. E., Cornell) M. Am. Soc. C. E., C. W. Melcher (M. E. Washington University.) M. Am. Soc. M. E., O. E. Hovey, (C. E., Thayer) Instructor in Civ. Eng., Washington University, and F. E. Turneaure (C. E., Cornell) Instructor in Civ. Eng., Washington University. Valuable assistance has also been rendered by Prof. Geo. F. Swain, M. Am. Soc. C. E. and by Mr. Clemens Herschel, M. Am. Soc., C. E. No simply clerical workers have ever been employed, and it is thought that the expression of opinion or of favor found in these notes may generally be relied on. The final preparation of this seven year summary for the press, its arrangement, classification, and the cross-references, has devolved upon Mr. Turneaure, who has been assisted by Mr. T. L. Condron C. E.

The value of a carefully prepared index such as is here offered can scarcely be overestimated. Even though the publications to which the references are made are not in one's private library, they generally are accessible and can be found in public or society libraries. On the other hand even though these volumes all stand on one's own shelves, if he does not know what they contain, or where to find an article which he dimly remembers to have seen, the task of making the search is so great as to forbid the effort, and the volumes remain unconsulted. It is the writer's firm belief that an index, such as this, without a library is of more value to a student than the library if he has no index to its contents.

If the volume now put forth should be appreciated for what it contains rather than criticised for what it lacks, and if it should contribute somewhat to the cause of good engineering in America, those to whom its usefullness is due will feel fully repaid for what has been largely a labor of love.

J. B. Johnson,
Manager Index Department.

Washington University, St. Louis, Apr. 15, 1892.

LIST OF PERIODICALS, ETC., INDEXED.

WITH GENERAL ABBREVIATIONS.

Age of Steel (*Age of Steel*), weekly, Equitable Building, St. Louis, Mo.

American Architect (*Am. Arch.*), weekly, Ticknor & Co., 211 Tremont Street, Boston, Mass.

American Engineer. (*Am. Eng.*)

American Journal of Railway Appliances (*Am. Jour. Ry. Appli.*), monthly, World Building, New York.

American Machinist (*Am. Mach.*), weekly, 96 Fulton street, New York.

American Manufacturer and Iron World (*Am. Mfr.*), weekly, Pittsburg, Pa.

Annales des Ponts et Chaussees (*Annales des P. & C.*), monthly, Vve. Ch. Dunod, 49 Quai des Augustins, Paris, France.

Annales des Travaux Publics, Paris, France.

Building, (*Building*), weekly, New York City.

Cassier's Magazine (*Cassier's Mag.*), New York City.

Centralblatt der Bauverwaltung.

Der Civil Ingenieur.

Deutsche Bauzeitung (*Deutsche Bauzeitung.*)

Electrical Review (*Elec. Rev.*), weekly, 22 Paternoster Row, London, England.

Engineering (Lon. *Eng.*), weekly, London, England.

Engineering and Mining Journal (*E. & M. Jour.*), weekly, 27 Park Place, New York.

Engineering News (*Eng. News*), weekly, Tribune Building, New York.

Engineering Record (*Eng. Rec.*), weekly, formerly The Sanitary Engineer, (*San. Eng.*), and the Engineering and Building Record (*Eng. & Build. Rec.*), 227 Pearl street, New York.

Fire and Water, (*Fire & Water*), New York City.

Illinois Society of Engineers and Surveyors, (*Rep. Ill. Soc. Eng. & Surv.*), Champaign, Ill.

Indian Engineering (*Ind. Eng.*), weekly, Calcutta, India.

Iron, (*Iron*), weekly, London, England.

Journal of the Association of Engineering Societies (*Jour. Assn. Eng. Soc.*), monthly, 51 Lakeside Building, Chicago.

Journal, Engineering Society, Lehigh University, (*Jour. Eng. Soc. Lehigh University*), Bethlehem, Pa.

Journal of the Franklin Institute (*Jour. Frank. Inst.*), monthly, Franklin Institute, Philadelphia, Pa.

Journal fur Gasbeleuchtung und Wasserversorgung, (*Jour. f. Gasbel. u. Wasserversorgung.*)

Journal of the New England Water Works Association (*Jour. N. E. W. W. Assn.*), quarterly, 113 Devonshire St., Boston, Mass.

Journal Royal Society of New South Wales (*Australia.*)

Journal of the Society of Arts (*Jour. Soc. Arts*), weekly, London, England.

Locomotive Engineering, (*Loc. Eng.*), monthly, 9-12 Temple Court, New York.

Manufacturer and Builder (*Mfr. & Build.*), monthly, New York City.

Mechanics (*Mechanics*), monthly, 907 Arch Street, Philadelphia, Pa.

Mining and Scientific Press, (*Min. & Sci. Press*), weekly, San Francisco, Cal.

Nouvelles de la Construction.

Popular Science Monthly, (*Pop. Sci. Monthly*), New York City.

Power (*Power*), monthly, 113 Liberty Street, New York.

Proceedings American Institute of Mining Engineers (*Proc. A. I. M. E.*), 13 Burling Slip, New York.

Proceedings, Cleveland Institute of Engineers, (*Proc. Cleveland Inst. Engrs.*), England.

Proceedings of the Engineers' Club of Philadelphia (*Proc. Eng. Club, Phila.*), quarterly, 1122 Girard St., Philadelphia, Pa.

Proceedings Indiana Association, Surveyors and Engineers, (*Proc. Ind. Assn. Surv. & Eng.*), Rensselaer, Ind.

Proceedings of the Institution of Civil Engineers (*Proc. Inst. C. E.*), 25 Great George St., Westminster, S. W., London, Eng.

Proceedings Master Car Builders' Association, (*Proc. M. C. B. Assn.*)

Proceedings Michigan Engineering Society (*Proc. Mich. Eng. Soc.*), Climax, Mich. Formerly Michigan Association of Surveyors (*Mich. Assn. Surv.*)

Proceedings, Nebraska Society of Associated Engineers and Surveyors, (*Proc. Neb. Assn. Eng. & Surv.*)

Proceedings of the Society of Arts (*Proc. Soc. Arts*), Mass. Institute of Technology, Boston, Mass.

Proceedings of Society of Civil Engineers, Paris, (*Proc. Soc. Civ. Eng., Paris*), Paris, France.

Proceedings of the United States Naval Institute (*Proc. U. S. N. I.*), quarterly, United States Naval Institute, Annapolis, Md.

Railroad and Engineering Journal (*R. R. & Eng. Jour.*), monthly, 45 Broadway, New York.

Railroad Gazette (*R. R. Gaz.*), weekly, 73 Broadway, New York.

Railway Review (*Ry. Rev.*), weekly, The Rookery, Chicago Ill.

Railway World, (*Ry. World*), Philadelphia, Pa.

Reports of Chief of Engineers, U. S. A., Washington, D. C.

Reports of Ohio Society of Engineers, Columbus, Ohio.

Revue des Mines.

Sanitary News. (*San. News.*)

School of Mines Quarterly (*Sch. Mines. Quart.*), Columbia College, New York.

Science, (*Science*), monthly, New York City.

Scientific American (*Sci. Am.*), weekly, 361 Broadway, N. Y.

Scientific American Supplement (*Sci. Am. Sup.*), weekly 361 Broadway, New York.

Selected Papers, Rensselaer Society of Engineers, Troy, N. Y.

Stevens Indicator (*Stevens Indicator*), Stevens Institute of Technology, Hoboken, N. J.

Street Railway Journal (*St. Ry. Jour.*), monthly, World's Building, New York.

Street Railway Review (*St. Ry. Rev.*), monthly, 334 Dearborn st., Chicago, Ill.

Technology Quarterly (*Tech. Quart.*), Mass. Inst. Technology, Boston, Mass.

The Electrical Engineer (*Elec. Engr.*), monthly, 11 Wall Street, New York.

The Electrical World (*Elec. World*), weekly, 177 Times Building, New York.

The Electrician and Electrical Engineer, (*Electrician & Elec. Eng.*)

The Engineer (Lon. *Engineer*), weekly, London, England.

The Engineering Magazine (*Eng. Mag.*), monthly, World Building, New York.

The Iron Age (*Iron Age*,) weekly, New York City.

The Locomotive (*Locomotive*), monthly, Hartford, Conn.

The Mechanical World (*Mech. World*), weekly, Manchester, England.

The National Car and Locomotive Builder, (*Nat. Car & Loco. Build.*), monthly, New York City.

The Polytechnic, (*Polytechnic*), Troy, N. Y.

The Practical Engineer, (*Practical Engineer.*). London, England.

The Progressive Age (*Progressive Age*), Philadelphia, Pa.

The Railway Engineer (*Ry. Eng.*), monthly, 8 Catherine St., Strand, W. C., London, Eng.

The Railway Master Mechanic (*Mast. Mech.*), monthly, "The Rookery," Chicago Ill.

The Sanitarian, (*Sanitarian*), New York City.

The Technograph, (*University of Illinois Annual*), Champaign, Ill.

The Street Railway Gazette (*St. Ry. Gaz.*), monthly, 8 Lakeside Building, Chicago.

The Technic (*Technic*), University of Michigan, Ann Arbor, Mich.

The Transit (*Transit*), University of Iowa.

Transactions of the American Institute of Electrical Engineers, (*Trans. A. I. E. E.*), Temple Court, New York City.

Transactions American Society of Civil Engineers (*Trans. A. S. C. E.*), 127 East Twenty-third street, New York.

Transactions American Society of Mechanical Engineers (*Trans. A. S. M. E.*), 12 West 31st Street, New York.

Transactions Arkansas Society of Civil Engineers Architects and Surveyors, (*Trans. Ark. Soc. C. E., Arch. & Surv.*), Little Rock, Ark.

Transactions Canadian Society of Civil Engineers (*Tran. Can. Soc. C. E.*), Sec'y McGill University, Montreal, Canada.

Transactions Engineers Society of Western Pennsylvania, (*Trans. Eng. Soc. W. Penn.*) Pittsburgh, Pa.

Transactions Liverpool Engineering Society, (*Trans. Liverpool Eng. Soc.*), Liverpool, England.

Transactions of the Technical Society of the Pacific Coast (*Trans. Tech. Soc. Pac. C.*), Rooms 14-15, 408 California street, San Francisco, Cal.

Van Nostrand's Engineering Magazine (*Van Nos. Eng. Mag.*), New York City.

Van Nostrand's Science Series, (*Van Nos. Science Series.*), New York City.

Wochenschrift des Oesterreichischen Ingenieur und Architekten Vereins, (*weekly*), Vienna, Austria.

Zeitschrift des Architekten und Ingenieurs Vereins (*Zeitsch. des Arch. u. Ing. V.*) Hanover, Germany.

Zeitschrift fur Bauwesen, (*Zeitsch. Bauwesen.*)

Zeitschrift des Oesterreichischen Ingenieur und Architekten Vereins, (*Zeitsch. des. Oesterr. u Arch. V.*), Vienna, Austria.

Zeitschrift des Vereins Deutscher Ingenieur.

Zeitschrift fur Vermessungwesen, (*Zeitsch f. Vermessungwesen.*)

There are a number of other references which are not catalogued above, for the reason that they are self explanatory and not of such frequent occurrence.

ASSOCIATION OF ENGINEERING SOCIETIES.

BOSTON SOCIETY OF CIVIL ENGINEERS.
S. E. TINKHAM, Secretary, 65 City Hall, Boston, Mass.

ENGINEERS' CLUB OF ST. LOUIS.
ARTHUR THACHER, Secretary, 1227 Spruce Street, St. Louis, Mo.

WESTERN SOCIETY OF ENGINEERS.
JOHN W. WESTON, Secretary, 51 Lakeside Bld'g., Chicago, Ill.

CIVIL ENGINEERS' CLUB OF CLEVELAND.
PROF. CHAS. S. HOWE, Secretary, Case School of Applied Science, Cleveland, O.

ENGINEERS' CLUB OF MINNEAPOLIS.
F. W. CAPPELEN, Secretary, 1628 Hennepin Ave., Minneapolis, Minn.

CIVIL ENGINEERS' SOCIETY OF ST. PAUL.
C. L. ANNAN, Secretary, City Engineers' Office, St. Paul, Minn.

ENGINEERS' CLUB OF KANSAS CITY.
WATERMAN STONE, Secretary, Kansas City, Mo.

MONTANA SOCIETY OF CIVIL ENGINEERS.
FRANCIS D. JONES, Secretary, Helena, Montana.

WISCONSIN POLYTECHNIC SOCIETY.
M. G. SCHINKE, Secretary, Asst. City Engineer, Milwaukee, Wis.

BOARD OF MANAGERS FOR THE ASSOCIATION.

BENEZETTE WILLIAMS, Chairman, 171 La Salle Street, Chicago, Ill.
L. P. MOREHOUSE, 78 Michigan Ave., Chicago, Ill.
A. GOTTLIEB, Major Block, Chicago, Ill.
JOHN NICHOL, 171 La Salle Street, Chicago, Ill.
For the Western Society of Engineers.

S. E. TINKHAM, City Hall, Boston, Mass.
W. E. MCCLINTOCK, 23 Court Street, Boston, Mass.
FREDERICK BROOKS, 31 Milk Street, Boston, Mass.
For the Boston Society of Civil Engineers.

PROF. J. B. JOHNSON, Washington University, St. Louis, Mo.
JOHN A. LAIRD, 77 E. May Street, St Louis, Mo.
For the Engineers' Club of St. Louis.

PROF. CHAS. H. BENJAMIN, Case School of Applied Science, Cleveland, O.
CHAS. H. STRONG, 35 Atwater Building, Cleveland, O.
For the Civil Engineers' Club of Cleveland.

ELBERT NEXSEN, 1620 S. E. Fourth Street, Minneapolis.
For the Minneapolis Society of Civil Engineers.

C. J. A. MORRIS, P. O. Box 2,544, St. Paul, Minn.
For the Civil Engineers' Society of St. Paul.

EDWARD BUTTS, City Engineer, Kansas City, Mo.
For Engineers' Club of Kansas City.

J. S. KEERL, Helena, Montana.
For Montana Society of Civil Engineers.

W. F. GOODHUE, Loan & Trust Building, Milwaukee, Wis.
For Wisconsin Polytechnic Society.

SECRETARY OF BOARD.
JOHN W. WESTON, 51 Lakeside Building, Chicago.

DESCRIPTIVE INDEX.

Abrasive Processes *in the Mechanic Arts.* A paper by John Richards read before the Tech. Soc. Pac. C., describing various cutting and grinding processes, and machines employed. *Ry. Rev.*, Sept. 12, 1891, pp. 603-4, *et seq.*

Accidents *in Mines.* A paper by Sir Fred. A. Abel, giving general review of legislation on, in England, and of causes and remedies which now obtain. *Proc. Inst. C. E.*, Vol. XC., p. [illegible].

See *Railroad Accidents, Mines, Bridge Accidents.*

Addresses.

Baker, L. O., President Illinois Society Engineers and Surveyors. Points out desirable changes in engineering practice of building roads, bridges, etc. *Rep. Ill. Soc. Engrs. & Surveyors*, [illegible], pp. 16-21.

Babcock, G. H. President's address before the American Society of Mechanical engineers, on "The Engineer, His Commission, and His Achievements." *Trans. A. S. M. E.*, Vol. IX., ([illegible]), pp. 23-37.

Bayles, J. C. President's address before the Inst. of Min. Engrs. at Pittsburgh, Feb. 1886. By J. C. Bayles. A forcible presentation of many of the temptations and moral pitfalls in the way of engineers, arising from their professional practice, and of the honorable course to pursue. Valuable for young engineers. *Sci. Am. Sup.*, Feb. [illegible] 1886.

Becker, M. J. Annual Address of Pres. Am. Soc. C. E. Delivered at Seabright, N. J., p. 12. *Trans. A. S. C. E.*, June, 1889, Vol. XX., p. 113.

Bruce, G. B., to the Institution of Civil Engineers. Gives his address on assuming the President's chair. A general review of engineering. *Engineer*, Nov. 11, 1887.

Chanute, Octave, Pres. Am. Soc. C. E., at the Annual Convention at Chattanooga, Tenn., May 22, 1891. Reviews some of the leading engineering works of the past year, and discusses new engineering proposals. *Trans. A. S. C. E.*, Vol. XXIV, May, 1891, pp. 397-419.

Chanute, Octave, President of the A. S. C. E., at the American Patent Celebration, on "The Effect of Invention Upon the Railway and Other Means of Intercommunication." A review of the progress in transportation, economic importance of the railway and probable future improvements. Condensed and printed in *Eng. News*, May 2, 1891, pp. 419-20.

Cleemann, T. M., Annual Address to the Engineers' Club of Philadelphia. By T. M. Cleemann, retiring President. Gives a comparison of the growth of engineering societies and a brief review of the work Philadelphia and Pennsylvania have accomplished. *Proc. Eng. Club. Philadelphia*, Vol. VI., pp. [illegible], (Feb., [illegible]).

Addresses, continued.

Cooley, L. E., Retiring Address of President L. E. Cooley to the Western Society of Engineers, on "The Modern Spirit of the Engineering Profession." *Jour. Assn. Eng. Soc.*, Feb., 1891, pp. 63 8.

Corthell, E. L., on Retiring from Presidency of the Western Society of Engineers. Jour. Assn. Eng. Soc., May, 1890, pp. 208-233.

FitzGerald, Desmond, President Boston Soc. of C. E. Delivered March 20, 1889. *Jour. Assn. Eng. Soc.*, May, 1889, Vol. VIII., p. 231.

Flad, Henry. One Year of Engineering Progress. The annual address before the Denver meeting of the American Society of Civil Engineers. *Eng. News*, July 17, 1886.

Herschel, Clemens, Pres. Boston Soc. C. E., "On the Advancement of the Profession of the Civil Engineer." *Eng. News*, Apr. 11, 1891, pp. 355-6, *et seq.*

Jones, Washington. A review of the great projects of the year. By the retiring President of the Philadelphia Engineers' Club. *Proc. Eng. Club Phila.*, Vol. VI., p. 81.

Meier, E. D., on Retiring from the Presidency of the St. Louis Engineers' Club. Jour. Assn. Eng. Soc., Feb., 1890, Vol. IX., pp. 43-50.

Potter, W. B., Retiring President, Engineers' Club of St. Louis. Gives brief history of the club and discusses its work and relations with other societies. *Jour. Assn. Eng. Soc.*, Jan., 1888, pp. 22-28.

Preece, W. H., Pres., to the Mechanical Science Section of the British Association, Bath, 1888. Reviews the developments of the practical applications of electricity. *Engineer*, Sept. 7, 1888; *T. J. and Elec. Rev.*, Sept. 7, 1888; *Jour. Soc. Arts.*, Sept. 14, 1888; *Sci. Am. Sup.*, Sept. 29, 1888.

Searles, William H., President Civil Engineers' Club of Cleveland. "The Outlook for Local and General Engineering Societies." *Jour. Assn. Eng. Soc.*, Apr. 1891, pp. 194-7.

Shinn, Wm. P., President A. S. C. E., at Cresson meeting. *Eng. News*, July 5, 1890, pp. 9-11.

Warner, W. R., President C. E. Club of Cleveland, at the annual banquet, March 29, 1890. *Jour. Assn. Eng. Soc.*, July, 1890, pp. 353-6.

Worthen, Wm. E., Before Am. Soc. C. E. Gives resume of the work of engineers during the past year. *Eng. News*, July 9, 1887; *San. Eng.*, July 16, 1877.

Wright, A. W., to the Western Society of Engineers. Gives brief review of some of the principal engineering achievements of the year. *Jour. Assn. Eng. Soc.*, Vol. VI., p. 131.

Adjutages, *Submerged, Experiments with.* By C. W. Clark. Gives details and results of experiments made at the University of Illinois. *Jour. Assn. Eng. Soc.*, Vol. VI., p. 308.

Aerial Navigation.

Extended discussion of the subject by Wm. Pole, F. R. S., in a paper before the *Inst. of Civ. Engrs.*, Vol. LXVII. Also, by same author, "Some Further Data on Aerial Navigation." *Am. Engr.*, Aug. 6, 1885. The former article is one of the most valuable that has ever appeared.

Flying Machine Memoranda. By L. Hargrave. Figure of 8 wing movement, and engine. Screw driven flying machine. Three cylinder air engine, and flying model. Theory of air rollers. *Royal Society of New South Wales Journal.* Vol. XXIII., pp. 70-74. 8 plates.

Gen. Thayer's System of Dirigible Balloons, either moving on wire "balloonways," or wholly disconnected from the earth. Fully illustrated in *Sci. Am.*, Dec. 26, 1885.

Improvements in. By Prof. W. Le Conte Stevens. An historical article, giving the recent improvements made in France up to 1885. Illustrated. *Pop. Sci. Monthly*, July, 1885.

Aerial Navigation, continued.

Interesting illustrated lecture delivered by Mr. O. Chanute, C. E., at Cornell Univ. *R. R. & Eng. Jour.*, July, 1890, *et seq.*, pp. 306 8.

Mechanical Flight. A practical criticism of Prof. Langley's article in the *Century* Magazine, by Wm. H. Harrison. *Am. Mach.*, Oct. 1, 1891.

Mechanics of Flying. By Ludwig Kargl. Object is to investigate under what circumstances it is possible for a machine to raise itself into the atmosphere and at the same time guide its motion in any given direction. *Van. Nos. Eng. Mag.*, Vol. IV., pp. 326-330.

Military Ballooning in France. An account of the most recent results attained. Illustrated. *Sci. Am. Sup.*, Nov. 6, 1886.

Power Required. By Hiram S. Maxim. Explains the principle of the aeroplane and describes his apparatus for testing it, giving general results of tests. Illus. *Century Mag.*, Oct., 1891, pp. 849-56.

Present State of, giving most modern devices, with mathematical discussion of principles, by M. DeBruignac. *Proceedings of Society of Civil Engineers*, Paris, October, 1884.

Problem of Air Navigation. A popular article by Prof. R. H. Thurston, discussing the history of the problem and experiments and observations upon birds. *The Forum*, Vol. VIII., pp. 542-51.

Progress in. A review by O. Chanute. Illus. *Eng. Mag.*, Oct. 1891, pp. 1-13.

Progress in, (continued). Article by Mr. O. Chanute containing illustrations of various flying machines from 1500 to 1879. *R. R. & Eng. Jour.*, Nov. 1891.

Prospects of Successful. Extract from letter in *New York Times*, by Hiram S. Maxim, stating how he is at present working on this problem. *R. R. & Eng. Jour.*, January, 1891.

Short sketch of its history and a description of a new form of motor for an air ship. *Am. Eng.*, May 22, 1885.

Vacuum vs. Inflation. A paper by Dr. A. de Bausset, Boston, Mass. *Am. Engr.*, Nov. 29, 1890, pp. 278-9, *et seq.*

Aerodynamics. *Experiments in.* By Prof. S. P. Langley, being experimental demonstrations of certain propositions in aerodynamics which prove that "flying" under proper direction is practicable. Very extensive experiments with "planes." Detailed description, results, etc. No. 801. Vol. XXVII., *Smithsonian Series*, 1891, pp. 115, pls. X.

Air.

Flow of, through Orifices in a Thin Plate. By A. Fliegner. Gives formula derived from experiments made with orifices from 3.17 to 12.36 millimeters in diameter. *Van. Nos. Eng. Mag.*, Vol. XXV., p. 217.

Flowing in Pipes, Coefficient of Friction of. By Prof. W. C. Unwin, M. I. C. E.

In Large Towns. A paper by William Thomson on the injurious effect of the air in large towns on animal and vegetable life, and the methods of securing a salubrious air. *Van. Nos., Eng. Mag.*, Vol. XX., p. 458.

Test of its Purity as to Carbonic Acid. A simple and exact quantitative test that may be applied by any one to test the fitness of air for breathing. *Abs. Proc. Inst. C. E.*, Vol. LXXXI., p. 384.

Air Resistance. *An Experimental Study of.* A valuable paper read by Mr. O. T. Crosby before the West Point branch of the Military Institute. Whirling experiments with velocities between 12 and 130 miles per hour seemed to show clearly a recti-linear relationship between pressure and velocity. Results plotted. Mr Crosby deduces the novel equation $P = .011 V$, P being in pounds per sq. ft. and V in miles per hour. *Lon. Eng.*, May 31, June 6 and 13, 1890. *Eng. News*, May 31, June 7, 14, 21, 1890, pp. 502, 530, 560, 589. Abstract in *Ry. Rev.*, June 7, 1890, pp. 326-8. Abstract and comments, *Lon. Engineer*, June 6, 1890, p. 461.

Air resistance, continued.

Laws of. A criticism on Mr. Crosby's experiments by Prof H. A. Hazen of the U. S. Signal Office. *Eng. News*, July 5, 1890, p. 5. Editorial, p. 14. Another editorial and letters from Prof. Hazen and Mr. Crosby discuss the best methods of making experiments, by using falling bodies and railroad trains. *Eng. News*, July 12, 1890, pp. 37-8. An excellent letter from Mr. C. F. Marvin, *id.* p. 61, discusses: (1) What we know of the laws of air resistance. (2) How we can further determine those laws.

Air Compressors.

Air Compressor, built in sections for transportation over mountainous countries on mule back. Illustrated. *E. & M. Jour.*, June 1, 1889.

Clayton. Description with indicator diagrams. *Eng. News*, Sept. 21, 1881, pp. 383-4.

Efficiency of Air Compressors Practically Considered. A paper of considerable value by Mr. Joseph Williams, read before the Inst. of Marine Eng'rs. *Mech. World*, March 28, 1891, pp. 124-5, *et seq.*

Notes on a Water Column. By Louis Brunin. Describes an invention for utilizing the power lost in allowing water to flow from the upper levels of mines to the lower level to be pumped. *Van. Nas. Eng. Mag.*, Vol. XXI., p. 1.

Plant at the Washington Aqueduct Tunnel. Illustrated description. *Eng. News*, Feb. 16, 1884, pp. 74-6.

See *Compressed Air. Lubrication.*

Air-Locks *and Shaft Tubes. The Construction and Testing of.* Advocates the periodic testing of all chambers. Air-locks not to be used over 1,500 days. Shaft-tube good for 5,000 days. By B. L. Brennecke. Lon. *Engineer*, May 7, 1886.

Alaska, *The Resources of.* By Fred'k Schwatka. With especial reference to timber and minerals. From *Bradstreet's* in *Sci. Am. Sup.*, July 4, 1885. Also *E. & M. Jour.*, June 27, 1885.

Allen, *Horatio.* A brief account of the life and work of the late Horatio Allen, one of the most prominent of American engineers. *Eng. News*, Jan. 11, 1890, p. 29.

Allowable Stresses. See *Dimensioning.*

Alloys *and Metalloids.* Address by C. Wood. Treats of Aluminum alloys, and value of silicon to iron founders. With discussion. *Proceedings Cleveland Inst. Engr's.*, 1888-9, No. 1, pp. 17-41.

Copper-Tin. A preliminary experimental research upon the mechanical properties of small castings of the alloys of copper and tin. Transverse, tension, torsion and compression tests in detail, with 81 plates and diagrams. By R. H. Thurston, Chairman. *Report Board of Testing*, etc., 1881, Vol. I., pp. 271-431.

Manganese Bronze. A paper of some length describing the properties, uses and manufacture of this alloy. Illus. Lon. *Engineer*, July 25, 1890, pp. 61-4.

Of Nickel and Steel. A paper read before the Iron and Steel Institute. By James Riley, of Glasgow. *Sci. Am. Sup.*, August 24, 1889.

See *Aluminum. Bronzes. Metals and Alloys.*

Aluminum.

And its Extraction on a Commercial Basis. By F. J. Seymour. *Sci. Am. Sup.*, Aug. 24, 1889.

And Other Metals Compared and *Aluminum in Wrought Iron and Steel Castings.* Two valuable papers by Mr. W. J. Keep, read at the Washington meeting of the A. I. M. E., Feb. 1890. Tables and diagrams, pp. 17 and 24. *Trans. A. I. M. E.* An abstract of the latter paper is in *E. & M. Jour.*, Aug. 30, 1890, pp. 245-7.

Influence upon Cast-Iron. By W. J. Keep, Prof. C. F. Maybery and L. D. Voice, before the American Association for the Advancement of Science. A valuable paper, giving details of experiments made to determine the effects of aluminum

Aluminum, continued.

on cast-iron. Good results were obtained by its use. *Sci. Am. Sup.*, Sept. 8, 1888.

Lecture delivered at the Royal Institution of Great Britain treating of the properties, manufacture and uses of this metal. By Sir Henry Roscoe. *Sci. Am. Sup.*, August 24, 1889.

Manufacture of. An illustrated description of the various processes, and a table giving the results of tests. Reprinted from *Le Genie Civil*, *Sci. Am. Sup.*, No. 753, June 7, 1890, pp. 12023-5.

Processes. A concise account of the several processes used for the extraction of this metal. From a census bulletin. *Eng. Record*, Sept. 5, 1891, pp. 223-4.

Aluminum and its Alloys, in the Electric Furnace. A lecture before the Franklin Institute describing the methods used by the Cowles Electric Smelting and Aluminium Company at Cleveland, O. Illustrated. *Jour. Frank. Inst.*, Feb. 1886. Description of the same. *Engineer*, July 9, 1886.

Electricity in the Production of Aluminum. Abstract of a paper by Alexander S. Brown read before the Am. Inst. E. E., giving a history and description of the various processes employed. *Elec. Eng.*, Apr. 29, 1891, pp. 501-2. *Elec. World*, May 2, 1891, pp. 306-8.

Grabau's Method of Making. *Annales des Mines*, Vol. XVI, p. 534. The Heroult process is described in *Le Genie Civil*, May, 1890. Abstracts of both of these papers are in *Proc. Inst. C. E.*, Vol. CL, 1890, p. 379. Former paper in *E. & M. Jour.*, Aug. 16, 1890, pp. 191-2.

Heroult Process. Describes the method of producing aluminum alloys; also gives a table showing tensile strength and elongation obtained from a series of tests made at Zurich. *Eng. News*, Sept. 8, 1888.

Heroult Process. Full description with illustrations of the plant used. *E. & M. Jour.*, Dec. 1, 1888; *Am. Mfr. and Iron World*, Nov. 23, 1888.

Its Production by Means of the Cowles Electric Furnace. Thomas D. West. *Am. Mach.*, Oct. 16, 1886.

Its Production by the Hall Process. A new electrolytic process by which aluminum is produced continuously, directly from alumina, at a single operation. *Elec. World*, Sept. 21, 1889.

Lecture delivered before the Franklin Institute by Hanford Henderson, Professor of Chemistry and Physics, Philadelphia Manual Training School. An interesting account of the various processes used to obtain metallic aluminum. *Sci. Am. Sup.*, Jan. 5, 1889.

Process of Dr. Kleiner, of Zurich, for the production of the metal direct from cryolite. Lon. *Eng.*, March 25, 1887.

Production of, with Special Reference to the Electrical Method. By Dr. Vander Weyde, before the Am. Inst. *Elec. World*, Jan. 15, 1887.

Production of the Alloys of Aluminum and Silicon in the Electric Furnace. (Cowles' process). A paper before Sec. B. British Association, by J. H. J. Dagger, F. C. S., F. Q. C., tracing the development of the industry from its inception, and giving concise descriptions of the new Cowles furnace and its operation; also giving a statement of the principal properties of aluminum alloys, as strength, elasticity, elongation, etc. *Sci. Am. Sup.*, Nov. 30, 1889.

Recent Development of the Cowles Process. By R. E. Crompton, before the Bath meeting of the British Association. Gives a description of the new plant for the production of aluminum at Milton, Eng. *T. J. and Elec. Review*, Sept. 14, 1888.

See *Aluminum, Properties.*

Properties. Abstracts from a report of Prof. Unwin on the results of the tests of a bar aluminium bronze produced by the Cowles process. Gives breaking weight,

Aluminum, continued.

30 tons per square inch: elongation, 30 per cent.; elastic limit, 15 tons per square inch. *Lon. Engineer*, Jan. 7, 1887; also see *Lon. Engineer*, Jan. 21, 1887.

Notes by Messrs. Cowles, with tables of physical properties. *R. R. Gaz.*, Oct. 21, 1887.

Properties of Aluminum. With some Information Relating to the Metal. An article by A. E. Hunt, J. W. Langley and C. M. Hall, giving much valuable information. Chemical analyses of metals, as made by several companies. Specific gravity, strength and other physical properties are fully given. *Am. Mfr.*, Feb. 21, 1890, p. 8, *et seq.*, *Eng. News*, Feb. 22, 1890, Vol. XXIII., p. 183. *Iron*, July 18, 1890, *et seq.*

Properties, Uses and Processes of Production of Aluminium. A lecture by A. E. Hunt, before the Boston Society of Arts. *Eng. News*, Feb. 28, 1891, pp. 200-2, *et seq.*

Transverse Strength of Aluminum. Record of transverse tests made by A. E. Hunt. Good deflection and permanent set are given for two tests. *Eng. News*, March 15, 1890, Vol. XXIII., p. 250.

See *Aluminum, Processes.*

Prospects of the Use of Aluminum in Railroad Work. A paper by A. E. Hunt, read before the N. Y. R. R. Club, setting forth the varied uses to which this metal may be put in this line. *Ry. Rev.*, March 28, 1891, p. 207. *R. R. Gaz.*, March 27, 1891, pp. 213-14. *Am. Mfr.*, March 27, 1891, p. 19. *Iron Age*, April 2, 1891, pp. 635-6.

Aluminum Bronze *and Brass as a Suitable Material for Propellers.* A paper read by Eugene H. Cowles at meeting of the American Institute of Mining Engineers, Feb. 21, 1890, at Washington. An interesting discussion, giving also several tests. *Sci. Am. Sup.*, April 12, 1890, p. 11890.

For Ordnance and Armor Plate. Defects in present steel guns and probable advantages to be gained by substituting aluminum bronze. by R. C. Cole, in *E. & M. Jour.* for Jan. 22, 1887.

See *Alloys.*

Aluminum Bronzes, *Physical Qualities of.* An article giving some very interesting facts concerning the physical properties of Aluminum Bronzes as brought out by certain tests made by Prof. L. Von Tetmajer of Zurich Polytechnic School. *Eng. News*, Feb. 8, 1890, Vol. XXIII., p. 122.

See *Aluminum.*

Ammonia, *Properties of.* Abstract of a paper by De Volson Wood, presented at Convention of Am. Soc. Mech. Eng., at Erie, May, 1889. *Mechanics*, June, 1889.

Ammoniacal Gas *as a Motive Power.* By Emile Lamm. Gives short history of ammonia and the method employed to convert it into a liquid. Also gives details of experiments made with a motor using ammonia gas instead of steam. *Van. Nos. Eng. Mag.*, Vol. V., p. 280.

Ammunition Cart, *Notes on Experimental.* Description of a cart designed for supplying infantry fighting line with ammunition. Two single-wheeled hand-carts make one two-wheeled horse-cart. Illus. *Proc. U. S. N. I.*, Vol. XVII., No. 1, 1891.

Anchor Gear. A description of Baxter's system for working cables, as applied to cargo and passengers steamers. *Lon. Eng.*, Feb. 20, 1885.

Anchor Ice, *Stoppage of Flow in a Water Main by.* By James B. Francis. Gives detail of the stoppage of flow at the Carleton, N. B., water works by anchor ice. With discussion in *Trans. A. S. C. E.*, Vol. XVI., p. 171; *Am. Eng.*, July 13, 1887; also editorial *Am. Eng.*, July 29, 1887.

Anemometers. *Experimental Investigations and Description of the Hagemann Anemometer.* By G. A. Hagemann. Translated by G. E. Curtis from the

Anemometers, continued.

"Annuaire Meteorologue" of the Danish Meteorological Institute, Copenhagen, 1877. *Journal of the Franklin Institute*, Sept., 1887, Vol. CXXIV., No. 741.

Whirled. Gives results of experiments made on the grounds of the Crystal Palace with whirled anemometers. *Van. Nos. Eng. Mag.*, Vol. XXV., p. 265.

Aneroid. See *Barometer*.

Anchoring Bolts. See *Cements, Adhesive Strength*.

Angle Bars. See *Railroad Track*.

Angle Irons. *Drifting Test for Steel Angle Bars.* Illustrated article. *Eng. News*, April 27, 1889.

Tests of Full Sized Angle Irons. Abstract of a paper by Charles F. Loweth, M. Am. Soc. C. E., and given in *Jour. Assn. Eng. Soc.*, May, 1889, *q. v.*, *Eng. & Build. Rec.*, Feb. 1, 1890, Vol. XXI., p. 133.

Angle Prisms. Discusses the construction and uses of angle prisms. *Eng. News*, May 12, 1888.

Appliances designed and used in connection with the Tennessee River Improvement, viz: *Derrick Winch* that lowers rapidly; *Derrick Grapple*. *Portable Drill*, for use on iron lock gates; *Iron Canal Lock Gates*, with solid and with trussed girders; *Balanced Wickets* designed to prevent leaking; *Inexpensive Switching Locomotive; Tests of Wooden Beams*, to show effects of notching and mortising, with important results; *Apparatus for testing the Strength of Explosives*, being simple and inexpensive. All these described and illustrated. *Rep. Chf. of Engrs., U. S. A.*: 1883, Vol. II., p. 1453.

Aqueducts.

Boston Water-Works. The construction of, on high embankments. By Desmond FitzGerald. *Eng. News*, May 3, 1884, p. 213.

Bridge Across the Potomac at Georgetown, D. C. Report of S. T. Abert, U. S. C. E., Ex. Doc., No. 156, House of Rep., 47th Congress, 1st Session; also in Ex. Doc. 169, House of Rep., 48th Congress, 1st session.

Cleaning of. Account of the cleaning of, on the Boston Water-Works. By Desmond FitzGerald. *Eng. News*, Apr. 26, 1884, p. 194.

Construction of the New Nadrai (Kali Nadi) Aqueduct, Ganges Canal. Consists of 15 arches of 60 ft. span. Well-foundations with concrete filling. Folding plate of illustrations. By William Good. *Proc. Inst. C. E.*, Vol. CV., 1891, pp. 147-60.

New Aqueduct for Paris. The new line is to be about 6 ft. in diameter and 62 miles long. Description with map. *Le Genie Civil*, Vol. XVI, 1890, p. 372. Abstract in *Proc. Inst. C. E.*, Vol. C., 1890, p. 448.

New Croton. Accident on. Gives detailed description of the method of timbering employed in the tunnel, and shows how the accident happened. Illustrated. *San. Eng.*, Oct. 7, 1887; also *R. R. Gaz.*, Sept. 16 and 30.

Cement Tests on. See *Cement Tests*.

Compressed Air Plant at. Illustrated article in *Eng. & Build. Rec.*, Dec. 28, 1889, Vol. XXI., p. 53.

Construction of arches of the new Croton aqueduct, including description of centres used in building. *Eng. & Build. Rec.*, Dec. 14, 1889, Vol. XXI., p. 24.

Description of the new aqueduct. By A. Fteley, Executive Engineer. *Science*, June 19, and *Eng. News*, July 11, 1885.

Design and engineering features are described in a valuable paper read by A. C. Chenoweth, Engr. in charge, before the Franklin Inst., *Jour. Frank. Inst.*, Feb., 1890. Abstract with map and profile, in *Sci. Am.*, July 19, 1890, pp. 40-1. The Harlem River siphon and pumping apparatus is described and illustrated, *ibid.*, pp. 31, 36. A summary of data concerning the old and new

Aqueducts, continued.

aqueducts is given in *Eng. News*, July 19, 1890, p. 50. For description of the reservoir see *Sci. Am.*, July 11, 1893, p. 21.

Illustrated description of drilling, blasting and timbering. *Eng. & Build. Rec.*, May 17, 1890, p. 376.

Inverts and Side Walls. Method of constructing some forms of invert, template used, etc. Illustrated. *Eng. & Build. Rec.*, Nov. 16, 1889.

Method of Detecting Bad Work. Gives a brief description of the methods employed to detect and repair the bad work on the Croton Aqueduct. *Eng. News*, Oct. 13, 1888.

Methods of Ventilating the new Croton Aqueduct. Illustrated article in *Eng. & Build. Rec.*, January 25, 1890, Vol. XXI., p. 117.

Notes on the Excavation of. A valuable paper by J. P. Carson describing the Geology of the section, methods of excavation, lighting and ventilation, and the methods used in the tunneling through some very bad ground. Fully illustrated. *Trans. A. I. M. E.*, 1890, pp. 56. Abstract from the portion relating to difficult tunneling in *E. & M. Jour.*, Apr. 25, 1891. pp. 494-6.

Plant at Shafts on the New Croton. Gives description of the plants at the different shafts, with plans of arrangement. *San. Engr.*, Nov. 12, 1887.

Pumps and Methods used for Grouting. A paper by P. F. Breadlinger. *Proc. Engrs. Club of Phila.*, Dec., 1889. Vol. VII., pp. 241-4.

Sanitary Inspection of the New Croton. A long article abstracted from the report of the New York State Board of Health. *San. Engr.*, May 27, 1886; also *Eng. News*, May 29, 1886, gives interesting data relating to accident in the tunnel.

Series of illustrated articles in *Sanitary Engineer*. Also see *Sci. Am.*, Nov. 7, 1885, for map, profile, and the interior views of tunnels.

Short illustrated article describing the dipping apparatus for emptying the Harlem River siphon. *Eng. & Build. Rec.*, Aug. 16, 1890, p. 163.

Specifications for Section 9, with many cross-sections shown. *Eng. News*, March 7, 1885, *et seq.* See also Croton.

Tunnel Excavation. A contractor's side of the tunnel excavation question. *Eng. News*, Oct. 20, 1888.

The various forms of the new Croton aqueduct for different materials traversed. Seven illustrations. *San. Eng.*, May 21, 1885.

Of Segovia. A brief description of the Roman aqueduct at Segovia. *Van. Nos. Eng. Mag.*, Vol. XXI., p. 298.

Plate Girder. Illustrated description of the new structure carrying the Delaware and Raritan Canal over four tracks of the Penn. R. R. at Trenton, N. J. *Eng. & Build. Rec.*, Sept. 27, 1890, p. 260.

The Weaver Crossing of the Vyrnwy Aqueduct. Description and full-page illustration showing the method of subaqueous crossing in steel pipes. *Sci. Am. Sup.*, July 19, 1890, pp. 12125-7.

Washington. Profile and sections of the new Washington Aqueduct Tunnel, with progress report. *An. Rpt. Chf. Engrs., U. S. A.*, 1888. Vol. IV., p. 2,706.

Report of experts appointed to examine and report on the condition of the Washington Aqueduct Tunnel. *Senate Report* No. 2,686, 50th. Cong., 2d Sess.

Zempoala, Mexico. Gives an illustrated description of Zempoala aqueduct, supplying the city of Otumba, which was built during 1553-7. *Eng. News*, July 7, 1883.

See *Pipe, Aqueduct*.

Arc, *Length of an.* Gives a new formula, which is simple and gives results near enough for practical purposes. It also admits of graphic construction. *Van. Nos. Eng. Mag.*, Vol. VII., p. 464.

Arches.

Application of the new method of graphical statics to the solution of stone and iron arches, by A. J. Du Bois. *Van. Nos. Eng. Mag.*, Vol. 13, p. 341.

Arched Ribs and Voussoir. By Mr. Martin, before a students' meeting of the Institution of Civil Engineers. Gives a mathematical discussion of arched ribs and voussoir arches. *Proc. Inst. C. E.*, Vol. XCIII., pp. 462-477.

Brick. See *Viaducts.*

Cabin John. A brief illustrated description of the above arch, having a span of 220 feet. *Eng. News*, June 25, 1887.

Concrete. Built in 12 hours, by 65 men, 39 ft. 4 in. span, 6 ft. 6⅔ in. rise of arch. Cost in Switzerland, complete, about $610. Abstracts of Papers, *Proc. Inst. C. E.* 1885.

Connemaugh, Reconstruction of. Illustration showing principal dimensions and details of new Connemaugh viaduct, with brief notice. *Eng. News*, March 29, 1890, Vol. XXIII., p. 291.

Description and complete statistics of three bridges of 140, 170 and 203 feet span, with a multitude of facts concerning the construction of other such large stone arches. One of the most sensible, practical articles that has appeared in engineering literature for a long time. *Annales des P. & C.*, October, 1886.

Elastic, Graphical Analysis of Stresses in, after Culmann. By R. H. Graham. Illustrated. Lon. *Engineer*, July 17, 1885.

Elastic, Theory of the. By Muller, Breslau. Investigation of arches of various uncommon shapes. *Zeitschr. Bauwesen*, 1886, pp. 274-304.

Elkader, Ia. Stone Arch Highway Bridge over the Turkey River. Description with plan, elevation, section, and general views of this two span arch bridge. Spans, 84 ft. each, and 27.9 ft. rise. *Eng. News*, Apr. 11, 1891, pp. 358-9.

Elyria, O. This fine stone highway arch is briefly described and illustrated. Clear span 130 feet. Rise 27 feet. *Eng. News*, May 31, June 21, 1890, pp. 506-579.

Employment of Mathematical Curves as the Intrados of. An article based upon the theory as given in Rankine's civil engineering. *Van Nos. Eng. Mag.*, May, 1884.

Error in a Common Theory of the Masonry Arch. A valuable discussion of the line of resistance in arches subjected to earth pressure. By Prof. I. O. Baker. *R. R. Gaz.*, May 31, 1889.

Fall of Beton. A short description of the beton arch on the Piræus & Corinth R. R., Greece, which fell on the removal of the centering. Illustrated. *R. R. Gaz.*, June 25, 1886.

Hinged. See *Viaduct, Garabit.*

Investigation of the theory. By W. H. Baker. *Van Nos. Eng. Mag.*, Vol. XXII., p. 33.

Iron. By W. Airy. Examines the conditions for equilibrium of a voussoir arch and then investigates the strains of a continuous iron arch. *Van Nos. Eng. Mag.*, Vol. III., pp. 450 and 641.

Masonry Arches for Railway Purposes. A paper by Mr. W. Bell Dawson, M. A., with discussion by Prof. Green, and others. *Trans. Can. Soc. C. E.*, Vol. II., p. 110. 1888.

Monier Method of Constructing. In this the arch is built of concrete, and has wire netting imbedded in it, near the soffit. Account of the construction of such an arch, with tests imposed. Illus. *Eng. News*, May 23, 1891, pp. 499-500.

Arches, continued.

From an article by F. Holzer, in *Wochenschrift des oesterr. Ing. u. Arch. Vereins.*

Of Brick and Masonry. A very good paper of 28 pp. By A. Wharton Metcalfe. Gives new tables for computing the form of arch suitable for any fixed load. Illustrated. *Trans. Liverpool Eng. Soc.*, Vol. VII., p. 19 (1886).

Practical Theory of Voussoir. By Wm. Cain. Treats of the effects of the compressibility of voussoirs, discusses different formula and extends the theory to groined and cloistered arches and domes. *Van Nos. Eng. Mag.*, Vol. XX., pp. 58, 97 and 201; also Van Nostrand's "Science Series," No. 42.

Proportion of. A careful digest by Mr. E. S. Gould of the actual practice of the French engineers in regard to the proportion of arches. *Van Nos. Eng. Mag.*, Vol. 28, p. 449.

Recent Construction in Germany of Masonry Arches with Joints at Crown and Point of Rupture. Lead joints are inserted at these points. Description of the method, with analysis of stresses, and application to four German bridges. Four folding plates. Paper by G. LaRiviere in *Annales des P. & C.*, June, 1891, pp. 899-949. Abstract, *Eng. News.*, Oct. 24, 1891, pp. 378-80.

Simple method of tracing the joints in elliptic arches. *Eng. News.*, Feb. 19, 1887.

Skew, Construction of. By M. P. Paret. Gives a history of interesting points on the construction of a skew arch on the Cincinnati & Richmond R. R., near Red Bank, O. *Eng. News*, Oct. 20 *et seq.*, 1883.

Description of a *skew* arch built at Harrisburg, Pa. *Van Nos. Eng. Mag.*, Vol. XIV., p. 351.

False Skew. A communication relating to the practice of building false skew arches, by an engineer who has constructed a number of them. *Eng. News.* March 26, 1886.

Geometry of the Oblique. By A. Ewbank. Developes the geometrical theory of the oblique arch. *Ind. Eng.*, April 30 *et seq.*, 1887.

Helicoidal Oblique (Skew) Arches. By J. L. Culley. A simple exposition of the theory and method of designing the templets and twist rules. *Van Nos. Eng. Mag.*, April, 1886. Also details of an actual construction of such an arch in *Proc. Eng. Club, Phila.*, Vol. V., p. 212.

Oakley, O. An oblique extreme skew arch, to carry a railway embankment over another double track railway. This valuable paper was read by Mr. J. F. Crowell at the Cresson meeting of the A. S. C. E. *Trans. A. S. C. E.*, Vol. XXIII., 1890, paper No. 431, pp. 155-76. Discussion, pp. 176-92. Abstract in *Mechanics*, July, 1890, pp. 152-4.

St. Paul, Colorado St. Description of this bridge of especial interest on account of method of coursing voussoirs parallel to lines of skew-backs, instead of in helices. *Eng. & Build. Rec.*, Nov. 23, 1889.

St. Paul, Minn. Plan, elevation, description, cost, etc. The cost was about twice that of corresponding plain arches. *Eng. News.*, Oct. 17, 1885.

Seventh Street Improvement Arches, St. Paul. By W. A. Truesdell. A description of these arches, the helicoidal method being used. Illustrated. *Jour. Assn. Engr. Soc.*, Vol. V., p. 317.

South Street, Boston & Providence R. R. Gives description, with plan, elevation and Sections, of the stone arch of 40 feet span to replace the Bussey bridge. *R. R. Gaz.*, Dec. 30, 1887.

Stability and Strength of the Stone. By G. F. Swain. Endeavors to show the fundamental principles, on which the stability and strength of arches depend. *Van Nos. Eng. Mag.*, Vol. XXIII., p. 265.

Stability of. By L. E. Ware. Gives Mr. Durand Claye's method of proving by

Arches, continued.

geometrical construction that any given arch is, or is not, in a good condition of stability. *Van Nos. Eng. Mag.*, Vol. XV., p. 33.

Stability of Loaded Masonry Arches. by Arthur S. C. Wurtele. *Trans. A. S. C. E.*, Vol. XXIII., 1890, pp. 1-13. Discussion, *ibid*, pp. 13-16.

Stone. Discusses the problem of the stone arch as designed from the catenary curve. *Eng. News.*, Nov. 19, 1887; *Mech. World*, Dec. 17, 1887.

Stone Arch at Falls of Schuylkill; Philadelphia & Reading R. R. 80 ft. span, with a rise of 26 ft. Segmental arch. Brief description with drawings. *Eng. News*, Jan. 24, 1891, p. 80.

Stresses in, See *Bridge Stresses*.

Theory of the Solid and Braced Elastic. By Wm. Cain. Gives a purely graphical method of treatment. *Van Nos. Eng. Mag.*, Vol. XXI., pp. 265, 353 and 443.

Transformed Catenary as a Figure for. By W. H. Booth. Advocates the use of the inverted catenary for arches, and gives methods of computing the same. *Van Nos. Eng. Mag.*, July, 1884.

Trussed. See *Viaduct, Blossom-Krantz*.

Voussoir. Stability of, determined graphically. *Proc. Inst. C. E.*, Vol. LXXXVI., p. 317.

Wheeling, W. Va., 135 Ft. Stone Roadway Arch. Illustrations of this handsome bridge, with description condensed from the specifications. *R. R. Gaz.*, Aug. 15, 1890, p. 565.

See *Aqueduct Construction. Aqueduct, New Croton. Bridges, Arches.*

Arch Curves. A paper presented by Mr. H. H. Suplee to the A. S. M. E. A description of the method of designing equilibrium arch curves as patented by T. J. Lovegrove. These curves are particularly adapted for sewers and subways. Diagrams. *Trans. A. S. M. E.*, Vol. XI., 1890, pp. 903-18. Abstract in *Mechanics*, June, 1890, pp. 126-129.

Architect, *Supervising.* History, organization and functions of the office of the Supervising Architect of the Treasury Department. U. S., 54 pp., 1886. *Treas. Dept. Doc. No. 817. Supervising Architect.*

Owner and Builder before the Law. Several papers relating to contracts, architects' duties, powers, etc. *Am. Arch.*, Aug. 8, 1891, *et seq.*

Architecture *of the United States.* Three papers illustrating and discussing the different classes. By M. Brincourt in *Encyclopédia de l' Architecture et de la Construction.* *Am. Arch.*, Nos. 812-14, July 18, 25 and Aug. 1, 1891.

See *Design, Elements of. Mill Architecture. Engineering and Architecture.*

Architectural Engineering. By J. N. Richardson. A plea for more art in engineering, and more engineering in architecture. *Jour. Assn. Eng. Soc.*, Vol. VI., p. 173.

Architectural Styles. *Sketches of* By Prof. N. C. Ricker. A fully illustrated series of articles, beginning with the Egyptian style. *Building*, Jan. 16, 1886, *et seq.*

Arctic Geography. By Lieut. A. W. Greely, U. S. A. A brief abstract, with map, of the work of the Greely Expedition. *Science*, February 27, 1885.

Armatures,

and Commutators. The Mechanical Design and Construction of. A practical paper treating of the mechanical parts only. Illus. *Mech. World*, Apr. 25, *et seq.*

Reactions. A theoretical investigation by F. C. Wagner, in *The Technic*. Reprinted in the *Elec. World*, Aug. 22, 1891, pp. 129-31.

Winding. Compound Winding for Constant Potential. A paper by Dr. Louis Bell, read before the A. I. of E. E., giving practical formulas for compounding. *St. Ry. Gaz.*, Jan., 1891, p. 9, *et seq.*

Armor.

Aluminum in. See *Aluminum Bronze.*

Belted and Internal. By J. H. Biles. A paper before the Institution of Naval Architects on the comparative effects of belted and internal protection upon the other elements of design of a cruiser. Lon. *Eng.*, April 15, 1887.

Manufacture of Iron and Compound. Translated from the Russian. Being a description of the methods employed by Chas. Cammell & Co. and John Brown & Co. Illustrated. *Mechanics*, Aug., 1885, *et seq.*

Modern Armor. An article by F. R. Brainerd, U. S. N., giving a brief description of the methods of manufacture, and chief requisites. Illus. *Sci. Am. Sup.*, No. 803, May 23, 1891, pp. 12829-30.

On the protective value of armor plates as proved in actual warfare. Lon. *Eng.*, May 15, 1885.

Armor Tests.

Brief account of tests on deflective armor plates, July 24, 1891. *Eng. News.*, Aug. 1, 1891, p. 95.

Comparative Trial of Steel and Steel-Faced armor at St. Petersburgh. Description of tests on three plates. Illustrated. Lon. *Engineer*, Nov. 21, 1890, pp. 409-11.

Competitive Trial. Report of Committee of Ordnance Dep't U. S. N., on firing test of three plates, respectively, of steel, nickel steel, and compounds. Illustrated fully by photographs. *Proc. U. S. N. I.* Vol. XVI., No. 3, 1890.

French and German Experiments against Chilled Cast Iron Armor. Results of extensive experiments. By J. Von Schutz. *Eng. News.*, Nov. 15, 1884, pp. 229-32.

Harvey Armor Plate. Account of trials on plates treated by the Harvey process, which shows a large gain in hardness over the non-treated ones. Illus. *Iron Age*, March 19, 1891, pp. 523-4.

Results of experiments with Gruson's chilled iron armor. Lon. *Engineer*, May 8, 1885.

Steel Armor Tests at Annapolis. Description of these tests which seem to indicate superior results from nickel steel. *Eng. News.*, Oct. 4, 1890, pp. 293-6. Reprinted from *N. Y. Sun* in *Sci. Am. Sup.*, No. 773, Oct. 18, 1890, pp. 12355-6. *Manfr. & Bldr.*, Oct., 1890, pp. 230-31. Some tests on nickel steel at a temperature 25 degrees below the normal, are described in *Eng. News*, Nov. 20, 1890, p. 475.

Trial of American Heavy Armor Plate. Partial results of trials of low and high carbon nickel-steel and Harvey steel. *R. R. Gaz.*, Nov. 6, 1891, pp. 780. Results of further trials, *ibid.*, Nov. 20, 1891, p. 813. *Am. Eng.* Nov. 21, 1891. Fully illustrated description, *Eng. News*, Nov. 23, 1891, pp. 510-13.

Armored Turrets. Results of the experiments at Bucharest on two model turrets, one French and the other German. Fully illustrated. Lon. *Eng.*, March 19, 1886. Also in Lon. *Engineer* for March 5, 1886, where the two systems are fully described.

Artesian Well *at St. Augustine, Fla.*, 1,390 feet deep; capacity, 10,000,000 gallons in 24 hours. Pressure of discharge sufficient to force the jet 42 feet above mouth of well. *Eng. News*, April 6, 1889.

Practice in the United States. A valuable series of articles, being a digest of a government report upon artesian wells in the United States and their relation to irrigation. Illus. *Eng. News*, Feb. 28, 1891, pp. 172-3, *et seq.*

Artesian Wells.

In Iowa. Cites a number of wells in Iowa, and then gives Prof. Leverett's discussion of the requisite conditions for a true artesian well. *San. Engr.*, Oct. 1, 1887.

In New England. A table of physical data regarding 34 wells is given in the re-

Artesian Wells, continued.

port of the Water Commissioners of Taunton, Mass., for 1889. It is reprinted in the *Eng. & Build. Rec.*, June 14, 1890, p. 20.

Irregularity of the Flow of. A paper by J. D. Caton giving some observations on this point. *Eng. News*, Oct. 17, 1878, pp. 330-1.

Materials and Methods of Boring. By C. W. Darley, of New South Wales. Fully illustrated. Lon. *Eng.*, June 19 and 26, 1885. Also *Eng. News*, July 25, and Aug. 8, 1885.

Memphis, Tenn. Description of this system said to be the most extensive in the world. *Sci. Am. Sup.*, Jan. 11, 1890, p. 11,697.

Requisite and Qualifying Conditions. By T. C. Chamberlain. This is a 50-page appendix to the Fifth Annual Report (1883-4) of the U. S. Geological Survey, J. W. Powell, Director. It is well illustrated and has considerable value to persons contemplating such enterprises. Apply to the Director, or to member of Congress.

World's Use of. A review of the subject giving considerable information. From the Government report on artesian wells, by Mr. R. J. Hinton. *Eng. News*, May 16, 1891, pp. 468-9.

Artillery.

See *Guns, Ordnance.*

Artillery Experiments *at Buchau and Tougerhuette.* Gruson's shielded mountings and quick firing guns. Lon. *Engineer*, Sept. 26, pp. 257-9, *et seq.*; *Iron*, Oct. 17, 1890, p. 339.

Asbestos. By S. A. Rogers, before the Chemists' Assistants' Association. Reviews the history, occurrence and properties of asbestos. *Sci. Am. Sup.*, June 16, 1888.

And its Applications. A lecture by James Boyd before the Soc. of Arts, London. *Jour. Soc. Arts*, April 16, 1886.

Its Occurrence and Mode of Working in Austro-Hungary. By E. L. Von Ebengreuth. Treats of Mineralogical character, analyses, and deposits in Europe, Asia and America. *Berg un Huttenmannisches Jahrbuch*, 1890, pp. 87-148.

Methods used in the mining and manufacture of asbestos. Lon. *Eng.*, March 6, 1885.

Ashtabula Disaster.

See *Bridge Accidents, Ashtabula.*

Asphalt.

and Concrete Foot Pavements. By George R. Strachan, before the Association of Municipal and Sanitary Engineers. A valuable paper containing much information relating to asphalt. *San. Eng.*, Aug. 6 and 13, 1887.

and its Application to Street Paving. By E. B. Ellice-Clark, describing methods of quarrying, manufacture, application and durability. *Eng. News*, May 1, 1880, pp. 154-7.

and Its Uses. By F. V. Greene, New York City. General descriptions, where found, chemical composition, methods of using, durability, etc. Illustrated, pp. 19. *Proc. A. I. M. E.*, Buffalo meeting, 1888.

Bituminous Rocks of California. This rock is composed of sand 85.3 per cent. asphaltic material 14.4 per cent. and water and loss 0.4 per cent. as shown in official report of the San Luis Obispo Rock. It is easily disintegrated by action of steam and used in the ordinary method for building asphaltic pavements, compressing about 10 per cent. under rollers. *Eng. News*, November 30, 1889, Vol. XXII., p. 506.

Description of deposits of asphaltum in California. *Eng. News*, Jan. 8, 1887.

for Engine Beds. Gives method of laying asphalt masonry for foundation of engines. It is said to do away with noise and jar. *Sci. Am. Sup.*, Feb. 19, 1887.

Asphalt, continued.

for Engine Foundations. An account of such foundations for a 100 horse-power engine in France. Is elastic, and does not settle or disintegrate. Illustrated. *Eng. News*, Dec. 11, 1886.

On the Use of, and Mineral Bitumen in Engineering. By W. H. Delano. Gives description of certain works with their cost, and such information concerning quality and preparation of material as may be useful to engineers. *Van Nos. Eng. Mag.*, Vol. XXIII., p. 460.

Recent Experience with Asphalt in Washington, D. C. Abstract from report of Mr. Clifford Richardson, inspector, giving considerable information as to the quality of asphalt. *Eng. Record*, Oct. 31, 1891, pp. 348-9.

Two articles treating respectively of the excavation and treatment of "asphalt comprimé" and the antiquity of the use of asphalt. Abstracted from memoir published by "Compagnie des Asphaltes de France." *Eng. News*, Jan. 25, 1890, Vol. XXIII., pp. 75 and 81.

Asphalts. By Dr. L. Meyer. Abstracted from the most complete and impartial account of the asphalts hitherto published. *Van. Nos. Eng. Mag.*, Vol. VIII., p. 74, etc.

Abstract of a portion of a census bulletin, giving considerable information on the production of the various bitumens. *Eng. Record*, Aug. 1, 1891, p. 144.

Asphaltum, Grahamite, Albertite, and Uintaite described and compared with observations on bitumen and its compounds. A paper by Wm. P. Blake, New Haven, Conn., *Trans. Am. Inst. Mining Eng.*, Washington meeting, February, 1890.

See *Pavements*.

Associations. See *Master Car Builders'*. *Master Car and Locomotive Painters'*. *Master Mechanics'*.

Astronomical Day. *Its Proposed Abolition*. The arguments stated, *pro* and *con* for beginning the astronomical day at midnight, according to the recommendation of the International Conference on Prime Meridian, 1884. *Van Nos. Eng. Mag.*, March, 1886.

Astronomical Instruments. See *Telescopes*.

Astronomy. *Method of Simultaneous Culmination of Polaris with Alioth and Adjacent Stars in the Handle of the Big Dipper*. Formulas and table, with examples of their use. By W. L. Marcy. *Eng. News*, Sept. 5, 1891, pp. 206-7.

See *Engineering and Astronomy*.

Atmosphere. *A Study of the Movements of the*. This article relates chiefly to the great continental air currents and winds. By Lieut. E. Fournier, of the French navy. Illustrated and published in *Proc. U. S. N. I.*, 1890, Vol. XVI., No. 4, pp. 557-68.

The Motions of, and their effect on the weather. *Deutsche Bauzeitung*, 1885, pp. 473, 490, 506, 538, 593.

See *Air*.

Avalanches *in Austria and Switzerland*. Abstract of a paper by V. Pollack, in an Austrian journal, describing various avalanches and methods of protection. *R. R. Gaz.*, Sept. 25, 1891, p. 671.

Axles.

Car Axles, Bearings and Lubricants. Summary of the discussion of the above subjects by the New England Railroad Club. *Railroad Gazette*, Nov. 18, 1887; also *Nat. Car and Loco. Builder*, December, 1887.

Effects of Temperature on the Strength of Railway. By Thomas Andrews. A valuable paper, giving experimental research on the effect of varying temperature on the resistance to impact of railway axles. *Proc. Inst. of C. E.*, Vol. LXXXVII, p. 340. Abstract, *Eng. News*, Feb. 18, 1885. A continuation o

Axles, continued.

these experiments given in *Proc. Inst. C. E.*, Vol. XCIV., p. 180, also *ibid.*, 1891, pp. 261-76.

For 60,000-lb Cars. A paper read before the Western Railway Club by Mr. G. W. Rhodes. *R. R. Gaz.*, Feb. 22, 1889. *Mast. Mech.*, March, 1889.

Fracture and the Remedy. By W. B. Adams. Treats of wheels and axles, longitudinal shocks, haulage resistance, torsion, etc. *Van Nos. Eng. Mag.*, Vol. IV., p. 174.

Locomotive Crank Axles. Experiments with a Steel Crank Axle. Experiments were made on an axle which had shown a small crack, to determine the effect of vibrations on the further developments of the crack. The experiments seemed to show that comparatively light blows on the axle when suspended had more effect than regular service. Illustrated description by Mr. H. V. Ivatt, in a paper read before the Inst. C. E., in Ireland., Lon. *Engineer*, Apr. 3, 1891, pp. 260-2; *Ry. Rev.*, Apr. 25, 1891, pp. 262-3.

Standard for 60,000-lb. Car. By H. C. Meade, before January meeting Western Railway Club. Gives comparison between the Johann and M. C. B. axles. *R. R. Gaz.*, Feb. 10, 1888; *Mast. Mechanic*, Feb. 1888.

Standard for 60,000-lb. Car. A paper by A. Forsyth, presented to January meeting of Western Railway Club by the committee on axles as their report. It discusses dimensions and loads, factor of safety and friction. *Mast. Mechanic*, Feb., 1888.

Steel Car. A paper by John Coffin before the Philadelphia meeting of the American Society of Mechanical Engineers. Discusses the treatment of the axle after it is forged. *Trans. Am. Soc. Mech. Engrs.*, Vol. IX, (1888), pp. 133-160; abstracted *R. R. Gaz.*, Dec. 23, 1887.

See *Car Wheels and Axles.*

Back Water, *Caused by Bridge Piers* and other obstructions. A theoretical investigation and some comparisons with actual tests, by S. W. Frescoln. *Jour. Eng. Soc. Lehigh University*, Feb. 1889.

In Streams Produced by Dams. By De Volson Wood. After a general discussion gives results of two cases occurring in his practice. *Trans. Am. Soc. C. E.*, Vol. II., p. 135.

Baldwin, *Loammi.* Said to be the Father of Civil Engineering in America. A biographical sketch, by Professor Vose, with portrait; published in pamphlet form by the Boston Society of Civil Engineers, 1885.

Ball-Bearing. *For Screw in Nut.* Description of a novel method of relieving screws of friction, by means of ball bearing nut. Illus. *Elec. World*, Nov. 1, 1890.

Balloons.

See *Aerial Navigation.*

Barbour, *William S.* A memoir prepared by a committee of the Boston Society of Civil Engineers, and read Sept. 18, 1889. *Jour. Assn. Eng. Soc.*, October, 1889. Vol. VIII., p. 506.

Barometer, *Aneroid.* Gives details of the construction and use of the aneroid; also contains tables for facilitating computations. *Van Nos. Eng. Mag.*, Vol. XVIII., pp. 104 and 202.

Aneroid Profile. By T. W. Floyd. Describes the method of observation and computation adopted with the aneroid barometer on Western surveys. *Van Nos. Eng. Mag.*, Vol. XXVI., p. 273.

Discussion of formulæ, and application to measurements of heights by Prof. C. Koppe. *Zeitschrift des Arch. u. Eng. Vereins* in Hanover, 1888, pp. 551-572. [Discussion of the different formulæ and methods of observation and computation. Quite complete.]

Barometer, continued.

Goldschmid Aneroid. By Geo. J. Specht. A new instrument wherein the movement of the corrugated cover is magnified by levers and read by a micrometer screw. Results of observations given, showing it to be a very accurate instrument. Illustrated. *Trans. Tech. Soc. Pac. Coast*; also in *Eng. News*, April 3, 1886; also description with cut showing construction, *Van Nos. Eng. Mag.*, Vol. XXV., p. 300.

Bartholdi Statue. Plans, elevations and sections of the pedestal and its foundations. *San. Engr.*, Oct. 23, 1886.

Barrels. See *Cement Barrels*.

Base-Line Apparatus. A new design, by the U. S. Coast Survey, consisting of a compensating apparatus, five metres long, composed of steel and zinc bars. Description of apparatus, comparisons and method of measuring the base. Report U. S. C. and G. Survey, 1882.

Description of the various apparatus with the theory on which they are constructed. By H. Breen. Illustrated. *Van Nos. Eng. Mag.*, Vol. XXVII., p. 59.

Base-Line Measurement *by Steel Tape*. Methods and results of such work on the survey of the Missouri River. By O. B. Wheeler. The results show that astonishing accuracy is attainable in this way. Illustrated. *An. Rep. Missouri Riv. Com.*, 1886. *H. Rep. Ex. Doc.*, *No.* 18, *49th Congr.*, *2d Sess.*

Aarberg base line, Switzerland. Fully illustrated description. *Eng. News*, March 22, 1884, pp. 134-7.

See *Tapes*.

Basic Open Hearth Furnace. See *Furnace*.

Batteries. See *Electric Batteries*.

Beams *and Girders*. *New Practical Formulas* for the strength of. By Prof. P. H. Philbrick. The results given in terms of the *dimensions* of the beam for various forms and loads. *Van Nos. Eng. Mag.*, Oct. & Nov., 1886.

Character of Position of Neutral Axis, as shown by experiments on glass, by polarised light. L. Nickerson. *Trans. A. S. C. E.*, Vol. III (1874), p. 31; also, by J. G. Barnard, Vol. III., p. 123. Discussion on these, Vol. IV (1875), pp. 177-189.

Concrete and Iron Beams. Tests of beams made of concrete and iron rods. Table of results. Lon. *Eng.* May 1, 1891, p. 521. Brief note, *Eng. Record*, May 16, 1891, p. 394.

Diagram, Giving, by Inspection, the Dimensions of Wooden Beams for a Given Span and Load. By J. M. Michaelson. *Trans. A. S. C. E.*, Vol. XXV., Aug., 1891, pp. 231-3.

Experiments Relating to the Theory of Beams. Some valuable experiments on various forms of beams of cast and wrought iron, and steel, tending to show that the same relation exists between stress and strain in cross bending as in tension and compression; and accounting for the variation of the modulus of rupture. By Jerome Sondericker. *Technology Quarterly*, Vol. II., 1885, No. 1, pp. 12-87.

Flexure and Transverse Resistance of. By Chas. E. Emery. The relation of the cross-breaking and tension moduli of in use discussed, and new formulæ given, dependent on actual tests in cross-breaking. *Trans. A. S. C. E.*, Vol VIII., (1879), p. 149, also Vol. IX., p. 353-361.

General Formula for the Normal Stress in. of any shape. By G. F. Swain. *Van Nos. Eng. Mag.*, Vol. XXIII., p. 63.

Laws of the Deflection of. By Prof. W. A. Norton. Gives results of experiments made for the purpose of testing the theoretical laws of the deflection of beams exposed to transverse strains. *Van Nos. Eng. Mag.*, Vol. III., p. 70.

Beams, continued.

Of Cast Iron and Steel. The details of an elaborate set of experiments on beams to discover the relation of the modulus of rupture in cross breaking with those in tension and compression and cause therefor. Deflections and distortions of extreme fibers measured. Observations poorly discussed. Illustrated. *Proc. Inst. C. E.*, Vol. XCVIII., pp. 3o8-316.

Practical Strength of. By Benj. Baker. A paper before the Institution of Civil Engineers. *Van Nos. Eng. Mag.*, Vol. XXIII., p. 444.

Strength of Cast-Iron Beams of Various Sections. Some experiments by students in Prof. Kennedy's laboratory. For beams of equal area of section the breaking modulus varies inversely with the stiffness. Solid, square and round, and also hollow box and circular forms, together with double T and unsymmetrical forms tested. *Proc. Inst. C. E.*, Vol. LXXXVI., p. 235.

Wooden. See *Wooden Beams. Appliances.*

See *Concrete and Iron. Flexure.*

Belts.

Adhesion of Leather. Results of experiments with iron pulleys. By Samuel Webber. *Trans. A. S. M. E.*, Vol. II., p. 224.

And Rope Gearing. The velocity of maximum efficiency. Theory, formula, and plotted curves. *Lon. Engineer*, Nov. 13, 1885.

Estimation of Power of. Formulæ for estimating the transmitting power of belts of any size. *Mechanics*, August, 1889, Vol. XI., p. 192.

Formulas for the Horse-Power of. By A. F. Nagle. Results diagramed. *Trans. A. S. M. E.*, Vol. II., p. 91.

Leather. Article by Joshua Rose, M. E., presenting several practical points on manufacture and use. Illustrated. *Eng. News*, Dec. 6, 1879, pp. 393-4.

Leather. Tractive Force of on Pulley Faces. An article containing an introduction to the subject and abstract of several papers, also a paper on the same subject by Scott A. Smith, of Providence, R. I., before Am. Soc. Mechanical Engineers. *Am. Eng.*, April 2, 1890, p. 150, *et seq.*

Link. A practical paper on leather belting made in links. *Sci. Am. Sup.*, July 9, 1887.

Origin and Progress of Leather Belting. With description of leather link belting. Paper read before the National Electric Light Association, Pittsburgh, by Charles A. Schieren. *Elec. World*, March 3, 1888; *Elec. Eng.*, March, 1888; *Age of Steel*, March 10 and 17, 1888.

Some New Points about Belts and Pulleys. By Fred J. Miller. Shows effect of varying thickness of belt at the edges, in connection with crown of pulley. Illus. *Am. Mach.*, Aug. 15, 1889.

Testing Machine for. A New Belt Testing Machine. Abstract of a paper by Geo. I. Alden, read before the Am. Soc. M. E., describing a special machine designed for the Engineering Laboratory of the Western Polytechnic Institute. Illus. *Mechanics*, July, 1891, pp. 157-8; *Am. Mach.*, July 9, 1891.

Transmission of Power by. Experiments at the Mass. Inst. of Tech. and of the firm of Wm. Sellers & Co. Probably the most thorough investigations of the efficiency of belting yet published. *Trans. A. S. M. E.*, Vol. VII., pp. 347 and 519.

Bench Marks. *Experiments on Stability of.* Paper by Geo. W. Cooley, M. Am. Soc. C. E., giving table of experiments on stability of bench marks as ordinarily made on trees; with discussion. *Trans. Am. Soc. C. E.*, Feb., 1889, Vol. XX., p. 73.

Beton *in Combination with Iron as a Building Material.* By W. E. Ward. Of interest as a study in fire-proof construction. Describes the construction of a dwelling-house at Port Chester, N. Y. *Trans. A. S. M. E.*, Vol. IV., p. 388.

Beton, continued.

Tensile Strength of Beton. An article giving valuable data. *Wochenschrift des Oesterr. Ingenieur Vereins.* A series of tests made during 1888-9. *Eng. News*, May 17, 1890, Vol. XXIII., p. 460.

Use of "En Masse." Condensed from report of Chas. K. Graham to the Department of Docks. Cites a number of works where used, and then gives method employed in the dock walls at New York City. *Van Nos. Eng. Mag.*, Vol. XIII., p. 203.

Beton Arch. *Fall of.* See *Arches, Fall of.*

Bitumen. See *Asphalt.*

Blast Furnace Wool, not a fit substance for steam-pipe covering when containing sulphur. Accidents caused by it at Columbia College. *Trans. A. S. C. E.*, Vol. XII., p. 253.

Blast Furnaces.

By Prof. M. L. Gruner. Treats of the chemical reactions, heat absorbed and discharged, weight of gases at the mouth, sampling, etc. *Van Nos. Eng. Mag.*, Vol. VIII., pp. 363, 450, etc.

A Lecture by J. M. Hartman before the Franklin Institute. An elementary but scientific analysis of the methods and operations of such furnaces, accompanied by five colored plates. *Jour. Frank. Inst.*, May, 1886.

American Blast-Furnace Practice. A discussion suggested by the paper of Mr. James Gayley on "The Development of American Blast-Furnaces," *Trans. A. I. M. E. XIX*, 932. *Trans. A. I. M. E.*, 1891, pp. 20.

American Practice, Notes on. Abridgement of a paper read before the South Staffordshire Institute of Iron and Steel Works Managers, Nov. 24, 1888. By Wm. John Hudson. *Am. Mfr.*, Dec. 14, 1888.

By-Products. See *Tar and Ammonia.*

Charges. By R. H. Richards and R. W. Lodge. Before the Duluth meeting of the American Institute Mining Engineers. Gives experiments illustrating the descent of charges in an iron blast furnace. Lon. *Eng.*, Jan. 10, 1888.

Charges, Calculation of. By A. Ledebur. Translated from *Handbuch der Eisenhuttenkunde.* *E. & M. Jour.*, Dec. 13, 1890, pp. 672-4.

Cost of. By B. Samuelson. Gives notes on the construction and cost of blast furnaces in the Cleveland District, England, in 1886. Lon. *Eng.*, June 17, etc., 1887.

Development of American Blast Furnaces with special reference to Large Yields. Very complete paper by Mr. James Gayley, read at the N. Y. Meeting of the Brit. Iron & Steel Inst. *Am. Mfr.*, Oct. 10, 1890, pp. 10, 18, 22. Abstract in *E. & M. Jour.*, Oct. 4, pp. 385-91. *Sci. Am. Sup.*, No. 776, Nov. 15, 1890, pp. 12397-7.

Distribution and Proportions of, in America. By John Birkinbine, before the Halifax Meeting of the *Am. Inst. of Min. Engrs.*, 1885.

Experiments on Blast-Furnace Gases. Paper by Jasper Whiting, giving results of a continuous series of analyses extending over several days. Curves showing variation drawn. *Trans. A. I. M. E.*, 1891, p. 11.

Gas. *The Mendheim Continuous Gas Furnace* for burning fire-brick and porcelain. A greatly improved device, well described and illustrated by Prof. Egleston, in *Sch. of Mines Quarterly*, Oct., 1884.

Hoists for. See *Hoisting Machinery.*

Irregularities of the Blast-Furnace Process, and how to avoid them. Paper read at a meeting of the American Institute of Mining Engineers, St. Louis, Oct., 1886. *E. & M. Jour.*, Nov. 6, 1886.

Improved Gas Heating. Gives estimate of the quantity of fall required for cast

Blast Furnaces, continued.

steel with the Siemens furnace, and the process of partial elimination of nitrogen of M. Ch. Schinta, of Strasburg. *Van Nos. Eng. Mag.*, Vol. I., p. 348.

Modern Blast Furnace Construction. An illustrated paper by J. L. White, *Iron Age*, Sept. 11, 18, 1890, pp. 426-8, 515-8.

From a Scientific Point of View. A series of lectures by C. R. Alden Wright before University College, London. Gives reactions and much other valuable information. *Am. Mfr.*, Jan. 17, 1890, p. 8, *et seq.*

See *Hot Blast, Theory of.*

Blasting.

Blowing up of Basin Wall at the Royal Albert Docks. An interesting paper by Mr. Joseph E. Thomas, M. I. C. E., read before the Inst. of Marine Engrs. Illustrated. Reprinted in *Eng. News*, Sept. 2, 1890, pp. 260-61.

Method of Determining the Charges of Dynamite for Sloping Rocks. Formulas with tables for use. Translated from the *Bulletin of the Soc. Austrian Engrs.*, for *Eng. News*, May 14, 1881, pp. 192-3.

Note on the Form of Crater Produced by Exploding Gunpowder in a Homogeneous Solid. A brief paper giving results of a series of experiments made by Frank Firmstone, shows crater to have in general a conoidal-shaped fracture rather than a true conical form. *Trans. A. I. M. E.*, Ottawa meeting, October, 1889.

Of Rock Under Water. Methods used on Welland Canal at Port Colborn, Ont. Giant powder proving ineffectual, nitro-glycerine was used with great success. *Eng. News*, July 11, 1885.

On Igniting Blasts by Means of Electricity. By Julius H. Striedinger. A very complete article on the subject by an engineer of large experience in such work. Illustrated. *Trans. A. S. C. E.*, Vol. VII., 1878, pp. 1-13.

On the Simultaneous Ignition of thousands of mines, and the most advantageous grouping of fuses. By J. H. Striedinger. Illustrated. *Trans. A. S. C. E.*, Vol. VI., 1877, pp. 177-194.

Operations at Hell Gate. A general account of the methods, cost, and results, with many cuts. L. F. Vernon-Harcourt. *Trans. Inst. Civ. Engrs.*, Vol. LXXXV., p. 264.

Subaqueous. A paper on the improvement of navigable waters. *Wochenschrift osterr. Ing. u. Arch V.*, Aug. 22, 29, 1890. Nos. 34-35, pp. 287-97, 299-306; discussion, Nos. 35, 36, pp. 298-7, 325-7.

Tamping Blast-Holes with Plaster of Paris. By F. Firmstone. *Trans Am. Inst. Min. Engrs.*, Vol. XII., *Eng News*, June 7, 1884.

Theory of. By Prof. H. Hoefer. Translated from the German and published by the Engineer Corps, U. S. Army, as a separate pamphlet of 80 pages. Theoretical equations are derived which agree with practice in homogeneous rock. Most effective relation between depth of hole and size of charge determined, both for fracture and for projection.

With Dynamite on a large scale in Italy. An account of some very successful blasts, in one of which 120,000 cubic meters were removed. *Proc. I. C. E.*, Vol. LXXXII., p. 423.

Blower, *Centrifugal Fan.* An investigation of some experiments on the relations of the velocity of discharge, pressure and velocity of fan blades in a centrifugal blower delivering air into the atmosphere at large. By Chief Engineer Isherwood, U. S. N. *Jour. Frank. Inst.*, April, 1889.

Blowers, *Experiments and Experience with.* By H. I. Snell, before the Philadelphia meeting of the American Society of Mechanical Engineers. *Trans. A. S. M. E.*, Vol. IX. (1888), pp. 51-73; *Am. Engr.*, Nov. 23, 1887; *Mech. World*, Dec. 17, 1887.

See *Ventilation.*

Blowing Engine.

See *Engines, Steam, Blast Furnace.*

Blue Printing. *Blue Print Frame.* Detailed drawings of the frame used in the Water Department, Philadelphia. *Eng. News*, Feb. 19, 1887.

Blue Process of Heliographic Printing. By Channing Whitaker. A thorough description of the chemicals, apparatus and methods. *Jour. Assoc. Eng. Soc.*, Vol. I., p. 349. *See* also *Proc. Phil. Engrs'. Club*, Vol. V., p. 151.

Giving white on blue ground, as well as blue on a white ground. Detailed instructions from the *Eng. Mech. and World of Sci.*, reprinted in *San. Engr.*, Aug. 29, 1885. *See* also *Proc. Engrs'. Club, Phil.*, Vol. V., p. 152.

Rapid Blue Printing Paper. Solution giving a more rapid printing paper and better prints. *Eng. News*, May 23, 1891, p. 486.

Boats. *Experiments with Lifeboat Models.* An elaborate paper read by Mr. J. Corbitt before the Institute of Naval Architects. Illustrated. Reprinted from Lon. *Eng.* *Sci. Am. Sup.*, No. 753, June 7, 1890, pp. 11032-4.

Full description of the U. S. Lighthouse Tender "Azalia". Illustrations of the compound engines, etc. *Iron Age*, Dec. 11, 1890, pp. 1029-34.

Navy Boats for life saving service. A paper in two parts by Ensign A. A. Ackerman, U. S. N. Folding plates. *Proc. U. S. N. I.*, Vol. XVI, No. 3, 1890, pp. 279-332.

Resistance of. See *Canal Traction.*

See *Keels.*

Boat Railway. *The proposed Boat Railway Around the Dalles Rapids of the Columbia River.* General description of the project with extracts from the report of the U. S. Engrs., giving details of car and hydraulic lifts, capacity and estimate of cost. Inset of details. *Eng. News*, Sept. 12, 1891, pp. 228-31.

Boiler Construction.

Boiler Making. A series of articles of considerable practical value. By a designer of boilers. Fully illustrated. *Mechanics*, Jan., 1886, *et seq.*

Boilers Adapted to Forced Draught. A paper by A. F. Yarrow, read before the Inst. of Naval Architects, discussing mainly the problem of avoiding leaky tubes. Illustrated. *Ry. Rev.*, April 4, 1891, pp. 222-3, *et seq.* Lon. *Eng.*, March 20, 1891, pp. 335-9, Lon. *Engineer*, March 20, 1891, pp. 225-8.

Horizontal Tubular. A new method of making horizontal tubular boilers by making the shell in two plates of steel the full length of the boiler. *Trans. A. S. M. E.*, Vol. VI., p. 110.

Joints. A new joint, consisting of double butt straps, with three rows of zig-zag rivets, the straps being cut with a wavy or sinuous outline, there being only half as many rivets in outer row as in the two inner rows. Develops 90 per cent. of the solid plate. Lon. *Eng.*, Oct. 10, 1884.

Latest Progress in. Report of the Boiler Committee of the Master Mechanics' Association, prepared by J. Davis Barnett, with discussion. *R. R. Gaz.*, Sept. 4, 1885, *et seq.*

Modern Practice in. An extended paper, with discussion, before the Inst. of Civ. Engrs., London. Gives present English practice in great detail. Illustrated. *Proc. Inst. C. E.*, Vol. LXXX., pt. 2.

On the Longitudinal Riveted Joints of Steam-Boiler Shells. An article by John H. Cooper, of Philadelphia, Pa., giving discussion of above question, formulæ for use in dimensioning boiler shells, etc. *The Locomotive*, July, 1889, Vol. X., p. 101.

Riveting. A paper by Prof. R. H. Smith, containing a careful investigation of the necessary relation of pitch, diameter of rivets and thickness of plates for tightness as well as for strength. Formulas derived and tables given. Lon. *Eng.*, Aug. 21, 1885.

Boiler Construction, continued.

Small Vertical Boilers. An article giving many practical hints concerning the proper designing of small vertical boilers, from *Am. Mach.*. *Am. Eng.* April 9, 1890, p. 156.

Steam Boilers, their construction, setting, and management. By J. C. Burwell, with discussion. *Trans. Engrs. Soc. West Penn.*, Pittsburgh, 1884.

Steel Rivets for Boilers. A short paper by Oliver Bros. and Phillips. An appendix gives the U. S. Government requirements and a table of shearing tests. *Ry. Rev.*, June 28, 1890, p. 433. *E. & M. Jour.*, July 19, 1890, p. 71.

Thick Steel Plates. Strictures on their use, derived from experience. *Van Nos. Eng. Mag.* Sept., 1885.

Thick Steel Plates in. By W. Parker. Investigation of a large steel boiler that burst on the test trial. *Iron Age*, May 7, 1885; Lon. *Eng.*, April 3, 1885.

See *Boiler Specifications. Boilers, Defects. Boilers, Marine.*

Boiler Corrosion.

Corrosion in Steam Boilers. By W. J. Norris, M. I. N. A. An examination into the various theories offered in explanation of the "pitting" of boilers and steam-drums, with arguments in favor of the theory of oxidation by free oxygen in the water. Reprinted from the *Trans. Inst. Naval Architects*, by the Bureau of Navigation, Navy Department. Naval Professional Papers, No. 16.

By F. J. Rowan. A paper read before the Glasgow meeting of the British Association. Points out some of the causes of corrosion and gives their remedies. *Van Nos. Eng Mag.*, Vol. XVI., p. 21.

Around Stay Bolts. Illustrated article showing corroded stay bolts and plates around stay bolt holes. *The Locomotive*, Jan., 1890. Vol. X., p. 1.

Relative of Iron and Steel. By William Parker, Chief Engineer-Surveyor of Lloyd's Register, London. A paper giving a brief analysis of the experiments of the Admiralty Boiler Committee, together with an account of further experiments made by the writer for the committee of Lloyd's Register. Reprinted from *Trans. Inst. Naval Architects*, Bureau of Navigation, Navy Department. Naval Professional Paper, No. 14.

See *Boilers, Marine.*

Boiler Designing. See *Boiler Construction.*

Boiler Deposits. A paper by Prof. V. B. Lewes, read before the Inst. of Naval Architects, investigating the nature and manner of formation of the oily deposits in high pressure boilers. *Mech. World*, April 4, 1891, pp. 138-9, *et seq.* Lon. *Eng.*, April 3, 1891, pp. 418-9.

Boiler Experiments *and Fuel Economy.* By J. Holliday, before the students' meeting, Institution of Civil Engineers. Gives details of experiments made to increase the efficiency and economy of a certain boiler. *Proc. Inst. C. E.*, Vol. XCII., pp. 336-352.

Boiler Explosions.

"*A Terrible Catastrophe.*" An account of the boiler explosion which caused the wreck of the Park Central Hotel, at Hartford, Conn., Feb. 18, 1889; with illustrations and description of exploded boiler. *The Locomotive*, March, 1889, Vol. X., p. 33.

Boiler Explosions. By Zerah Colburn. Very valuable. Treats of overheating, electricity, decomposed steam, overpressure, etc. *Van Nos. Eng. Mag.*, Vol. IX., p. 209.

Boilers as Magazines of Explosive Energy. A valuable paper by Prof. Thurston, showing the source of the explosive energy to be mostly in the heated water, and the amounts of such energy in different styles of boilers. *Trans. A. S. M. E.*, Vol. VI., p. 199.

Cause and Remedy. By Thos. Kays. Adopts the Lawton theory of explosions.

Boiler Explosions. continued.

Gives results of some experiments. Advocates the Lawton diaphragm. *Van Nos. Eng. Mag.*, November, 1884.

Cause of. Abstract of paper before the Scientific Society of Bridgeport, Conn., by F. G. Folwer, describing experiments indicating the possibility of sudden increase of pressure due to air and gases. *R. R. & Eng. Jour.*, May, 1890.

Cause of. Diagrams showing causes of 1,079 boiler explosions classified under fifteen heads. *The Locomotive*, Sept., 1891.

Dangers in the Employment of Boilers with Movable Covers. Paper by MM. Polonceau and Orly, describing accidents occurring in industrial establishments, and discussing causes. Three folding plates of illustrations. *Annales des P. & C.*, July, 1891, pp. 79-111.

Danger of Employing Certain Disincrusting Agents in Small Boilers. Full account of two explosions due to this cause. Details illustrated. By —— Orly. *Annales des P. & C.*, March, 1891, pp. 434-40.

Failure of a Firmenich Boiler. By C. F. White, before the Engineers' Club of St. Louis. Gives results of examinations as to the cause of explosion of a Firmenich boiler. *Jour. Assn. Eng. Soc.*, Aug. 1888, Vol. VII., pp. 319-335. *R. R. Gaz.*, Sept. 7, 1888.

In 1884. Summary published by the Midland Steam Boiler Assurance Company. Lon. *Eng.*, March 13, 1885.

In 1888. Gives a list of explosions in the United Kingdom during 1888, with illustrations in many cases, a statement of cause, results, etc., and a tabulated comparison with previous years. Lon. *Eng.*, July 19, 1889. Vol. XLVII., p. 59.

In 1890. Annual report by Mr. E. B. Marten. This is a valuable paper giving a very brief account of 77 explosions in Great Britain. Tables and illustrations. Lon. *Eng.*, July 18, 1891. pp 83-6.

See *Boilers, A Lecture, etc.*

Boiler Incrustation.

Automatic Apparatus for Preventing Boiler Incrustation. Illustrated description of an apparatus for injecting a disincrusting fluid. *Sci. Am. Sup.*, No. 813, Aug. 1, 1891, pp. 12990-1.

Incrustation. Circular prepared by the Bavarian Boiler Inspection Society, containing statements of what is most important for boiler owners to know concerning feed water and incrustation. *R. R. Gaz.*, Jan. 21, 1887.

Boiler Incrustation. By F. J. Rowan. Treats of the incrustation of land boilers using natural fresh water, and of marine boilers using sea water. Gives tables of analysis of boiler crusts and deposits. *Van Nos. Eng. Mag.*, Vol. XV., p. 345.

Danger from Disincrusting Agents. See *Boiler Explosions.*

Scale in Steam Boilers. By G. S. King, before the Royal Society of Arts. *Van Nos. Eng. Mag.*, June, 1884.

Use of Kerosene Oil to Prevent. A paper by L. F. Lynes before the American Society of Mechanical Engineers. Gives practical experience in using kerosene oil in steam boilers to remove and prevent scale. Advocates its use. *Trans. A. S. M. E.*, Vol. IX., (1888), pp. 247-258. *Am. Engr.*, Nov. 30 and Dec. 7; *Power*, December, 1887; Abstract in *Eng. & Build. Rec.*, Dec. 31, 1887; *Sci. Am. Sup.*, Feb. 11, 1888.

See *Boilers, Defects. Locomotive Boilers.*

Boiler Plates. *Tests for Ship and Boiler Plates.* Physical and chemical, with discussions. By P. G. Salom. *Trans. A. I. M. E.*, Vols. XII. and XIII.; also in *Iron*, Nov. 14, 1884.

Boiler Plants. See *Smoke Consumers, Smoke Prevention. Heating and Ventilation.*

Boiler Setting. Description and discussion of merits of the different forms of bridge walls and combustion chambers in use. *Lon. Engineer*, Feb. 20, 1885.

Setting Boilers over a Single Furnace. Illustrated description of a method of suspending two or more boilers over a single setting or furnace. *The Locomotive*, April, 1889. Vol. X., p. 49.

See *Boilers, Lancashire*.

Boiler Specifications.

A paper by C. G. Darrach, with a discussion of considerable interest by several prominent engineers. *Proc. Eng. Club Phila.*, December, 1887. Vol. VI., pp. 179-224.

In the Manhattan Co's. Bank Building, New York. Illustrated. Giving portions of the specifications. Some new and commendable features in the matter of internal bracing and dome. *San. Engr.*, January 29, 1885.

Materials and Tests. Report of a committee to the American Boiler-makers Association at Pittsburgh meeting, Oct. 14, giving specifications proposed and tests recommended for boiler materials, with discussion. *The Iron Age*, October 21, 1886.

Specifications for Boilers. By C. G. Darrach, before the Philadelphia Engineers' Club. Gives general specifications for boilers which require the bidder to state not only the price for the entire work, including boiler, setting, fixtures, etc., but also the economy and capacity he will guarantee. Discussed. *Proc. Eng. Club. Phila.*, December, 1887, Vol. VI., pp. 179-206.

Tests of Material and Manufacture prescribed by the Steam Users' Association of Manchester, Eng. From *Iron* in *San. Engr.*, Nov. 27, 1886.

Boiler Tests.

At Nashville, Tenn. By John W. Hill. A Sibley College lecture describing one of the most elaborate tests on record. Time of test 72 hours. Five boilers under test. *Sci. Am. Sup.*, Dec. 18, 1886.

At the Phila. Electrical Exposition. Final Report of the Com., giving methods and results. *Jour. Frank. Inst. Sup.*, July, 1885.

Babcock & Wilcox Boilers. Full account of the trial of. Diagrams and table. *Lon. Eng.*, July 11, 1890, pp. 43-6.

Steam Boilers with Hot Blast Apparatus. Probably the most careful, elaborate and valuable series of boiler trials on record, being a nine weeks' test of the Marland Warm Blast Apparatus at the Pacific Mills, Lawrence, Mass., by J. C. Hoadley, in 1881. The heat of the escaping gases returned to the furnace and the draught created by a blower in the flue. From 10 to 20 per cent. of the heat saved, but singularly enough the cost of operating the blower was not found. Tables, diagrams, apparatus used, etc., fully stated, the report occupying 177 pp. in *Trans. A. S. M. E.*, 1884, p. 676.

Boiler Trials. Wm. P. Trowbridge. Treats of the evaporative efficiency of steam boilers. *Van Nos. Eng. Mag.*, Vol. VII., p. 546.

Boiler Trials. Many short topical articles by the different members of the committee which was appointed by the American Society of Mechanical Engineers to report on the subject. These articles cover the whole ground of boiler trials for various purposes giving many practical hints and precautionary measures. *Mechanics*, May and June, 1885.

Coil Boilers for the U. S. Navy. Description with account of test of a Ward coil boiler. Also a discussion of the advantages of this type of boiler. Illus. *Eng. News.*, Aug. 29, 1891, pp. 184-5.

Comparative Tests of. A report of tests made at the boiler house of the Brush Electric Light Co., Philadelphia. The plants consisted of 10 boilers, 4 Babcock & Wilcox, and 6 return tubular boilers. Tests gave 77 in favor of former. *Van Nos. Eng. Mag.*, Vol. XXVIII., p. 265.

Corliss Boilers. Evaporation of. By James B. Francis. Gives details of experiments made on 4 Corliss boilers *Van Nos. Eng. Mag.*, Vol. III., p. 175.

Boiler Tests, continued.

Dupuis Pattern. Tests in Silesia of boilers of this type showing an evaporation of 4½ pounds of water per sq. ft. of heating surface per hour, and using from 60 to 70 per cent. of the heating power of the coal. Consists of two horizontal boilers joined by one vertical tube in front and by a vertical fire tubular boiler in the rear. A new and most efficient type. Abstracted from the German, *Proc. Inst. C. E.*, Vol. LXXXVI., p. 452.

Efficiency of Using Waste Gas as Fuel. By D. S. Jacobus, before the Birmingham meeting of the American Institute of Mining Engineers. Gives the results of trials made to determine the efficiency of a water-tube boiler with waste gas from a blast furnace as fuel. *Am. Eng.*, Aug. 15, 1888; *Eng. News*, Aug. 18, 1888.

Experimental. By Prof. R. H. Thurston. Gives details of the experiments made at Hoboken and Staten Island, on full-sized boilers. *Van Nos. Eng. Mag.*, Vol. VI., p. 473.

Evaporation, Performance of. By D. K. Clark. A valuable article; gives results of a large number of experiments with boilers. *Van Nos. Eng. Mag.*, Vol. XVI., p. 243.

Galloway Boilers. A table giving record of a series of tests of Galloway boilers made by Edge Moor Iron Co. *Am. Eng.*, March 16, 1887.

Harvard Boilers. Report of the Judges on the tests of the Harvard boilers at the fair of the American Institute. *Van Nos. Eng. Mag.*, Vol. XIII., p. 166.

Heating Feed-water with Live Steam. Account of a test showing a remarkable gain in economy by this method of heating. Lon. *Engineer*, July 3, 1891. Reprinted in *Eng. News*, July 25, 1891, pp. 82-3.

Hydraulic Tests. A brief article discussing the value of such tests, reprinted from the *Practical Engineer*, (Lon.) *Am. Eng.*, Dec. 13, 1890, pp. 258-9.

Standard Methods of Making Trials. Being the Report of the Committee of the Am. Soc. Mech. Engrs., consisting of Messrs. Kent, Hoadley, Thurston, Emery and Porter. Containing a Code of Rules and methods of treating the several problems involved. Also form of final report. *Van Nos. Eng. Mag.*, March, 1885. *San. Engr.*, June 18, 1885, *Am. Eng.*, Nov. 14, 1884.

Steam Boiler Experiments. By Bryan Donkin, Jr., and Prof. Kennedy, F. R. S. These papers record the result of a splendid series of comparative experiments upon some of the chief types of boilers in general use in England. Lon. *Eng.*, July 4, 1890, *et seq.*

Tables for Facilitating the Calculation of. The tables are for reducing the evaporation under the actual conditions to the equivalent evaporation from and at 212°. More complete than any heretofore published. By Wm. Kent, in *Trans. A. S. M. E.*, Vol. VI., p. 84.

Water Tube, Trial of a. By R. H. Thurston. Gives very full details of a test of a Babcock-Wilcox water-tube boiler at Sibley College, Cornell University. *Sci. Am. Sup.*, April 14, 1888.

Boiler Tubes.

Resistance to Collapsing. A series of formulas, rules and tables of the same. *Iron Age*, Dec. 11, 1884.

Serve Boiler Tube. Account of experiments made at Sheffield, Eng., with these tubes, indicating a gain of 12 to 14 per cent. in fuel economy. Lon. *Engineer*, Oct. 24 and 31, 1890, p. 361. *Eng. News*, Nov. 8, 1890, pp. 410-11.

Serve Ribbed Boiler Tube. Results of tests made with both forced and natural draught, show a remarkable gain in economy of fuel and capacity of boiler. Tests were made on a vertical tubular boiler. *Eng. News*, May 9, 1891, p. 442. Abstract, *R. R. Gaz.*, May 8, 1891, p. 320; *Ry. Rev.*, May 9, 1891, p. 302.

See *Boiler Construction*.

Boilers *and Engines for high Pressure.* By L. Perkins. Gives plans for working steam with pressure of from 250 to 1,000 lbs. per sq. in. With comments from Lon. *Eng.* *Van Nos. Eng. Mag.*, Vol. XVII., p. 143.

A lecture by J. M. Allen in the Sibley College course. Gives an elaborate discussion of the subject, and defines a theory of explosions. *Sci. Am. Sup.*, Feb. 19, 1887.

A lecture at Cornell University, by J. M. Allen. The subject adapted to young mechanical engineers. *Sci. Am. Sup.*, March 20, 1886.

A practical paper on boilers, engines, condensers, etc., by E. E. Roberts, M. E. *Am. Eng.*, July 2, p. 1891, *et seq.*, pp. 2, 3.

A special number of *The Locomotive* treats of boilers, pipes, steam fittings, nozzles and their connections. A very practical paper, fully illustrated. *The Locomotive*, Aug. 1890, Vol., VI., No. 8, pp. 113-28. Published by the Hartford (Conn.) Steam Boiler Inspection and Insurance Co.

An exhaustive paper and discussion on materials used in boilers, joints and seams, stays, general construction and special types, covering 68 pages in *Proc. Inst. C. E.*, Vol. LXXX., p. 100.

Circulation of Water in Steam Boilers. Extracts from a lecture by Mr. G. A. Babcock before the students of Cornell University, discussing and illustrating this question. *Eng. Rec.*, May 10, 1891, pp. 377-8.

Condenser for. See *Condensers.*

Construction of. See *Boiler Construction.*

Coil. See *Boiler Tests, Coil.*

Comparison of the Belpaire and the Radial Stay Boiler. Brief descriptions and sections showing details. *R. R. Gaz.*, Aug. 1, 1890, p. 538. Editorial, *ibid* pp. 533-4.

Corrosion of. See *Boiler Corrosion.*

Defects. Some Land Boiler Defects and how Prevented. A paper of some practical value, by Geo. Walker, read before the Aberdeen Mech. Soc. *Mech. World*, Oct. 4, 1890, p. 138, *et seq.* Reprinted in *Am. Eng.*, Nov. 8, 1890, pp. 204-5, *et seq.*

Defects and Remedy. By A. C. Engert. An article treating of incrustations, smoke and unconsumed gases and priming. *Van Nos. Eng. Mag.*, September, 1884.

Deposits in. See *Boiler Deposits, Boiler Incrustation.*

Deterioration of. By J. M. Allen. A Sibley College lecture, treating of errors in boiler construction, and of the natural cause of their deterioration. Illustrated. *Sci. Am. Sup.*, June 9, 1888.

Deterioration of. Some interesting examples given. *R. R. Gaz.*, Nov. 6, 1885.

Deterioration of. See *Boilers, Preservation.*

Dupuis. See *Boiler Tests, Dupuis.*

Economy. Relative Economy of Sectional and Shell Boilers. By R. H. Buel. Gives tabular performances of many types, with conclusions drawn therefrom. *Am. Eng.*, July, 1885.

Effect of Oil in Bulging. Illustrated, described, and explained. *San. Engr.*, May 21, 1885.

Essex Vertical. Describes a new internal arrangement adopted for small vertical boilers. Lon. *Engineer*, Dec. 9, 1888.

Experiments on. See *Boiler Tests, Boiler Experiments.*

Explosion of. See *Boiler Explosions.*

Failures. See *Boilers, Marine, Steel.*

Feed Water Heating. See *Boiler Tests, Heating.*

Feeding of. Danger from feeding at bottom. *Am. Eng.*, March 20, 1885.

Boilers, continued.

For Electrical Installations. Paper by W. H. Booth and F. B. Lee, discussing the merits of various forms, and general requirements. *Elec. Rev.*, Sept. 18, 1891, pp. 328-39. Reprinted in *Mech. World*, Oct. 9, 1891, *et. seq.* *Am. Eng.*, Oct. 17, 1891, pp. 155-6.

Forced Draught fog. Blast System for Steamships, Mains. A description, with illustrations, of a system of hot blast for forcing the fires, and at the same time cooling the smokestack and ventilating the ship, used by the U. S. and B. M. S. S. Co., with marked saving in fuel. *Am. Mach.*, March. 7, 1889.

A paper read before the Institute of Marine Engineers, Oct. 19, 1889, by J. Williams, Birmingham, Eng., discussing subject of forced draught at some length. Lon. *Eng.*, Nov. 22, 1889, Vol. LXVIII., p. 433, *et seq.*

A paper before the N. E. Coast Inst., of Scotland, by James Patterson and Magnus Sandison. An important paper going to show the advantages of forced draught. Experiments given. Lon. *Eng.*, March 12 and 19, 1886.

An account of an experiment made on the Clyde by James Howden, showing great advantages. Lon. *Eng.*, Oct. 24, 1884.

As Applied on Certain English Steamers, by connecting the fan directly with the mouth of the furnace by a separate passage. This connection is broken when doors are opened. Accomplishes same result as by increasing air pressure in boiler-room. May be applied to stationary boilers. Cut given. Lon. *Eng.*, April 3, 1885, and Lon. *Eng.*, April 17, 1885.

A paper by H. Williams, Chief Inspector of Machinery, R. N. Reprinted from *Naval and Military Mag. Mech. World*, July 26, 1890, *et seq.*

Closed Stokehold System. By Thos. Soper, before the Institution of Naval Architects. Gives a discussion on the use of forced draught under boiler in the closed stokehold system. Contains the experience gained with vessels in the British Navy. Lon. *Engineer*, April 6, 1888. Lon. *Eng.*, April 6. 1888.

Experiments on the Use of Steam Jets for Producing Forced Draught in Boiler Furnaces. Tests made at the N. Y. Navy Yard. Results with illustrations of nozzles used. *Eng. News*, Sept. 19, 1891, pp. 266-7.

Forced by a fan in the flue. Successfully applied by Martin & Co., London, Description and cuts. Method of keeping bearings cool shown. A draught equal to 3½ inches of water obtained. *Iron*. Nov. 6. 1883.

By J. R. Fothergill, before the Institution of Naval Architects. Gives the results of trials made with forced draught on the steamers "Marmora," "Dania" and "Elna." The summary shows a slight reduction in speed, with a saving of from 20 to 30 per cent. in fuel. Lon. *Engineer*, April 6, 1888; Lon. *Eng.*, April 6. 1888.

Report of a series of trials of an apparatus for transferring part of the heat of the escaping flue gases back to the furnace. Read by J. C. Hundley before Am. Soc. Mech. Engrs. *Am. Mach.*, June 13. 1885. *See* also Forced Draught.

Report of a series of trials of a warm-blast apparatus for transferring a part of the heat of escaping flue gases to the furnace, by J. C. Hundley, to the Am. Soc. Mech. Engrs. Results in saving 10 to 18 per cent. of the fuel. Illustrated. *San. Engr.*, July 16, 1885.

Stationary Boilers. A plea for their greater use. By Chas. W. Baker. A theoretical view of the matter. *Eng. News*, Jan. 2, 1886.

For Steam Boilers, by the use of a fan in the chimney flue. A short account of some experiments at the N. Y. Navy Yard, which have resulted in doubling the consumption of coal per square foot of grate surface. Probably the only practicable solution of the forced-draught problem. *Proc. U. S. N. I.*, Annapolis, Md., Vol. XI.

Steam Trials of the "Satellite" and "Conqueror." By R. J. Butler, M. I. N. A. A paper describing the experiments made with closed boiler-rooms and fan-blast on board H. M. S. "Satellite" and H. M. S. "Conqueror," whereby the

Boilers, continued.

steaming power of those vessels was increased 60 per cent. over that with natural draught; with deductions from the experiments as to the advantages of this system, particularly with reference to its utility as a form of emergency power for war vessels. With discussion. Reprinted from *Trans. Inst Naval Architects* by the Bureau of Navigation, Navy Department. Naval Professional Papers, No. 16.

vs. Natural. Tests on two steamers making identical voyages, at almost the same time—one using forced draught and the other natural, the machinery, steamers and other conditions being practically alike. London *Eng.*, Jan. 4, 1889.

Working of the Howden system on the steamer "New York City." Advantages shown to be: Increased power from smaller boilers combined with large saving of fuel; reduced wear and tear of boilers and fittings; perfect control of combustion; coolness of stoke-hole and absence of smoke. Lon. *Engineer*, July 10, 1881.

See *Boiler Construction*, *Blower*, *Combustion*.

Gas Fired. Gives a description of Frederick Siemens' improvement in generating steam with gaseous fuels. *Sci. Am. Sup.*, March 31, 1888; *Am. Mfr.*, May 11, 1888.

Heat Transmission in. An article by John G. Hudson, M. Inst., C. E., investigating the laws of heat transmission and boiler efficiency. Lon. *Eng.*, Dec. 5, 1890, pp. 449-50, *et seq.* Abstract, *Ry. Rev.*, Jan. 10, 1891, p. 22.

High Pressure. See *Boilers, Marine*

Incrustation in. See *Boiler Incrustation.*

Influence of position on the value of heating surface. Lon. *Eng.*, March 13, 1885.

Lancashire. By L. E. Fletcher. Describes the construction, equipment and setting of the Lancashire boilers, and calls attention to some points which are sometimes overlooked. *Van Nos. Eng. Mag.*, Vol. XV., p. 210.

Low Water in a Steel Boiler. By J. E. Codman. Gives details of a severe test of the material of an internally fired boiler. Drawing shows position of crown sheet before and after the overheating. *Proc. Eng. Club, Phila.*, Vol. VI., p. 21.

As Magazines of Explosive Energy. A paper read before the Am. Soc. of Mech. Engrs. Stored Energy computed, the explosive energy determined for various patterns of boilers, with conclusions as to the best types. By Robt. H. Thurston, in *Jour. Frank. Inst.*, Dec., 1884.

Marine. Application of Zinc in Boilers. A paper by Mr. John Green read before the Inst. of Marine Engrs., discussing the use of zinc to prevent corrosion, and giving several cases from practice, with discussion. *Iron*, March 27, 1891, pp. 267-70. *Am. Eng.*, Apr. 18, 1891, pp. 157-9.

Connected Arc Marine. A demonstration of the principles of their construction. By Charles E. Emery. Illustrated. *Trans. A. S. C. E.*, Vol. VI. (1887), pp. 169-176. Discussion, p. 274.

Construction, Lloyd's Rules as Affecting. By J. T. Milton, Asst. Chief Engineer and Surveyor to Lloyd's Register, M. I. N. A. An exposition of the principles upon which Lloyd's Rules are based, and a defense of the Rules, with discussion concerning both Lloyd's and the Board of Trade Rules. Reprinted from *Trans. I. N. A.* by the Bureau of Navigation, Navy Department. Naval Professional Papers, No. 16.

Construction of Marine Boilers, with a view to the Use of Higher Pressure. A paper by C. B. Casebourne, suggesting constructing marine boilers with two shells; the intermediate space to be charged with steam of a lower temperature than within the inner shell, thus subjecting the inner to only the difference of pressure. *Mech. World*, Jan. 11, 1890, p. 14.

Boilers, continued.

Corrosion and Pitting in Marine Boilers. An article stating results of experience, and giving composition of boiler scale; also suggesting methods of preventing corrosion, etc. *Mechanics*, September, 1889.

Efficiency of. A paper read before Inst. N. A., by J. T. Milton. Lon. *Engineer*, April 10, 1885; *Am. Eng.*, May 1, 1885.

Efficiency of Some Types of. A discussion of this subject by W. M. Chambers. *Eng. News*, Jan. 31, Feb. 7, 1878.

Forced Draught in. See *Boilers, Forced Draught.*

For High Pressure. Description of two new forms of marine boiler, one tubulous and one tubular, the first applied by Yarrow & Co., to a torpedo boat, the second proposed for general marine service. Illus. *R. R. & Eng. Jour.*, March, 1891. *Eng. News*, Feb. 7, 1891, p. 139.

Illustration and description of the boilers of the Cromwell steamer, Louisiana, which had been fitted with brick-enclosed furnaces. By Miers Coryell, Lon. *Eng.*, March 6, 1885.

Kemp's High and Low Temperature Marine. Gives illustrated description of the boiler put in the steel ship "Bleville." Lon. *Eng.*, Jan. 21, 1887.

A lecture delivered recently in both Brooklyn and New York by Mr. J. W. Walters. *Power*, December, 1889, p. 15.

Normal Indicated Horse-Power of Marine Engines and Boilers. Five tables showing the relation of pressure, heating surface, etc. *Mech. World*, June 14, 21, 1890, pp. 235, 243.

Of the S. S "County of York." The two boilers are 12 ft. in diameter by 6½ long, are single ended and each contains three of Fox's corrugated furnaces, 2 ft. 9 in. in diameter internally, 181 tubes 3¼ in. external diameter: heating surface, 2691 sq. ft.; grate area, 93 sq. ft. Detailed drawings showing arrangement of stays, tubes, etc. Lon. *Eng.*, April 1, 1887.

Of S. S. "Earl." Illustrated description of the boiler placed in the tug and salvage steamer "Earl." Lon. *Eng.*, Oct. 14, 1887.

Pressure in. By Richard Sennett, before the Institute of Naval Architects. Discusses working and test pressures for marine boilers. Lon. *Eng.*, March 30, 1888.

Steel. Some astonishing and apparently unaccountable failures of steel marine boilers from cracking into pieces when not in service. Illustrated. Lon. *Eng.*, Dec. 11, 1885.

United States Government Rules for Marine Boiler Pressures. Pressure allowed for various thicknesses and qualities of plates, flues, etc. *Mechanics*, January, 1888.

Water-Tube Steam Boilers for Marine Engines. A paper by John Isaac Thornycroft, read before Inst. of C. E., 19th Nov., 1889. Three folding plates. *Proc. Inst. C. E.*, Vol. XCIX, paper No. 2404, pp. 41-56; appendix with tables, pp. 57-70; discussion of the paper, pp. 71-121; correspondence, including tables, pp. 121-28.

Preservation of. By Rear-Admiral Aynsley. Gives results of experiments on the prevention of deterioration of boilers. *Van Nos. Eng. Mag.*, Vol. XXIII., p. 293.

Purification of Water for. A paper read at the Louisville Convention of the American W. W. Assn., April, 1889, by F. W. Gerické. *Am. Eng.*, April 24, 1889.

Pumps for Boiler Feeding. Illustrated article in *The Locomotive*, Sept., 1889, Vol. X., p. 129.

Purifying Feed Water for Steam Boilers. An article of some value describing and illustrating three different devices. *Am. Eng.*, April 9, 1890, p. 155.

Rating of, by Horse-powers, for Commercial Purposes. A paper before *Am. Soc.*

Boilers, continued.

Mech. Engrs., by W. P. Trowbridge and C. B. Richards. A valuable discussion of the whole question. *Am. Eng.*, Nov. 11, 1885, *et seq.*

Report of Mr. Fletcher, Engineer to the Manchester Steam Users' Association. Treats of the overheating of furnace crowns and other boiler plates when covered with water. *Van Nos. Eng. Mag.*, Vol. I., p. 901.

Setting of. See *Boiler Setting.*

Specification for. See *Boiler Specifications.*

Strength of. Experiments on the Strength of. Results. Paper by J. C. Spence before the N. E. Coast Inst. of Engrs. and Shipbuilders. Experiments on strength as affected by length. Illus. *Mech. World*, May 30, 1891, p. 316, *et seq.*

Strength of Short Boilers. Theoretical discussion as to the effect on strength due to shortness of the boiler. Lon. *Eng.*, April 17, 1891, pp. 468-9.

Temperature of the Fire Surface of Boilers. Summary of a recent paper in *Annales du Conservatoir des Arts et Metiers.* By M. Hersch, in *R. R. Gaz.*, Dec. 13, 1889, Vol. XXI., p. 815.

Tests of. See *Boiler Tests.*

Tubulous. A valuable paper read by Asst. Engr. S. H. Leonard, U. S. A., before Am. Soc. of Naval Engrs. Reprinted from the *Journal* of the Society. Tables and illustrations *R. R. & Eng. Jour.*, July, 1890, *et seq.*, pp. 310-11.

Warming Dwelling Houses. A paper before the Am. Soc. Mech. Engrs. By J. W. Anderson. Illustrated. Gives general requirements and illustrates the design and workings of a new form. *Am. Eng.*, June 5, 1885.

Water Tube. Discussion on their uses and drawbacks. Lon. *Engineer*, Oct. 7 and 28, 1887.

Water Tube. See *Boilers, Marine.*

See *Feed Water. Fire Grates. Heat. Heating. Locomotive Boilers. Stokers. Engines, Steam. Notes.*

Bolts and *Screw Threads.* Experiments to determine the advantage of increasing the number of threads. Results show 10 per cent. increase of static strength and 3 or 4 times strength to resist impact, by doubling the number of threads to the inch. Maj. Wm. R. King, before Am. Inst. Min. Engrs., Chattanooga meeting, 1885. *Eng. News*, Oct. 17, 1885, and *Sci. Am.* of same date.

Bore-Hole Testing Appliances, for giving a complete survey of a bore-hole, so that its exact position at any depth may be accurately determined. Designed by E. F. Macgeorge, of England. A scientific, simple and efficient device. Lon. *Eng.*, March 13, *et seq.*, 1885.

Boring Machines *for Mines.* By M. Alfred Geyler. Extracts from a chapter of the report of U. S. Commissioner to the Paris Exposition. Treats of percussion and rotary drills; also of boring by steam and compressed air. *Van Nos. Eng. Mag.*, Vol. III., p. 10.

By Theo. Allen. An illustrated description of boring machine worked by hydraulic pressure. *Trans. A. S. C. E.*, Vol. II., p. 33.

Brake Beams. *Manner of Flanging.* Discussion at a meeting of the N. W. R. R. Club. *Ry. Rev.*, Oct. 11, 1890, p. 591. Editorial *ibid.*, p. 606.

Pressed Steel Brake Beam and Shoe-head. New form of steel brake beam. Described and illustrated. *R. R. Gaz.*, May 22, 1891, pp. 352-3.

Tests of. Results of tests of brake beams of different patterns and different woods. *R. R. Gaz.*, Feb. 8, 1889; Feb. 22, 1889.

Tests of the Schoen Pressed Steel Brakebeam. Tables giving results of tests of this and other beams, show a much greater strength in the former. *R. R. Gaz.*, June 5, 1891, p. 393.

Trussed. The Strength, Stiffness and Resistance of. By Prof. J. B. Johnson. Formulas, and table of the result of tests *R. R. Gaz.*, May 30, 1890, pp. 373-4.

Brake Beams, continued.

The table of tests is also given in *Mast. Mech.*, June, 1890, p. 94. *Ry. Rev.*, May 31, 1890, p. 310.

Brake Rigging. *Its Care and Operation.* A paper by R. A. Parker, read before the N. Y. Railroad Club. *R. R. Gaz.*, Dec. 26, 1890, pp. 890-2. *Ry. Rev.*, Dec. 27, 1890, pp. 778-9.

M. C. B. Committee Report. Gives complete details of brake gear recommended by the Master Car Builders' Committee. Illustrated. *R. R. Gaz.*, June 28 and July 5, 1889; *Proc. M. C. B. Ass'n.*, Saratoga, N. Y., 1889, p. 109.

See *Car, Ore.*

Brake Shoes. *The Best Metal for Brake Shoes.* A paper before Western Railway Club, by E. C. Case, with an extended discussion by other members. *Mast. Mech.*, December, 1889, Vol. IV., p. 217.

Metals for Brake Shoes. Preliminary report of a sub-committee of the M. C. B. A. Diagrams. *R. R. Gaz.*, May 30, 1890, pp. 374-5. *Eng. News*, May 31, 1890, p. 510. *Ry. Rev.*, May 31, 1890, p. 304.

Report of Committee of M. C. B. Assoc., giving results of some laboratory tests on wear and friction on various metals. Diagrams drawn. *Ry. Rev.*, June 20, 1891, p. 400. *R. R. Gaz.*, June 19, 1891, pp. 429-3. *Eng. News*, June 20, 1891, p. 587.

See *Car Wheels, Steel.*

Brake Shoe Tests. Circular issued by the committee appointed by the Master Car Builders' Association and illustrative of methods used in making friction tests. *R. R. Gaz.*, Jan. 24, 1890, p. 55; *Ry. Rev.*; Jan. 25, 1890, p. 46.

Editorial review of the proposed brake shoe tests by the Master Car Builders' Ass'n., with some data derived from tests already made. *Eng. News*, March 17, 1890, Vol. XXIII., p. 470.

Brake Tests.

Apparatus for. Gives illustration of the apparatus used in the brake tests of the Berlin Elevated Railroad, and of a form of apparatus proposed by Mr. Albert Kopteyn for similar tests. *R. R. Gaz.*, March 4, 1887.

Burlington, Ia. The most elaborate series of tests of freight car brakes ever made. Full description of tests, discussions, and reports and illustrations in *R. R. Gaz.*, July 23, 1886, *et seq.* Also synopsis of data of the above tests in *Ry. Rev.*, Aug. 7, 1886.

A comparison of the Burlington brake tests of 1886 with those of 1887. *Eng. News*, June 25, 1887.

Resume of the results attained in the above tests. Also editorial on same subject. *R. R. Gaz.*, May 20 and 27, 1887; also *Eng. News*, May 21, 1887. A very good analysis of. By H. Hollerith. *Sch. Mines Quart.*, October, 1887, pp. 43-61. Illustrated. An account of. By W. W. Hanscom. *Trans. Tech. Soc. Pac. C.*, Vol. III., p. 73.

Review of the tests at Burlington, Ia., during July and Aug., 1886. By W. F. Hall. *Proc. Eng. Club, Phila.*, Vol. VI., p. 4.

Text of the report of the sub-committee on Electrical Appliances. *R. R. Gaz.*, June 17, 1887; also *Eng. News*, July 9, 1887.

Galton-Westinghouse Experiments. Details of the experiments by Capt. D. Galton and Mr. Westinghouse, to determine the effects of brakes. Lon. *Eng.*, May 2, 1879; also *R. R. Gaz.*, June 10, 1887.

In New South Wales. Report of board of experts giving full details of tests, with table of results. *R. R. Gaz.*, Oct. 16, 1891, pp. 722-4. Extracts from this report, *Eng. News*, Oct. 24, 1891, pp. 376-7.

Park Electric Freight Train. Gives details of the above brake made at Burlington, Ia., Dec. 30, 1886. *R. R. Gaz.*, Jan. 7, 1887.

Brake Tests, continued.

Westinghouse. Gives details of the test made at Weehawken with a train of 50 empty freight cars. *Eng. News*, Nov. 25, 1887; *R. R. Gaz.*, Nov. 25, 1887; *Nat. Car and Loco. Builder* for December, 1887. Gives table of the tests at various places.

In South Australia. Official reports containing results, with diagrams, etc., of the tests made in October, 1890. *Eng. News*, Aug. 1, 1891, pp. 97-9.

New Quick Acting Apparatus. Tests made at Carlsruhe, on the Baden State Railways. Illustrations of new automatic time, pressure, speed and distance recorder used in making the tests. *Sci. Am. Sup.*, May 11, 1889.

Quick-Acting Brakes applied to Freight Trains on the Belgian State Railways. Diagrams of speed, cylinder pressure and valve operation. *Iron*, Oct. 31, 1890, pp. 379-82. Lon. *Eng.*, Oct. 17, pp. 461-62. *Ry. Rev.*, Nov. 28, 1890, pp. 683-4.

Table showing results obtained with an experimental 50-car freight train during October and November, 1887. *R. R. Gaz.*, Nov. 11, 1887.

Brake Valve. *The Eames Pattern*. Acts the same as the Westinghouse, but much simpler. Described and illustrated. *R. R. Gaz.*, Oct. 15, 1886.

See *Brakes for Freight Trains*.

Brakes.

Action of Some Unconsidered Elements in. Analysis of the forces acting, and effect of different methods of suspending the beam. By R. A. Parke. *R. R. Gaz.*, Oct. 24, 1890, pp. 730-1.

Automatic Friction for R. R. Trains. The Heberlein System. Awarded a gold medal at the Inventions Exhibition and adopted largely on the continent. Fully described and illustrated in Lon. *Engineer*, April 2, 1886.

Buffer. A brief article explaining, with formulæ, the nature and action of buffer-brakes. *Mast. Mech.*, April, 1878.

Carpenter Electro Air. A description, with illustrations, showing details of the above brake. The power is compressed air with valves worked by electricity. *R. R. Gaz.*, May 3, 1887.

Classification of Continuous Railroad. By A. W. Metcalf, before the Students Institution of Civil Engineers. Gives a classification of railroad brakes based upon the general principles of action. *Proc. Inst. C. E.*, Vol. XCII., pp. 313-335.

Coefficient of Friction of Railroad. By Capt. D. Galton. Gives details of experiments made on the London, Brighton & South Coast R. R. *Van Nos. Eng. Mag.*, Vol. XIX., p. 519.

Eames' Automatic Vacuum. A very full illustrated description of the above brake, with cuts showing details. *R. R. Gaz.*, May 27, 1887.

Efficiency of. Discussion of methods determining their efficiency, and its ultimate practical limit; with tables to facilitate computations, deduced from experiments on the North British Railway. Editorials in *R. R. Gaz.*, May 15, 22 and 29, 1885.

Electro-Magnetic Railroad Brake. Account of some tests of a new form of electric railway brake recently made in England. *Elec. Rev.*, Dec. 29, 1889, p. 715.

For Freight Trains. Air-Brake Standards and Lease of Air-Brakes on Freight Cars. Report of Committee of M. C. B. Assoc., giving rules for the guidance of employees, with drawings of valves, etc. *Ry. Rev.*, June 13, 1891, pp. 385-7. *R. R. Gaz.*, June 12, 1891, p. 407. *Eng. News*, June 13, 1891, pp. 564-5.

Air Brakes in Freight Service. A paper by Mr. Angus Sinclair before the N. Y. R. R. Club, treats mainly of inspection and maintenance. Discussion. Also a discussion on this subject at a meeting of the Central Railway Club. *R. R. Gaz.*, Apr. 17, 1891, pp. 261-2. *Eng. News*, Apr. 18, 1891, pp. 379-80. Abstracts. *Ry. Rev.*, Apr. 18, 1891. *Mast. Mech.*, May, 1891.

Brakes, continued.

Air Brakes in Freight Service. A paper by G. W. Rhodes before the Western Ry. Club, giving a discussion of the new M. C. B. rules. Abstract. *R. R. Gaz.*, Oct. 10, 1891, pp. 707-11.

Equipment of Freight Cars with the Westinghouse Automatic Brake. Map showing progress of this improvement, with statistics from the various railroads. *Eng. News*, May 2, 1891, pp. 430-2. Editorial discussion, pp. 425-6.

Paper by H. H. Westinghouse before the New York Railroad Club. Describes the construction, operation and maintenance of brakes. With discussion by the Club. *Mast. Mech.*, February, 1888; *R. R. Gaz.*, Jan. 27, 1888.

Paper by Mr. Lauder, and discussion that followed it at the December meeting of the New England R. R. Club. *R. R. Gaz.*, Dec. 23, 1887, also *Mast. Mech.*, January, 1888.

Power of. By Wm. P. Shinn. A valuable contribution to the desirableness and efficiency of such appliances. Has especial reference to the Westinghouse and American systems. Includes many experiments. Illustrated. *Trans. A. S. C. E.*, Vol. XIV., p. 405.

Friction Brake of the Mount Pilatus Railroad. The automatic brake in use on this very steep rack railroad is described and illustrated in *R. R. Gaz.*, Feb. 13, 1891, p. 110.

Friction, some new forms of, from a new French work on steam engines. Several different forms given with cuts, in *Mechanics*, New York, March 1885.

Manomatik. Gives a description of the Manomatik lever momentum brake which is operated by power transmitted from the drawheads through buffer springs. Illustrated. *R. R. Gaz.*, March 23, 1888.

Paper by Mr. James Howard, with discussion, at meeting of New York R. R. Club. Giving mechanical principles which control the construction of main brakes, with special reference to locomotive brakes. *R. R. Gaz.*, Nov. 23, and *Mast. Mech.*, December, 1888.

Recent Improvements in. A paper by H. H. Westinghouse, read before the Am. Ry. Sup. Assoc., Oct., 1890. *R. R. Gaz.*, Oct. 10, 1890, pp. 615-720. *Ry. Rev.*, Oct. 11, 1890, pp. 607-9. *Ry. Age*, Oct. 11, 1890, p. 716-18.

Suggestions of Radical Changes in Automatic Brakes, especially for freight trains. The main feature of improvement suggested is that the power of the brake should increase with the *load* in the car. By A. K. Mansfield. *R. R. & Eng. Jour.*, January, 1888.

Tests of. See *Brake Tests.*

Thermo-Dynamics of the Automatic Vacuum Brake. A paper by Leonard Archbutt, F. C. S., and R. M. Deeley, M. I. M. E. Illustrated. Lon. *Engineer*, June 27, 1890, *et seq.*, pp. 509-11, 21-2.

Vacuum automatic, for railroad trains. Fully described and illustrated. *Iron*; also *Sci. Am. Sup.*, Jan. 9, 1886.

See *Car Couplers. Standard, etc.*

Breakwaters. See *Harbor Improvement Breakwater. Jetties. Lake Currents.*

Brick Industry, About New York. By Calvin Tomkins. Gives a good review of the condition of the brick industry about New York. *Trans. A. S. C. E.* Abstracted in *Eng. News*, June 11, 1887.

Brick-Making *in England.* A paper by Henry Ward, with discussion, giving methods used in the plastic, semi-plastic, and semi-dry processes, by hand and machinery, with various kinds of clays. Cost, strength, etc., given. The whole occupying 58 pp., with many cuts, in *Trans. Inst. C. E.*, Vol. LXXXVI., p. 1.

Brickmaking Machinery. A paper by M. Powis Bale, M. Inst. C. E., giving an account of machinery and methods used in brick manufacture. *Sci. Am. Sup.*, Feb. 8, 1890, p. 11 840.

Brick Pavements. See *Pavements.*

Brick Walls, *The Efflorescence on.* Its causes and remedies discussed by C. J. Anderson, before Ill. Assoc. of Architects. *Sanitary News*, Dec. 12, 188[illegible], and *Am. Eng.*, Dec. 24, 188[illegible].

Brick Masonry.

Crushing Strength of. Experiments on. Abstracts of Papers, 188[illegible]. *Proc., Inst. C. E.* Also *Iron Age*, Nov. 26, 188[illegible].

Experiments on, at Alwa. Experiments made by H. Leonard, M. I. C. E., with brick piers having different forms of footings to test the strength of foundations for heavy buildings. *Ind. Eng.*, Dec. 15 and 22, 188[illegible].

Permissible Stress upon. From the German. Various qualities of brick and mortar. Table given. *Van Nos. Eng. Mag.*, March, 1886.

Vs. Stone Masonry. A Paper by Prof. I. O. Baker, entitled "A Plea for Brick Masonry in Engineering." In *An. Rep. Ill. Soc. of Engs. & Surveyors*, 18[illegible].

Strength of. A good article, containing formulæ deduced from many experiments. *Van Nos. Eng. Mag.*, Vol. VII., p. 481.

Working loads on, as determined by experiments in Berlin. Abstracted in *Proc. Inst. C. E.*, Vol. LXXIX., p. 376.

Bricks *and Brickwork.* By Prof. T. R. Smith. Gives a good historic and technical review of the subject. *Sci. Am. Sup.*, July 9, 188[illegible].

Laws of Crushing Strength of. By Prof. Ira O. Baker. Experiments made to determine effect upon the apparent strength of a variation in the size of piece tested. *Eng. News.*, Feb. 2, 18[illegible].

Paving, Experiments on. A paper by J. F. Fisher giving the results of laboratory experiments. *Selected Papers, C. E. Club, Univ. of Ill.*, 18[illegible], pp. [illegible].

Record of some tests made by Mr. E. Butts, City Engr., Kansas City, on crushing strength, absorption and abrasion. Results plotted. *Eng. News*, Apr. 11, 1891; p. 352. *Eng. Record*, May 2, 1891, p. 367.

Paving Bricks and Brick Pavements. A valuable paper by Chas. P. Chase, C. E., describing the manufacture and use of paving bricks. Illus. 7 tables. *Eng. News*, July 19 and 26, 1890, pp. 55, 70-3.

Permeability of, Experiments on. A paper by A. W. Hale, Asst. Engr. New Croton Aqueduct, giving methods of testing and results. Illus. *Eng. News*, Jan. 3, 10, 1891.

Slag Brick Manufacturers in Montana. Description and illustration of simple apparatus. By Thos. Egleston. *Sch. Mines Quart.*, Apr. 1891, Vol. XII., pp. 189-93. *Eng. Record*, June 13, 1891, pp. [illegible]. Brief notice in *Eng. News* June 20, 1891, p. 601.

Tests of, for Hoosac Tunnel, as to absorption and porosity. *Jour. Assn. Eng. Soc.*, Vol. I., p. 143.

For Vosburg Tunnel. Tests for crushing and cross-breaking, as well as for absorption, of eleven different makes, and over 350 samples. *R. R. Gaz.*, Oct. 14, 1886.

Resistance to a Crushing Force. By Geo. S. Greene, Jr. Gives details of the crushing of six bricks. The force required for crushing varied from 1,734 lbs. to 3,322 lbs. per sq. in. *Trans. A. S. C. E.*, Vol. II., p. 185.

Summary of Results and Conclusions on a Series of Tests on brick for absorption, compression and cross breaking strength; effects of plaster beds and of testing lengthwise, edgewise and flatwise. The most valuable generalizations yet given. By Prof. I. O. Baker. *Clay Worker*, (Indianapolis) March, 18[illegible].

Tabulated results of tests of several kinds of brick. *Eng. News*, June 21, 1890, p. 174.

See *Fire Brick.*

Bridge Accidents.

By G. H. Thomson, before the 52th meeting of the British Association. Discusses bridge failures and their causes, and details of experiments made on various types of bridges. *R. R. Gaz.*, Sept. 28, 1883. Editorial discussion. *Eng. & Build. Rec.*, Dec. 1, 1883.

Ashtabula Disaster. A careful description of the structure, and an account of the probable method of failure. By Charles Macdonald. Illustrated. Discussion. *Trans. A. S. C. E.*, Vol. VI., 1877, p. 74-87. Discussion continued on pp. 193-222, also p. 231-234.

Bussey. See *Bridges, Bussey*.

Fall of a Suspension. Gives details of the failure of the suspension over the Ostrawitza at Ostraw. Illustrated. *San. Eng.*, Sept. 24, 1887.

Fall of a Bridge Over the Big Otter River. An illustrated description of a bridge that gave way under a loaded coal train, killing seven men. Plan and elevation shown. *R. R. Gaz.*, April 8, 1887.

Flat Creek Trestle. Illustration and description of the scene of this disaster, and discussion of causes, etc. *Eng. News*, August 31, 1889, Vol. XXII., p. 198.

Forest Hills Disaster. A full discussion of the Forest Hills disaster from an engineering point of view, with plan and elevation of the truss of the bridge and view of the wreck. A full-page view of the structure before it fell. *Eng. News*, March 19, 1887; see also *R. R. Gaz.*, March 18, 1887.

Munchenstein Bridge Disaster. Full account of this disaster with illustrations of various details of the bridge. *Eng. News*, Aug. 1., 1891, pp. 90-94. *R. R. Gaz.*, July 17, 1891, pp. 498-9. Brief account, *Eng. Record*, July 25, 1891, p. 119.

On the Means of Averting. Report of a committee of the Am. Soc., consisting of James B. Eads, C. Shaler Smith, Thos. C. Clarke, J. W. Adams, A. P. Boller, Chas. Macdonald, and Theo. G. Ellis. The Committee brought in four reports. *Trans. A. S. C. E.*, Vol. IV., (1875), p. 122-135. Discussion, p. 218-221. *Van Nos. Eng. Mag.*, Vol. XIII., p. 305.

Tay Bridge Disaster. Many articles on this subject in *Eng. News*, 1880, *i. e.*: Description of bridge, Jan. 10; statement of main facts, Jan. 17; lengthy discussion as to cause, from Lon. *Eng.* Illus., Jan. 24; details of piers, Jan. 31; wind-resisting surfaces, Feb. 14, 21; illustrations of the wrecked piers, Feb. 28; abstract from report of Board of Inquiry, July 31.

Bridge Designing, and Economy.

American vs. English. As coming from Prof. Kernot, an English engineer in Australia, through Prof. Waddell, an American engineer in Japan. The comparison very favorable to American structures. *Eng. News*, March 27, 1886.

Arching Wooden Howe Trusses to increase their strength. By Gustav Lindenthal. Illustrated. *Jour. Assn. Eng. Soc.*, Vol. I. p. 83.

Braces. By P. H. Philbrick. Gives practical formulæ for angle blocks and braces of bridges. *Van Nos. Eng. Mag.*, Vol. XXIV., p. 377.

Brazilian-British-American Bridge. Designed by an English engineer after an American type. Contains several peculiar details. Illustrated description, also editorial discussion of the design. *Eng. News*, July 25, 1891, pp. 79, 80, 83-4

Details of Chicago, Kansas City & Texas R. R. Bridge. The connections are designed with especial care in reference to avoiding difficult shop work, commanding a minimum pound price. Inset showing details. *Eng. News*, Jan. 24, 1891, p. 87.

Details in Ordinary Iron. By J. A. L. Waddell. Illustrated. *Jour. Assn. Eng. Soc.*, Vol II., p. 39.

Construction. Discusses the proper inclination of tie and strut. *Van Nos. Eng. Mag.*, Vol. I., p. 230.

Economical Height of Trusses for a Given Panel Width. By John Lundie. *Jour. Assn. Eng. Soc.*, Vol. VII., pp. 101-103 (March, 1888).

Bridge Designing and Economy, continued.

Economy of Bridge Building. By A. D. Ottewell. Endeavors to show that to secure true economy in bridge building, beside providing for future increase in rolling load, we must adopt a new method of dimensioning. *R. R. Gaz.*, March 25, 1885.

Economy in Bridging Rivers. An article by Alfred D. Ottewell, in which he deduces a formula showing the number of spans giving great economy for any case. *Eng. News*, Dec. 14, 1889. Vol. XXII.

French Pier Masonry and Bridge Detail. Details of piers, and arch decorations of the bridge over the Rhone at Lyons. From *Le Genie Civil*. *Eng. News*, Sept. 5, 1891. pp. 220-1.

General Design of Iron Bridges of Very Large Span for Railway Traffic. Paper by T. C. Clarke before the Inst. C. E. Treats mainly of American practice. Many details illustrated, also experiments on wrought iron columns. *Eng. News*, Jan. 15, Feb. 12, 1881.

Wrought Iron Girder Work. Paper by C. G. Smith before the Liverpool Engr. Soc., describing method of designing and discussing many features. *Eng. News*, Sept. 29, Oct. 6, 1877; also Nov. 17, Dec. 22, 1877.

Lateral Systems for Iron Highways. By J. A. L. Waddell. Illustrated. *Jour. Assn. Eng. Soc.*, Vol. III., p. 17.

Notes on the most economical depth of straight girders and trusses. By Emil Adler. *Van Nos. Eng. Mag.*, Vol. XVI., p. 61.

Paper by H. J. Lewis before the Eng. Soc. West. Penn., discussing briefly the designing of the various parts. With discussion. *Eng. News*, Oct. 17, 1891. pp. 367-9.

Pins. By Chas. Bender. A valuable paper. Gives a mathematical deduction of the proportions of pins *Van Nos. Eng. Mag.*, Vol. IX., p. 289. A criticism on the above article by Prof. De Volson Wood. *Van. Nos. Eng. Mag.*, Vol. IX., p. 504.

Pins and Eye Bars, Proportion of. By C. F. Stowell. Discusses the present state of pin calculation and gives formula for computing the stress in the side of the head of eye-bars. *Eng. News*, March 31, 1888.

Pins, Bending Moments on, by Graphics. By Chas. E. Greene. Illustrated. pp. 3. *The Technic*, University of Michigan, No. 2, New Series, p. 36. 1889.

Relative Quantities of Material in, of different kinds and heights. By Charles E. Emery. Two papers and discussion. Plates. *Trans. A. S. C. E.*, Vol. VI, pp. 134-49, also pp. 277-88, and Vol. VII., p. 192.

Rivets Bearing Value of. A series of three graphical diagrams or tables giving bearing value of rivets or pins for 15,000 lbs. per sq. in. and 12,000 lbs. sq. in. for diameters from ½ in. to 6-ins. and thickness of plates from ⅛ in. to 1½ in. See article "Safe Building." *Am. Arch.* Feb. 22, 1890, p. 121.

See *Bridge Stresses, Maximum, etc. Bridge Members.*

Bridge Details, See *Bridge Designing*.

Bridge Erection.

Appomattox, High. Illustrated description of the method of replacing 21 spans of 112 feet each, of Fink truss by Pratt truss, without stopping traffic. *Eng. News*, Oct. 22, 1887.

At Binghamton, N. Y. An account of the manner of erecting a plate-girder bridge of 85-foot span without interruption of the street traffic. *Eng. & Build. Rec.*, March 8, 1890. Vol. XXI., p. 213.

Erecting the St. Paul High Bridge. An interesting illustrated description of the traveler used in erecting this high bridge. *Eng. & Build. Rec.*, Dec. 21, 1889, p. 38. *Eng. & Build. Rec.*, Dec. 28, 1889, p. 58.

Erection of a Bridge over the Danube. By W. Howard White. Describes the manner in which four spans of lattice truss, each 223 feet long, were built upon

Bridge Erection, continued.

shore and shoved out over the piers into their proper position. *Trans. A. S. C. E.*, Vol. II., p. 291.

Erection of Iron Bridges. Cleveland & Mahoning Valley Railway, second track. Paper by James Ritchie. Mem. Civ. Eng. Club of Cleveland. Read May 14, 1889. *Jour. Assn. Eng. Soc.*, June 1889. Vol. VIII., p. 339.

Niagara Cantilever. Traveler used on. Illustrated description. *Eng. News*, March 1, 1884, pp. 98-9.

Plant. Details of the traveler used for erecting columns and girders of the Kings Country Elevated Railroad are given in *Sci. Eng.*, Jan. 4, 1887.

Traveler, A Viaduct Erecting. Illustrated description of traveler designed by the King Bridge Co., for erecting a viaduct mainly of plate and lattice girders up to about 80 ft. span. *Eng. & Build. Rec.*, Dec. 23, 1888.

Temporary Bridge Support. A paper read by M. J. Becker before the Eng. Soc. of Western Penn., May 25, 1890. The paper describes the renewal of the Ohio River bridge at Steubenville. Discussion. *Proc.* of the Soc. Reprinted with illustrations in *Eng. News*, July 12, 1890, p. 32.

Torkham, India. Describes the method employed to launch three short spans of riveted girder of Torkham bridge. Illustrated. *Ind. Eng.*, Oct. 1, 1887; *Eng. News*, Nov. 19, 1887; Lon. *Eng.*, Jan. 13, 1888.

See *Bridges, Cantilever, Forth. Hydraulic Appliances.*

Bridge Floors.

Comparative Weight and Cost. Tables showing weight and cost of various types of bridge floors. By E. Olander. *Mech. World*, April 2, 1887.

Concrete Floors. By R. K. Field. Describes the usual forms in which such bridges are built. *Report Conn. Assoc. C. E.*, 1887, p. 15.

Design, Strength and Cost. By Edmund Olander, before the Society of Engineers. Gives a comparison of weight, strength and cost of various designs of bridge floors. Four plates. *Trans. Soc. Engrs.*, 1886, pp. 27-67.

For Railroad Bridges. By W. H. White. A model system described and illustrated, having rerailing arrangement, guard rails, etc. *Trans. A. S. C. E.*, Vol. XII., (1883), p. 431.

Of Buckle Plate. Description of a very shallow but substantial floor used on the C. M. & N. R. R. Illustrated. *Ry. Rev.*, Aug. 30, 1890, pp. 508-9.

Of Old Rails. Description of viaduct constructed from old rails. *R. R. & Eng. Jour.*, October, 1889, Vol. LXIII., p. 452.

Old Rails. Illustrated description of the use of old rails for floors, as a relief to failing arches of short spans. Used on Indian railways. *R. R. Gaz.*, March 10, 1891, p. 155.

Of Old Steel Rails. Economically used on the N. Y. C. & H. R. R. R. *R. R. Gaz.*, Oct. 8, 1890, pp. 678-9.

Railroad Bed for Bridge Structures. Abstract of a paper by O. C. Woolson, before the Philadelphia meeting of the American Society of Mechanical Engineers describing an elastic floor system which has been tried on the elevated roads of New York. Illustrated. *Trans. A. S. M. E.*, Vol. IX., (1888), pp. 276-285; *R. R. Gaz.*, Dec. 30, 1887.

Solid. Editorial discussions, on. *R. R. Gaz.*, Oct. 31, 1890, p. 746.

Solid Bridge Floors on the N. Y. Central & Hudson River R. R. Brief illustrated article, showing method of treating comparatively small openings, giving an unbroken roadbed. *Eng. News*, Nov. 16, 1889. Vol. XXII., p. 465.

Solid Bridge Floor, Oriskany Bridge. New York Central & Hudson R. R. R. An inset plate illustrating an excellent example of the solid bridge floor system as applied to riveted bridge of 60 feet span. *Eng. News*, Dec. 7, 1889. Vol. XXII., p. 537.

Bridge Floors, continued.

Street Bridges. By Carl Gayler, before the St. Louis Engineers' Club. Discusses the different kinds of floors in use and gives cost of the different classes used in St. Louis. *Jour. Ass. Eng. Soc.*, May, 1888: *Eng. & Build. Rec.*, June 30, 1888.

See *Bridges, Cantilever, Forth. Bridges, Arches, Illinois and St. Louis.*

Bridge Guards.

Guard Rails on Bridges. A circular issued by the Massachusetts Board of Railroad Commissioners to all of the railroads in that state, recommending a certain form of guard rail on bridges. *R. R. Gaz.*, Dec. 30; *Eng. News*, Dec. 31; *Eng. & Build. Rec.*, Dec. 31.

An illustrated description of the *Latimer Improved Re-railing* Bridge Safety-Guard. *Eng. News*, Feb. 19, 1887.

Location of on Bridge Structures. A discussion of the necessary requirements. By Archibald A. Schenck. Illustrated. *R. R. Gaz.*, Feb. 6, 1891, pp. 96-7.

Plain Tires Letter calling attention to the action of broad plain tires on guard rails, and editorial comment on same. Illustrated. *R. R. Gaz.*, Aug. 23, 1889, Vol. XXI., p. 549.

Standard Bridge Guard on Savannah, Florida & Western Ry. Illustrated description of a good form of re-railing bridge guard. Plant system. *R. R. Gaz.*, Jan. 17, 1890, p. 38. Experiments on. *Ibid.*, Jan. 31, 1890, p. 69.

Used on the State Railways of Holland. Illustrations of re-railing device, taken from the *Organ des Eisenbahnwesens*. *Eng. News*, March 7, 1891, p. 224.

See *Bridge Floors*.

Bridge History.

American Railroad Bridges. An extensive paper treating this subject historically and in detail by Mr. Theodore Cooper, M. Am. Soc. C. E. Fully illustrated by plates and diagrams. Pp. 60: plates, 27. *Trans. A. S. C. E.*, July, 1889. Vol. XXI., p. 1.

Complete historical treatment of the development of bridge building, by Joseph M. Wilson, retiring President of Engineers' Club of Philadelphia. *Proc. Eng. Club, Phila.*, Vol. VII. No. 1, Feb., 1889, p. 63.

Details of Whipple Truss built in 1852. *Eng. News*, April 7, 1883, pp. 160-1.

Development of the Iron Bridge. Notes by Squire Whipple, "the father of iron bridge building in America," with sketches. *R. R. Gaz.*, April 19, 1889.

Evolution of the Modern Bridge. A popular article by Prof. Charles D. Jameson, giving the various steps in the development of modern bridges of various types. Illustrated by numerous diagrams. *Pop. Sci. Monthly*, Feb. 1890. pp. 461-481.

Illustrated, Historical Description of all Expired Patents, which under the law are now Public Property. By F. B. Brock. A series of articles of historical value. *Eng. News*, Oct. 28, 1882, July 21, 1883.

Review of Bridge Building. By Prof. W. P. Trowbridge. Of the development of bridge construction, with notices of some remarkable historic bridges. *Sci. Am. Sup.*, March 17, 1887.

Sketch of a bridge constructed by an uneducated Mexican laborer from his own design. *Trans. A. S. C. E.*, May, 1886, also *Sci. Am. Sup.*, July 17, 1886.

See *Bridges, Cantilever, History*.

Bridge, Inspection.

And Maintenance. A discussion at the Annual Convention of the American Society of Civil Engineers on the inspection and maintenance of railway structures. The discussion, by many prominent engineers, covers 30 pages in the *Trans. A. S. C. E.*, Vol. XVII., Dec., 1887.

And Inspectors. By S. T. Wagner, before the Annual Convention of the American Society of Civil Engineers. Discusses the characteristics and work of

Bridge Inspection, continued.

bridge inspectors and makes suggestions for their work. Discussion. *Trans. A. S. C. E.*, Vol. XVII., December, 1887. pp. 319-329.

Complete set of rules for inspection of bridges, by Col. H. S. Haines, General Manager of Plant system. These rules present the method of conducting inspections in a concise manner. *R. R. Gaz.*, May 2, 1890, Vol. XXII. p. 303.

Iron Bridges and Viaducts. Paper by B. L. Marsteller, read before the Engineers' Club of Kansas City, Feb., 16, 1889, with discussion. *Jour. Assn. Eng. Soc.*, January, 1889.

On the Erie R. R. Gives description of the blank employed and the general orders for the inspection of bridges. *R. R. Gaz.*, July 29, 1887.

Orders to bridge inspectors in use on the Buffalo, Rochester & Pittsburgh Railroad. *Eng. News*, May 11, 1888.

Some Points on Bridge Inspection. A Paper by Henry Goldmark, read before Engineers' Club of Kansas City. *Jour. Assn. Eng. Soc.*, February, 1890. Vol. IX., pp. 62-65.

Bridge Loads and Weights.

Dead Loads of Iron Highway. By J. A. L. Waddell, accompanied by an extended table. *Jour. Assn. Eng. Soc.*, Vol. II., p. 277.

Comparative weights of different designs and for different loads. Both English and American practice discussed. *R. R. Gaz.*, Dec. 3, 1886.

Diagrams for Determining Weights, used on Austrian Railroads. Curves given for various types of bridges. *R. R. Gaz.*, July 17, 1885.

Formulas for the Weight of. By A. J. DuBois. Presents a series of rational formulas, which accommodate any specifications and design, and gives accurately the weight of iron for any given style. Followed by 38 pages of discussion by prominent bridge engineers. *Trans. A. S. C. E.*, Vol. XVI., p. 191, *et seq.*

Formulas for the weight of iron and steel railway bridges under standard specifications. By Geo. H. Pegram. Five classes of loading discussed, for which constants are evaluated. Results tabulated. Discussion by Whittemore, Wilson, Hutton, Hughes, Cooper, Macdonald, DuBois, Elliot and Burr. *Trans. A. S. C. E.*, Vol. XV., p. 85.

Live Loads and Weights. Diagram giving total weight of iron for single spans for various rolling loads, and spans up to 210 ft., with accompanying formula, and formula for longer spans. The diagram gives also the equivalent uniform load per foot for chord stresses. *Eng. News*, Jan. 10, 1891, pp. 28-9. Editorial discussion on rolling loads, pp. 37-8.

Weights of Iron Railroad. A series of curves showing weights of modern iron bridges for various types of locomotives, mostly taken from Pegram's Paper before the American Society of Civil Engineers. *R. R. Gaz.*, Sept. 17, 1886.

See *Bridges, Railroad. Bridges, Girders.*

Bridge Members.

Austrian Tests of Mild Steel for Bridge Building. A lengthy report of a committee of the Austrian Eng. and Arch. Soc., on the qualities of wrought iron and soft steel, and their relative value for bridge building. The report contains the results of many tests, several being on full size girders. These latter are briefly described and illus. in *R. R. Gaz.*, Nov. 27, 1891, pp. 834-5. For full report see the *Zeitschrift* of the above Society.

Compression Members.

Full-sized members tested to failure at Keystone Bridge Co.'s Works. Especially valuable as showing failure of columns at the ends due to cutting away the flanges of the channels. Illustrated. *Jour. Assn. Eng. Soc.*, Vol. I., p. 117.

Full-sized Phœnix Columns tested to failure at the Watertown Arsenal by the Government. Full account of the tests, with discussion by many engineers as to correct formula. Illustrated. The whole filling 110 pp. in *Trans. A. S. C. E.*, Vol. XI., (1882), p. 1.

Bridge Members, *Compression Members*, continued.

Strength of. By J. G. Dagron before the annual convention of the American Society of Civil Engineers, 1887. Gives the results of the testing of eight full-size compression members made of the steel used in the B. & O. bridge over the Susquehanna River. *R. R. Gaz.*, Jan. 6, 1888.

Eye Bars.

Manufacture of Eye Bars for Pin Connected Structures. A valuable paper by William Sellers, describing process of manufacture at Edge Moor Bridge Works, for both iron and steel eye bars. Tables of dimensions and strength are given. *Proc. Eng. Club, Phila.*, June, 1886, Vol. VI., pp. 3-6.

Eye Bars, Manufacture of. Kloman's Improved Process and Machinery for. Illustrated description. The ends of the bar are left thicker than the body. From *Iron Age*, *Eng. News*, May 17, 1879. Also paper on same, by C. Macdonald before the *A. I. M. E.*, discussing also the various methods of manufacture and their defects. Gives etchings of upset bars. *Eng. News*, May 24, 1879, pp. 163-4.

Proportions of Heads and Pins. By C. Shaler Smith. Determinations from experiments. Illustrated. *Trans. A. S. C. E.*, Vol. VI. (1877), pp. 263-267.

Stresses in Heads of. By Wm. H. Burr. An elaborate theoretical investigation, but rather too highly mathematical to prove of much practical value. *Trans. A. S. C. E.*, Vol. VI., (1877), p. 137, *et seq.* Discussion, Vol. VII., p. 189.

Steel Eye-Bars. Tensile Strength of Full-sized Members, tested by the Watertown machine, for the Keystone Bridge Co., together with some notes on steel in construction. By Gus. C. Henning. *Trans. A. S. C. E.*, Vol. IV., p. 410.

Wrought Iron and Steel. Paper by Carl Gayler, read before the Engineers Club of St. Louis, Jan. 16, 1889. *Jour. Assn. Eng. Soc.*, April, 1889.

Floor-beam. Test of a. By A. P. Boller, before the American Society of Civil Engineers. Gives details of the testing of a full-sized wrought-iron double track floor beam. Three plates. Discussion. *Trans. A. S. C. E.*, Vol. XVIII., May, 1888, pp. 119-130; Lon. *Engineer*, April 27, 1888; *Sci. Am. Sup.*, June 2, 1888.

Pine Stringers and Floor-Beams for Bridges. A paper by Onward Bates, giving tables of tests on the strength of pine timber, also drawings of standard pile and trestle bridges, C. M. & St. P. Ry. *Trans. A. S. C. E.*, Vol. XXIII., 1890, paper No. 455, pp. 261-76.

Tests of. By J. D. Steel. Gives results obtained by testing full-sized bridge members. Discussed. *Trans. A. S. C. E.*, Vol. II., p. 223.

See *Columns, Experiments. Bridge Designing.*

Bridge Piers.

Barrel Pier. Detailed drawing and description of. Used by the C. B. & Q. R. R. in renewing pile piers and wooden bridges. *R. R. Gaz.*, Feb. 11, 1887.

Backwater from. See *Backwater.*

Construction of Pier 4 of the Kansas City. Gives details of the method employed and the obstacles met with. *Van Nos. Eng. Mag.*, Vol. III., p. 404.

Concrete Piers, Red River. By C. D. Purdon. Gives details of the construction of concrete piers for the St. Louis & San Francisco R. R. bridge over Red River, Texas. *Eng. News*, June 2, 1888.

Notes on River Crossings and Piers. Papers by M. Oingawa, Tokio, Japan. Discusses the subject very thoroughly, taking especially the considerations of dimensions of waterway, effect of obstructions to discharge, height of eddy caused by piers, form and thickness of piers, and gives some examples from Japanese practice. *R. R. & Eng. Jour.*, Nov. and Dec., 1888, and Jan., 1889.

As obstructions to River Discharge. Paper by Q. A. Gilmore describing a

Bridge Piers. continued.

special case and investigating this subject. Abstract with discussion by J. W. Adams. *Eng. News*, June 10, 1882, pp. 190-3.

Piers of the Soulevvre Viaduct, France. A common parabola with horizontal axis was taken as the generating curve of each face. An illustrated description abstracted from *Le Genie Civil*. *Eng. News*, June 28, 1890, p. 606.

Repairing a Bridge Pier's Foundation. A paper by Conway B. Hunt, describing the repairing of a pier of the Aqueduct Bridge over the Potomac River. Concrete was deposited in bags. *Proc. Eng. Club Phila.*, Dec., 1889, Vol. VII., pp. 239-41. Abstracted in *Eng. Rec.*, Dec. 27, 1890, pp. 60-1.

Repairing of the "Pont-Neuf," Paris. Part of one pier of this arch bridge was replaced, and other pier foundations repaired. Described by M. Guiard. Three folding plates. *Annales des P. & C.*, June, 1891, pp. 185-98.

Repairing under Water. An article by M. Rossignol, translated from *Les Annales des Ponts et Chaussees*. Illustrated. *R. R. Eng. Jour.*, Sept., 1890, pp. 393-5.

Stability of. Investigation of stability of masonry piers. From Lon. *Eng.*, *Eng. News*, Feb. 14, 1880, pp. 59-60.

Underpinning. By A. P. Boller. Describes methods used for supporting feet of iron piers while masonry was renewed. Illustrated. *Trans. A. S. C. E.*, Vol. XI. (1882), p. 130.

See *Bridges*.

Bridge Pins. See *Bridge Designing*.

Bridge Rollers. *Strength of Solid Metallic to Resist Crushing.* By J. B. Johnson. A new formula for computing the strength, with demonstration. *Jour. Assn. Eng. Soc.*, Vol. IV., p. 110.

Bridge Shops. See *Bridge Works*.

Bridge Specifications.

German Specifications for Iron Construction in Bridges and Buildings. Gives specifications adopted by the German Architects and Engineering Societies in co-operation with the German Society of Engineers and the Society of German Iron Manufacturers, 1885. *Eng. News*, Aug. 6, 1887.

General Specifications for Highway Bridges. By J. A. L. Waddell, of Kansas City, Mo. In addition to general specifications and specifications for iron by bridge companies, there are articles given on Highway Bridge failures, Bridge Lettings, How Bridges are, and how they should be, built, etc. *Gen'l Spec. for Highway Bridges*, J. A. L. Waddell, Kansas City, June, 1889.

Iron Bridges. By I. O. Baker. Gives specifications relating to ultimate strength, elongation and fractured area. *Rpt. Ill. Soc. Engrs. & Surv.* 1888, pp. 51-57.

Iron Bridges. By Joseph M. Wilson, with a discussion by Messrs. Cooper, Wright, Bouscaren, Burr, Robinson, Vose, Davis, Swain, Moulton, Boller, Merriman, Dagron, Sellers, Thatcher, Pegram, Schneider and Clarke, the whole covering over 100 pages. This is the latest, fullest and best source of information on the subject in print. *Trans. A. S. C. E.*, June, 1886, Vol. XV, p. 389. See also Dimensions of Iron Structures by Prof. Krohn, abstracted in *Proc. Inst. C. E.*, Vol. LXXXIV., p. 461. By J. A. L. Waddell. A criticism on the specification of Mr. Joseph M. Wilson. Also advocates the use of general specifications for ordinary cases and special specifications of extraordinary cases. *Trans. A. S. C. E.*, Vol. XVI., p. 13. By J. M. Wilson. A reply to the above article. *Trans. A. S. C. E.*, Vol. XVI., p. 35.

Iron bridges in use on Erie Railroad. *Van Nos. Eng. Mag.*, Vol. IX., p. 378.

Railway. A paper by Edwin Thatcher before the Engrs. Soc. of West Penn. Treats of the general specification of the Keystone Bridge Company. *Am. Eng.*, April 22, 1887.

Bridge Specifications, continued.

St. Lawrence Bridge near Montreal. Total length, 1¾ miles. Published in full in the *Eng. News*, Oct. 10, 1885.

Summary of the regulations drawn up by the Saxony Association of Engineers and Arch'ts, relating to the construction of iron bridges. *Van Nos. Eng. Mag.*, Nov., 1884.

Youghiogheny Bridge, McKeesport, Pa. Complete specifications. *Eng. News*, Feb. 18, 1882, pp. 51-55, Feb. 25, pp. 59-60.

Bridge Stresses.

By Wm. H. Booth, M. Am. Soc. C. E. *Trans. A. S. C. E.*, April, 1889. Vol. XX., p. 127.

By Wm. Cain. Gives notes on decomposition of resultant on a cross section, strength of long posts and safe strains for different members of a bridge. *Van Nos. Eng. Mag.*, Vol. XVII., p. 459.

And Deflection. Application of the Principle of Virtual Velocities to the Determination of the Deflection and Stresses of Frames. A paper by Prof. George F. Swain, of Mass. Inst. of Technology, Boston. *Jour. Frank. Inst.*, Feb., 1883, Vol. CXV., p. 102, *et seq.*

And Deflection. Deflection of Framed Structures and Distribution of Stresses over Redundant Members. By Prof. J. B. Johnson. *Jour. Assn. Eng. Soc.*, May, 1890, pp. 240-251.

Arch, Calculation of Strains in. By Charles Pfeifer. Treats of the resistances of abutments and deflections of arch produced by movable loads. *Van Nos. Eng. Mag.*, Vol. XIV., p. 481.

Arches, Maximum Strains in. By E. A. Werner. *Van Nos. Eng. Mag.*, Oct., 1884.

Arch Truss. A valuable paper by Mr. F. Gilman, giving formulas and tables of strains for trusses where the upper chord is any curve. *Eng. News*, Aug. 16, 1890, pp. 142-3.

Computations. Two communications of value by Bridge Engineers Geo. H. Pegram and J. P. Snow, and also an editorial on more simple methods of computing stresses in bridge members. Both favor a uniform load with a single concentrated excess. Examples given. Illustrated. *R. R. Gaz.*, Dec. 10, 1887.

Experiments Upon an Iron Bridge of 35 Metres Span. Paper by M. Cuenot. The bridge, a riveted Warren girder, was subjected to various loadings, both stationary and moving, and the resulting stresses obtained by measurement of elongations and compressions of the various members. The stresses due to a moving load were about 25 per cent. greater than those due to the same stationary load. *Annales des P. & C.*, July, 1891, pp. 5-36.

Girder Bracing, Stresses on. By A. C. Elliot. Treats of the maximum stresses in the bracing of large railway girders. Lon. *Eng.*, July 1, 1887.

Graphical Evolution of Stress in Lattice Girders. By Wm. Robertson. Gives a comparison between the values of the stresses in the flanges of various forms of latticing as determined by their numerical evolution and the ordinate to the parabolic curves of moments, and deduces rules for graphical solution. Lon. *Eng.*, March 16, 1883.

Graphical method for the determination of strains in bridge trusses with parallel chords. By C. H. Tutton. *Van Nos. Eng. Mag.*, Vol. XVII., p. 385.

Graphical Method of Calculating Bridge Stresses for Concentrated Loading. A paper by Ward Baldwin, giving a convenient method of finding maximum bending moments and shears, from a graphical diagram for any span, length and loading. *Eng. News*, Sept. 28, 1889, Vol. XXII., p. 295.

Graphical Method of Obtaining Stresses in Lattice. By Wm. Robertson. Gives a new graphical method of computing strains in lattice bridges. Lon. *Engineer*, Dec. 30, 1887. *Sci. Am. Sup.*, March 24, 1888.

Bridge Stresses, continued.

Highway, Computation of Strains in. By C. M. Brown. A paper showing county commissioners and surveyors how to compute strains in highway bridge structures. *Rpt. Ohio Soc. Surv. & Engrs.*, 1888, pp. 193-203.

Maximum Stresses. By Wm. Cain. Investigates the maximum strains that can occur from live loads; also treats of the most economical height of trusses. *Van Nos. Eng. Mag.*, Vol. XIX., p. 71 and 43; also *Van Nos. Sci. Series*, No. 38.

Maximum. A method for determining the maximum bending and shearing stresses in girders or trusses. Diagrams. A paper by H. E. Vautelet. *Trans. Can. Soc. C. E.*, Vol. III., pt. II., 1889, pp. 247-56. Discussion on ditto, pp. 257-71.

Maximum. A new graphical construction for determining the maximum stresses in the web of a bridge truss. *Van Nos. Eng. Mag.*, Vol. XIX., p. 538.

New Graphical Solution of the Problem,—What Position of Concentrated Loads will cause the Greatest Stress in any given part of a Truss or Girder. A valuable paper by Henry T. Eddy, Ph.D. Folding plate and 26 figures. *Trans. A. S. C. E.*, May, 1880, No. 437, pp. 259-353.

Portal Bracing. Paper by Prof. Wm. Cain, Charleston, S. C. Gives a satisfactory discussion of the method in ordinary use in bridge offices of computing stresses in portal bracing. *Eng. News*, July 13, 1889, Vol. XXII., p. 26.

Post Truss. An article by C. H. Tutton, comparing results and showing the discrepancies in the computation by different methods of the Post truss. *Van Nos. Eng. Mag.*, Vol. XVIII., p. 303.

Rail Joints, Experimental Effects of, on Bridge Strains. Gives results of the studies of M. Considère, a Chief Eng. of the French Department of Roads and Bridges. *R. R. Gaz.*, Jan. 28, 1887.

Slide Moment Diagram for Computing. By J. E. Greiner, before the Engineers' Society of Western Pennsylvania. Gives a description of a slide moment diagram, which has been in use in the Baltimore & Ohio office for three years, and is considered the best method of finding shears and moments in bridges. Abstracted in *Eng. News*, April 14, 1888.

Tables for finding Stresses in Parallel-Chord Bridges, Uniformly Loaded, Having Verticals and Equal Panels. Fourteen tables of factors computed for uniform loads and for one or two excesses. By Prof. H. A. Hitchcock. *Eng. News*, May 13, 1892, pp. 503-4, with supplement.

Trestles. The Stresses in. By Ch. F. Stowell. Determined analytically and graphically in *Eng. Era*, March 18, 1885.

Whipple Truss. Method of Calculating the Stresses in, for a concentrated Loading. A valuable paper by Ward Baldwin. The method is partly graphical and partly analytical. *Eng. News*, Aug. 30, 1890, pp. 180-1.

See *Columns. Bridges, Cantilever, Theory. Bridges, Arches, Trussed. Bridge Designing. Graphics.*

Bridge Tests. See *Bridge Members.*

Bridge Trusses.

A New Truss. By Geo. H. Pegram. Proposes a new form of truss. Gives formulas and applies them to a through span of 533 ft., etc. Valuable. *Eng. News*, Dec. 10, 1887.

Burr. See *Bridges, Highways.*

Missouri Pacific Railway Bridges. Inset showing details of a 100 ft., 150 ft., and a 200 ft. span bridge with trusses of the Pegram type. *Eng. News*, Feb. 14, 1891.

Plan proposed by J. Freeman Clark for a truss bridge for very long spans. *Jour. Assn. Eng. Soc.*, Dec. 1886.

See *Bridges, Bridge Stresses, Bridge Designing. Bridges, Howe.*

Bridge Vibrations.

Sukkur Bridge Oscillation. Some interesting and somewhat remarkable facts concerning observed oscillation. *Ry. Rev.*, February 22, 1890, p. 118.

Vibration of Bridges. By S. W. Robinson. Presents facts and figures obtained by the application of an indicator to railway bridges to investigate the cumulative vibration. Describes the indicator and its application and mentions some of the conclusions toward which the figures point. *Trans. A. S. C. E.*, Vol. XVI., p. 42.

Vibration of Bridges Under Passing Trains. Gives six diagrams taken from the same bridge, showing vertical and lateral vibrations. Abstracted from advance sheets. Report of R. R. Com. of Ohio. *R. R. Gaz.*, June 25, 1:66.

See *Bridge Stresses. Rail Joints. Bridge Stresses, Experiments. Cranes, Traveling.*

Bridge Weights. See *Bridge Loads.*

Bridge Works.

New Shop of the Edge Moor Bridge Works. A complete and fully illustrated description. *Eng. & Build. Rec.*, April 19, 1890, Vol. XXI., p. 312.

The Plant of the Berlin Iron Bridge Co. at East Berlin, Conn. Description with view of the new shop. *Eng. News*, Oct. 3, 1891, pp. 316-18. *R. R. Gaz.*, Oct. 2, 1891, p. 697.

Bridges, *Accidents.* See *Bridge Accidents.*

Alexandria Bridge, Punjab Northern State Railway. Sixty-four spans, 142 ft. c. to c. upon brick piers. Detailed description by Henry Lambert before the Inst. C. E. Reprint, *Eng. News*, Oct. 31, Nov. 7, 1875.

Arches. See *Bridges, Arches* below.

Avon, Design and Construction of. By R. F. Uniacke. One Folding plate. *Trans. Can. Soc. C. E.* Vol. III., Pt. II., 1890, pp. 272-280. Discussion on ditto, pp. 281-6.

Battersea. Detailed drawings of the new bridge to be erected over the Thames from designs of Sir Joseph Bazalgette. Lon. *Engineer*, July 9 and 16, 1886.

Battersea Cast Iron Bridge. This work, designed by Sir Joseph Bazalgette, M. I. C. E., is briefly described; a double-page supplement shows the elevation, plan and some details. Lon. *Engineer*, July 23, 1890, p. 65.

Helsia Bridge. Samara-Ufa Railway, Russia. Double plate, showing details of the truss, and other illustrations of this interesting work. Lon. *Eng.*, July 11, 1890, pp. 40-1. Description and additional illustrations *ibid.*, July 25, pp. 90-1, 95, 95. Brief notice in *Ry. Rev.*, Aug. 9, 1890, p. 463.

Big Warrior River. Gives a brief description of a 300-foot through span over the Big Warrior River, near Cordova, Ala., with full detailed drawings. *R. R. Gaz.*, June 29, 1888; *Sci. Am. Sup.*, July 21, 1885.

Bismarck. Traveler, for Erecting 400 ft. span of. Illustrated description, with bill of particulars. *Eng. & Build. Rec.*, Nov. 3, 1888.

Blair Crossing. Gives full description, with plates showing details, of the bridge over the Missouri River at Blair, Neb. Plates reproduced from the Chief Engineer's report. Lon. *Eng.*, Sept. 2, 1887.

Blauw Kraats. A third paper, with many illustrations, showing it in process of erection. Details and strain sheet. Lon. *Engineer*, April 17, 1885.

Boylston Street Bridge, Boston. A valuable article, giving complete dimensions strain sheets, details, method of erection, etc. Illustrated. *Eng. & Build Rec.*, June 8, 1889, *et seq.*

Brooklyn. See *Bridges, Suspension.*

Bussey, Design and Strains of. A communication by H. S. Prichard, showing strains in the Bussey Bridge. *Eng. News*, April 2, 1887.

Cairo. By S. F. Balcom. Gives brief description of the Cairo bridge, and

Bridges, *Cairo*, continued.

describes some of the details of construction and progress of the work. *Rep. Ill. Soc. Engrs. & Surv.*, 1888, p. 75-84; and *R. R. Gaz.*, June 1, 1888.

Fast Erecting on, 518 *ft.* 6 *in.* Channel span erected in four days. With illustrations, showing daily progress. *R. R. Gaz.*, Jan. 11, 1889.

Substructure of the. A valuable paper by Mr. Edward H. Connor. Thirteen tables of quantities and tests. Six plates. The paper concludes with the specifications. *Jour. Assn. Eng. Soc.*, June, 1890, pp. 270-323. Abstract *Eng. News*, Feb. 7, 1891, pp. 122-4.

Cantilever. See *Bridges, Cantilever*, below.

Cast Iron. See *Bridges, Arches, Belfast. Bridges, Battersea.*

Channel Bridge of Messrs Schneider and Hersent. Proposed. Brief illustrated description. *Eng. News.*, Nov. 2, 1889, Vol. XXII., p. 421; *Sci. Am. Sup.*, Nov. 2, 1889, Vol. XXVIII., p. 53.

Channel Bridge. Preliminary Designs for. A paper of some length, by Messrs. Schneider & Co. and H. Hersent, giving general description of the bridge, plans for foundations and methods of constructing them, etc. Illustrated. *Lon. Eng.*, Sept. 27, 1889, *et seq.*; *Iron* (London), Oct. 18, 1889, *et seq.*

Chittravati Bridge. A paper by E. W. Stoney, giving an account of the construction of this large bridge, including the sinking of eighteen cylinder-piers. 2 plates. *Proc. Inst. C. E.*, Vol. CIII., 1891, pp. 135-150. Discussion and correspondence, pp. 151-170.

Chittravati Bridge, Madras Ry. An article by Edward W. Stoney, describing briefly the construction of this bridge and giving in detail an exhibit of results of final test of same, including a plate of deflection diagrams. *Ind. Eng.*, April 5, 1890, p. 276.

Cincinnati and Covington Elevated Railway. Transfer and Bridge Company. River Spans of. Description, specifications and tables of tests. 8 folding plates. Paper by Wm. H. Burr. *Trans. A. S. C. E.*, Vol. XXIII., 1890. Paper No. 445, pp. 47-94.

City. See *Bridges, Selection, etc.*

Conestoga Replacing the. Gives method employed to shift the old bridge on to the temporary trestles. Cuts show details of trestle pier. *Eng. News*, Oct 22, 1887.

Creeping of Bridge Spans on the Louisville Bridge, over the Ohio River. By M. J. Becker, Chief Engineer P., C. & St. L. R. R. Fully illustrated. Ends similarly supported, not on rollers, with movements not more than may be attributed to changes in temperature. *Eng. News*, March 28, 1885.

Deflection. See *Bridge Stress and Deflection.*

Designing of. See *Bridge Designing.*

Detroit River, Proposed. Plan and elevation of the proposed Winter bridge across the Detroit River, at Detroit. *Eng. News*, March 5, 1887.

Development of. See *Bridge History.*

Draw. See *Bridges. Draw* below.

Dufferin. Construction of the Dufferin Bridge over the Ganges, at Benares. An interesting paper by F. T. G. Walton, C. I. E. *Proc. Inst. C. E.*, Vol. CI., 1890. Paper No. 2398, pp. 13-24. Two folding plates.

Duluth. Project for a Combined Railway and Highway Steel Bridge over the ship canal, Duluth, Minn. By Alfred P. Boller, C. E. *An. Report of the Board of Public Works, Duluth, for the year ending Feb. 28, 1890*, pp. 97-104.

Economy in. See *Bridges, Long Span.*

Over the Elbe at Hamburg and at Harburg. Built 1869-72. Spans of about 318 feet of "fish-belly" form, built of wrought iron, the upper chord consists in itself of two chords and panel bracing, as does likewise the lower chord, and the

Bridges, continued.

two are connected by vertical suspension rods only, these suspension rods holding up the track construction. The constructive analysis of the bridge makes it an arched bridge, with parallel chords and panel bracing, supported on towers erected on the piers, the height of the towers equal to the deflection of the stiffened suspension bridge structure (above spoken of as the lower chord of the "fish-belly" bridge), and whose sole office is to neutralize the thrust of the arch. The bridge platform, or track construction, is tangent to this double lower chord, and is supported from the arch by suspension members. A comparison of the weight of this bridge with others of like span and strain on the materials of construction, shows it to be a favorable form of bridge-truss, as far as own weights are concerned. *Zeitschrift f. Bauwesen*, 1885, pp. 76-178.

Erection of. See *Bridge Erection*.

Failures. See *Bridge Accidents*.

Fall River. Description of the building of the cast iron cylinder piers. By E. N. Winslow. *Eng. News*, April 9, 1881, pp. 146-7.

Floors of. See *Bridge Floors*.

Foot. *River Ouse, Bedford, Eng.* Gives plan, elevation and cross-section of a foot-bridge of 100 ft. span, practically without abutment. *Sci. Am. Sup.*, Sept. 8, 1888.

Fort Madison. By W. W. Curtis. Gives a good description of location and construction of Chicago, Sante Fe & California railroad bridge across the Mississippi River at Fort Madison, Ia., with cuts showing details of caisson and piers. *Eng. News*, June 2 and 9, 1888. See also *Sci. Am. Sup.*, Aug. 31, 1889.

Forth. See *Bridges, Cantilever*.

Girders. See *Bridges Girders* below.

Guard Gates. *Description of Guard gates* at Point street bridge at Providence, R. I. Paper by Wm. D. Bullock, M. Am. Soc. C. E., giving description and complete illustrations of gates and operating machinery. *Trans. A. S. C. E.*, Feb., 1889, Vol. XX., p. 78.

Guard Rails on. See *Bridge Guards*

Hannibal. Gives description of the combined highway and railroad bridge across the Mississippi River at Hannibal, Mo. *Van Nos. Eng. Mag.*, Vol. V., p. 366.

Hawkesbury, Australia. Drawings of the fourteen competitive designs, three American, ten English and one French, with the corresponding bids. All the American bids, being far below the others, the reward was made to the Union Bridge Co. Lon. *Eng.*, *Eng. News*, May 8, 1886; *R. R. Gaz.*, April 30, 1886; also *Ry. Rev.*, May, 8, 1886. Small map and elevation showing depth of foundation and details of caissons given. These to be sunk to a depth of 185 feet below high water. Proposed method described. *Eng. News*, Feb. 13, 1886. Method to be employed in sinking piers to the depth of 185 feet. *Ry. Rev.*, May 15, 1886.

A two-page plate showing general design and details. Lon. *Eng.*, Feb. 25, 1887.

A description of the above bridge, with plates showing strain sheet, method of loading and various details. Lon. *Eng.*, April 8-22, 1887.

Hawkesbury, New South Wales. Illustrations and description of the method of erecting on pontoons and floating to place. *R. R. Gaz.*, Aug. 10, 1888; *Indian Engineer*, July 28, 1888; *Sci. Am. Sup.*, Aug. 11, 1888; Lon. *Eng.*, Sept. 7, 1888.

Motions, Observed in. An article by W. Ewald, Hamburg, advancing the idea that the material immediately surrounding the piers and underneath them is mud which would present but slight resistance to lateral motion. Lon. *Eng.*, Feb. 6, 1891. Reprinted in *Eng. News*, March 14, 1891, pp. 247-8.

Bridges, continued.

Highway. See *Bridges, Highway*, below.

History of. See *Bridge History.*

Howe Trusses.

Mountain Creek, on the Can. Pac. Ry. A wooden Howe truss on timber piers 140 feet high. Total length, 1,071 feet. Sustains a 10-degree curve on a 100-foot grade. Illustrated. *Eng. News.*, Sept. 26, 1885.

Oregon Pacific Ry. An article giving short description and details as well as complete strain sheets of Howe Truss Bridges on above road for spans of 30 to 150 feet. *Eng. News*, April 26, 1890, Vol. XXIII., p. 402.

Willamette River, Oregon. Gives elevation, cross section and details of a timber Howe truss across the Willamette River, Oregon. It has two spans 175 feet long, and a draw span 280 feet in length. Lon. *Eng.*, Jan. 6, 1888; *Sci. Am. Sup.*, March 17, 1887.

See *Bridge Designing, Arching, etc. Howe Truss.*

Iciuy River, Brazil. Total length, 3,933 feet, 60 spans of 78 feet, and 4 spans of 230 feet. Two-page plate showing details. Lon. *Eng.*, Feb. 14, 1887.

Inspection of. See *Bridge Inspection.*

Keokuk and Hamilton. Statement of the public test of the bridge over the Mississippi River at Keokuk, Ia. *Van Nos. Eng. Mag.*, Vol. V., p. 374.

List of. See *Bridges, Long Span.*

See *Bridge Loads, Locomotive Wheel Weights.*

Long Span, Discussion of. By Gustav Lindenthal. Gives a discussion of cantilever, general features of arch bridge and suspended arches. *Eng. News*, March 3, 1888.

Long Span. The economic conditions of long-span bridges, with special reference to the proposed North River Bridge at New York City. Address before Am. Assn. for the Advancement of Science (No. 1) at Toronto meeting, by Gustav Lindenthal, C. E. *Eng. News*, Nov. 9, 1889, Vol. XXII., p. 436.

Long Span Bridges. A list of long span, 400 feet or over, bridges in the world. The name, locality, purpose and number of spans are given. *R. R. Gaz.*, May 2, 1890, Vol. XXII., p. 379.

Louisville and New Albany. A short description by the engineers, T. C. Clarke and Chas. Macdonald. Read before the British Assn. for the Adv. Science, 1886. With full page illustration. *Eng. News*, Nov. 27, 1886.

Manhattan.

Description of the elevated railroad bridge over Harlem River, New York, with cuts showing details. *Sci. Am. Sup.*, May 14, 1887.

Caisson for Pier II of. Good description of a caisson 104.8 ft. long, 54.4 ft. wide and 13.3 high, recently placed in position. *Eng. News.*, May 26, 1887.

Mannheim. Gives brief illustrated description of five competitive designs for a bridge at Mannheim. Lon. *Eng.*, Dec. 16, 1887.

Memphis. Abstract of specifications for superstructure, progress, profile of pier sinking for Feb. 7, 1890, and other information. *R. R. Gaz.*, Feb. 21, 1890, pp. 122-128.

Memphis Bridge Erection. A series of illustrated articles describing this work. *Eng. Record*, July 25, 1891, pp. 120-2, *et seq.* Statement of progress. *R. R. Gaz.*, June 26, 1891, p. 442.

Mississippi River. Extract from the report to the Secretary of War by the Board of U. S. Engineers on the bridging of the Mississippi River at St. Louis, Mo. They report against a low bridge below the mouth of the Missouri River and emphasize the difference of the Mississippi above and below the mouth of the Missouri. *San. Eng.*, April 28, 1887.

Missouri River. A diagram giving number and length of spans, elevation of

Bridges, continued.

lowest point of superstructures, standard low and standard high water elevations, is given in *Report of Missouri River Commissioners*, 1889, or *Appendix W. W., of Report of Chief Engineers*, 1889.

Monongahela River at Pittsburg, Rebuilding of the. A suspension bridge replaced by Pauli trusses. The work described by G. Lindenthal, engineer of the bridge. Illustrated. *Trans. A. S. C. E.*, Vol. XII., p. 353. *Eng. News*, May 24, 31, 1884.

New Orleans, Proposed. Description with map showing location. *R. R. Gaz.*, July 11, 1890, pp. 489-90.

New Orleans Bridge Projects. A review of the project proposed by Mr. T. C. Clarke, for a high level bridge over the Mississippi River at New Orleans. *R. R. Gaz.*, May 16, 1890, p. 336.

Newer Bridges in Switzerland. By O. Riese. Interesting description, with detailed drawings. *Zeitschr. f. Bauwesen*, 1886, p. 214-228, 352-367.

Ohio Connecting at Pittsburgh. Double inset with description. The method of erection proposed by C. L. Strobel, C. E., is also described. *R. R. Gaz.*, June 20, 1890, pp. 430-1.

Details of Top and Chord Floor Beams. *Eng. Rec.*, Jan. 3, 1891, pp. 96-7.

Erection of. No. CVI., of a series of articles on Erection Plant. *Eng. & Build. Rec.*, Dec. 13, 1890, p. 25. Description with detail drawings, *ibid.*, Dec. 20, 1890, pp. 41, *et seq.*

Floating the Channel-Span. An article giving details of this engineering feat. Inset plate. *Eng. & Build. Rec.*, Sept. 13, 1890, pp. 226-30. *R. R. Gaz.*, Aug. 22, 1890, pp. 579-586, Sept. 12, 1890, p. 631. *Eng. News*, Sept. 20, 27, 1890, pp. 253-276.

Ottawa, River, Canada. General description of the Chaudiere bridge and account of its erection. *Eng. News*, June 11, 1881, pp. 235-8.

Paso. By M. T. Seyrig. Interesting account of this great bridge, 150 m. in span; with plates. *Mem. de la Soc. des Ing. Civils*, January, 1886; pp. 36-79.

Pegram. See *Bridge Trusses.*

Pile and Trestle. See *Bridge Members. Bridges, Trestle.*

Pontoon Railway Bridge across the Miss. Riv. at Prairie du Chien. Description, with plates and discussion. *Trans. A. S. C. E.*, Vol. XIII., p. 67.

Pony Lattice, W. S. R. R. Gives plan, elevation and cross section, with dimensions of a pony lattice bridge truss built at Normanskill, N. Y., on the West Shore Railroad. Span, 86 ft.; clear width, 14 ft.; height, 10 ft., and weight, 30 tons. *R. R. Gaz.*, Sept. 11, 1885.

Preservation of Iron. By E. Paschen. Gives a scheme for periodic inspection. *Van. Nos. Eng. Mag.*, Nov., 1881.

Quindaro. Caissons and Cribs for. Very complete drawings. *Eng. News*, April 4, 1891, p. 321. (With inset.)

Railroad. See *Bridges, Railroad* below.

Ravi Bridge, Punjab Northern State Railway. Thirty-three 90 ft. spans supported on brick piers. Foundations were of brick cylinders. Description by R. T. Mallet before the Inst. C. E. Reprint, *Eng. News*, Oct. 11, 1878, pp. 328-9.

Rope. See *Bridges, Suspension.*

St. Charles. A history of the enterprise and notes on the construction of the bridge. *Van. Nos. Eng. Mag.*, Vol. V., p. 186.

St. Louis Merchants Bridge.

Description, details, and general specifications. *Lon. Eng.*, June 5, 1891, pp. 685-7, *et seq.*

Bridges, *St. Louis Merchants'*, continued.

An illustrated description of this bridge, including approaches and proposed terminals. *Eng. News*, Dec. 21, 1889, Vol. XXII., p. 578.

Sault Ste. Marie.

A description of the bridge building over the present ship canal and river at Sault Ste. Marie. *R. R. Gaz.*, July 15, 1887.

View of this structure with extract from a paper by G. H. Massey, published in *Proc. Can. Soc. E.* *Eng. News*, Oct. 18, 1890, p. 334.

Schuylkill River, on the B. & O. R. R. A series of articles describing the bridge and its foundations. By Col. Wm. M. Patton, Engr. in Charge. *Eng. News*, Feb. 6, 1886, *et seq.*

Selection and Maintenance of. By D. W. Mead. Gives hints relating to the selection and maintenance of bridges for cities. *Rep. Ill. Soc. Engrs. & Surv.*, 1885, pp. 65-68.

Sibley. An account of the construction of this bridge, including general plan, foundation, and superstructure, with tables. By O. Chanute, John F. Wallace and W. H. Breithaupt, M. Am. Soc. C. E. Pp. 33; plates, 14. *Trans. A. S. C. E.*, Sept., 1889. Vol. XXI., p. 97.

Sin Ho, China. Brief description, with elevations, cross section and half plans showing bracing of the Sin Ho bridge. Lon. *Engineer*, Dec. 9, 1887.

Sioux City Bridge. Description condensed from a report by George S. Morison, Chief Eng. Illustrated. *R. R. & Eng. Jour.*, May, 1891, pp. 219-22.

Skew.

Boylston Street Bridge, Boston, Mass. Illustrated description of the interesting skew bridge, with photo engravings showing process of erection. *Rep. City Engineer* of Boston, Mass., 1888, p. 19.

Central Avenue, at Newark, N. J. By A. P. Boller. Describes a novelty in bridge engineering. The bridge crosses the Morris Canal at a very sharp angle, and is also intersected by two streets. *Trans. A. S. C. E.*, Vol. II., p. 379.

On the Illinois Central Railway. Description and illustrations of a skew span crossing an 82-foot canal at an angle of 22° 10', four tracks, two independent spans of two tracks each. *Ry. Rev.*, Jan. 15, 1890, p. 32.

South St., Phila. Failure of. Brief description of method of building foundations, piles, etc. *Eng. News*, Feb. 28, 1878, p. 67.

Specification for. See *Bridge Specifications.*

Stresses in. See *Bridge Stresses.*

Suspension. See *Bridges, Suspension* below.

Susquehanna River Bridge at Harrisburg, Pa. Brief description with illustration of coffer dam and diagram showing the system of triangulation used for locating the piers. *R. R. Gaz.*, Feb. 13, 1891, pp. 108-9.

Swing. See *Bridges, Draw.*

Tay at Caputta Ferry. Brief description, with illustrations of a few details of construction, both trusses and caissons, showing present tendency of English practice. Lon. *Eng.*, Sept. 13, 1889. Vol. XLVIII., p. 306.

Tay.

By Wm. Inglis, before the Institution of Civil Engineers. Gives details of the construction and difficulties overcome of the Tay viaduct. *Sci. Am. Sup.*, June 16, 1888; *R. R. Gaz.*, June 29, 1888.

Condensed description of the new Tay bridge. *Eng. News*, Aug. 6, 1887; also see *Sci. Am. Sup.*, July 16, 1887. Paper by F. F. S. Kelsey, *R. R. Gaz.*, Sept. 2, 1887.

Erection of the Superstructure of the Tay. By A. S. Biggart. Interesting ac-

Bridges, continued.

count of the methods used and full details of the operations. *Sci. Am. Sup.*, Feb. 12, 1887.

Report of Mr. Henry Law to the Commissioners for the Casualty. *Van Nos. Eng. Mag.*, Vol. XXIII., p. 63.

Two papers with discussion and plates by the engineers of the bridge. *Proc. Inst. C. E.*, Vol. XCIV., pp. 87-131.

Tay. See *Viaduct.*

Temperature Movements. See *Bridges, Cantilever, Forth.*

Thames, New Railway Bridge on the Thames at Blackfriars. A descriptive paper by G. E. W. Cruttwell. Folding plate. *Proc. Inst. C. E.*, Vol. CI., 1890. Paper No. 2401, pp. 23-37. Discussion on railway bridges. *ibid*, pp. 38-58; correspondence, pp. 59-72.

Transfer Bridge. Harsimus Cove, Jersey City, N. J. Paper by J. A. Bensel Jr., Am. Soc. C. E., pp. 2. A brief description of a somewhat novel counterbalanced bridge for transferring cars from the shore to a floating vessel. Illus. *Trans. A. S. C. E.*, December, 1888, Vol. XIX., p. 309. Well illustrated description. *Eng. News.* Jan. 18, 1890, p. 67.

Traveling. A brief illustrated description of a traveling bridge or rolling ferry, between St. Malo and St. Servan, in north France. Docks of masonry are built out from either side into the harbor leaving a waterway of 301 ft.; they are 33 ft. high, tide being 33 ft. The bottom is bare at low tide. The truck runs on two rails placed on the bottom and will carry about 100 passengers. *San. Eng.*, April 23, 1887.

Trestle. See *Bridges, Trestles* below.

Van Buren, Ark.

Construction of the Bridge over the Arkansas River at Van Buren, Ark. Monograph by Mr. C. D. Purdon, M. Am. Soc. C. E. Illustrated, pp. 43, plates 10. *Trans. A. S. C. E.*, May, 1889, Vol. XX., p. 191.

Locating the Piers of the Van Buren, Ark., Bridge. St. Louis & San Francisco Ry. Abstracted from a paper by C. D. Purdon before Am. Soc. C. E. in *Eng. & Build. Rec.*, Jan. 11, 1890, Vol. XXI., p. 86.

Piers of. Descriptions and dimensions of caissons, piers, etc., with experiments on cement, mortar, etc. Illustrated. *Eng. & Build. Rec.*, Feb. 8, 1890, Vol. XXI., p. 149.

Vibration of. See *Bridge Vibration.*

Victoria Tubular. History and description. *Eng. News*, July 12, 1879, p. 221. From Lon. *Eng.*

Walnut St. Bridge at Chattanooga, Tenn. A 5-span iron and steel bridge over the Tennessee river. Description with folding inset showing elevation, design of masonry, strain sheets and details of superstructure. *Eng. News*, May 16, 1891, pp. 462-3.

Weights of. See *Bridge Loads.*

Winner Bridge at Kansas City, Mo.

Descriptive article with a very valuable inset plate giving the strain diagram and ordering sheet as actually used for one of the 253-feet spans of the Winner Bridge, by Mr. Frank D. Moore, C. E. the Chief Engineer. *Eng. News.* March 15, 1890, Vol. XXIII., p. 249.

A description of the bridge with illustrations. A folding inset shows the strain diagram and ordering sheet for one of the 428 ft. trusses. *Eng. News*, July 19, 1890, pp. 51-2.

Wainganga River, Bengal-Nagpur Ry., India. Illustrated. Showing a novel method of erection. *Ind. Eng.*, April 6, 1890.

Bridges. See *Viaducts.*

Bridges, Arches.

Belfast. New Lagan Cast Iron Bridge. Illustrated description of this handsome arch bridge. Lon *Engineer*, Aug. 22 and 29, 1890.

Ben Rhydding, Eng. Gives brief description with two-page plate of detailed drawings, of two lattice arches, with suspended roadway, over the River Wharfe near Ben Rhydding, Yorkshire. Lon. *Engineer*, May 25, 1883.

Brunswick, Eng. Gives two-paged plate showing elevation and details of a hinged-arch foot bridge, spans 79 feet, over the River Oker at Brunswick, England. Lon. *Eng.*, Aug. 17, 1885.

Bow Girders (three-hinged Arches) which do not change the Signs of the Moments, the Load Running over the Bow. A theoretical investigation, by E. W. Welker. *Eng. News*, Oct. 2, 1880, p. 331.

Cedar Avenue Bridge—Baltimore, Md. A braced arch of 150 ft. span. General view and page of details. *R. R. Gaz.*, Sept. 18, 1891, pp. 649-51.

Clermont, Ia. Three-hinged Iron Highway Trussed Arch, of 100 feet span. Illus. *Jour. Assn. Eng. Soc.*, Vol. III., p. 24. Lon. *Eng.*, Aug. 12, 1887.

Driving Park Avenue Bridge, Rochester, N. Y. A three-hinged iron arch of 428 ft. span. Full description with detail drawings. *Eng. Record*, July 18, 1891, *et seq.* Brief description, *R. R. Gaz.*, July 17, 1891, pp. 452-3.

Gruenenthal. See *Canals, Ship, North Baltic.*

Harlem River. Designs for the proposed, at One Hundred and Eighty-first street, which were awarded the first and second prizes. *Eng. News*, also *Sci. Am.*, March 6, 1886. The accepted design, as modified from those presented. Cut given in *Eng. News*, April 10, 1886. Detailed drawings and specifications for the new Harlem River bridge; also, illustrated description of the design for the Harlem River bridge submitted by Messrs. Vaux and Radford. *Eng. News*, July 31, 1886. Designs submitted for, *Eng. News*, Dec. 27, 1890, p. 561.

Plan and elevation showing the arrangement of the plan and the condition of the work just before the last segments of span No. 2 were closed. *Eng. & Build. Rec.*, Jan. 21, 1888. False Works, skewback segment and hinges are shown in *Eng. News*, Feb. 4, 1888.

A series of articles describing the erection of the Harlem River bridge, with details of contractors' plant, staging, etc. *Eng. & Build. Rec.*, July 14, *et seq.*, 1888.

Sinking the Foundation for the Harlem River Bridge at 181st. street, New York City. Description of caisson, and method of sinking in a bed of part earth and part rock, with an account of the comparative advantages of dynamite and rackarock for blasting under such conditions. *Sci. Am.*, April 16, 1887.

Javroz Bridge in Freiburg, Switzerland. Short description, with illustrations, of this iron arch of 285 feet span, without hinges. *Deutsche Bauzeitung*, 1885, p. 546-548.

Illinois and St. Louis.

By Theo. Cooper. Gives notes on the mode of setting and adjusting the skew backs on the insertion of the centre tube of the different spans, and the tests of the completed bridge. *Trans. A. S. C. E.*, Vol. III, (1874), pp. 239-254.

Reconstruction of the Floor of. By N. W. Eayrs. Gives details, with drawings, of the plan adopted in the reconstruction of the railroad floor of the St. Louis bridge. *R. R. Gaz.*, Aug. 31, 1883.

Minneapolis Steel Arch Bridge. Brief and well illustrated description of main features of this structure. *Eng. & Build. Rec.*, May 10, 1890, p. 358.

Paderno, Italy. Gives brief description, with elevation and cross section, of a bridge to be built over the River Adda, at Paderno, Italy. Length of main arch, 493 ft.; rise, 123 ft.; lattice truss spans, 109 ft.; total length, 997 ft. *R. R. Gaz.*, Sept. 14, 1888.

Bridges, Arches, continued.

Putney. Brief description of the bridge over the Thames River at Putney, with double-page drawing of the entire structure and detailed drawing of the centring, etc. Lon. *Eng.*, July 23, 1886.

Raising a Masonry Bridge in France. Account of the raising of an existing arch by cutting through at the springing line and lifting by screws. From *Annales des P. & C.* *Eng. News*, Oct. 3, 1878, pp. 315-16.

Ravine, Lowestoft. Description, with elevation and details, of a wrought-iron arched bridge. The arch ribs are made of ⅝-inch web plate and angle iron. Lon. *Engineer*, Sept. 2, 1887.

Stone Bridge, St. Anthony's Falls (*Miss. River*). Description. Illus. *R. R. Gaz.*, Nov. 23, 1883, p. 772.

Tower Bridge across the Thames, London. Cuts and description of design offered. A trussed arch, with crown, 150 feet above high water, supporting low rail and wagon bridges, with lifting draws. Lon. *Eng.*, May 8, 1885.

Trussed Arches, hinged at crown and springing, maximum strains in. By Emmerich A. Werner. Graphically and analytically treated. *Van Nos. Eng. Mag.*, Vol. XXXI., p. 280 (Oct., 1884).

Upright Arched. By J. B. Eads. Endeavors to show that upright arched bridges can be more economically constructed than is possible by any other method. *Trans. A. S. C. E.*, Vol. III., 1874, pp. 195-235.

See *Arches.* *Bridge Stresses.*

Bridges, Cantilever, Forth.

Erection of. A full page illustration and brief description of the erection of the cantilever pier superstructure. Lon. *Engineer*, Feb. 4, 1887, also see *R. R. Gaz.*, March 11, 1887.

Erection of. By A. S. Biggart. A paper before the British Association, treating of the problems that occurred during the erection of the Forth bridge and methods of overcoming them. Illustrated. Lon. *Engineer*, Nov. 25, 1887. *Sci. Am. Sup.*, Dec. 31, 1887.

South Pier. A full-page illustration showing the erection of the south pier of the Forth bridge. *Eng. News*, April 2, 1887.

Erection of Superstructure. By A. S. Biggart, before the Scotland Institution of Engineers and Shipbuilders. Describes briefly the principal features of the erection of the superstructure of the Forth bridge. *R. R. Gaz.*, May 18, 1888.

Fife Cantilever Pier. A two-page plate of the Fife cantilever pier of the Forth bridge, showing all of the main tubes and connections, including junction girders completed to the full height of 360 feet, the north cantilever carried out 170 feet, the first struts and braces to a height of 220 feet, and 130 feet of the viaduct completed. Lon. *Engineer*, Feb. 3, 1888. A small view of the same in Lon. *Eng.*, Jan. 27, 1888, also *Eng. News*, March 10, 1888.

Floor system of the Forth Bridge. Illustration and detail description of manner of laying rails, etc. Rails are laid in a trough. *Ry. Rev.*, March, 1890; Vol. XXX., p. 151.

Inspection of. The final inspection report on the Forth bridge by C. S. Hutchison and F. A. Marindin, of the Royal Engineers. *Eng. News*, March 29, 1890, Vol. XXIII., p. 304.

Machinery at. By Wm. Arrol, before the Institution of Mechanical Engineers. Gives description of the machinery employed in the erection of the Forth bridge. Illustrated. Lon. *Eng.*, Sept. 9, 1887.

Observations on. A paper read by Dr. Charles E. Emery. Elevation and sections. Tables of temperature and diagrams showing temperature movements. *Trans. A. S. C. E.*, Vol. XXII., June, 1890, pp. 609-24.

Paper by W. Westhofen of great value and interest, fully illustrated and de-

Bridges, Cantilever, continued.

scribing this great bridge in a reasonably complete and satisfactory manner. Lon. *Eng.*, Feb. 28, 1890, pp. 231-283.

Popular article describing several interesting features and giving a table of existing cantilever bridges. By Thos. C. Clarke. *Eng. Mag.*, Apr., 1891, Vol. I., No. 1, pp. 65-79.

Putting in the Connecting Spans of the Forth Bridge. These spans were erected as temporary prolongations of the cantilevers: then connected at centre and wedges at ends released when the span became independent. *R. R. Gaz.*, Nov. 15, 1889. Progress on to Dec. 12, 1888. Lon. *Eng.*, Jan. 11, 1889.

Raising one of caissons which accidently sank out of position. Described and illustrated. Lon. *Eng.*, Oct. 30, 1885.

Sixteenth quarterly report of progress made to the Board of Trade. *Am. Eng.*, May 18, 1887, also Lon. *Eng.*, April 15, 1887.

History of First Cantilever in America. Eng. News, Dec. 29, 1883, pp. 617-8.

Jubilee, Hooghly River, India. By Sir B. Leslie. A paper before the Institution of Civil Engineers, giving details of the construction of the Jubilee bridge carrying the East Indian railroad over the Hooghly River at Hooghly. It has a central double cantilever 360 feet long by 52 feet high, and side spans 420 feet long and 47 feet deep. *Proc. Inst. C. E.*, Vol. XCII., pp. 73-141: abstract Lon. *Eng.*, Jan. 27, 1888; *Mech. World*, Feb. 4, 1887; Lon. *Engineer*, Feb. 10, 1888; *Eng. & Build. Rec.*, Feb. 4, 1888, also Oct. 4, 25, 1890.

Kanawha River Bridge. Very complete details of this cantilever bridge. *Eng. & Build. Rec.*, Aug. 17, 1889, Vol. XX., p. 161, *et seq.*

Kentucky River or Tyrone Cantilever Bridge, on the Lexington extension of Louisville Southern Ry., built by Union Bridge Co. A general description is given with several illustrations, including details of adjusting machinery, and an inset plate giving complete strain sheets and detail drawings of structure, including method of anchoring shore arm of cantilevers. *Eng. News*, April 5, 1890, Vol. XXIII., p. 319.

Kentucky and Indiana. By Mace Moulton. A paper before the American Society of Civil Engineers, containing a full account of the construction, with extracts from specifications, tables showing tests of materials, etc., of the bridge over the Ohio river at Louisville. Plates show design, locations, strain sheet and details. *Trans. A. S. C. E.*, Vol. XVII., September, 1887, pp. 111-168, abstract in Lon. *Eng.*, Jan. 27, 1887.

Lachine.

Gives description with a two-page plate, with details of the bridge across the St. Lawrence River. Lon. *Eng.*, April 13, 1888.

The Foundations. By G. H. Massy. And *the Superstructure.* By J. W. Schaub. Illustrated. These are the first authoritative descriptions of this important structure. 35 pp. and 2 plates, including discussion. *Trans. Can. Soc. C. E.*, Vol. I., 1887, p. 36. Abstract, *Eng. News*, Oct. 1, 8, 1887.

List of, See Bridges, Cantilevers, Forth.

Long Island, Proposed. A description of the proposed bridge between New York and Long Island, over Blackwell's Island. To consist of two cantilever bridges of 850 feet in the clear, resting on braced steel piers 110 ft. high. General elevation and some details in *Eng. News*, April 23, 1887.

Market Street Cantilever Bridge over the Schuylkill River, Philadelphia, Pa. A paper by Fred. J. Amweg, describing the substructure, and asphalt pavement, with table of tests of iron. *Proc. Eng. Club Phila.*, Dec. 1889, pp. 301-18.

Market Street Bridge, Philadelphia. A deck-cantilever highway bridge, 77 ft. wide and 198 ft. long. Plans, elevations and details. *Eng. News*, Oct. 30, 1886

Niagara River. By C. C. Schneider, the engineer of the bridge. A most satis-

Bridges, Cantilever, continued.

factory article on this unique design, with numerous cuts, details, specifications, deflections, etc. *Trans. A. S. C. E.*, Vol. XIV., p. 499.

A paper by C. F. Findlay, M. A., A. M. Inst. C. E. This paper gives when taken with discussion thereon, a valuable treatment of cantilever bridges. *Trans. Can. Soc. C. E.*, Vol. III., pp. 54-110, 1889.

Poughkeepsie.

A series of articles on the erection of the Poughkeepsie bridge. *Eng. & Build. Rec.*, May 5, *et seq.*, 1888.

By Thomas C. Clarke. The Second Sibley College lecture describing the erection of bridge over the Hudson at Poughkeepsie. *Sci. Am. Sup.*, May 19, 1888.

For an historical sketch of the project, general dimensions, description, etc., as well as many fine illustrations of the bridge and its vicinity, see *Poughkeepsie Daily Eagle*, Souvenir Edition, 1889: also *E. & M. Jour.*, Feb. 2, 1889.

Foundation and Substructure of the Poughkeepsie. Gives description of the methods used for founding the pier, and some details of the difficulties met and overcome. Illustrated. *Eng. News*, Oct. 29, 1887: see also Lon. *Eng.*, Aug. 26, 1887, and *R. R. Gaz.*, July 1, 1887.

Quebec. Proposed Cantilever over the St. Lawrence. Illustrated. *Am. Eng.*, May 1, 1885.

Red Rock, Col.

A description of the foundations and superstructure of the largest cantilever bridge in the U. S. A valuable folding inset shows the details. *Eng. News*, Sept. 27 and Oct. 4, 1890, pp. 274, 303-5.

An account of this bridge now under construction across the Colorado River, including outline profile, details of caissons and inset sheet of photo-reproduction. *R. R. Gaz.*, April 25, 1890, Vol. XXII., p. 278.

St. John's River, New Brunswick. Central Span 447 feet. Full description and cuts showing stages in erection. *R. R. Gaz.*, Oct. 30, 1885. Also Lon. *Eng.*, Aug. 6, 1886, *et seq.*

Sukkur, India.

Span, 820 feet. Several new features. Cut given. From *Mech. World*, London; in *Ry. Rev.*, Dec. 12, 1885: also Lon. *Engineer*, July 9, 1886.

Brief description with three full-page illustrations. Lon. *Eng.*, Jan. 4, 1889.

Illustrated article showing the various proposed crossings of the Indus, and a general view of the bridge. *Ind. Eng.*, April 20, 1889.

By Wm. Parsey. Gives a description of staging and temporary erection of the Sukkur cantilever bridge at the bridge works. A two-page plate gives details of staging, etc. Lon. *Eng.*, March 2, 1888.

A brief description, with large colored plate, of the staging for the main pillars and guys of the 820-foot cantilever span. *Ind. Eng.*, Nov. 5, 1887.

A paper by F. E. Robertson, before the Inst. C. E., describing the erection. Abstract with elevation and plan of superstructure. *Ry. Rev.*, Apr. 11, 1891, pp. 230-1.

Theory of Cantilever Bridges. Notes on. A paper by Charles McMillan, C. E., Professor of Civil Engineering at Princeton College. Methods of computing stresses, both analytically and graphically, are given. Illustrated by two plates. *Selected Papers of Rensselaer Soc. of Engrs.*, Vol. II., No. 2, p. 61, June, 1889.

See *Bridges, Long Span.*

Bridges, Draw

Arthur Kill. Gives a brief description, with plan and details of the drawbridge recently constructed between Staten Island and New Jersey. Total length of draw, 496½ feet; clear water-way 2×6+212 feet. *R. R. Gaz.*, June 22, 1888.

Bridges, Draw, continued.

By Clemens Herschel. Treats on the principles of construction of and the calculation of the strains in revolving drawbridges having two spans as openings and built as continuous girders, more especially as continuous panel girders. *Trans. A. S. C. E.*, Vol. III., (1874), pp. 395-448.

Carrying a Cable R. R. Across. See *Cable Railroads.*

Clarence Bridge at Cardiff. A steel structure 464 ft. long with a 190 ft. draw span. Details of superstructure, caissons, and fenders are given with a general description of the bridge. Lon. *Eng.*, Feb. 13, 1891, pp. 155-8, *et seq.*

Coteau.

The Swing Span and Pivot Pier of the Coteau Bridge; Method of Floating Spans into Place. Illustrated description with inset showing details of machinery for operating. *Eng. News*, May 30, 1891, pp. 521-5.

An illustrated description of the process used in sinking the foundations of this Canadian bridge over the St. Lawrence. Open caissons filled with concrete were used. *Eng. News*, April 12, 1890, Vol. XXIII., p. 338.

Electric Motor for Drawbridges. Illustrated description of the new electric turning gear of a drawbridge at Bridgeport, Conn. *St. Ry. Jour.*, May, 1889, Vol. V., p. 119.

Erection of a, without false works. By C. S. Maurice. Describes the erection of a pivot-span 260 feet long over the Tombigbee River, near Demopolis, Ala. *Trans. A. S. C. E.*, Vol. II., p. 150.

Florence Draw Bridge. 422 ft. span with both railroad and wagon roadways. Has special hand turning gear. Details of bridge and turning gear illustrated. *Eng. Rec.*, Feb. 28, 1891, pp. 206-12.

Folding-Floor Drawbridge. A novel type of drawbridge recently completed at Chicago, Ill. Brief description with illustrations. *Eng. News*, May 23, 1891, pp. 496-7.

Glamorganshire Canal Bridge. Fully illustrated description of a highway swing bridge. Entire weight rests on the pivot. Designed by Max am Ende. Lon. *Engineer*, Aug. 7, 1891, pp. 105-7.

Hackensack Draw. Gives description of new draw-bridge recently built by the Erie Railroad over the Hackensack River, with drawings showing details of girders, turn-table, wedges and foundations of draw span. *R. R. Gaz.*, July 20, 1888.

Harlem River. A brief description of the draw-bridge over the Harlem River, with illustrations of details. Lon. *Eng.*, April 1, 1887.

Harlem River, New York. Designed by Theo. Cooper. Operated by hydraulic pressure. Illustrated and described in *Sci. Am.*, Jan. 2, 1886.

Hydraulic Swing. By W. E. Armstrong. Describes the Armstrong Swing Bridge over the Ouse. *Van Nos. Eng. Mag.*, Vol. I., p. 498.

Lifting, Tarante, Italy. Description of a bridge at Tarante, Italy, with plates showing details. It consists of two half arms meeting in the center when closed: each half has a rising and rotating movement, and is worked by hand or turbines. Distance between axes of rotation, 220 feet. Lon. *Eng.*, Oct. 26, 1889, *et seq.* Brief description, illustrated. *Sci. Am. Sup.*, Jan. 14, 1888.

Lifting, Utica, N. Y. By Squire Whipple. Gives description, with elevation and cross-section, of a "lift-draw-bridge" over the Erie Canal at Utica, N. Y. *Trans. A. S. C. E.*, Vol. III., pp. 190-1.

Masnedsund. Description of the swing bridge over the Masnedsund, between the Isle of Falster and Seeland, Denmark, with plan, elevation and some of the more important details. Lon. *Engineer*, March 4, 1887.

Milwaukee. Description of a 203-foot span double-track draw-bridge at Milwaukee. Cut showing details. *San. Engr.*, Nov. 26, 1887.

Milwaukee. Specifications for Draw-Bridge over Canal, for the C. M. & St. P.

Bridges, Draw, continued.

Ry. General specification for superstructure. *Eng. News*, Jan. 18, 1879, pp. 21-3.

Newark Bay Drawbridge. Gives illustrated description of the old draw-bridge on the New Jersey Central road now being replaced by a new bridge. Also shows the construction of the temporary drawbridge now in use. *R. R. Gaz.*, June 17, 1887.

Pamunkey. The erection of the Pamunkey River Drawbridge, a 150 ft. span is described and fully illustrated in *Eng. Rec.*, Feb. 14, 1891, pp. 176-7.

Petaluma Draw. Gives brief description, with general view and plan and elevation of the central pier of the Petaluma drawbridge on the San Francisco & North Pacific Railroad. *Eng. News*, Oct. 13, 1888.

Rolling bridge over the —— lock in the harbor of Antwerp. *Wochenschrift des Oest. Ing.-u. Arch.-V.*, 1886, pp. 37-9.

Swinging Drawbridges by Electricity. Brief article describing arrangement of motor and gearing of a Chicago drawbridge. *R. R. Gaz.*, Aug. 21, 1891, p. 578.

St. Mary's Falls Canal Bridge and Dam. An illustrated description of this draw span, with attached movable dam. Includes illustrations of principal details and complete strain sheet. *Eng. & Build. Rec.*, March 22, 1890, Vol. XXI., p. 246.

Thames River Bridge. Complete general description. Abstract from the report of Alfred P. Boller, Chief Eng., *Eng. & Build. Rec.*, Oct. 11, 1890, pp. 295-7. Details of floor beams, pedestals and piers, *ibid*, Oct. 18, 1890, pp. 310-11. Drawings of turn-table, rail-lifts and locks, *Ry. Rev.*, Oct. 25, 1890, pp. 634-5. Piers and superstructure, *id.*, Nov. 29, 1890, pp. 713-15. Foundations, *R. R. Gaz.*, Nov. 7, 1890, pp. 763-4.

Tower Bridge. London. A series of illustrations of various details of this new and interesting structure, foundations and superstructure. *Lon. Engineer*, Vol. LXVIII., pp. 138, 140, 170, 202, 309, 336, 356, etc., 1889.

Tower Bridge at London. Short general description with illustration. A novelty in draw bridges. *R. R. & Eng. Jour.*, Oct., 1890.

Wells Street, Chicago, Removal of. Gives details of the moving of the Wells street draw-bridge, bodily, to its new position on Dearborn street. *Eng. & Build. Rec.*, April 14, 1888.

And their Turn-Tables. By C. Shaler Smith. A paper for "non-specialists" showing method of computing the strains; also, gives table of draw-bridge tests to obtain co-efficient of rolling friction. *Trans. A. S. C. E.*, Vol. III., pp. 129-41.

Willamette River. Bridges of the Oregon Railway & Navigation Co., over the Willamette River at Portland, Oregon, with illustrations of "Draw" in detail. *R. R. Gaz.*, April 19, 1889.

Winona Bridge. Brief description of the entire structure, with details of trusses, turning and lifting machinery of the swing span, illustrated. Specifications as to quality of steel also given. *Eng. News*, Oct. 17, 1891, p. 370.

See *Bridges, Arches, Tower.*

Bridges, Girders.

Bengal-Nagpur Railway Bridges. Specification for triangulated girder bridges, with insert, giving complete details. *Ind. Eng.*, May 18, 1889.

Continuous.

Application of the Theory of, to Economy in Bridge Building. By Chas. Bender. Reviews at length the many objections to the use of continuous girders, derives working formula, and compares weights with those of equal discontinuous spans. Illustrated. *Trans. A. S. C. E.*, Vol. V. (1876), pp. 147-198.

Bridges, Girders, continued.

Calculation of Continuous Girders, analytically and graphically. By M. Bertrand de Fontviolant. A lengthy paper, containing a tolerably full discussion of the subject, including, besides the well-known theories, some investigations regarding effect of deformation of piers, and temperature. The paper, however, contains little, especially in the graphical treatment, which has not been given in previous discussions of the subject. *Memoires de la Soc. des Ing. Civ.*, Sept., 1885, pp 255-348, with 2 plates.

Discussion of continuous girders, with examples, by Mr. M. S. Hodgins. *Van Nos. Eng. Mag.*, Vol. XIX., p. 553.

International across the River Minho (Spain and Portugal). Detailed description of foundations and superstructure, lattice continuous girder, 5 spans, 61.5 to 69 meters, progress of the work, tests of materials, etc. *Revista de Obras Publicas*, 1886, January to June.

Paper by Chas. Bender on the application of the theory of continuous girders to economy in bridge building, with discussion. *Trans. A. S. C. E.*, Vol. V., p. 146.

Stress produced in continuous girders during launching. Said to be the first published solution of the problem. Illustrated. Lon. *Eng.*, Nov. 27, 1853.

Theory and Construction of. By Mansfield Merriman. The object of the paper is to present some of the main principles and laws and to illustrate this application to the practical designing of continuous bridge. *Van Nos. Eng. Mag.*, Vol. XV, pp. 141 and 193. Also *Van Nos. Sci. Series*, Vol. XXV. A criticism on the above by Chas. Bender. *Van Nos. Eng. Mag.*, Vol. XV., p. 269.

Variable Moment of Inertia. Moment and Load Coefficient. Graphical method. Illus. By C. H. Lindenberger. *Jour. Frank. Inst.*, Jan., 1891.

Chenab, India. Gives two pages of detailed drawings and abstracts from the specifications of the Chenab bridge, India state railroads. It is composed of 17 spans, of 206 feet each, of riveted triangular girder. Lon. *Engineer*, Sept. 14, 1888.

Dalmarnock Bridge. Fully illustrated description of this plate girder bridge at Glasgow. Details of piers are given. Lon. *Eng.*, March 28, 1890, p. 381.

Flanges. The Lateral Stability of. Theoretical discussion including the relation of wave length to stability. By Max am Ende. Lon. *Engineer*, May 29, 1891, pp. 419-20.

Formulas for the Weight of Girder. By Max am Ende. A discussion of the various formulas. *Proc. Inst. C. E.*, Vol. LXXXVII., p. 386.

Lattice Girder Overhead Crossing, Chicago, Santa Fe & California Ry. Paper by W. H. Breithaupt, read before the Engineers' Club of Kansas City, June 18, 1888. *Jour. Assn. Eng. Soc.*, Jan., 1889.

Napier Bridge Madras. An old lattice girder of ten 50 ft. spans on screw piles. Illustrated description. *Ind. Eng.*, May 2, 1891, p. 342.

On some Points for the Consideration of English Engineers, with reference to the Design of Girder Bridges. A paper before the Br. Assoc. Adv. Sc. Opposes the Board of Trade Rules and favors the American practice. Lon. *Engineer*, Sept. 10, 1886. Also *Eng. News*, Oct. 30, 1886.

Plate.

Highway Bridge; Brookline, Mass. Sections of abutments and wingwalls, with inset of details of girders. Owing to lack of head room the water pipe is carried through the triangular openings in the floor beams. *Eng. News*, Sept. 19, 1891, pp. 257-8.

Paper with formulas, plates and practical suggestions, by M. J. Becker, Chief Engr., P., C. & St. L. Ry. *Report Ohio Soc. Surv.*, 1885.

Web. Thickness of. By J. J. Webster. A sensible discussion of the subject. *Trans. Liverpool Eng. Soc.*, Vol. III., p. 49.

Bridges, Girders, *Plate*, continued.

Pin Bearing Plate Girder Railroad Bridge. A 3-span girder, lower flange continuous, pin bearings, one end of the bridge being fixed and the other bearings on rockers. Illus. description. *Eng. Rec.*, Nov. 21, 1891, p. 399.

See *Aqueduct, Plate Girder.*

Riveted. The Designing of, with investigation as to distribution of web stress and discussion on reamed rivet holes. By A. F. Hill. *Eng. News*, Apr. 8, 1888, p. 109.

Types of Iron Girder, Indian Midland R. R. A series of plates giving elevations, plans and details of types of iron girder in use on the Indian Midland R. R., India. *Ind. Eng.*, Aug. 25, *et seq.*, 1888.

Bridges, Highway.

Building of. By J. O. Wright. Discusses the present practice of building highway bridges and gives hints for improvements. *Rpt. Ill. Soc. Eng. & Surv.*, 1888, pp. 60-65.

Highway Bridges, Construction, Maintenance and Repair. By S. A. Buchanan. A very valuable paper, which points a way to the solution of the highway bridge problem. Supervision by a local engineer, with detailed plans and specifications, aided by a superintendent of repairs. A rational and efficient scheme which has been in successful operation for many years. *An. Rpt. of Ohio Soc. Eng. & Surv. for* 1888. C. N. Brown, Secretary, Columbus, O.

Design of. By M. J. Butler. Attempts a practical treatment, and endeavors to make plain some of the principles of design. *Proc. Prov. Land. Sur. of Ontario for* 1887.

By James Owen. A paper of much value to the young engineer, since the practical results from an experience with some 500 bridges of all sizes is given in the paper. Illustrated. *Trans. A. S. C. E.*, Vol. XI, (1882), p. 177.

Improved. By J. H. Burnham. Discusses the improvements made in highway bridges. The discussion on the paper relates mostly to the use of brick in place of stone. *Rpt. Ill. Soc. Eng. & Surv.*, 1888, pp. 47-54.

Iron Viaducts for Highways. A valuable paper by J. A. L. Waddell, C. E., of Kansas City, Mo. *Published in pamphlet form*, by Selden G. Spencer, Kansas City, Mo., June 1889.

Nova Scotia. By M. Murphy. Illustrated article, *R. R. Gaz.*, May 3, 1888.

Old Wooden Highway at Waterford, N. Y. A typical Burr truss span, 200 feet built in 1804, and still in use. Fully illustrated. *Eng. News*, June 1, 1889.

Overhead, N. Y. C. & H. R. R. Gives details of the 60-ft. span overhead highway bridge erected in New York City. *R. R. Gaz.*, Nov. 9, 1888.

Short Span. By S. A. Buchanan. Discusses the construction, maintenance and repairs of short span highway bridges. *Rpt. Ohio Soc. Eng. & Surv.*, 1888, pp. 184-194.

Supervision for New York. Proposed system of supervision. Paper by Chas. F. Stowell, M. Am. Soc. C. E. *Eng. News*, March 22, 1890. Vol. XXIII., p. 269, *et seq.*

Types. Drawing, with dimensions of two types of highway bridges, built by the N. Y., N. H. & H. R. R. *R. R. Gaz.*, Oct. 28, 1887.

See *Bridge Designing. Bridges, Draw.*

Bridges, Railroad.

Bridges and Culverts. Detailed drawing of a number of bridges and culverts recently constructed on the Eastern and Midland R. R., England. Lon. *Eng.*, Jan. 14 and 21, 1887.

Iron. A description of those recently erected on a new English road, which may be taken as modern English practice. Two plates. *Proc. Inst. C. E.*, Vol. LXXXII., p. 348.

Bridges, Railroad, continued.

Recent Construction of. A paper by James Ritchie. Includes a table of weights per foot of actual bridges of various spans. *Jour. Assn. Eng. Soc.*, Nov., 1890, pp. 511-17.

See *Bridge Floors.*

Bridges, Suspension.

By C. Bender. Traces the successive improvements in their mode of construction. *Trans. A. S. C. E.*, also *Van Nos. Eng. Mag.*, Vol. IV., p. 594.

Of Any Degree of Stiffness. By C. B. Bender. Gives the theory of equilibrium of a loaded elastic beam suspended from an elastic parabolic catenary. *Van Nos. Eng. Mag.*, Vol. XXV., p. 399.

Brooklyn Bridge.

Chatham Street Extension of the New York and Brooklyn. A description, with plan, elevation and details, of the New York and Brooklyn Bridge. *R. R. Gaz.*, May 13, 1887.

Enlarging the Capacity of the. Gives the report of the Board of Experts on the plans for enlarging the capacity of the Brooklyn bridge; also the report submitted to the Board by Mr. A. M. Wellington. *Eng. News*, March 17, 1888.

Foundations of. A paper before the Am. Soc. of C. E., by F. Collingwood, giving notes on the caissons of the East River Bridge. *Trans. A. S. C. E.*, paper, No. XXX., also *Van Nos. Eng. Mag.*, Vol. VII., p. 399.

Masonry of. By F. Collingwood. A valuable paper, giving details of the masonry work on the towers and anchorages of the East River Bridge. *Trans. A. S. C. E.*, Vol. VI., p. 7.

Progress of Work at, June, 1879. By F. Collingwood. Gives numerous details of methods used. *Trans. A. S. C. E.*, Vol. IX. (1880), pp. 161-172.

Railroad on. See *Cable Railroads.*

Report of the committee on Terminal Facilities, and the adopted plans for the terminals. *Eng. News*, April 21, *et seq.*, 1888; *R. R. Gaz.*, April 27, 1888; *Eng. & Build. Rec.*, Apr. 21, 1888.

Second annual report of the Chief Engineer of the New York Bridge Co. *Van Nos. Eng. Mag.*, Vol. V., p. 381.

A series of articles compiled from official reports and observation taking up the description of this work from the beginning in 1867. *Eng. News*, Apr. 30-Dec. 17, 1881.

Grand Avenue Bridge, St. Louis, Mo. General view and brief description of this handsome suspension bridge. *Eng. News*, June 27, 1891, pp. 610-11. Details of trusses, floor system, and anchorages, *ibid.*, July 18, 1891. General description and detail drawings are also given in *Eng. Record*, June 6, *et seq.*

Hammersmith. A two-page plate showing plan and elevation; also cuts showing details of towers, chain saddles and anchorages. Lon. *Engineer*, April 22 and 29, 1887. Details of construction of towers and roadway suspension connectors, also details of pier foundations, chains and anchorage. Lon. *Engineer*, May 20, 1887.

Iron Suspension Bridge at Harpers Ferry. Illus. *C. E. & Arch. Jour.*, 1857, p. 156.

Minneapolis Suspension. By T. M. Griffith. Gives brief description of the old and new bridges. *Van Nos. Eng. Mag.*, Vol. XVIII., p. 248.

Minneapolis. A paper by F. W. Cappelen, read before the Engrs. Club, Minneapolis, describing in detail this early structure. *Jour. Assn. Eng. Soc.*, Aug., 1891, pp. 400-26.

Monongahela. Detailed description with historical sketch. By Col S. M. Wickersham before the Engrs. Soc. West. Pa. *Eng News*, May 26, 1883, pp. 243-4.

Bridges, Suspension, continued.

Niagara.

The reinforcement of the anchorage and renewal of the suspended superstructure of the Niagara Railroad Suspension Bridge. By L. L. Buck. An elaborate paper of 34 pp. fully illustrated, describing a very important work. *Trans. A. S. C. E.*, Vol. X., (1881), p. 195.

Replacing Towers of. By L. L. Buck, before the American Society of Civil Engineers. Gives details of the work of replacing the stone towers of the Niagara suspension bridge with iron towers. *Trans. A. S. C. E.*, Vol. XVII. (Oct., 1887), pp. 204-212. Lon. *Engineer*, Dec. 9, 1887; Lon. *Eng.*, Dec. 9, 1887; abstracted *Proc. Inst. C. E.*, Vol. XCIII., pp. 510-512. *R. R. Gaz.*, Nov. 4, 1887. *Eng. News*, Dec. 10, 1887.

North River, Proposed.

By G. Lindenthal. Gives brief description of the proposed bridge, also gives a full page plate comparing the bridge with four of the greatest bridges in the world. *Eng. News*, Jan. 14, 1888, and *Eng. & Build. Rec.*, Jan. 14, 1888.

By Gustav Lindenthal, before the American Society of Civil Engineers. Gives very full details of the proposed bridge over Hudson River, at New York. Proposed dimensions are: River span, 2850 feet; two shore spans, 1,800 feet; width, 68 feet, with six railroad tracks; height above water, 145 feet. Abstracted in *Eng. News*, Jan. 28, *et seq.*, 1888.

Pittsburgh, Repairing the Cables of. By F. Collingwood, M. I. C. E. A detailed account of the cause and extent of the rusting of the wires, and the methods adopted to repair them. *Proc. Inst. C. E.*, 1884, Vol. LXXXVI. *Eng. News*, Aug. 2, 1884.

Point Bridge, Pittsburg. Description and account of test. *Eng. News*, Apr. 14, 1877, pp. 89-91.

Replacing the cable by links on the Albert Bridge, London, the cable having rusted until the bridge is considered unsafe. Cuts showing the design; also four other systems. Lon. *Engineer*, Dec. 4, 1885.

Rope Bridges and Their Military Applications. Brief article discussing the best form of military bridge and illustrating one form of rope bridge. From *Le Genie Civil*. *Sci. Am. Sup.*, No. 824, Oct. 17, 1891, pp. 13163-4.

St. Ilpize and at Lamothe. By M. Nicou. *Annales des P. & C.*, 1885, Oct., pp. 660-683.

Theory of Modern American. By Prof. C. Clericetti. Gives the mathematical theory. *Van Nos. Eng Mag.*, May, Vol. XXIII., p. 221.

Vishwamitri River. Short description and abstract from specifications of a chain suspension bridge of 190 feet span, with two large plates showing elevation and details. *Ind. Eng.*, Dec. 10, 1887.

See *Bridge Accidents*, *Cables*.

Bridge Trestle.

Approaches to Arthur Kill. Description and detail illustrations of the trestle approaches. The approaches consist of about 6,600 feet of framed trestles and some 2,000 feet of pile trestle. *R. R. Gaz.*, July 26, 1889. Vol. XXI. p. 489.

Cluster Bent. By J. A. Hanlon. Gives details of a high trestle near Flushing, O., constructed on the cluster bent plan; shows plan and cross-sections. *Eng. News*, Dec. 31, 1887.

In Deep Water in Halifax Harbor. Water 80 ft. deep. Trestle weighed down by rock. Fully illustrated. *R. R. Gaz.*, April 9, 1886.

High Trestles of the Esquimalt & Nanimo Railway. On 10 degree curve. Drawing of a center bent, 150 ft. high, with details. *R. R. Gaz.*, Feb. 6, 1891, p. 89.

Iron Joint Plates in. Gives detailed drawings showing the use of iron plates instead of tenons, etc., as used in the construction of a trestle on the N. Y. L. E. & W. R. R. *Eng. News*, Nov. 5, 1887.

Bridge Trestle, continued.

McCoy's Creek, Fla. Description of pile trestle for which piles 130 ft. long were driven. Method of driving described. *Eng. News*, pp. 133-4.

Novel Form. Brief description, with plan, elevation, and cross-section, of a novel form of trestle used on the L. & N. R. R. *Eng. News*, Oct. 29, 1887.

Pile and Trestle. By A. F. Robinson. Discusses the use of pile and trestle bridges, and gives design of the standard trestle of the Chicago, Burlington & Northern Railroad Company. *Eng. News*, April 7, 1888.

Pile and Trestle Bridges, Construction of. A paper read by A. Amos before the N. W. Track & Bridge Assoc. Abstract giving arrangement of details. *Ry. Rev.*, Oct. 4, 1890, p. 583.

Pile and Trestle Bridges. Description and drawings of standard pile and trestle bridges of various railroads. Condensed from W. C. Foster's forthcoming book on this subject. *R. R. Gaz.*, Oct. 10, 1890, p. 699, *et seq.*

Pile and Trestle Bridges on the Minneapolis & St. Louis Railway. Description and illustrations of a pile bridge and a high wooden trestle. By W. H. Hixon, Supt. Bridges and Bldgs. *R. R. Gaz.*, April 17, 1891, pp. 260-1.

Railway. By E. A. Hill. Treats of permanent and temporary trestle work, sustaining power of piles, review of railroad practice, etc. *Sec. An. Rep. Ill. Soc. Engrs. & Surv.*

Special Features in Wooden. Discusses the use of mortise and tenon-cap bolting, and bearings, etc. *Eng. News*, July 21, 1888.

Standard Plans. Give plans and details of the standard pile trestle in use on the Chicago, Burlington & Northern Railroad. *Eng. News*, June 9, 1888.

Standard Wooden. Gives plan, elevation, sections and bills of material for the standard wooden trestle of the Kansas City & Omaha R. R. *Eng. News*, Aug. 13, 1887.

Trestle Bridges and Modes of Maintaining them. By J. H. Curtis. An article on the erection of temporary structures. *Van Nos. Eng. Mag.*, Vol. XXVI., p. 383.

Wooden. Designs for Canadian Pacific Ry. Drawings of two standard trestles. *Eng. News*, May 2, 1891, pp. 418-9.

Wooden Trestle Bridge over the Huron River, at Ann Arbor, Mich., with plate. By Chas. E. Greene. *Report Mich. Assoc. Surv.*, 1881.

Bronze Casting. A lecture by George Simonds before the Society of Arts, London. Gives a detailed description of the several steps in producing a bronze statue or artistic model of any kind. Illustrated. *Sci. Am. Sup.*, March 20, 1886.

Bronzes.

Manganese. A paper read before the British Assoc. by P. M. Parsons. Gives characteristics, tensile strength, etc., of the different alloys. *Van Nos. Eng. Mag.*, Jan. 1884.

Strength of. A contribution to the subject from English experiments. *Trans. Liverpool Eng. Soc.*, Vol. V., p. 31.

Strength of. Prof. R. H. Thurston constructed a surface, showing in relief the relative strength of alloys of copper, zinc and tin; and Lieut. Pitman, U. S. A., drew a contour chart of this model. Cuts of both and a very clever editorial discussion given in *R. R. Gaz.*, Aug. 7, 1885.

Strongest of the. The method described of finding this combination; also strength shown graphically for all combinations. By Prof. Thurston. *Trans. A. S. C. E.*, Vol. X., (1881), p. 1; also *Van Nos. Eng. Mag.*, Vol. XXV., p. 126; also supplemental paper in *Trans. A. S. C. E.*, Vol. X., p. 309.

See *Alloys.*

Buffer. See *Car Buffer.*

Buffer Stop. *Hydraulic.* Designed and described by A. A. Langley, Eng. Furnishes a constant resistance till train stops. Fully illustrated. *Lon. Eng.*, Feb. 19, 1886.

Building Construction.

Anchoring of Buildings. Present Defects and How Improved. A paper read before the National Association of Fire Engineers, Sept. 11, 1889, at Kansas City, Mo., by Mr. Henry A. Goetz, New Albany, Ind. *Sci. Am. Sup.*, Dec. 21, 1889. p. 11,615.

Chicago, Tall Buildings in. General description of design, methods of erection and framing. Details of the last illustrated. *R. R. Gaz.*, Oct. 30, 1891, pp. 759-60.

Chicago Construction, or Tall Buildings on a Compressible Soil. Paper by W. L. B. Jenney, before the Am. Inst. Arch., giving general design, methods of founding, framing, etc. Illustrated. *Eng. Record*, Nov. 14, 1891, pp. 388-90.

In Earthquake Countries. A valuable paper on this subject by John Milne, read before the Inst. C. E., giving the precautions and the peculiar methods of construction adopted. Condensed and printed in *Eng. News*, Jan. 18, 1891, pp. 33-4.

Flooring. Brief description of some German experiments on double flooring, with and without asbestos paper between. *Am. Arch.*, No. 820, Sept. 19, 1891 p. 157.

High Unbraced Wrought-Iron Columns. Illustrated description of the construction of an electric lighting station, New York, in which unbraced Phœnix columns 35 ft. long were used. *Eng. News*, May 30, 1891, pp. 514-5.

Lessons from the Fire. A paper by Mr. John Fox, at a meeting of the Boston Society of Architects, containing a good discussion of the contributing causes of this fire, and several valuable and practical suggestions concerning the construction of large business houses. *Am. Arch.*, Dec. 14, 1889, p. 281.

Lessons of the Park Place Disaster. An article by Edward Atkinson discussing the subject of factory construction. Contains details of various constructions and a table of floor weights from various materials, from C. J. H. Woodbury. *Eng. Mag.*, Nov. 1891, pp. 157-80.

Elevators and warehouses. 9 folding plates. *Zeitschrift des oester. Ing. u. Arch. V.* (Vienna.) Vol. II., 1890. pp. 55-62.

Roman. A Prize Essay, Edinburgh University. Discusses principles and forms, processes and methods, materials and workmanship. *Van Nos. Eng. Mag.*, September, 1883.

Slow Burning. A series of articles by W. H. Dabney in *Building*, Feb. 6, 1886. *et seq.*

Slow Burning Construction. Specifications and requirements of the Continental Fire Insurance Company. *Am. Arch.*, No. 815, Aug. 8, 1891. pp. 80-90.

Slow Burning Construction of Buildings. Paper by H. B. Chess read before the Eng. Soc. West. Pa., describing and discussing various methods of construction. Fully illus. *Iron Age*, Aug. 13, 1891. pp. 247-50. *Ry. Rev.*, Oct. 10, 1891, pp. 657-8.

Standard Storehouse Construction. Abstract from a pamphlet prepared by C. J. H. Woodbury. Gives detailed description with drawings of typical storehouse. *Eng. Record*, Sept. 19, 1891, pp. 246-8.

Steel Plate. Detailed plans and descriptions of a method of constructing buildings with embossed galvanized steel plates. *Sci. Am. Sup.*, Nov. 5, 1887.

See *Columns. Concrete Buildings. Factory Construction.*

Building Laws. *Comparative Municipal Building Laws.* Classified synopses of the building laws of the principal American cities. *Am. Arch.*, Aug. 1, Sept. 5, 1891.

Proposed National. Draughted at a meeting of the Combined Committee on Building Ordinances, held at New York, Apr. 2, 1891. *Eng. Record*, April 25, 1891, p. 339.

Building Material, *Strength of.* Abstracted from a paper by Capt. Siddon before the Inst. of British Architects. *Van Nos. Eng. Mag.*, Vol. VII., p 254.

Building Materials *and their use in Fire-Proof Construction.* By S. E. Loring. An illustrated series describing the best and latest practice in the construction of fire-proof buildings. *Building*, Dec. 17, 1887, *et seq.*

Buildings.

Chicago Auditorium. An illustrated description of this new building. Many details of construction are given. *Eng. & Build. Rec.*, April 19, 1890, Vol. XXI, p. 296, *et seq.* *Wochenschrift des. Oest. Ing. u. Arch. V.*, March 28, 1890, p. 103.

Drexel Building, Philadelphia. Complete description, with illustrations. *Eng. & Build. Rec.*, May 4, 1889, *et seq.*

Erection of. See *Cranes, Traveling Jib.*

Failure of. Abstract of a lecture by Prof. T. R. Smith, giving several cases of failure and discussing causes. *The Architect.* Reprinted in *Am. Arch.*, April 11, 1891, pp. 28-9.

Foundations of. See *Shoring of Buildings. Foundation for Heavy.*

Height of. French law of 1884 in full. *San. Eng.*, Sept. 24, 1885.

Masonic Temple, Chicago. Description of this 20-storied building with illustrations of foundation, roof trusses, etc. Lon. *Eng.*, Aug. 7, 1891, pp 126-7-50-52.

Metropolitan Opera House, St. Paul, Minn. Illustrated general description and details of. *Eng. Rec.*, June 13, 1891, pp. 23-4, *et seq.*

Observations on Heavy. A paper read at the Buffalo Convention of the Am. Inst. of Architects, October, 1888, by R. W. Gibson, of New York. *Eng & Build. Rec.*, Nov. 3, 1888.

See *Heating and Ventilation.*

Bulkhead Construction *on North River, New York City.* Description, with illustrations. *Eng. News*, May 25, 1889.

Bulkheads. Illustrated description of a proposed system of water-tight bulkheads. Proposed to use piles made of metal bent U-shaped, fitted together and filled with concrete. *Eng. News*, Jan 8, 1887.

Buoys. *Electric Lighted, in New York Harbor.* Illustrated description of the apparatus and the means of generating and transmitting the current. *The Elec. World*, Feb. 2, 1889.

Cable Conduit Yokes.

Their Strength and Design. Paper by J. B. Johnson, member St. Louis Engineers' Club, with discussion. Read Feb. 6, 1889. *Jour. Assn. Eng. Soc.*, June 1889. Vol. VIII., p. 307.

The relative strength of different forms of cast-iron yokes, as found by actual tests, including the "Johnson solid web yoke." *St. Ry. Gaz.*, March, 1889.

Detail drawing of the cable yokes and connections on Third Ave., and Broadway, N. Y., lines. *St. Ry. Jour.*, Nov., 1890, pp. 54-5, 557.

Yokes and Subway. A valuable illustrated article. The details are well shown and described. *St. Ry. Jour.*, July, 1890, pp. 346-52, *et seq.*

Cable-Driving Drums. *Wear of Grooves of Cable-Driving Drums.* Tables showing actual wear under varying conditions are given. *St. Ry. Jour.*, March, 1890, Vol. VI., p. 147.

An illustrated description of Walker's differential cable drums, designed to prevent the irregular wear in grooves of cable drum and obviate excessive wear of cable. *Am. Eng.*, Dec. 3, 1888.

Cable Grips.

Life of Grip-faces and Cables on Cable Roads. A brief article by F. W. Richart,

Cable Grips, continued.

giving the results of experience on several roads. *Technograph*, (Univ. of Ill. Annual), 1890-91. Reprinted in *Eng. News*, July 11, 1891.

Underground Cable Clip. In use in English collieries. Illustrated description. Lon. *Eng.*, Feb. 1, 1889.

Various forms of grips are fully described and illustrated. An interesting paper by C. B. Fairchild. *St. Ry. Jour.*, Aug., 1890, pp. 367-74.

Cable Railroad Systems.

Agudio System. Description of the system proposed by Agudio in 1863; also of a road from Torino to the Superga, built by Agudio and opened in 1884. *Eng. News*, July 19.

Agudio. *Worked by Endless Ropes for Steep Inclines*. By T. Agudio. Proposes to use turbine wheel for power to run a double system of wire ropes which will communicate power to a "locomotor" with pinion gearing. *Van Nos. Eng. Mag.*, Vol. XXIII., p. 230.

Burgenstock Mountain Railway. This new and interesting line is fully described by A. Sommerguth. The average grade is 53.2 in 100. The line is operated by two Thury dynamos worked by turbines. *Zeitschrift des Vereines deutscher Ingenieure*, 1890, p. 61. Abstract in *Proc. Inst. C. E.*, Vol. C., 1890, pp. 434-6.

Improved. Illustrated description of new form of cable road, in which the cable is supported on trucks attached to it, and each truck frame carries a projection intended to engage with the driving device carried by the car. *Mechanics*, January, 1889.

Inclined, of Hoboken, N. J. Illustrated description of its peculiarities, especially the grip and brakes. *Sci. Am.*, Feb. 20, 1886.

Otto System. Describes the installation prepared for the New Castle Exhibition, with illustrations showing details. Lon. *Eng.*, April 6, 1886.

Rasmussen. A description of this system, which is about to be put down at Sioux City, Ia., with cuts showing the construction of the cable-tube, sprocket-wheel, etc., in *Eng. News*, May 7, 1887.

A summary of the special features of all the cable roads in America, with discussion. By Wm. H. Searles. *Jour. Assn. Eng. Soc.*, Vol. VI., p. 10.

Vogel system, in which the conduit is composed of rolled steel bars and is but ten inches deep, is illustrated and described in *St. Ry. Jour.*, Jan., 1891, p. 28. *Eng. News*, Jan. 31, 1891, pp. 104-5.

Cable Railroads.

Belleville, Paris. Illustrated description. Abstract from *Le Genie Civil*. *Sci. Am. Sup.*, Sept. 6, 1890, No. 766, p. 12231.

By D. J. Miller. A valuable paper on cable roads. Treats of conduits, pulley vaults, drainage, curves, cables, and gives an estimate of cost. Illustrated. *Jour. Assn. Eng. Soc.*, Vol. VI., p. 201. *Eng. News*, June 4 and 11, 1887. Abstracted in *R. R. Gaz.*, June 4, 1887.

Birmingham, Eng. Gives a brief description, with a two-page detailed drawing, of the Birmingham cable road. Lon. *Engineer*, June 22, 1888.

Broadway Cable Road, New York. General construction and methods of operation, plant, etc., described. Illustrated. *Eng. Record*, July 25, 1891, *et seq.*

Brooklyn Bridge.

By G. Leverich, before the American Society of Civil Engineers. Gives a very complete description of the road, plant and particulars of traffic and operation, details of wear, renewals and changes, with 28 plates showing details. Very valuable. *Trans. A. S. C. E.*, Vol. XVII., (March, 1888), pp. 67-100.

Terminal Plans. Sketch and description of general plan for proposed terminals. *Eng. News*, January 11, 1890, p. 26, 37.

Cable Railroads. *Brooklyn Bridge*, continued.

New Terminals for. Report of the Board of Experts. *Eng. News*, Jan. 17, 1891, pp. 54-5. *Eng. Record*, Jan. 17, 1891, pp. 105-6. *R. R. Gaz.*, Jan. 16, 1891, p. 42. *Ry. Rev.*, Jan. 24, 1891, pp. 54-5.

Traffic Capacity of. An exhaustive and valuable paper discussing this interesting problem quite completely. Much valuable data is given. Illustrated *R. R. Gaz.*, March, 28, 1890. Vol. XXII., p. 200, *et seq.*

Carrying a Cable Railway Across a Drawbridge. Accomplished by means of a conduit beneath the river. Illustrated description. *Eng. News*, Sept. 19, 1891, pp. 268-9.

Chicago City Railway. Power Station. An exhaustive description of this new and complete power station. Illustrated. *St. Ry. Jour.*, August, 1889.

Chicago. North and West Chicago St. Ry. Co's. Cable Lines. An account of the extensive Chicago system of cable railways, showing location, tunnel sections, variation of H. P. exerted, etc. *St. Ry. Jour.*, May, 1890, p. 213.

Construction and Operation of Cable and Electric Railways. A paper by H. M. Kebby, before the Eng. Club of St. Louis, comparing cost of construction and of operation of the two systems. In the paper and following discussion, considerable data is given on the subject. *Jour. Assn. Eng. Soc.*, February, 1891. pp. 79-94. Abstract in *Ry. Rev.*, March 14, 1891. pp. 166-7.

Edinburgh Northern Cable Tramways. Abstract of a paper by W. N. Colam, Assoc. M. Inst. C. E., describing this system. The conduit is only 19 inches deep by 9 inches wide, with recesses every 90 feet to receive the carrying pulleys. Details illustrated. *St. Ry. Jour.*, Jan. 1891, pp. 13-17. Lon. *Engineer*, Aug. 8, 1890. pp. 102, 116.

Efficiency Diagram. A diagram and explanation by Mr. Frank Allen, made for Metropolitan Street Ry. Co., of Kansas City, Mo. *Eng. News*, May 3, 1890, Vol. XXIII., p. 416.

Efficiency of. Its variation with length of cable and other elements of construction. Limiting length at which the efficiency would become zero, with formula and curves. Read at the February meeting of the Engineers' Club of St. Louis, by James A. Seddon, *Jour. Assn. Eng. Soc.*, Vol. VI., No. 4, February-April, 1887; *Eng. News*, June 18 and 19, 1887. An abstract from the Philadelphia *Press*. Contains table giving a comparison of the costs, expenses and profits compared with horses. *Eng. News*, March 12, 1887.

Highgate Hill, London. The peculiar features described of this, the first cable road in Europe, opened 1884, by W. N. Colam, before the Soc. of Engrs. *Van Nos. Eng. Mag.*, Aug. 1885.

History and Development. Report of the committee of the Am. St. Ry. Assn., giving the history and developement of the system, short accounts of the systems of various cities, and some practical considerations in regard to construction and operation. *St. Ry. Jour.*, Nov., 1891, pp. 578-87. *St. Ry. Rev.*, Oct., 1891, pp. 427-31.

Jersey City New Elevated Cable Railway. Brief description with illustration. *Sci. Am.*, Aug. 16, 1890, p. 95.

Kansas City. Table showing present condition and mileage of Kansas City cable roads. *Eng News*, Dec. 8, 1888.

Los Angeles, Cal. Viaduct for. Details of a cable railway viaduct, presenting some novel and unusual features. Designed by Samuel G. Artingstall, C. E., of Chicago. *Eng. News*, Jan. 25, 1890. Vol. XXIII, p. 82.

Los Angeles, Cal. An illustrated description of tracks, engines and driving plant and boilers. Endless rope transmission of power is used in the driving of machinery and "Hazelton Tripod Boilers" of 500 H. P. are illustrated and briefly described as they appear in actual operation. *St. Ry. Jour.*, March, 1890, Vol VI., p. 97. Sept, 1890, pp. 409-11.

Cable Railroads, continued.

Melbourne, Australia. Description with map. *St. Ry. Jour.*, Feb., 1890, p. 60.

Oakland, Cal. The conduit is formed entirely of concrete. *St. Ry. Jour.*, Feb., 1890, p. 583-5.

Operating Expenses on Cable Railways. How to Reduce. Practical suggestions from an operator's standpoint. Gives considerable information as to operating expenses, best practice and various details of operating. By M. K. Bowen. *St. Ry. Jour.*, June, 1891, pp. 304-5.

Philadelphia. Causes of the failure of the system found in the cheap and unscientific character of the yokes. Paper by Prof. L. M. Haupt, in *Proc. Eng. Club, Phila.*, Vol. V., p. 226.

St Louis, Winding Plant at. Illustrated. *Am. Eng.*, July 30, 1890.

San Diego, Cal., Cable Tramway. Description of power plant and rolling stock on this light cable road. By Frank Van Vleck. *St. Ry. Jour.*, July, 1891, pp. 837-42.

Third Avenue, N. Y. Track Details of. Folding plate. *Eng. News*, Dec. 13, 1890, pp. 523-4, *Cable Pulleys and Lifts, ibid*, Dec. 20, p. 541-2.

Washington, D. C. Description of system and power plant, with illustrations of conduit, tension apparatus, rope drive, etc. *St. Ry. Jour.*, Dec., 1890, pp. 615-8.

West Point, New York. Cable railway constructed by C. W. Hunt & Co., to convey coal from the wharf to the military academy, a distance of 1,200 feet horizontally and 130 feet vertically. Illus. *Am. Mfr.*, Jan. 24, 1890, p. 7.

Yokes. See *Cable Conduit.*

See *Electric Railroads, Street.*

Cable Towing.

In France. A description of the methods and apparatus used in France for towing boats by continuous cables carried over pulleys on shore. *The R. R. & Eng. Jour.*, April, 1889.

On the Volga. Description of the capstan navigation by horse and steam power, the cable being anchored a few miles ahead continually by auxiliary tug *Am. Eng.*, Feb. 27, 1885.

Theoretically and experimentally considered as employed on the Rhine. By Prof. K. Teichmann, Stuttgart. Abstract given in *Proc. Inst. C. E.*, 1886, Vol. LXXXVI., p. 407.

Theoretical investigation by J. B. Johnson on the relative amounts of work performed when propelling the same boat at different speeds, with or against currents of different velocities, by cable and by paddle-wheel or screw. *Van Nos. Eng. Mag.*, Vol. XXIII., p. 169.

See *Canal Traction.*

Cables.

Breakage of a Cable on the Brooklyn Bridge. A short article including a tabulated statement of number of days service, miles hauled, passengers hauled, tons hauled and ton miles of four cables. *R. R. Gaz.*, April 4, 1890, Vol. XXII., p. 231.

Chain. A full discussion of the determination of the character of iron best adapted for chain cables, the best form and proportions of links, with details of the testing of a large number of specimens. *Report U. S. Board of Testing*, Vol. I., 1881, pp. 1-298.

And Driving Machinery. An excellent paper by C. B. Fairchild. Fully illustrated. *St. Ry. Jour.*, Sept., 1890, pp. 423-31.

Experiments on the Wire Rope of the Superga Railway. A paper by A. Galassini in *Il Politecnico*, 1890, p. 69. Three short tables giving the breaking weight and elongation. Abstract in *Proc. Inst. C. E.*, Vol. C., 1890, pp. 433-4. *Eng. & Build. Rec.*, July 19, 1890.

Cables, continued.

Hoisting. A formula for calculating the section of hoisting cables, adapting them to a given load, and allowing for their own weight, is given. *Eng. News*, March 5, 1887.

Manufacture for Suspension Bridges. As exemplified in the East River Bridge. By W. Hildenbrand. A detailed description of the cable-making machinery and of the process employed on the East River bridge. *Van Nos. Eng. Mag.* Vol. XVII., pp. 171, 193 and 289. Also *Van Nos. Sci. Series*, No. 32.

Manufacture of Street Cable at San Francisco, where they are made at the power house of the company. *Sci. Am. Sup.*, Oct. 30, 1886.

Steel. Abstract from the specifications for the steel cables for Birmingham cable roads. Lon. *Eng.*, Aug. 12, 1887.

Submarine Cable Industry. By E. Vlasto. I. Historical; II. Statistical; III. The Workshop at Calais. *Societe des Ingenieurs Civils; Memoires*. March, 1891, pp. 267-93.

Tests of Wire Ropes. A valuable paper by A. S. Biggart. *Proc. Inst. C. E.*, Vol. CI., 1890. Paper No. 2447, pp. 231-41; appendix containing 8 tables, pp. 242-8. *Eng. News*, Jan. 17, 1891, pp. 50-51.

See *Submarine Cable*, *Cable Grips*.

Caissons. See *Foundations*, *Compressed Air*, *Coffer Dams*.

Calculating Machines. *A Few Practical Uses of the Logarithmic Curve*. By Wm. Cox. Several examples of its use explained. *Eng. News*, Aug. 29, 1891, pp. 195-6.

A brief description of the remarkable machine invented by Leon Bollee. Illustrated. *Mfr. & Build.*, July, 1892, p. 156.

Edmondson's. Described and cut given in Lon. *Eng.*, and *Iron Age*, July 2, 1885.

Invented by A. P. Leschorn, M. E., with Phoenix Bridge Co. Has special application to truss calculations. Brief illustrated description. *Eng. News*, Nov. 24, 1888.

Calorimeters. *A New Form of Calorimeter*. Determines amount of water in steam by observing the fall in temperature of a given quantity of superheated steam necessary to re-evaporate it. *Am. Eng.*, Dec. 3, 1885.

For Testing Fuels on a Small Scale, with Notes on Fuel Testing Stations. By Bryan Donkin and John Holliday. *Proc. Inst. C. E.*, Vol. CII., 1890. Paper No. 2466, pp. 290-304.

Notes on. Paper by R. C. Carpenter, before the Am. Soc. M. E., describing several forms and their uses, especially of the throttling calorimeter. Abstract. *Mechanics*, Aug., 1891, pp. 181-4.

Steam. A paper by G. H. Barrus, read at the meeting of the A. S. M. E. Illustrated. *Eng. News*, May 31, 1890, pp. 520-2. *Trans. A. S. M. E.*, Vol. XI., 1890, pp. 790-828. Abstract in *Mechanics*, June, 1890.

See *Gas Calorimeter*. *Locomotive Tests*.

Cams. A brief treatise on the Steamboat Cam, by Louis Johnson. Illustrated. *Trans. A. S. M. E.*, Vol. II., p. 121.

Canal Congress.

Brussels. Synopsis of the proceedings of the Brussels Inland Navigation Congress; details and illustrations of the principal canal work in Europe. Lon. *Eng.*, July 9, 1886; also *Van. Nos. Eng. Mag.*, Aug. 1886.

International Canal Congress at Manchester, July 28th to August 1st, 1890. Report of proceeding of section A, dealing with inland canals. Illus. Lon. *Eng.* August 8 and 15, 1890, pp. 166, 181-4. Brief abstract in *Eng. News*, August 30, Sept. 6, 1890.

Canal Conference. *Society of Arts.* At a conference recently held under the auspices of the Society of Arts, fifteen papers on canals and inland navigation were presented. They cover the use, history, progress and present condition of canals, their influence on railroads, and a comparison of the cost of traffic on each. *Jour. Soc. Arts*, May 28, *et seq.*, 1888.

Canal Construction. *Mechanical Appliances for.* Illustrated description of machines, chiefly dredges, used in the construction of the Manchester canal. By E. L. Williams before the Inst. M. E. *Lon. Eng.* *Eng. News*, Oct. 31, 1891, pp. 418-20.

Canal Engineering. By L. F. Vernon-Harcourt, before the Society of Arts Canal Conference. Treats of the past, future aims and the prospects of canal engineering in the future. *Jour. Soc. Arts*, May 25, 1888.

Canal Lift.

Fontinettes, France. Gives a discussion on canal lifts *vs.* locks, and a description, with view, of the hydraulic lift at Fontinettes, France. *R. R. Gaz.*, Sept. 21, 1888, *E. & M. Jour.*, Sept. 29, 1888.

Hydraulic Elevator, replacing five locks by a single lift of 43 ft., at Fontinettes, France. Many cuts and full description. Translated from the French. *Sci. Am. Sup.*, Feb. 21, 1885; also Lon. *Eng.*, July 10, 1885.

Hydraulic Lift, at La Louviere, Belgium. Illustrated description. Lon. *Engineer*, Dec. 14, 1888; 21, 1888; Jan. 25, 1889.

Hydraulic Lift. A description, with illustrations, of the hydraulic lifts used to overcome large differences of level in the canal of the Centre, in Belgium. *R. R. & Eng. Jour.*, March, 1889.

Inclines Elbing Hiland, Holland. The locks replaced by inclines operated by turbines. Abstract *Inst. C. E.*, Vol. LXXXIII., p. 470; also *Zeitschrift f. Bauwesen*, 1885, p. 63.

Inclined Plane to serve instead of Locks. The boat floated in water on moving trucks running on a railway. Illustrated. From the French. *Sci. Am. Sup.*, July 4, 1885.

Vertical Lifts for. An address by Oscar W. Petri, Magdeburg. The relative advantages of various forms of locks are discussed including vertical lifts, and a detailed description of illustrations given of a proposed design for a large lift intended for use on the German canals. *Eng. News*, May 9, 1891, pp. 442-4. Lifts for ocean vessels discussed; *ibid.* May 23, pp. 500-2.

Canal Locks.

Description of the Aubois canal lock, on the lateral canal of the Loire River. Method of supplying and filling is by means of oscillating liquid columns. *Van Nos. Eng. Mag.*, Vol. XIX., p. 85.

Gates, the Strength of. By C. H. Romanes. A series of articles beginning Nov. 21, 1884, in Lon. *Engineer*.

Gates. See *Appliances, Lock Gates.*

Project for a Canal Lock of 20 Meters (65.6 feet) Lift. Proposed by Mr. Fontaine, chief engr. Ponts et Chaussees. A masonry dam takes the place of the lower gate, boats entering the lock through an arched opening. Illustrated description. *Eng. News*, Sept. 19, 1891, pp. 250-3.

Ship, to be operated by Steam. Describes those in use on the Delaware and Raritan Canal. Designed by Ashbel Welch. Illustrated. *Trans. A. S. C. E.*, Vol. IX. (1880), pp. 293-318.

See *Hydraulic Lifts. Lock Gates, Canals, Ship. Panama.*

Canal Traction.

A paper by J. B. Johnson on the relative amount of work done in propelling boats by paddle-wheels and by cables. *Van. Nos. Eng. Mag.*, Vol. XXIII., p. 309.

Abstract of a paper before the International Congress on Inland Navigation,

Canal Traction, continued.

Brussels, 1885. Describes the various methods of chain or cable towage, and particularly a new system which is analogous to a cable tramway. Endless chains succeed each other on the bank, the boat being fastened to them by grip. *Proc. Inst. C. E.*, Vol. LXXXII., p. 419.

In Belgium. The various methods of navigation and cost on the different canals is described in a paper in *Revue Universelle des Mines*. Abstract. *Eng. News*, March 3, 1883, pp. 98-100.

Relation of Resistance to Speed on narrow and shallow water ways. By R. F. Conder, M. I. C. E., with discussion by Lloyd, Hartley, Williams, Gordon, Latham, and many others, making over 100 pp. of matter in *Proc. Inst. C. E.*, 1884, Vol. LXXVI.

Resistance of vessels in confined channels and its effect on the dimensions of navigation canals. By M. Graevell. Interesting to students of theoretical hydraulics. *Der Civil Ingenieur*, 1887, pp. 91-110.

Steam on Canals. Propulsion by a fixed wire rope and clip-drum. Experiments in Belgium. *Van Nos. Eng. Mag.*, Vol. I., p. 744.

Steam on Canals. By G. E. Harding. Gives some of the most prominent experiments which have been designed to improve the construction of vessels adapted to inland navigation and the application to them of mechanical means of propulsion. *Van Nos. Eng. Mag.*, Vol. V., p. 84.

Steam Power on Canals. Several articles describing its use on various canals. *Proc. Inst. C. E.* Vol. XXVI, pp. 1.

See *Canals, Ship.*

Canals, *and Inland Navigation.*

By W. J. C. Moens, before the Society of Arts Canal Conference. Gives much information relative to inland navigation in France, Belgium and Holland. *Jour. Soc. Arts*, June 8, 1888.

By Gen. Randall, before the Society of Arts Canal Conference. Advocates the control of canals by the Government as national works. *Jour. Soc. Arts*, June 1, 1888.

European Inland Navigation. Extensive extract from a report by Lieut. Col. William E. Merrill. *Eng. & Build. Rec.*, Nov. 22, 1879, p. 392. Tables of canals and canalized rivers of the U. S. *Ibid.*, Nov. 29, p. 411.

Transport by Canals and Railroads. By G. Lester, before the Society of Arts Canal Conference. *Jour. Soc. Arts*, June 1, 1888.

Transportation in the 19th Century. By F. R. Condor, before the Society of Arts Canal Conference. Discusses transportation in England by land and water, and shows how the canals have been taken in hand by the railroad at a loss. *Jour. Soc. Arts*, June 1, 1888.

Birmingham and Bristol. Gives details of the proposed canal between Birmingham, Gloucester and Bristol. Estimated cost is about $10,000,000. *Lon. Engineer*, June 24, 1887.

Of Canada. A historic sketch of the Lachine, Beauharnois, Williamsburg, Welland, St. Peter's and other canals. *Van Nos. Eng. Mag.*, Vol. V., p. 154.

Cape Cod. Description of the above canal, with map showing location. *San. Eng.*, Nov. 5, 1887.

Closing Breaks Under Difficulties. By O. T. Whitford. Gives details of the closing of a break in the guard lock of Chenango Canal at Binghamton, N. Y., *Trans. A. S. C. E.*, Vol. II., p. 161.

Coast Line, between Norfolk, Va., and Wilmington, S. C. Report by T. Thayer Abert in regard to the physical features of the route. Contains historical information regarding the shape of the coast as well as results of late surveys. Ex. Doc., No. 34, 44th Congress, 1st session.

Canals, continued.

Cornwall Canal. By S. Keefer. A history of the work from 1833. Three folding plates. *Trans. Can. Soc. C. E.* Vol. III., Pt. II., 1890, pp. 297-313.

Deterioration of. A paper treating of the causes of ruin to canals and irrigation works, also on the estimated losses from deterioration of irrigation canals. *Van Nos. Eng. Mag.*, Vol. XXI., p. 381.

Erie. Engineering problems involved in the proposed improvement of the Erie canal by increasing the depth of the channel one foot. By E. Sweet, Jr., *Trans. A. S. C. E.*, Vol. IX., pp. 99-110. Discussion on p. 257.

On the Radical Enlargement of the, to permit the lake boats to pass directly to New York. By E. Sweet, with discussion by Corthell, North, Pope, Poe, Drake, Boller, Walker, Chesbrough, Evans, Clarke, Edwards, Merrill, Van Buren, Henry, Chanute, Herschell, Cooper, Collingwood, and others. The proposition is to increase the depth to 18 feet, width to 100 feet on the bottom, length of locks to 450 feet, and make lake Erie the summit. Discussion brought out a mass of statistics and opinion. Estimate of cost varies from $115,000,000 to $140,000,000. *Trans. A. S. C. E.*, February and March, 1885.

Slope of Water Surface on. By W. H. Searles. Results of a careful determination, with description of levels. *Trans. A. S. C. E.*, Vol. VI., (1877), pp. 189-293.

Great Britain. By M. B. Cotsworth, before the Society of Arts Canal Conference. Gives the history, use and progress of canal and river navigation in England and Ireland. *Jour. Soc. Arts*, May 25, 1888.

Hennepin, and Illinois and Michigan. Surveys and estimates, comparison of routes, etc. Report of Chief of Engrs., 1883, Vol. II, p. 1763.

Hennepin and Ill. & Mich. canals considered as engineering and commercial problems, treated by an engineer acquainted with the facts and surveys. Map and profile. *Eng. News*, May 1, 1886, and Chicago *Morning News*, April 7, 1886.

Irrigation.

Head Gates and Boom of Main Canal, Dolores, Cal. Description, with complete details of head gate and boom of the main canal of the Colorado Water Supply Co. W. H. Wells, Chief Engineer. *Eng. News*, January 4, 1890, Vol. XXIII., p. 7.

Large Irrigation Canals. Condensed from a paper read before the Denver Soc. of C. E., by G. G. Anderson, C. E. *R. R. & Eng. Jour.*, July, 1890, pp. 303-9

Pecos Valley Canals. Article by H. M. Wilson, M. Am. Soc. C. E., describing the irrigating canal system of the Pecos valley, and giving details of the principal one. Illus. *Eng. News*, Oct. 17, 1891, pp. 350-1.

The Sooukesala. By S. A. Latham. The object of this paper is to give a general description of the canal between Sooukesala and Cuddapah, constructed for irrigating purposes, and direct attention to the mode of safely constructing high banks for canals and tanks. *Van Nos. Eng. Mag.*, Vol. XV., p. 385.

See *Irrigation.*

Improvement of, between London and Birmingham. By Henry J. Marten. Gives details of the methods proposed, for improving the efficiency and economy of the canals between London and Birmingham. *Jour. Soc. Arts*, June 1, 1888.

Improvement of. By Sam. Lloyd, before the Canal Conference of the Society of Arts. *Jour. Soc. Arts*, July 6, 1888.

Lake Michigan and Mississippi. A paper by T. T. Johnston, discussing the great water way to connect Lake Michigan with the Mississippi River, and its influence on floods in the Illinois River. *Jour. Assn. Eng. Soc.*, Vol. VI., p. 182.

Canals, continued.

By R. E. McMath, before the Engineers' Club at St. Louis. Discusses the proposed waterway from a St. Louis point of view in respect to its physical, sanitary, economical and political consequences. *Journ. Assn. Eng. Soc.*, August, 1888, Vol. VII., pp. 313-19.

Levels of the Lakes as affected by. By Geo. Y. Wisner and others. Read before the Western Soc. of Engrs., Sept. 5, 1888. *Jour. Assn. Eng. Soc.*, March, 1889.

Laws of. By A. B. Kempe, before the Society of Arts Canal Conference. Object of the paper is to give a concise statement of the existing laws relating to canals in England. *Jour. Soc. Arts*, July 8, 1888.

Maintenance of. By G. R. Jebb, before the Society of Arts Canal Conference. Discusses the work of a canal, method of maintaining them, with remarks on the special difficulties to be overcome in mining districts. *Jour. Soc. Arts*, May 25, 1888.

Mussel Shoals Canal. Brief description with illustration, map and profile of this important improvement on the Tennessee River. *Harper's Weekly*, Oct. 18, 1890, pp. 813-14. *Eng. & Build. Rec.*, Nov. 22, 1890, p. 400.

In New York State. Report of State Engineer John Bogart, Albany, 1889. Cloth pp. 251. The description of the present condition of the Erie canal and its improvements is quite full. Tables and plates.

North Sea. A full description of the North Sea Canal and improvement of the navigation from Rotterdam to the sea. *Van Nos. Eng. Mag.*, Vol. VIII., p. 176.

North Sea Canal of Holland. An official report by Gen. J. G. Barnard to the Chief of Engineers, U. S. A. A quarto of 60 pp. and 11 plates. Includes also the improvement of navigation from Rotterdam to the sea. *Prof. Papers Corps of Engrs., U. S. Army*, No. 22.

Tancarville, France. Gives brief description of the canal being constructed between Havre and the Seine. *Sci. Am. Sup.*, Sept. 15, 1888.

See *Inland Navigation*.

Canals, Ship.

Birmingham and Liverpool Ship Canal. Brief description, map and profile of this projected work. Lon. *Engineer*, Sept. 26, 1890, pp. 247-8. *Eng. News*, Oct. 25, 1890, p. 370. *Eng. & Build. Rec.*, Oct. 18, 1890, p. 310. *Iron*, Sept. 19, 1890, p. 251.

Chesapeake and Delaware Ship Canal. Report on surveys giving estimates. *Eng. News*, March 22, 1879, p. 93.

Corinth. A paper read before the Vandois Society of Engineers and Archts., giving the history of the enterprise, the methods of excavation employed, the cross-sections, etc.; with many cuts. *Am. Eng.*, Nov. 28, 1884.

Corinth. Proposed Methods of Blasting and Removing the Material. Brief description of canal and methods of work. Map, profile and sections. By G. T. Specht. *Eng. News*, June 7, 1884, pp. 270-81.

Harlem River. Gives plan and brief description of the proposed Harlem River Ship Canal. *Eng. News*, Aug. 13, 1883.

In 1889. A paper by R. E. Peary, M. Am. Soc. C. E., reprinted from *Trans. A. S. C. E.*, in *Sci. Am. Sup.*, Feb. 1, 1890; p. 11,736 *et seq.*

Interoceanic Communication by way of the American Isthmus. A series of papers by Lieut. H. H. Barroll, U. S. N., giving full account of work done and various routes proposed. Illustrated. *R. R. & Eng. Jour.*, April, May, June, July, August, 1890.

Interoceanic Communication. American Isthmian Routes. A lecture by Daniel Ammen, Rear-Admiral U. S. N. Historical review and comparison, showing superior practicability of the Nicaragua Route. *Jour. Frank. Inst.*, December, 1889.

Canals, Ship, continued.

Interoceanic Projects. By A. G. Menocal. The problem as presented in 1879. Includes a great deal of valuable data. *Trans. A. S. C. E.*, Vol. VIII., (1879), p. 311-326. Discussion in Vol. IX., pp. 1-88, 117-160; also pp. 429-436.

Lake Erie and Ohio River Ship Canal. A pamphlet issued by the Canal Commission contains estimates of cost and considerable information as to speed of ships in canals, cost of dredging, etc., by J. M. Goodwin. Brief Abstract. *R. R. Gaz.*, April 10, 1891, p. 269.

Manchester. Construction of. A series of articles describing in detail the progress made and methods employed in the construction of the Manchester ship canal. Lon. *Eng.*, May 18, *et seq.*, 1888. Abstracted *Eng. News*, June 30, *et seq.*, 1888. *Eng. & Build. Rec.*, Sept. 29, *et seq.*, 1888.

Description of the locks, gate sluices, bridges, viaducts and other works connected with the canal. Lon. *Eng.*, May 31 and June 6, 1890.

Engineering Features. A succinct account of the alternative schemes and of the accompanying engineering studies. By A. G. Lyster, *Trans. Liverpool Eng. Soc.*, Vol. VII., p. 47, (1886).

General account of the canal as a whole. Map and profile. Lon. *Engineer*, July 11, 1890, *et seq.* See also *Sci. Am. Sup.*, No. 763, Aug. 16, 1890, p. 12189.

History of the Bridgewater navigation, with details of cuttings and disposal of earth, and details of the estimates for different items of the work on the ship canal, with cuts showing different types of sections, geology, etc. Lon. *Eng.*, Sept. 2, *et seq.*, 1887.

Manchester Docks Section. Descriptions and illustrations showing progress on these important docks. Lon. *Engineer*, Aug. 2, 1889.

Plant and Machinery. By L. B. Wells, before the 58th meeting British Association. Gives a brief description of the principal machinery now in use on the Manchester Ship Canal. Lon. *Engineer*, Sept. 21, 1888.

Popular description of this work as to its political and physical aspect. By G. R. Dunell, C. E., *Eng. Mag.*, Nov., 1891, pp. 204-12. Part II., *ibid.*, Dec., 1891.

Summary of the principal features of the canal from Manchester to Liverpool. Lon. *Eng.*, July 16, 1886. Also, by H. S. S. Smith, gives a short history and details of the scheme. *Van Nos. Eng. Mag.*, Nov., 1884.

Nicaragua.

Description of the proposed work, with map. By Commander H. C. Taylor, U. S. N. Paper read before Franklin Institute, Oct. 15, 1888. *Jour. Frank. Inst.*, Jan. and Feb., 1889.

Earthquake Shocks. Possible Effect on the Structures of. Report of Maj. C. E. Dutton on this subject. Discusses the question on the basis of a personal exploration, and data of previous earthquakes. Abs., *Eng. News*, Oct. 31, 1891, pp. 402-4.

Engineering features of, with estimates, etc., by J. F. Crowell. *Proc. Eng. Club, Phila.*, April, 1886.

Estimates, etc. A paper by Lieut. Henry H. Barroll, U. S. N. Maps and profile. *R. R. & Eng. Jour.*, August, 1890, pp. 351-5.

Historical paper on the Nicaragua canal, by Rear Admiral Daniel Ammen, *Lippincott's Mag.*, Sept., 1890, No. 273, pp. 349-59.

Location. Notes on the survey and new location made in 1885, with colored profiles and map. By Ensign W. I. Chambers, member of the expedition. *Proc. U. S. Naval Institute*, Annapolis, Md., Vol. XI., p. 807.

Location of, 1885. Gives maps showing results of the survey of the Nicaragua Canal during 1885, with full description of the work to be done. *Eng. News*, July 14, 1888.

Location of proposed canal with sections, profiles, etc., and description of proposed works. *Eng. News*, May 10, 1890, Vol. XXIII., p. 451.

Canals, Ship. *Nicaragua*, continued.

Notes on. Account of inspection tour of officers of the company and others. Various interesting features described, progress of the work noted. Illus. *Eng. News*, May 30, 1891, pp. 510-12, *et seq.*

Pamphlet describing work, project, etc., and giving plates showing location and general character of work. *The Maritime Ship Canal of Nicaragua.* Pamphlet issued by Canal Company, 41 Wall street, N. Y.

Report on Prospective Tonnage of Traffic, in 1891, and as Estimated for 1897. By Thomas B. Atkins, Sec. and Treas. of the Company. Pamphlet, pp. 27. N. Y., 1890. Also by the same, an account of the explorations and surveys from 1500 to the present time, and the relations thereto to the Government of the United States. Pamp. pp. 56, N. Y., 1890.

Surveys of. By R. E. Perry, before the American Association for the Advancement of Science at Cleveland. Gives details of the surveys and their results. Abstracted in *Eng. News*, Aug. 18, 1888.

North Sea and Baltic Canal.

Description with data as to tonnage, etc. From a U. S. Consular report. *Eng. News*, Oct. 17, 1891, pp. 369-70.

Interesting description of the methods of constructing sand walls through marshes, of the bank protection, and of the large arch bridge near Gruenenthal. Illus. From *Centralblatt der Bauverwaltung*. *Eng. News*, Aug. 1, 1891, pp. 104-6.

Panama.

Actual Status of the. Gives a carefully prepared article, with official profile and cuts from photographs, showing the actual condition of the work. *Eng. News*, June 2, *et seq.*, 1888.

A series of controversial letters by Capt. Bedford Pim, Capt. Henry Gorringe, Capt. Jas. B. Eads, Admiral Ammen, and C. Colne, Secretary Am. Committee, all appearing in the files of the New York *Sun* for November. Also, a paper by C. Colne before the Franklin Institute, treating the subject historically, and describing the present state of the works, with profiles and maps. *Jour. Frank. Inst.*, Nov., 1884.

A very complete statement of present condition of work. Map and profile. *Eng. News*, Aug. 4, 1883. Geological survey of, *ibid.*, Aug. 11, 1883.

Brief account of the history of various projects and statement of work done. *Eng. News*, May 24, 1879, pp. 161-2.

And Its Rivals. By J. S. Jeans, before the Society of Arts. Gives a brief review of the historical, engineering and commercial aspects of the Panama and Nicaragua canals. *Jour. Soc. Arts.* April 6, 1888.

Estimates. By Lieut. Henry Barroll, U. S. N. *R. R. & Eng. Jour.*, July, 1880, pp. 298-300.

In 1887. By Lieut. C. C. Rogers, before the American Society of Civil Engineers. Gives details of the condition of the canal as seen during an inspection trip of nearly three weeks during March and April, 1887. Abstracted *Eng. & Build. Rec.*, Jan. 28, 1888.

Intelligence Report by Lieut. Chas. C. Rogers, U. S. N., giving valuable statistics and other information concerning progress of this work. Illustrated by numerous process plates, maps and profiles. Pp. [illegible]. *House of Representatives, 50th Cong., 1st Sess., Mis. Doc., No. 599.*

Notes on. By Chas. D. Jameson. This is a historical and descriptive article by an engineer for engineers. It is the most satisfactory presentation of the whole subject the Index Manager has yet seen. Accompanied by map, plans, etc. A very readable and valuable contribution by one who takes a hopeful view of the enterprise. Salaries, expenses, etc., given. *Jour. Assn. Eng. Soc.*, Vol. V., p. 367.

Canals, Ship, *Panama*, continued.

Paper by Dr. Wolfred Nelson before the American Society for the Advancement of Science on some of the difficulties to be overcome in making the Panama Canal. *Sci. Am. Sup.*, July 23, 1887. By F. G. Corning. A review of the French company, the work accomplished, and the future prospects of the undertaking. *Sci. Am. Sup.*, Oct. 8, 1887. Abstracts from the report of Lieutenant Kimball, United States Navy. Gives account of the plan for the construction of the Chagres barrage. *Eng. News*, July 30, 1887. The company's side of the scheme is given in *Eng. News*, Aug. 13, 1887. By R. N. Boyd. A paper before the Civil and Mechanical Engineers' Society. An important paper on the present condition of the enterprise and probable time of completion, being conclusion taken from notes on the ground. *Lon. Engineer*, April 15 and 22, 1887; *Sci. Am. Sup.*, June 4, 1887.

Plant and Material of the Panama Canal. A paper by W. P. Williams, Jr., Am. Soc. C. E., p. 36. Fully illustrated. *Trans. A. S. C. E.*, December, 1888, Vol. XIX., p. 273. *Sci. Am. Sup.*, May 25, 1889, *et seq.* *Eng. News*, Aug. 18, 1888.

Proposed Locks on the. Gives a description, with general view, of the proposed locks on the Panama Canal. There are to be four locks, two of 8 m., and two of 11 m. lift on the Atlantic side, and three of 11 m. and one of 8 m. lift on the Pacific side. They are to be 18 by 18 m. *Le Genie Civil*, Feb. 18, 1888; *Eng. News*, March 10, 1888. *Eng & Build. Rec.*, March 10, 1888; *Sci. Am. Sup.*, March 31, 1888.

Reports to the Secretary of the Navy of Rear-Admiral G. H. Cooper, U. S. N., and of Lieut. R. P. Rodgers, U. S. N., with numerous illustrations, including thirty-three elegant phototypes (full page) from photographs taken in January 1884; also maps, colored profiles, etc. An exceedingly valuable report. Senate Ex. Doc. 123, 48th Congr., 1st Session.

San Blas Canal vs. Panama Canal, and Sea Level vs. Locks. Paper by Wm. W. Redfield. Read before the Minneapolis Society of Civil Engineers, May 8, 1889. *Jour. Assn. Eng. Soc.*, June, 1889, Vol. VIII., p. 345.

Work on the. Gives a good statement of what has been done up to the present time. Illustrated. *Sci. Am. Sup.*, March 10, 1888.

San Blas. General description, map of route, tunnel section proposed, etc. *Eng. News*, Sept. 13, 1879, pp. 290-1. Discussion, *ibid.*, Oct. 25, 1879.

Sault Ste. Marie.

Described and well illustrated in *Van. Eng.*, March 25, 1876.

Abstract from the report of Colonel Poe to the War Department. Gives comparison of tonnage of the St. Mary's Canal lock with the Suez Canal. The tonnage is now about 4,000,000 tons per year. *R. R. Gaz.*, March 4, 1887.

A set of 13 drawings, 18 in. × 23 in., together with a description, and 12 photographs, showing details of construction, stages of progress, sections, plans, etc. A most valuable set of plates issued by the *Engr. Dept. of U. S. Army*. Apply to Chief of Engrs. U. S. A., Washington, D. C., or to Mem. Congr.

Locks, Accident to. A short descriptive article containing statistics. Details of the lock illustrated. *R. & M. Jour.*, Aug. 30, 1890, pp. 241-2. For a brief description of the canal see *Sci. Am. Sup.*, Sept. 6, 1890, No. 766, p. 12231.

Suez.

Engineering, Commercial and Political Aspects. By Lieut.-Gen. Rundall, before the Society of Arts. *Van Nos. Eng. Mag.*, January, 1884.

Enlargement of. From the German. Gives a historical account and the scheme of enlargement. *Eng. News*, Sept. 26, 1885.

History and description from 1869 to 1884. Illustrated, 260 pp. Prepared under orders of the Bureau of Navigation, Navy Dept., by Prof. J. E. Nourse, U. S. N., and issued as Senate Ex. Doc. No. 198, 48th Congr., 1st Session.

Report of Commission on building a parallel canal, or increasing the cross-

Canals, Ship. *Suez*, continued.

section of the present canal. Describes the currents in the canal, etc. Valuable. *Annales des. P. & C.*, Oct., 1885.

Resume of the report of the International Commission (of 1884) on the Suez Canal. By M. Flamant. *Annales des P. & C.*, October, 1885, pp. 714-760.

Tehuantepec Railway and Canal. Report of a commission appointed on the artificial waterways of Europe, with a view of applying the knowledge to the railway and ship canal across the Isthmus of Tehuantepec. *Van Nos. Eng. Mag.*, Vol. VI., p. 129.

Canon. *First Trip Through the Big Horn Canon.* Brief paper by E. Gillette describing his experience in this previously unexplored canon. *Trans. A. S. C. E.*, Vol. XXV., July, 1891, pp. 8-14.

Car Axles. See *Axles, Car Wheels and Axles.*

Car Buffer. *Westinghouse Friction Buffer.* Elevation, plan and section, with description of the latest form of Westinghouse Friction Buffer. *Ry. Rev.*, Feb. 15, 1890, p. 83; *Mast. Mech.*, March, 1890, p. 42.

See *Buffer.*

Car Cooling. *Notes on Experiments made with a Cooled Carriage on the East Indian Railway.* By Horace Bell, M. I. C. E. The air supply by passing over ice was cooled 12 to 16 degrees, F. Results of two trials given. *Ind. Eng.*, June 27, 1891, p. 513.

Car Construction. *Aluminum in.* See *Aluminum, Prospects.*

Car Couplers.

By Prof. S. W. Robinson, R. R. Inspector of Ohio. The essential features of a good coupler considered, seriatim with much force and profit. Is deserving of a careful reading by inventors and railroad managers. From the forthcoming report of the R. R. Commissioner of Ohio. *Sci. Am. Sup.*, May 8, 1886; also *Ry. Rev.*, April 17, 1886.

Automatic. An account of the trial of 42 couplers, at Buffalo, by the Exec. Com. of the Master Car-Builders' Assoc., with tables of particulars and results, and cuts of all. *R. R. Gaz.*, Sept. 25, 1885. Also editorials giving estimates of R. R. employes killed and injured in coupling accidents. *R. R. Gaz.*, Sept. 18 and 25, 1885.

C. B. & Q. R. R., Couplers for Freight Cars on. Tables of fractures and photographic record of broken couplers. *R. R. Gaz.*, June 6, 1892, pp. 391-6, 420. The same paper without illustrations, *Ry. Rev.*, June 7, 1892, p. 322.

Chicago Vertical Plane, Tests of the. Both malleable iron and cast steel couplers tested. *R. R. Gaz.*, Oct. 3, 1890, p. 675.

Disadvantages of Mixed Unlocking Gear for Vertical Plane Couplers. Illustrations of various forms, with discussion. *R. R. Gaz.*, Nov. 13, 1891, pp. 793-5.

Dowling. Gives detailed drawing of above coupling and a record of tests made at Columbus, O. *R. R. Gaz.*, Aug. 19, 1887.

Janney.

Life of. Tables showing in detail the breakage of couplers for over twelve months. Taken from the records of Manufacturers. Editorial discussion. *R. R. Gaz.*, Nov. 7, 1890, pp. 794-5, 770-1.

Service Record. A complete and valuable report on the durability and record of many couplers in actual service. A number of illustrations showing manner of failure are given. *Eng. News*, Feb. 15, 1890, Vol. XXIII., p. 146. *Ry. Rev.*, Feb. 15, 1890, p. 90.

Master Car-Builders Standard.

At Interchange Points. Illustrated. *R. R. Gaz.*, July 25, 1890, *et seq.*

Contour lines length of draw-bar and arrangement of dead-block for the automatic coupler, as established by the committee of the Master Car-

Car Couplers, continued.

Builders' Association. *R. R. Gaz.*, April 20, 1888; *Mast. Mech.*, May, 1888; *Nat. Car & Loc. Build.*, May, 1888.

Equipment of Freight Cars with. Map showing progress in this improvement with table of statistics for various railroads. Also editorial discussion. *Eng. News*, April 25, 1891, pp. 401-2, 406-8.

Gauges and Limits for Automatic Coupler. Description and drawings of gauges proposed by the M. C. B. Committee. *Ry. Rev.*, Sept. 19, 1891, pp. 610-11. *R. R. Gaz.*, Sept. 18, 1891, p. 646. *Eng. News*, Sept. 19, 1891, pp. 267-8.

Maintenance of the Standards of the Association for. Report of Committee of M. C. B. Assoc. Gives contours of the various couplers on the market as compared with M. C. B. standard. *Ry. Rev.*, June 13, 1891, pp. 387-9. *R. R. Gaz.*, June 12, 1891, pp. 407-9. *Eng. News*, June 13, 1891, pp. 564-5.

On Curves. Diagrams and tables showing relative positions of couplers connecting cars of different lengths, when on curves. Editorial discussion. *R. R. Gaz.*, Oct. 3, 1890, pp. 678, 684.

Wear and Breakage of. A discussion of the causes of by an Engineer of Tests. Illus. *R. R. Gaz.*, Jan. 2, 1891, p. 2-3.

Standard Contour Lines for Vertical Plane Couplers; and the Lateral Displacement of Couplers on Curves. A paper by Mr. D. L. Barnes before the Western Railway Club on Vertical Plane Couplers and Automatic Air Brakes. Abstract with diagrams, *Eng. News*, Jan. 2, 1891, pp. 3-5 contains abstracts from the entire paper. *Ry. Rev.*, Jan. 3, 1891, pp. 6-8.

Car Couplings.

For Steam Heating. Discussion of the subject before the New England Railroad Club. *R. R. Gaz.*, Feb. 22, 1889; *Mast. Mech.* March, 1889.

For Steam. Brief description, with drawings of, nearly all the different forms of coupling now in use for continuous heating of cars by steam from the locomotive. *R. R. Gaz.*, Dec. 2, 1887; Lon. *Engineer* Jan. 27, 1888.

For Steam. A very full discussion of the subject by the New York Railroad Club. *Mast. Mech.*, May, 1888.

Car Heating.

By C. P. Kerr. Treats of the twofold continuous or independent method. *Sci. Am. Sup.*, June 4, 1887.

By C. P. Kerr. Treats of the amount of heat required for heating the air and to supply waste; mathematical treatment of the steam supply question. Illustrated. *Sci. Am. Sup.*, April 9, 1887.

Electric Car Heater. The Burton Electric Heater for Electric Railway Cars. An illustrated description of this heater with selections from a report on same by Committee on Science and Arts of the Franklin Institute, recommending the award of the "Scott Legacy Medal" to its inventor. *St. Ry. Jour.*, Feb., 1890, p. 48

Coupling for. See *Car Coupling*.

Emerson System. An illustrated description of a system of steam heating that has been in use for five years on the Connecticut River road. Illustrated. *Am. Eng.*, March 30, 1887.

Gold System.

As applied on Long Island Railroad. *R. R. Gaz.*, Oct. 21, 1887.

Illustrated description of the Gold system adapted to the Baker heater. *R. R. Gaz.*, Dec. 16, 1887.

Water in radiators is heated by steam from the boiler on down grades, the water acting as a temporary storage. Illustrated. *Sci. Am. Sup.*, Dec. 4, 1886.

Heat Tender for Trains on the C. M. & St. P. Ry. Full description with plan

Car Heating, continued.

and sections. Electric light is supplied from the same tender. Indicator tests are tabulated. *R. R. Gaz.*, June 13, 1890, pp. 420-2.

Hot-Air System. Gives diagrams and description of the Pearce system of heating cars by hot air furnished from the engine. *Am. Eng.*, Sept. 21, 1887.

McElroy System. Test of the. Gives details of test of McElroy system of continuous heating made on the Hudson River Railroad. *R. R. Gaz.*, April 6, 1888.

On the Heating of Railway Trains. By E. Bellroche. A continuation of previous papers by the said author. Decription of authors system of hot water reservoirs. *Revue des Mines*, 1891, Vol. XIII, pp. 1-24.

Pennycuick System. An illustrated description of the system of heating and ventilation invented by J. G. Pennycuick. It is a hot water system, the water being heated by steam injected into it from the engine. *R. R. Gaz.*, Oct. 28, 1887.

Present Status of Car Heating in the United States. Very complete statistics concerning about 96 per cent of the entire equipment of North American roads. Number of cars heated by steam, by hot water and by stoves, number of engines equipped for steam heating, total equipment, etc., is given. *Ry. Rev.*, Apr. 11, 1891, pp. 228-29. Editorial discussion, *id.*, pp. 237-8.

Process of continuous steam heating of passenger trains, with inset map and tables. It appears that about 25 cent. of the passenger cars of this country are equipped for steam heating. *Eng. News*, June 7, 1890, pp. 541-4.

Sewall System. By C. P. Kerr. A very full description of the system, with drawings showing details, coupling, traps, etc. *Sci. Am. Sup.*, Dec. 24, 1887; *R. R. Gaz.*, Jan. 20, 1888.

Steam.

Experiments on. A paper by Prof. Lanza, before the April meeting of the New England Railroad Club. Giving details of experiments made to determine the amount of steam used in heating passenger cars, with discussion. *R. R. Gaz.*, April 20, 1888; *Mast. Mech.*, May, 1888; *Nat. Car & Loco. Build.*, May, 1888.

Gives tabulated results of experiments in continuous heating, from reports collected by the Committee of the Master Car-Builders' Association. *R. R. Gaz.*, June 22, 1888; *Eng. News*, June 30, 1888.

Notes. By W. F. Baldwin. A paper before the American Society of Mechanical Engineers. Gives experience gained while making experiments on the Long Island Railroad. *Eng. & Build. Rec.*, May 12, 1888; *Am. Eng.*, June 20, 1888; *Eng. News*, Aug. 18, 1888.

In Germany and Sweden. Gives good description of the practice of heating cars by steam in Germany and Sweden. *Mast. Mech.*, Nov., 1887.

On the Boston, Revere Beach & Lynn R. R. Description of the system of continuous steam heating being used on the the Boston, Revere Beach & Lynn R. R. *R. R. Gaz.*, Nov. 4, 1887.

On the C., M. & St. P. R. R. Gives description and detailed drawings of the couplings of the Gibbs system of steam heating now being tested on the C. M. & St. P. R. R. *R. R. Gaz.*, Jan. 13, 1888; *Eng. & Build. Rec.*, Jan. 21, 1888.

Pennsylvania R. R. Full description of system employed, by Theo. N. Ely. Illustrated. *Jour. Frank. Inst.*, Nov., 1890; *Ry. Rev.*, Nov. 8, 1890, pp. 668-9; *Eng. News*, Nov. 15, 1890, pp. 445-6.

Review of the different systems for heating cars by means of steam from the locomotive. By W. A. Smith, before the December meeting of the Western Railway Club. Illustrated by cuts of the different styles of couplings. *Mast. Mech.*, Jan. 1888, also *Am. Eng.*, Dec. 28, 1887.

Car Heating, continued.

Steam Heated Water. Abstract of a paper by W. S. Johnson, read before the Western Ry. Club, advocating a system in which steam heated water is used. *R. R. Gaz.*, June 5, 1891, p. 385. *Eng. News*, June 6, 1891, p. 549.

Steam Loop. Illustrated description of a new and valuable invention for automatically returning condensed steam to the boiler. *Eng. News*, Aug. 2, 1890, p. 106; *Am. Eng.*, Aug. 20, 1890, p. 79.

Winter's Lesson. By Geo. Gibbs, before the April meeting Western Railway Club. Discusses steam heating in the light of the experience of the past winter. *Mast. Mech.*, May, 1888; *Nat. Car & Loc. Build.*, May, 1888; *R. R. Gaz.*, April 20, 1888.

Temperature Regulator. Tests of the Consolidated Car Heating Co.'s Automatic Temperature Regulator. Brief account of a very satisfactory test. *Eng. News*, May 16, 1890, p. 469.

Theory and Practice. Report of 25 pp., by Prof. Lanza, giving theory and present practice. An important paper. *An. Rpt. R. R. Commissioners of Mass.*, 1889.

Timlin Heidinger System. Illustrated description of the system in use on the Ill. & St. L. R. R. The Timlin-Heidinger system has been in operation on this road since October, 1887. With temperature at 10 degrees below zero is said to heat cars to 70 degrees in fifteen minutes. No perceptible increase in fuel consumption on locomotive has been noticed, or any decrease in the motive power. *R. R. Gaz.*, Dec. 7, 1888.

Ventilating and lighting. Paper by J. D. Barnett. Covers 31 pp., including discussion. *Trans. Can. Soc. C. E.*, Vol. I. (1887), p. 72.

Westinghouse Car Heating System. Illustrated description of the. *Eng. News*, Jan. 26, 1887.

Car Journals. See *Car Lubrication.*

Car Lighting.

Comparative Merits of Various Systems of Car Lighting. A series of articles taking up the three systems of lighting, oil lamps, compressed oil gas, and air gas from gasoline—and investigating the nature and efficiency of illuminant, cost of construction and operation of plant, etc., and safety of the system. *Eng. News*, April 18, 1891, pp. 362-3, *et seq.*, running up to April 16, 1892.

Description of Various Systems. Abstract of a paper by Geo. Gibbs, read before the Western Railway Club describing various systems and discussing questions of safety, cost and relative advantages. A valuable paper. *R. R. Gaz.*, March 6, 1891, pp. 158-60. Paper given in full in *Ry. Rev.*, March 7, 1891, pp. 152-3, 6-7. *Eng. News*, March 7, 1891, pp. 234-5, *et seq.* *Mast. Mech.*, April, 1891, pp. 47-8, *et. seq.*

Electric.

Electric. By T. P. Carswell. Describes a proposed system of electric lighting for trains. *Lon. Eng.*, Sept. 23, 1887.

Paper read before the American Institute of Electrical Engineers at its June meeting in New York, by G. W. Blodgett, Electrician of the Boston & Albany Railroad. Describes various methods, especially that in use on the B. & A. Railroad, employing storage batteries. *R. R. Eng. Jour.*, Oct. 1887. *R. R. Gaz.*, April 15, Sept. 9, 1887.

A paper by Mr. Charles Selden, Supt. of Telegraph, Baltimore & Ohio. The paper reviews three systems, and gives a detailed statement of the cost of equipment and subsequent operation. *R. R. Gaz.*, October 25, 1889. Vol. XXI., p. 694.

By Wm. Stroudley. Describes a method in use in England, where the motor is geared by belting to the axle of a car. When the train stops the current is supplied from accumulators. Seems to be very simple and successful. *Trans. Inst. C. E.*, Vol. LXXXIII., p. 370.

Car Lighting, continued.

A paper by J. A. Timms before the Brit. A. A. S., discussing conditions and requirements. *Elec. Rev.*, Sept. 11, 1891. pp. 315-17.

Treats of lighting cars by electricity, first, by batteries; second, by dynamo machines, with costs of the same. *R. R. Gaz.*, April 15, 1887. A full discussion of the subject by the members of the New England Railroad Club. *R. R. Gaz.*, April 22, 1887.

C. M. & St. P. Ry. Description of the light and heat tender designed by G. W. Gibbs, M. E. Plan and sections. Indicator tests. *R. R. Gaz.*, June 13, 1890, pp. 411-2. *Elec. World*, June 21, 1890, pp. 419-20.

Electric Light Car, Pennsylvania R. R. Brief description with folding inset. *Eng. News*, Aug. 23, 1890, p. 174.

On Railway Train Lighting. A paper by William Langdon describing the general features of electric lighting plants on several English railway trains, and giving some statistics regarding first cost and maintenance. Folding plate showing arrangements. *Proc. Inst. C. E.*, Vol. CVI. 1891, pp. 106-49. Discussion, pp. 150-77.

Pintsch System.

In France. Gives details of the cost of lighting cars by the Pintsch system in France. *R. R. Gaz.*, Nov. 11, 1887.

On the Chicago & Northwestern. Illustrated description of the gas plant at Chicago, and of the method of manufacture. *R. R. Gaz.*, July 24, 1891, pp. 507-8.

Some Interesting Facts About the Way the Gas is Made. Brief descriptive article. Illus. Details of compressor shown. *Mast. Mech.*, June, 1891, pp. 82-3.

Lighting and Ventilation. Discussion of the above subjects by the New York Railroad Club. *R. R. Gaz.*, Nov. 25, 1887.

Car Lubrication.

English Car Journal Lubricants and Methods of Testing. An article giving English practice. By an engineer of tests on an English railway. *R. R. Gaz.*, Feb. 14, 1890, Vol. XXII., p. 103.

Extracts from a forthcoming book by Willis E. Hall. Treats of the relation of friction to the size of bearings, and of distribution of pressure. *Eng. News*, June 27, 1891, pp. 621-3.

A paper by W. E. Hall before Engineers' Club, Phila., Nov., 1886. *Proc. Eng. Club, Phila.*, Dec., 1886, Vol. VI., pp. 37-42.

Car Starters.

For Grades. Flood's. Illus. *Elec. Eng.*, Oct. 29, 1890.

Theory of. By J. B. Johnson. Gives mathematical demonstrations for sizes of steel springs and air chambers designed for storing the energy of a moving car. Shows the futility of all steel spring devices, and the unprofitableness of all such schemes. *Jour. Assn. Eng. Soc.*, Vol. IV., p. 293.

Car Transfer Apparatus, *of Robert H. Ramsay*. Report of a committee of the Franklin Institute on its merits, who recommend awarding to the inventor the Elliott Cresson medal. The apparatus described and illustrated. Its purpose is to transfer cars from tracks of one gauge to those of another. *Jour. Frank. Inst.*, September, 1886.

Car Trucks.

Barber Roller Bearing Truck. A description with drawings of a very satisfactory truck in use on the Northern Pacific. A lateral movement of one inch is allowed. *R. R. Gaz.*, Feb. 20, 1891, pp. 124-5.

Fox Solid Pressed Steel Freight Truck. Description of this new form of truck. *R. R. Gaz.*, March 27, 1891, pp. 210-11.

Car Trucks, continued.

Passenger Car Truck. Gives detailed drawings with dimensions of a passenger truck with eight brake shoes. *R. R. Gaz.*, Dec. 9, 1887.

Six-Wheel Trucks for Freight. By J. M. Barr, before the March meeting of the Western Railway Club. Discusses the use of the collarless axle, and advocates the use of six-wheel trucks for freight cars of 60,000 lbs. capacity. *Mast. Mech.*, April, [illegible]; *R. R. Gaz.*, March 23, [illegible]; *Nat. Car. & Loc. Build.*, April, [illegible].

Swing Beam Trucks. Discussion at the Oct. meeting of the Western Railway Club. *Ry. Rev.*, Nov. 6, 1890, pp. 664-5. Abstract in *Eng. News*, Nov. 8, 1890, p. 419. *Ry. Rev.*, Nov. 15, 1890, pp. 681-2. *R. R. Gaz.*, Nov. 14, 1890, pp. 781-2.

Swing Motion and Rigid Trucks. Discussion at Nov. meeting of New Eng. R. R. Club. *R. R. Gaz.*, November 21, 1890, pp. 802-3. *Ry. Rev.*, Nov. 22, 1890, p. 696.

Car Vestibule. *Barr's Vestibule, Chicago, Milwaukee & St. Paul Railway.* Description with detail drawings. *R. R. Gaz.*, July 17, 1891, pp. 492-3.

Car Wheels.

And Axles for 60,000-lb. Freight Cars. Discussion of the subject before the New York Railroad Club, at the December meeting. Discussion opened by a paper on the subject by Mr. G. N. Barr. Also editorial notes on same. *R. R. Gaz.*, Dec. 28, [illegible]; *Mast. Mech.*, Jan. 1889.

And Axles. Their Relation to the Track. A discussion by the members of the New England Railroad Club at its February meeting. Relates mainly to the relative merits of steel and cast-iron wheels. *Mast. Mech.*, March, [illegible]. *R. R. Gaz.*, Feb. 17, [illegible]; *Nat. Car & Loco. Build.*, March, [illegible].

And Contracting Chills. A discussion before the Northwest Railroad Club, in which this subject is thoroughly discussed. *Mast. Mech.*, Dec., 1889. Vol. IV., p. 211.

And Tires. By C. F. Allen, before the March meeting of the New England Railroad Club. Discusses the question of safety in the use of wheels and tires. Followed by discussion. *Mast. Mech.*, April [illegible]. *Nat. Car & Loco. Build.* April, [illegible]; *R. R. Gaz.*, March 23, [illegible].

Best Diameter of. A paper by Samuel Porcher, discussing the subject and giving the results of his experience. Read before the New York Railroad Club. *R. R. Gaz.*, Feb. 27, 1891, pp. 139-40; *Ry. Rev.*, Feb. 28, 1891, p. 136.

Conical Tires on Railway Rolling Stock a cause of resistance to traction (6 to 20 per cent.) and a supposed cause of the creeping of rails. Abstracted from the German. Highly Mathematical. *Proc. Inst. C. E.*, Vol. LXXXVI., p. 410.

Cushion Car Wheel. Description with drawings of wheel having a cushion of rubber between the tire and wheel center. *R. R. Gaz.*, Sept. 4, 1891, p. 613.

Cylindrical Wheels and Flat Topped Rails for Railways. A paper by D. J. Whittemore, M. Am. Soc. C. E., objecting to the usual form of rail, and proposing to substitute a flat topped rail exactly filling the tread of the wheel. Pp. 19; plates, 7. *Trans. A. S. C. E.*, Sept., 1889. Vol. XXI., p. 133.

See *Iron, Car Wheel, Microscopic Structure of.*

Guarantees.

Article by P. H. Griffin discussing report of M. C. B. Committee. *R. R. & Eng. Jour.*, Sept., 1891, pp. 584-85.

Three valuable papers presented to the February meeting of the New York Railroad Club, on the guarantee for car wheels, mileage of steel-tired wheels and the safety of cast-iron wheels. *Mast. Mech.*, March, [illegible]; *R. R. Gaz.*, Feb. 24, [illegible].

Lecture by Rob. W. Hunt at Cornell University. This is a valuable and complete discussion of American car wheels. Illus. *Sci. Am. Sup.*, Nos. 761, 762, Aug. 2 and 9, 1890, pp. 12166-8, 12171-3.

Car Wheels, continued.

Machine for Rolling. Illustrated description of a machine designed by J. R. Jones of the Pencoyd Iron Works. *R. R. Gaz.*, Oct. 9, 1891, pp. 704-5. *Ry. Rev.*, Oct. 31, 1891, p. 712. *Iron Age*, Sept. 24, 1891, pp. 493-4.

Machined. An interesting pamphlet describing the method of manufacturing machined wheels. Illus. Issued by the N. Y. Car Wheel Co., Buffalo, N. Y. Extended extracts in *Eng. News*, July 12, 19, 1890, pp. 40, 49.

Sections and Mechanical Conditions of. Abstract of a paper by P. H. Griffin read before the Am. Soc. C. E., discussing the question of proper action, methods of manufacture, and mechanical defects. Illus. *Eng. News*, June 27, 1891, pp. 612-14. *R. R. Gaz.*, June 12, 1891.

Specification for Cast-Iron.

Gives specifications for cast-iron car wheels, as proposed by Mr. Barr before the Western Railway Club. *R. R. Gaz.*, Dec. 23, 1887; *Mast. Mech.*, January, 1888.

M. C. B. specification for cast-iron car wheels. Specifications respecting both materials and testing are given. *Ry. World*, July 13, 1889; *R. R. Gaz.*, July 5, 1889; *Proc. M. C. B. Ass'n.*, Asratoga, 1889, p. 82.

Steel, Richards' Process for Casting. A special method of moulding producing a very sound wheel. Illustrated description. *R. R. Gaz.*, June 5, 1891, pp. 386-8.

Steel Tired and Chilled. Extract from the report of the Massachusetts Railroad Commissioners on the Haverhill Accident, showing the kind of wheels in use in Massachusetts. *R. R. Gaz.*, May 11, 1888; *Nat. Car. & Loco. Build.*, June, 1885; *Eng. News*, May 19, 1888.

Steel Tired Wheels and the Ross-Meehan Shoe. Discussion at New Eng. R. R. Club meeting. *R. R. Gaz.*, Oct. 18, 1890, p. 127; *Eng. News*, Oct. 18, 1890, p. 317; *Ry. Rev.*, Oct. 18, 1890, p. 616.

Table showing percentages of wheels removed in 1884 for various causes. 18,000 wheels removed out of a total of 350,000, from 24 different makers; with editorial discussion of the lessons to be learned from this remarkable exhibit. *R. R. Gaz.*, June 5 and 12, 1885.

Testing of. A paper by P. H. Griffin, read before the N. Y. R. R. Club. This paper gives an account of the methods of testing employed at the N. Y. Car Wheel Works, and some excellent results attained in the strength of their cast iron. *R. R. Gaz.*, March 6, 1891, pp. 160-1; *Ry. Rev.*, March 7, 1891, p. 150.

Tire Testing. Woehler's experiments on. An editorial abstract. *R. R. Gaz.*, Sept. 2, 1887.

"*Truing up old Car Wheels in Five Minutes.*" A description and illustration of a remarkable new machine for truing up both cast-iron and steel car or locomotive wheels. *Mast. Mech.*, Jan., 1890, p. 13.

See *Wheels*.

Cars.

American Cars for the Southeastern Railway, England. Description with inset of drawings of some handsome cars recently built for the above railway. *Eng. News*, Oct. 10, 1891, pp. 338-40, Oct. 17, p. 361.

Best Size of Freight. A good editorial on the subject. The conclusion reached is that "a short car cheaply handled at the termini will, on a short haul, convey freight more cheaply than a large car. On a long haul the terminal charges have little influence on the total cost, and the car that carries the greatest load appears the cheapest." *R. R. Gaz.*, April 8, 1887.

Canada's Cattle. Gives description, with plans, elevation and cross-section, of Canada's cattle cars. They are provided with hayracks, water-troughs and movable partitions. *R. R. Gaz.*, March 2, 1888.

Coal, 60,000 lbs. Capacity. Gives drawings of a 60,000 lbs. capacity coal car for the Georgia Pacific Railroad. *Nat. Car & Loco. Build.*, June, 1888.

Cars, continued.

Combination. Systems in use on railroads in Belgium and in Rhenish Prussia. *Annales des P. & C.*, 1884-2-635.

Derrick, 20-Ton. Baltimore & Ohio Railroad. Standard Derrick Car. Description and drawings. *R. R. Gaz.*, Jan. 9, 1891, pp. 23-4.

Dining. Gives plan, cross-sections and specifications for a dining car for the Michigan Central R. R. *R. R. Gaz.*, Aug. 19, 1887.

Draft Rigging for Freight Cars: Denver & Rio Grande Railroad. Continuous throughout the entire length of the car. Very serviceable. Illustrated description. *Ry. Rev.*, April 25, 1891, pp. 263

Draw-Bar Rigging for Freight Cars. Discussion before meeting of Western Railway Club. Also an illustrated description of the Graham draft rigging in use on the Pennsylvania R. R. *R. R. Gaz.*, March 29, 1889.

Dump.

Side Dump Car. Eighty thousand pounds capacity. New York, Lake Erie and Western R. R. A complete description of this novel and convenient form of side dump car, with inset drawings in detail of same. *Eng. News*, April 19, 1890, Vol. XXIII., p. 369.

Standard Dump Car. Boston and Albany Railroad. Plan, elevation, end elevation and section giving dimensions and details. *R. R. Gaz.*, March 7, 1890, Vol. XXII., p. 157; *Ry. Rev.*, March 8, 1890, p. 135.

Fifty-Thousand Pounds.

Standard. Gives brief description with drawings and bill of material of the standard 50,000-lb. freight car of the Lehigh Valley Railroad. *R. R. Gaz.*, June 8, 1888.

Standard 50,000 lbs. Freight Car, for Chicago & Northwestern Ry. Description, complete details and weights of parts are given. *R. R. Gaz.*, Jan. 10, 1890, Vol. XXII., p. 19.

Gondola. Standard 50,000-lb. Gives detailed drawing, with abstract from specifications for the standard 28-ton gondola car of the Newport News & Mississippi Valley Co. *R. R. Gaz.*, April 6, 1888

Twin Hopper 60,000-lb. Gives description, with bill of lumber and detailed drawing, with dimensions, of a twin hopper bottom gondola car having a capacity of 60,000 lbs. recently constructed for the Lehigh Valley Railroad. *R. R. Gaz.*, Sept. 14, 1888.

Freight. An essay by W. R. Brown on the construction of the best form of *freight cars* with special reference to economy in dead weight. *Van Nos. Eng. Mag.*, Vol. XVI., p. 138.

Freight Depreciation of. Gives a table showing the value of a freight car at any age, estimated at 6 per cent. per annum as per Master Car Builders' rules. *R. R. Gaz.*, April 27, 1888.

Grain Weighing and Transfer Car. Description and drawings. *Ry. Rev.*, June 6, 1891, pp. 358-9.

Iron. A short article making comparison between iron and wooden railway cars favorable to the adoption of the former. *Am. Mfr.*, Nov. 30, 1889.

Iron. Uses of in Freight Car Construction. A paper by G. W. Eitinger, read before Western Railway Club with discussion. *R. R. Gaz.* Nov. 30, and *Mast. Mech.*, December, 1888.

100,000-lb. Car. Penn. R. R. Gives drawing, showing details of a car of 100,000 lbs. capacity, designed for carrying cables for street railroads, and built for the Pennsylvania Railroad. *R. R. Gaz.*, May 11, 1888.

Ore.

Twenty-five Ton Iron Ore. Gives a two-page plate of detailed drawings of a twenty-five ton iron ore car used on the Swedish Railroad. *Lon. Engineer*, April 27, 1888.

Cars. *Ore*, continued.

Wisconsin Central Ore Cars—60,000 lbs. Capacity. Description with detailed drawings. *Ry. Rev.*, Oct. 17, 1891, p. 673. Brake rigging of these cars illustrated in *Ry. Rev.*, Oct. 24, 1891, p. 690.

Pullman Vestibule Sleeping. General and detail drawings and description. *R. R. Gaz.*, Nov. 23, 1888.

Thirty-Ton Combination Car, Chesapeake & Ohio Railway. Illustrated description. *R. R. Gaz.*, Feb. 1, 1889.

Sixty Thousand Pounds.

Union Pacific Railway. Description, principal dimensions, details of truck, etc. *Ry. Rev.*, April 12, 1890, Vol. XXX., p. 214.

Michigan Central Standard Car and Truck. Description with drawings. *Ry. Rev.*, Oct. 11, 1890, pp. 598-600.

Standard dimensions, forms of construction, etc., adopted by the Master Car-Builders' Association, with discussions on the same. See *Proc. M. C. B.* 20th annual convention.

Steel.

Harvey Steel Car. Description with drawings of this new car, built for the C. B. & Q. R. R. *Ry. Rev.*, Nov. 1, 1890, pp. 650-1, Dec. 13, p. 747.

Illustrated description of a new form of car built of steel, asbestos, iron pipe and wood. *R. R. Gaz.*, Jan. 4, 1889.

Schoen Pressed Steel Car. Description and details of this somewhat novel car. *R. R. Gaz.*, May 23, 1890, p. 352.

Stock Car. Detail drawings with description of the new Harvey Steel Car. *Eng. News*, July 11, 1891, pp. 24-5.

Thirty Ton Steel Cars—Belfast and Northern Counties Railway, Ireland. Designed for ore carrying. Both car and truck have steel frames. Description and drawings. *Ry. Rev.*, Aug. 15, 1891, pp. 534-5.

Cast Iron. See *Iron, Cast*.

Castings.

Defects in Structural Castings. By Thos. D. West. Gives many valuable facts concerning the design and execution of such work. *Jour. Assn. Eng. Soc.*, Vol. II., p. 247.

New Process of Making Ornamental Castings. Consists in lining the inside of the mould with carbonized lace or other textile fabric. Abstract of remarks made at the meeting of the Franklin Institute, April 20, 1887. By A. E. Outerbridge, Jr., *Jour. Frank. Inst.*, June, 1887, Vol. CXXIII., No. 738. Report of Franklin Institute Committee on same. *Jour. Frank. Inst.*, Nov., 1887, Vol. CXXIV., No. 743.

Sound. A paper by Thos. D. West, with discussion. Furnishes considerable valuable information to engineers as to the causes of defects in castings and their remedies. *Trans. A. S. M. E.*, Vol. VI., p. 91.

Cattle Guards. See *Railroad Structures*.

Cement Barrels. *Cubic Contents of*. Table exhibiting the cubical contents of the barrels used by nine cement manufacturers. *Eng. News*, Feb., 22, 1890, Vol. XXIII., p. 185.

Cement Laboratory *of the St. Louis Water-works Extension*. An article by S. Bent Russell, describing apparatus used and methods of testing. Illustrated. *Eng. News*, Jan. 3, 1891, pp. 2-4.

Cement Tests.

Carried on in the Department of Engineering, State University of Iowa. Many tests are recorded, and comparisons made between specimens hardening in air and those hardening in water. *The Transit*, Dec., 1890, pp. 7-40.

Different Forms of Briquettes. By J. E. Codman before the Philadelphia Engi-

Cement Tests, continued.

neers' Club. Gives results of testing cement in different forms of briquettes. *Proc. Eng. Club, Phila.*, December, 1887, Vol. VI., pp. 168-73.

Experiments with Testing Appliances. By Alfred Noble. On Sault Ste. Marie Canal Locks, with discussion. Illustrated. *Trans. A. S. C. E.*, Vol. IX. (1880), pp. 186-201.

Hints on. A paper read before the Engrs. Club of St. Louis. Discusses relative importance of various tests, especially that of fineness. *Jour. Assn. Eng Soc.*, Sept., 1891, pp. 431-[illegible].

How to Test the Strength of Cements. By J. Sondericker. Gives a description of an apparatus for testing cements, and presents some of the results obtained. *Jour. Assn. Eng. Soc.*, June, 1888, Vol. VII., pp. 207-[illegible]. Also *Trans. A. S. M. E.*, Vol. IX. (1888), pp. 172-84.

Long Time Tests. An answer, by experiments extending over six years, for Boston Main Drainage Works, of many questions relating to the use of cement. *Trans. A. S. C. E.*, Vol. XIV., p. 141. Also in *Eng. News*, July 4, 1885. Illustrated.

Gives data derived from breaking 200 briquettes at dates varying from 7 to 2,319 days of gauging. *San. Eng.*, July 2, 1887.

Methods.

By Edmund Yardley. Gives the methods employed and results obtained in testing cements for the Pennsylvania Railroad. *Trans. A. S. C. E.*, Vol. II. p. 143.

Answers to eight queries by the government chemist on methods of testing (should be tested wet) relative strength of Portland and light burned cements, effects of alumina, magnesia, alkali, sulphuric acid. *Eng. News*, Dec. 26, 1885.

Microscopic Method. Account of a microscopical examination of some cements, with discussion as to the value of the method. Illustrated. By Alden H. Brown, in *The Transit*, of the Univ. of Iowa. Reprint, *Ry. Rev.*, Nov. 7, 1891, pp. 726-7. *Eng. News*, Nov. 21, 1891.

Neat Tests vs. Sand Tests for Portland Cement. A paper by S. Bent Russell, discussing the results of a large series of experiments made for the St. Louis Water Works. Describes also a machine for mixing briquettes. Illustrated. *Trans. A. S. C. E.*, Vol. XXV., Sept., 1891, pp. 293-300. Discussion, pp. 300-4.

New Croton Aqueduct. Gives profile showing strength of cements used in the construction of the new Croton Aqueduct. *Eng. & Build. Rec.*, Aug. 18, 1888.

Notes and Experiments on the Use and Testing of Portland Cement. By Wm. W. Maclay. An exhaustive paper of great merit, giving principles for both testing and using. Illustrated. Received the Norman medal. *Trans. A. S. C. E.*, Vol. VI. (1877), pp. 311-64. Discussion, Vol. VII., p. 274, where cement tests for East River Bridge are given; also, p. 180.

Portland Cement Testing. By I. J. Mann, before the Institution of Engineers. Treats of the color, weight, pulverization, and tensile strength of Portland cement. *Van Nos. Eng. Mag.*, Vol. XVII., p. 17.

By T. Guillain. Gives details of the specification for the supply and testing of Portland cement used for harbor work at Calais and Boulogne. *Nouvelles de la Construction.* Vol. III., p. 59. *San. Eng.*, July 9, 1887. *Eng. News*, Sept. 3, 1887.

By H. Faija, before the American Society of Civil Engineers. Gives details of the method employed by himself for a number of years; shows details of apparatus for mixing and testing cements. *Trans. A. S. C. E.*, Vol. XVII., November, 1887), pp. 218-28.

On the Mechanical Examination and Testing of Portland. By Henry Faija.

Cement Tests, continued,

Considers details of manipulation and other matter affecting results in a cement test, and then the properties which a good cement should show. *Van Nos. Eng., Mag.*, Nov., 1884.

Very elaborate, both as to times allowed for setting and as to composition. Results plotted. Very valuable record. *Rep. Chf of Engrs.*, 1883, Vol. II., p. 1849.

Standard.

German Specifications for Standard Portland Cement Tests. Translated for the Testing Laboratory of Cornell University. Many of the requirements will appear novel to American readers. *Eng. News*, Nov. 13, 1886.

New German (1887) *Standard Rules* in. *Proc. Inst. C. E.*, Vol. XC., p. 474.

Report of a committee of the Am. Soc. of Civ. Engrs., prescribing a series of "standard tests" for fineness, checking, or cracking, and tensile strength. It would seem to be well for all tests in America to be made by the methods here prescribed in order to enable intelligent comparisons to be made. *Trans. A. S. C. E.*, Vol. XIV., p. 475, also in *Eng. News*, Dec. 19, 1885.

Tensile. An appliance for more accurate determinations. By D. J. Whittemore. With discussions. Illustrated. *Trans. A. S. C. E.*, Vol. IX. (1880), pp. 378-47.

Wet and Dry Sand. A table showing the great increase in bulk of sand, due to the addition of a small amount of water, and hence the variation in the quality of mortar from using wet sand. From Lon. *Engineer*. *Eng. News*, Feb. 21, 1891, p. 173.

Cement-Testing Machine. Home-made, at a cost of ten or twelve dollars. Seems to work satisfactorily. Full detail drawings given. *Proc. Eng. Club, Phila.*, Vol. V., p. 198.

Cement Works.

On the Lehigh. Brief description of the cement (Portland and Anchor) works and the deposits, with tables of analyses and results of tests. p. 9, Chap. XII. of Report of Frederick Prime, Jr., *Second Geol. Survey of Pa.*, *D. D.*, 1878.

Louisville. An account of the location and capacity of the mills on the Ohio River, manufacturing Louisville cements, with records of tests of most of the brands, made by the St. Louis inspector, Thos. D. Miller. *Jour. Assn. Eng. Soc.*, Vol. V., p. 187.

Porta Cement Works at Bremen. Description of plant and method of manufacture. Illustrations of buildings. Lon. *Eng.*, July 17, 1891, pp. 60-2.

Portland Cement Grinding Mill House. Illustrated description of, Lon. *Eng* January 30, 1891, pp. 130-2. *Eng Rec.*, February 18, 1891, p. 187.

Cement. *Action of Hydraulic Cements upon Embedded Metals.* Numerous experiments described and results given. By J. C. Trautwine. *R. R. Gaz.*, Jan. 27, 1882, p. 51.

Adhesive Strength of Portland. A paper by I. J. Mann, before the Inst. C. E., which refers principally to the adhesive strength of cement, and a description of improved method of determining its quality and value. *Van Nos. Eng. Mag.*, Vol. XXIX., p. 233.

Adhesive Strength of Sulphur, Lead and Portland Cement for Anchoring Bolts. Short article reprinted from *The Polytechnic*. Careful experiments showed that cement is stronger and more reliable than either lead or sulphur, and that its resistance is from 400 to 500 lbs. per sq. in. of surface exposed. *Eng. News*, July 19, 1890, p. 53. *Ry. Rev.*, Nov. 1, 1890, p. 650.

Adhesive Strength of, with Special Reference to an Improved Method of Testing. By I. J. Mann before the Inst. C. E. An elaborate series of tests. Reprint. *Eng. News.*, May. 19, 1883, p. 230-3.

American Material, with especial reference to Rosendale cement. By F. O.

Cement, continued.

Norton, manufacturer. Illustrated. *Trans. A. S. C. E.*, Vol. IX., (1880), p. 278-286.

American vs. Portland. An article by F. Collingwood going to show that for most purposes for which cement and concrete are used in this country, our own natural cements are quite "good enough." *San. Eng.*, Nov. 5, 1885.

And Mortar, Selection, Inspection and Use of. By S. F. Burnett, before the Engineers' Club of St. Louis. Gives practical hints in regard to the selection, inspection and action of cement and sand, and the methods of mixing and using, to produce a good mortar. *Jour. Assn. Eng. Soc.*, July, 1888. Vol. VII., pp. 255-264.

Compressive Strength of. Progress report of the American Society of Civil Engineers' Committee on the compressive strength of cements and the compression of mortar and settlement of masonry, with five plates. *Trans. A. S. C. E.*, Vol. XVII. (November, 1887), pp. 213-218. Abstracted, *Proc. Inst. C. E.*, Vol. XCIII., p. 526.

Formula for a number of useful cements. *Van Nos. Eng. Mag.*, Vol. IX., p. 265.

From Waste Product Lime. By J. S. Rigby, before the Society of Chemical Industry, Liverpool University. A valuable paper on the utilization of waste lime from chemical process for the manufacture of cement. *Sci. Am. Sup.*, June 26, 1888.

Hardening of. A series of articles embodying the results obtained in the most recent and important investigation on the hardening of cements. The subject is treated from a chemical point of view. Lon. *Engineer*, Sept. 21, *et seq.*, 1883.

Hydraulic, Chemically Considered. Extracts from the annual report of the Engineer Department of the District of Columbia. Seven brands of native and foreign cements compared in strength and chemical composition. *Eng. News*, Dec. 5, 1885.

Hydraulic Lime, Roman Cement, Portland Cement. Process of manufacture, chemical analyses, best proportions for ingredients and effect of variation in these proportions upon the strength of the cement. By Prof. Engler in Ladenburg's *Handwörterbuch der Chemie.* Translated by Alden H. Brown and published in *The Transit*, December, 1890. Reprinted in *Eng. News*, Feb. 7, 1891, pp. 123-7, *et seq.*

Influence of Sand upon. From the German. Results of experiments, showing remarkable differences in ultimate strength, due to different qualities of sand used. *Eng. News*, July 11, 1884.

Influence of Sugar upon. By H. DeB. Parsons, before the American Society of Mechanical Engineers. Gives details of tests of cements mixed with from ¼ to 2 per cent. of sugar and molasses. Molasses greatly retarded the setting, but at the end of two months the strength was rapidly increasing. With Portland cement one per cent. of sugar gave maximum strength. With Rosendale cement one-fourth per cent. gave maximum results. With large per cent of sugar the cements were too soft to take from the molds in 48 hours. The effect appears to be mechanical. *Trans. A. S. M. E.*, Vol. IX., pp. 286-291; *Eng. News*, Dec. 24, 1887; *Eng. & Build. Rec.*, Dec. 24, 1887; *Am. Eng.*, Jan. 18, 1888; Lon. *Engineer*, Jan. 27, 1888; Lon. *Eng.*, Jan. 28, 1888; *Mech. World*, Jan. 14, 1888.

Lime and Hydraulic. By F. W. Gibbs. Treats of the sources, composition and properties of limes and cements. *Trans. Arkansas Soc. C. E. Arch. & Surv.*, November, 1887, pp. 38-46.

Magnesia Cement, The Practical Application of. A paper by C. O. Weber, Ph. D., read before the Society of Chemical Industry, describing the properties of various cements, with several tests showing great strength. *Sci. Am. Sup.*, May 16, 1891, No. 802, pp. 12810-13.

Cement, continued.

Manufacture of.

And Properties of the Cements of the Isere. An abstract of an article from *Annales des P. & C.*, Vol. XVII., 1889, p. 318: in *Proc. Inst. C. E.*, Vol. XCVIII., p. 415, 1889.

Notes on the hydraulic or other cements at the Philadelphia exhibition. By Gen. Gillmore. Gives methods of manufacture and tables of crushing and tensile strength. *Van Nos. Eng. Mag.*, Vol. 16, p. 261.

Statement of the Industry. A statement showing the extent of the industry in the United States. *Eng. & Build. Rec.*, March 17, 1888.

In Ulster Co., N. Y. A description of the method of quarrying the rock and preparing the cement. Illus. *Eng. News*, Sept. 15, 1883, pp. 436-7.

Of the Gate of France. Nature of deposits, methods of manufacture, workings of an aerial cable for automatic transportation, the final uses and products. From *Le Genie Civil.* 19 illustrations. *Sci. Am. Sup.*, Nov. 28, 1885.

Paper read before the British Association by Gen. H. Y. D. Scott on the conversion of lime into cement by the Selenitic method, and the economy and other advantages of the process. *Van Nos. Eng. Mag.*, Vol. VII., p. 512.

Permeability of Cements, and their Decomposition under the Action of the Sea. By MM. Leon Durand-Claye, Eng. in Chief of Bridges and Roads, and Paul Debray, Ass't Eng. Translated by Prof. L. M. Haupt. An account of some valuable experiments on the above subject. Illustrated. *Jour. Frank. Inst.*, March, 1889.

Permeability of Portland Cement Mortar and its Decomposition by Sea Water. Abstract from *Annales des Ponts et Chaussees. Eng. & Build. Rec.*, Nov. 23, 1889.

Portland.

And Its Use. A series of articles by Hormosji Nowroji, giving a practical description of the manipulation and treatment of cement, mortar and concrete. *Ind. Eng.*, March 14, 1891, *et seq.*

For Engineering Works. Paper by Capt. W. W. Maclay, C. E., with an appendix containing abstracts of the new German rules for setting cements. Published and distributed gratuitously by James Brand, 81 and 83 Fulton St., New York.

Improvement in the Manufacture of. By Fred Ransome before the Manchester meeting of the Association for the Advancement of Science. Gives a description of a revolving furnace in which the fine cement material is calcined by passing through gas flames. *Am. Eng.*, Sept. 21, 1887.

Manufacture and Use of. By Dr. W. Michaelis. Gives analysis of Portland and Roman cements, and treats of selecting and mixing the materials, calcination, etc. *Van Nos. Eng. Mag.*, Vol. I., p. 306.

Manufacture of Portland cement described in a lecture by Calvin Brown, before the Tech. Soc. Pacific Coast. *Eng. Record*, May 30, 1891, pp. 415-6.

Ransome's Improvements in the Manufacture of. By R. J. Friswell. Reprinted from the London *Engineer*, March 4, 1887. *Jour. Frank. Inst.*, June and July, 1887, Vols. CXXIII., CXXIV., Nos. 738, 739.

Sources of strength. By Henry Faija. *Van Nos. Eng. Mag.*, Vol. XXII., p. 463.

Specifications for. Gives specifications for the supply and delivery of Portland cement for paving portions of streets known as King's road and Pont street, in Parish of Chelsea. *Eng. & Build. Rec.*, Dec. 31, 1887.

Rate of Hardening of. By Prof. W. C. Unwin. The increase in tensile strength found to follow the law of a cubical parabola, and of compressive strength the law of a common Parabola. Tests, equations and curves given for Portland cement. *Proc. Inst. C. E.*, Vol. LXXXIV., p. 399.

Cement, continued.

Relative Value of. A paper by Prof. C. D. Jameson and Hubert Remley, discussing the relative value of artificial and natural cements and based on experiments carried out at the Iowa State University, which are given in full in *The Transit. Pop. Sci. Mon.*, March, 1891, pp. 663-70.

Rock and Gypsum Deposits in Buffalo. Paper by Julius Pohlman, Buffalo, N. Y. *Proc. A. I. M. E.*, Buffalo meeting, October, 1888.

Slag.

Hydraulic. By R. J. Friswell. Discusses the composition of slag cements. *San. Engr.*, Nov. 12, 1889.

Manufacture and Properties of. A paper by G. R. Redgrave stating the conditions essential for a good cement, and giving analysis of slags and cements, and results of tests showing great strength. *Proc. Inst. C. E.*, Vol. CV., 1891, pp. 215-30.

See *Slag Cement.*

Use of. By G. R. Redgrave. Before the Architectural Association advocates the use of cements instead of lime in all buildings, and gives tables of comparative strength and costs. *Van Nos. Eng. Mag.*, Vol. XXIV., p. 121.

See *Concrete. Mortars.*

Center of Gravity.

A new formula of general applicability. By J. W. Davis. Gives formula and determines the extent of its application. *Van Nos. Eng. Mag.*, Vol. XX., p. 467.

Of a Polygon. A new graphical method for finding. By J. W. Davis. *Van Nos. Eng. Mag.*, Vol. XXI., p. 327.

Chaining.

See *Surveying, Land.*

Chains.

Use and Care of. By Henry Adams. A paper before the Society of Civil and Mechanical Engineers, on the use and care of chains for hoisting and hauling. Contains much useful information. *Lon. Engineer*, July 8, 1887.

Weldless Steel. Illustrations of machinery used, and description of process of manufacture. From Lon. *Engineer. Sci. Am. Sup.*, No. 819, Sept. 12, 1891.

Weldless Steel. Description, with results of tests of these chains, which are cut from a solid piece of steel. Lon. *Eng.*, May 1, 1891, p. 524; *R. R. Gaz.*, May 15, 1891, p. 337; *E. & M. Jour.*, May 16, 1891, p. 587.

Chanute, *Octave, Pres. A. S. C. E.* Biographical sketch, with portrait. *Eng. News*, May 23, 1891, p. 504.

Chart. *The Pilot Chart of the North Atlantic Ocean.* Lecture before the Franklin Institute, by Everett Hayden, of the United States Hydrographic Office. Presents a copy of the pilot chart for March, 1888, and interesting description of it. *Jour. Frank. Inst.*, April and May, 1888.

Chesbrough, *Ellis Sylvester.* A memorial by a committee of the Western Society of Engineers. *Jour. Assn. Eng. Soc.*, Vol. VI., p. 130.

A Memoir. By Benezette Williams. A very readable and satisfactory account of one of the greatest of American Engineers. *Jour. Assn. Eng. Soc.*, Vol. VI., p. 1.

Chimneys.

And Flues. Formula and Notes on Designing. Tables of capacity in lbs. of coal per hour, temperature and velocity of air, etc. From a forthcoming book by Geipel and Kilgour. *Electrician*, June 12, 1891, pp. 163-5.

Circular, 440 feet high, near Cologne. Description of this recent work. *Trans. Inst. C. E.*, Vol. LXXXV., p. 383.

Construction of. Continuation of a series of articles on ventilating chimney

Chimneys, continued.

shafts, moving, straightening and taking down chimneys. *Eng. News*, Oct. 31 and Nov. 14, 1885.

Construction of several large chimneys in this country, at Lawrence and Lowell, Mass., at New York City and at Johnstown, Pa. *Trans. A. S. C. E.*, April, 1885, and *Eng. News*, July 11 and 18, 1885.

Description of the chimney of the West Cumberland Hematite Iron Works, abstracted from a paper by Prof. Rankine, before the Inst. of C. E. in Scotland. *Van Nos. Eng. Mag.*, Vol. I., p. 146.

Dimensions in height and cross-section for a given horse-power of engine and of boiler capacity. By Wm. Lowe, in *Am. Mach.*, March 27, 1886.

Empirical Formula for Chimney Area. Formula and graphical diagram, as deduced and used by Mr. E. M. Hugentobler, M. E., engineer of Abendroth & Root Manufacturing Company. *Eng. News*, Dec. 7, 1889. Vol. XXII., p. 533.

For Furnaces, Fire-Places and Steam Boilers. By R. Armstrong. *Van Nos. Eng. Mag.*, Vol. IX., p. 130.

For Horizontal Tubular Boilers. Paper by Frederic Stamm. *Proc. Eng. Club, Phila.*, Vol. VII., No. 2, Feb., 1889, p. 105.

For the Narragansett Electric Lighting Company, Providence, R. I. Detailed account of its erection, with five plates of illustrations. By John T. Henthorn. *Trans. A. S. C. E.*, Vol. XXVI., July, 1891, pp. 1-7. Reprint, *Sci. Am. Sup.*, Oct. 31, 1891, No. 826, pp. 13191-2; *Eng. Rec.*, June 20, 1891, pp. 40-1.

Notes on the construction of a large chimney, being an iron frame lined with brick. *Sci. Am. Sup.*, Dec. 10, 1887; *Lon. Engineer*, Jan. 6, 1887.

Proper Dimensions of. By Wm. Kent. *Eng. News*, Oct. 2, 1886.

Proper Size of. By William Kent. A paper read before the Am. Soc. of Mech. Eng.; gives formulas and table of sizes for given horse-power of boilers. *Am. Eng.*, Nov. 28, 1884. Also *Mechanics*, Jan. 1885.

Righting those which have become inclined. A description of a successful method employed by removing courses of brick. *Iron*, Nov. 21, 1884. Another method in *Van Nos. Eng. Mag.*, Jan., 1885.

Stability of. By R. J. Hutton before the Society of Engineers. Proposes to point out some errors which have crept into the theory of the stability of chimneys, and to offer some considerations as to the economical application of the theory in practical designing. *Trans. Soc. Eng.*, 1882, pp. 153-184.

Two Modern. By C. P. Kerr. A study of the designs and construction of the chimney at the Clark Thread Company's Works at Newark, N. J., and the stack of the Marshall & Company Thread Works, at Newark, N. J. *Sci. Am. Sup.*, Jan. 29, 1887.

Wind Pressure. By Jas. B. Francis. Discusses dimensions usually adopted and estimate of wind pressure upon. *Eng. News*, Aug. 28, 1880, pp. 288-90.

Chimney Draft.

Idiosyncrasies of Chimney Draft. A table of data as to dimensions of chimney, fuel burned, and observed draft of several chimneys. From a paper by W. F. Crane before the Am. Soc. M. E. *Eng. News*, Dec. 5, 1891, p. 531.

Measurement of. By H. W. Spangler. A description of various methods used for this purpose, by which the amount of the draft is magnified for direct measurement. *Proc. Eng. Club, Phila.*, Vol. IV., No. 5.

Several papers on this subject were read by Prof. De Volson Wood, J. Burkitt Webb and W. Kent, at the Cincinnati meeting of the A. S. M. E. *Trans. A. S. M. E.*, Vol. XI., 1890. Abstracts in *Mechanics*, July, 1890, pp. 168-72.

Theory and Design of Chimneys. Paper before the American Society of Mechanical Engineers by Horace B. Gale, bringing out a new theory of chimney draught, with some useful practical formulæ. *Trans. A. S. M. E.*, Vol. XI., 1890, pp. 451-91. Abstract in *Mechanics*, Dec., 1890.

See *Draught Gauge.*

Chimney Gases. *Analysis of.* By apparatus designed by Prof. Elliott, of Columbia, Cal. Cut of instrument, methods of use and chemical formulas. *Mechanics*, Jan., 1883.

Chlorine. *Manufacture of.* Brief outline of the Weldon-Pechiney process for the manufacture of. *E. & M. Jour.*, Nov. 24, 1888.

Chordel, *and its Application to the General Section of an Angle.* By J. B. Millar. *Van Nos. Eng. Mag.*, Vol. XXII., p. 206.

Circular Arc. See *Arc.*

City Engineering. See *Municipal.*

City Lots, *Recording of.* See *Registry Bureaux, Phila.*

City of Mexico. See *Mexico City.*

Civil Engineering.

Address of Sir J. W. Bazalgette, before the Institution of Civil Engineers. Contains a good description of the public works of London, also of Paris and New York City. *Van Nos. Eng. Mag.*, April, 1884.

And Surveying Progress, Report on, by a committee of the Civil Engineers' Club of Cleveland, read Oct. 9, 1888. *Jour. Assn. Eng. Soc.*, January, 1889.

In England, Notes on. Paper by William H. Searles, Member Civil Engineers' Club of Cleveland. Read September 10, 1889, p. 7. Discussion by members, p. 8. *Jour. Assn. Eng. Soc.*, Nov., 1889. Vol. VIII., pp. 553-62.

Societies. See *Engineering Societies.*

Civil Engineers.

Education of. By Thomas C. Clarke. With discussion. *Trans. A. S. C. E.*, Vol. III. (1874), pp. 255-66.

Training for Students. By Geo. L. Vose. *Van Nos. Eng. Mag.*, Vol. XXVIII., p. 191.

Clay *for Puddling, New Method of Using it.* It is dried, pulverized and compacted in the interior of the embankment. On swelling with moisture it becomes impervious to water. *Am. Eng.*, Oct. 13, 1886.

Varieties of, and their Distinguishing Qualities for Making Good Puddle. A paper by William Gallon, Assoc. Member Inst. C. E. *Eng. News*, Dec. 1, 1888.

Climate, *Relation to Health.* See *Health.*

Coal.

Anthracite. Analyses of large commercial samples, with full analytical data. *II. Geol. Survey of Penn.*, 1883, Vol. A. A.

Fields of Pennsylvania, a description of. By Charles A. Ashburner, of the Geol. Survey of Penn. A 30-page article, giving the Geography, History, Topography, Structural Geology, Stratigraphical Geology of the region, together with the Composition and Origin of this Coal, methods of mining, and production statistics. *Proc. Eng. Club. Phila.*, Vol. IV., No. 3.

Relative Value to the Consumer of certain Pennsylvania anthracites. By H. M. Chance. *Trans. A. I. M. E.*, Vol. XIII.

Table giving weights per cubic foot and specific gravity of anthracite broken to market sizes. *E. & M. Jour.*, June 1, 1889.

Bietrix Briquette Machinery for utilizing coal slack. Three and a half million tons made and used in Europe annually. Described and illustrated. *E. & M. Jour.*, Dec. 11, 1886. *Age of Steel*, Nov. 13, 1886.

Comparative Merits of Bituminous and Anthracite Coals. Fully considered for naval uses by special board of naval officers. *Iron Age*, May 21, 1885.

Consumption of, as Affected by Enginemen. Article by Geo. H. Baker, of considerable value. *R. R. Gaz.*, Aug. 2, 1889; *Mast. Mech.*, September, 1889.

Heating Power of Illinois Coal. Table of comparative tests. Thesis by R. B.

Coal, continued.

McConney and F. H. Clark. Abstract in *Technograph*, (Univ. of Ill. Annual), 1890-91, pp. 22-3.

How to Analyse. An article describing the methods for the determination of the various constituents in coal which are considered best. Lon. *Engineer*, April 20, 1888.

Tests of Bituminous for Steam Heating. Results of extended experiments by the Cincinnati Water-Works of four kinds, viz.: Youghiogheny No. 2, lump; Kanawha "Winefrede" lump; Kanawha Campbell's Creek lump; and "Peach Orchard," Ky., lump. Each test extending over 48 hrs. Complete results and coal analysis given. *Van Nos. Eng. Mag.*, Dec., 1884.

Testing the Relative Value of Different Kinds of Coal. By Wm. Kent, M. E. Table. *E. & M. Jour.*, July 19, 1890, pp. 66-7.

Washing. By Arthur Beckwith. Shows the different forms of machine and makes comparison of them. *Van Nos. Eng. Mag.*, Vol. II., p. 317.

Washing for Coke Production. The Bell-Ramsey machine described and illustrated. *Mechanics*, Nov., 1884.

See *Combustion. Minerals. Mining. Mines.*

Coal Handling Machinery.

Trade publication of the C. W. Hunt Co., N. Y., containing fully illustrated description of various machinery, including a steam coal shovel, many coal docks and data as to cost of handling, etc. Pamp., pp. 48.

Separating and Washing Plant. Illustrated description of a plant for cleaning very dirty coal. Lon. *Eng.*, February 13, 1891, pp. 184-5.

Simmerly's system of elevating derrick and buckets for reducing breakage loss. *Iron Age*, Oct. 8, 1885.

The Iron Breakers at Drifton, with a description of some of the machinery used for handling and preparing coal at Cross Creek collieries. A valuable paper by Eckley B. Coxe, describing at some length the machinery used, means employed for elevating the coal, the breaker, and the preparation of the coal. Forty-two plates. *Trans A. I. M. E.*, 1890, pp. 77.

Coal Tar.

Distillation. A lecture before the Franklin Institute, by Prof. S. P. Sadtler. Gives several methods, the products, etc. Illustrated. *Jour. Frank. Inst.*, Feb. 1886, *et seq.*

Coast Defences. *Abstract of Lecture by Gen. H. L. Abbott*, before the Academy of Sciences in New York, March 31. Reprinted from the *New York Herald*, by *R. R. & Eng. Jour.*, April, 1887.

Of the United States. A summary of the reports to the Secretary of War, of General Duane, Chief of the Engineer Corps, and General Benet, Chief of Ordnance for the army. *R. R. & Eng. Jour.*, December, 1887.

See *Forts.*

Coffer Dam.

And Concrete Mixer at Mare Island Dry Dock, California. A brief description of this large coffer dam which stood about 12 years, is abstracted from a paper by Otto V. Geldern in the *Trans. Tech. Soc. Pac. Coast. Eng. News*, Jan. 31, 1891, p. 106.

And Crib of the Point Pleasant Railroad Bridge, W. Va. A coffer dam, with puddle-wall, inclosing an inner caisson or crib, inside of which the excavation is made. Illustrated. *Eng. News*, March 14, 1885.

And Floating Caissons. A valuable paper by Randell Hunt, C. E. Illustrated. Abstracted from *Proc. Tech. Soc. Pac. Coast. E. & M. Jour.*, July 19 and 26, 1890, pp. 77-9, 98-100.

Stopping the Leak in the Coffer-Dam at the Sault Ste. Marie Canal. A very large leak stopped by stock-ramming. *Eng. News*, May 9, 1891, pp. 450-3. Short note describing the break. *ibid.*, Apr. 11, 1891, p. 357.

Coffer Dams.

In the Connecticut River. By W. W. H. Bunall. Gives description of structure built in the river, and the difficulties experienced. *Van Nos. Eng. Mag.*, Vol. XIV., p. 366.

By J. J. de Kinder and E. F. Smith. Two short papers, with drawings. *Proc. Eng. Club, Phila.*, Vol. IV., No. 4.

Coke.

As a Fuel for Steam Boilers. By A. Ehreadorfer. Concludes that fuel may be advantageously used in almost all cases, with proper arrangements, and is to be recommended especially where smoke is a nuisance. *Journal fur Gasbeleuchtung und Wasserversorgung*, 1886, pp. 35-41.

Bauer's Ovens. Describes a group of ovens so arranged that they can be worked continuously with or without condensing apparatus. *Lon. Eng.*, Nov. 11, 1887. *Lon. Engineer*, Jan. 20, 1888.

Best oven for coking coal for furnace use. Discussion of the subject with tables of cost. By John Fulton, E. M. Pp. 18. Appendix, *Report 11, Second Geological Survey of Pa.*, 1877.

Manufacture of. Being part of Vol. X. of U. S. Census, 1880, 4to. 110 pp. Gives practice in U. S. and Europe with statistics. *H. Rep. Misc. Doc.*, 42, pt. 10, 47th *Congr.*, 2d *Sess.*, 1884.

Manufacture of, from Illinois Coal. By Henry Leubbers. Describes the process as carried out in Western Illinois. The coke is made from the slack crushed and washed to free it from impurities. It weighs about 38 lbs. per bushel, and the yield is about 60 per cent. of the weight of washed coal. *Trans. A. S. C. E.*, Vol. II., 163.

Simon-Carve's Coking Process. Illustrated description of this process for coking inferior coals by utilizing the waste products. *Iron Age*, Dec. 11, 1890, pp. 1036-8.

Value of. Discussion of a paper on the use of coke in blast-furnaces. Read before the Iron and Steel Institute by I. Lowthian Bell. *Lon. Eng.*, May 15, 1885.

See *Coal Washing.*

Cold Storage.

See *Ice Machines. Freezing Mixtures.*

Columns.

Comparison of Formulæ for the Strength of. By Th. H. Johnson, C. E., of Pittsburgh, Pa. *Eng. News*, Dec. 22, 1888.

Experiments on the Strength of Bessemer Steel Bridge Compression Members, giving results and discussion of a series of eight tests on full sized compression members, made by Mr. James G. Dagron, M. Am. Soc. C. E. Pp 8, 4 plates. *Trans. A. S. C. E.*, June, 1889, Vol. XX., p. 254.

Formulæ. A highly mathematical paper by L. M. Haskins, taking exceptions to the common theory of flexure. *Van Nos. Eng. Mag.*, Nov., 1886.

Formula for Strength of. By Th. H. Johnson. The formula is Euler's for very long columns and the equation of a tangent to this curve for short lengths. They give results agreeing remarkably well with the recorded experiments, and are likely to come into use. Formulas given for wrought iron, steel and wood; for flat, hinged and round ends. Also curves and plots of observation. *Trans. A. S. C. E.*, Vol. XV., p. 517. Also theoretical derivation of column formulas. By R. Krohn, p. 537.

High Unbraced. See *Building Construction.*

Notes on the strength of wrought-iron columns. By C. E. Moore. *Am. Eng.*, April 17, 1881.

Of Cast and Wrought Iron, Brick, Concrete and the various building stones. Their behavior in case of fire. Results of experiments in Berlin. Cast-iron

Columns, continued.

shown to be much better than wrought-iron, not so good as cement or brick, but much better than the stone. *Proc. Inst. C. E.*, Vol. LXXXII., p. 394.

Practical Strength of Long Columns. By Thos. C. Fidler. Recommends a modified form of Euler's formula, using the elastic limit instead of the ultimate strength of the material. Illustrated. *Proc. Inst. C. E.*, Vol. LXXXVI., p. 261.

Resistance of the Action of Fire and sudden cooling by water. Many materials experimented on by Prof. Bauschinger, and cast iron found to be the best. *Van. Nos. Eng. Mag.*, Vol. XXXV., p. 455, Dec., 1886.

Strength of. By E. Hauel. Derives formulæ for long columns and applies results to cast and wrought iron and stone columns. *Van Nos. Eng. Mag.*, Vol. 17, p. 461.

Strength of Iron. Part LII. of "Safe Building," by Louis DeCoppet Berg, has several useful diagrams giving strength of various sizes of hollow cylindrical columns and also factors for use in designing wrought iron columns. *Am. Arch.*, No. 793, March 7, 1891, pp. 151-5.

Tables of the Strength of Cast Iron. By Edwin Thacher. Gives tables to facilitate the use of Hodgkinson's formulæ. They give the breaking weight in pounds per square inch of section of cylindrical or octagonal columns from 4 to 60 diameters, and for all thicknesses of metal from solid to one-sixteenth the external diameter. *Trans. A. S. C. E.*, Vol. II., p. 294.

Tests. See *Viaduct, Antofagasta Ry.*

Tests of Pin-Ended Columns. Completion and discussion of results of tests made at the Watertown arsenal. By Theodore Cooper. Feb. 3, 1883, pp. 35-6.

Theory of the Strength. By Ward Baldwin. Derives and discusses Gordon's and Hodgkinson's formulæ. *Van. Nos. Eng. Mag.*, Vol. XXII., p. 353.

Wooden. See *Wooden Columns. Tests of full sized*

Wrought Iron. The Strength of. By G. Bouscaren. The results of the tests made for the Cincinnati Southern Railway brought together and discussed. This, perhaps the most valuable series of tests of large columns ever made. Plates given showing tests in detail. *Trans. A. S. C. E.*, Vol. IX. (1880), p. 447-54.

Z-Iron. Experiments on. By C. L. Strobel before the American Society of Civil Engineers. Gives details of the testing of 15 columns made of four "Z"-shaped iron bars. 5 plates. *Trans. A. S. C. E.*, Vol. XVIII., April, 1888, pp. 103-16. Abstracted *R. R. Gaz.*, July 13, and *Eng. & Build. Rec.*, Dec. 3, 1887.

See *Bridge Members. Fire Effect on. Pillars. Struts.*

Combustion.

Heat of. By R. H. Buel. Treats of the heating powers of fuels as compared with that obtained from the results of chemical analysis. Gives tables of experiments on the heat of combustion of various compounds and coals. *R. R. Gaz.*, July 27, 1888.

Forced. A paper read before the Inst. of Naval Arch. By James Howden. Gives results of forced combustion on the "City of New York." Shows an increase in the number of revolutions and decrease in amount of fuel used. Illustrated. Lon. *Eng.*, May 7, 21 and 28, 1885.

Forced Draught. By James Howden. Advocates forced draught. Gives description of different methods in use. *Van Nos. Eng. Mag.*, Sept., 1884.

Lecture before the engineering students of Cornell University on *Fire.* By J. C. Hoadley. A graphic description of the mechanical and chemical changes and methods of most economic use of fuel. Illustrated. *Sci. Am. Sup.*, May 8, 1886.

Natural and Forced Draught. A description of a great many experiments made in England, with results and conclusions. By James Howden. Reprinted from *Iron*, in *Van Nos. Eng. Mag.*, Vol. III., p. 248. (Sept. 1884).

Combustion, continued.

Natural and Forced Draught. By W. G. Spence before the Northeast Coast Institution. A valuable paper giving the results of experiments with forced and natural drafts. Lon. *Eng.*, Feb. 10, *et seq.*, 1888. An editorial on the above paper, Lon. *Engineer*, Feb. 10, 1888 and the article Feb. 17, 1888.

Notes on. By C. Chomienne, engineer. Composition of Coal. The Grate, the Ash-Pan, etc. *R. R. & Eng. Jour.*, August 11, 1891, pp. 357-60, *et seq.*

With especial reference to Practical Requirements. A paper by Fredk. Siemens before the Iron and Steel Inst. An able discussion of the essential conditions of perfect combustion for heating purposes. *Van Nos. Eng. Mag.*, Vol. XXXV., p. 437, Dec., 1886.

Compass.

Deviation on Shipboard. A paper by H. C. Pearsons, in *Proc. Mich. Eng. Soc.*, 1885. Illustrated.

Deviation of vessels and the conversion of compass courses. By H. C. Pearsons. *Report Mich. Assn. Surv.*, 1882.

Marine. Sir William Thompson's new compass. Fully illustrated. Lon. *Engineer*, Jan. 1, 1886.

See *Pocket Compass.*

Compressed Air.

Applied to Mining. A paper before the Liverpool Society of Engineers. A general discussion of the advantages derived from its use. *Eng. News*, April 11, 1885.

Efficiency of. By Prof. C. M. Woodward. A development of the fundamental equations, with illustrations showing effects of jacketing and compounding. *Jour. Assn. Eng. Soc.*, Vol. III., p. 101.

Flow of, Through Long Pipes. By E. Stockalper. Gives details of experiments made at the St. Gothard tunnel. *Van Nos. Eng. Mag.*, Vol. XXIV., p. 96.

For Blast in Cupola Furnaces. Brief notice of the successful application of compressed air direct to a cupola furnace for the melting of iron for casting purposes. *E. & M. Jour.*, Nov. 24, 1888.

Living Force of. By M. Perrigault. *Van Nos. Eng. Mag.*, Vol. VI., p. 274.

Motor, Mekarski's Compressed Air. Gives an illustrated description of the Mekarski compressed air motor for street railroads, several of which are now working on a London road. Lon. *Eng.*, March 23, 1883.

Motors. Latest Developments in Compressed Air Motors for Tramways. A paper by D. S. Jacobus, describing the railroads at Nantes, and from Vincennes to Nogent, with records of tests, etc., and estimating the cost of a line in the United States. *Trans. A. I. M. E.*, 1890, pp. 19.

Popp System of Distribution of Power by Compressed Air in Paris. Article giving the substance of a recent paper by Prof. Alexander B. W. Kennedy before British Association. *Eng. News*, Dec. 14, 1889, Vol. XXII., p. 456.

Physiological Effects of. By C. M. Woodward. Gives full account of all matters relating to the use of compressed air during the building of the St. Louis Bridge. *Van Nos. Eng. Mag.*, Vol. XXVI., p. 89.

Plant. See *Aqueduct, New Croton.*

Production. A lecture by Wm. L. Saunders, C. E., before the students of Sibley College, Cornell University. The theory of air compression is briefly taken up, after which the various types of compressors, details of construction and necessary requirements in design are discussed and fully illustrated. *Sci. Am. Sup.*, No. 799, April 25, 1891, pp. 12763-8.

Results of the experiments made by order of the Italian Government at the Mt. Cenis tunnel. *Van Nos. Eng. Mag.*, Vol. VI., p. 63.

See *Air Compressors, Street Railways.*

Compression Members. See *Bridge Members.*

Concrete.

And Iron to Resist Transverse Strains. By G. W. Percy before the Technical Society of the Pacific Coast. Gives details of experiments made on compound iron and concrete beams. *Eng. News*, Sept. 8, 1888.

Arch. See *Arches, Monier Method.*

As a Substitute for Masonry in Bridge Work. Paper by M. Murphy before Canadian Society of Civil Engineers. *Trans. Can. Soc. C. E.*, Vol. II., pp. 79-111, 1888.

Beams. See *Beams.*

Best Materials for. Treats of the best materials for it, the proper proportion of lime or cement; its strength and loss of bulk. *Van Nos. Eng. Mag.*, Vol. V., p. 875.

Blocks of Large Size made by manual labor for Department of Docks, New York. Details of process. Size from 13 to 60 tons weight. Tests given of crushing strength. Illustrated. *Trans A. S. C. E.*, Vol. IV. (1875), pp. 93-103. Discussion, pp. 312-317.

Building. By J. H. Owens, before the Arch. Asso. of Ireland. Gives detailed instructions for constructing building of concrete. *Van Nos. Eng. Mag.*, Vol. VIII., p. 113.

Building. An account of the erection of the Government offices of concrete, at Simla, India. By Walter Smith. Illustrated. *Trans. Inst. C. E.*, Vol. LXXXIII., p. 390.

Cost of. By O. E. Michaelis. Gives cost of concrete used in constructing the foundations for the new plant of the Troy Steel and Iron Company on Breaker Island; 9,803 cubic yards of concrete, costing $3.52 43-100 per cubic yard, were used. Also contains a large number of cement tests. *Trans. A. S. C. E.*, Dec., 1887.

Effect of Low Temperature on. By P. M. Bruner before the Engineers' Club of St. Louis. Discusses the effect of low temperature on Portland cement concrete. *Jour. Assn. Eng. Soc.*, April, 1888.

Erection. A new adaptation of Portland cement concrete to buildings. The walls formed of molded blocks on the two outer sides of the wall, with concrete filling. Well illustrated. *Iron*, Sept. 25, 1885.

Floors. See *Bridge Floors.*

In Harbor Works. Some data as to use in existing structures. Lon. *Eng.*, Apr. 17, 1891, pp. 467-70

In Sea Water. An abstract from the report of P. J. Messent to the Aberdeen Harbor Board. Gives as a cause for the failure of some of the concrete work at the Aberdeen Graving dock, injudicious specification for the cement or improper method of mixing it. Lon. *Eng.*, Jan. 28, 1886.

Mixer. See *Coffer Dam.*

Mixing and Handling. Abstract from a paper by W. T. Learned before the New England Water Works Association. Gives details of the methods employed at the Ashland Basin No. 4 Boston Water-Works. *Eng. News*, Dec. 24, 1887.

Notes on. By John Lundie. Gives notes on the selection of material, mixing depositing concrete in place. *Jour. Assn. Eng. Soc.*, December, 1887, Vol. VI., pp. 437-440. *Sci. Am. Sup.*, April 28, 1888. *Mech. World*, May 12 and 19, 1888.

Piers, Failure of in Scotland from too small a proportion of cement. *Eng. News*, Dec. 11, 1886.

Piers. See *Bridge Piers.*

Concrete, continued.

Portland Cement.

Porosity of, as determined by experiments on absorption of different mixtures. *Trans. Liverpool Eng. Soc.*, Vol. VI., p. 36.

Ingredients and graduation of sizes for maximum strength, crushing resistance, methods of mixing, conditions of economic use in place of masonry, etc., by Welfrid S. Boult. A standard article by an experienced engineer. *Trans. Liverpool Eng. Soc.*, Vol. I., p. 61.

Crushing Strength of in large blocks. Some experiments in, in *Trans. A. S. C. E.*, Vol. IV. (1875), p. 98.

Under Water.

New plant for depositing concrete under water. *Annales des P. & C.*, April, 1885. Also *Eng. News*, June 13, 1885.

By W. R. Kinipple. Gives the results of thirty years experience in the use of concrete. Describes the method employed at various works. *Proc. Inst. C. E.*, Vol. LXXXVII., p. 65.

Some experiments, by grouting ballast previously thrown in. By H. F. White. *Ind. Eng.*, Nov. 1, 1890, p. 352. Abstracted in *Eng. News*, Dec. 20, 1890, p. 548.

Used to build a harbor pier. Scow-loads sunk in a body by lining scow with sail cloth, then sewing upper edges of cloth together over the load, and sinking the whole through movable scow-bottom. Also concrete mixer for this work. Haven harbor, in England. *Zeitschrift f. Bauwesen*, 1884-300.

Work for Harbors. 100 pages of discussion and correspondence on the above subject by a large number of engineers. *Proc. Inst. of C. E.*, Vol. LXXXVII., p. 134.

See *Beton. Foundations. Harbor Improvements. Asphalt.*

Condensers.

Air-Surface. Explains their principles and their proportions. *Van Nos. Eng. Mag.*, Vol. L. p. 257.

By W. B. Coggswell. Gives details of the use of a surface condenser in connection with a set of blast furnace boilers at the Franklin Iron Works, Oneida, N. Y. Illustrated. *Trans. A. S. C. E.*, Vol. II., p. 41.

For Steam Engines, by J. H. Kinealy, read before the Engineers' Club of St. Louis, Dec. 5, 1889. *Jour. Assn. Eng. Soc.*, Jan., 1890.

Conduits. *Proposed, New York.* Gives substance of a report to the Commissioner of Public Works of New York as to feasibility and cost of removal to subways under sidewalks of all pipes, conduits, wires, etc., now buried under street pavements. *Eng. & Build. Rec.*, Aug. 4, 1888.

Sections of many forms illustrated. *Eng. News*, Feb. 8, 1879, pp. 44.

Connecting Rods. See *Locomotive, Connecting Rods.*

Continuous Girders. See *Bridges, Girders.*

Contractor's Plant. *Mechanical Appliances for Civil Engineering Works in India.* Articles describing various appliances used in prosecuting engineering works, No. I. Portable Engines and Hoisting Engines. By W. A. Francken, M. I. M. E. *Ind. Eng.*, Sept. 12, 1891, pp. 207-8. Shafting, Belting, etc., Centrifugal Pumps, Mortar Mills and Inspectors, *id.*, Sept. 19, pp. 227-8.

Contracts.

Instruction to Bidders and General Conditions used on municipal works in Washington City, D. C. *Eng. News*, Oct. 23, 1886.

Regulation of. Two essays on how far men should now be left free to make their own bargains. By A. T. Hadley and W. G. Sumner in *Science*, March 5, 1886.

Standard Building. Gives text of a standard building contract, the adoption of which is advised by the Committee of Conference of the American Institute o

Contracts, continued.

Architects, the Western Association of Architects and the National Association of Builders. *Eng. & Build. Rec.*, Sept. 15, 1888.

See *Architect. Engineers.*

Copper.

Analysis of. By A. A. Blair. A paper giving the methods used for the analysis of copper. *Report of Board on Testing, etc.*, 1881, Vol. I., pp. 247-265.

Effect of Heat on. See *Heat and Cold.*

Elmore Process of Depositing. This process, developed in England, is by electrically depositing the copper on revolving mandrels and at the same time rubbing or burnishing it by agate rubbers. This product has a tensile strength as high as 41 tons per sq. in., and a greatly increased electrical conductivity. Briefly described. *Eng. News*, Feb. 14, 1891, pp. 147-8.

Industry, Review of for 1888. "Never before has information of such enormous practical value and of such serious import been published for the benefit of the copper industry." *E. & M. Jour.*, Jan. 12, 1889.

Joints. The Strength of Riveted Copper Joints. Account of a test on a copper vessel, tested to destruction; also, tests on a rolled copper plate. *Lon. Eng.*, Apr. 24, 1891, pp. 497-9.

Resources of the United States. A paper by James Douglas, reviewing the state of the present supply from various districts, and the probable source of future supply. *Trans. A. I. M. E.*, 1890, pp. 26.

Tempered Copper. Report of committee of Frank. Inst., with table of tests. *Ry. Rev.*, Dec. 17, 1892, pp. 749-50.

Treatment of Copper Slates at Mansfeldt. A thorough description of the processes used at this mine in Germany. The ore is worked under great disadvantages, but the amount of copper produced is very large. Paper by T. Egleston, Ph. D. *School of Mines Quar.*, January, 1891, pp. 85-117, *et seq.*

Corrosion.

And Fouling of Steel and Iron Ships. An exhaustive paper by Prof. V. B. Lewis. *Sci. Am. Sup.*, Aug. 3, 1889.

Effects of Steel on Iron in Salt Water. By J. Farquharson, M. I. N. A. A paper giving an account of experiments by the author. Reprinted from *Trans. Inst. Naval Architects*, by Bureau of Navigation, Navy Department. Naval Professional Papers No. 14.

Investigation of causes under various circumstances and in various metals, with discussion of the efficiency of preservatives. *Mech. World*, May 30 and June 13, 1891.

Of an Iron Water Main. By P. D. Borden, Jr., and Wm. Ripley Nichols. An exhaustive account, with chemical analyses, of a peculiar case of corrosion, wherein the iron is reduced to a state similar to plumbago. *Jour. Assn. Eng. Soc.*, Vol. IV., p. 274.

Of Cast-Iron Pipes. Causes, methods of action, effects, etc. By Sam'l McElroy. *Jour. Assn. Eng. Soc.*, Vol. II., p. 287.

Of Metals during long exposure to Sea Water. The results of numerous experiments extended over several years. *Proc. Inst. C. E.*, Vol. LXXXII., p. 281.

Corrugated Iron, *The Strength of.* By J. E. Hart. Contains experiments on the transverse strength of corrugated iron. *Bombay Builder*, Aug. 1868; Lon. *Engineer*, Nov. 13, 1868; *Van Nos. Eng. Mag.*, Vol. I., p. 129.

Counterbalancing. See *Engines, Steam, Counterbalancing. Locomotive Counterbalancing.*

Cranes.

As Labor-Saving-Machines. By C. J. Appleby, with discussion. Illustrated by many cuts. *Trans. A. S. C. E.*, Vol. XV., p. 369.

Cranes, continued.

Atlas, 100-Ton Floating. Illustrated description of the cranes constructed for the Mersey Dock and Harbor Board. *Eng. News*, Jan. 15, 1887.

Derrick Plant Used in the Erection of the Equitable Building at Denver, Col. Description of this plant of six Norcross derricks, with full detail drawings. *Eng. News*, Aug. 22, 1891, pp. 156-9.

Description of a crane, 30 meters high, used in the construction of the church of Montmartre. *Mem. de la Societe des Ing. Civ.*, August 1883, pp. 208-221. (With 3 plates.)

Electric.

Application of Electricity to Cranes. Paper by H. D. Wilkinson, describing briefly their construction. *Electrician*, Oct. 31, 1890, pp. 736-7.

Brief description of a portable, self-propelling electric crane to lift 15 cwt., recently constructed and now in operation. Lon. *Eng.*, *Elec. Review*, Nov. 30, 1888.

Full description of the machine used by the Compagnie des Entrepots et Magasins, Paris, for handling bales of wool. Illustrated. *Sci. Am. Sup.*, May 22, 1886.

Erecting Crane Used in the Hudson River Tunnel. Described and illustrated in *Eng. & Build. Rec.*, October 25, 1890, p. 327; *Eng. News*, November 15, 1890, p. 439.

Five Ton Portable Steam Crane. Account of a crane of considerable value as in handling heavy freight at stations. *Eng. News*, April 5, 1890, Vol. XXIII., p. 321.

Floating. Calculations for Constructing. Full analysis. From *Annales des Travaux Public.* *Eng. News*, Nov. 5, 1881, pp. 444-6.

For Railroad Use. A series of papers by Wm. L. Clemens, M. E., describing recent improvements in the application of cranes for railroad uses. Illustrated. *R. R. Gaz.*, Aug. 14, 1891, pp. 558-9, *et seq.*

"*Goliath,*" *Twelve Ton Steam.* Gives a two-page plate showing details of a twelve-ton steam travelling crane. It has a span of 60 feet, and a clear height of 28 feet. Lon. *Eng.*, Jan. 13, 1888.

High-Speed Rope-Driven Cranes. Adapted to any width of span and any length of travel. Illustrated description. *R. R. Gaz.*, July 10, 1891, p. 470.

Hydraulic.

Hydraulic Charging Crane. Description with inset of drawings. *Eng. News*, Feb. 7, 1891, p. 138.

Ridgeway Balanced Steam Hydraulic Crane. Can be operated by water from city water works, also by a combination of steam and water. Illustrated description. *R. R. Gaz.*, July 31, 1891, p. 527.

Traveling Cranes. A paper by Erwin Graves read before the Am. Soc. M. E. at the Richmond meeting. *Rr. Rev.*, Dec. 6, 1890, p. 732.

1½ and 5-Ton Portable Cranes. Complete computations with stress diagrams. Translated from *Der Praktische Maschinen Constructeur*, *Mech. World*, Oct. 11, 1890, pp. 146-8, *et seq.*

Sixty-Ton Derrick Crane at Port Glasgow, described and cut given. Lon. *Eng.*, Dec. 11, 1885.

Thirty-five Ton Accident and Construction Crane. Illustrated description of this powerful and convenient form of crane for railway purposes. *R. R. Gaz.*, March 28, 1890, Vol. XXII., p. 209.

Traveling.

Compound Trusses for Traveling Crane. Description of a novel method of constructing trusses for traveling crane, when the head room limited the depth to 8½ feet. By Mr. Henry B. Seaman, M. Am. Soc. C. E. Illus. *R. R. Gaz.*, August 23, 1889, Vol. XXI., p. 550.

Cranes, *Traveling*, continued.

Description of the largest traveling crane in the world, built for the U. S. Navy Dept., with capacity of 246,400 lbs. *R. R. Gaz.*, June 20, 1890, pp. 431-4.

Gives a description, with full details, of a six-ton universal traveling crane for the erection of the Union Elevated Railroad, Brooklyn. *Sci. Eng.*, Dec. 17, 1887.

Illustrated description of a traveling crane, 30 feet span, to lift three tons. Lon. *Engineer*, Sept. 9, 1887.

Jib Crane. Illustrated description of crane designed for use in the erection of high buildings. *R. R. Gaz.*, Sept. 4, 1891, p. 614.

Tre-ton Electric Traveling Cranes at Paris Exposition. Description and illustrations of this new application of electricity to ordinary traveling cranes. *Sci. Am. Sup.*, Sept. 21, 1889.

Water, for Indian Railroads. Gives brief description and full detailed drawing of water cranes to be used on the Indian State Railways. Lon. *Engineer*, Oct 7, 1887.

Wharf. Drawings of a twenty-five-ton wharf crane and a three-ton locomotive crane in Lon. *Engineer*, April 6, 1888.

Crank Shafts. See *Engines, Steam, Crank Shafts.*

Crank Transmission. Results of a series of tests made by H. J. Crandall & Son. Illustrated. Given in *Power*, Sept., 1889.

Cranks. See *Engine Cranks.*

Crematorium. *The Manchester Crematorium.* Discussion of the question of cremation, with description and illustration of the Manchester crematorium. Lon. *Eng.*, July 31, 1891, pp. 123-6.

Cribs. *Cribwork in Canada.* Description of the various cribs. Illustrated. *Eng. News*, April 16, 1881, pp. 154-5.

New Arrangement of Superstructure, for increased cheapness and strength. Rpt. Chf. of Engrs., 1883, Vol. II., p. 1812.

Crib Sinking *at Chicago.* The method pursued on the Great Lakes to build and hold wooden cribs against undermining and lateral displacement. A folio pamphlet, with plates, issued by the Engr. Dept., U. S. A.

Croton Aqueduct. See *Aqueduct, New Croton.*

Crushers.

By Prof H. Fischer. A valuable discussion of all the different types of machines. Illustrated. *Zeitschrift des Vereins deutscher Ingenieure*, 1886, pp. 175-180, 189-194, 210-213, 229-231, 278-284, 333-338, 353, 358, 390-403.

Gates. An illustrated description of the Gates Stone Breaker and Ore Crusher, *Am. Mfr.*, Dec. 14, 1888.

Morris Ore Crusher and Pulverizer. Illustrated description of a new form of jaw crusher. *Am. Mfr.*, Feb. 22, 1889.

Culverts. *Size of.* See *Floods in Stony Brook;* also *Washouts.*

See *Railroad Culverts.*

Current Meters.

Description, use, methods of rating and accuracy of results. By F. P. Stearns. Applies to the meter with helicoidal vanes and to its use in small channels and conduits. *Trans. A. S. C. E.*, Vol. XII., p. 301.

Measurements in the Rhine. Gives description of method employed and results obtained. Floats gave a slightly higher value of the mean velocity than the meter. *Van Nos. Eng. Mag.*, Vol. XXIX., p. 131.

Observations. Report on Experiments on Mississippi River at Burlington, Ia., in October, 1879. Illustrated with one sketch and 41 plates. Issued separately by Engr. Dept., U. S. A.

Current Meters, continued.

Testing. By Robert Gordon. Paper read before the British Inst. of Mech. Engineers. Illus. *Mechanics*, New York, July, 1884.

See *Jets of Water on Curved Vanes*. *Gauging*.

Current Observation *in Lake Erie*. Preliminary report by W. P. Rice, City Engineer of Cleveland, O. This is a valuable paper on the effects of wind on currents and sewage disposal. Reprinted in *Eng. & Build. Rec.*, June 14, 1890, p. 21.

Cutting and Grinding. See *Abrasive Processes*.

Currents. See *Littoral Movements*, *Lake Currents*.

Cutters. See *Milling Cutters*.

Cutting Speeds *in Turning and Boring*. See *Machines*.

Cyclones. See *Meteorology*.

Cylinder Condensations. See *Engines*, *Steam Condensation*.

Damp. An extract from a paper by Mr. M. G. Phillippe on the cause, effects, prevention and cure of damp in buildings. *Van. Nos. Eng. Mag.*, Vol. XXV., p. 476.

Dams.

By David Gravell before the Society of Civil and Mechanical Engineers. The paper brings together, in a convenient form, the sections and salient facts concerning many dams. Illustrated by profile of many dams. Lon. *Engineer*, March 11, 18, 1887; also, *Sci. Am. Sup.*, May 28, 1887.

For a complete report on the various dams and reservoirs in the State of Rhode Island, including sketches showing principal characteristics, dimensions and manner of construction, etc., see *Annual Report of the Commissioner of Dams and Reservoirs* for 1885, 1886, 1887, 1888 and 1889.

Some important notes on their construction, from recent experiments and failures. From the Italian. *Van Nos. Eng. Mag.*, Vol. XXXV., p. 445, December, 1886.

Across the Strug in Galicia. Built of wood and founded on piles in soft ground. *Wochenschr. d. Oesterr. Ing. u. Arch.-V.*, 1886, pp. 9-14.

An. Rep. of the Commissioner of Dams and Reservoirs in the State of R. I. By L. M. E. Stone. Pamphlet; 84 pp., January, 1890. 433 reservoir structures are reported and 13 dams are illustrated.

Backwater Produced. See *Backwater*.

Bear Valley. A satisfactory description of the proposed new Bear Valley Dam, giving principal dimensions, strain sheets, plan, details of outlet gates and gate well; also specifications for foundations, superstructure, manner of laying both faces and the rubble filling; proportions for mortar, strength required for Portland cement, etc., *Eng. News*, Nov. 23, 1889, Vol. XXII., p. 484, *et seq*.

Building in Navigable and other Streams. Paper by Edwin F. Smith, discussing the "best and cheapest forms of timber dams, as well as some details of their construction, which have by long experience been found to give the best results." Illustrated by plates. *Proc. Eng. Club Phila.*, Aug., 1888, Vol. VII. pp. 7-13.

Cambridge Water-Works Dam on Stony Brook. A paper by Wm. S. Barbour, read before the Boston Society of Civil Engineers, giving methods of construction used and difficulties encountered in building a dam on sand and gravel foundation. *Eng. & Build. Rec.*, June 5, 1889.

Colorado River Dam at Austin, Tex. Description with estimates of cost of this large masonry dam. Illustrated. *Eng. News*, July 11, 1891, p. 24.

Concrete Dam, Bettalow Water Works. Brief description with plan, section, and

Dams, continued.

photographic views of this dam 100 ft. in height. Lon. *Engineer*, May 8, 1891, pp. 362, 376. Brief note. *Eng. Record*, May 23, 1891, p. 411.

Cost of. A tabular exhibit of the bids opened Feb. 5, 1890, for building the Tilghman River "Reservoir M" Dam. New York City supply. The detailed bids of twelve contracting firms are given. *Eng. News*, Feb. 15, 1890. Vol. XXIII., p. 169. *Eng. & Build. Rec.*, Feb. 8, 1890, p. 159.

Davis Island Dam on the Ohio River, Pittsburgh, Pa. Illustrated description of this movable dam. *Sci. Am. Sup.*, No. 813, Aug. 1, 1891, pp. 12983-4.

Defective Construction. By W. B. Rider. Attention called to the effects of frost and ice on reservoir embankments and walls. Illustrated. *Eng. News*, June 6, 1884.

Failure. Description of the broken mud pond dam at East Lee, Mass. *Eng. News*, May 1, 1886.

Mill River. Gives report of a committee appointed to report upon the failure of a dam on Mill River, at Williamsburg, Mass. *Trans. A. S. C. E.*, Vol. III., pp. 118-92.

Failure of 100 ft. high. See *Water Supply, Genoa, Italy.*

Staffordville Dam. Account of failure of a dam 26 ft. in height built of masonry, loose rock and earth. *Eng. News*, April 7, 1877, p. 83.

Worcester, Mass. Investigated and reported on by a committee of the Am. Soc. C. E. Caused by pervious stratum underneath, which was not removed. *Trans. A. S. C. E.*, Vol. V. (1876), pp. 244-50.

South Fork Dam. Very complete report, containing numerous photographs and maps. *Trans. A. I. C. E.*, Vol. XXIV, June, 1891, pp. 431-69. Abstract, *Eng. Record*, Aug. 29, Sept. 19, 1891.

South Fork Dam. Discussion on the cause of failure. *Proc. Engrs. Soc. of W. Penn.*, Vol. V., 1889, pp. 89-90. See also a paper by Arthur Kirk on *Removing the Drift Jam at Johnstown by the use of Dynamite.* *Ibid.*, pp. 123-34. Discussion, pp. 133-48.

The Folsom Dam. Illustrated description of this large masonry dam built for irrigation and water power. *Eng. News*, Oct. 17, 1891, pp. 364-5. From *Mining & Scientific Press.*

Gileppe. By A. Marichal. Gives a brief description of the curved masonry dam near Verviers, Belgium. Illustrated. *Proc. Eng. Club. Phila.*, Vol. VI., pp. 243-40.

The Gileppe Dam, near Verviers, Belgium. By G. Grugnola. Section, plan and general discussion. Translated for *Eng. News*, Dec. 25, 1886.

High Masonry.

A valuable synopsis of research on high masonry dams made by A. Fteley, for the Chief Engineer of the Croton Aqueduct for the purpose of determining the form, dimensions, etc., of the proposed Quaker Bridge dam. Reprinted from the Report to the Aqueduct Commission in *Eng. News*, Feb. 4 and 11, 1888.

By James B. Francis, Past. Pres. Am. Soc. C. E. A very valuable discussion of the problem of high masonry dams under water pressure. With discussion. Illustrated, pp. 24. *Trans. A. S. C. E.*, Vol. XIX., p. 147.

Profile of. By Isaac Morley. Derives a formula for determining the profiles of high masonry dams and discusses its application. *Eng. News*, Aug. 11, 1888.

Theory of. A reprint of the report of Mr. B. S. Church to the Aqueduct Commission on the design of the Quaker Bridge dam. Gives a comparison of all the great masonry dams of the world, with plates of cross-sections and plans of the same; also table of data, etc. A valuable article. *Eng. News*, Jan. 7 and 14, 1888; Lon. *Engineer*, March 30, 1888.

Dams, *High Masonry*, continued.

Holyoke, Mass.

An account of the repairs now making to the Holyoke dam on the Conn. Riv. under the direction of Clemens Herschel. The dam is composed of crib-work, first built in 1849, now being made impervious to water by filling in with gravel, well washed or "puddled" with streams of water. Illustrated. *Am. Eng.*, Oct. 15, 1885.

Preservation of. A paper by Clemens Herschel, the engineer, giving the history of the dam across the Connecticut River, which is made of timber, 30 feet high, its sources of weakness, and the method used for preserving it. Also a study for a stone dam to replace the timber one at a future day. Many illustrations and photographic plates. *Trans. A. S. C. E.*, Vol. XV., p. 343.

Lumber Dam, Pennsylvania. Fully illustrated description from a report by J. J. R. Croes, C. E. *Eng. News*, Sept. 5, 1891, p. 205.

Masonry. By J. W. Hill. A paper before the American Society of Civil Engineers. Gives description of the masonry dam at Eden Reservoir, Cincinnati, and shows the methods of computation used, with discussion and three plates. *Trans. A. S. C. E.*, Vol. XVI., pp. 261-282, June, 1887.

Masonry, Memoir of the Construction of. By J. J. R. Croes. Gives details of the construction of a masonry dam on a branch of the Croton River, in Putnam County, N. Y., by the Croton Aqueduct Board. *Trans. A. S. C. E.*, Vol. III., (1874), pp. 337-367.

Masonry. Theoretical derivation of profiles for. By W. B. Coventry. *Trans. Inst. C. E.*, Vol. LXXXV., p. 281.

Movable Dam.

At Davis Island, Ohio River. A valuable article, giving description of the construction of a Chanoine dam composed of 305 wickets, having a total length of 1,223 ft. By William Martin, Assistant Engineer in charge. Illustrated. *Eng. News*, May 15, 1886.

The Kanawha. Gives views of the movable dams, taken during construction. Sections of dam and details of tripping bars. *San. Eng.*, Oct. 22, 1887.

Pochet's. Account of a new form of movable dam after a system devised by M. Leon Pochet. Illustrated. *Sci. Am. Sup.*, May 24, 1890, p. 11992.

On the River Rhone. Full illustrated description. *Eng. News*, Apr. 14, 1883, pp. 171-3.

See *Bridges, Draw, St. Mary's.*

New Croton Dam. Comparison of three proposed sites, with inset, illustrating the proposed earth and masonry dam. *Eng. News*, Nov. 15, 1890, pp. 428-9.

New York Water Supply. A short description of the dam, with a very complete description of the machinery used in its construction. With illustrations. *Eng. News*, Dec. 1, 1888.

Notes on European Dams. Abstracted article from communication by Dr. P. Kreisnik in *Wochenschrift des Osterr. Ing. u Arch. V.* *Eng. News*, March 15, 1890, Vol. XXIII., p. 244.

Philadelphia, At Fairmount. Notes on the early history of the employment of water power for supplying the city with water, and building and rebuilding of the dam at Fairmount, Philadelphia. Illustrated. *Proc. Eng. Club, Phila.*, August, 1886, Vol. V., pp. 372-379.

Paris Dam across the Seine. This dam is composed of folding shutters of slats which roll up like a Venetian blind. Illustrated from *La Nature*. *Sci. Am. Sup.*, No. 758, July 12, 1890, p. 12111.

Quaker Bridge

Gives full text of the report of the Board of Experts on the plans of Quaker Bridge dam. *Eng. News*, Nov. 3, 1888.

History of the. By E. E. R. Tratman. Gives a good review of early history

Dams. *Quaker Bridge*, continued.

of Quaker Bridge dam and reasons for its adoption. Illustrated by maps, etc., from the Report of the Aqueduct Commissioner. Lon. *Engineer*, Jan. 27, 1888.

Paper by Wm. A. Pike, read before the Minneapolis Society of Civil Engineers, Jan. 2, 1889. *Jour., Assn. Eng. Soc.*, April, 1889.

Plan Formation of. By A. Marichal before the Philadelphia Engineers' Club. Discusses the question whether the dams should be built with a curved or straight line and advocates the former. *Am. Eng.*, Jan. 18, 1888.

Plan, section and general view, with description. *Sci. Am. Sup.*, Sept. 11, 1886.

Report of Chief Engineer A. Fteley as to the question of its construction. Gives three possible sites for a dam and advocates the construction of a dam at a point much above the Quaker Bridge site. *Eng News*, Oct. 11, 1890, pp. 327-8. Report with profiles and maps in *Eng. & Build. Rec.*, Oct. 18, 1890, pp. 312-13. Editorial discussion, *ibid.* Oct. 25, 1890, p. 323.

Reservoir. By W. J. McAlpine. Gives copy of specifications for restoring the earthen dam at Worcester. *Van Nos. Eng. Mag.*, Vol. XVI., p. 54.

Reservoir, Safety of. Measures adopted for the safety and service of reservoir dams. This paper, by Dr. P. Kresnik, refers mainly to numerous examples of dams constructed across valleys in France and Spain. *Eng. News*, May 10, 1890, Vol. XXIII., p. 435.

Reservoir, of earth, 80 feet high. Very interesting. *Annales des P. & C.*, September, 1887.

Rock Fill Dam across the Pecos River. Description by Mr. H. H. Cload, Chief Eng. Illustrated. *Eng. News*, May 17, 1890, Vol. XXIII, p. 459.

Rock Fill. An editorial discussing the use of rock fill dams. *Eng. News*, July 28, 1888.

Sodom Dam and Reservoir. Description and outline of specifications for proposed Sodom dam by Mr. Harold Brown. Illustrated. *Sci. Am.*, Aug. 17, 1889.

South Fork. See *Johnstown Disaster*.

State supervision of dams and reservoirs. Editorial article. *Eng. News*, Aug. 24, 1889, Vol. XXII., p. 162.

Strains in.

Curved. Correspondence. Curved vs. Straight Dams. Relative to proposed design of Quaker Bridge Dam. *Eng. News*, Dec. 1 and 29, 1888, *Am. Eng.* Nov. 28, 1888.

Curved Dams. A short mathematical investigation of the curve of bending moments in a circular caisson or dam of uniform section. *Eng. News*, Aug. 16, 1890, p. 149.

High Masonry. By E. S. Gould. *Van Nos. Eng. Mag.*, April, 1884.

Sweetwater.

Description of the dam now in course of construction by the San Diego Land and Town Company. Gives profile of dam and cost of masonry work in detail. *San. Engr.*, June 4, 1887.

By J. D. Schuyler, before the American Society of Civil Engineers. Gives details of the construction of the Sweetwater dam, San Diego, Cal. *Eng. News*, Oct. 27, 1888.

Construction of. By Jas. D. Schuyler, M. Am. Soc. C. E. A complete description of this dam. Dimensions: Base, 46 feet; top thickness, 12 feet; height, 90 feet; dam on curve; radius of arch, 220 on line of face at top. Illustrated, pp. 18. *Trans. A. S. C. E.*, Nov. 1888, Vol. XIX., p. 201.

Timber Dam, with navigable pass, closed by a hydraulic lifting apparatus, called the American bear trap system. To be used on the Kentucky River at Beatty-

Dams, continued.

ville, Ky. Report Chief of Engineers U. S. A., 1884, Part III., p. 2731. Illustrated.

Villar Reservoir Dam on the River Lozoya. Description with profile of this high masonry dam. *Eng. News*, Feb. 24, 1883, p. 93.

Vyrnwy.

A description, with illustrations, of the dam on the Vyrnwy River in Wales built in connection with the Liverpool Water Works. Gives dimensions of structure, composition of the concrete, with results of tests on same, etc. *R. R. & Eng. Jour.*, Jan. 1889.

Description and sketch on the Lake Vyrnwy masonry dam, constructed for increasing the water supply of Liverpool. Character of stone, binding material and method of laying given, cross section and diagram of stresses. Rubble masonry, using very large blocks, filling spaces with concrete. Area of lake 1,115 acres; available contents, 12,000,000,000 gallons; length of embankment, 1,255 feet; height, 139 feet. Abstracted from Lon. *Eng.*, Jan. 8, 1886; *Eng. News*, Jan 30, 1886; also *San. Eng.*, Jan., 1886.

Description illustrated in section and perspective of the Vyrnwy Dam now building for Liverpool, Eng. *Eng. News*, Dec. 29, 1888.

Full page engraving of this beautiful masonry dam. Lon. *Engineer*, June 27, 1890, p. 548; brief description, p. 514. *Eng. & Build. Rec.*, July 12, 1890, pp. 8-9.

Report of the experts, Gen. Clark and Russel Aitken, appointed by the city to examine whether or not the construction of the masonry embankment has been carried out in a substantial and workmanlike manner. Also, report of Mr. A. H. Holmes, a councilman. *Eng. News*, June 19, 1886.

Walnut Grove. Gives views, sketch plan and details of construction of the Walnut Grove "rock fill" dam, near Prescott, Ariz. *Eng. News*, October 20, 1888.

Washington, D. C., Potomac River. By S. H. Chittenden, before the American Society of Civil Engineers. Gives description of the work of constructing a dam across the Potomac River for increasing the water supply of Washington, D. C. *Trans. A. S. C. E.*, Vol. XVIII., Feb., 1888, pp. 70-79.

Wildbach Retaining. By Paul Grueber. Gives a sketch of Sarrell's study of torrential streams, and an account of one of the dams built to carry out his method. The dam is built to retain the detritus from the upper part of the river Gall. *Eng. News*, June 5, 1886.

Woodbridge, Conn. Description of the New Haven Water Company's Dam at. An earth dam, 1,000 ft. long, and 46 ft. high at highest point. Very irregular rock formation was encountered. Described and illustrated by L. A. Taylor. *Jour. N. E. W. W. Assn.*, Sept., 1891, pp. 51-56.

See Water Supply. *Water-Works.*

Dams and Reservoirs. A brief abstract of a paper by W. W. Follett, read before the Denver Soc. Civil Engrs. and Arch., having special reference to the use of these in irrigation. *Eng. Record*, March 14, 1891, p. 242.

Connecticut Law Concerning Dams and Reservoirs. This law while failing at one critical point is an advance on the almost universal neglect of all regulations. *Eng. News*, April 12, 1890, Vol. XXIII., p. 351.

Embankments for Water-Works in U. S. and Canada. Plans, elevations and sections selected from drawings sent with official reports for publication in the *Manual of American Water-Works*, *Eng. News*, June 21, 1890, pp. 578-9, July 12, 1890, p. 37.

Dead. See *Sanitary Disposition of the Dead.*

Declination of Needle, as reported by various observers. Table of one page. *Rpt. Mich. Assoc. Surv.*, 1882. *See* also Magnetic Declination.

Deflection of Bridge Trusses. By Harry Sheridan. Some valuable simple formulæ for finding the deflection under loads. *Van Nos. Eng. Mag.*, Vol. XXXV., p. 457, Dec. 1886.

Derrick, lately built by the American Shipbuilding Company, of Philadelphia. H. H. Gorringe. *Proceedings U. S. Naval Institute*, No. 32.

Depot. See *Railroad Station*.

Derrick. See *Appliances, Car, Derrick, Cranes*.

Derrick Car, *Fifteen-Ton.* Illustrated description of a well-designed derrick car on the Pennsylvania railroad. Dimensions and bill of material given. *R. R. Gaz.*, Oct. 18, 1889.

Design, *Elements of Architectural.* A course of four lectures, by B. H. Statham, under the auspices of the Society of Arts. Treats of architectural decoration functional and applied; influence of roofing in; influence of the constructive principles of the beam and arch; mouldings, carvings, etc. *Jour. Soc. Arts*, Dec. 23 and 30, 1887, Jan. 6, 1888; *Eng. & Build. Rec.*, Jan. 21, *et seq.*, 1888.

Determination of Time, Longitude, Latitude and Azimuth by means of the transit instrument. A very exhaustive treatment of the subject. U. S. C. and G. Sur. Rep., 1880.

Diamond Mining. See *Mining*.

Diary *and Reference Book for Engineers.* This is a substantial volume containing considerable information, besides many advertisements. Some account is given of all the principal English engineering societies; also tables for the conversion of thermometer and hydrometer scales, weights and measures, and money. Cloth, 4o, 3octs. Spon & Co., London and New York.

Dikes *of Isle de Re.* A short description of the dikes, with profile. *Van Nos. Eng. Mag.*, Vol. XXVII., p. 279.

Screen Dike and wet Pile Sinking. Missouri River Commission. Abstract from report of Samuel H. Yonge, describing their construction. Illus. *Eng. News*, Dec. 6, 1890, p. 498.

Dimensioning.

Prof. Wm. Cain. Gives criticism on the paper of Mr. Alfred D. Ottewell. *Eng. News*, July 16 and 23, 1887.

Discussion of Prof. Cain's formula, by Mr. Ottewell; also a communication by Professor Cain. *Eng. News*, Aug. 20, 1887, and Sept. 3, 1887.

New Method of. By Alfred S. Ottewell. Discusses the use of the formulas of Laundhardt and Weyrauch and their modifications. Gives diagram showing the effect of the various formulas to reduce the working strain per square inch as the range of stress became greater. Derives formula and considers application to compression members. *Eng. News*, May 14 and 21, 1887.

Various Methods of Determining. By Dr. J. Weyrauch. Gives a short demonstration and comparison of the methods of determining dimensions based on the assumption of variable strength. A number of examples and tables given. *Van Nos. Eng. Mag.*, Vol. XXIX., pp. 310 and 385.

Discharge *of Gas and Steam from Orifices.* By Gustav Beuner. An investigation to show that new results and extensions can readily be introduced into common theory. *Van Nos. Eng. Mag.*, Vol. VII., pp. 10 and 210.

Of Streams in Relation to Rain-fall. By A. G. Coghlan before the Inst. of C. E. A valuable paper discussing the results of seven years' observation on streams in Australia. Area of water-shed, 344 sq. miles. Rain-fall from 16 to 64 inches; average, 34. Per cent. of discharge varied from 9 to 68 inches, average, 44 inches. *Van Nos. Eng. Mag.*, July, 1884.

Of Water Main, as determined by the Pressure Gauge. Part of a paper by G. A. Ellis, before the New England Water-Works Association. *Eng. News*, June 26, 1886.

Discharge *in Times of Flood from Catchment Areas.* See *Flood Discharge.*

Disinfection. Describes the machines used for disinfecting clothing, bedding, etc., by means of hot air and steam, acting separately and in combination. Lon. *Engineer*, May 11, *et seq.*, 1888; *Sci. Am. Sup.*, June 30, 1888.

Disinfector. *Aero Steam.* A disinfector in which all articles are subject to a moist heat of at least 217° Fah. Lon. *Engineer*, July 29, 1887.

Distance Measuring Micrometers. Paper by W. F. King before Association of Dominion Land Surveyors; treating of fixed and movable wire. Rochon and Sugeol Micrometers. Pp. 6. *Proc. Asso. D. L. S.*, Ottawa, Feb., 1889, p. 171.

Distribution of Power. See *Power Transmission.*

Ditch Apportionment. *Report of a Committee* of the Ohio Association of Engineers and Surveyors, very fully covering the ground. *Report of the Society* for 1890, Benj. Thompson, Sec'y., Urbana, O.

Ditches. *Flow of Water in.* A paper read before the Technical Society of the Pacific Coast by Aug. J. Bowie, Jr. Gives a full discussion of the value and methods of measuring the miner's inch, and methods of determining the discharge of ditches.

Drainage. *Formulas for cross-section and slope,* for maximum discharge under given conditions. By R. F. Hartford. *Am. Eng.*, Feb. 27, 1885.

By L. S. Alter. Discusses size, depth, location and slope. *Proc. Ind. Assoc. Surv. & Engrs.* L. S. Alter, Sec'y, Rensselaer, Ind.

See also *Drainage.*

Diving. *Appliances for Working Under Water or in Irrespirable Gases.* History of diving and description of various dresses. By W. A. Gorman before the Inst. M. E. *Eng. News*, June 3, 1882, pp. 181-2.

Dock Gates. *On the New Steel Dock Gates of Limerick Floating Dock.* Paper by Wm. J. Hall, B. E., Assoc. M. Inst. C. E. Pp. 7. *Proc. Inst. C. E.*, Vol. XCVII., p. 331. London, 1889.

Docks.

Alexandra, Hull. By A. C. Hurtzig before the Institute of Civil Engineers. Gives details of the construction of the Alexandra Dock, 1881-5. The work included a dock of 46½ acres, two miles of dock wall, two graving docks, a lock 550 × 85 feet; embankment, 40 feet high and 6,000 feet long, and dredging an artificial channel. *Proc. Inst. C. E.*, Vol. XCII., pp. 144-56. Abstracted Lon. *Eng.*, Feb. 10, 1888. Abstracted *Mech. World*, Feb. 18, 1888. Abstracted Lon. *Engineer*, March 2, 1888.

Description of the deep-water dock at Tilbury, England. *Sci. Am. Sup.*, May 29, 1886.

A brief description of the docks and railroad now in course of construction at Barry Island, near Cardiff. The work will consist of a 40-acre and 7-acre basin, also about 18 miles of railroad. Lon. *Engineer*, March 4, 1887.

The Barry Dock Works, including the Hydraulic Machinery and the Mode of Tipping Coal. Interesting descriptive paper by John Robinson. 3 folding plates. *Proc. Inst. C. E.*, Vol. CI., 1890. Paper No. 2413, pp. 129-46; discussion, pp. 152-72; correspondence, pp. 173-83.

Blowing up of. See *Blasting.*

Dry. See *Docks Dry* below.

Esquimalt Graving Dock Works, British Columbia. Paper by W. Bennett, Can. M. Soc. C. E., with discussion and plates. *Trans. Can. Soc. C. E.*, Vol. III., p. 210-36, 1889.

Esquimalt. Gives description of the new dock at Esquimalt, British Columbia, with two two-page plates showing plans and details of the work. The dock is 451 feet long, 65 feet wide at the entrance, and has 27 feet of water on the sills. Lon. *Eng.*, July 20 and 27, 1888.

Docks, continued.

Esquimalt. Illustrated description of the Esquimalt, British Columbia, dock. Extreme length, 475 feet; entrance width, 65 feet; top width, 90 feet; bottom, 40 feet; depth, 30 feet. Lon. *Eng.*, Feb. 18, 1887.

Extension at Liverpool, with a Description of the Port of Liverpool, and the River Mersey. Paper read by Geo. F. Lyster, 14th Jan., 1890. 4 plates and appendix of statistics. *Proc. Inst. C. E.*, Vol. C., paper No. 2431, pp. 2-39; discussion, pp. 40-81; correspondence, pp. 84-114.

Floating. A brief illustrated description of a new English dry dock and its arrangements for taking on listed vessels. Lon. *Eng.*, Sept. 9, 1887. *Sci. Am. Sup.*, Oct. 15, 1887.

Floating Docks and Pontoons. By Alex. Taylor, before the Northeast Coast Institution of Engineers and Shipbuilders. Gives approved practice for pontoons and floating docks. *Sci. Am. Sup.*, July 7, 1888.

Glasgow Harbor, New Docks. Illustrated article, descriptive of Glasgow docks and details of construction of the new docks. Lon. *Eng.*, August 9, 1889.

And Harbors, England. Description of several. From *Proc. Inst. C. E.*, *Eng. News*, Dec. 5, 1878, pp. 387-8.

New York City Freight. Being an account of the terminal facilities of the Penn. Ry. at New York. Fully illustrated. Shows, also, section of new river wall on pile foundation on North River. *R. R. Gaz.*, Feb. 20 and 27, 1885. *See* also Terminal Facilities.

Locks for. See *Locks.*

And Machinery. See *Ore Docks.*

Ore Dock of the Duluth, Southshore & Atlantic Railway Company of Marquette, Mich. This dock, 1300 ft. long, is fully described and details illustrated in *E. & M. Jour.* Jan. 18, 1891, pp. 72-3, *et seq.*

Ore Dock of Wisconsin Central, at Ashland, Wis. Illustrated. *R. R. Gaz.*, Oct. 7, 1887.

Preston and River Ribble. Gives history of the work of constructing a 40 acre dock at Preston, Eng., and the improvement of the river Ribble. Lon. *Engineer*, Sept. 30, 1887.

Pumping and Hauling Machinery. Description with double-page drawing of the machinery at the Windsor Slipway, Lower Grange Tower, Cardiff. Lon. *Eng.*, July 9, 1886.

Tidal Dock at Southampton. Illustrated description with plans and sections. Lon. *Eng.*, Sept. 12, 1890, pp. 308-10.

See *Canals, Ship, Manchester, Harbors.*

Dock Dry.

For the Broadwise Docking of Ships. Description and method of operation of a dock of this system. Illustrated. *Sci. Am. Sup.*, Aug. 8, 1891, pp. 13000.

Calliope Dry Dock at Auckland, New Zealand. Paper by Mr. John H. Swainson, before the Inst. C. E. Abstracts in *Eng. & Build. Rec.*, Nov. 22, 1890, pp. 389-90.

Colonial Government Dry Docks at St. Johns, Newfoundland. Paper by H. C. Burchell, giving a description of these docks and reviewing their construction. With discussion. *Trans. Can. Soc. C. E.*, Vol. III., pp. 222-19, 1889.

Compressed Air Pier Foundations for the New Harbor of La Pallice at La Rochelle. The paper is accompanied by 5 plates giving very very full information on all parts of the subject. *Annales des P. & C.*, Nov. 1889, p. 435. Abstract in *Proc. Inst. C. E.*, Vol. C., 1890, pp. 430-2.

Government Timber Dry Dock, St. Johns, Newfoundland. A paper by H. C. Burchell, before Canadian Society of Civil Engineers. Illustrated. *Eng. News*, Oct. 12, 1889, Vol. XXII., p. 338.

Dock, Dry, continued.

Havre. Full description, with plan, transverse section and views of the large dry-dock being built at Havre, France. *Le Genie Civil*, Oct. 29, 1887; *Sci. Eng.*, Dec. 3, 1887.

Hydraulic Dry Dock—Union Iron Works, San Francisco. In this the vessel rests upon steel girders which are raised by hydraulic lifts. Fully illustrated description. *R. R. Gaz.*, Nov. 27, 1891, pp. 836-7.

Hydraulic Lift, Bombay. Brief description of the hydraulic lift dock at Bombay. It is the largest hydraulic structure in the world: is 350 ft. long, 65 ft. clear width, and docks vessels drawing 30 ft. Lon. *Eng.*, Nov. 25, 1887.

League Island Dry Dock. Brief description of this timber dry dock, with illustrations showing sections. *Eng. News*, March 14, 1891, p. 249.

Simpson's Timber Dry-Docks. Description and comparison with stone-dry-docks. Illustrated. Paper read before U. S. Naval Institute, Feb. 2, 1887, by Lieut.-Comdr. C. H. Stockton, U. S. N. *Proceedings of the U. S. Naval Institute* for 1887, Vol. XIII., No. 2.

See *Shipyard.*

Dock-Wall. Twenty-three feet of water at low tide in front of them, built in form of hollow piers 31′ × 23′ in plan, by 27 ft. high, the well-hole 19′ × 8′ in plan, sunk by pumping out the water and excavating inside the well-hole, then filled with concrete. The piers set three feet apart and sunk at intervals: first every other one, then the missing ones. Cost of excavation, of net measurement of exterior of the completed piers, including cost of pumping, stagings, lighting, fuel, cost of plant, with cost of repairs and charging whole original cost of plant, inclusive also of all supplementary work, and excluding only cost of masonry laid in cement above ground, but by tide-work, was about $2.05 per cubic yard: of which price about 63 cts. represents the cost of the plant. *Annales des P. & C.*, 1885-1-96. *See* also works on the Plantation quay at Glasgow, described in the same journal, February, 1876, and on the Rochefort wet-dock, Feb., 1834.

Failure and Construction of the Limerick Dock Walls. A paper by W. J. Hall describing this difficult work. Illustrated. Abstract from Proc. Inst. C. E. *Eng. Record*, March 28, 1891, pp. 273-4.

Repairing the Kidderpore Dock Walls, India. Account of the failure of the walls, with illustrations of the plans adopted for repair. *Eng. News*, July 11, 1891, p. 28.

Stability of. By R. Allen. Gives description of new method of computation. *Van Nos. Eng. Mag.*, Vol. XX., p. 36.

Wet Dock at the Port of Havre, France. Interesting description of the method of construction, with illustrations, as given in *Le Genie Civil*, May 17, 1890. Abstract in *Eng. News*, June 28, 1890, pp. 602-3.

Dock Yards. *Report on European Dock Yards.* By Naval Constructor Philip Hichborn, U. S. N. A valuable and exhaustive report, containing descriptions of vessels, and giving details of all leading European dock yards. *Washington Govt. Print.* 4to. pp. 68. 63 plates.

Dome. *Description of the Wooden Dome of the Temple Beth Zion, Buffalo.* Plan and sections. *Eng. News*, June 28, 1890, p. 609.

Investigation of a Ribbed Dome. Mathematical thesis by R. G. Manning. *The Technic*, Univ. of Mich., pp. 35-51. 50 cts.

See *Arches, Practical Theory of.*

Drainage. A paper by George Wilson giving details of the drainage and improvement of 1,500,000 acres of desert land in the west of France. *Van Nos. Eng. Mag.*, Vol. XX., p. 522.

By J. P. Frizell. Treats of the reclamation of the submersible lands of the Mississippi Valley. A criticism on the report of the Commission authorized by Congress in 1874. *Van Nos. Eng. Mag.*, Vol. XIII., p. 18.

Drainage, continued.

And Sewerage of Cicero. Description of the system adopted for this township, which is adjacent to the city of Chicago. By J. W. Alvord. Rept. 6th annual meeting Ill. E. & S., 1891, pp. 64-73.

And Sewerage of Hyde Park, Ill. Report upon. By Benezette Williams and John A. Cole, Civ. Engrs., Chicago. A 40-page pamphlet, with large map. The report based on a careful topographical survey. The project involves many new features. Apply to the authors.

A Water Supply of Chicago. See *Water Supply.*

Of Charleston, S. C. See *Tidal Drains.*

Chicago. Gives extracts from a special report of the Chicago Drainage and Water Supply Commission on the diversion of the flood water of the Desplaines River and the north branch of the Chicago River. *San. Eng & Const Rec.*, July 30, 1887.

Chicago Drainage Problem. The Overflow of the Desplaines River. Paper by Ossian Guthrie, member Western Society of Engineers; read September, 1889. A valuable paper. P. 17. *Jour. Assn. Eng. Soc.*, March, 1890, Vol. IX., p. 77.

Chicago. See *Water Supply, Chicago.*

For Cities. The Separate vs. Combined System. By S. A. Bullard. Presents the favorable points of both systems, and makes an examination of the faults of both systems. *Soc. An. Rep. Ill. Soc. Eng. & Surv.*, p. 37.

City of Mexico. Account of the present plan for draining the city and lowering Lake Texcoco several feet. Estimate of quantities of work and prices are also given. *Eng. News*, April 5, 1890, Vol. XXII., p. 319. *Eng. News*, April 12, 1890, Vol. XXIII., p. 339.

Fens and Low Lands by Steam Power. By W. H. Marshall. A series of papers giving a general description of the pumps and engines used in draining low land. Lon. *Engineer*, Feb. 4, 11, 18, 1887.

Hillsides by Tile Drains. By John L. Cully. An account of a successfull drainage of a springy slope. Illustrated. *Jour. Assn. Eng. Soc.*, Vol. V., p. 100.

Hints on Field Works in. By A. H. Bell. *Soc. An. Rep. Ill. Soc. Engrs. & Surv.*, p. 33.

House. See *House Drainage.*

House of Parliament. Gives sketch of old system and its defects, and describes the system now being put in. Lon. *Engineer*, Feb. 4, 1887, also Lon. *Eng.*, Feb. 4, 1887.

Improvements, Just Apportionment of the Cost of. By S. J. Stanford. Discusses the best methods of estimating the proportion of taxes each tract should pay. *Soc. An. Rep. Ill. Soc. Engrs. & Surv.*, p. 58.

Lake Copais. A plan being carried out in Greece for the purpose of drainage of about 60,000 acres of land. Illustrated. Lon. *Engineer*, July 9, 1886.

Lake Fucino, Italy. Description of the conduit and other drainage works. Memoire by M. A. Durand-Claye, Pont et Chaussees. Reprint, *Eng. News*, July 26 and Aug. 15, 1878.

Lake Fucino. A description of the work and methods employed. *Van Nos. Eng. Mag.*, Vol. XVIII., p. 433.

Land.

A paper by E. B. Opdycke, on the construction, comparative cost and efficiency of tile drains and open ditches. *Rpt. Ohio Soc. Surv. & Engs.*, 1888, pp. 148-151.

Paper by R. C. Carpenter. *Rep. Mich. Assn. Surv.*, 1880.

Prairie Drainage. A paper by M. H. Messer treating of the drainage of land in the prairie country of Illinois. *Eng. News*, March 28, April 11, 1875.

Drainage, *Land*, continued.

Report of committee upon. Also a paper upon, by D. W. Pampel. *Report Ohio Soc. Surv.*, 1884.

In Illinois on a large scale by means of dredges. Cut of dredge shown. *Eng. News*, Dec. 19, 1885.

Large Marshes. By C. E. Hollister. Gives description of the draining of several marshes in Michigan. *Van Nos. Eng. Mag.*, Oct., 1884.

Lough Erne Drainage. Paper describing this important work by Jas. Price, Jr. Folding plate. *Proc. Inst. C. E.*, Vol. CI., 1890. Paper No. 2451, pp. 73-91; discussion and correspondence, *ibid.* pp. 95-117.

New Brighton, S. I. A discussion of problems connected with the drainage and sewerage of New Brighton, Staten Island. By Arthur Hollick: pp. 10. *School of Mines Quart.*, Columbia College, July, 1889. Vol. X., p. 322

Newhaven. Brief description, with details, of the drainage work at the Port Newhaven, Eng. Lon. *Engineer*, July 29, 1887.

New Orleans. Brief abstract of scheme devised by B. M. Harrold, City Surveyor of New Orleans. *Eng. News*, Dec. 15, 1888.

Of New Orleans. See *New Orleans*.

New York, Twenty-Third and Twenty-Fourth Wards. Report of E. B. Van Winkle giving general description of the region, common theories of malaria and of drainage and sewerage and recommendations. *Eng. News*, Aug. 13, 20, 1881.

Oak Orchard and Tonawanda Swamps, New York. The reclamation of 25,000 acres of swamp land found perfectly feasible at a small cost. The scheme outlined accompanied by maps. *N. Y. State Survey Report*, 1883.

Problem of Just Apportionment of cost (of construction of among landholders benefited thereby.) A statement of legal principles involved. By E. B. Opdycke. Also a shorter paper on same subject by J. L. Geyer. *Report Ohio Soc. Surv.*, 1885.

Specifications for the *Rangoon*, India, draining project. *Ind. Eng.*, Feb. 5, 12 and 17, 1886.

Specifications. See *Pumping and Drainage*.

Stickel Valley. Lawrence, Mass. By Arthur D. Marble. Consists of straightening channel and providing sewers. Fully illustrated. *Jour. Assn. Eng. Soc.*, Vol. V., p. 401.

St Louis, The Future. By R. E. McMath. Gives a brief description of the ground covered by St. Louis, and outlines a plan for the drainage of a portion of the unsewered portion of the city; also gives a map of the city showing the 25 feet contours. *Jour. Assn. Eng. Soc.*, Vol. VI., p. 133.

Swamp Land in Illinois. By I. O. Baker. Some facts as to extent, laws and methods. *Jour. Assn. Eng. Soc.*, Vol. V., p. 426. Also *Eng. News*, Oct. 30, 1886.

Swamp Lands. Two papers of considerable practical value to country surveyors and civil engineers. *Trans. Ind. Assoc. Surv.*, L. S. Alter, Sec'y, Remington, Ind.

Underground Tile and Open Ditch. Capacity of Tile Drains and Open Ditches. Some working formulas, by Prof. Baker. *Eng. News*, Sept. 4, 1886.

Valley of Mexico. By Francisco de Garay. A very instructive historical account of the anomalous condition of this valley, the evil effects on health of a want of proper drainage, and of the various attempts that have been made to drain the basin. Also a description of the author's plan for accomplishing this result, with large map. *Trans. A.S.C.E.*, Vol. XII. (1883), p. 15.

Valley of Mexico. By Richard E. Chism. An elaborate project for the drainage of the Valley of Mexico, with map. Gives geology of the valley, rainfall and evaporation. *E. & M. Jour.*, Dec. 8, 1888.

Drainage, continued.

Valleys of Mystic, Blackstone and Charles Rivers, Mass. A report by a State Commission upon, with Eliot C. Clarke as Chief Engineer. A book of 300 pp. and 48 plates. The recommendations indorsed by Jos. P. Davis and Rudolph Hering as consulting engineers. The net results of an expenditure of some $3[illegible]. Intermittent downward filtration recommended in most cases. The book is a substantial addition to the literature on sewage disposals. Apply to *Eliot C. Clarke*, Chf. Engr., 15 Brimmer street, Boston, Mass. A three-column abstract given in *Eng. News*, Jan. 23, 1886.

Wandle Valley Main Drainage. Description of the extensive works draining a district of [illegible] acres, in England. Illustrated. *Eng. News*, Oct. 11, 18, 1884.

See *Plumbing, Pavements, Railroad Construction.*

Drainage Canal. *The Rungpur Drainage Canal.* Description with map, of this canal, which is 7 miles long, 29 feet wide at base, and slopes, two to one. *Ind. Eng.*, Oct. 18. pp. 312-3.

Drainage Machinery *in the Valley of the Po.* The water drawn off by ditches and pumped into the river. Pumps of three kinds, centrifugal, turbines, and lifting wheels. The latter most efficient, giving 80 per cent. efficiency. *Proc. Inst. C. E.*, Vol. LXXXI., p. 400.

Drainage Main, *of Marseilles, France.* The new scheme that has been adopted to prevent the fouling of the port. *Engineering Era*, (Cleveland, O.), Oct. [illegible], 1885.

Drainage Tables. By G. H. Johnson. Gives tables showing the diameters of circular pipes of given length which will discharge given volumes of water per second under a given head. *Eng. News*, May 5 and 12, 1888.

Drainage Tunnel *for City of Mexico.* Map and profile of proposed drainage tunnel, with description of present conditions and proposed improvements. *Eng News*, Aug. 10, 1889. Vol. XXII., p. 122. *Sci. Am. Sup.*, Aug. 10, 1889.

Drainage Works *of the City of Boston.* By Eliot C. Clarke. Full description of the new main works, their origin and working. Fully illustrated: 31 plates. Text, 110 pages, with Appendix giving results of a very extended and painstaking series of cement tests, the latter reprinted from a paper presented to Am. Soc. Civ. Engrs., Boston, 1885. Valuable.

Near Markdorf, in Bavaria. Short description of work executed at a cost of $25,000, consisting of the drainage of swamp lands covering about 865 acres. *Deutsche Bauzeitung*, 1885, pp. 497-9 and 521-3.

Drain Pipes, *Willow Roots in.* An account of the obstruction of glazed tile drains, 8 inches in diameter, with mortar joints, by willow roots, these having grown through the mortar. A remedy offered. F. A. Calkins, in *Eng. News*, Dec. 20, 1884.

Drains, *Tile Construction. Comparative Cost and Efficiency,* as compared to open ditches. A good paper by E. B. Opdycke in *An. Rep. Ohio; Soc. Engrs. & Surv.*, 1885. C. N. Brown, Columbus, O.

And Traps. See *House Drains.*

Testing of. An unpatented appliance, readily procured, devised by Charles Hawksley, M. Inst. C. E. Illustrated by cut. Tests soundness of all traps and joints in a house system. *San. Eng.*, March 26, 1885.

Ventilation and Trapping of. Abstracted from a paper before the Society of Arts, by James Lovegrove. *Van Nos. Eng. Mag.*, Vol. I., page 731

Draught Gauge, *a Simple Form of*, for measuring the draught of chimneys. Designed by Prof. J. Burkitt Webb. Consists of an ordinary pair of pan scales, with vacuum chambers sealed with mercury and connected with the flue. *Jour. Frank. Inst.*, June, 1885. *See* Chimney Draught.

Drawing Boards. *Recent Improvements in.* By Theo. Bergner. A ruler moves parallel to itself over the board by means of pulleys and an endless cord. In

Drawing Boards, continued.

accuracy is practically perfect. Highly recommended by those who have used it. Not patented. Details shown. *Proc. A. S. M. E.*, Vol. VI., p. 224.

Drawing Table. Illustration of a cheap and practical form of adjustable drawing table. *Mast. Mech.*, Dec., 1889, Vol. IV., p. 210.

Drawings.

Duplication of. By J. M. Bradford. Describes the "blue prints" method, with formula. *Rep. Ohio Soc. Surv. & Engrs.*, 1882, pp. 150-61.

Office System for. By Henry R. Towne. *Trans. A. S. M. E.*, Vol. V., p. 193. *Mechanics*, June and July, 1884.

Reproduction of, by Actinic Processes. By Benj. H. Thwaite. Describes many methods, with the necessary mixtures, apparatus and manipulations, including two giving white lines on blue ground; three giving blue lines on a white ground; two giving white lines on a black ground; two giving black lines on a white ground; two giving brown lines on a white ground; also the zincographic process. *Proc. Inst. C. E.*, Vol. LXXXVI., p. 312. See *Heliograph Black Process and Copying Drawing.*

In black on white ground. Sensitised paper used, and process similar to the blue process, but no chemicals needed. *Eng. News*, Nov. 28, 1885.

Black Process. A process of reproducing drawings which gives black lines on a white ground. *Building*, March 5, 1887, also *Am. Eng.*, March 16, 1887; *Eng. News*, Dec. 25, 1886.

By Chemical Processes. By Robt. Marshall. Gives the ferro-prussiate and cyani-ferric processes of making negative and positive blue prints, with formulas and mode of application. *Sci. Am. Sup.*, Feb. 19, 1887.

Cyanotype Process of Reproducing. Gives notes compiled in the Photographic Office Survey of India Department, Calcutta, on the positive cyanotype process of reproducing drawings with dark lines on a clear ground. *Ind. Eng.*, Aug. 4, 1888.

See *Blue Printing.*

Section Lining, Standard. By T. Van Vleck. Recommends the adoption of standard section to represent the conventional sections of iron, steel, etc., on all engineering drawings. Cuts show the proposed sections. *Eng. News*, Dec. 31, 1887.

Shades and Shadows in Architecture. Valuable series of papers by Prof. A. D. F. Hamlin. Illustrated. *Am. Arch.*, Aug. 30, 1890. No. 766, *et seq.*, pp. 130-3. Continued from No. 759, p. 155.

Workshop. By Alfred D. Ottewell. Two short articles giving many valuable practical instructions. *Am. Eng.*, Nov. 7 and 21, 1884.

Dredging.

Bristol, England and the West Indies. Description of a steel, four-screw and single-ladder hopper dredge, 218 feet long, built for the Dock Committee of the Bristol Corporation. Also of a twin screw dredge "Dolphin" for the West Indies. Illustrated with full page drawings. Lon. *Engineer*, Jan. 7, 1887.

Cost of. See *Canals, Ship, Lake Erie.*

Hydraulic, in New York. Brief description of method of dredging by pumping up the material. *R. R. Gaz.*, Aug. 28, 1891, pp. 593-4.

Mersey Dock Estate. Description of the dredges and method of work. Fully illustrated. Lon. *Eng.*, May 31, June 6, 1890.

Oakland Harbor, Cal., being an account of extensive operations where the material is pumped up and forced through long lines of pipe to the deposit beds. With plate. *Trans. A. S. C. E.* Vol. XIII., p. 9.

Ocean Bars. By Gen. A. Q. Gilmore. Gives details of the removal of bar at the mouth of St. John's River by means of sand pumps. *Van Nos. Eng. Mag.*, Vol. VII., p. 311.

Dredging, continued.

Operations and Appliances. By John J. Webster. Gives historic sketch o dredging machines, and a great deal of valuable matter relating to dredging. Discussion of the above paper by prominent engineers. *Inst. C. E.*, Vol. LXXXIX, p. 2; *Eng. News*, July 16 and 23, 1887; Lon. *Eng.*, March 4, 1887.

Dredging Contracts, *Prices of.* A list of prices of the accepted bids on dredging contracts let between Feb. 1888, and Feb., 1889. *Eng. News*, March 23, 1889.

Dredging Machines.

Hydraulic. By C. R. Hunt. Refers briefly to one or two typical machines, and then describes the Von Schmitt system of hydraulic dredging in detail. *Proc. Eng. Club. Phila.*, Vol. VI., p. 124.

Recent Improvements in. A paper before the Am. Soc. Mech. Engrs., by A. W. Robinson. Mainly confined to the chain-bucket or elevator system in comparison with others. Illustrated. *Eng. News*, Dec. 4, 1886.

Dredging Plant.

Gives description, with detail drawing, of the Von Schmidt dredge used at Oakland, Cal. *San. Eng.*, Oct. 8, 15, 1887.

Dredges.

Description, with large detailed drawing, of a twin screw dredge, having a capacity of 400 tons per hour in a depth of 32 feet of water, constructed for Auckland Harbor. Lon. *Eng.*, June 11, 1886.

Double Ladder, Swansea Harbor Trust. Gives brief description, with two-paged plate, showing plan, sectional elevation and sections, of a double ladder dredge recently constructed for the Swansea Harbor Trust. Dimensions, 150 × 42, with 12 hold, capacity 920 tons per hour from a depth of 38 feet. Lon. *Eng.* July 13, 1888.

Ejector, using compressed air as power to set column of water in motion. *Annales des P. & C.*, June, 1888.

For Use in Shaft Sinking underwater or for pier foundations. From the French. Illustrated. Buckets on an endless chain; and also by jet operated by forcing air into the elevating tube near the bottom. *Eng. News*, March 6, 1886.

Hercules Dredges for the Panama Canal. Illustrated description of these dredgers. *Eng. News*, Feb. 3, 1883, pp. 49-51.

Hopper Suction Dredge for Port of Liban. Illustrated description. At test, hopper was filled in 35 minutes with 800 tons of sand and clay, then steamed out to sea and unloaded in a few minutes through bottom doors. Lon. *Eng.*, Jan. 11, 1889.

Hydraulic. Works to a depth of 15 feet and excavates 90 tons per hour. Sectional views shown. Bruce and Bathos's system. Lon. *Eng.*, June 18, 1884.

Illustrated description of various machines. *Eng. News*, May 10, 1879, pp. 148.

Lockwood. Gives description of the dredge designed and built by F. A. Lockwood for work on the Cape Cod Canal. Drawings show details. *San. Engr.* Nov. 5, 1887.

Lobnitz Rock-Breaking. Illustrated description of a dredge for the removal of rock under water by first breaking it into blocks of convenient size, and then lifting these to the surface by a strong chain of buckets. *Eng. News*, Jan. 26, 1889.

New Canal Excavator. The dredge is placed upon a truss, the ends of which rest upon trucks. These trucks run on lines of track built along the bank of the canal. Illustrated description. *R. R. Gaz.*, Sept. 21, 1881, p. 669.

Pumping. The Badger dredge, as operated at Coney Island, New York, described and illustrated. Seems to be very efficient. *Eng. News*, Jan. 30, 1886.

Rock, Suez Canal. An illustrated description of the sub-aqueous rock dredger "Derocheuse," built for the Suez Canal. Its dimensions are 180×40×12 ft. It

Dredges, continued.

has ten chisel bars 40 feet long, weighing four tons each, and dredging machinery to remove the broken rock. Its capacity is about 40 tons per hour. Lon. *Engineer*, March 9, 1888.

Steam Dredge at Hamburg. By G. Vogeler. Large bucket dredge. Length of bucket ladder over 80 feet. Capacity averaged about 800 cu. yds. per hour for over a year. Cost about $80,000. Described and illustrated in *Zeitschrift des V. Deuts. Ing.*, 1886, pp. 441-443, 455-470.

Twin Screw Hopper, "Pholas." A brief description, with two-page plate showing plan and sections of the dredge boat "Pholas," built for the Bombay Port Trust. The boat is 159 feet long, 32 feet beam, and 10 feet depth molded, built of steel, with 8 compartments; capacity of hoppers, 10,000 cubic feet, with displacement for 500 tons of dredged material. Lon. *Eng.*, June 24, 1887.

Used on the Panama Canal, of American Manufacture. A large cut, with description. Capacity, 1,000 cubic yards per hour deposited on the bank. *Sci. Am. Sup.*, Feb. 14, 1885; also *Sci. Am.*, Aug. 15, 1885.

On Schmidt Dredge. Description of this form of sand dredger or suction-dredger. Folding plate. By George Higgins. *Proc. Inst. C. E.*, Vol. CIV., 1891, pp. 191-5.

And Soil Transporter at the Manchester Canal. Illustrated description of this very efficient machine. From Lon. *Eng.* *Eng. News*, Sept. 5, 1891, pp. 221-2.

See *Canal Construction.*

Drift Bolts. *Experiments on the Adhesive Power of.* By J. H. Tscharner. *Selected Papers, C. E. Club, Univ. of Ill.*, 1889-90, pp. 33-8. *Eng & Build. Rec.*, Nov. 29, 1890, pp. 413-14.

Holding Power of. A summary of three series of experiments, which comprises all the information accessible on this subject. *Eng. News*, Feb. 28, 1891, pp. 169-170.

Holding Power of Drift Bolts and Spikes. Table of results of a large series of tests made in 1874-7. Extracted from a government report. *Eng. News*, Sept. 26, 1891, pp. 262-3.

Square Drift Bolts. Results of tests on one inch square bolts in various sized holes. By J. H. Powell and A. E. Harvey. *Technograph* (Univ. of Ill. Annual), 1890-91, pp. 39-41.

Drift Mining.

See *Mining.*

Drill, *Cheap Portable.* See *Appliances.*

Drilling, *Relative Economy of Hand and Machine.* By W. A. Wheeler. Gives a valuable comparison of the cost of hand and machine work from a purely economical standpoint. Shows there is but little difference in the cost. *Jour. Assn. Eng. Soc.*, February, 1886.

See *Boring.*

Drilling Machine. *The Ainsley-Oakes Square Hole Drilling Machine.* Description with inset of working drawings of this ingenious machine. *Eng. News*, June 20, 1891, p. 579.

Drills, *Diamond.* Shows results obtained by the use of the diamond drill. *Van Nos. Eng. Mag.*, Vol. IX., p. 434.

Githens. An illustrated description of a direct acting percussion drill. Its distinguishing features are the valve motion, rotation, front head and feed screw connection. *Eng. News*, March 26, 1887.

Trials of Rock Drills. A paper by Mr. E. H. Carbutt and Mr. Henry Davy, read before the Inst. of Mech. Engrs., giving the results of some recent trials, and describing the mechanism of each drill, seven power drills and two hand drills were tested, all being of the percussive kind. *Iron*, Apr. 3, 1891, pp. 289-91.

Drills. *Twist.* An inquiry into the requirements of the cutting edges of twist drills. A paper before the Am. Soc. Mech. Engrs. by Wm. H. Thorne. *Am. Eng.*, Nov. 19, 1885.

See *Mining Machinery.*

Drop Press. *Pressure Attainable by the Use of the.* By Prof. R. H. Thurston. Obtained an efficiency of 99%. Gives table of pressures obtained by using drops weighing from 50 pounds to 2,000 pounds, falling from 3 inches to 5 feet. *Van Nos. Eng. Mag.*, Jan. 1884.

See *Foundations for.*

Dye Works. See *Ventilation of.*

Dynagraph. *Dudley's.* Illustrated description of an instrument invented by P. H. Dudley for making a complete record of power exerted by a locomotive. *Eng. News*, May 15, 1891, pp. 167-69.

Dynamics. See *Parallel Motions.*

Dynamometers. A description of several styles, with the principles on which they act. By C. F. White. *Jour. Assn. Eng. Soc.*, Vol. IV., p. 313. See also *Indicators.*

A continuous registering belt dynamometer, wherein the friction of the apparatus is eliminated. Described and illustrated in *Sci. Am. Sup.*, December 4, 1886.

Atchison, Topeka & Santa Fe Railway. This simple form consists essentially of a heavy locomotive spring, whose deflection under the pull of the engine is indicated on the dial. Full drawing. *Ry. Rev.*, Jan. 1891, pp. 8-9.

Construction of a Large Prony Brake. By R. H. Thurston. Specifically described and worked out. *Jour. Frank. Inst.*, April, 1886.

French Transmission. Gives a description of a dynamometer constructed and used to determine the efficiency of a plant for the transmission of power by electricity. *Lon. Engineer*, Feb. 17, 1888.

Friction Brake. By W. W. Beaumont. A paper and discussion covering 77 pages, and covering the whole ground of appliances, principles and practice. Illustrated. *Proc. Inst. C. E.*, Vol. XCV.

Friction of a Transmitting Dynamometer. Abstract of paper by Samuel Webber, Charlestown, N. H., describing tests on the friction of a Webber transmitting dynamometer at different loads and speeds. *Mechanics*, June, 1889.

The Tatham. Used for testing motors at the Phil. Elec. Ex., capacity 120 H. P. Tested by churning water and getting thus the mechanical equivalent for 1° Fahr., which was found to be 772.8 ft.-pds. thus proving its absolute accuracy within exceedingly small limits. Illustrated. *Jour. Frank. Inst.*, Dec., 1885.

Testing the Force of the Hammer Blow of Locomotive Drivers. Designed by a joint committee from the Master Mech. Assn. and the Franklin Inst. Described and illustrated. *Ry. Rev.*, Oct. 9, 1886.

Wallace's Graphic Dynamometer. Brief description of a dynamometer for measuring amount of power consumed by any machine or part of any machine while in motion. *Ind. Eng.*, Jan. 11, 1890, p. 30.

See *Engines, Steam. Indicators.*

Dynamometer Car. *Western Railway of France.* Description of an car designed to make the following researches: (1) Measurement and registration of tractive efforts; (2) registration of total work done; (3) registration of speed; (4) registration of number of revolutions of wheels; (5) registration of time at ten-second intervals; (6) marking kilometres run and other points of interest; and (7) the analysis of the products of combustion from locomotive. Illustrated. *Ry. Rev.*, Oct. 26, 1889, Vol. XXIX., p. 617.

Dynamometric Brake. *Brauer's.* Described and illustrated in *Proc. Inst. C. E.*, Vol. LXXX., p. 366.

Dwellings. See *Houses.*

Dynamos. See *Armatures. Electric Dynamos.*

Eads, *James B.* An exhaustive biography, with portrait. *Sci. Am. Sup.*, May 7, 1887.

A short sketch of his life and achievements. *R. R. & Eng. Jour.*, April, 1887.

Eulogy by E. L. Corthell. *Jour. Assn. Eng. Soc.*, September, 1890, pp. 456-61.

With Portrait. Mechanics, April, 1887; *R. R. Gaz.*, March 18, 1887.

Earth, *The Figure of.* By Frank C. Roberts, C. E. An historical and mathematical paper, showing the various ways in which the spheroidal form is determined. *Van Nos. Eng. Mag.*, March, 1884.

Earth Pressure. By Wm. Cain. Gives a comparison between results obtained from a formula and from experiments. *Van Nos. Eng. Mag.*, Vol. XXVI., p. 69.

Actual Pressure of. By Benj. Baker. A paper of great value to all who are likely to be confronted with the question of high retaining walls. *Van Nos. Eng. Mag.*, Vol. XXV., pp 333-353 and 492; also No. 56. *Van Nos. Science Series.*

New Investigation on Earth Pressure, and the most economical shape of retaining walls. By M. L. Leygue. This very interesting paper contains the results of an extended series of experiments on the cohesion of earth, the direction, point of application, and amount of earth pressure, and the best form of retaining wall, with empirical formulas deduced therefrom, and all compared with results given by ordinary theories and with other experiments. *Annales des P. & C.*, November, 1885, pp. 788, 1004.

Sand, The Thrust of. Gives a new experimental method of studying the reactions between a mass of earth and its supporting wall, and details some experiments made with the method. *Sci. Am. Sup.*, Sept. 3, 1887.

Earthquake Countries. *Building in.* See *Building Construction.*

Earthquakes.

A paper by Prof. J. S. Newbury. Gives the commonly received theories. *Sch. Mines Quar.*, October, 1886.

Charleston, Origin of, The. By Emil Starek. Treated with equations to determine point of origin. Illustrated with a map. *Sch. Mines Quar.*, October, 1886.

Grenada in 1884. A description which is of interest in view of the recent earthquake at Charleston. *Trans. Inst. C. E.*, Vol. LXXXV. p. 275.

Spanish. By C. G. Rockwood, Jr. With sketch map and views. Also, four other papers on the same subject. *Science*, March 6, 1885.

Studying, by Electricity. A good description of methods and apparatus. Illus. *Elec. World*, March 12, 19, 26 and April 2 and 9, 1887.

Their Effect on the alignment of railroads in the South. Described and photographic cut shown. *R. R. Gaz.*, Oct. 29, 1886.

1886, Their influence on light-houses. *An. Rpt. Light-House Board*, 1886.

See *Canals, Ship, Nicaragua.*

Earthwork. A paper by George J. Specht, before the Tech. Soc. of the Pac. Coast. Treats of hull construction, and also the results of the extended investigation of the relative volumes in cut and fill. Also gives the German method of computing haul, with a full page diagram. A valuable paper. Reprinted in *Eng. News*, Aug. 22, 1885.

Borrow-Pit Excavation at Portland, Oregon, for N. Pac. R. R. terminal yards. The earth loosened on bluff banks by Judson powder, and wheeled to cars on elevated platforms. Proved more economical than steam shovels. Illus. *Eng. News*, Nov. 28, 1885.

Calculation of. By N. B. Putnam. Discusses Simpson's rule and the prismoidal formula. *Van Nos. Eng. Mag.*, Vol. XVI., p. 161.

Earthwork, continued.

Calculation of, by the use of the slide rule. By M. Le Brun. Considerable discussion of the methods of calculating excavations and embankments, with use of graphical scales of various kinds, and the slide rule. *Mem. de la Soc. des Ing. Civils*, Aug., 1886, pp. 157-216.

Center of Gravity of. By J. W. Davis. Proposes a method for simplifying the computation of haul, etc. *Van Nos. Eng. Mag.*, Vol. XVII., p. 83.

Consolidation of treacherous slopes in heavy cuts by means of rubble spurs perpendicular to face of slopes. *Annales des P. & C.*, 1888, p. 5; Abstract in *Proc. Inst. C. E.*, Vol. XCV., p. 465.

Computation by Diagrams. Gives some diagrams illustrating practically the method used in Mr. Wellington's treatise. *Eng. News*, Sept. 3, 1887.

Computation of. By E. W. Hyde. Investigates the application of the prismoidal formula and compares it with Weddle's formula. *Van Nos. Eng. Mag.*, Vol. XV., p. 227.

Computations. *The legal status of the "end-area" earthwork computations*. An editorial article of value in *Eng. News*, Nov. 9, 1889, Vol. XXII., p. 445.

Estimating. By J. R. Gillis. Gives method of estimating the work from profile when lack of time prevents more accurate methods. *Van Nos. Eng. Mag.*, Vol. II., p. 132. *Trans. A. S. C. E.*, 1869.

Estimating Overhaul in Earthwork by Means of the Profile of Quantities. This method is described and an example given by S. B. Fisher. *Eng. News*, Jan. 31, 1891, pp. 48-9.

Filling South Boston Flats. By F. W. Hodgdon. Gives details of the methods employed by the Commonwealth of Massachusetts to fill 120 acres of South Boston Flats from two feet below to thirteen feet above mean low water. *Jour. Assn. Eng. Soc.*, January, 1888, pp. 5-9.

Formula for. Gives a new formula, derived from the prismoidal formulas, for computing railroad earthwork. It also has a graphical representation. *Eng. News*, July 28, 1888.

Graphical Estimates for Railways. Gives method consisting essentially of three diagrams, viz.: for section-areas of through-cut sections, for areas of hill-side sections, and for a correction to be applied to end-area solidities. *Van Nos. Eng. Mag.*, Vol. VI., p. 84.

Graphic method for measuring cross-sections. By M. H. Willette. Gives simple method of constructing profile of cross-section and obtaining its area. *Van Nos. Eng. Mag.*, Vol. XXIV., p. 153.

Overhaul. A paper by G. M. Walker, Jr., of University of Michigan, '90, reviewing methods of calculating "average overhaul" and "overhaul." p. 9 *The Technic*, Univ. of Mich. No. 3, new series, p. 61, 1889.

Profile of Quantities. From the German. Illustrated. A graphical method of determining grade line and haul for given relations of cut and fill. *Am. Eng.*, Oct. 1, 1885.

Shrinkage of. By P. J. Flynn, before the Tech. Soc. Pac. Coast. A very valuable paper, giving a summary of all published rules and experiments on the subject. Attention called to the error made in allowing, in embankments, a correction only for ultimate shrinkage from cut to fill, without reference to first increase in embankment. *Van Nos. Eng. Mag.*, Dec., 1884. *Eng. News*, May 1 and 8, 1886.

Specifications. Suggested improvements and changes in the ordinary methods of classifying materials in excavations by a "Contractor." *Eng. News* April 5, 1890, Vol. XXIII., p. 325.

Tables. By W. B. Ross. Describes some simple tables that are universal in their application. *Van Nos. Eng. Mag.*, Vol. V., p. 594.

Tables for finding end areas for three-level sections. A description of the tables

Earthwork, continued.

used by W. W. Redfield, whereby the labor is greatly reduced. *Jour. Assn. Eng. Soc.*, Vol. V., p. 217.

Theory of Calculating Slopes. By Prof. Merriman. Wherein the cohesion of the material is taken into account. *Eng. News*, March 14, 1885, *et seq.*

See *Excavation. Railroad Construction. Railroad Earthwork.*

Economy of Structures, *Comparison of the.* By Prof. G. F. Swain, before the New England Water-Works Association. Discusses the proper method of comparing the economy of structures of different classes. *Jour. N. E. W.-Works Assoc.*, March, 1888, Vol. II., pp. 31-34.

Comparative, or other engineering and architectural works. By Arthur Cobb. The subject investigated by means of algebraic equations, all the functions entering in with their proper values. An interesting and valuable discussion. *Van Nos. Eng. Mag.*, December, 1886.

Education.

An address by Robert H. Thurston, delivered at the seventeenth annual commencement of the Worcester Polytechnic Institute on technical training, considered as a part of complete and generous education. *Sci. Am. Sup.*, July 16, 1887.

Report of the Commissioner of the Society of German Engineers in regard to the preparatory schools for higher scientific training. Discussion of the "Gymnasien" and "Realschulen." Published with *Zeitsch d. V. Deutscher Ing.*, No. 16, 1886.

Of the Mechanical Engineer. A paper by C. H. Benjamin, read before the Civil Engr. Club of Cleveland, giving a comprehensive and practical discussion of the subject. *Jour. Assn. Eng. Soc.*, July, 1891, pp. 344-53.

Technical, on the Baltimore & Ohio R. R. A long extract from the principle of the B. & O. school, with long comments on the same. *R. R. Gaz.*, April 22, 1887.

See *Technical Education.*

Effects of Heat and Cold *on Metals.*
See *Heat and Cold.*

Eiffel Tower. See *Tower.*

Elastic Extension. By R. H. Graham. Gives a mathematical treatment of the subject of elasticity. Considers it a form of motion and subject to the laws of velocity and acceleration. Lon. *Engineer*, Aug. 19, 1887.

Elastic Limit. By J. Bauschinger. On the change of the elastic limit and strength of iron and steel by drawing out, by heating and cooling and by repitition of loading. *Am. Eng.*, Jan. 26, 1886.

Of Iron and Steel. On a new method of detecting overstrain in iron and other metals, and on its application in the investigation of the causes of accidents to bridges and other constructions. Shown by the permanent elevation of the elastic limit by overstraining. By R. H. Thurston. *Trans. A. S. C. E.*, Vol. VII., p. 13. See Iron and Steel.

See *Iron and Steel, Increase in Elastic Limit.*

Electrical Apparatus. See *Dynamos; Galvanometer; Gramme Generator and Motor and Governors.*

Electric Balance, *Thomson Composite.* By Thomas Gray. Full description of Sir Wm. Thomson's new balance, available as volt, ampere, or watt-meter. *Sci. Am. Sup.*, July 14, 1888.

Electric Batteries. A paper by Desmond Fitz Gerald before the Society of Telegraph Engineers and Electricians on reversible lead batteries and their use for electric lighting. Lon. *Engineer*, March 25, *et seq.*, 1887.

Causes of variation of Clark Standard Cells. Abstract of a paper by J. Swin-

Electric Batteries, continued.

burne before the British Assn., giving the results of a series of experiments. *Electrician*, Sept. 4, 1891, pp. [illegible].

For Light and Power. By A. H. Bauer, before the Am. Inst. of Elec. Engrs. Reports result of three years' work with the Faure Battery at Baltimore. *Elec. Eng.*, June, 1886.

The Possibilities and Limitations of Chemical Generators of Electricity. A paper read before the Am. Inst. of Elec. Engrs., May 16, 1888, by Francis B. Crocker. Shows what can be expected of any given combination of materials and gives table of the cost per horse-power per hour of voltaic battery energy, with different materials. *Elec. World*, May 26, 1888; *Elec. Eng.*, June, 1888.

Primary, for Illuminating Purposes. By Perry F. Nursey, before the Society of Engineers. Treats briefly the principles of the primary battery, outlines its history and then describes in chronological order the various batteries brought out. *Trans. Soc. Eng.*, 1888, pp. 165-223.

Electric Batteries, Secondary. An historical and descriptive paper by F. G. Howard, Stud. Inst. C. E. *Elec. Rev.*, June 20, 1885.

By Oliver Lodge. Gives description of the Faure battery. Also methods and instruments for measuring the various quantities involved in storage by secondary batteries. *Van Nos. Eng. Mag.*, March, 1884.

By William Henry Preece, F. R. S. *Jour. Soc. Arts*, May 3, 1889, Vol. XXXVII., p. 540.

Description of a large one for commercial distribution, by E. P. Roberts, M. E., before the Am. Inst. Elec. Engrs. *Elec. Eng.*, New York, July, 1893.

Faure's. Illustrated description. *Elec. Rev.*, Nov. 16, 1889.

Notes on the Chemistry of Secondary Cells. A paper by Prof. W. E. Ayrton, C. G. Lamb and E. W. Smith, read before the Inst. of Elec. Engrs. *Electrician*, Nov. 28, 1890, pp. 1003 *et seq.* *Elec. Rev.*, Nov. 28, 1890, pp. 652-53 *et seq.*

On the Charging of, for obtaining an efficient, uniform and constant source of currents for electric lighting. By Wm. H. Preece, F. R. S. Illustrated. *Elec. Eng.*, New York, July, 1885.

The Use of in Telegraphy, giving estimate of relative cost. By W. H. Preece. *Elec. Rev.*, Nov. 22, 1884.

Electric Batteries, Storage.

By Oliver J. Lodge. *Electrician & Elec. Eng.*, Jan., 1883, from *Nature*.

By E. Raynier. N. Y. *Electrician*, March, 1883.

Developement and Progress of. By Wm. Bracken, at Niagara meeting of National Electric Light Association. *St. Ry. Gaz.*, Aug., 1889.

Discharge of. By E. Frankland. Gives conditions of successful working and shows that currents of 25,000 amperes may be obtained from 100 cells. *Elec. Rev.*, Oct. 25, 1884.

In Electric Lighting. An excellent article, fully illustrated, read as a paper before the Am. Inst. of Electrical Engrs. By George B. Prescott, Jr. Describes methods of using and regulating accumulators in various classes of electric lighting work. *Elec. Eng.*, Nov., 1889.

For Electric Locomotion. By A. Reckenzaun. A paper read before the Electric Light Association. Gives historical facts and working figures of expense, etc. *Sci. Am. Sup.*, Nov. 5, and 19, Dec. 24, 1887.

Montaud. Full account of the construction and power of the above battery. *Sci. Am. Sup.*, June 18, 1887.

Practice in Europe as seen by an American.—VI. A description of accumulators at the Frankfort Exhibition. Illus. *Elec. World*, Oct. 3, 1891.

Railroad Motors. Illustrated description of the system used in Birmingham, England. Abstracted from Lon. *Eng.*, *St. Ry. Jour.*, Sept., 1890, pp. 411-3

Electric Batteries, Storage, continued.

Report of Prof. Henry Morton on the storage batteries of the Electrical Power Storage Co. *Van Nos. Eng. Mag.*, Vol. XXVIII, p. 470.

Street Cars at Paris. Some results of experiments as to electrical work required. Abstracted from *Electrician*. *Eng. & Build. Rec.*, October 4, 1890, p. 274.

Street Car Motors. An abstract of statements from a circular of the "River and Rail Electric Light Company," 45 Broadway, N. Y., which will prove of interest. *Eng. News*, Dec. 7, 1889. Vol. XXII., p. 535.

Systems of different kinds are described and illustrated in *St. Ry. Jour.*, July, 1890, pp. 314-6, 318-9, 376.

System of the Electric Storage Battery Co. Embodies latest improvement in battery construction. Illus. *Elec. Eng.*, Oct. 8, 1890.

Working Efficiency of Secondary Cells. A very valuable paper by Messrs. Ayrton, Lamb, Smith and Woods. Read at the Edinburgh meeting of the I. of E. E., July 16, 1890. Tables and diagrams. *Electrician*, July 25, Aug. 1, 8 and 15, 1890. *Elec. Rev.*, (N. Y.) Aug. 9, 16, 1890. *Elec. World*, Aug. 9, 16, 23, 30, pp. 89, 107, 124, 138.

See *Electrical Railroads*.

Electric Block Signals. See *Railroad Signals*.

Electric Brakes. See *Electric Railroads*. *Florence*. *Brakes*.

Electric Buoys. See *Buoys*.

Electric Calculating Machines. Description of the Hollerith machines used in tabulating census returns. *Elec. Eng.*, Nov. 11, 1891, pp. 521-30. Illus.

Electric Calculations. Articles by T. O'Connor Sloan, Ph. D., giving a concise review of the principal electrical terms, formulas and computations. Article of a popular character. *Elec. Rev.*, April 11, 1890, p. 404.

Electric Capstans. An article from *La Nature*, by M. Max De Nasonty, describing several applications of electric power to driving capstans and windlasses, for handling freight, etc., giving a good illustration of the wide applicability of electricity to ordinary purposes. *Elec. Rev.*, Nov. 22, 1889.

Electric Car. *Reversing Gear. Baldwin's*. Allows motor to run at constant speed, while car is stopped, or run forward or back. Illus. *Elec. Eng.*, Oct. 22, 1890.

Electric Car Heating.
See *Car Heating*.

Electric Car Truck. *The new Stephenson*. Built partly of wood. Has some valuable features. Illus. *Elec. Eng.*, Oct. 22, 1890.

Electric Cars, *Hydraulic Gear for*. See *Hydraulic Gear*.

Electric Cars.
See *Electric Railroad*. *Street Cars*.

Electric Circuit, *Lines for*, the construction of. By Th. D. Lockwood. A series of valuable articles giving working methods. *Elec. Eng.*, New York, May, June, July, *et seq.*, 1885.

An account of the successful working of an 80 kilometer circuit containing arc and incandescent lamps. By M. Gaulard. *Elec. Rev.*, Oct. 18 and Nov 8, 1884.

Electric Cranes. See *Cranes*. *Railroad Shops*.

Electric Currents. A paper by C. W. Siemens, on certain means of improving the steadiness of dynamo-electric currents. *Van Nos. Eng. Mag.*, Vol. XXVI., p. 63.

Alternating. An article by Prof. Elihu Thomson on the phenomena of alternating currents. *Sci. Am. Sup.*, July 9, 1887.

Excellent discussion of the main problems in the distribution and care of alter-

Electric Currents, *Alternating*, continued.

nating currents by T. Carpenter Smith, before the National Electric Light Association. *Elec. World*. Feb. 28, 1891.

Working Capacity and Self Induction in. A paper by Gisbert Kapp. *Electrician*, Dec. 10, 1890, pp. 1979 *et seq*.

Experiments upon Apparatus, to determine the effect of heating the core of a transformer. Paper read before the A. I. E. E. by Prof. H. J. Ryan. Tables and diagrams. *Elec. Eng*., July 2, 1890, pp. 10-15. *Elec. Rev*., (N. Y.), July 5, 1890, pp. 9-10.

Experiments with. An illustrated paper by Prof. Elihu Thomson, giving some interesting experiments on induced currents. *Proc. Soc. Arts*, 1890, pp. 108-18.

Novel Phenomena of. Paper read before the American Institute of Electrical Engineers, May 18, 1887, by Elihu Thomson. Treats of the repulsion of a closed circuit by an alternating current. A valuable contribution to electrical science. *Electrician & Elec. Eng*., June, 1887.

Some Effects of Flow in Circuits having Capacity and Self Induction. A paper by Dr. J. A. Fleming read before the Inst. E. E. *Electrician*, May 8, 1891, pp. 17-20, *et seq*.; *Elec. Rev*., May 8, 1891, pp. 583-9, *et seq*.; *Elec. World* May 23, 1891, *et seq*.

Calculations on Electric Shocks from Contact with High Pressure Conductors. A paper by P. Cardew, read before the Inst. E. E., giving calculations of discharge of current under various conditions, with discussion. *Electrician*. May 8, 1891, pp. 11-17

Continuous Currents from Alternating Currents. Describes briefly several methods of transforming alternating into continuous currents, and vice versa, without mechanical commutators. By E. E. Ries. *Elec. World*, No. 23, 1889.

The Determination of Currents in Absolute Electro-Magnetic Measure. A series of articles giving an account of some experiments on quantitative chemical effects of currents. By G. F. C. Searle. *Electrician*, Sept. 11, 1891, *et. seq*.

Distribution in Networks of Conductors A valuable paper by Messrs. J. Herzog and L. Stark. Theorems and diagrams. *Elec. World*, Aug. 23, 1890, pp. 118-9.

Distribution by Transformers. A valuable paper by Dr. S. A. Fleming, describing different systems. Fully illustrated. *Electrician*, July 4, 1890, *et seq*.; *Elec. World*, July 26, 1890, *et seq*.

The Heating of Aerial Conductors. By J. G. White. Gives details of experiments made to determine the effects of heating and consequent changes of resistance of aerial conductors carrying electrical currents. *Elec. Eng*., Aug., 1886.

Heating of Conductors. Account of an excellent experimental investigation into the laws of the heating of conductors by electric currents under the conditions of practice. Results plotted as curves. By E. A. Kennelly. *Elec. World*. Nov. 23 and Nov. 30, 1889.

High Tension Underground Currents. *English Authorities on*. A resume of the opinions of many English authorities in electrical matters concerning safety of transmitting currents of high voltage underground. *Elec. Rev*., April. 18, 1890, p. 411.

Induction of Electric Currents and Induction Coils. A lecture by Prof. Elihu Thomson, before the Franklin Inst. Illustrated by experiments many principles concerning induction, especially of alternating currents. Illus. *Jour. Frank. Inst*., Aug., 1891, pp. 81-101.

Modern Views as to Electric Currents. Abstract of an address at anniversary of John Hopkins University, by Prof. Henry A. Rowland. *Elec. World*. March 9, 1889.

Electric Currents, continued.

Originating in ordinary and Aerial Telegraph Conductors. Paper by Mr. A. R. Bennett, read before the Inst. of E. E., July 15, 1890. Diagrams. Lon. *Elec. Rev.*, July 25, 1890, pp. 105-8.

Prices and Conditions for the Supply of, at Monaco. Brief article giving prices rent of meters, etc. *Elec. Rev.*, Nov. 13, 1891, pp. 557-8.

Relation to the Human Body. Alternating vs. Continuous Currents. A paper by H. N. Lawrence and A. Harries, M. D., read at the Leeds meeting of the British Asso. Adv. Sci., 1890. Tables. Lon. *Elec. Rev.*, Sept. 12, 1890, pp. 301-3.

Relation to the Nature and Form of Conductor; the Self-Induction. Discussion by the Soc. Tel. Engrs. and Elect. Engrs. *Elec. Eng.*, June, 1884.

Rupture of Copper by. A remarkable rupture of copper wire used in construction of "Bernstein lamps." *Elec. Rev.*, Nov. 22, 1888.

Strength, Measurement of. With a shunted galvanometer. By Carl Hering. *Electrician & Elec. Eng.*, Jan. 1885.

Traction Increaser. By E. E. Ries. Results of experimental examination of the electric current as a means of increasing the tractive adhesion of railway motors and rolling contacts. *Sci. Am. Sup.*, Dec. 10, 1887.

Paper by Elias E. Ries, reprinted from *Elec. World. St. R. Gaz.*, July, 1891, pp. 106-7. See also a paper by C. Sheldon, *id.*, pp. 111.

Brief summary of the discussions and experiments on the alleged effect of the electric current to increase adhesion: also a description of the Ries machine. *Eng. News*, Aug. 9, 1890, p. 117.

See *Electric Lighting.*

Electric Dynamos.

A lecture by Prof. Forbes, at the Electrical Exhibition, developing the subject from the simple and elementary facts to the ordinary generators and motors now in use. *Jour. Frank. Inst.*, Dec., 1884.

An account of elaborate experiments, with cuts of apparatus used by Alabaster, Gatehouse & Co., London, copied from Lon. *Eng.*, in September No. of *Mechanics*, New York.

A new form of engine dynamo and motor, invented by Mr. Henry Joel, and made interchangeable in all its parts. Perspective view and details. *Sci. Am Sup.*, July 3, 1886.

By J. Hopkinson, M. A., D. Sc., F. R. S., and E. H. Hopkinson, M. A., D. Sc. From the Philosophical Transactions of the Royal Society. Theoretical construction of characteristic curve, efficiency experiments, etc. *Electrician & Elec. Eng.*, Jan. and Feb. numbers, 1887.

By G. M. Hopkins. Gives a full and detailed account of a dynamo capable of supsupplying a current to 8 sixteen candle-power lamps, with elaborate instructions for constructing the same. *Sci. Am. Sup.*, July 2, 1887.

Alternating Current Machinery. A paper by Gisbert Kapp, Assoc. M. Inst. C. E., with discussion and correspondence on same by various members. Pp. 104. *Proc. Inst. C. E.*, Vol. XCVII., p. 1. London, 1889.

Alternating Current System. A brief description of the new dynamo and converter of Chas. F. Brush. Illustrated. The dynamo has a fixed armature and revolving field. *Elec. Eng.*, Aug., 1889.

And Their Engines. By Gisbert Kapp. Considered theoretically and practically, and fully illustrated. This, with the discussion, covers 150 pp. in *Trans. Inst. C. E.*, Vol. LXXXIII., p. 123, and *Elec. Eng.*, Aug., 1886. *See also* Engines, High Speed.

And Motors, Designing and Calculating. The most practical method of designing and proportioning dynamos yet published. Francis B. Crocker. *Elec. World*, April 28, 1888.

Electric Dynamos, continued.

And Motors, Low Speed, for Continuous Currents. By Frank H. Perret. *St. Ry. Gaz.*, Dec. 1890, pp. 229-31.

Ball Unipolar. C. W. Raymond. *Electrician*, (N. Y.) Feb., 1884. Gives details and results of tests.

Building, for Amateurs. A series of illustrated articles intended to enable any mechanic, with the ordinary shop conveniences, to build a dynamo of any desired size. From the *English Mechanic* in *Elec. Eng.*, (N. Y.), May, 1886, *et seq.*

Considerations Governing the Choice of a Dynamo. By E. P. Roberts before the C. E. Club of Cleveland. Analizes and discusses the subject. *Jour. Assn. Eng. Soc.*, Sept., 1891, pp. 445-54.

The Consideration of the Chief Features which Regulate their Application. A valuable and practical paper by A. B. Blackburn, read before the Manchester Assoc. Engrs. *Elec. Rev.*, Jan. 30, 1891, pp. 141-3, *et seq. Mech. World*, Feb. 7, 1891, pp. 52-3, *et seq.*

Construction and Pole Pieces. By H. F. Watts. *Elec. World*, Jan. 22, 1886.

Coupling and Control of. Paper by James Swinburne. *Industries*, Aug. 15, 1890, *et seq.* Abstract in *Electrician*, Aug. 22, 1890, *et seq.*, *Elec. Rev.*, (N. Y.) Sept. 6, 1890, *et seq.*

Design of Armatures. Paper by A. Isenthal. Rules are given for the number of layers of armature wire, depth of winding, number of coils, and the dimensions of commutator, spindle and pulley. *Elektrotechnische Zeitschrift*, Feb. 1890, p. 83.

Eickemeyer's. A new departure in dynamo construction, presenting apparently some marked advantages. The field coils surround the armature, and a new and ingenious method of winding the latter is described. *Elec. Eng.*, March, 1888; *Tel. Jour. & Elec. Rev.*, March 23, 1888.

Experiments on the Speed of. Paper by M. Rechniewski, read before the International Soc. of Electricians. Diagrams and tables. *Elec. World*, Sept. 13, 1890, pp. 189-90.

Exploration of the Magnetic Fields Surrounding Dynamos. Experiments made at the Philadelphia Electrical Exhibition in 1884, by Carl Hering. *Electrician & Elec. Rev.*, August, 1887. Reprinted in the *Elec. Rev.*, August 19 and 26, 1887.

Foundations for. See *Foundations Elastic.*

Frankfort Exhibition. Illustrated description of the various types. By W. B. Esson. *Elec. Rev.*, Sept. 18, 1891, *et seq.*

The Lauffen-Frankfort Transmission of Power. Detailed description of Alternator at Lauffen. Illus. Lon. *Engineer*, Oct. 23, 1891, pp. 337-8.

Low Speed Multipolar Dynamos and Motors for Continuous Currents. By Frank A. Perret. *Elec. Rev.*, (N. Y.) Nov. 29, 1890, p. 173.

Mordey's Alternating Current. Gives an illustrated description of Mordey's new alternating current dynamo. It has fixed armatures and revolving magnets. *Tel. Jour. & Elec. Rev.*, June 1, 1888; *Sci. Am. Sup.*, July 21, 1888.

Notes on the Deposition of Copper by. By E. T. Delafield. *Elec. Eng.*, August, 1886.

Notes on the Performance of Some. Reprint of table published in April, 1887, with corrections and additions. Prof. Jackson, Cornell University. *Electrician & Elec. Eng.*, June, 1887.

Notes on the Design of Multipolar Dynamos. A practical article by W. B. Esson, presenting the subject in a concise manner. *Electrician*, Apr. 10, 1891, pp. 702-3; *Elec. Rev.*, Apr. 10, 1891, pp. 456-61.

On the establishment of a common basis for the tests and efficiency percentages. By Ch. F. Heinrichs. *Elec. Eng.*, N. Y., Feb. 1886.

Electric Dynamos, continued.

On the Theory of Alternating Currents, particularly in reference to the alternate current machines connected by the same circuit. Being elaborate experiments to determine the effects of coupling them in parallel circuit and in series, with the mathematic discussion. By J. Hopkinson, F. R. S., in Lon. *Engineer*, Dec. 5 and 12, 1884.

Practical Deductions from the Franklin Institute Tests of. By Carl Hering. The most important and valuable tests ever published. Valuable for dynamo designers. *Electrician & Elec. Eng.*, Nov., 1886; also *Jour. Frank. Inst.*, Dec., 1886.

Predetermination of the characteristics of dynamos. By Gisbert Kapp. Paper read before the Society of Telegraph Engineers and Electricians, London, England. *Elec. Rev.*, Nov. 19, 1885.

Relation between the Cross-section of the iron in Armature and Field. Paper read before the American Inst. of Elec. Engrs., May 18, 1887, by Dugald C. Jackson, and discussion of the same. Gives results of the author's experiments on a Gramme dynamo at Cornell University, showing the effect of varying the section of the armature. *Electrician & Elec. Eng.*, June, 1887.

Relation of the Air Gap and Shape of Poles. An article by Prof. H. J. Ryan giving experimental data read before Am. Inst. Elec. Eng. *Elec. World*, Oct 3, 1891, illus., pp. 252-3; *Elec. Eng.*, Sept. 30, Oct. 7, 1891.

Regulation, by means of a third brush. Revised abstract from graduating Thesis, Cornell University, by Edward Caldwell. Contains some interesting experiments on constant potential regulation by this means. *Elec. Rev.*, Jan. 4, 1889.

Some New Forms of. An illustrated article by F. L. O. Wadsworth. *Elec. World*, Sept. 13, 1890, pp. 183-4.

Some Points in Dynamo and Motor Design. Paper read by Mr. W. B. Esson before Inst. of Electrical Engineers Feb. 20, 1890. Illustrated by diagrams. *Elec. Rev.*, Feb. 20, 1890, p. 243.

A series of lectures by Prof. S. P. Thompson before the Society of Arts. Treats of the construction in detail, and describes special machines as typical of a class. *Van Nos. Eng. Mag.*, Vol. XXVIII., pp. 211 and 265; also No. 66 of *Van Nos. Sci. Series*.

Stanley Alternate Current Arc. Account of very thorough test of the performance and self-regulating qualities of this machine, by W. B. Tobey and G. H. Walbridge. Paper before Am. Inst. Elec. Eng. Illustrated. *Elec. World*, Nov. 8, 1890; *Elec. Rev.*, (N. Y.) Nov. 8, 1890, pp. 134-5.

Synthetic Study of. A series of articles giving a full synthetic study of dynamo machines, including a good exposition of induction. *Tel. Jour. & Elec. Rev.*, Aug. 3, 1883.

Systems, Construction and Operation of. By S. D. Field. Treats of the proportions, amount of iron in armatures, brush lead and other items. *Sci. Am. Sup.*, May 7, 1887.

Testing their Commercial Efficiency. A description of the application of Hopkinson's method. Illustrated. Lon. *Engineer*, March 5, 1886.

Tests of, at the Phil. Elec. Exhib. Report of Committee. Illustrated, covering 60 pp., in Sup. to *Jour. Frank. Inst.*, Nov. 1885.

Tests of an Edison-Hopkinson. Giving efficiency and results of a continuous run for 10 days. *Elec. Rev.*, Jan. 24, 1885.

Theoretical construction of characteristic curve. Effect of the current in the armature, etc., By J. Hopkinson, M. A., D. Sc., F. R. S., and E. Hopkinson, M. A., D. Sc. From the Philosophical Transactions of the Royal Society. Published in the *Elec. Rev.*, Nov. 12 and 19, 1886.

Theory of. By Prof. Rowland, at National Conference of Electricians. *Elec. Rev.*, Nov. 1, 8, 15 and 22, 1884.

Electric Dynamos, continued.

Thomson's new Alternating. No revolving wire. Illustrated. *Elec. Eng.*, July. 30, 1890.

The Thomson-Houston, with detail drawings. *Elec. Rev.*, Nov. 15 '90.

Winding of Fields and Armatures. A concise and clear statement of the methods of determining the proper winding for a dynamo which is to produce a given E. M. F. and current at any given speed, by Prof. Francis E. Nipher. *Jour. Assn. Eng. Soc.*, May, 1889. Vol. VIII., p. 271.

See *Electric Lighting Station. Electro-Magnetic Reciprocating Engine.*

Electric and Dynamo Machinery. *Design and Construction of.* Read before the Engineers' Club of St. Louis, Feb. 26, 1887, by Dr. Wellington Adams. Applies the formula of Kapp to the design of an efficient dynamo. *Electrician & Elec. Eng.*, Dec., 1887.

Electric Drill. *Edison Electric Percussion Drill.* Explanation of the various methods of operating. Illus. *Iron Age*, June 11, 1891, pp. 1112-13.

Electric Energy. *Distributed by Constant Current.* Abstract of paper read before the Electrotechnical Society of Berlin, by Alexander Bernstein, describing what is likely to become one of the leading systems of electrical distribution and regulation. *Elec. World*, Jan. 25, 1890.

Electric Elevator. *The Otis Electric Elevator.* Illustrated description of the various appliances used with the Otis elevator. *Elec. World*, July 4, 1891, p. 7.

Electric Furnaces. By E. J. Houston. Describes some of the early electric furnaces. *Tel. Jour. & Elec. Rev.*, April 6, 1888.

Electric Galvanometer.

A New. J. Rosenthal. *Elec. Rev.*, Dec. 27, 1884, from Wiedemann's Annalen. A simple and sensitive instrument.

Electric Generators, *Danger from.* A. D'Arsonville. *Elec. Rev.*, London. Feb. 14, 1885, recommends the use of storage battery of proper E. M. F. to abolish the extra currents.

Influence Machines. By James Wimshurst before the London Royal Institution, giving a full account of recent forms of generators of static electricity. Illus. *Tel. Jour. & Elec. Rev.*, May 26, 1888.

Novel. New method of generating electricity from heat, direct, by utilizing expansion and contraction. Edison's patent. Illus. *Elec. Eng.*, Sept. 3, 1888.

System used at the Erie Colliery, near Scranton, Pa. Illustrated. *Am. Mfr.*, June 20, 1890, p. 15.

Electric Haulage *in the Anthracite Mines.* By A. W. Sheafer, from Proc. Eng. Club of Philadelphia. Read June 2, 1888. Gives comparative costs by mule, steam and electricity. *Elec. Rev.*, Nov. 30, 1888.

Electric Hoist. *Mining.* Paper by F. G. Bulkley describing the construction and use of an electric hoist at Aspen, Col. The power is obtained from a Pelton hurdy-gurdy water-wheel. Illustrated, pp. 9. *The Technic*, Univ. of Mich., No. 2, new series, p. 24, 1889.

Electric Horse Power. *Electrical Horse Power Characteristics.* A paper by W. F. D. Crane, giving a graphical diagram showing at a glance the electrical horse power corresponding to any given number of volts and amperes. *Elec. World*, July 13, 1889.

Electric Installations, *Regulations for the Safety of.* Issued by the Electrotechnical Association of Vienna. Contains many good suggestions for those making out specifications for electric installations. *Elec. Rev.*, Jan. 12, 1889.

Safety and Devices in. Paper before the National Electric Light Assoc., giving thorough and intelligent discussion of the subject, with some valuable suggestions, by Prof. Elihu Thomson. *Elec. World*, Feb. 22, 1890.

Electric Lamps. Commercial efficiency of the different systems used in America, all things considered. From Frank. Inst. Tests. By Carl Hering in *Elec. Eng.*, (N. Y.), Jan., 1886.

Electric Lamp Arc.

Nine Years with the. Paper before the Nat. Elec. Light Assoc. by M. D. Law, of Philadelphia. Gives results of experiments as to care, adjustment, etc. *Elec. World*, Feb. 22, 1890.

Electric Lamps, Incandescent.

An account by Count Du Moncel of the Edison system as displayed at the Paris Exhibiton. *Van Nos. Eng. Mag.*, Vol. XXV., p. 439.

Abstract of a paper by Thos. G. Grier, read before the Chicago Electric Club, discussion, life, cost, and efficiency, in connection with a recent test of an incandescent lamp. *Elec. Rev.*, May 8, 1891, pp. 580-1. *Elec. Eng.*, April 29, 1891, pp. 477-8. Full paper in *Elec. World*, Apr. 25, 1891, pp. 311-12.

Efficiency of. By W. E. Ayrton and J. Perry. Before the Physical Society of London. Treats of the efficiency of incandescent lamps with direct and alternating currents. The mean of 75 experiments gave same results for both. *Tel. Jour. & Elec. Rev.*, June 1, 1888.

Efficiency and Duration of. Report of the Special Committee on, appointed by the President of the Franklin Institute. The report fills 187 pages, and forms a supplement to the *Jour. Frank. Inst.* for Sept., 1885.

Fittings of the Weston System. Illustrated by 16 engravings. Lon. *Engineer*, Aug. 21, 1885.

Highest Economy in the Use of. A clear and concise statement of the requisite conditions for greatest economy. By H. Ward Leonard. *Elec. Rev.*, Apr. 24, 1891, pp. 527-9. *Elec. Eng.*, Apr. 15, 1891, pp. 448-9. *Elec. World*, Apr. 18, 1891, pp. 288-9.

Maximum Efficiency of. A paper read before the Am. Inst. of Elec. Engrs., April 10, 1888, by John W. Howell. A valuable contribution to electric lighting literature. Illustrates how to determine at what candle-power it is most economical to operate any given lamp, and determines for a particular Edison lamp that it is working at its maximum efficiency when the cost of the lamp is 15 per cent. of the total cost of operation. *Elec. World*, April 14, 1888.

Method of Testing. Data and plots of various incandescent lamps showing effect upon nominal efficiency of different methods of measuring candle-power. By W. L. Puffer. *Elec. Eng.*, January and February, 1890. *Tech. Quart.*, Vol. III, 1890, No. 1.

Report by Messrs. Morton, Mayer and Thomas on the measurement of an incandescent paper-carbon horseshoe lamp constructed by T. A. Edison. *Van Nos. Eng. Mag.*, Vol. XXIII., p. 1.

Report on, at the Paris Exposition. By the Experimental Committee. Gives methods employed and conclusions arrived at. *Van Nos. Eng. Mag.*, Vol. XXVII., p. 372.

Their Use and Manufacture. By Gen. C. E. Webber, R. E., C. B. A very complete paper, describing in detail the methods of manufacture. *Jour. Soc. Arts*, London, Dec. 10, 1886.

Electrical Lamp, Safety *for Mines*, designed by F. Wilson Swan. Weighs, with its portable battery, about six pounds. *Engineering Era* (Cleveland, O.), Oct. 21, 1885.

Electric Launches. *Eng. News*, Aug. 23, 1890, p. 174.

Review of what has been done, with estimates of operation and first cost. By Fred Reckenzaun. *Elec. Eng.*, August 13, 1890. *Eng. News*, August 23, 1890, p. 174.

Electric Light.

A description of the means now used for that purpose. By N. S. Possons, Cleveland. *Jour. Assn. Eng. Soc.*, Vol. V, p. 463.

Compared with other Sources of Light. See *Light*.

Cost of Arc Lights. A brief article containing much valuable information con-

Electric Light, continued.

cerning the actual cost of arc lights in many American cities. The results are taken from report of investigation by Engineer Commissioner Raymond of Washington, D. C. *Sci. Am. Sup.*, Dec. 14, 1889, p. 11,629.

Danger from Fire. Some curious experiments recently made in the laboratory of the Paris Soc. of Electricians as to the danger of fire from electric lighting. *Elec. Rev.*, Nov. 23, 1889.

Distribution. A short description of the method employed by the Edison Electric Illuminating Company of Boston. *San. Eng.*, Jan. 8, 1887.

Economical Generation of Steam for. Paper before the National Electric Light Association by George H. Babcock. *Elec. World*, Feb. 22, 1890.

The Electric Arc and its Use in Lighting. A paper by Elihu Thomson, read before the National Elec. Light Assoc., describing the action of the current on the carbons, and explaining many points in the practical working of an arc lamp. *Elec. Rev.*, March 20, 1891, pp. 367-70, *et seq.*

Experiments with Alternate Currents of Very High Frequency, and Their Application to Methods of Artificial Illumination. A lecture by Nikola Tesla before the Am. Inst. E. E., in which many interesting and valuable experiments are given. Fully illustrated. *Elec. World*, July 11, 1891, pp 19-27.

In Works and Factories. By Prof. J. A. Fleming before Iron and Steel Institute. Gives a great deal of valuable information relating to choice of cost, durability and efficiency of apparatus required. Lon. *Eng.*, Sept. 30, *et seq.*, 1887; *Am. Eng.*, Oct. 19, 1887.

Life and Efficiency of Arc Light Carbons. An elaborate paper by L. B. Marks, M. E., read before the Am. Inst. of Elec. Eng., Boston, May 21, 1890. Twenty-nine tables and numerous illustrations. *Elec. Eng.*, May 28, 1890, pp. 401-8. *Electrician*, June 13 and 20, 1890.

New Cruisers. Paper read before the National Electric Light Association, Boston, Aug. 10, 1887, by Lieut. J. B. Murdock, presenting in a general way the views of the Bureau of Ordnance of the Navy Department on the peculiar conditions to be met in the lighting of ships by electricity, and the best methods of meeting them. *Electrician & Elec. Eng.*, Sept., 1887; *R. R. Eng. Jour.*, Oct. 1887.

Problems in the Physics of an Electric Lamp. A very interesting lecture delivered before the Royal Inst. by Dr. J. A. Fleming. Fully illustrated. Reprinted in *Sci. Am. Sup.*, Aug. 23, 1890, No. 764, pp. 12204-6.

Electric Light and Gas.

Comparative Cost of. Reprint from *Gas and Water Rev.* in *Elec. Rev.*, Sept. 13, 1889.

Electric Light Association. Full report of the convention at Cape May, Aug. 19-21, 1890; *Elec. World*, Aug. 30, 1890, pp. 133-51; *Elec. Eng.*, Aug. 27, 1890, pp. 207-42. *Elec. Rev.* (N. Y.), Aug. 30, 1890, pp. 1-3, 6-14.

The National Convention at Chicago. A full account of the proceedings, including an interesting discussion on underground wires, ownership by municipalities of electric light plants, use of fuel oil, etc. *Elec. World*, March 2, 1889.

Meeting of the National Electric Light Association at Philadelphia, February, 1887. Full report. Papers on "Standard Wire Gauge," "High Insulation" of lines, "The U. S. Patent System," "Belting," "Electric Motors," "Steam Plant," "Incandescent Lights on High Tension Circuits," "Secondary Generators," "Alternating Current Distribution," etc., with discussions. *Elec. World*, Feb. 26, 1887; also in *Electrician & Elec. Eng.*, March, 1887.

Electric Light-house. See *Light-houses*.

Electric Lighting. A very elementary article by David Solomons. *Lippincott's Mag.*, Oct. 1890, No. 274, pp. 528-37.

Electric Lighting, continued.

Abstracted from a work of W. E. Sawyer. Treats of the division and regulation of the current. *Van Nos. Eng. Mag.*, Vol. XXIV., p. 230.

See *Incandescent Lamps*.

Applied Upon the Suez Canal. By R. Percy Sellon before the Mechanical Science Section of the British Association for the Advancement of Science. Describes the use of the electric light on the Suez canal, and gives text of the regulation issued by the canal company. *Tel. Jour. & Elec. Rev.*, Sept. 14, 1888.

The Brush System. By Chas. F. Brush. Explains the principal features of the system. *Van Nos. Eng. Mag.*, Vol. XXI., p. 393.

By incandescence, The Economy of. By John W. Howell. Treats of the economy of the generator; conductors and lamps. *Van Nos. Eng. Mag.*, Vol. XXVI., page 51.

Candle-Power of. By P. Higgs. A paper before the Inst. of C. E., with full discussion. *Van Nos. Eng. Mag.*, Vol. XXVII., p. 33 to 105.

For Cars. See *Car Lighting*.

Cost of Arc Lighting. A paper read by C. M. Keller before the Western Gas Association, giving the running expenses of a number of cases for Thompson Houston and American arc lights. *Progressive Age*, June, 1888.

Domestic. By W. H. Preece. Gives details of an isolated plant of incandescent lights, operated by storage battery. *Elec. Rev.*, Oct. 25, 1884.

Economy in Incandescent. *Van Nos. Eng. Mag.*; *Elec. Rev.*, London. By W. D. Weaver.

Fire Insurance Rules. Paper by Mr. Wilson Hartnell, read before the Brit. Asso. Adv. Sci., giving experiments on the heating effect of the current. *Mech. World*, Sept. 27, 1890, p. 128, *et seq.*

Frankfort-on-the-Main, Germany. A commission of experts reported upon (1) the comparative merits of direct and alternate current supply; (2) safety and economy; (3) the best system to be adopted. The report is published in *Elektrotechnische Zeitschrift*, Vol. V., 1890, p. 109. Abstract in *Proc. Inst. C. E.*, Vol. C., 1890, pp. 492. *Eng. & Build. Rec.*, July 19, 1890, p. 109.

Hoosac Tunnel. Brief description of system. By John A. Grier. *Mechanics*, June, 1891, pp. 141-2.

How to make Arc Light Circuits Safe. Paper before National Electric Light Convention by J. E. Lockwood, of Detroit. *Elec. World*, Feb. 23, 1891.

In America. By Prof. Geo. Forbes, before the Mechanical Science Section of Brit. Assoc. Adv. Sci. Compares the present state of the central system of lighting with its condition four years ago. Also compares the state of the system in the United States with that of England. *Tel. Jour. & Elec. Rev.*, Sept. 14, 1888.

In America. W. H. Preece in *Elec. Rev.*, Dec. 6, 1884.

Electrical Measurement. *See* Ampere, etc.

In Berlin. An illustrated paper by F. Uppenborn, describes the 70-ton dynamos, *Elektrotechnische Zeitschrift*, Vol. V., 1890, p. 53. See also an article by F. Kovarik; double plate, showing details. *Wochenschrift des esten. Ing. u. A. V.*, July 18, 1890, p. 250. *Sci. Am.*, June 28, 1892, p. 407.

Incandescent and arc lamps on the same circuit. By Hermann Lemp. *Electrician & Elec. Eng.*, Nov., 1886.

Incandescent Lamps on Arc Circuit. Diehl's induction system. An interesting description of a very ingenious device. Illustrated. *Elec. World*, February 12, 1887.

Independent Engines. A paper before the Nat. Elec. Light Assoc., by William L. Church, discussing the advantages of independent engines over the system of concentrated power for incandescent lighting. *Tel. Jour. & Elec. Rev.*, April 13, 1888.

Electric Measurement, continued.

In London. Details of plant described and illustrated. Lon. *Engineer*, Sept. 5, 12, 1890.

Management of Large Arc Lighting System. Paper by J. I. Ayer describing operation of Municipal L. & P. Co.'s System, St. Louis, in detail. *Elec. World*, Sept. 19, 1891, pp. 217-19.

In Mills. Abstracted from Report of C. J. H. Woodbury, Inspector of Boston Mfrs Mutual Fire Ins. Co. Deals with expense. Experience with light in mills and the safeguards adopted. *Van. Nos. Eng. Mag.*, Vol. XXVIII., p. 228.

By Municipal Authorities. A paper by Henry Robinson, M. I. C. E., giving calculations on the relative cost of gas and electricity. Reprinted in *Eng News*, Sept. 12, 1891, pp. 231-2.

Notes on Central Station Electric Lighting. Paper by C. H. Yeaman, read before the Liverpool Eng. Soc., discussing the different systems. *Elec. Rev.*, Nov. 21, 1890, pp. 619-23.

Progress in London. A paper by F. Bailey, read before the Society of Arts, giving a brief description of the various plants. *Jour. Soc. Arts*, Dec. 12, 1890, pp. 51-63. *Electrician*, Dec. 19, 1890, pp. 208-12. *Elec. Rev.*, Dec. 19, 1890, pp. 747-52.

Power expended in. Lon. *Elec. Rev.*, March 7, 1885. By Gilbert Kapp.

Railway Service. A valuable paper read by Mr. M. B. Leonard, before the Niagara Falls meeting of Ry. Tel. Superintendents. *Elec. World*, July 12, 1890, pp. 28-30. *Ry. Rev.*, July 12, 1890, pp. 400-1. Extracts in *R. R. Gaz.*, July 11, 1890, p. 487 and *Eng. News*, July 12, 1890, p. 39.

Railroad Service. A paper by W. H. Markland, read before the National Elec. Light Ass'n, describing its use in R. R. stations, yards, etc., and noting some essential features in such work. *R. R. Gaz.* Feb. 27, 1891, pp. 142-3.

Report of Committee on Safe Wiring of Nat. El. Lt. Ass. Rules for central stations, conductors, electric railways, etc. *Elec. World*, September 19, 1891, pp. 214-15.

Rules for Electric Light Installations. By Phenix Fire Office. *Elec. Rev.*, London, March 14, 1885.

From a Sanitary Point of View. By J. A. Fahie. *Elec. Rev.*, Oct. 25, 1884.

Silvertown. Gives description of the electric plant at the works of the India Rubber Company, at Silvertown, Eng. Lon. *Eng.*, June 11 and 18, 1886.

Steamships. A valuable illustrated paper, with discussion, in *Proc. Inst. C. E.* Vol. LXXIX., pp. 1-95.

Some Points on. A lecture before the Institution of Civil Engineers, by Dr. John Hopkinson, treating on some points relating to the conversion of mechanical energy into electrical energy. *Van Nos. Eng. Mag.*, Vol. XXIX., p. 394. Also *Van Nos. Sci. Series*, No. 71.

Test of Plant, Edison. Paper by John W. Hill, read before Am. S. C. E. A very full account of test of an Edison isolated incandescent electric light plant at the Union Central Passenger Depot at Cincinnati consisting of nominal capacity of 800 16 candle-power lamps. *Elec. Rev.*, Dec. 21 and 28, 1888. *Trans. A. S. C. E.*, June, 1888.

Tower System. Describes the system in practice at Detroit. Gives detailed drawing of the 150-foot towers, with dimensions. *Eng. & Build. Rec.*, Dec. 24, 1887.

Universal System of Accounts for Central Station. Paper before the National Electric Light Association by Mr. T. Carpenter Smith. *Elec. World*, Feb. 22, 1890.

Validity of the Incandescent Patent. Gives text of the decision of the High

Electrical Measurement, continued.

Court of Justice, England, in the matter of the Edison & Swan United Electric Light Co. vs. Holland and others. *Eng. & Build. Rec.*, Aug. 4, 1888.

Of Warships. A paper by Lieut. H. Hutchins, U. S. N. Enumerates special requirements, and describes various forms of projectors. Illustrated. *Eng. News*, Dec. 5, 1891, *et seq.*

Westinghouse System by Alternating Currents and Converters. By F. L. Pope. An excellent description of the system and apparatus. Fully illustrated. Undoubtedly one of the best systems in use for central station lighting. *Electrician & Elec. Eng.*, Sept. 1887.

See *Lighting*.

Electric Lighting Stations.

Breakdowns. By R. F. Jones, before the old Students' Association, Finsbury Technical School. Gives a classification of breakdowns in electric plants; then gives actual cases, with their symptoms, causes and cures. *Tel. Jour. & Elec. Rev.*, June 22, 1888.

Development of Generating Stations for Incandescent Light and Power. Paper before the Nat. Elec. Light Assoc. by C. J. Field, giving a full description of the new Edison station at Brooklyn, with consideration of cost, earning capacity, operating expenses and economy. Illustrated. *Elec. World*, Feb. 22, 1890.

Distribution of steam from. Paper before the National Electric Light Convention by F. H. Prentiss, showing how the distribution of exhaust steam for heating may add another source of revenue to the business of electric companies. *Elec. World*, Feb. 28, 1891. *Am. Eng.*, Feb. 18, 1891. pp. 83-5.

Installation, "Kaiser Gallerie," Berlin. Gives description of the plant at the King's Gallery, Berlin, with two-paged plate showing plan and section of engine room. *Lon. Engineer*, Sept. 21, 1888.

Installation of Plants. A series of articles by J. M. Ordway, Superintendent of Bridgeport Electric Light Co. Discusses the economical features of the problem. *Elec. Engr.*, (N. Y.), May, 1886.

Installation of Plants. By J. W. Deane. Abstract of paper before the Elec. Club. *Elec. World*, Jan. 15, 1887.

Keswick, Eng. Paper by Messrs. W. P. Fancus and E. W. Cowan read before the Inst. C. E. A 50-H. P. turbine furnishes the power. Full description with tables of tests and inset plate. *Eng. News*, Sept. 13, 1890, pp. 227-9. *Proc. Inst. C. E.*, Vol. CII., 1890. Paper No. 2459, pp. 151-64.

Largest in the World. A full description with illustrations of the central station of the Municipal Elec. L. & P. Co., of St. Louis, together with details of the organization and method of conducting the business. Contains many features of interest. *Elec. World*, Jan. 3, 1891.

London. A paper by Mr. Frank M. Gilley, describing the various stations. *Proc. Soc. Arts*, 1892, pp. 118-28.

Natick, Mass. Some novel features are described and illustrated. *Elec. Eng.*, Aug. 6, 1890, pp. 127-8.

New Station of the Edison Electric Light Company at Philadelphia. Illustrated description. *Power*, May, 1887.

New uptown stations of the Edison Illuminating Company, New York. Illus. Gives details of building and arrangement of steam and electrical apparatus *Elec. World*, Jan. 19, 1889.

Electric Lighting Arrester. *The Wood. Double Break*. Illus. *Elec. Eng.*, Nov. 5, 1890.

Electric Locomotive.

Field. Gives a description of the above locomotive recently tested on the New York Elevated Railroad. *R. R. Gaz.*, Nov. 11, 1887.

Electric Measurement.

Combination Voltmeter and Ammeter. An illustrated paper by Mr. Anthony C. White. *Proc. Soc. Arts*, 1890, pp. 60-76.

For Commercial Purposes. By J. N. Schoolbred. Treats of the supply of electricity, and describes numerous machines for measuring the quantity. *Van. Nos. Eng. Mag.*, Vol. XXIX., p. 319; also *Van Nos. Sci. Series*, 71.

Resistance of Electric Arc. W. Peukert. *Elec. Rev.*, London, March 14, 1885, from *Zeitschrift fur Electrotechnik*.

Of Supply. By W. Lowrie before the Bath meeting of the Brit. Assoc. Describes the system of measurement of house-to-house supply of electricity in use in Eastbourne, Eng. *Tel. Jour. & Elec. Rev.*, Sept. 21, 1888.

Electrical Measuring Instruments.

See also *Electric Balance*.

Electric Meters.

A brief review of the requirements to be met by an electricity meter, with an account of the principal advances made in this direction, and elementary descriptions of meters in use. *Elec. World*, Nov. 22, 1890.

For Central Stations. A paper read before the Society of Arts, Jan. 23, 1889, by Prof. George Forbes, F. R. S. Gives description of the different forms of meters now in use, with full discussion of the subject. *Jour. Soc. Arts*, Jan. 25, 1889.

Chemical Meter. The Edison, six years practical experience with. Paper read before the Am. Inst. E. E., New York, Dec. 18, 1888, by W. J. Jenks, with illustrations and discussion. A very complete and valuable paper. Gives full account of development of the Edison meter, its degree of accuracy, cost of operation, etc. *Elec. World*, Jan. 5, 1889. *Elec. Eng.*, Jan., 1889.

Current meter, the invention of Prof. George Forbes, described by him before the Am. Inst. E. E., Oct. 11, 1887. Illustrated. With interesting discussion. The meter is very simple, and works on the principle of a smoke jack by a current of heated air rising from a coil. *Electrician & Elec. Eng.*, Nov., 1887.

Current Meter for Direct and Alternating Currents. A paper by Prof. Wm. E. Geyer, read before the Am. Inst. E. E., Nov. 13, 1888. *Elec. Rev.*, Dec. 21, 1888. Discussion of the above paper. *Elec. Rev.*, Dec. 28, 1888.

Edison. By Francis Jehl. Gives complete exposition of the principle upon which the Edison electric light meter is operated. *Van Nos. Eng. Mag.*, Vol. XXVII., p. 197.

Forbes. Description of various forms and analysis of patent claims. By Prof. Edwin J. Houston. Also paper by Prof. George Forbes, describing his electrical current meter, read before the Franklin Institute, Oct. 19, 1887. *Jour. Frank. Inst.*, Dec., 1887. Vol. CXXIV., No. 744.

Lippmann. *Elec. Rev.*, London, Feb. 21, 1885. An ingenious and simple recording meter.

Meylan-Rechniewski. Illustrated description. *Elec. Rev.*, Feb. 13, 1891, pp. 197-201.

Thomson's. An electric current meter for continuous or alternating currents. Invented by Prof. Elihu Thomson. The vaporization of a volatile liquid by the heat of the current is employed to effect a reciprocating motion, which is registered by a train. *Elec. World*, April 28, 1888.

Thomson Self Recording Watt Meter. Description of instrument and test of accuracy. Illus. *Elec. Eng.*, No. 5, 1890.

Sir William Thomson's new Electricity Meter. Adjustable to various capacities. Description and illus. Paper read at the Leeds meeting of the Brit. Assoc. Adv. Sci. Lon. *Elec. Rev.*, Oct. 10, 1890, pp. 423-4.

See *Electric Wave Measurer*.

Electric Mining Machinery.

Coal Mining Machine. The Jeffrey. Illustrated description. *E. & M. Jour.*, July 6, 1889.

Edison Electric Mining Pump. Table of results of tests, showing efficiencies at various speeds. *Elec. Eng.*, July 22, 1891, p. 87.

With Special Reference to the Application of Electricity to Coal-Cutting, Pumping, and Rock-Drilling. By L. B. Atkinson, and C. W. Atkinson. Shows how the special conditions have been met, and describes some of the machinery. Two large folding plates. *Proc. Inst. C. E.*, Vol. CIV. 1891, pp. 83-113. Discussion, pp. 114-84.

See *Electric Railroads in Mines. Electric Power Transmission.*

Electric Motor Power. See *Street Railways, Motive Power for.*

Electric Motors.

A paper by Henry Morton on the theory and construction of the leading forms of electro-motors and their influence in the production of electric light. Illus. *Van Nos. Eng. Mag.*, Vol XXII., p. 397-441.

A series of illustrated articles describing the various motors on the market. *Eng. & Build. Rec.*, Oct. 4, 1890, p. 283, *et seq.*

Applied to the Propulsion of Street Cars. Paper by Nathan S. Possons, member of Civil Engineers' Club of Cleveland. Read October 8, 1889. Discussion by members. *Jour. Assn. Eng. Soc.*, Nov. 1889, Vol. VIII., pp. 563-569.

Alternating Current.

Paper read before the American Institute of Electrical Engineers, May 16, 1888. By Nikola Tesla. A new principle in electric motors, apparently the most practical scheme yet proposed for alternating current work; also some remarks on transformers. *Elec. World* June 1, 1888. *Elec. Eng.*, June, 1888; *Tel. Jour. & Elec. Rev.*, June 15 and 22, 1888.

A paper by Dr. Louis Duncan, giving a brief discussion of two types of motors. *Elec. World*, May 9 and 16, 1891.

A new and ingenious form of alternating current motor, described by Lieut. F. J. Patten, U. S. A., in a paper before the American Institute of Electrical Engineers, Sept. 10, 1889. Illus. *Elec. World*, Sept. 28, 1889.

Charges for Services. A paper presented to the Electric Light Convention, showing that there is a general average controlling the use of machinery which is safe for power companies to follow in making charges for electric motors. *Sci. Am. Sup.*, Sept. 22, 1888.

Comparative Tests of an Electric Motor and Steam Locomotive on the Manhattan Ry. A valuable paper presented by Mr. Lincoln Moss at the Cresson meeting, A. S. of C. E. Diagrams. Mr. Moss concludes from his experiments that the cost of direct electric propulsion would be four times that of steam locomotion. *Trans. A. S. C. E.*, Vol. XXIII, 1890, pp. 193-216. *R. R. Gaz.*, July 11, 1890, pp. 488-9. *Eng. & Build. Rec.*, July 19, 1890, pp. 101-3. *Ry. Rev.*, July 26, 1890. Editorial review discusses the same experiments and concludes that they show a decided economy in favor of electric propulsion. Diagrams. *R. R. Gaz.*, July 11, 1890, pp. 491-3. Mr. Moss replies in the same Journal, July 18, 1890, p. 503.

Construction. A New Departure in. A description by H. W. Leonard of what his recently invented motor is intended to accomplish, and his method of applying it to various uses. He claims it will be possible to "vary the voltage as the speed desired" and "vary the amperes as the torque required." *Eng News*, Nov. 28, 1891, pp. 521-3.

Description of Brown's 20 h. p., three-phase alternate current motor. Illus. *Elec. Eng.*, Nov. 4, 1891.

Designing. By T. Wahu. Discusses the best practical method of proportioning and the proper winding of motors, and gives practical experience in the con-

Electric Motors, *Designing*, continued.

struction of special motors. *Mech. World*, Feb. 18, 1888. *Tel. Jour. & Elec. Rev.*, Feb. 24, 1888.

Discussion of the new principle of constructing alternating current motors advanced by Mr. Nicola Tesla, by C. O. Mailloux. *Elec. World*, Dec. 28, 1889.

And Gramme Generator, Mechanical Efficiency of. By F. E. Nipher. Original experiments from which curves and equations for relative economy of various speeds are obtained. *Jour. Assn. Eng. Soc.*, Vol. IV., p. 15.

Governing of. By W. E. Ayrton and J. Perry, before the Physical Society. *Tel. Jour. & Elec. Rev.*, July 20, 1888.

How to make a Simple. By G. M. Hopkins. Gives full instructions by which a motor can be made with ordinary tools. Illus. *Sci. Am. Sup.*, April 14, 1888. *Tel. Jour. & Elec. Rev.*, April 13, 1888.

Principles. An exposition of, with special reference to the Sprague system. Presented to the U. S. Naval Institute, May 16, 1887, by F. J. Sprague. *Proc. U. S. Naval Inst.*, Vol. XIII., No. 3.

Proper Basis for Determining Rates. Paper by Mr. H. L. Lufkin, at the Cape May Convention of the Electric Light Association. 28 diagrams show the variation of load in actual service. *Elec. World*, August 30, 1890, pp. 145-8. *Elec. Rev.*, (N. Y.), August 30, 1890, pp. 11-14. *Elec. Eng.*, August 27, 1890, pp. 214-22.

Short Gearless. Illustrated description, with brief statement of the principles of construction of this and other motors. *Eng. News*, Sept. 5, 1891, pp. 213-4.

Single Reduction Gear Motor. Description of a new slow speed motor for street railway work. Illus. *St. Ry. Jour.*, Feb., 1891, pp. 55-6.

Speed Regulation of a Motor in the Electrical Transmission of power. Paper by M. Marcel Deprez, in Comptes Rendus. *Elec. Rev.*, April 26, 1889.

In Workshops. Paper by Chas. F. Jenkins, Stud. Inst. C. E. Illustrated. *Am. Mfr.*, Sept. 26, 1890, pp. 507-8. *Proc. Inst. C. E.*, Vol. CII., 1890, Students' Paper No. 270, pp. 307-28.

See *Bridges, Draw, Electric Motor for.*

Electric Parcel Exchange System. Small underground railways for carrying parcels in London. Illustrated description of proposed plan. By A. R. Bennett read before the Brit. A. A. S. *Elec. Rev.*, Sept. 4, 1891, pp. 271-3. *Mech. World*, Sept. 18, 1891, pp. 120-1, *et seq.* Brief notes, *Eng. Record*, Oct. 17, 1891, p. 318.

Portelectric System of Transportation. Description, with illustrations, of a system of electric transportation adapted for letters and small parcels in cities. *Elec. World*, May 4, 1889.

Electric Position Indicator, *Pointing Guns by*. Description of the instruments and methods invented by Lieut. Bradley A. Fiske for indicating the position of an object, such as that of a ship in a harbor at a point distant from the observation station, so that the guns of a fort for example, may be correctly aimed when the target is hidden from the gunners by smoke or otherwise. *Elec. World*, March 8, 1890. *Elec. Eng.*, Oct 1, 1890.

Range Finding by Electricity. The Enemy's Distance. By Park Benjamin, Ph. D. *Harper's Mag.*, June, 1890, 6 pp. Illustrated.

Electric Potential, *Energy and Work*. By Lieut. B. A. Fiske. *Van Nos. Eng. Mag.*, Vol. XXVIII., p. 380.

Electric Power Stations.

Central Electric. Paper by C. J. Field, M. E., giving a complete account of the latest developments in Edison central station practice at Brooklyn, N. Y. Illus. *Elec. World*, Dec 21, 1889.

Central Station of the Narragansett Electric Lighting Company, Providence, R. I. A paper read before the National Electric Light Association, Aug. 7, 1889.

Electric Power Stations, *Central, etc.*, continued.

by J. T. Henthorn and C. R. Remington. Illustrated. Gives a very complete account of the arrangement of steam and electrical apparatus, shafting, etc., for a first class plant. *Elec. World*, Aug. 17, 1889; *Elec. Eng.*, Sept., 1889; *Power*, Oct., 1889.

Deptford Central Station. Description and illustrations of this, as yet uncompleted station, being built under the plans of Mr. Ferranti. *Elec. Rev.*, May 8, 1891, pp. 591-4.

Largest in the World. Description of work on West End Railway Station, Boston. Illustrated. *Elec. Eng.*, Oct. 15, 1890.

Metropolitan Supply Company. Illustrated description of these large stations, by Dr. J. A. Fleming. *Electrician*, Oct. 24, 1890, pp. 703-7, *et seq.*

Modern Development of. Includes descriptions and drawings of stations of Edison Electric Illuminating Co., of Brooklyn, embodying most recent improvements, with figures on cost, operating expenses, etc., by C. J. Field. *Elec. Eng.*, March, 1890.

At Providence, R. I., Knoxville, Tenn., and Indianapolis, Ind., are described and illustrated. *St. Ry. Jour.*, Aug., 1890, pp. 379-83.

Standard Type of. Requirements of the Electric Mutual Insurance Company, of Boston, for the construction, equipment and maintenance of electric light or power stations to secure safety from fire. *Elec. Eng.*, April 16, 1890.

Station Combining the Advantages of Both the Continuous and Alternating Current Systems. A paper by H. Ward Leonard read before the Nat. Elec. Light Assoc., outlining such a possible station. *Elec. Rev.*, Oct. 2, 1891, pp. 391-3, *et seq.*

See *Electric Lighting Station*.

Electric Power Transmission.

A comparison of cost with Steam and Air. By W. S. Schulz. *E. & M. Jour.*, Nov. 15, 1884.

A valuable and comprehensive paper read Feb. 19, 1890, by Mr. Eugene Griffin, Boston. *Jour. Assn. Eng. Soc.*, Sept. 1890, pp. 419-48; discussion, pp. 448-56.

By C. J. Van Depoele. Gives results of experience with electrical motor on street railways. *Ry. Rev.*, June 12. By Geo. and Wm. E. Gibbs. Gives the result of some experiments. *Van Nos. Eng. Mag.*, Vol. XXVII., p. 247.

By Prof. Ayrton, before the Bath meeting of the British Association. Discusses the advantages of electrical transmission of power, and tells what is being done. *Tel. Jour. & Elec. Rev.*, Sept. 21, 1888; *Am. Eng.*, October 3, *et seq.*, 1888.

By Alternating Currents Differing in Phase (Rotary Currents). By Von Dolivo-Dobrowolsky in *Elektrotechnische Zeitschrift*. Explains general principles and describes various types, giving results of tests on one system. *Electrician*, Aug. 7, 1891, pp. 389-91; *Elec. World*, Oct. 10, 1891.

Alternating Currents for. Addresses by Prof. Fr. Vogel giving a clear explanation of general principles. Abs. *Eng. News*, Aug. 21, 1891, pp. 161-2.

Bessbrook and Newry. By E. Hopkinson, before the Institute of Civil Engineers. Describes the construction and discusses the working of the Bessbrook and Newry electrical tramway, designed for freight and passenger traffic. Gives full details. Experiments show the electrical efficiency to be 72 per cent. Lon. *Engineer*, Dec. 16; Lon. *Eng.*, Dec. 9, 1887. Abstract in *R. R. Gaz.*, Feb. 24, 1888.

Cantor lectures by Gisbert Kapp. Three valuable papers giving an exhaustive review of the subject. *Jour. Soc. Arts.*, July 3, 10, 17, 1891. Reprint *Mech. World*, July 25, 1891, *et seq.*; *Elec. Rev.*, July 17, 1891, *et seq.* A review comparing cost of above method with cost of steam power. *Eng. Rec.*, March 7, 1891, pp. 225-6.

Electric Power Transmission, continued.

Cost of Long Distance Electrical Power Transmission. Abstract of a paper lately read at a meeting of the New England Cotton Manufacturers' Association by W. S. Keller, giving results and conclusions arrived at in paper. *Power*, Dec., 1889, p. 10.

Description of the Lauffen-Frankfort Plant. Article by Carl Hering. Illus. *Elec. World*, Sept. 26, 1891, pp. 232-33.

Difficulties in the way of, by alternating current motors. Papers read before the Elektrotechnichen Verein of Berlin. By Alard Du Bois Raymond. *Elec. Rev.*, Feb. 1, 1889.

Discussion of economy of plants now in operation, showing a marked increase over steam. Abstract from report of Chief Inspector of Mines in Ohio. *E. & M. Jour.*, Oct. 18, 1890, pp. 456-7.

Electrical Practice in Europe as Seen by an American. IV. Paper by Carl Hering containing description of rotary alternating current motors. Illustrated. *Elec. World*, Sept. 19, 1891, pp. 193-5.

Electrical Transmission and Conversion of Energy for Mining Operations. A paper by H. Ward Leonard read before the Assoc. Mining Engr's. of Quebec. Three kinds of electric drills are described, and an estimate of a plant given, with formulas and diagrams to aid in the design. *Eng. News*, May 16, 1891, pp. 470-1, 473-4. Editorial comparing these drills with others, *ibid.*, pp. 473-4. Mr. Leonard's paper is also given in (N. Y.) *Elec. Rev.*, May 23, 1891, pp. 178-9. Abstract in *Elec. World*, May 16, 1891, pp. 361; *Elec. Eng.*, May 13, 1891, pp. 552-3.

Gives further tests of the electrical transmission of a water-power of 50-horse maximum by four Brown dynamos at Kriegstetten, Switzerland. Lon. *Eng.*, April 20, 1888.

Gives results of the electrical transmission of work from Kriegstetten to Solotharn. Is described by Prof. H. F. Weber, Reporter to the Commission of Measurement. *Tel. Jour. & Elec. Rev.*, Feb. 17, *et seq.*, 1888; from the *Schweetz Bauzeitung*, Vol. XI., Nos. 1 and 2; condensed account in *Sci. Am. Sup.*, March 3, 1888.

Gives details of the experiments made in November, 1887, at the Oerlikon Works at Zurich. The dynamos were about 50 horse-power, and were to be used over a distance of five miles. Illustrated. Lon. *Engineer*, April 13, 1888.

Lecture before the Franklin Institute, Nov. 19, 1888, by Frank J. Sprague. Gives some new and useful formulæ for designing systems of electrical power transmission. *Jour. Frank. Inst.*, March, 1889.

In Mining Operations. Valuable paper by H. C. Spaulding, before the A. I. M. E., Sept. 1890. *Elec. Rev.*, Oct. 11, 1890, pp. 84-5; *Elec. Eng.*, Oct. 8, 1890, pp. 391-3. Abstract in *Elec. World*, Oct. 11, 1890, p. 264.

Multiphase Alternating Current. Article by R. O. Heinrich describing above. Illus. *Elec. Eng.*, Sept. 9, 1891, pp. 273-5.

From Niagara. By Benj. Rhodes. Discusses the economic problems involved. *Trans. A. S. C. E.*, Vol. XIV., p. 205.

Preliminary report of the experiments between Paris and Creil, a distance of 35 miles. An industrial efficiency of fifty per cent. obtained. *Elec. Rev.* (N. Y.), Dec., 1885. Also an account of the many experiments made to transmit power of large amounts and to long distances given, including the Paris-Creil experiments. From *La Lumiere Electrique*, in *Sci. Am. Sup.*, January 23, 1886, *et seq.*

Present Status of. A paper read before Am. Inst. Min. Engrs., Oct., 1883, by Richard P. Rothwell. *E. & M. Jour.*, Jan. 5, 1884.

Recent Improvements in. Paper by N. S. Possons, read before the Civil Engineers' Club of Cleveland, Nov. 13, 1888. *Jour. Assn. Eng. Soc.*, April, 1889.

Electric Power Transmission, continued.

Theoretical consideration of the subject. By M. Levy. *Van Nos. Eng. Mag.*, Vol. XXVII., p. 341.

See *Electric Railroads, and the Transmission of Power. Tunneling Plant. Electricity, Distribution of.*

Electric Pumps.

See *Pumping, Electric.*

Electric Power Transmission.

A comparison of cost with steam and air. By W. S. Schultz. *E. & M. Jour.*, Nov. 15, 1884.

Electric Railroads.

Albany and Elsewhere. Fully described and illustrated. *St. Ry. Jour.*, June, 1890.

Application of Electricity to Street Railways. Valuable paper by Frank J. Sprague before the National Electric Light Association, giving account of progress made, methods now in successful use in the United States, economy of operation, electrical and legal relations to telephone business and prospects of application to trunk lines. *Elec. World*, Feb. 22, 1890. *St. Ry. Gaz.*, March, 1890. p. 40.

Application of Storage Batteries to Street Car Propulsion. A paper E. W. Hesine, describing a system invented by W. L. Stevens. *Proc. Soc. Arts*, 1890. pp. 99-108.

Bentley-Knight System. A short illustrated description of the proposed electric railway to be constructed in New York City on the Bentley-Knight system Lon. *Engineer*, Jan. 28, 1887.

Bessbrook and Newry Tramway, Eng. A successful Electric Railway. *Proc. Inst. C. E.*, Vol. XC., 193-381

Birmingham (Eng.) Electric Railway. Description of this storage battery line in considerable detail. *St. Ry. Jour.*, Dec. 1891. pp. 669-71.

Birmingham Bristol Road Electric Railway. The Elwell-Parker storage batteries are used. The plant is fully described. Lon. *Engineer*, July 18, 1890, p. 48.

Blackpool Tramway, England, as now in operation. Siemens system with underground conductors. Lon. *Engineer*, Jan. 1, 1886. *Sci. Am. Sup.*, Jan. 8, 1887.

Buffalo—The Latest Developed Type of Electric Street Railway Practice. A general description of the plant, with illustrations of power house. By C. J. Field. *St. Ry. Jour.*, Feb. 1891, pp. 66-70.

Cincinnati, System. Description of system with discussion of question relating to single and double trolley systems. *St. Ry. Gaz.*, Aug. 1891, pp. 160-2.

Comparison of the Single and Double Trolley Wire Systems. Paper by Geo. W. Mansfield, read before the Boston Electric Club. *St. Ry. Gaz.*, June, 1890, p. 96.

Construction and Operation, and a Consideration of their Connection with Central Station Interests. Paper by C. J. Field, before the National Elec. Light Assn., giving some details of practical problems in this connection. *St. Ry. Gaz.*, October, 1891, pp. 208-10. *Elec. Eng.*, September 16, 1891. *Elec. World*, Sept. 19, 1891.

Dependent Overhead or Underground—System of Electric Motive Power. A paper by Geo. W. Mansfield giving a general discussion. Paper read at the meeting of the Am. St. Ry. Assn. *St. Ry. Jour.*, Nov., 1891, pp. 487-91. *St. Ry. Rev.* Oct., 1891. pp. 434-37.

Development of. A paper by Eugene Griffin read before the National Electric Light Assoc. Reprinted from *Elec. World.* *Sci. Am. Sup.*, No. 823, Ocober 10, 1891. pp. 13143-6. *St. Ry. Gaz.*, October, 1891. pp. 203-8. *Elec. Eng.*, Sept. 16, 1891.

Electric Railroads, continued.

Direct Electric Co's. Storage Battery Traction System. A system possessing several promising features. Description and account of trial trip. *Elec. World*, May 16, 1891, pp. 354-5. *Elec. Eng.*, May 13, 1891, pp. 542-4.

Edison's proposed system. A short description of high potential transmission to motor generators which convert current to low potential—the rails being the working conductors. *Elec. Eng.*, Nov. 18, 1891, pp. 545-6.

Efficiency Test of Syracuse, N. Y. By G. D. Hulett. Diagrams. *St. Ry. Jour.*, Sept., 1890, pp. 417-9.

Evolution of the. By Dr. Wellington Adams. An historical account, together with a description of the author's own devices. Illustrated. *Jour. Assn. Eng. Soc.*, Vol. III., p. 285.

Florence and Fiesole Electric Railway. A paper by C. P. Shelbner describing this railway and its equipment. Folding plate of details of track, cars, motors, etc. A maximum grade of 8 per cent is overcome. Electric brakes used. *Proc. Inst. C. E.*, Vol. CVI., 1891, pp. 248-63. Abstract, *Elec. Rev.*, September 5, 1891, pp. 370-3.

Hamburg. By J. L. Huber, before the Institution of Civil Engineers. Gives details of the trial trips made on the Hamburg electric road with the Julien system. *Proc. Inst. C. E.*, Vol. XCII., pp. 304-311.

High Railroad Speed with an Electric Motor. Abstract of a paper read before the Am. Inst. C. E., by O. T. Crosby, reviewing the experiments made in 1889 on high speed electric cars, and discussing a design for a proposed plant and four mile road for further experiments, also giving calculations as to cost of such a railroad, speed of trains to 125 miles per hour. *R. R. Gaz.*, March 13, 1891, pp. 184-5. Full paper with illustrations in *St. Ry. Gaz.*, March, 1891, pp. 48-52. Abstract, *Ry. Rev.*, April 4, 1891, pp. 216-17.

Independent—Storage or Primary Battery—System of Electric Motive Power. Paper by Knight Neftel before the Am. St. Ry. Assn., giving a brief discussion of the application of secondary batteries. *St. Ry Jour.*, Nov., 1891, pp. 593-5. *St. Ry. Rev.*, Oct., 1891, pp. 473-4.

In Mines. Electric Locomotives in German Mines. A paper by Karl Eilers describing and illustrating these locomotives and giving comparative statement of cost of haulage. *Trans. A. I. M. E.*, 1891, pp. 12.

Lineff System of Underground Conductors, Report on. This system, described by Gilbert Kapp, consists in a conductor placed in a subway without any slot; connection being made by magnetic action through a third rail which is laid in short insulated pieces. Lon. *Elec. Rev.*, July 25, 1890, pp. 102-3. Lon. *Eng.*, Aug. 15, 1890, p. 129. *Elec. Eng.*, Oct. 20, 1889.

London City and South London Railway. Description of the line, methods of tunnelling, locomotive, and horse power station. Illustrated. Lon. *Engineer*, Nov. 7, 1890, pp. 382-4. Lon. *Eng.*, Nov. 7, 1890, pp. 531-2. *Elec. Rev.*, Nov. 7, 1890, pp. 534-8. *Ry. Rev.*, Nov. 29, 1890, pp. 716-17. *Sci. Am.*, Nov. 29, 1890, pp. 342-3.

Motive Power for Street Railways. Paper by J. N. Beckley before the N. Y. St. Ry. Assn., discussing the adaptability of the trolley system and giving some data on cost of operating. Discussion on snow plows, motors, etc. *Eng. News*, Sept. 26, 1891, pp. 279-81.

Motive Power for Street Railways. Report of committee on above to Street Railway Association of the State of New York, and discussion. *Elec. World*, Sept. 26, 1891, pp. 237-9.

New York. The Daft system described and illustrated. *Science*, Aug. 21, 1885. Also *Elec. World*, Sept. 5, 1885.

Of the United States. By T. C. Martin. Advocates the use of electricity, and shows what is being done by giving the performances of the various motor companies, seriatim. *R. R. Gaz.*, April 29, 1887.

Electric Railroads, continued.

Of the World. Two tables showing location, length of line, system of conductors, cost of operation, etc., of electric roads of the world. *Eng. News,* May 28, 1887.

Operation of.

A paper by O. T. Crosby. The observed coal consumption, attendance expenses, repairs, interest charges, etc., etc., are given for electric roads in Washington, Richmond, Cleveland, and Scranton, forming a valuable collection of data gathered from actual experience. *St. Ry. Gaz.,* Feb., 1890, p. 18. *Elec. World,* Dec. 21, 1889.

Operation of, Results and deductions from statistics gathered from 85 railway lines. *Elec. World,* Oct. 18, 1890, p. 274.

Paper prepared by Capt. Eugene Griffin, U. S. A., before Soc. of Arts of Mass. Inst. of Tech., pp. 12. *Proc. Soc. Arts* (Mass. Inst. of Tech.), 1888 and 1889, p. 102.

Paper by George W. Mansfield, read at Niagara meeting of National Electric Light Association. *St. Ry. Gaz.,* Aug., 1884.

Power Tests. By Irving Hales. Tabulated results of experiments: diagrams. Reprinted from *Elec. World. Elec. Rev.,* May 31, 1890, pp. 606, 610.

Practical Side of. Article by J. H. Bickford giving a consideration of the details of track construction. *St. Ry. Jour.,* Nov. and Dec., 1891.

Problem in Electric Traction. A paper by Dr. Louis Bell, Perdue University, giving results and discussion of a series of experiments made on an electric railway at Lafayette, Ind. *Elec. World,* June 22, 1889; *Elec. Rev.,* July 12, 1889.

For Rapid Transit Between Cities. An interesting paper by Mr. Carl Zipernousky read at Frankfort Electrical Congress, Sept. 1891, describing the projected high speed electric road between Vienna and Buda-Pesth. *Elec. World,* Oct. 10, 1891.

By A. Reckenzaun. Describes the details of several electric tramways in Europe and gives cost of working them. *Jour. Soc. Arts* April 22, 1887.

Schlesinger System described by the inventor. The conductor carried in a conduit between the tracks. One such in operation in Philadelphia. *Jour. Frank. Inst.,* Nov., 1886.

Solution of Municipal Rapid Transit. By Frank J. Sprague, before the American Institute of Electrical Engineers, June 19, 1888. Gives valuable data as to efficiency of different methods of rapid transit in cities, a description of the Sprague Electric Railway in Richmond, and proposes a plan for solving the problem in New York. *Elec. Eng.,* August, 1888. Discussion in November number. *R. R. Gaz.,* Nov. 2, 1888.

Some recent Electrical Work on the Elevated Roads, and its bearing on the rapid transit problem. Paper by Mr. Leo Daft read before Institute of Electrical Engineers, June 25, 1889. *Elec. Rev.,* July 6, 1889; *Elec. World,* July 6, 1889.

Sprague System. General description. *Elec. Rev.,* Oct. 29, 1886.

Standard Potential for. Report by a committee of the National Electric Light Association, giving data and recommending 500 volts. *Elec. World,* Feb. 22, 1890.

Street. By G. W. Mansfield. Gives comparison of the cost of construction and operation of cable, horse and electric street railroads. *Sci. Am. Sup.,* Sept. 17, 1887.

Street Cars, Methods of Gearing for. Paper read before the American Institute of Electrical Engineers, Sept. 20, 1887, by A. Reckenzaun, C. E. Compares different methods of gearing in use, and advocates the use of worm-gearing. *Elec. World,* Oct. 1, 1887.

Electric Railroads, continued.

St. Paul. Gives a brief sketch, with drawing, of an electric railroad in St. Paul. The cars are suspended from an overhead rail. *Eng. & Build. Rec.*, Aug. 4, 1888.

And the Transmission of Power by Electricity. A paper before the Society of Arts, by Alex. Siemens, with discussion. *Van Nos. Eng. Mag.*, Vol. XXV., p. 45.

Underground. Brief description of the methods of operating the tunnels of the City and South London Railway by Electricity. Illustrated. *Elec. Eng.*, March 4, 1891.

Zipernowsky Electric Tramway with a Vertical Rail. Illustrated description of a new form of electric tramway having a single duplex central surface rail with an auxiliary balancing rail below the surface in a conduit which may be adapted for either cable or electric traction. *Elec. Rev.*, Nov. 15, 1889. Experiments on. *Elec. Rev.*, June 19, 1891, pp. 780-2.

Electrical Art *in the U. S. Patent Office.* By C. J. Kintner, a patent examiner. A review of the patents issued previous to 1876, with many interesting examples of worthless applications. *Jour. Frank. Inst.*, May, 1886. *Sci. Am. Sup.*, June 5, 1886.

Electrical Commutators. Description of the Hartnell's Brush contact and compound switches. Illustrated. Lon. *Eng.*, Oct. 21, 1887.

Electrical Conductors. *Andrew's System of Concentric Wiring.* In this system one conductor is placed within the other, with insulating material between. Illustrations of joints and various details. *Elec. Rev.*, May 15, 1891, pp. 616-18.

Cables, Influence of Electric Tension in the Insulation of. Paper by A. Paley, in *La Lumière Electrique*, giving results of a series of experiments. *Elec. Rev.*, Oct. 31, 1890, pp. 481-2.

Cables in the St. Gothard Tunnel. A full account of the laying of new electric cables is given in *Elecktrotechnische Zeitschrift*, Vol. XVI., p. 85. Abstract. *Eng. Record*, Aug. 22, 1891, p. 186.

Cable Testing. A practical paper by Henry W. Fisher, describing methods of determining the working conditions of the cable, location of faults, etc. Illus. *Elec. World*, June 27, 1891, pp. 481-2.

Capacity, Self-Induction and Mutual Induction Measurements on Overhead Lines. By M. Massin in *Comptes Rendus*, July 13, 1891. Gives description of methods and results obtained. Reprint, *Elec. Rev.*, July 31, 1891, pp. 129-30. *Elec. World*, Aug. 29, 1891, p. 145.

Diagram for the Application of the Law of Heating as it Affects Insulated Electrical Conductors. Based on a given increase of temperature over that of the surrounding atmosphere. By Thos. J. Fay. *Elec. World*, July 4, 1891, p. 5. Reprinted in *Eng. News*, July 11, 1891, pp. 32-3.

Economy in. London *Elec. Rev.*, Feb. 28, 1883, from *Science*. Balancing of heat waste and interest on capital.

Effects of High Temperature upon the Insulation Resistance and Inductive Capacity of Vulcanised India-Rubber. Results of tests on three cables. By Wm. Maver, Jr. *Elec. Eng.*, Aug. 12, 1891, pp. 159-60.

Flad's Overhead System for Cities. Proposed method of electric distribution by overhead wires supported on high towers. By Col. Henry Flad, of St. Louis. *Electrician & Elec. Eng.*, July, 1887.

Formulas for Dimensioning conductors for carrying power when the power to be delivered at a given distance is known. Results very different from those obtained by Thomson's formulas. By Prof. Ayrton and Perry. Lon *Eng.*, March 19, 1886.

High Insulation; Commercial Testing of. A clear and practical exposition of the subject by Carl Hering. Illus. *Elec. World*, March 7, 1891.

Electrical Conductors, continued.

Law of Economy of. An interesting paper by Georges Santarelli. Translated from *La Lumière Electrique*. *Elec. World*, July 26, 1890, pp. 56, 57.

Mechanical and Electrical Tests of, at the Phil. Elec. Exhib. Report of Committee in Suppl. to *Jour. Frank. Inst.*, Nov., 1885.

The Most Economical Loss in Conductors. A paper by E. P. Roberts, discussing the question from a financial point of view. *Elec. World*, July 18, 1891, pp. 42, *et seq.*

Safety of Overhead. Condensed from report of evidence before the Massachusetts Committee on Street Railways during February and March. *St. Ry. Jour.*, May, 1889, Vol. V., p. 125.

Size of. By H. W. Leonard. Treats of the size of conductors for incandescent lighting when lamps are in multiple arc, as determined by a required loss of electromotive force. *Elec. Eng.*, Aug., 1886.

Sizes of, for uniform heating: determined mathematically by Carl Hering. *Elec. Eng.*, N. Y., Feb., 1886.

Some Points Connected with Mains for Electric Lighting. A paper by Wm. H. Preece read before the Inst. E. E., setting forth the advantages of the concentric system and discussing the question of insulation, etc., *Electrician*, May 29, 1891, pp. 108-11, *et seq.*

Wiring Buildings for the Electric Light. Article giving "amended standard for electric equipments" adopted by the New York Board of Fire Underwriters. Feb., 1890. *Eng. & Build. Rec.*, Jan. 11, 1890, Vol. XXI., p. 70.

See *Electric Lighting, Report, etc.* *Wiring Chart.*

Electrical Conductors Underground.

A lecture by Stuart A. Russell, describing several kinds of cables and discussing the merits of various insulating substances. Relating chiefly to English practice. *Elec. Rev.*, March 13, 1891, pp. 323-4, *et seq.*

Cables. Includes an abstract of a paper, read by Prof. George Forbes, and quotations from eminent authorities. *Sci. Am. Sup.*, Aug. 10, 1889.

Cables in Germany. Methods of laying described and illustrated by full page cut. *Sci. Am. Sup.*, Nov. 14, 1885.

Edison System. Rep. of Exam. of Phila. Ex. *Electrician* (N. Y.), March, 1885.

For Electric Lighting. Full discussion of the subject at the National Electric Light Convention in Chicago. *Elec. World*, March 2, 1889.

By W. W. Jacques. Difficulties, precautions and cost. *Science*, July 3, 1885.

By W. W. Leggett, before the National Electric Light Association. Discusses the difficulties of putting the arc light wires underground. *Tel. Jour. & Elect. Rev.*, April 6, 1888.

Report of Sir Wm. Thomson and Prof. Jenkin. New York *Elec. Rev.*, Jan. 17, 1881.

Wires in Washington, D. C. Report containing discussion and recommendations of the Electrical Commission. *Eng. News*, Nov. 7, 1891, pp. 441-2. *Eng. Record*, Nov. 14, 1891, pp. 388-9.

Electrical Conduits.

The current is supplied through sections of a surface conductor which are automatically connected with a main conductor. By C. K. Harding. *St. Ry. Gaz.*, Nov. 1890, pp. 210-11.

Denver, Colo. An account of the 8 mile tile conduit system recently laid in Denver for telephone cables. Capacity, 100-pair cables. *Eng. News*, March 28, 1891, p. 290.

Edison System of Underground Electric Tubes. This ingenious system, which has been in successful operation for ten years, is fully described and illustrated. *Elec. Eng.*, July 2, 1890, pp. 4-8.

Electrical Conduits, continued.

For Telegraph Wires used in England. By D. E. Lain, London. *Elec. Eng.* (N. Y.), January, 1886.

Mains of Paris. A brief description of the systems used by the four principal companies. Lon. *Eng.*, Jan. 30, 1891, pp. 119-20, *et seq.*

Report of examiners on the exhibit in this section at the recent electric exhibition at Phila. A pamphlet of 51 pp. issued as a supplement of the *Jour. Frank. Inst.* for February, 1885. The report describes thirteen systems, and is illustrated by many cuts.

Underground Conduits and Electrical Conductors. A paper read before the Inst. of E. E., by John B. Verity. *Elec. Rev.*, April 26, 1889.

See Electric Subways.

Electrical Distribution. A series of lectures by Prof. Geo. Forbes before the Soc. of Arts, London. They are designed to assist engineers in drawing specifications and in selecting the system to be adopted. Accompanied with valuable tables. *Jour. Soc. Arts*, London, Oct. 2, 1885, *et seq.* *Van Nos. Eng. Mag.*, April, 1886. *Elec. Eng.* (N. Y.), April, 1886.

Magnetic Circuit of Transformers; closed vs. open. A discussion of relative merits, taking up the question of the various losses in each system, and describing a method of diagraming these quantities. By S. Evershed. *Electrician*, Feb. 20, 1891, pp. 477-80, *et seq.*

Notes on Economy in Conductors. A paper by Hamilton Kilgour. *Elec. Rev.*, Dec. 12, 19, 1890.

System of. A paper read before the Society of Telegraph Engineers and Electricians, by Henry Edmunds, Nov. 22, 1888. Illustrated, with discussion. *Elec. Rev.*, Nov. 30, 1888.

See *Electric Power Transmission.*

Electrical Energy. *Cost of the Generation and Distribution of Electrical Energy.* A paper by R. E. B. Crompton discussing the effect of various factors on the cost of production and distribution. *Proc. Inst. C. E.*, Vol. CVI, 1891, pp. 1-32. Discussion, pp. 33-113.

Electrical Engineering. *Course of Study for.* By F. R. Hutton. Should be "Mechanical Engineering with an Electrical Engineering Attachment." *Sch. of Mines Quar.*, Oct., 1887, p. 69.

Edinburgh Exhibition. An interesting article, fully illustrated, describing the light and telpherage plants. Lon. *Engineer*, July 11, 1890, pp. 23-5. Aug. 15, 1890, p. 125.

Practice in Europe as Seen by an American. Interesting papers of considerable value. By Carl Hering. *Elec. World*, Aug. 8, 1891, *et seq.*

Recent Advancement in. By James Ritchie, before the C. E. Club of Cleveland. Gives some recent improvements. *Jour. Asso. Eng. Soc.*, Sept., 1891, pp. 433-7.

Several papers on branches of this subject were read at the Leeds' meeting of the British Asso. Adv. Sci., Sept., 1890: Tables and diagrams. Lon. *Elec. Rev.*, Sep., 12, 19, 1890, pp. 299-316, 340-9. Abstracts of a number of other papers on the same subject are published in *Proc. Inst. C. E.*, Vol. CI, 1890, pp. 387-403.

Electrical Industries *in St. Louis.* Address by Francis E. Nipher, Retiring President of the Eng. Club of St. Louis. A brief account of the progress and present condition of the electrical industries is given, also a short review of their development. *Jour. Asso. Eng. Soc.*, Jan., 1891, pp. 1-8.

See *Railroads, Operating by Electricity. Railroads, Report on Application of Electricity to. Cable Railroad.*

Electric Traction. Railroads Elevated. Street Railways.

Electric Resistance Governor, *Sir David Salomon's*. Brief illustrated description of an electrical machine intended for automatically introducing or withdrawing resistance from a circuit in case of an alteration of E. M. F. *Elec. Rev.*, Nov. p. 188.

Electric Signalling Apparatus. See *Railroad Signalling.*

Electric Sparking.

A Few Points About. By L. W. Serrell, Jr., "with reference to that troublesome occurrence which makes the commutators and brushes on our dynamos so short-lived." *Elec. Rev.*, Nov. 23, 1888.

Electric Steering Apparatus *for Boats*. See *Torpedo Boat, Grimold's.*

Electric Subways.

In Europe. Facts obtained by President Plympton of the Brooklyn Subways Commission in a recent visit to Europe. Only three cities in all Europe have telephone wires under ground. *E. & M. Jour.*, Sept. 18, 1886.

Letter and report of Major of Engineers, Chas. W. Raymond. A valuable paper reporting on the desirability of underground wires for city of Washington, with numerous letters from electric companies and electricians, also including report on proposed subways for New York by Julius W. Adams, etc. Illus. *Senate Mis. Doc.*, No. 15, 50th Congress, second session.

New York. Description of the Dorset system and tests of the material, this system having been adopted by the Commission. Illus. *Eng. News*, Sept. 11, 1886.

Annual Report of the New York Board of Electrical Control. Recommendations in regard to subways, underground conduits, etc., for distribution of electricity in New York City. *Elec. World*, Jan. 12, 1889.

Full report of the New York Commission on placing wires under ground giving result of their work for the past year. *Elec. Eng.*, Aug., 1886.

Gives a good history of the Board of Electrical Control of New York City and its work. *Eng. News.* April 21, *et seq.*, 1888.

The Practical Working of. Paper before the Am. Inst. of E. E. by Wm. Meyer, Jr. Illus. *Elec. World*, March, 1890.

Paris. Description of cement conduit adapted for use in Paris, the cables being supported by glass insulating hooks. The conduits are directly under the surface paving, and have manholes at each street intersection. *Eng. News*, Jan. 25, 1890, Vol. XXIII. p. 74.

Electric Tabulating System. *The Hollerithe.* Description of electrical method of tabulating the census returns, etc. Illus. *Jour. Frank. Inst.*, April, 1890.

Electric Telegraph and Telephone.

See *Telegraph. Telephone.*

Electric Traction. *A New System of.* The River and Rail system is described and illustrated. *Elec. World*, June 28, 1890, pp. 54-5. *Elec. Eng.*, Aug. 6. 1890, pp. 132-6. *St. Ry. Gaz.*, Sept., 1890, pp. 145-7.

Data. Paper by A. Reckenzaun, read in London before the Old Students' Assn., April, 1890. Diagrams and tables. *Elec. Eng.*, June 18. 1890, pp. 480-3. *Elec. World*, June 7, 1890, pp. 389-90. *Elec. Rev.*, May 9, 1890, p. 527, *et seq.*

Waller-Manville System of Electric Traction. Some novel features are shown in the details of the system. *Elec. Rev.*, May 9, 1890, p. 513.

See *Electric Current. Electric Railroad.*

Electric Transfer Table. See *Transfer Table.*

Electric Transmission of Power.

See *Electric Power Transmission.*

Electric Voltmeters, *Portable, Ayrton and Perry's*. These instruments are an improvement upon the Cardew voltmeters, and depend on the expansion of a fine wire due to the heating effect of a current. *Elec. World*, Jan. 28, 1888.

Electric Wave *and Phase Indicator*, for alternating and undulatory currents. A diaphragm is made to move a light mirror in harmony with the current vibrations, and the form of the waves is indicated by the movement of a spot of light. In this way photographs of the wave forms may be made. Elihu Thomson in *Elec. World*, Jan 28, 1888.

A very delicate machine for measuring minute and rapid fluctuations of current. Illustrated description, from *Electrotechnische Zeitschrift*, by C. Grawinkel and K. Streckner. *Electrician*, Feb. 13, 1891, pp. 459-61.

Electric Weighing Machine. The Snelgrove Electric Weighing Machine. An ingenious self-balancing weighing device, described and illustrated in *Elec. Rev.*, Aug. 16, 1889.

Electric Welding.

Applied to the Manufacture of Projectiles. Paper by Lieut. W. M. Wood, U. S. N., read before the Soc. of Arts, Boston, describing the method of manufacture. *Sci. Am. Sup.*, No. 777, Nov. 22, 1890, pp. 12413-14.

Benardos System. Illustrated description. *E. & M. Jour.*, Jan. 5, 1889.

Bramwell on. The Rheostat is fully described and table of tests on iron and steel are given. *Proc. Inst. C. E.*, Vol. CII, 1890, paper No. 2460, pp. 1-37. Discussion and correspondence, pp. 37-72. Abstracted in *Eng. News*, Nov. 15, 1890, pp. 435-6.

Arc Welding. An illustrated description of the method and apparatus. *Elec. World*, Feb. 25, 1888.

Electricity as a Blacksmith. An article by Ralph W. Pope, describing several special applications of electric welding. *Eng. Mag.*, July, 1891, pp. 475-83.

Electricity in Welding and Metal-Working. Brief paper by A. B. Wood, describing the methods of the arc-welding system. With discussion. *Trans. A. I. M. E.*, 1891, p. 6.

Fibrous structure of an electric weld, printed from an etched specimen of a butt welded bar. *Eng. News*, Sept. 21, 1889, Vol. XXII, p. 273.

Machines. A paper by Hermann Lemp, read before the meeting of the Am. Inst. of E. E. Boston, May 21, 1890. Illus. *Elec. World*, June 7, 1890, pp. 351-3.

Practical Aspect of. A paper by F. A. C. Perrine, read before the Am. Inst. E. E. Abstract in *Eng. News*, June 6, 1891, pp. 539-40. *Eng. Rec.*, June 20, 1891, pp. 39-40. *R. R. Gaz.*, May 19, 1891, pp. 365-6.

On Ship-Board. Report of U. S. Naval Board on Thomson system for use on ship-board. *Jour. Frank. Inst.*, July, 1890.

Thomson, Prof. Elihu. An interesting description of this new process, illustrated by cuts of the apparatus used, and a collection of specimens of the work done. *Elec. World*, Dec. 25, 1886, also *Am. Eng.*, Feb. 2, 1887.

Thomson, Prof. Elihu. A full account of the art of electrical welding; with full description of the apparatus. *Sci. Am. Sup.*, May 7, 1887.

Thomson, Prof. Elihu on. Recent Development and Improvements in. Elec. Eng., Oct. 9, 1891. *Elec. World*, Oct. 11, 1890, pp. 264-5.

Thomson Electric Welding Process. An article stating the manner in which this process was invented and perfected, giving the requisites for an electric welding plant, and showing results obtained as indicated by tension tests, etc. *Eng. News*, Jan. 18, 1890, Vol. XXIII, p. 65.

Thomson's. Illustrations showing sections of bars welded by the Thomson process. *E. & M. Jour.*, Feb. 9, 1889.

Thomson Electric Welding Process. A paper read before the Iron and Steel Institute, by Mr. W. C. Fish, of Boston, Mass. Describes essentials of a welding plant, and gives a few results of tests of welded specimens of metals. Lon *Eng.*, Oct. 4. Abstract in *Iron Age*, Oct. 24, 1889.

Electric Welding, continued.

Vs. Hand Welding. Brief note giving comparison of cost in the Victoria Works, Eng., Electric welding much cheaper. *Eng. News*, May 2, 1891, pp. 426. *R. R. Gaz.*, May 1, 1891, pp. 302-3.

Woodbury, C. J. H., before the Scranton meeting of the American Society of Mechanical Engineers. Reviews the principles upon which electric welding is based, describes the apparatus used and then considers its practical applications. *Am. Eng.*, Oct. 17, 1888; *R. R. Gaz.*, Nov. 2, 1888.

Woodbury, C. J. H. Paper read before the A. S. of M. E. Table is included showing the results of tensile tests of electrically welded bars made by Lieut. Col. F. H. Parker, U. S. A. Lon. *Eng.*, June 27, 1890, *et seq.*, p. 811. Discussion on the paper. *id.* July 4, 1890, p. 9.

Electric Wires.

See *Electrical Conductors*.

Electrical Induction. A paper by J. E. Taylor discussing the subject in a popular way. *Elec. Rev.*, July 31, Aug 7., 1891.

See *Electric Current Induction*.

Electrical Measuring Instruments. *Sir William Thompson's New.* A new system of standard electrical measuring instruments in which the electrical force is balanced by gravity. *Elec. World*, Feb. 25, 1888.

Standardizing. By A. W. Meikle, before the Physical Society of Glasgow University. Gives a description of an application of the electrolysis of copper sulphates which has been employed for standardizing purposes in the Physical Laboratory for the last two years. *Tel. Jour. & Elec. Rev.*, March 23, *et seq.*, 1888.

Electrical Plants.

Construction of. By Elihu Thomson, before the Am. Nat. Elec. Light Assn. Discusses the insulation and installation of wires and the construction of plants. *Tel. Jour. & Elec. Rev.*, March 23, 1888.

Boilers for. See *Boilers for Electrical, etc.*

Insulation of Industrial Electric Installations. A communication to the Société Internationale des Electriciens, Nov. 27, 1888. By R. V. Picou. *Elec. Rev.*, Nov. 23, 1889.

Rules for the installation of electric plants as adopted by the Boston Fire Underwriters' Union. Pamphlet. 9 pp. Reprinted in *Eng. News*, Aug. 9, 1890, pp. 119-20. *Eng. & Build. Rec.*, Aug. 16, 1890, pp. 171-3. *Am. Arch.*, Aug. 30, 1890, No. 766, pp. 135-7. *Elec. Rev.*, (N. Y.), Aug. 23, 1890, pp. 10-11.

Standard Specifications for. The specifications published in the catalogue of the United Edison Mfg. Co., are reprinted in the *Elec. Rev.*, July 4, 1890, pp. 9-11.

Steam Plants for Electric Service. See *Steam Plants.*

Electrical Resistance, *Compensated Standards of.* Paper read before the Am. Inst. of E. E., May 16, 1888. By Edward L. Nichols. Describes a method of making a standard of resistance unaffected by temperature, by combining copper and carbon. *Elec. Eng.*, June, 1888. *Elec. World*, June 9, 1888.

Influence of Temper on the Electrical Resistance of Steel. By H. Le Chatelier. Experiments show that the measurement of electrical resistance determines roughly the condition of the carbon in the iron. *Comptes Rendus*, Vol. CXII., 1891.

Electrical Speed Indicator. *Elec. Rev.*, Dec. 13, 1884.

Electrical Streams. By A. W. Rucker and C. V. Boys, before the Society of Telegraph Engineers and Electricians. An interesting paper on some phases of static electricity. Illustrated. *Sci. Am. Sup.*, May 19.

Electrical Theories. *A Review of Modern.* By Prof. Wm. A. Anthony, before Am. Inst. E. E., New York, Jan. 21, 1890. *Elec. Eng.*, Feb., 1890. *Elec. World*, Feb. 1, 1890.

Electrical Transformers.

Electric Light. Description of the "transformers" on the Zipernowsky-Deri-Blathy system, which is being successfully worked in Europe. *Elec. World.* Jan. 15, 1887.

Alternating Electric Currents. Paper before the Am. Inst. E. E., New York, by Prof. Harris J. Ryan, of Cornell University, giving results of tests of the efficiency of a commercial transformer. Illustrated by curves, etc. *Elec. World* Dec. 28, 1889.

Alternate Currents. By Gisbert Kapp, before the Society of Telegraph Engineers and Electricians. Gives a full treatment of alternate current transformers, with special reference to the proportion of iron and copper. A valuable paper. *Tel. Jour. & Elec. Rev.*, Feb. 18, *et seq.*, 1888: Lon. *Engineer*. Feb. 10, *et seq.*; Lon. *Eng.*, Feb. 16, 1888, *et seq.*, *Elec. World*, March 17, 1888.

Efficiency of. Paper by Messrs. Calvin Humphrey and W. H. Powell. Tables and plates. *Elec. Eng.*, July 2, 1890, pp., 16-9. *Elec. Rev.*, (N. Y.) July 5, 1890, pp. 11-2. *Electrician*, July 18, 1890, pp. 281-2.

Electrical Distribution by Alternating Currents and Transformers. By Rankin Kennedy. Describes and compares the various systems that have been patented, and explains briefly the principles involved. *Elec. Rev.*, March 18, 25; April 1, 8, 15 and 22, 1887. Reprinted by *Electrician & Elec. Eng.*, May, July and August, 1887. See Westinghouse System of Electric Lighting.

Historical Development of the Induction Coil and Transformer. A valuable series of articles by Dr. J. A. Fleming. *Electrician*, January 30, 1891, pp. 396-7 *et seq.*

Vs. Accumulators. By R. E. Crompton, before the Society of Tel. Engineers and Electricians. A valuable paper, presenting facts and figures relating to the distribution of electricity by accumulators as transformer vs. transformers. Discussion. *Tel. Jour. & Elec. Rev.*, April 20, *et seq.*

Prof. Elihu Thomson's New Method of Regulating Current, or, Potential in the Secondaries of Transformers. Illustrated paper. *Elec. Rev.*, April 26, 1889. *Wochenschrift esterr. Ing. u Arch. V.*, Sept. 5, 12, 1890.

Electrical Units. A paper by Prof. F. E. Nipher, explaining the origin and meaning of the terms, volt, ohm and ampere. *Jour. Assn. Eng. Soc.*, Vol. VII., pp. 83-89 (March, 1888); *Tel. Jour. & Elec. Rev.*, April 27, 1888.

Units of Measurement. By Sir Wm. Thomson, before the Inst. of Civil Engrs. *Van Nos. Eng. Mag.*, May, 1884.

Of the Present and Future. An address by Prof. Francis B. Crocker before the Electric Club, N. Y. *Elec. Rev.*, (N. Y.), Jan. 3, 1891, p. 228-9.

Report of meeting of International Congress of Electricians held at Paris, 1889, session of Aug. 29, 1889. Several valuable reports abstracted in *Elec. Rev.*, Sept. 13, 1889, *et seq.*

Theory of. By G. Szarvady. A good paper on electrical units. Treats of the equations between concrete magnitudes, change of units, absolute units, history and criticism of the C. G. S. system. *Sci. Am. Sup.*, Aug 13, 1887.

The Volt, the Ohm, and Ampere. A mathematical exposition of the methods employed in fixing the value of these units. Read before the Engineers' Club of St. Louis, by Prof. F. E. Nipher, of Washington University. *Jour., Assn. Eng. Soc.*, March, 1888. *Tel. Jour. & Elec. Rev.*, April 27, 1888.

Watt and Horse-Power. A comparison of the ordinary and electrical units of power, with a plea for making the electrical standard universal. *Van Nos. Eng. Mag.*, December, 1885.

Electricity.

And Gas, Light and Energy of. By J. T. Sprague. Examines some of the questions asked and answered before the Committee of Parliament on lighting by electricity. *Van Nos. Eng. Mag.*, Vol. XXI., p. 394.

Electricity, continued.

Application of. See *Telpherage, Telephones, Street Railways, Railroads.*

Applied to Engineering. By Wm. Geipel before the Institution of Mechanichal Engineers. An exhaustive view of the practical application of electricity to industrial uses. Treats of the electrical transmission and distribution of power, of locomotion, lighting and metallurgy. Lon. *Eng.*, Oct. 24; *Sci. Am. Sup.*, March 10, 1888.

Ampere, Ohm and Volt. Direct Measurement of. J. Kessler. Lon. *Elec. Rev.*, April 11, 1885.

Atmosphere. See *Lighting.*

Direct from Fuel. A proposed new process of generating electricity, which promises to be practical and economical on a large scale. By Olaf Dahl. *Elec. World*, Dec. 14, 1889.

Distribution of. Geo. Forbes. *Elec. Rev.*, London, Feb. 7, 14 and 21, 1885. Contains tables of sizes of conductors for currents from two to two thousand amperes and of net annual cost per electrical horse-power.

Two lectures by Prof. Geo. Forbes, London. Gives the latest knowledge to principles and practice in distributing electricity. *Elec. Rev.*, New York., June and July, 1885.

By J. K. D. Mackenzie before the Society of Telegraph Engineers and Electricians. Discusses the use of secondary generators as transformers for the distribution of electricity. Gives points on the practical working of the system. *Elec. Rev.*, Feb. 17, *et seq.*, 1888.

Edison's Three-Wire System. Compiled from *Science* and *Am. Eng.* in *Elec Rev.*, Jan. 10, 1885. Advantages of the system shown.

By the Ferranti System. Paper before the National Electric Light Association, by Caryl D. Haskins, describing the remarkable features of this system as employed in London, etc. *Elec. World*, February 28, 1891.

By Induction Apparatus. By Richard Ruhlmann. *Zeitschrift des V. deutscher Ing.*, 1886, pp. 68-71.

By Secondary Generators. By J. Dixon Gibbs. Describes the evolution and working of a secondary generator, whereby the energy delivered from the primary generator is *transformed* for all purposes, as for various kinds of electric lights, motive power, etc. A grand prize of 10,000 francs granted the invention by the Italian Government. *Van Nos. Eng. Mag.*, June, 1885.

Systems. By Elihu Thomson, Sibley College, Lecture No. V. Gives a general classification of the systems of distribution, and describes each in detail. *Sci. Am. Sup.*, July 22, 1887.

Dynamic. A series of elementary articles by Carl Hering, a theoretical and practical electrician. *Elec. Eng.* (N. Y.), April, 1886, *et seq.*

Elementary Text Books on. List of some elementary books on electricity in the library of the Gramme Society, of Kansas City, Mo. *Elec. Eng.*, September, 1885.

Ether, Electricity and Ponderable Matter. Presidential address before the London Inst. of Electrical Engineers, by Sir Wm. Thomson. A mechanical explanation of electrical phenomena. *Elec. World*, April 27, 1889.

Exhibition at Philadelphia. General Report of the Chairman of the Committee on Exhibitions. An historical and descriptive account, accompanied by many full-page solar prints of special exhibits. *Jour. Frank. Inst. Sup.*, July, 1885.

Fire Hazards from. See *Fire Hazards.*

Flashing Carbon Filaments at Different Temperatures. By L. S. Powell. Gives details and results of experiments made to obtain a clearer insight of what really goes on by employing different temperatures in flashing. *Tel. Jour. & Elec. Rev.*, May 4, *et seq.*, 1888.

Electricity, continued.

From Carbon. A brief description of method recently patented by Mr. Edison consisting of an iron melting pot which contains a soluble carbonaceous electrode and oxides, the whole being heated by fuel burned on a grate. Illus. *Elec. Eng.*, Oct. 7, 1891.

Galvanism, Magnetism and the Telegraph. Chronological History of from B. C. 2637 to A. D. 1888. By P. F. Mottelay. *Elec. World.* April 18, 1891, p. 286 *et seq.*

Index to Current Electrical Literature, extending from October, 1887, to July, 1888, inclusive; 172 pages, with some 10,000 references. *Trans. A. I. E. E.*, Vol. V., 1889 (Published by the Institute, Temple Court, New York.)

In the Art of Tanning. A paper by Messrs. Rideal and Trotter read before the Soc. Chem. Indus., giving results of a series of experiments. Extended abstract. *Elec. Rev.*, June 26, July 3 and 10, 1891.

Kirchoff's Laws and their application. By E. C. Rimington. Gives a good description of the application of Kirchoff's laws to the finding of currents in network of conductors. Illustrated with many diagrams. *Tel. Jour. & Elec. Rev.*, March 2, *et seq.*, 1888.

For Light and Power, with Particulars of Canadian Installation. Abstract of a very practical and comprehensive paper read by Mr. A. J. Lawson, before the Canadian Soc. of C. E. Several statistical tables are included. *Elec. Rev.*, June 27 and July 4, 1890, pp. 731-6, 836.

In Mills, Various Uses for. By C. J. H. Woodbury. Mentions briefly various applications of electricity to purposes other than lighting. *Electrician & Elec. Eng.*, November and December, 1887.

In Mining, as Applied by the Aspen Mining and Smelting Company, Aspen, Col. Paper by M. B. Holt, containing a brief record of what has been accomplished since 1888 by this company. Illus. *Trans. A. I. M. E.*, 1891, p. 8. Abs., *Ry. Rev.*, Oct. 17, 1891. Full paper in *Eng. News*, Oct. 10, 1891.

In the French Navy. A statement of achievements and results in ten years. From *Le Genie Civil. Electrician*, Feb. 13, 1891, pp 458-9.

Measurement of. See *Electric Measurement.*

Modern Views of. Extracts from lectures by Dr. Oliver Lodge in London and Birmingham. Not published before. Lon. *Eng.*, Nov. 4, 1887.

As a Motive Power, conditions necessary to a financial success in use of. A paper before Am. St. Ry. Assn. By Thos. C. Barr, discussing the financial side of the question mainly. *St. Ry. Gaz.*, October, 1889.

By Prof. Geo. Forbes. Discusses the results which have already been obtained, and shows what steps are immediately possible with our present knowledge and experience. *Van Nos. Eng. Mag.*, Vol XXIX., p. 161.

By Wm. Wharton, before the Philadelphia Street Railway Convention. Gives a full review of electricity as applied to the propulsion of street cars, with figures of expense and practical details. *Sci. Am. Sup.*, Dec. 3, 1887.

Past, Present and Future. Lecture by R. W. Pope, Sec'y, Am. Inst. Elec. Engr's. *Jour. Frank. Inst.*, Jan. and Feb., 1891.

Physiological Bearing of, on Health. By Dr. W. H. Stone. *Elec. Rev.*, Oct. 18, 1884. Continued from a previous number. Discusses conditions of danger to life.

Production Direct from Fuel by Edison's Pyromagnetic Dynamo. Paper read before the American Association for the Advancement of Science, New York By Thomas A. Edison. *R. R. & Eng. Jour.*, October, 1887.

Projection of Lines of Force. By J. W. Moore. Gives an illustrated description of the use of the lantern in obtaining direct optical projections of electrodynamic lines of force and other phenomena. *Sci. Am. Sup.*, April 21 and 28, 1888.

Electricity, continued.

Purification of Sewage by. See *Sewage, Purification of.*

Report of Board of Control, N. Y. Gives an abstract of the report of the Board of Electrical Control, addressed to the Governor and Legislature of New York State. Illus. *Eng. & Build. Rec.*, Jan. 14, 1888.

Report of Boards of Examiners on *Telegraphs, Electro-Dental Apparatus* and *Applications of Electricity to Warfare*, as shown in the recent Electrical Exhibition in Philadelphia. Making 44 pp., well illustrated, in *Jour. Frank. Inst.*, March and April, 1885.

Report on the International Exhibition at Paris in 1881. By Major D. P. Heap, U. S. A. A book of 270 pp., issued by the Engr. Dept., U. S. A. The work is mainly confined to electrical engineering subjects.

Self-Induction. A report of the experiments performed by Professor Hughes before the Royal Society of Arts. Gives methods, apparatus and general conclusions and their practical importance. *Sci. Am. Sup.*, July 24, 1886.

Storage of. A paper by Samuel Sheldon, Ph. D., giving briefly the development of the storage battery. *Pop. Sci. Mon.*, Jan. 1891, pp. 355-63.

Study of Earthquakes by. See *Earthquakes.*

Traction Force, as a. By Roman Baron Gostkowski. Gives formulæ, with special reference to the accumulator battery. *Elec. Rev.*, Dec. 28, 1888.

Transportation, in. Compared with Steam. A valuable economic paper by Mr. O. T. Crosby, read at the annual meeting of the Am. Inst. of Elec. Engrs. Six tables. *St. Ry. Gaz.*, June, 1890, pp. 69-92. *Elec. Eng.*, May 28, 1890, pp. 395-400. Abstract and comments in Lon. *Engineer*, June 13, 1890, p. 471.

Theories of Electrical Action. A review of various theories, by Prof. H. S. Carhart. Abstract. *Jour. Frank. Inst.*, Nov., 1889.

The transformation of heat into electrical energy without batteries, thermopiles and dynamo machines: also the relative expense of several sources of energy. *Sci. Am. Sup.*, July 31, 1886.

In Transitu. From Plenum to Vacuum. Address of Wm. Crookes, Pres. Inst. C. E., describing many phenomena accompanying the passage of electricity through various media. Illus. *Elec. Rev.*, Jan. 16, 1891, pp. 76-81, *et seq.*

In Warfare. A Lecture before the Frank. Inst., by Lieut. B. A. Fisk, U. S. N. Gives an account of all recent applications of electricity in both land and naval warfare. *Van Nos. Eng. Mag.*, December, 1885.

In Warfare. A lecture by Lieut. B. A. Fisk, U. S. N. *Jour. Frank. Inst.*, Sept., 1893. *Eng. News*, Sept. 20, 27, 1893, pp. 250, 271-2.

Electricians', National Conference, of Philadelphia. A full report of this important conference is given in the *Elec. Rev.*, Oct., 18, 1884, and the four following numbers.

Electro Dynamics *of Stationary Bodies.* A valuable mathematical paper by Prof. H. Hertz. Maxwell's formulas are deduced without the aid of quaternions. *La Lumiere Electrique*, July, 1890. Abstract in *Electrician*, Aug. 29, 1890, *et seq.* *Elec. World*, Sept. 20, 1890, *et seq.*

Electrolysis. *Verification of Faraday's Law*, with reference to silver and copper. Paper read before the M. & P. S. Section of the British Association at Birmingham, Sept. 7, 1886, by W. N. Shaw, M. A. Gives results of author's experiments. *Elec. Rev.*, April 8, 1887. The *Electrician & Elec. Eng.*, April, 1887.

Electro-Magnetic, *Induction Experiments of Prof. Elihu Thomson.* A paper read by Prof. J. A. Fleming before the Society of Arts. Thirty-eight figures. *Elec. Rev.*, May 23, 30, 1890, pp. 577-618. *Elec. World*, June 14, 21, 1890, pp. 406-8, 422-3. *Jour. Soc. Arts*, May 16, 1890, p. 626.

Electro-Magnetic Mechanisms. Presidential Address to the Junior Engineering Society, by Prof. Silvanus P. Thompson. *Electrician*, Dec. 26, 1890, pp. 235-41, *et seq.*

Electro-Magnetic Reciprocating Engine. Paper by Charles J. Van Depoele, before the A. I. M. E., at N. Y., Sept., 1890. Abstracted in *Elec. Eng.*, Oct. 1, 1890, p. 390; *Eng. News*, Oct. 4, 1890, pp. 291-4.

Tesla Alternating. In mining work. Illus. *Elec. Eng.*, Aug. 13, 1890.

Electro-Magnetic Repulsion. A discourse delivered at the Royal Inst. of Great Britain, by Dr. J. A. Fleming, giving many experiments, illustrating various phenomena. Illus. *Electrician*, March 13, 1891, pp. 567-71, *et. seq.*

Electro-Magnets *Applied to the Working of Railway Signals and Points.* A paper read by I. A. Tummis, of London, before the Institution of Mechanical Engineers, the peculiar feature being the great power of the magnet which is exerted through long range. *Iron*, Nov. 7, 1884.

The Cantor lectures by Prof. S. P. Thompson, D. Sc. Very valuable and comprehensive papers. Illus. *Elec. World*, Aug. 30, 1890, *et seq.* *Sci. Am. Sup.*, No. 777, Nov. 22, 1890, *et seq.*

The Construction of. Translated from the French of Th. Du Moncel. Gives mathematical theory (algebraically) based upon patient and skillful experiments. The laws are concisely translated, and the limits of applicability tested *Van Nos. Eng. Mag.*, Vol. XXVIII., pp. 51, and 89.

Electromotive Forces *in the Voltaic Cell, Theory of.* O. J. Lodge. *Phil. Mag.* London, March, April and May, 1885. An important paper.

Electro-Pneumatic Block Signalling. See *Railroad Signalling.*

Electrostatic Machine. *Wimhurst's Electrical Review*, Jan. 14, 1885. Gives details of the largest machine yet made, the diameter of the plates being 84 inches.

Electrotechnics. A valuable series of papers, being a translation of *Uppenborn's Kalender fur Electrotechniker.* By Chas. M. Sames. Begun in *Mechanics*, Oct., 1890, pp. 219-20.

Elevated Railways. See *Railroads, Elevated.*

Elevation of Outer Rail. See *Railroad Track.*

Elevations Above Sea Level *in New York.* Determined by the *N. Y. State Survey. See* Reports for 1883.

See also Primary Triangulation of the U. S. Lake Survey. Prof. Papers, Corps of Engineers, U. S. Army, No. 24, p. 610, for descriptions and elevations above sea level of permanent bench marks between Albany, N. Y., and Oswego; also between Port Colborne and Port Dalhousie, Can.; also between Lakes Erie and Huron and between Escanaba and Marquette, Mich.

With description of permanent bench marks on a line from Sandy, N. J., to St. Louis, Mo., together with map showing location of principal bench marks. Report U. S. C. and G. Survey, 1882.

At Various Points on the Missouri River, referred to the St. Louis City Directrix. A table giving extreme high and low water, standard high and low water, as adopted by Missouri River Commission, etc. *Report of Missouri River Commission, 1889*, p. 2757. *Appendix W. W. of Report of Chief of Engineers*, 1889.

Elevator. *Balanced Hydraulic Ram.* A description of the Edoux Lift in the Eiffel Tower, Paris. *Am. Mach.*, Aug. 8, 1889.

Bins, pressure of grain on sides and bottoms of. Results of experiments on a large scale in England. *Sci. Am. Sup.*, Dec. 18, 1886, p. 9,108.

Evolution of, in America. A paper by Mr. R. M. Sheridan, read before the Engr's Club of Kansas City. *Jour. Assn. Eng. Soc.*, Dec., 1890, pp. 582-8. Abstract in *Eng. News*, Jan. 24, 1891, p. 77. *Eng. Record*, Jan. 24, 1891, p. 129.

Grain. A general article on the construction of grain elevators. Illustrated by details from various structures. *San. Eng.*, Nov. 12, 1887.

Alexandra Docks, Liverpool. Description of the warehouse and machinery for the storage and transit of grain. By Wm. Shapton before the Inst. M. E. Illus. Lon. *Eng.*, Sept. 11, 1891, pp. 296-8.

Elevator, *Grain*, continued.

Construction of those on the Can. Pac. Ry. By Stewart Howard. Illustrated. *Trans. Can. Soc. C. E.*, Vol. I., p. 24, 1887.

Pressure of Stored Grain. Paper by Isaac Roberts before the Brit. Assn., describing experiments made with wheat and deriving formulas. *Eng. News*, Apr. 7, 1883, pp. 158-9.

The Helicoidal. Gives description of the system to be applied to the Tower in Paris. *Sci. Am. Sup.*, Sept. 10, 1887.

Most Economic. By E. E. Magovern. Gives results of tests made on the elevators supplied by the New York Steam Company. *Steven's Indicator*, January, 1887; *San. Eng.*, Dec. 3, 1887.

Passenger Elevators on the North Hudson County Ry. Description with inset of illustrations of these large hydraulic elevators, which are designed to carry 20,000 lbs. *Eng. News*, Jan. 17, 1891, p. 50.

And Reloader for Coal. The Rondout coal yard machinery described and illustrated. *Sci. Am.*, June 7, 1890, p. 360.

See *Hydraulic Elevators. Hydraulic Lifts. Hydraulic Machinery.*

Elliptical Curves. *The False Ellipse Reduced by Equation of Condition.* Paper by Arthur S. C. Wurtele, deducing equations from which false ellipses may be constructed of any number of centers. Gives also special constructions. *Trans. A. S. C. E.*, Vol. XXIV., June, 1891, pp. 540-55.

Ellipsograph, *Arbter's.* Description and theoretical discussion in *Wochenschrift d. Oesterr. Ing. u. Arch. V.*, 1886, pp. 168-170.

Embankments *made by Water Power.* By Ed. B. Dorsey. Describes how a reservoir embankment 80 ft. high could be built by hydraulic excavation and water-carriage, the materials being deposited in place by the water. Illustrated. *Trans. A. S. C. E.*, May, 1886, Vol. XV., p. 348.

On Slopes. The lesson to be learned from the West Deerfield accident. An embankment having stood 22 years, gives way suddenly under a passenger train. Editorial in *R. R. Gaz.*, April 20, 1886. With illustrations.

Rapid Construction. By J. A. Smith. Describes the method adopted for filling Hall street, St. Louis, in a short space of time. *Jour. Assn. Eng. Soc.*, Vol. VII., pp. 103-105 (March, 1888). *Eng. & Build. Rec.*, March 24, 1888. *R. R. Gaz.*, June 1, 1888. *Eng. News*, Aug. 18, 1888.

Stability of Swamp. By Samuel McElroy. Gives experience in dealing with embankments over swampy ground. *R. R. Gaz.*, Aug. 31, 1888.

See *Canals, Ship, North Baltic.*

Emery Wheels. A valuable paper on the comparative economical value of emery wheels and files. *R. R. Gaz.*, May 27, 1887.

And Emery Wheel Machinery. A Miller Prize paper before the Inst. of Civil Engineers. *Proc. Inst. C. E.*, Vol. LXXVII., p. 272. Reprinted in *Van Nos. Eng. Mag.*, February, 1885.

And Files. Comparative amount of work done by each. Illustrations showing cuts made in same piece of metal by emery wheel and file in same length of time. *R. R. Gaz.*, Feb. 15, 1889.

Report by Coleman Sellers, J. E. Denton and A. R. Wolff, on "The Comparative Value of Fifteen Varieties of Solid Emery Wheels." *R. R. Gaz.*, Oct. 9, 1891, p. 705. *Am. Eng.*, Oct. 17, 1891, pp. 171-2. *Ry. Rev.*, Oct. 17, 1891, p. 681.

Solid. The Manufacture and Use of. By P. Dunkin Paret. Illustrated. *Sci. Am. Sup.*, April 24, 1886.

Energy. A paper on losses of energy other than frictional losses. It includes a computation of the amount lost by vibration in shafting, belting and bearings. By Prof. R. H. Smith. *Lon. Eng.*, April 24 and May 1, 1885.

Engineer. *Some Often-Neglected Duties of the.* By Francis Collingwood, before the Rensselaer Polytechnic Institute. A most excellent review of many important matters. *Van. Nos. Eng. Mag.*, Aug., 1886. *Eng News*, June 19, 1886. *San. Eng.*, June 24, 1886.

The Bridge Engineer. A paper by J. A. L. Waddell in *The Polytechnic*, outlining a course to be pursued by the would-be bridge engineer. *Eng. Record*, Oct. 17, 1891. pp. 316-18.

Engineering.

Agricultural, in India. A series of articles on irrigation, with the side issues, geological, social and financial, which must be considered in an extensive scheme. Lon. *Eng.*, April 6, *et seq.*, 1888.

Advancement of the Profession. See *Addresses, Herschel*,

A descriptive account, with plates, of the various railways in the Alps. By Vernon Harcourt. *Proc. Inst. C. E.*, Vol. XCV.

American, A Memoir of. By John B. Jervis. A valuable account of the early engineering in this country. *Trans. A. S. C. E.*, Vol. VI, (1877), pp. 39-67.

America at the Paris Exhibition, 1876. The Am. Soc. of Civ. Engrs., exhibit described, and many plates given of designs there shown. *Trans. A. S. C. E.*, Vol. VII., (1878), pp. 380-422.

And Architecture. Art and Utility in Building. An article by Barr Ferree, Univ. of Penna., discussing early methods, and advocating a better union of engineering and architecture in modern construction. *Eng. Mag.*, Aug., 1891, pp. 605-12.

Artistic. Article by H. J. Haber discussing briefly various points in designing. *Eng. News*, May 26, 1877, pp. 134-6.

And Astronomy. Recent Progress in Civic and Mechanical Engineering and Astronomy. Report of Committees of the Civil Engineers' Club of Cleveland presenting a review of the progress made in these branches. *Jour. Assn. Eng. Soc.*, June, 1890, pp. 319-32.

And Surveying. By Chas. Latimer. *Jour. Assn. Eng. Soc.*, Vol. VI., p. 416.

And Technical Literature, Handy List of. A very useful catalogue of books printed in English from 1880 to 1888 inclusive, arranged alphabetically by authors and subjects; 2145 titles and thousands of cross-references. Compiled and published by H. E. Haferkorn. Milwaukee, Wis., 8vo. Cloth, pp. 168. ($2.75.)

As a Profession. An Address to the Alumni Assoc. of Stevens Inst. of Tech., by Wm. Kent, M. E. A vigorous plea in behalf of this as one of the learned professions. *Van Nos. Eng. Mag.*, Aug., 1883.

Beginnings of. A paper by J. Elfreth Watkins, tracing the history of the various departments of engineering from the oldest times, including the evolution of engineering instruments with discussion. *Trans. A. S. C. E.*, May, 1891, Vol. XXIV, pp. 309-84.

Canal. See *Canal Engineering.*

Centennial Exhibition. Report on, by Maj. D. P. Heap, U. S. A. A carefully prepared document of 450 pp., amply illustrated; giving a complete description of all engineering models and devices, together with much valuable related matter. Issued separately by the Engr. Dept., U. S. A.

Century of Progress. An editorial article with "diagrams showing growth of railways, water-works and population of United States and Canada. *Eng. News*, April 27, 1882.

Civil, Father of. See *Baldwin.*

Electrical, Review of Development. See *Addresses, Preece.*

Estimates, Costs, Accounts, etc. A series for young engineers showing the methods of making estimates, etc., with a discussion of the underlying principles. *Mech. World*, Jan. 6, 1877.

Engineering, continued.

Finance. Costs—Estimates—Prices. An article discussing the methods of computing these elements. Lon. *Engineer*, Dec. 19, 1890, pp. 493-8.

Inventions Since 1837. An address by Sir Fredk. Bramwell. Prest. Inst. Civ. Engrs. Gives a valuable review of the various new appliances and methods in all departments of engineering in the past 25 years. *Iron*, Jan. 26, 1885.

Laboratories. Abstracted from a paper by Prof. B. W. Kennedy before the Inst. of Civil Engrs. *R. R. Gaz.*, Jan. 28, 1887.

Laboratories. Notes on the Engineering Laboratories of the Mass. Inst. Tech. A description of the laboratories with methods of conducting various tests. By Gaetano Lanza. From *Proc. Inst. C. E.*, Vol. XCI., Part I. *Technology Quarterly*, Vol. I., No. 4, pp. 356-74. Description of the new laboratory, *id.*, Vol. III., 1890, No. 2, pp. 144-50.

Laboratory. See *Cement Laboratory.*

Literature. An American Society for the writing of Engineering Works. By Prof. J. A. L. Waddell. Advocates the formation of such a society. *Jour. Assn. Eng. Soc.*, Vol. IV., p. 419.

Literature. See *Diary.*

Modern Spirit of the Engineering Profession. See *Addresses, Cooley.*

Progress in Civil and Mechanical Engineering and Architecture for 1889. Report of Committee on Civil Eng. and Surv., by Cyrus G. Force, Jr., on Mechanical Eng'r'g by Walter Miller; and on Architecture by John H. Richardson; all of Engineers' Club of Cleveland, p. 9. *Jour. Assn. Eng. Soc.*, March, 1890, Vol. IX., p. 117.

Proposal for an Academy of. A paper read before the Am. Assoc. Adv. Sc., by Wm. Kent. A rather visionary proposition to establish an exclusive body, at once honorary and executive, in America. *Van Nos. Eng. Mag.*, October, 1886. Also *Eng. News*, Oct. 16, 1886, and *Am. Eng.*, Oct. 6, 1886.

Review of for 1887. A long editorial in Lon. *Engineer*, for Jan. 6, 1888, gives a good review of the engineering progress and practice for the year 1887.

Review of. See *Addresses; Brace, Chanute.*

School of Application at Willets Point, N. Y. Its present organization and work done by West Point Cadets graduating into the engineer corps of the army. *An. Rep. Chief of Engrs.*, U. S. A., 1884, pp. 417.

Societies (Civil) in the U. S. A. A list of such societies, with names of officers. *Eng. News*, July 11, 1885. See also *Addresses, Clemens.*

Societies. St. Louis Club. History of. See *Addresses, Potter.*

Students. The Training of. By Prof. Geo. L. Vose, in Vol. II., p. 1; and *A Course of Study for*, by John D. Crehore, Vol. I., p. 380. *Jour. Assn. Eng. Soc.* *See* Technical Education; *also* Mechanical Engineering.

Structures, Destructive Agencies in. A series of articles discussing the agencies tending to destroy engineering structures and their remedies. *Am. Eng.*, Aug. 15, 1888.

Steam at the Philadelphia Edison Electric Station. Illustrated description. *Elec. Eng.*, Oct. 8, 1890.

Steam. River Practice in the West. By J. M. Sweeney, before the Nashville meeting of the American Society of Mechanical Engineers. *Am. Eng.*, June 13, 1888.

Steam. Introduction to the Study of. By R. H. Buel. A series of articles for practical men with a limited education. *Am. Eng.*, May 30, *et seq.*, 1888.

Engineering. See *Miscellaneous Notes, Mechanical Engineering, Municipal Engineering, Civil Engineering, Sanitary Engineering, Electrical Engineering, Mining.*

Engineers.

And Wage-Earners. The responsible position of Engineers in the delicate relations of employers and employed, which is coming to have such ominous import in all civilized countries. Clearly and forcibly presented by J. C. Bayles. *Trans. A. I. M. E.*, Halifax meeting, 1885.

Civil vs. Military. A controversy on the subject of employing graduates of West Point on government engineering works, to the exclusion of civilians from positions of responsibility. Both sides presented. *E. & M. Jour.*, March 7, 14, and 28, April 4, 1885. Also *Am. Eng.*, April 3, 1885. *See* also River and Harbor Bills.

Contractors and the Public. Article discussing their relations and general conditions governing public works. By C. F. J. *Eng. Record*, Oct. 3, 1891, pp. 281-2. Also on same by Julius W. Adams. *Eng. Record*, Nov. 7, 1891, p. 358. R. E. McMath. *Eng. Record*, Dec. 5, 1891, p. 5.

Royal, of England. Difficulties experienced in filling the corps. Lon. *Engineer*, Dec. 11, 1885.

Status of. See *Addresses.*

Their Relations and Standing. By R. E. McMath. Gives a short history of the development of the Engineering Department of the U. S. Army. *Jour. Assn. Eng. Soc.*, Jan., 1887.

Engine Room Practice. A paper by Wm. W. Wilson giving an account of recent improvements in construction, arrangement and management of marine engines. Lon. *Eng.*, May 29, 1891, pp. 658-60.

Engine Fly Wheels. See *Engines, Steam. Fly Wheels.*

Engine Governors. See *Engines, Steam. Governors.*

Engines, *Air*. See *Aerial Navigation, Flying Machine.*

Ammonia. See *Ammoniacal Gas.*

Aqua Ammonia, Theory of. By E. E. Magovern, Jr., Am. Soc. C. E. Illustrated, pp. 10. *Trans. A. S. C. E.*, Oct., 1888, Vol. XIX., p. 127.

Bisulphide of Carbon, Propelled by. The fallacy of claiming any advantage for this fluid over steam shown by W. P. Trowbridge. An entirely superfluous argument, however, to one acquainted with the fundamental principles of efficiency in heat engines. *Sch. of Mines Quar.*, April, 1886.

Foundations for. *See Foundations, Elastic, Asphalt.*

Ether, Lusini's. Illustrated description of this engine in which ether vapor replaces steam. From *Les Inventions Nouvelles.* *Sci. Am. Sup.*, Nov. 14, 1891. No. 828, pp. 13223-4.

Friction of. *See Friction.*

Heat. See *Engines, Heat*, below.

Petroleum. See *Engines, Petroleum*, below.

Pumping. See *Pumping Engine.*

Shafts of. See *Shafts, Shafting.*

Steam. See *Engines, Steam*, below.

Water vs. Air. By L. Trasenster. A mathematical treatment of air and water engines. Concludes that compressed air possesses advantages in certain places; for ordinary service high pressure water engines are preferable on score of efficiency and economy. *Van Nos. Eng. Mag.*, Vol. XIX., p. 429.

Without Fire, for Rail and Tramways. Full account, with cuts and certificates, of performance for the past year of two such engines. Invention of Mr. Honigmann. Works with a solution of hydrate of soda; seems to be economical and efficient. Lon. *Engineer*, Jan. 23, 1885. See also *Locomotive, Caustic Soda.*

Engines, Gas.

Atkinson Engine. Fully described and illustrated. Lon. *Eng.*, December 18, 1885.

Engines, Gas, continued.

Atkinson's and others. Report of the Committee on Science and Arts of the Franklin Institute upon Atkinson's Gas Engine in comparison with others. Cuts of the principal gas engines, awards, premium and medal to Atkinson's as superior in simplicity, economy and regulation. Ten horse-power engines consumes 22½ cubic feet of gas per hour per brake horse power. *Jour. Frank. Inst.*, June, 1889.

Atkinson's. Abstract of a report from Prof. Unwin on the trial of the gas engine invented by Mr. James Atkinson. The tests gave a consumption of 22 cu. ft of gas per hour per horse power. Lon. *Engineer*, May 6 and 13, 1887.

Beck's. A report of an exhaustive series of experiments made by Prof. A. B. W. Kennedy on Beck's gas engine. Illus. Lon. *Engineer*, May 4, 1888. Abstracted *Eng. News*, June 9, 1888.

Combustion in the. By Prof. R. Schöttler. *Zeitschrift des Deutscher Ing.*, 1886, pp. 210-211, 234-240. Also by Prof. Slaby. *Id.*, pp. 325-329.

Controversy in England. Suits for damages for infringement of the Otto patents. An injunction sustained after a four weeks' trial. Full proceedings in the cases. Lon. *Eng.*, Nov. 28, 1885, *et seq.*

Daimler Gas and Petroleum Motor. Illustrated description of a two-cylinder gas engine. *Am. Mfr.*, March 6, 1891, p. 201.

Economical Use of, for the *Production of Electricity.* An abstract from a lecture delivered at the Paris Electrical Exhibition, by W. E. Ayrton. *Van Nos Eng. Mag.*, Vol. XXVI., p. 294.

Efficiency of. A paper read before the Liverpool Eng. Soc. by Thomas L. Miller, M. E. Lon. *Engineer*, June 13, 1890, *et seq.* *Am. Mfr.*, June 27, 1890, *et seq.*

Elementary Principles of. By D. Lane. *Van Nos. Eng. Mag.*, Nov., 1885.

Experiments. By Mr. Benj. Davies. Indicator cards and diagrams, and methods of ignition. Lon. *Engineer*, Sept. 26, 1890, pp. 251-2.

Griffin. Gives details of experiments with a Griffin gas engine. Showed a consumption of 13.86 cubic feet per hour per horse-power. Lon. *Eng.*, April 13, 1888; Lon. *Engineer*, May 25, *et seq.*, 1888.

Lecture by Dugald Clerk before Lit. and Phil. Soc. Oldham. History, principles and prospects. *Sci. Am. Sup.*, April 11, 1885.

Otto. A full report of tests made at the Stevens Institute on a 10 horse-power gas engine. It also gives the relative economy of the gas, steam and hot air engines. *Van Nos. Eng. Mag.*, February, 1884.

Paper on gas engines, with a description of the Simplex engine. By Edward Delamon-Deboutteville, of Rouen. Lon. *Eng.*, Aug. 2, 1889, *et seq.*

Paris Exhibition. By M. Armengaud. Presents explanations of the structure and mode of working of the different engines exhibited. *Van Nos. Eng. Mag.*, Vol. XX., p. 148.

Report of the committee on, at the Electrical Exhibition. *Jour. Frank. Inst.*, October, 1885.

Sturgeon. Brief description, with drawing, of the cylinders of the Sturgeon gas engine. Lon. *Engineer*, June 15, 1888.

Theory of. By Dugald Clerk, before the Inst. of Civ. Engrs., with discussion. Gives results of the author's experiments. A good paper. *Van Nos. Eng. Mag.*, Vol. XXVII., pp. 351 and 443; also *Van Nos. Sci. Series*, No. 62.

Theory of. Published in 1862 by M. Beau de Rochas. An important document pertaining to the legal contest of *Otto vs. Steel*. Translation given in Lon. *Engineer*, Dec. 4, 1885.

See *Heat.*

Engines, Heat.

Action of Heat Engines as Explained by the Molecular Theory. Presidential address to the Liverpool Polytechnic Soc., by H. S. Hele-Shaw, M. I. C. E. *Iron*, May 1, 1891, pp. 383-5.

An illustrated description of a new and practical form of hot air engine. A test, of which a detailed account is given, showed that 2.98 pounds of coke was burned per brake horse power, and 1.91 pounds per indicated horse power. *Eng. News*, Sept. 14, 1889. Vol. XXII., p. 242.

Benier's Hot Air. Description of the above engine, with plan, sectional elevations and sections. Lon. *Engineer*, Nov. 4, 1887; *Sci. Am. Sup.*, Dec. 10, 1887. *Polytechnische Journal*, Vol. CCLXVII., 1888, p. 193; Abstract, *Proc. Inst. C. E.*, Vol. XCII., pp. 488-9.

Of the Future. A discussion of the relative merits of the steam boiler and engine compared with the gas generator and motor. Illus. Trials by Mr. Aime Witz showed a considerably greater economy from the latter. From *La Nature*. *Sci. Am. Sup.*, No. 799, April 25, 1891, pp. 12761-2. Discussion from *Gas World*, ibid., p. 12769. Lon. *Eng.*, April 17, 1891, pp. 453-4.

Hargreaves. Gives a description, with sectional elevation, of Hargreaves thermo-motor, which, at 102 revolutions per minute, indicated 40 horse-power. It consumes 20¾ lbs. of coal tar per hour. The highest available efficiency is 73 per cent. Lon. *Engineer*, Jan. 27, 1888; *Power & Steam*, March, 1888.

Other than Steam. A series of lectures by H. Graham Harris, M. Inst. C. E. *Jour. Soc. Arts*, July 12, 1889, *et seq.* Vol. XXXVII., p. 687, *et seq.*

Operated by Other Mediums than Steam. By George H. Babcock. A paper on "Substitutes for Steam," covering with the discussion 60 pp. fully illustrated. Many kinds of substances are used which are here discussed, together with the forms of engines using them. *Trans. A. S. M. E.*, Vol. VII., p. 680.

Triple Thermic. By Chas H. Haswell before the American Society of Civil Engineers. Gives description, operation and results of a single expansion, non-condensing steam engine, supplemented by the evaporation of bisulphide of carbon and expansion of its vapor. *Trans. A. S. C. E.*, Vol. XVII., pp. 193-199, October, 1887; *Sci. Am. Sup.*, April 14, 1888; Lon. *Engineer*, June 1, 1888.

Review of the Thermic of the Paris Exhibition. By V. Dwelshauver-Dery. Compares the economy that it is possible to obtain in the employment of heat in the engine with that which is possible to obtain in the production of heat in the fire-place. *Van Nos. Eng. Mag.*, Vol. XXII., p. 89.

Engines, Petroleum.

A description of the action of Spiel's engine, which works very much like the gas engine. Fully illustrated. Lon. *Eng.*; also *Sci. Am. Sup.*, April 3, 1885. For tests of, see *Engineer*, Jan. 21, 1887.

A paper by Prof. Wm. Robinson before the Brit. Assn., stating the general principles, and describing various apparatus. Tables giving properties of various oils given. Lon. *Engineer*, Sept. 11, 18, 25, 1891. Extracts in *Am. Mfr.*, Oct. 30, 1891, *et seq.*

Gives a series of interesting diagrams from a petroleum spirit engine. Lon. *Engineer*, June 15, 1888.

Engine, Steam.

Description of a novel form of steam engine: dispensing with all eccentrics, rods, steam chests, valves, etc., the piston itself by its peculiar construction acting as its own slide valve. Illustrated. *Eng. News*, October 26, 1889, Vol. XXII., p. 389.

Translation of M. Hallauer's experimental study of the influence of expansion in simple and compound engines. *Jour. Assn. Eng. Soc.*, Vol. I., pp. 228, 329, 345, 386, and Vol. II., pp. 61 and 109.

Engine, Steam, continued.

Translation of M. Keller's Report upon M. Hallauer's Memoir on the expansion and efficiencies of simple and compound engines. *Jour. Assn. Eng. Soc.*, Vol. I., p. 18.

Allen, Report on. Gives details of the experiments conducted at the Exhibition of the American Institute to ascertain the economy and effective power of the Allen engine. *Van Nos. Eng. Mag.*, Vol. IV., p. 5?4.

Applied to Bicycle and Tricycle. A description of a small, light, steam engine and boiler, invented by L. D. Copeland. *Am. Mach.*, March 3, 1888.

Blast Furnace Blowing Engine. A paper by Fred. W. Gordon, read before the Am. Soc. M. E., describing in detail a carefully designed engine. Illustration of engines and valves, and indicator diagrams given. *Eng. News*, July 6, 1889, pp. 6-8.

And Boiler. A very clear and elementary statement of the scientific principles involved in the steam engine. By A. R. Wolff, M. E. A series of articles. *Eng. News*, Feb. 20, 1886, *et seq.*

Calculations, Simple Formulas for. By Prof. W. C. Unwin. Some of the most important theoretical and empirical formulas brought together, and tables given to assist in evaluating them. Lon. *Engineer*, April 9, 1886. Also *Am. Eng.*, May 6, 1886.

"*Climax.*" Gives an illustrated description of the Climax steam motor, and detail of a trial run of five hours. The boiler is composed of two coils of pipe. Lon. *Eng.*, April 1, 1887.

Compound.

A very complete description of Willan's receiver compound engine, for which an economy of 24 lbs. of steam per indicated horse-power per hour is claimed when indicating only eight horse-power. Lon. *Eng.*, April 17, 1885.

By H. Turner. Gives a description of a new form, tandem compound, of steam engine. Lon. *Engineer*, July 20, 1888.

Calorimetric Investigation of the Performance of a Compound Engine. A paper by Mr. W. H. Nauman describing a test made on a compound Engine at Augsburg, Bavaria, and discussing results. *Proc. Eng. Club. Phila.*, Dec. 1887, Vol. VI., pp. 313-320.

Corliss. Gives a brief description with two-page plate and other engravings of a compound Corliss engine. Cylinders 40 × 70 in., stroke 72 in., 2,500 indicated horse-power. Lon. *Eng.*, April 16, 1888.

Cut-off in the Large Cylinder. By Prof. R. H. Smith. A theoretical discussion of considerable value. Graphical solutions given. Lon. *Engineer*, Nov. 27, 1885.

At Dublin. A short description, with a two-page plate showing plan, elevation, cross-section and valve gear of a compound condensing engine fitted with Collman's valve gear. Cylinders are cast steel, jacketed, 14 and 20 inches in diameter, 28 inches stroke; speed, 87 revolutions; pressure, 110 lbs.; horse-power 150. Lon. *Eng.*, Feb. 3, 1888.

Horizontal. Gives an illustrated description of a compound horizontal, 2,000 indicated horse-power engine. Cylinders 38 and 66 in. in diameter, 6 feet stroke, pressure 95 lbs., piston speed 600 feet per minute. Lon. *Eng.*, Jan. 20, 1888; *Sci. Am. Sup.*, Feb. 25, 1888.

Horizontal, with Frikart valve gear. Description and double plate. Lon. *Eng.*, June 18, 1890, p. 739.

Investigates a formula which shall determine the pressure at any part of the stroke, etc. *Van Nos. Eng. Mag.*, Vol. IX.

For Manufacturing Purposes. By Chas. S. Main, before the Scranton meeting of the Am. Soc. M. E. Discusses the use of the compound engine for manufacturing purposes, the relative areas of cylinders and the regulation

Engine, Steam, *for Manufacturing Purposes*, continued.

of pressure in the receiver. *Am. Eng.*, Oct. 24, 1888; abstracted *R. R. Gaz.*, Oct. 19, 1888.

Mill Engine. An illustrated description, with details of the performance of an engine provided with double-beat equilibrium valves. The engine was constructed under a guarantee of economy in steam consumption and in regularity of speed. Lon. *Eng.*, May 27, 1887.

Multicylinder, or Compound; its Theory and its Limitations. Paper read before Am. Soc. M. E., by Robert H. Thurston, from advance sheets of Transactions. *Jour. Trans. Soc.*, Dec., 1889, and Jan., 1890. Abstract in *Mechanics*, Dec., 1889. *Trans. A. S. M. E.*, Vol. XI., 1890, pp. 135-87.

Non-Condensing, Cost of Power in. Experiments and theoretical discussion showing relation of cost of power to cut-off, etc., by Charles E. Emery. *Trans. A. S. M. E.*, Vol. X., 1889.

And Non-Condensing Steam Jackets, etc. By Chas. E. Emery. Presents tabular statement showing the results of experiments made in 1874 on a number of steamers to ascertain the best means of securing economy of fuel. *Trans. A. S. C. E.* Vol. III., (1874), pp. 367-394.

Relative Economy of Simple and Compound Engines. An argument in favor of the simple engine for mills and factories. *Am. Eng.*, April 24, 1889.

Paper by S. W. Robinson treating of the two-cylinder compound engine in which the strokes are simultaneous, or co-initial and coterminal with receiver, cushion, clearance, etc. *Van Nos. Eng. Mag.*, Vol. XXIX, pp. 129 and 313.

Specifications of a Single Tandem Compound Corliss Condensing Engine. *Mech. World*, Nov. 1, 1890, p. 176, *et seq.*

Tandem. Brief description with two-page plate of a pair of compound tandem engines having cylinders 30 and 42 inches in diameter and 6 feet stroke. Lon. *Eng.*, Aug. 26, 1887.

Tandem. Brief description, with two-page plate showing plan, elevation and sections of exhaust valves of a tandem compound horizontal engine, constructed at Rouen. Lon. *Eng.*, Nov. 11, *et seq.*, 1887.

"*Triumph*" *Compound Engine.* Illustrated description of a form of compound engine having three high and three low pressure cylinders, each set of pistons controlling the steam supply of the set next preceding; so serving as a perfect valve device. *Ry. Rev.*, Feb. 8, 1890, p. 81.

Vertical. Detail drawings, fully dimensioned, of all the parts of the engines for the Indian State Railway. Lon. *Engineer*, Sept. 2, etc., 1887.

Condensation in Cylinders.

By M. E. Couste. Gives an analysis of condensation from the physical and mathematical theory of the condenser. Compares the surface and injector systems, etc. *Van Nos. Eng. Mag.*, Vol. I., pp. 613, 801, etc.

Determined from the law of cooling of iron in contact with a gas. By Prof. L. D'Auria. *Jour. Frank. Inst.*, May, 1886.

Effect of the Steam-Jacket on. Paper by W. W. Bird before the Am. Soc. M. E., giving results of some experiments. Abs., *Mechanics*, Oct. 1891, pp. 109-11.

Emery, Chas. E. A paper before the Am. Soc. of M. E., on cylinder condensation, and the reduction of the same by the use of the compound engine. *Am. Eng.*, June 8, and 15, 1887.

English. A paper by Major English, detailing experiments made to directly measure the condensation taking place in a steam cylinder. Lon. *Eng.*, July 6, 1887. *R. R. Gaz.*, Oct. 28, 1887.

Engine, Steam, *Condensation, etc.*, continued.

An exhaustive investigation on a 350 horse-power engine. The paper was presented to the Ann Arbor meeting of the Am. Assoc. Adv. Sc., by Professor Thurston. *Jour. Frank. Inst.*, October, 1885, *et seq.*

Experiments on the Steam Jacketing and Compounding of Locomotives in Russia. A paper read by Alex. Borodine, of Russia, before the Inst. Mech. Engrs., London, 1886. This is a series of very careful and important experiments, carried out in a most thorough and scientific manner, Mr. Hirn having been consulted in regard to them. *Am. Eng.*, Sept. 8, 1886, *et seq.*

Final Improvement of the Steam Engine. Paper by Dr. R. H. Thurston giving results of experiments in which cylinder walls were treated with acid, varnished, etc. *Proceedings of the U. S. Naval Institute*, Vol. XVII., No. 3, pp. 417-528.

Thermo-Electric Method of Studying. A paper by Prof. Edwin H. Hall, read before the A. I. E. E., describing a method for determining the temperature of the cylinder at various depths in the metal, and at different parts of the stroke. Curves showing results of experiments given. *Elec. World*, June 20, 1891. Abstract, *Elec. Eng.*, June 24, 1891.

Condensers. See *Condensers.*

Corliss Engine at the Late Paris Exposition. An illustrated description of this unusually large engine, especial attention being directed to novel features. The design was by M. Joseph Farcot, Saint Ouen. *Mech. World*, March 15, 1890, p. 106, *et seq.*

Corliss Engine. The Bullock. Illustrated. Description of new design of Corliss engine built by the M. C. Bullock Mfg. Co., Chicago. *E. & M. Jour.*, Nov. 21, 1888.

Corliss, Horizontal. Gives two-page plate and short description of an 18 by 48 Corliss engine. Lon. *Eng.*, April 20, 1888.

Corliss. See *Engines, Steam, Compound Corliss.*

Counterbalancing of Engines and other Machinery. By Prof. S. W. Robinson. The paper, with discussion, covering 40 pp. Illustrated. *Trans A. S. M. E.*, Vol. II., p. 213.

Crank Pin Pressures and Steam Distribution. Communication from J. B. Strube, discussing mainly the relative advantages of the old Stephenson link valve motion and that of Mr. Wm. Wilson. Gives several interesting indicator cards. *Mast. Mech.*, Dec. 1889, Vol. IV., p. 206.

Crank Shafts.

Bent. An account of the processes of making crank shafts of large size by bending to the required shape as practiced in England. By W. H. Booth. *Am. Mach.*, June 27 and Aug. 22, 1889.

Graphical Method of Finding the Moment in. By I. N. Hollis, Asst. Engr. U. S. N. Illustrated. *Am. Eng.*, Dec. 8, 1886.

Strength of. A review of Mr. Longridge's report to Engine and Boiler Insurance Co., of Manchester, Eng. Lon. *Engineer*, Oct. 24, 1884. *See* also Shaft and Steamboat.

Cushioning in. By George L. Morton, M. E. Formula deduced and working tables computed, of value in engine designing. *Van Nos. Eng. Mag.*, June, 1884.

Cushioning.

The effect of cushioning and deduction of formulas for determining the work done by a given volume of steam. Tables are given for finding conditions of maximum efficiency. By George L. Morton. *Van. Nos. Eng. Mag.*, June, 1884.

Cut-off in. Most Economical Point of. By Alfred R. Wolff. A paper which, with the discussion, occupies 19 pp. in *Trans. A. S. M. E.*, Vol. II., p. 147 and 251. Graphical and analytical solutions given.

Engine, Steam, continued.

Cylinders.

Proportioning of. By R. H. Thurston, before the American Society of Mechanical Engineers. *Trans. A. S. M. E.*, Vol. IX., (1888), pp. [illegible].

Ratios of Triple Expansion Engines. Abstract of a paper by J. M. Whitman, Fayetteville, Ark., read before the Am. Soc. M. E., May, 1889. *Mechanics*, June, 1889.

Steam Jackets. By H. Hildebrant. Treats of its action, advantages and disadvantages. *Van Nos. Eng. Mag.*, Vol. VIII., p. 195.

Steam Jacketing. A description of the theory and use of the steam jacket, by H. L. Greene. *Jour. Frank. Inst.*, September, 1885.

Thermometers in and About the Cylinder Walls of Steam Engines, with some Experiments on the Temperature of the Metal under Different Conditions. By Bryan Donkin. Gives typical results of experiments made upon various parts of the cylinder. Folding plate showing apparatus used. *Proc. Inst. C. E.*, Vol. CVI., 1891, pp. 261-76. Abstract, *Ry. Rev.*, Sept. 5, 1891, pp. 581-2. Diagram of temperatures, Lon. *Engineer*, Sept. 25, 1891, pp. 262-3.

See *Lubrication*.

Dead Point, Simultaneous. Prof. C. W. MacCord. Gives an elaborate mathetical investigation of the subject. *Sci. Am. Sup.*, Oct. 8, 1887.

Design and Construction of Stationary Engines. A series of papers discussing the principles of design, and giving the best practice. *Mech. World*, Sept. 11, 1891, pp. 106-7, *et seq.*

Economy. A Uniform Basis for Comparison. By Chas. E. Emery. Based on the relation of heat units to foot-pounds of work. *Trans. A. S. C. E.*, Vol. VII. (1878), p. 60. Discussion on p. 193.

Economy. A Uniform Basis for Comparison. By C. E. Emery. The basis advocated is that the boiler is capable of absorbing 10,000 heat-units per pound of coal consumed. *Van Nos. Eng. Mag.*, Vol. XIX., p. 41.

Economical Limits of Steam Pressures. Tables and diagram giving amounts of water per horse power per hour consumed by engines under various conditions. Theoretical. By A. F. Nagle. *R. R. Gaz.*, Nov. 11, 1891, pp. 801-2.

Economical Use of Steam in Colliery Engines. Paper by John McLaren before the Federated Inst. Min. Engrs., discussing this question and giving results of a four-hours trial of a triple expansion engine with locomotive boiler. Lon. *Engineer*, July 17, 1891, pp. 45-6. *Mech. World*, June 13, 1891, *et seq.*

Efficiencies. The Ideal Engine Compared with the Real Engine. A paper by Prof. R. H. Thurston, read before the Am. Soc. M. E., giving comparisons between the two, for various ratios of expansion. Table of quantities and diagrams given. *Eng. News*, July 4, 1891, pp. 16-19.

Efficiency of Plant. By Prof. De Volsen Wood. A review of the recent steam engine efficiency. *Sci. Am. Sup.*, Sept. 1, 1888.

Fly Wheels.

A discussion of the requirements in various engines. Numerous diagrams. *Mech. World*, Nov. 22, 1890, pp. 207-8.

The Bursting of. Abstract of a paper made from official returns of a government investigation in Prussia. Causes and remedies. *Proc. Inst. C. E.*, Vol. LXXXIV., p. 518.

Strength of. By O. Kruger. Gives general formula and then investigates a special case. *Van Nos. Eng. Mag.*, Vol. VIII., p. 171.

Strength of. Gives calculation of the strength of wheels and the stresses to which they are subjected, the limit of safety as regards speed and the application of wrought-iron hoops. *Sci. Am. Sup.*, Jan. 1, 1887; also Lon. *Engineer*.

Engine, Steam, *Fly Wheels,* continued.

Theory of the. Equations and diagrams, with rules for proportioning for given conditions, by Prof. R. H. Smith. Lon. *Engineer,* Jan. 9, 1885.

Friction in.

Distribution of Internal Friction, and its relation to speed and work. Papers by R. H. Thurston, giving results of experiments. *Trans. A. S. M. E.,* Vol. IX., 1889.

Internal Friction, Variable Load, and Engine Speed and Work. An abstract of two papers by Prof R. H. Thurston, read at the Scranton meeting of the Am. Soc. M. E., with diagrams. *R. R. Gaz.,* Dec. 28, 1888.

Non-Condensing. Paper read at a meeting of the Am. Soc. M. E., New York, November, 1886, by Prof. R. H. Thurston, of Cornell University. Describes experiments showing that the friction of a non-condensing engine is practically the same for all loads. *Trans. Am. Soc. M. E.,* Vol. VIII., 1887; *Am. Eng.,* Dec. 14, *et seq.,* 1887; *R. R. & Eng. Jour.,* Feb. 1887. A paper before the Philadelphia meeting in *Trans. Am. Soc. M. E.,* Vol. IX., 1888, on the same subject. Abstracted *Eng. News,* May 26, 1888.

See *Friction.*

Governors.

Application of Governors and Fly-wheels to Steam Engines. A discussion of the requirements of a governor, and investigation in regard to the degree of perfection attainable. By Prof. V. A. E. Dwelshauvers-Dery. *Proc. Inst. C. E.,* Vol., CIV, 1891, pp. 196-208.

Ball Steam Engine Governor. Account of some recent improvements in making a steam engine governor to give uniform speed. Consists in the combination of an auxiliary spring and dash-pot with the ordinary shaft-governor. From a paper, by Mr. Frank H. Ball, before the Am. Soc. M. E. at Philadelphia. *Power,* Dec., 1887.

Compound Governor for the Westinghouse Compound Engine. New fly-wheel governor. Description and drawings, with indicator cards continuous throughout changes of load. *R. R. Gaz.,* June 12, 1891, pp. 402-3.

Improved Form of Shaft. By F. H. Ball before the Am. Soc. M. E. *Trans. A. S. M. E.,* Vol. IX. (1888), pp. 300-323.

New Method of Investigation of, wherein the frictional resistance to movement is taken into account. *Proc. Inst. C. E.,* Vol. XC., p. 60.

Paper by Mr. J. M. Smith, read at the Cincinnati meeting of the Am. Soc. M. E. *Trans. A. S. M. E.,* Vol. XI., 1890, pp. 1081-1100. Abstract, *Eng. News,* July 19, 1890.

Paper of considerable value, by C. F. Budenberg, read before the Manchester Assn. Engrs., giving a practical discussion of the subject. Illus. *Mech. World,* Apr. 18, 1891, p. 157, *et seq.*

Theory of. By J. F. Brunton, M. I. C. E., in Lon. *Engineer* for Jan. 6, 1885. The mathematical analysis, with discussion, for designing various patterns.

Theory of Centrifugal Governors. By Sven S. Ekman. *Mechanics,* No. 227. Nov., 1890, pp. 211-6.

Theory of the Conical Pendulum or steam governor. Illustrated. Lon. *Engineer,* Dec. 18, 1885.

Heat Expenditure in, Investigation of the, especially with reference to methods of diminishing cylinder expansion. A paper by Prof. Dwelshauvers-Dery, of Liege University, pp. 26. Translated in *Proc. Inst. C. E.,* Vol. XCVIII., p. 854, 1889.

Heat in the Steam. By Maj. T. English. Gives results of experiments made to ascertain the most economical method of working some direct-acting, high-pressure pumping engines for the Egyptian expedition. Lon. *Engineer,* July 1, 1887, *et seq.* Abstract, *R. R. Gaz.,* Oct. 28, 1887.

Engine, Steam, continued.

Heat in the Steam. Gives translation of the explanation of Prof. Dwelshauvers-Dery's diagrams of exchange of heat between metal and steam in a steam engine. Illus. Lon. *Eng.*, July 27, 1888.

For High Pressure. See *Boilers and Engines.*

High Speed.

A discussion of interesting features of rotary and reciprocating high-speed engines, and the proper utilization of the inertia of the reciprocating parts. By John Imray, in *Trans. Inst. C. E.*, Vol. LXXXIII., p. 106.

By Prof. R. R. Werner. Theoretical discussion and study of different engines. *Zeitschr. d. V. deutscher Ing.*, 1887, pp. 533, 538, 553, 557, 578-581, 606-609.

For Electric Lighting. By Prof. R. H. Thurston. *Elec. World*, Aug. 9, 1890, pp. 84-5. *Am. Mfr.*, Aug. 15, 1890, p. 8.

Obscure influence of Reciprocation in. By Arthur Rigg, before the Soc. of Engrs. Gives tables of proportions of reciprocating parts and numerous momentum diagrams. Lon. *Engineer*, June 4, 1886. Illustrated by 28 cuts Lon. *Eng.*, April 26, 1886.

Riggs'. Describes a new and ingenious type of four-cylinder revolving steam engine, capable of attaining 2,000 revolutions per minute. Illus. *Mech. World*, Feb. 18, 1888; *Sci. Am. Sup.*, March 10, 1888.

Rota. Gives brief description of a new type of high-speed engine. *Ill. Eng.* July 20, 1888.

Woodbury Automatic. Description of details of construction with illustrations. *Iron Age*, May 14, 1891, pp. 913-18.

History of, A Contribution to its. *Am. Eng.*, March 13, 1885.

History of, Early. Illustrated article. *Cassier's Mag.*, Vol. I., No. I, Nov., 1891, pp. 14-23.

Horizontal. Gives description, with large detailed drawings, of a horizontal engine, the special feature of which is the valves. Lon. *Eng.*, June 11, 1886.

How to Manage a Steam Engine. A popular article of some merit. Illus. *Sci. Am. Sup.*, April 26, 1890, p. 11927.

Indicators.

And Dynamometers. The principal points that have to be attended to in making a test with either. *Mech. Eng.*, April 4, 1885.

Averaging Machine for. A description of Coffin's machine for averaging, or measuring, the mean ordinate of an indicator diagram, thus giving at once the mean effective pressure. Illus *Trans. A. S. M. E.*, Vol. II., p. 314.

Comparison of Indicators. A paper read at the New York meeting of Mechanical Engineers, Nov., 1889. By Prof. J. Burkitt Webb, Hoboken, N. J. *Am. Eng.*, Nov. 27, 1889. *Trans. A. S. M. E.*, Vol. XI., 1890, pp. 311-26.

Construction and use. Series of articles. Illustrated. *Mech. World*, Aug. 10, 1889.

Description of. A paper by C. F. Budenberg, M. S. read before the Sheffield Soc. of Eng. All the principal forms of indicators are described and their merits discussed. *Iron*, June 20, 1890, pp. 538-40. Lon. *Ry. Engr.*, Sept., 1890, No. 128, *et seq.*

Diagrams. Two very exhaustive papers by Osborne Reynolds and A. W. Brightmore, which, with the discussion, covers over 100 pp. The errors to which these diagrams are subject fully investigated. *Trans. Inst. C. E.*, Vol. LXXXIII., p. 1.

Emery on Indicators. An interesting paper by Chas. E. Emery, before the American Society of Mechanical Engineers, showing the necessity of taking diagrams from the steam-chest at the same time as from the cylinder. Gives actual diagram. *Trans. A. S. M. E.*, Vol. IX., (1888), pp. 293-[illegible]; *Am. Eng.*, Dec. 7, 1887.

Engine, Steam, *Indicator*, continued.

Error Limit in Thompson & Tabor's by Assistant Engineers Windsor and Barry. *Eng. News*, Feb. 9, 1889. *Am. Eng.*, Jan. 30, 1889, *et seq.*

Le Van on. A series of articles on the steam engine indicator and its use. By Wm. Barnet Le Van. *Mech. Eng.*, Feb. and March, 1885.

A Novel. Designed for use especially on high-speed engines. The diagram is produced by a ray of reflected light from a mirror. Illustrated description. *Eng. News*, Sept. 5, 1891, p. 219.

Practical directions for applying and taking care of the instrument. *Van Nos. Eng. Mag.*, Vol. I., p. 412.

Revolving Steam Engine Indicator. An article advocating the use of a steam engine indicator, with a continuously revolving drum, instead of the reciprocating style. *Am. Mach.*, Dec. 24, 1887.

And Leaky Pistons. An illustrated paper by Mr. H. W. Spangler. *Proc. Eng. Club, Phila.*, Vol. VII., No. 2, Feb., 1890, p. 108.

Inertia of the Reciprocating Parts in Modifying the Force Transmitted to the Crank Pin. A graduation thesis by Francis E. Jackson, M. E., communicated by Prof. De Volsen Wood. A highly mathematical paper, with graphical interpretations given. *Jour. Frank. Inst.*, Sept., 1886.

Internal Waste of.

A Practical Method of Reducing the. A paper by Prof. R. H. Thurston, giving results of some experiments showing a considerable saving of heat by treating the interior surfaces of the cylinder, not rubbed, with acid and oil. *Trans. A. S. C. E.*, Vol. XXIII., 1890, pp. 39-43.

Reducing. Additional notes on this subject by Prof. R. H. Thurston, stating the results of some recent experiments. *Trans. A. S. C. E.*, Vol. XXV., July, 1891, pp. 15, 16.

Link. A new Reversing, for steam engines. By A. C. Smith. Reviews the objection to the common link motions and proposes a new design to overcome all difficulties. *Van Nos. Eng. Mag.*, Vol. XXI., p. 123.

Link Motion for. See *Link Motion.*

Long and Short Stroke. By Wm. H. Bryan. Discusses relative merits of each. *Jour. Assn. Eng. Soc.*, Vol. IV., p. 358.

Marine.

A paper by Mr. F. C. Marshall before the British Institution of Naval Architects, treating of the progress and development of the marine engine. *Am. Eng.*, Aug. 17, etc., 1887.

An Assistant Cylinder For. A paper by Mr. David Joy, read before the Inst. of Naval Architects. This cylinder is used to supply most of the power for moving the valves. Illustrated description with indicator diagrams. Lon. *Eng.*, April 10, 1891, pp. 430-2. Lon. *Engineer*, April 17, 1891, p. 301.

Auxiliary Engines in Connection with the Modern Marine Engine. A paper by William H. Allen, giving a brief history and description of the various auxiliary engines now in use. *Proc. Inst. C. E.*, Vol. CIV., 1891, Pt. II., pp. 1-23. Discussion, pp. 24-31.

Compound. A description and drawings of engines, boilers and feathering paddle-wheels of new steamer Normandy. Lon. *Eng.*, March 20, 1885.

Compound, Economy of. By W. Parker, Chief Engineer-Surveyor of Lloyd's Register, M. I. N. A. A paper setting forth that as steam pressure are increased, according to the tendency of the times, we must, for pressure beyond 80 or 90 lbs., "compound our compound engines," in order to gain the economy to be expected. With discussion. Reprinted from *Trans. Inst. Naval Architects*, by the Bureau of Navigation, Navy Department. Navy Professional Papers, No. 16.

Engine, Steam, *Marine*, continued.

Compound, For Service on the Great Lakes. By Walter Miller. A review of their introduction and advantages on lake boats. *Jour. Assn. Eng. Soc.*, Vol. V., p. 1.

Compound of the S. S. "Sidney" and "Wapping." Double page illustrations with description. They involve many new and valuable features. Lon. *Eng.*, Sept. 17, 1886.

Crank and Shafts. Flexible crank and propeller shafting in lieu of rigid shafting for marine propulsion. By J. F. Hall, before the Inst. Naval Archts. With many illustrations. Lon. *Eng.*, April 23, 1885.

Development of. A series of illustrated articles with the object of describing and illustrating the marine engines in the existing ships of the British Navy, and to trace the development of the engine from the type fitted in the oldest of them to that at present being fitted in the most modern. Lon. *Engineer*, March 23, *et seq.*, 1888.

Development of in the last Decades. Extracts from paper read before the Society of German Engineers in 1883, by Carl Busley, Esq., Engineer Corps, German Navy; translated by Asst. Engineer F. C. Bieg, U. S. N., for *Jour. Am. Soc. Naval Engineers*; *Mechanics*, July and August, 1889.

Development of the Marine Engine, and the Progress made in Marine Engineering during the Past Fifteen Years. A valuable paper by A. E. Seaton, M. I. C. E., read before the N. Y. meeting of the Brit. Iron and Steel Inst., Oct., 1890. *Sci. Am. Sup.*, No. 773, Oct. 25, 1890, pp. 12349-8, *et seq.*

Of Cruiser No. 12, the 3-screw fast cruiser, recently contracted for by U. S. Navy Dep't. Drawings of Engines, Boilers, etc., with description. *R. R. & Eng. Jour.*, December, 1890.

English Compound Side-Wheel Engine. Description and drawings. *R. R. & Eng. Jour.*, May, 1880.

Experimental Marine Engine in the Walker Engineering Laboratories of University College, Liverpool. Full description and illustrations. A paper by H. S. Hele-Shaw before the Inst. M. E. Lon. *Engineer*, Sept. 4, 1891, pp. 200-2.

First Century of the. By Prof. H. Dyer, before the Institution of Naval Architects. Gives a brief *resume* of the chief steps in the development of the steam engine and marine navigation. Lon. *Engineer*, Sept. 21 and 28, 1885.

Full illustrations of the engines and boilers, and other machinery of a new steamer on an African Line. Lon. *Eng.*, July 3, 10, 24, 1891.

Horse Power of. Normal Indicated and Boilers. Five tables showing relation of pressure, heating surface, etc. *Mech. World*, June 14, 21, 1890, pp. 235, 248.

Illustrated description of the compound engines of the twin-screw dredge "Dolphin," recently constructed for harbor work in the West Indies. Gives section and elevation of one pair of the engines. Lon. *Engineer*, Jan. 14, 1887.

Illustrations and description of diagonal engines for paddle-wheel steamer. Lon. *Engineer*, Feb. 20, 1885.

Increased Boiler Pressure and Increased Piston Speed for Marine Engines. A paper by W. R. Cummings discussing the advisability of these changes. *Mech. World*, Dec 27, 1890, pp. 257-8, *et seq.*

Kriebel Boat Engines. A concise description of this novel type of oscillating engine for use on small boats. Illustrated. *Am. Eng.*, July 10, 1889.

Protected Cruiser No. 12, U. S. Navy. Description, with folding inset of drawings, of the triple screws and engines of this ship. *Eng. News*, Aug. 15, 1891, pp. 134-5.

Quadruple-Expansion. See *Shipyard.*

Engine, Steam. *Marine*, continued.

Screw Propeller Engines. The new features of engines and boilers introduced in the compound beam engines of the new U. S. twin screw propeller, Chicago. Fully illustrated and described. *Mech. Eng.*, Nov. 28, 1885.

Steamship "Poplar." Gives illustrated description of a part of compound engines. Lon. *Engineer*, Feb. 4, 1887.

Triple Expansion.

A table giving tonnage, dimensions, size of cylinders, boiler pressure, proportions between cylinders, name of builders of engines, and date of building of steamships having triple expansion engines. By J. F. Redman. Lon. *Eng.*, Feb. 14, 1885.

And Quadruple-Expansion Marine. Abstract of paper before the Inst. of Engrs. and Shipbuilders in Scotland. Gives performance of quadruple expansion engine with only two piston valves. Illustrated. Lon. *Eng.*, April 2, 1886.

Experiments made upon a Horizontal Triple-Expansion Engine (Sulzer Type), by Prof. M. Schröter, of Munich. Lon. *Eng.*, Dec. 9, 1890, pp. 669-70.

Ferryboat "Bergen." Description with illustrations in perspective and section. *Mech. World*, Dec. 8, 1888.

H. M. S. "Orlando" and "Undaunted." Two two-page plates, showing vertical and horizontal sections through engine and boiler-rooms, two cross sections, and three perspective views of the triple expansion engines of steamers "Orlando" and "Undaunted.". Lon. *Eng.*, Nov. 4 and 11, 1887.

Sectional elevation and perspective view of the triple expansion engines of the Italian cruiser "Dogali." Lon. *Engineer*, Aug. 5, 1887.

Steamship "Aberdeen." By A. C. Kirk, M. I. N. A. An article setting forth the considerations which led to the adoption of the triple expansive type of compound engine in the design of the machinery for the steamship Aberdeen, with arguments in favor of the general introduction of this type of engines for very high pressures. Reprinted from the *Trans. Inst. Naval Arch.*, by the Bureau of Navigation, Navy Department. Naval Professional Papers, No. 16.

Steamship "City of New York." View and brief description of. The cylinders are 45, 71 and 113 inches; stroke, 5 feet; indicated horse power, 20,000 at 150 lbs. pressure. Lon. *Eng.*, Aug. 3, 1888.

S. S. "Coot." Full plate, general and detail drawings. Lon. *Engineer*, April 16, 1886.

S. S. "Courier." Brief description, with two-page plate, giving two perspective views of the triple expansion engine of the steamship "Courier." Lon. *Eng.*, Jan. 6, 1888.

S. S. "Indria." Description with sections of the vessel showing arrangement of machinery. Also illustrations of boilers. Lon. *Eng.*, May, 1891, pp. 581-8.

S. S. "Monowai," which was converted from a compound to triple-expansion engine, and shows an increase of speed of 9¼ per cent., and 5 per cent. decrease in fuel, or the distance through which a ton of coal will propel the loaded ship has been increased 15 per cent. Lon. *Engineer*, May 20, 1887.

S. S. "Victoria" and "Domira." Illustrated description. Lon. *Engineer*, Jan. 4, 1889.

S. S. "Violet." Description, with elevation, of the 4,000 H. P. engines, adapted to a paddle steamer. Lon. *Engineer*, April 17, 1891, pp. 598-9.

Wylie, Robert. Paper read before the Institute of Mechanical Engineers, at Leeds, England, October 18, 1886. *Am. Eng.*, Nov. 17, and 24, 1886.

Engine, Steam. *Marine*, continued.

See *Ferry-boat*.

Trials of. Contract trials of U. S. cruisers *Philadelphia* and *San Francisco*. Official reports, with full details of dimensions and performance of engines, boilers, and steam apparatus, and indicator cards. *Jour. Am. Soc. Naval Eng.*, November, 1890.

Trials. Paper read by Prof. Alexander Kennedy, before the Inst. Mech. Engrs. giving complete results of a trial of the triple expansion engines of the steamship "Iona." Diagrams of feed water and coal consumed. temperature of chimney gases and indicator cards given. Lon. *Eng.*, Mar. 8, 1891 pp. 261-2. *Mech. World*, May 9, 1891. p. 183, *et seq*.

Works, Plan and Construction. By Thomas Mudd. Gives sketch of the construction and equipment of the Central Marine Engine works at West Hartlepool, but gives prominence to features that are important to all engine works. A valuable paper. Illustrated. Lon. *Eng.*, March 25, 1887.

See *Boilers, Marine. Engine Room Practice. Ferry-boat. Marine Engineering. Steamboat. Steamship*.

Mill. By B. H. Thwaite. A lecture before the Textile Society of Yorkshire College, Leeds, giving retrospective history of the transition in development of the steam engine. Lon. *Eng.*, July 13, *et seq.*, 1885.

Mill. See *Engines, Steam, Plant*. Also *Quadruple Expansion*.

Multi-Cylinder, Philosophy of the. Article by Prof. R. H. Thurston. *Cassier's Mag.*, Vol. I., No. 1, Nov., 1891, pp. 1-7.

Multiple-Expansion, Application of Hirn's Analysis to. Paper by C. H. Peabody, before the Am. Soc. M. E. Abstract in *Mechanics*, Aug., 1891, pp. 179-81.

Non-Condensing, Output of the, as a function of Speed and Pressure. A valuable contribution to the scientific treatment of the steam engine, including some new laws, by Prof. F. E. Nipher. *Trans. St. Louis Acad. Science*, Vol. V., No. 3; also *Am. Jour. of Science*, Oct., 1889.

Notes on the Action of Furnaces, Boilers and Steam Engines. Abstract from various sources. I. II on steam; III. fuel and combustion; IV. chimney, furnace and boiler; V. VI. engines. By P. H. Philbrick, *Eng. News*, Jan. 15, April 2, 1881.

Performance. By J. W. Hill. The advantages of high expansion, steam-jacketing and compounding discussed, together with their limiting conditions. *Van Nos. Eng. Mag.*, Feb., 1886.

Piston-Packing, A Rational System of. By Prof. S. W. Robinson. A theoretical investigation of value. Illustrated. The packing ring pressed out by the fluid-pressure on the compression side of piston-head, which connects with space under bottom of ring. *Trans. A. S. M. E.*, Vol. II., p. 19.

Piston-Packing, New Principle in. By John E. Sweet, before the Philadelphia meeting of the Am. Soc. M. E. *Trans. A. S. M. E.*, Vol. IX., pp. 91-99.

Piston Valve, Child's Patent. An entirely new form of valve for steam engines, invented by F. D. Child, Manager of the Hinckley Locomotive Works, in which the ports run entirely around the cylinder, and are opened and closed by annular pistons working inside the cylinder itself. *R. R. & Eng. Jour.*, Jan. 1889.

Piston Velocity in Crank. By R. Werner. Find formula for the maximum velocity of piston. Illus. *Van Nos. Eng. Mag.*, Vol. VII., p. 325.

Pistons. See *Friction of P. Packing Rings. Piston Springs. Engines, Steam. Indicators*.

Plant of the Washington Mills at Lawrence, Mass. Short description, with illustration of plant and engines designed by E. D. Leavitt, guaranteed to devel-

Engine, Steam, *Plant of Washington Mills*, continued.

ope an indicated horse-power on 1·7 lbs. of coal at 140 lbs. steam pressure, and against a back pressure of 8 lbs. above the atmosphere. *Power*, March, 1889.

Portable Straw-Burning Compound, Tests of; with full illustrations, sections, indicator diagrams, etc. Lon. *Engineer*, Sept. 4, 1885.

Portable, Recent Patterns of. *Sci. Am. Sup.*, March 14, 1885.

Practical Treatise on. By Arthur Rigg, M. E., London, Eng. Fully illustrated. Reprinted in *Eng. News*, Sept. 13, 1879. July 3, 1880.

Process of Steam in its Development of Power by means of a Steam Engine. A lecture before Sibley College, Cornell University, by Chief Engr. Isherwood, U. S. N., Dec. 13, 1889. *Jour. Frank. Inst.*, Vol. C., pp. 301-17, *et seq*.

Pumping. See *Pumping Engines*

Quadruple Expansion.

A brief description, with drawing showing arrangement of details. Lon. *Eng.*, March 18 and April 1, 1887.

Designed by Mr. Thomas Mudd. Illustrated description. Lon *Engineer*, Jan. 18, 1890.

Disconnective of the steam yacht "Rionnag-na-mara." A very interesting solution of a special problem. Two full-page plate and indicator diagrams, and over a page of description. Lon. *Eng.*, April 9, 1886.

Mill Engine. Description and drawings of a 2,000 horse-power quadruple expansion mill engine at the mills of the Warren Mfg. Co., Warren, R. I. *Power*, June, 1889.

Quarrel. A brief illustrated description of an engine which can be worked with economy. Experiments show that its consumption of steam can be reduced to about 14 lbs. per horse-power. Lon. *Engineer*, Jan. 28, 1887.

Reciprocating, Dynamics of. By Prof. M. E. Cooley. A very careful and thorough study of the forces acting on the crank pin, etc., with diagrams. *The Technic*, Univ. of Michigan, 1888.

Rotary, Inclined Shaft. A valuable description of many types of this very interesting and ingenious class of motors. Illus. Lon. *Engineer*, Dec. 18, 1885.

Rotary. Their theory, advantages, disadvantages, construction, use, types, etc. Given in series of articles in Lon. *Engineer*, Oct. 16, 1885, *et seq*.

Rotary, With Movable Partitions. A description of several such types. Lon. *Engineer*, Dec. 25, 1885.

Shaft. *Graphic Representation of the Strains on*. *Mech. World*, December 29, 1883.

Shafts. See *Shafting, Shafts*.

Single-acting. An historical and descriptive article. By J. Richards. *Mechanics*, May, 1885.

Small Engines, Economy of. Trials of three small portable engines. Lon. *Eng.*, Nov. 14, 1890, pp. 577-8. *Mech. World*, Nov. 22, 1890, pp. 208-9.

Stationary Engine Practice in America. A series of illustrated articles by James B. Stanwood, describing the principal features of the various types. Lon. *Eng.*, Jan. 2, 1891, pp. 1-4, *et seq*.

Steam Consumption of, at Various Speeds. Paper by J. E. Denton and D. S. Jacobus, giving results of experiments. *Trans. A. S. M. E.*, Vol. X., 1889.

Steam Consumption of Engine at Various Speeds Abstract of a paper by James E. Denton and D. S. Jacobus, Hoboken, N. J., read at meeting of Am. Soc. Mech. Engrs., at Erie, May, 1889. Gives results of experiments on an engine at various speeds, pressures and points of cut-off, showing effect of each on economy. *Mechanics*, June, 1889

Steam Pipes, Copper. By W. Parker, before the Institution of Naval Architects. Gives a summary of investigations and results of experiments made to ascer-

Engine, Steam, *Steam Pipes, Copper*, continued.

tain the behavior of different kinds of commercial copper under various treatments and temperatures. Lon. *Eng.*, August 3, 1888; *Am. Eng.*, Aug. 22, *et seq.*, 1888; *Sci. Am. Sup.*, Sept. 1, 1888.

Steam Pipes Copper for Modern High Pressure. By W. Parker, Chief Engineer Surveyor to Lloyd's Register. Gives results of tests on strength of brazed, solid-drawn and electro-deposited copper pipes, to the advantage of the latter; also an account of the method of manufacture, and causes of failure in certain cases. *The Locomotive*, November and December.

Tests.

Application of Hirn's Analysis to Engine Testing, and a Method of Measuring Directly the Quality of the Steam in the Clearance Space. A paper by Prof. R. C. Carpenter before the Am. Soc. M. E. Abstract in *Mechanics*, Aug. 1891, pp. 176-9.

At the Philadelphia Electrical Exhibition. Comparative tests of the Porter-Allen, Buckeye and Southwark engines. *Jour. Frank. Inst.*, February, 1886.

Condensing Compound. An account of the experiments made on such an engine by a committee of the Industrial Society of Mulhouse, in Alsace, Germany. By Chief Engineer Isherwood, U. S. N. A most exhaustive set of experiments. *Jour. Frank. Inst.*, October, 1884.

Discussion of a test of a Corliss engine at Creusot, with tables. Object of test; to determine value of condenser and value of jacket. Lon. *Engineer*, March 6, 1885.

Gives conditions of tests of steam engines prepared for use at the Royal Agricultural Show at Newcastle; also gives copy of blank, showing what data will be collected. Lon. *Engineer*, June 24, 1887.

Of a Compound Engine at the Howland Mill, New Bedford, Mass. A very valuable article, by George H. Barrus, M. E. of Boston. Gives complete description of boilers and engines, etc., also exact methods used.

Of a 300 Horse-Power Triple Expansion Engine. Iron Age. Nov. 13, 1890, pp. 843-4.

At the Mass. Inst. of Tech. They were made to show the water per H. P. per hour actually used, and the amount shown by the indicator; also the condensation and re-evaporation in the cylinder. Methods and results given. *Trans. A. S. M. E.*, Vol. VII., p. 328.

By M. A. Quernel. Account of test, with tables. *Mem. de la Soc. des Ing. Civils*, Oct., 1885, pp. 461-482.

Tests of several types of engines under conditions found in actual practice. A paper by Prof. R. C. Carpenter before the Cincinnati meeting of the A. S. M. E. *Trans. A. S. M. E.*, Vol. XI., 1890, pp. 723-60. Abstract in *Eng. News*, Aug. 2, 1890, p. 108.

Of a Triple-Expansion Engine. A paper by J. T. Henthorn, read before the Am. Soc. M. E., giving description of the engine and details of the trial. Fully illustrated. Abstract in *Mechanics*, July and Aug., 1891.

Of a Triple-Expansion Engine. Paper by J. T. Henthorn before Providence meeting of the A. S. M. E., giving account of test of engine of Narragansett Electric Lighting Co. Abstract, *Am. Mach.*, July 9, 1891.

Report of the Steam Engine Tests made at the Franklin Institute Electrical Exhibition at Philadelphia in 1884. Very complete tests of Porter-Allen, Southwark and Buckeye engines. *Jour. Frank. Inst.*, February, 1886. Vol. CXXI., No. 722.

See *Pumping Engine. Engines, Steam, Marine.*

Theory.

A paper by Prof. R. H. Thurston on Hirn and Dwelshauvers Theory of. *Trans. A. S. M. E.*, Vol. XI., 1890, pp. 685-99. Abstract in *Mechanics*, June, 1890, pp. 129-31.

Engine, Steam, *Theory*, continued.

Contribution to a Rational. A series of articles intended to supply a description of the phenomena attending the performance of work in a steam engine and certain deductions logically following the phenomena. Lon. *Engineer*, July 6, *et seq.*, 1883.

Development of. By R. H. Thurston. A paper read before the British Assoc. at Montreal, being an historical outline sketch. *Iron*, Nov. 21, 1884.

Notes on. Part of the course of instruction given in the Royal School of Naval Architecture and Marine Engineering. *Van Nos. Eng. Mag.*, Vol. VI., p. 613.

Traction.

Bissell's Road Locomotive. Described and illustrated. Lon. *Eng.*, June 20, 1890, p. 779.

European Experience with. Full abstract of a paper read before the Inst. C. E., by Jno. McLaren on "Steam on Common Roads." The paper discusses the subject fully, giving a description of the engines of various makers and illustrating special features. *Eng. News*, April 4, 1891, pp. 317-21.

For Common Roads. Brief illustrated description of an English steam engine with a crane. *Eng. News*, June 28, 1890, p. 606.

In France. By D. Sienard. Treats of the engines at the Exposition and gives results of experiments. *Van Nos. Eng. Mag.*, Vol. I., p. 209.

McLaren's High-Speed. Describes a compound 12 horse-power road engine working with a pressure of 175 lbs. Lon. *Engineer*, Dec. 16, 1887.

Road Locomotives. By Prof. R. H. Thurston. Gives a short history of traction engines and then gives details of some experiments made with them. *Van Nos. Eng. Mag.*, Vol. VIII., p. 351.

Trials.

Engine, Economy Trial of an Automatic. By J. W. Hill. Details of the testing of Harris-Corliss engine as a condensing and non-condensing engine, at an Indianapolis flouring mill. *Van Nos. Eng. Mag.*, Vol. XVII., p. 544.

Economy Trials of the Non-Condensing. By P. W. Williams, before the Inst. of C. E. Gives details of a series of economy trials made on a triple expansion engine used as a simple, compound and triple engine. *Proc. Inst. C. E.*, Vol. XCIII., pp. 128-243; *R. R. Gaz.*, April 6, 1888; *Mech. World*, March 24, 1888; Lon. *Engineer*, April 6, 1888; *Sci. Am. Sup.*, May 26, 1888.

For Economy of non-condensing Simple Compound Engines, with steam pressures from 40 to 170 pounds and with revolutions from 110 to 450 per minute. A very valuable series of experiments, containing 25 pages of tabulated results, with paper and discussion, covers 170 pages, in *Proc. Inst. C. E.*, Vol. XCIII.

Experiments on a Pair of Compound Steam Pumping Engines at the Copenhagen Waterworks. By F. Olgaard. Results of a trial with and without steam jackets. Abs., *Proc. Inst. C. E.*, Vol. CIV., 1891, pp. 389-90.

Trials of Compound Horizontal Tandem-Engines; cylinders 26 and 52 inches. The object of experiments were: (1) The consumption of fuel and water per indicated horse power per hour; (2) The advantage resulting from the application of expansion gear to large cylinders. Lon. *Engineer*, July 3, 10, 17, 1885.

Newcastle. Gives full particulars of the engines and boilers, with results of trial, tested by the Royal Agricultural Society at Newcastle, England. Lon. *Eng.*, Nov. 18, *et seq.*, 1887; Lon. *Engineer*, Nov. 18, *et seq.*, 1887.

Newcastle. Gives a report of the competitive trials of agricultural engines at the Newcastle Exhibition. Contains description of the engines with table showing the principal dimensions; also table giving result of brake trial. Lon. *Eng.*, July 15, 1887.

Engines, Steam, *Trials*, continued.

Results of Trial of Boiler and Engine of the Temple Street Cable Railway. Los Angeles, Cal. By F. W. Wood, Gen. Man. *St. Ry. Jour.*, August, 1890, pp. 403-4.

Society of Arts Engine Trials. A valuable paper giving the results of a series of tests of various forms of hot air, gas and steam engines. Lon. *Engineer*, March 1, 1889, *et seq.*

Triple Expansion.

And Engine Trials at the Whitworth Engineering Laboratory, Owens College, Manchester. An elaborate paper read by Prof. Osborne Reynolds, L. L. D., F. R. S., before the Inst. of C. E., 10th Dec., 1889. Plate. *Proc. Inst. C. E.*, Vol. XCIX., Paper No. 2407, pp. 152-7; discussion, including tables and diagrams, pp. 198-253; correspondence, pp. 254-63.

Experimental Engine. Detailed description of the new experimental engine at the Mass. Inst. Tech. Lon. *Eng.*, Aug. 14, 1891, pp. 174-5, 910.

Non-Condensing. Gives a description, with dimensions, of a triple expansion non-condensing engine that gives an indicated horse-power per hour for 1.43 lbs. of coal. *R. R. Gaz.*, Aug. 3, 1888; *Eng. News*, Aug. 11, 1888.

Perspective view of a set of triple expansion engines and specifications to which they were built. Lon. *Engineer*, Aug. 26, 1887.

Vertical Condensing Engines. Illustrated description of an electric lighting engine with cylinders 31, 19, and 12 in. in diameter. Lon. *Engineer*, Feb. 6, 1891, pp. 115-16.

Underground at the Beaumt Colliery. Results of tests, and the conditions favorable to putting the engines under ground. From the French. *Abst. Inst. C. E.*, Vol. LXXXIII., p. 521.

Valve Gears, Radial. An extended study of this type of gear, with the necessary mathematics and cuts. Lon. *Eng.*, Sept. 17, 1886, *et seq.*

Valves, Modern Practice in Slide. A paper before the Cleveland (Eng.) Inst. of Engrs. Illus. Lon. *Engineer*, Feb. 5, 1886.

Valves. See *Valves.*

Water Consumption of Engines of Less than Fifty Horse-Power. Paper by Prof. J. E. Denton, giving results and discussion of experiments on a 7 in. × 24 in. engine, etc. *Steven's Indicator*, Jan., 1889, Vol. VI., p. 34.

Engines, Vapor.

Vapor, Efficiency of Fluid in. By H. L. Gantt and D. H. Maury, Jr. Experiments made with water, alcohol, ether, bisulphide of carbon and chloroform to find the best working fluid. *Van Nos. Eng. Mag.*, November, 1884.

Some Properties of Vapor and Vapor Engines. Abstract of a theoretical discussion, by DeVolson Wood, presented at convention of Am. Soc. Mech. Eng. at Erie, May, 1889. *Mechanics*, June, 1889.

Volatile Vapor. By A. F. Yarrow, before the Institution of Naval Architects. Discusses the possible advantages of using highly volatile liquids in lieu of water for the purposes of propulsion. Describes a launch propelled by a volatile hydrocarbon. Lon. *Eng.*, April 9, 1888.

Epidemic *at Plymouth, Pa.* An analysis of the causes, which were found in the water supply. *San. News*, May 16, 1885.

Equations *of third degree and of second degree.* Graphical solution of. *Annales des P. & C.* November, 1884.

Equilibruim Polygon. Method of drawing equilibrium polygon, so that three given strings may pass through three given points. By Huppner. [A beautiful method, identical, however, with that given by Levy in the new ed. of his "Statique Graphique," p. 67.] *Der Civilingenieur*, 1885, pp. 89-92.

Erecting Plant. See *Contractors Plant.*

Ericsson, John. An interesting account of the life of this great engineer by Wm. Conant Church, in *Scribner's Magazine*, February, 1890. Vol. VII., pp. 169, *et seq.*

Some of his Works. Brief biography with description of some of his inventions. *Eng. News*, Oct. 8, 1881, pp. 403-5.

The story of his life and achievements, with portrait. *R. R. & Eng. Jour.*, April, 1889.

Estuaries. *Tidal and the Bar of Mersey.* By W. H. Wheeler. Gives a general discussion of two papers read before the British Association. *Lon. Engineer*, Nov. 11, 1887.

Evaporation.

And Percolation. By Chas. Greaves. Discusses the maxima and minima of rainfall, rain Percolation through ordinary ground and sand evaporated from a water surface. *Van Nos. Eng. Mag.*, Vol. XVI., p. 48.

Investigation of the Laws of. By Desmond FitzGerald. The most elaborate, extended and scientific experiments ever made. The influence of heat, wind, sun and depth investigated. Tables, cuts, plates, etc., the whole covering 66 pp. in *Trans. A. S. C. E.*, Vol. XV., p. 581.

From Water Surfaces. Abstract of a paper before the Am. Soc. Civil Engrs., by Desmond Fitzgerald. Gives the results of some very careful experiments at Boston with a new formula. *Eng. News*, April 10, 1886.

Of Water, The Latent Heat of. By R. H. Buel. Showing the various values now in use, and drawing conclusions from Regnault's experiments as to its value. *Van Nos. Eng. Mag.*, February, 1885.

Examination Questions for the position of Assistant Engineer on the Municipal Works, New York City. One hundred and seven questions. *Eng. News*, Aug. 7, 1886.

Excavation.

And Embankment by Water Power. By E. B. Dorsey. Gives details of cost and construction of earthen dams by the use of water jets and flumes. *Trans. A. S. C. E.*, May, 1886; also *Sci. Am. Sup.*, July 17.

Freezing Process. Paper by Mr. Edward L. Abbott, C. E., read before the Boston Society of Civil Engineers, describing the development and success of this method. pp. 4. *Jour. Assn. Eng. Soc.*, March, 1890, Vol. IX., p. 94. *Eng. & Build. Rec.*, May 17, 1890, p. 378.

As Applied at Iron Mountain, Mich., in sinking a shaft through quicksand. Interesting paper by D. E. Moran, C. E. Illustrated. *School of Mines Quar.*, April, 1890, pp. 237-54. Abstract in *Eng. & Build. Rec.*, June 14, 1890, p. 23. *R. R. Gaz.*, May 26, 1890, p. 350.

Poetsch, as applied at the Chapin Iron Mine. Description of this freezing method as actually applied at a shaft sunk for the Chapin Mining Company at Iron Mountain, Mich., by Sooy Smith & Co. The material excavated was sand, with boulders embedded in it. Illustrated. *Eng. News*, Aug. 3, 1889, Vol. XXII., p. 102.

Poetsch Shaft Sinking at the Housser Colliery, Belgium. Abstract from *Bulletin* de la Societe de l'Industrie Minerale, 1888, Vol. II., p. 21. The shaft met quicksand at a depth of 295 feet, which was frozen and excavated successfully. *Proc. Inst. C. E.*, Vol. XCVI., 1888, p. 432.

Shaft Sinking. Experiments with in France. *Annales des P. & C.*, September, 1887.

Soft Material. Paper read before the Society of Arts of the Mass. Institute of Technology, by Mr. Charles Sooysmith. *Proc. Soc. Arts, Mass. Inst. Tech.*, for 1886-1887.

Quicksand in. By Charles L. McAlpine, with discussion. Gives successful treatment of a large body of quicksand in a canal cutting. *Trans. A. S. C. E.*, Vol. X., (1881). p. 275-289.

Excavation, continued.

Steam Shovel. With table of details of cost on five different pieces of work. *Eng. News*, June 8, 1889.

Rock under water without the use of Explosives. A paper by Frederick Lobnitz. Assoc. M. Inst. C. E. A description is given of the dredger, the method of breaking up the rock being to drop a heavy cutter from a height. This method is shown to be both rapid and economical, as proved by experience on Suez Canal, pp. 13. *Proc. Inst. C. E.*, Vol., XCVII., p. 339. *Eng. News*, Oct. 19, 1889. Vol. XXII., p. 364. London, 1889. See Rock Removal.

Submarine Rock Excavation. An interesting application of rock drill to submarine work by Ingersoll-Sergeant Drill Co., in New York harbor. *Am. Mfr.*, Jan. 31, p. 7.

Submarine in Sand. Illustrated description of the method of submarine excavation in sand, used by the Dutch engineer, W. H. Ter Muelen, by means of the water jet *Sci. Am., Sup.*, May 18, 1889.

Wet Material by the Hydraulic Process. An account of work done on the Northern Pacific Railway by J. T. Dodge. *Jour. Assn. Eng. Soc.*, Vol. II., p. 67.

Water Jet in. See *Land Slide*.

Excavators.

Dunbar and Ruston's Steam. Capacity one thousand cubic yards per day of ten hours. Illus. *R. R. Gaz.*, Nov. 13, 1885.

Land Dredges. By Albert Millet. Describes the different forms of machines in use on the Panama Canal, and gives cost of excavating with them. *Trans. Soc. of Engrs.*, Vol. I., p. 152.

Steam. By W. L. Clements before American Society of Mechanical Engineers. Describes the general construction of steam excavators, and then gives details of a special machine. Abstracted. *R. R. Gaz.*, May 11, 1888; also *Eng. News*, May 26 *et seq.*, 1888, and supplemented by information from other sources.

Executive Department, *a New.* The May number of the *Forum* contains a good article of the above title, setting forth the necessity for re-organizing the several bureaus of a scientific character under one department, where engineers, both civil and military, will have ample scope and just rewards for their service.

Expansion of Metals. See *Metals*.

Exploration, *and Best Outfit for Such Work.* By Maj.-Gen. Fielding. Gives many practical hints as to methods and outfit for scientific engineering, military and geographical explorations. Reprinted in *Van Nos. Eng. Mag.*, May, 1885.

Explosives.

Gives composition and relative intensity of a large number of high explosives, and then discusses those best adapted to submarine mining. Lon. *Eng.*, Aug. 26. *et seq.*, 1887.

By M. P. E. Berthelot. A series of lectures delived before the College de France Paris. *Van Nos. Eng. Mag.*, Vol., XXIX., p. 100; also No. 70, *Van Nos. Sci. Series*.

W. H. Deering, F. C. S., F. J. C., read before the Society of Arts, London. Gives composition and characteristics of various kinds of explosives. *Sci. Am. Sup.*, May 18, 1889. *Jour. Soc. Arts*, Dec. 14, 1888. *Errata*, Dec. 21, 1888.

Ammonite, Experiments with. Account of experiments with this new explosive. Claimed to be very safe and have an explosive force equal to dynamite. Lon. *Engineer*, July 17, 1891, pp. 41-2.

Apparatus for Testing Force of, by forcing a lead cylinder into a conical opening. A German invention. Described and illustrated in *Sci. Am. Sup.*, Nov. 7, 1885. *Eng. News*, April 10, 1886.

Bellite, Description of Tests made with comparatively new explosive Bellite. *F. & M. Jour.*, Dec. 1, 1888.

Explosives, continued.

Composition of High. Gives interesting list of the composition of high explosives. *Le Genie Civil*, Oct. 22, 1887; *San. Eng.*, Nov. 26, 1887.

Composition of. By W. C. Foster. Gives a list of the explosives most commonly used, with the composition, and references to publications in which notes may be found. *Eng. News*, June 30 *et seq.*, 1888.

Compressed Gun Cotton. Describes process of manufacture, properties and uses. By Eissler in *Mining & Scientific Press.* Reprint, *Eng. News*, July 30, 1881, pp. 308-9.

Dynamite, Use of, in Mines. By G. M. Cerbeland. *Mem. de la Soc. des Ing. Civils*, Dec., 1893, pp. 791-806.

Dynamite. Removing the Drift Jam at Johnstown by the Use of Dynamite. See Dam, *South Fork.*

Emmensite. Gives some of the characteristics and properties of a new explosive called Emmensite, said to be better than dynamite. *Am. Eng.*, Nov. 23, 1887, *Sci. Am. Sup.*, Jan. 7, 1888.

Experiments made on various explosives of the Favier company, in the presence of firedamp and coal dust. By —— Larmoyeux. *Assoc. des Ing. de Liege*, 1891, pp. 14-82.

Explosive Gelatine. Gives descriptive properties and composition of this explosive. *Van Nos. Eng. Mag.*, Vol. XXIV., p. 160.

Gun Cotton.

Compressed for Military Use, with special reference to gun-cotton shells. By Max V. Forster, Berlin. Translated by J. P. Wisser. Illustrated. *Van Nos. Eng. Mag.*, Nov. and Dec., 1886.

History, Manufacture and Use. Paper by Lieut. Karl Rohrer, U. S. N. Reprinted from *Sci. Am.*, in U. S. Naval Institute, Vol. XV., p. 463, 1889, No. 50.

Tonite. Gives a short history of explosive and special properties of the form of gun-cotton known as tonite. *Van. NosEng. Mag.*, Vol. XIX., p. 321.

Gunpowder, as a Propellant. Lecture by Maj. F. W. J. Barker, R. A. Illustrated *Sci. Am. Sup.*, June 21, 1890, No. 751, pp. 12009-62.

Gunpowder, The Force of. By Prof. Dr. Rudolph Wagner. A review of Berthelot's essay on gunpowder and other explosives. *Van. Nos. Eng. Mag.*, Vol. VII., p. 62.

Heat Action of. See *Heat Mechanical Applications of.*

High Explosives in Warfare. Lecture before the Franklin Institute by Commander F. M. Barber, U. S. N. *Jour. Frank. Inst.*, Feb., 1891.

High, Some Experiments on the Use of. By Prof Chas. E. Munroe, U. S. N. A. Illustrated. Nitro-glycerine, gun-cotton and dynamite used without any premature explosions of the shells fired. *Van. Nos. Eng. Mag.*, Jan., 1885.

Mechanical Work of. By F. von Reihm. Compilation, for different explosives, of theoretical and effective work. *Zeitschrift d Oesterr Ing.-u. Arch.-V.*, 1886, pp. 19-22.

Mixtures and. Gaseous Properties of, giving pressures developed by forty-two different compounds. From the French. Lon. *Engineer*, Oct. 2, 1885.

Modern Explosives and their Practical Application on the St. Gothard R. R. and the Corinth Canal. Compiled by G. T. Specht. *Eng. News*, July 5, 1884, pp. 1-3, 8.

New Explosives, Favier, Bellite and Roburite. Description of these explosives, which are in considerable use in Europe in place of dynamite. From *Le Genie Civil.* *Eng. News*, Apr. 25, 1891, pp. 391-2.

Notes on Literature of. By Charles E. Munroe, in U. S. Naval Institute. Vol. XV., p. 491, 1889, No. 50.

Explosives, continued.

And Ordnance Material, Considered with Reference to some Recent Experiments with Emmensite, Gelbite and Aluminum Bronze. By S. H. Emmens. Contains tables giving composition, etc., of various explosives. *Proc. U. S. Naval Inst.* Vol. XVII., No. 3, pp. 393-445.

Past and Future. Smokeless Powder. An interesting article reprinted from *La Nature. Sci. Am. Sup.*, No. 756, June 28, 1890, pp. 12081-2.

Properties and Use of Certain High Explosives. Abstract and quotations from a set of "Regulations" for use of explosives issued by the New York City Fire Department. The directions for care and use in blasting are concise and careful. *R. R. Gaz.*, *Eng. News*, July 6, 1889, Vol. XXII., p. 2.

Rack-a-rock for use in tunnels, caissons, etc. See Bridges.

Reactions and Formulas. Gives formulas for polynitro cellulose and asphaline, with the reactions which occur in the latter. *Le Genie Civil*, May 7, 1887, also *Eng. News*, July 2, 1887.

Recent Developments in High Explosives. By Perry F. Nursey. A paper read before the Society of Engineers, London, giving history and development of high explosives, and the composition and uses of many kinds of explosives. *Sci. Am. Sup.*, July 20, 1889, *et seq.*

Regulations to Govern the Manufacture, Transportation, Sale, Storage and Use of, in New York City. *E. & M. Jour.*, June 22, 1889.

R ...r Strength of High Explosives. A brief summary of results obtained by Lieut. Willoughby Walker, published in *Jour. Am. Chem. Soc.*, is given in *Eng. News*, Jan. 31, 1891, pp. 105-6.

Roburite. Gives results of experiments in mining coal with roburite. Lon. *Engineer*, Oct. 28, 1887; *Sci. Am. Sup.*, Nov. 19, 1887.

Smokeless Gunpowder. By Hudson Maxim. Sets forth requirements and possibility of their being met. *Sci. Am. Sup.*, No. 821, Sept. 26, 1891, pp. 13114-16.

Submarine. How their efficiency is tested by the government. Illustrated. *Eng. News*, Oct. 31, 1884.

Tests of the Relative Strengths of nitro glycerine and other explosives. A paper by Frederick W. Clark, Chicago, Ill. *Trans. A. I. M. E.* Colorado meeting, June, 1890.

Testing Strength of. See *Appliances.*

Transportation of High. Considers the nature and character of the explosives giving rules for handling and the treats of legislation on the subject. *Van Nos. Eng. Mag.*, Vol. XVII., p. 300.

Undulatory Pressures Produced by the Combustion of Explosives, in a Closed Vessel. Experiments with a steel cylinder and automatic registering apparatus. By M. Vallie. *Comptes Rendus*, 1890, pp. 639-41.

See *Blasting. Foundations.*

Exposition, *International Inventions, of 1885.* The London *Engineer* of May 8, and *Engineering* of May 1 (double number), are devoted entirely to the description of the most noteworthy inventions exhibited. Lon. *Engineer* also contains a plan of the building and arrangement of the exhibits.

Paris, 1889. A very complete and valuable review of the Paris Exhibition 1889. Profusely illustrated. Lon. *Eng.*, May 30, 1889, Vol. XLVII., p. 615.

Paper by Mr. O. Chanute, read in Chicago, Feb. 5, 1890. *Jour. Assn. Eng. Soc.*, July, 1890, pp. 341-50.

World's Columbian Exposition.

Article by Joseph Frietag, Asst. Engr., describing briefly the various buildings. Illus. *Eng. Mag.*, Nov., 1891, pp. 161-75.

Discussion on the question of site by Members of the Western Soc. of Engrs. *Jour. Assn. Eng. Soc.*, Sept., 1890, pp. 261-71.

Exposition. *World's Columbian*, continued.

A paper by James Dredge read before the Soc. of Arts, London. Lon. *Eng.* Dec. 5, 1890, pp. 678-80, *et seq.* *Jour. Soc. Arts*, Dec. 5, 1890.

Extractors. *Centrifugal.* By R. F. Gibson. Gives the mathematics and mechanics of centrifugal extractors; also describes the different forms of machines. *Sci. Am. Sup.*, Sept. 24, etc., 1887.

Factory Construction.

Comparison of English and American Types of Factory Construction. A valuable paper by John R. Freeman, read before the Boston Soc. C. E. A comparison of many details is given, with efficiency and cost of various forms of fire-proof construction. Illustrated. *Jour. Assn. Eng. Soc.*, January, 1891, pp. 19-51.

See *Building Construction, Manufactories.*

Falling Bodies.

Velocity of Bodies of Different Specific Gravity Falling in Water. An interesting paper by R. H. Richards and A. E. Woodward, read at the February meeting of the Am. Inst. Min. Engrs. An original table and 4 plates showing the relation between the diameter of the particle and the velocity of fall. *Trans. A. I. M. E.*, 1892. *Tech. Quart.*, Vol. III., No. 2.

Falls of St. Anthony.

In the Mississippi River, Preservation of. By Col. F. U. Farquhar. An account of the means used to prevent the undermining of the rock bed and the retrogression of the falls. Illus. *Trans. A. S. C. E.*, Vol. XII., (1883) p. 303.

Preservation of the Apron at the. By A. Johnson, before the Engineers' Society at St. Paul. Gives description of the Falls of St. Anthony, the main work constructed for their preservation, and describes in detail the construction of a crib to protect the apron of the fall. *Jour. Assn. Eng. Soc.*, July, 1888, Vol. VII., pp. 271-279.

Water Power. By Jos. P. Frizelle. The present difficulties in the way of its complete development and preservation, and suggestions for making effective one of the greatest water powers on the continent. *Trans. A. S. C. E.*, Vol. XII., (1883), p. 418.

Fans and Blowers.

Efficiency of. Experiments by Wm. T. Trowbridge and Geo. A. Souter, to determine the volumes of air delivered under various conditions and the power required. From *Trans. A. S. C. E.*, Vol. VII., reprinted in *Eng. News*, Dec. 11, 1880.

Efficiency of. An editorial in *The Engineer*, Nov. 14, 1884, in which a new rule for computing the H. P., by Prof. Herschel, is given and discussed; also by W. C. Unwin. A short discussion, with formulæ and illustrations, very simply put. Lon. *Engineer*, Dec. 12, 1884; also by Robt. H. Smith in same for Dec. 19, 1884.

Testing of. To find (1) how much mechanical work is required to be done by a proposed fan not yet in place, in order to give a desired volumetric ventilation through a given mine; and (2) how much mechanical work is being done by a fan already in place and working under given conditions, and in connection with this test to calculate the mechanical efficiency of the fan. By Prof. R. H. Smith. Lon. *Engineer*, July 24, 1885.

See *Ventilation.*

Fatigue of Metals. An account of some new devices and methods used by Prof. Egleston in examining the molecular changes accompanying fatigue. Illus. Lon. *Eng.*, July 10, 1885. *See* also Iron.

Faucet. *An Indestructible Faucet.* Description, with section, of an ingeniously designed faucet. Proved very serviceable under thorough tests. Design of Sir Wm. Thomson. *Eng. News*, Sept. 19, 1891, p. 255.

Feed Water

By T. A. Paget. Treats of the chemical and physio-chemical effects. *Van Nos. Eng. Mag.*, Vol. I., p. 917.

Heating. See *Boiler Tests, Heating.*

Heaters. Their value in the saving of fuel. *Eng. News*, Dec. 18, 1886.

Ferro-Silicon, *and the Economy of its Use.* Paper by W. J. Keep, Detroit, Mich., and Edward Orton, Columbus, O., pp. 9. *Proc. A. I. M. E.*, Buffalo meeting, 1888.

Ferroid. *A new Artificial Stone.* See *Stone.*

Ferry Boat.

Description of new steel screw ferryboat "Bergen" built by Hoboken Ferry Co. Illus. *R. R. Gaz.*, Aug. 9, 1889.

Double Screw, Performance of. Account of tests of the performance of the screw ferry-boat "Bergen" and a comparison of the same with that of the wheel ferry-boat "Orange." Illustrated. Paper before the Am. Soc. of Mech. Engrs., by E. A. Stevens and J. E. Denton. *R. R. & Eng. Jour.*, March, 1890. *Trans. A. S. M. E.*, Vol. XI., 1890, pp. 370-431.

Improvements in. By Wm. Cowles. Has especial reference to the boats used in New York Harbor. *Trans. A. S. M. E.*, Vol. VII., p. 180.

Pennsylvania Railroad Co.'s Steam Ferry Boat "Cincinnati." Description with inset showing arrangement of machinery and engines. *R. R. Gaz.*, Nov. 6, 1891, pp. 776-7. *Eng. News*, Nov. 7, 1891, pp. 428-30. *Iron Age*, No. 12, 1891, pp. 821-4.

Railroad Ferry Steamer "Solano." A complete and fully illustrated review of this interesting boat by Robert L. Harris. *Trans. A. S. C. E.*, April, 1890, Vol. XXII., pp. 247-257.

Steel Ferry Steamer. Description and drawing of the new ferry-boat of the Michigan Central Railway, designed to ferry 21 loaded cars across the Detroit River, breaking its way through the ice in winter. *R. R. Gaz.*, Jan. 25, 1889. *R. R. & Eng. Jour.*, Feb., 1889.

Transfer, at Portland, Ore., as constructed in 30 days on a very difficult slope. Illustrated. *R. R. Gaz.*, Aug. 7, 1885.

"*Transfer" of the Canadian Southern Railroad Company.* An illustrated description of this new ferry boat, built by the Cleveland Shipbuilding Company. *Sci. Am. Sup.*, May 10, 1890, p. 11,980.

Ferry Landing.

A brief description, with drawing showing details of the inclines, etc., for the proposed ferry across the river Ganges. *Lon. Eng.*, March 11, 1887.

For a Car Ferry on the Mississippi River at St. Louis. By Robt. Moore. The general problem discussed, with full drawings of the device used. *Trans. A. S. C. E.*, Vol. XIII., p. 247.

For Railway Transports. The Isle of Wight Steam Ferry. A landing moving on inclined ways connects with steamer at different stages of tide. Illustrated. *Lon. Eng.*, Aug. 7, 1885. *See* Floating Landings.

See Landing.

Fiber. *Lime Sulphite Fiber Manufacture in the United States.* Paper by Maj. O. E. Michaelis, M. Am. Soc. C. E., and also Remarks on the Chemistry of the Process. By Martin L. Griffin, M. A., p. 23. *Trans. A. S. C. E.*, June, 1889, Vol. XX., p. 263.

Filters.

Analysis of Mechanical Action of Filters. A paper by Mr. Emil Geyelin. This paper discusses nearly all the more prominent filters in use and made for use of water works, and forms a valuable source of information. *Am. W.-W. Assn.*, Louisville meeting, 1889, p. 67.

Filters, continued.

Breyvis' Micromembran Filter. A short description of this filter, in which ground asbestos is used as filtering material, and which it is claimed will prevent the passage of bacteria. *Journ. f. Gasbel. u. Wasservers.*, 1886, pp. 415-416.

Describes a number of small filters used in different places. *Van Nos. Eng. Mag.*, Vol. XXIII., p. 152.

Piefke's Filter. Adapted for use on large or small scale. Filtering material is a specially prepared form of asbestos and other fibres. Short description without data, regarding cost, etc. *Journal f. Gasbel. u. Wasservers.*, 1886, pp. 781-783.

Warren. Description of the Warren water filter, with plan and sections. *Eng. News*, Nov. 19, 1887.

See *Water Filtration and Purification.*

Filtration,

Data of. Papers by Prof. Wm. T. Sedgwick in *Tech. Quart.* No. I, "Some Recent Experiments on the Removal of Bacteria from Drinking Water by Continuous Filtration through Sand," giving the results. Discussion of the experiments in Berlin. Vol. III., 1890. No. 1, pp. 69-75. No. II., "On Crenothrix Kuehniana (Rabenhorst), Zopf," giving an account of several cases of occurrence and its effect on the water. Vol. III., 1890, No. 4, pp. 138-65.

Experiments on. By Dr. Plagge. Report to the German Soc. of Naturalists. Experiments on filtration of germs or bacteria. Bischoff's spongy iron filter and all the charcoal filters were entirely unable to remove germs, filtered water often containing more than the unfiltered. The same was true of some paper and cellular filters. The Chamberland clay filter and that of Dr. Hesse were found to filter germs entirely at first, though after a time they were not effectual. Some asbestos filters (Breyvis' "micromembran filter") filtered perfectly at first, but allowed germs to pass in a short time. *Journal f. Gasbel. u. Wasservers.*, 1886, pp. 809-812.

Experiments on the Efficiency of Filtration Through Sand. By Dr. Carl Frankel and C. Piefke. This is a valuable report on the city water of Berlin. Tables. The proportion of bacteria in the unfiltered water to those in the filtered was, roughly speaking, 1,000 to 1. *Zeitschrift fur Hygiene*, Vol. VIII., 1890, p. 1. Abstracted in *Proc. Inst., C. E.*, Vol. C, 1890, p. 455.

Gerson's System. Describes a process in which water is passed through sponges and pumice stone filled with insoluble tannate of iron. Gives analysis of water before and after passing the filter. *Lon. Eng.*, Nov. 15, 1887.

Of Natural Waters. A paper by Mr. T. M. Drown, read in Boston, Jan. 15, 1890. *Jour. Assn. Eng. Soc.*, July, 1890, pp. 356-68. *Eng. & Build. Rec.*, Nov. 22, 1890, p. 394-6.

Practical Results of Mechanical. By W. S. Richards. A paper before the American Water-Works' Association, giving experience with Hyatt filters at the Atlanta water-works. *Proc. Am. W.-W. Assn.*, 1888, pp. 148-152. Abstracted *Eng. & Build. Rec.*, May, 1888.

Of Public Water Supply. Account of extensive experiments by the East London Water Company with the Atkins filter. Illustrated. *Iron*, Nov. 6, 1885.

Rate of Through Sand. In terms of the head and the average size of the grains of sand. A formula from the German. *Eng. News*, Oct. 24, 1885.

Or Subsidence. Abstract from a paper by J. D. Cook, before the American Water-Works Association. Treats of the most practicable and economical method of obtaining a supply of clear water from streams carrying considerable quantities of silt. *San Eng.*, July 22, 1887.

See *Sewage Disposal, Water Purification, Water Supply.*

Fire Apparatus.

Modern Fire Apparatus. A popular article by John R. Spears. Illus. *Scribner's Magazine*, Jan., 1891, pp. 51-61.

Fire Apparatus, continued.

About Shops and Buildings. A valuable paper by Mr. C. J. H. Woodbury before Am. Soc. of Mechanical Engineers. *Ry. Rev.*, Dec. 4, 1889, Vol. XXXI., p. 718.; *Eng. News*, Dec. 11, 1889, Vol. XXII., p. 568.

Fire Bricks.

As Manufactured at Mt. Savage, Md. A paper before the Am. Inst. Min. Engrs. at Pittsburg meeting, Feb., 1886, by Robert Anderson Cook. This is probably the best known manufactory of fire-brick in America. *Eng. News*, April 10, 1886; also *E. & M. Jour.*, March 13, 1886.

By Lieut. G. E. Glover. Extracts from a paper presented to Corps of Royal Engineers. *Van Nos. Eng. Mag.*, Vol. VI., p. 6.

Fire Boats. By William Cowles. A paper read before the Am. Soc. Mech. Engrs. Gives description of the fire boats at New York, Chicago and Cleveland. Fully illustrated. *Lon. Eng.*, Jan. 14, 1887.

Fire Bulkheads.

A description of such devices as are used to prevent the spread of fires in mines. By Chas. M. Rolker. *Eng. News*, May 2, 1885.

Fire-Damp *in Coal Mines.* Apparatus for indicating at the surface the exact amount of it in any part of the mine without having to visit the region personally. A very ingenious and apparently practicable invention. Fully illustrated. *Sci. Am. Sup.*, Feb. 11, 1882. Caused by falling barometer.

See *Coal Mine Explosions.*

Fire Grates. *Donnelley for Boilers.* Gives plan, elevation and section of the Donnelley fire grate and the results of experiments with different kinds of coal. *R. R. Gaz.*, Jan. 6, 1888.

Fires. *Prevention and Extinction of.* By A. Chatterton, before the students of the Institution of Civil Engineers. Discusses the causes of fires, fire-proof material, fire-proof construction, internal and external appliances for extinguishing fires. *Proc. Inst. C. E.*, Vol. XCIII., pp. 437-461.

Fire Hazards.

From Electricity. A lecture by C. J. H. Woodbury before the students of Sibley College. *Sci. Am. Sup.*, No. 830, Nov. 28, 1891, pp. 13267-9. *Elec. Eng.*, Nov. 18 and 25, 1891.

See *Electric Lighting in Mills.*

Fire Streams. *Hydraulics of Fire Streams.* A valuable paper by Mr. John R. Freeman, M. Am. Soc. C. E.; with discussion by Clemens Herschel, Mansfield Merriman, E. Kuichling, Geo. F. Swain, J. P. Church, E. B. Weston, R. Hering and John R. Freeman. Fully illustrated by plates and diagrams. *Trans. A. S. C. E.*, Nov., 1889, Vol. XXI., pp. 303 *et seq.*

Fires. Abstract of. *Eng. & Build. Rec.*, Sept. 7, 1889, Vol. XX., p. 221, *et seq.*

Effect of fire and sudden cooling on columns of cast iron, wrought iron and stone. An abstract of the experiments of Prof. Bauschinger on this subject (as detailed in the 12th Heft of his "Mittheilungen") is given in the *Deutsche Bauzeitung*, 1885, pp. 343-346.

Extinction. Apparatus for the automatic extinction of fires. A lecture before the Soc. of Arts, London, by Prof. S. P. Thompson. Discusses automatic sprinklers, automatic alarms, automatic fire-doors, and miscellaneous appliances. Illustrated. A valuable paper. *Jour. Soc. Arts*, Nov. 27, 1885.

Extinguishing, Means of. A lecture by C. J. Hexamer before the Franklin Institute. Discusses the chemistry of combustion, water, pumps, vertical pipes, hydrants, valves, hose, nozzles, tanks, sprinklers, steam jets, extinguishing fluids and powders, fire brigades, watchmen and fire alarms. *Jour. Frank. Inst.* Aug., 1885.

Hydrant. The Greathead Injector Hydrant for Fire Extinction. Two systems of pipes, one for service, the other small, with water at very high pressure

Fires. *Hydrant*, continued.

given to it by hydraulic accumulator. A connection established between these two systems at a hydrant working on the principle of the steam injector draws most of the water from the large service pipes with a very much higher pressure. Cut given. *Eng. News*, Feb. 20, 1886.

Loss, Method of Reducing. Paper by C. J. H. Woodbury before the American Soc. of Mech. Engs., at New York City, describing practical methods of reducing the fire loss on isolated manufacturing property. *Trans. A. S. M. E.*, Vol. XI., 1890. pp. 271-311. *Mechanics*, Dec., 1889.

Prevention of. See *Sprinklers.*

Prevention and Extinction of. A paper by A. Chatterton, describing methods of fire-proofing buildings and means of quenching fires. *Proc. Inst. C. E.*, Vol. XCIII., p. 437.

Protection of Mills, from. By C. J. H. Woodbury. A very complete paper of 35 pages, with many cuts, giving the appliances used and organization of the fire brigade as in vogue in the New England Mills. *Trans. A. S. M. E.*, Vol. II., p. 301.

Firearms. *Development of Automatic.* Gives a two-page plate and short description of the details of the Maxim gun. Lon. *Eng.*, Jan. 27, 1888.

Fireproof Arches. *Tests of.* Full account of tests made on three kinds of fireproof arching at Denver, Colo. Tests for strength under static load, resistance to fire and water and to continuous fire, were conducted, and the results are given in full. Illustrated. *Am. Arch.*, No. 796, March 25, 1891. pp. 195-201.

Fire-Proof Buildings. A letter from Mr. Edward Atkinson in *Am. Arch.* containing many valuable suggestions concerning fire-proof and slow-burning business buildings. *Am. Arch.*, Dec. 21, 1889, p. 293.

Fire-Proof Construction.

At Moderate Cost. By C. T. Aubin. Precautions that should be observed. Appliances to put out fires, etc. *Jour. Assn. Eng. Soc.*, Vol. IV., p. 158.

A series of articles of value, by F. Collingwood, Mem. A. S. C. E., and M. Inst. C. E., in *San. Eng.*, (New York), Feb. 11, *et seq.*

A very comprehensive paper by John J. Webster, describing the various fire proof systems. Considers separately the fire resisting properties of the various materials, their combination, and the general arrangement of structures. Gives some new experiments regarding the first. *Proc. Inst. C. E.*, Vol. CV., 1891. pp. 249-88.

By Theo. Rosenberg. Some practical suggestions of value. *Jour. Assn. Eng. Soc.*, Vol. V., p. 101.

Fire-Resisting Construction. A lecture by C. J. H. Woodbury before the Franklin Institute on "Conflagration in Cities" is abstracted in *Eng. Record*, Jan. 24, 1891, pp. 122-3.

Fire Proof Construction. See *Below*, *Building Construction*.

Fireproof Materials, *Tests of, at Boston.* Walls and sides of buildings treated with various materials were subjected to great heat. Details described and illus. *Eng. News*, Oct. 31, 1891, pp. 417-18. Abstract from official report giving conclusions, *ibid.*, Nov. 14, 1891, p. 460. Full report in *Am. Arch.*, Nov. 14, 1891, pp. 101-2. Abs. *Science*, Nov. 13, 1891. *Eng. Record*, Oct. 24, 1891.

Fish Plates.

Their Defects. An illustrated editorial of four columns, showing that fish-plates fail more from the bending moment on them than from shear, and suggesting that a form be used to resist shear only. *R. R. Gaz.*, Sept. 3, 1885.

Some Tests of Iron Fish-Plates at a Freezing Temperature. A paper by David Buel read before the Wisconsin Polytechnic Society, giving results of several tests. *Eng. Record*, July 4, 1891, p. 70.

Flexure. *Resistance of Beams to.* By J. G. Barnard. An abstract from a paper on the resistance of materials, by M. Decamble, with comments. *Trans. A. S. C. E.*, Vol. III., pp. 123-128.

Floating Deflectors. See *Harbor Improvement.*

Floating Landings. A description of such landings designed for passenger traffic. From the Dutch. Illustrated. *Eng. News*, Aug. 8, 1885.

Flood Announcements.

Treatise on, and on the Hydrology of the River Seine. (Manuel Hydrologique du Bassin de la Seine. De Preaudeau. 1884.) Reviewed in *Annales des P. & C.*, 1884-2-610. For several years the heights and times of maximum flood heights along the Seine at Paris, and at Points below, have been publicly announced several days in advance, to the great advantage and benefit of all interested inhabitants. The book in question shows how this result has been accomplished. The review gives suggestions as to still further improving this "hydrological service."

Valley of the Ohio. Plan for same. *Annales des P. & C.*, November, 1854.

Flood Discharge *from Catchment Areas.* A theoretical formula with empirical coefficients with special reference to India. *Proc. Inst. C. E.*, Vol. LXXX., p. 301.

Flood Gates, *Automatic.* Gives brief description, with cuts, of the Cvetkovics automatic flood gate. Lon. *Eng.*, July 13, 1883.

Flood Heights *in Rivers as Affected by the Change from Wild to Cultivated Conditions.* By Gustav v. Wex. Translations of two pamphlets on the subject, using European data. Issued by the Engr. Dept., U. S. A.

As Affected by the Destruction of Forests. By Thos. P. Roberts, with discussion and plates. A large mass of data, with important discussions. *Trans. Eng. Soc. West. Penn.*, Pittsburg, 1884.

Mississippi River. See River Improvements.

Flood Rock Removal.

A carefully written account of the entire work, by Lt. Geo. M. Derby, who has been officially connected with the work. An historical and descriptive account of permanent value. Illustrated. *San. Eng.*, Dec. 3, 1884.

See *Harbor Improvements, Hell Gate.*

Floods.

The Flood of June, 1889, in the West Branch of the Susquehanna River. Abstract of the report of Maj. Chas. W. Raymond U. S. Engrs., giving the main facts of the flood, reviewing the causes and giving several possible methods of protection from floods. *Eng. News*, Feb. 14, 1891, pp. 152-3.

Great Floods on the Lower Miss. Riv., as illustrated in the flood of 1882. By J. B. Johnson. Accompanied by map of the valley showing overflow area below Cairo. *Jour. Assn. Eng. Soc.*, Vol. II., p. 113.

Inefficiency of Reservoir to Prevent Inundation. An article by Mr. Gros, in *Annales des P. & C.*, II., 1890, giving as the result of several years study of some large rivers of France, the conclusion that floods cannot be appreciably diminished by this means. *Eng. News*, March 14, 1891, pp. 258-9.

Mississippi Flood of 1890. A discussion of this flood and the effect of levees on same; mainly by Mr. D. M. Harrod. The conclusion is that the levees have been beneficial. *Eng. News*, April 5, 1890 Vol. XXIII., p. 325.

Prediction of Floods in the Central Loire. Account of methods employed, with discussion of various formulas and diagrams. *Annales des P. & C.*, Oct. 1890, p. 410. *Proc. Inst. C. E.*, Vol. CIII., 1891. Foreign abstracts, pp. 495-6.

Prediction of High Water on the Elbe in Bohemia. The discharge of the various tributaries is measured. Account of method with curves of discharge given. By —— Holtz. *Annales des P. & C.*, April, 1891, pp. 477-84.

Floods, continued.

Report on the Protection of the City of Elmira, N. Y., Against Floods. A report by F. Collingwood to the Mayor of Elmira. *Report*, Feb. 12, 18.0.

In Stony Brook, Prevention of. A report of 29 pp. by a Commission, consisting of Jas. B. Francis, Eliot C. Clarke and Clemens Herschel, to the Mayor of Boston, accompanied by a map and six plates. The report is of value to engineers generally, as bearing on the question of the maximum instantaneous discharge from a given valley. Persons designing sewers, culverts, and water-ways under bridges would do well to consult it. *Boston City Document* 159. 1886.

See *Johnstown Disaster*.

Floor Construction.

A system of constructing fire proof flooring on a simple, economical and effective plan. Comparative cost of several fire proof systems of flooring are given. Illustrated. *Eng. News*, April 19, 1890, p. 367.

Abstract of a method used in "*Centralblatt der Bauverwaltung*." Iron girders are used as usual with "plaster beams" as filling. *Eng. News*, April 12, 1890, Vol. XXIII., p. 341.

Fireproof. Designs and estimates for a proposed combination of cement and iron, by F. H. Kindl. *Eng. News*, July 4 1891. pp. 2-3. Illustrations of a similar construction, *id.*, July 25, 1891, p. 81.

Floor Construction. See *Building Construction*. *Fire Proof Arches*.

Flooring, *Steel.* Gives description, with illustrations showing the application of Lindsay steel flooring. Lon. *Engineer*, Oct. 7, 1887; *Eng. News*, Nov. 26, 1887.

Floors.

Brick and Cement. Details of a number of experiments made to determine the strength of the different systems of brick and cement floors. *Van Nos. Eng. Mag.*, Vol. III., p. 45.

Experiments Upon the Resistance to Shock of Beton Floors and Arches in Brick Masonry. Illustrated description of several experiments on floors. *Eng. News*, July 12, 1884, p. 11.

In Mills. By C. J. H. Woodbury. Loads, formulas for designing, and practical details of construction. Illustrated. *Trans. A. S. M. E.*, Vol. II., p. 468.

Weights of. See *Building Construction*.

See *Lime Floors*.

Flour-Mill Analysis. A full account of some flour-mill tests to determine the power used, and where it was spent. Apparently a profitable kind of investigation for any mill. *Am. Eng.*, Dec. 3, 1873.

Flour Milling. *Modern Flour Milling.* A paper by J. Harrison Carter, describing the various details of flour manufacture, and machinery used, with discussion. *Jour. Soc. Arts*, March 6, 1891, pp. 299-313.

Flour Mills *and their Machinery.* A historical and descriptive article. Illustrated. *Proc. Inst. C. E.*, Vol. XC., p. 366.

See Milling.

Flow of Air.

And Other Gases. Experiments and formulæ especially adapted to blast-furnaces. By F. W. Gordon. *Trans. A. I. M. E.*, Chattanooga meeting, 1885.

In Sewers. See *Sewers*.

Flow of Metals *in the Drawing Processes used in the Arts.* A lecture by Oberlin Smith. Describes tube and wire drawing, stamping, etc. Illustrated. *Jour. Frank. Inst.*, Nov., 1886.

Flow of Solids. See *Forging*.

Flow of Water. See *Ditches. Hydraulics. Flow of Water*

Flues. *Experimental Investigations of the resistance of, to collapse.* By C. R. Roecker. Details the experiments made of flues and discusses the formulæ derived from the experiments by various writers. *Van Nos. Eng. Mag.*, Vol. XXIV., p. 308.

Fluid Friction. *Experiments on Rotation of Disks in Fluids.* By Prof. W. C. Unwin. *Proc. Inst. C. E.*, Vol. LXXX., p. 221.

Fluids. *Resistance of.* Abstract from "Sonnet's Dictionnaire des Mathematiques, Appliquees." *Van Nos. Eng. Mag.*, Vol. IV., p. 431.

Flushing. See *Siphon.*

Flushing Tanks.

Automatic, for urinals and water-closets. Causing a saving of 80 per cent. of water used, as compared with a small continuous stream. *Abstracts of Papers, Inst. C. E.*, 1885.

The Rosewater Flush. A description of an automatic flush-tank now in use on the Omaha Sewage Works. *Eng. News*, July 17, 1886.

Flushing Tunnel. *The Milwaukee.* A full illustrated description of the tunnel from Lake Michigan to the Milwaukee river, built for flushing the river from sewage contamination. Water is forced through the tunnel from the lake by a wheel operated by a 350 H. P. engine. *Eng. & Build. Rec.*, Dec. 8, 1888.

Fly Wheels. See *Engines, Steam, Fly Wheels.*

Forced Draught. See *Boilers, Forced Draught for.*

Forces. On the mutual forces acting between masses of matter, with application to mechanical, physical and chemical phenomena. By M. P. Berthot. Considers molecular forces as well as forces acting at finite distances, application to liquids and gases, capillarity, chemical phenomena, etc. *Mem. de la Soc. des Ing. Civils*, Nov. 1885, pp. 380-401.

Forests. *Their Influence on Rainfall.* A paper by Prof. Geo. F. Swain, giving an able and rational discussion of the subject, and including a synopsis of the known facts relating thereto. *Jour. New. Eng. W. W. Assn.*, Vol. I., No. 3.

Forging. *Hydraulic Forging and the Flow of Solids.* Notes suggested by alleged defects in certain forgings made by pressure. A paper by Coleman Sellers, E. D., Professor Engineering Practice at Stevens Institute of Technology. *R. R. Gaz.*, Feb. 7, 1890, Vol. XXII., p. 86. *et seq.*

Forging Machines.

Hydraulic Forging Machines and Steam Hammers. Reprint of paper by Prof. Coleman Sellers before senior class at Mass. Inst. Tech. From *Stevens Indicator. R. R. Gaz.*, Aug. 2, 1889.

Press of 4,000 tons capacity. A description, with many cuts, of this hydraulic press now building at Sheffield, Eng. Lon. *Eng.*, April 23, 1886.

Forging Machines. See *Hammers. Screw Forging Machines. Steam Hammers.*

Forts. *Protection of Heavy Guns for Coast Defense.* Extracts from an article by Capt. G. S. Clarke, R. E., taken from the *Proceedings of the Royal Artillery Institution*, Woolwich, Febr. 1887. *Proc. of the U. S. Naval Inst.* for 1887, Vol. XIII., No. 2.

Foundations

Lecture delivered by Francis Collingwood, before the students of the Rensselaer Polytechnic Institute. The specific subjects discussed are: examination of the soil, compressibility of masonry, bearing power of various soils and problems connected with building upon them, bearing power of piles, and the various methods of founding under water. Many examples from actual practice are given, also references to the literature bearing on the subject. *The Polytechnic*, Jan. 24, 1891, pp. 92-100. Reprinted in *Eng. News*, Feb. 14, 1891, pp. 161-2, *et seq.* *Eng. Record*, Feb. 21, 1891, pp. 196-9, *et seq.*

By W. C. Street. Treats of foundations in peat, sand and clay soils. *Van Nos. Eng. Mag.*, Vol. XXVI., p. 317.

Foundations, continued.

Bartholdi Statue. Elevations and sections. *Sci. Am.*, June 13, 1885.

Bartholdi Statue. See *Bartholdi Statue.*

Bridge.

And Piers of the W. & L. E. Railroad Bridge at Toledo. By Chas. E. Greene. *Report Mich. Assoc. Surv.*, 1882.

Brooklyn Anchorage of the East River Bridge. By F. Collingwood. Gives details of method of construction adopted. *Trans. A. S. C. E.*, Vol. III., pp. 142-146.

Empress Bridge over the Sutlej, India. The piers are of stone, resting on metallic tubes, three to each pier, which extend to a depth of 102 feet below low water. From the French. Illustrated. *Sci. Am. Sup.* Aug. 8, 1885.

Forth Bridge. The Great Caissons; Their structure, building and founding. By H. S. Biggart, before the Inst. of Engrs. and Shipbuilders, Scotland. Lon. *Eng.*, Nov. 27, 1885, *et seq.* See also *Caissons* below.

Hawkesbury Bridge. *Recent Progress in Sinking Deep Foundations for Engineering Works.* By Charles Ormsbee, Engineer in charge of Hawkesbury Bridge. From *Proc. of Australian Soc. for the Advancement of Science. Eng. News*, Feb. 1, 1889, Vol. XXIII, p. 114.

Hawkesbury River Bridge. Detailed drawing of the plan proposed by the Union Bridge Co. *R. R. Gaz.*, June 18, 1886.

New London, Conn., R. R. Bridge. A combination of crib, piles and open caisson. A. P. Boller, Engineer. Description, with illustration of bridge, in *Sci. Am.*, June 8, 1889.

New Tay Bridge. Description and cuts showing the work in progress, from the French, in *Sci. Am.*, Jan. 9, 1886.

New Tay Bridge. Illustrated and described by C. Barlow before the British Assoc. *R. R. Gaz.*, Dec. 4, 1885.

New Tay Bridge. *Sinking of the Cylinders by Pontoons.* By Andrew S. Biggart. A paper before the Inst. of Engrs. and Shipbuilders, Scotland. Illustrated. Lon. *Engineer*, July 10, 1885; also *Eng. News*, Aug. 1, 1885; also Lon. *Eng.*, June 26, 1885.

Schuylkill B. & O. Bridge at Philadelphia. Description of caisson, crib, and cofferdam used on the pivot pier. Sooysmith & Co., contractors; Wm. M. Patton, engineer in charge. Well illustrated. *Eng. News*, March 27, 1886. Also by A. Brooks Cuthbert. A series of articles in *Eng. Era*, Dec. 31, 1885, *et seq.* Illustrated.

Tensas River R. R. Bridge, Ala. The Cushing system of piles inside iron cylinders was used. Fully illustrated and described by Col. Wm. M. Patton, in *Eng. News* of June 20, 1885.

See *Piles*, *below.* *Also Bridge Foundations.*

Caissons. *Cribs and Coffer Dams* used on the foundations of the new Havre de Grace bridge on the B. & O. Ry. By Col. Wm. M. Patton, engineer in charge. This work is under contract by Gen. Sooysmith & Son, and the rate of progress made has been remarkable. Plans, elevations and sections given of the caissons. First article in *Eng. News* of Feb. 7, 1885.

Forth Bridge. A full-page cut in perspective; also sectional illustrations. Lon. *Engineer*, Feb. 6, 1885; also Lon. *Eng.*, same date.

With lateral opening against a rough masonry wall. Description of how the joint was made by means of tarpaulin, to be readily set and removed. Lon. *Eng.*, June 12, 1885, also *Iron Age*, July 2, 1885.

Coffer Dams and Floating Caissons. A paper by Randall Hunt, read before the Tech. Soc. Pacific Coast. Several actual examples are given and illustrated. *Ry. Rev.*, Jan. 3, 1891, pp. 3-5.

Foundations, continued.

Coffer Dams. See *Coffer Dam.*

Concrete. By J. E. Blackwell. Gives specifications for and cost of concrete foundations; also the relative strength of different cements. *Sci. Am. Sup.*, Oct. 8, 1883.

Concrete, Brickwork and Stonework. A description of an attempt to render concrete, brick and stone more easily available for cylindrical foundations. *Van Nos. Eng. Mag.*, Vol. VIII., p. 335.

Cooper Institute. Why they failed and how the faults have been remedied. An excellent example of very common faults in buildings. Fully described and illustrated by 11 cuts in *San. Eng.*, New York, Nov. 5, 1885.

Deep. By Col. Wm. M. Patton, C. E. Discusses timber piles, open caissons, pneumatic process, suspension of caissons while sinking, and concrete. *Eng. News*, July 18, 1885.

For Drop Presses. Illustrated account of manner of constructing the foundations for a drop forging plant for the Gorham Manufacturing Co. at Providence, R. I. *Eng. News*, Feb. 22, 1890, Vol. XXIII., p. 186.

Elastic, for Dynamos, Steam Engines etc. Describes and illustrates a method of preventing the vibration of such machinery being communicated to the ground and buildings. *Elec. World*, Jan. 5, 1887.

For Engines. See *Asphalt.*

Equalizing Pressure on. By J. H. Apjohn, M.I.C.E. With especial reference to foundations in the alluvial soil of Bengal. *Ind. Eng.*, January 5, 1889, *et seq.*

And Floors for the World's Fair Buildings. Abstract of a paper by A. Gottlieb before the West. Soc. Engrs., describing methods of determining supporting power loads adopted, and plan of construction. *Eng. Record*, December 5, 1891, pp. 15-16.

Of Garfield Monument. A paper by James Ritchie and a lengthy discussion. *Jour. Assn. Eng. Soc.*, Vol. V., p. 410.

Of a Heavy Fire-proof Building on Compressible Soil. An account of the methods and results used for such a building in Chicago, by W. L. B. Jenney. *San. Eng.*, Dec. 10, 1885. *San. News*, Chicago, Jan. 2, 1886.

For Heavy Buildings. A short article in the London *Architect*, discussing the best manner of distributing the weight upon the foundations. *Eng. Record*, Jan. 31, 1891, pp. 140-2.

House. See *House Foundations, Healthy.*

Of a Lighthouse renewed and lighthouse removed a distance of 300 feet. Foundation of creosoted piles. J. W. Putnam. Illustrated. *Trans. A. S. C. E.* Vol. X., (1881), p. 14.

Of Lighthouse in Delaware Bay. Pneumatic process used for sinking a cast-iron cylinder, made in sections and bolted together. Illustrated. *Eng. News* Oct. 31, 1885.

See Pile.

Lighthouse. See *Lighthouse.*

Of the New Capitol at Albany. By W. J. McAlpine. A description of the requirements for the foundation and details of the experiments made to ascertain the sustaining power of the blue clay upon which it is built. *Van Nos. Eng. Mag.*, Sept., 1884.

Of the New Capitol at Albany, N. Y. By W. J. McAlpine. Gives brief description of soil and experiments made to ascertain its bearing power. *Trans. A. S. C. E.*, Vol. II., p. 177.

Open Brick Walls. See *Viaduct.*

Foundations, continued.

Pile.

By Julian Griggs. Describes the common methods of managing pile foundations for bridge masonry and trestles. *Report Ohio Soc. Surv. & Engrs.*, 1888, pp. 209-216.

Fall of the Western Arched Approach to South Street Bridge, Philadelphia. Foundation of piles in soft ground gave way. D. McN. Stauffer. Illus. *Trans. A. S. C. E.*, Vol. VII., (1873) p. 261. Also Discussion in Vol. IX. p. 349.

Method Pursued in Replacing a Stone Pier on a Pile. By J. A. Monroe. Gives description of the method employed to replace on the pile foundation a grillage with two courses of masonry, which had broken loose and settled 14 inches out of place. *Trans. A. S. C. E.*, Vol. III., p. 59.

Reinforcements of. Brief account of the re-enforcing of the pile foundation of a draw bridge by means of auxiliary piles, a portion of the weight on the old piles being transmitted to these by means of cast-iron brackets and set screws. Illus. *Eng. News*, March 7, 1891, p. 224. *Am. Arch.*, No. 794, March 14, 1891, pp. 175-6.

Repair of Bridge Pier at Kansas City, Mo. Supports to piles had washed out. Illustrated. *Eng. News*, Dec. 25, 1886.

Repairs to the Foundations of the Chestnut Street Bridge, Philadelphia. The stone pier, resting on piles, prevented from lateral movement by inclined pillars of concrete put in by the use of compressed air. With illustrations. *Eng. News*, Jan. 3, 1885, and *Sci. Am.*, Jan. 10, 1885; also *Proc. Eng. Club, Phila.*, Vol. V., No. 1.

Placed beneath the Pumps and Engines at the Newton, Mass., Pumping Station. By Albert F. Noyes. *Jour. Assn. Eng. Soc.*, Vol. IV., p. 213.

Pneumatic.

By A. Heinerscheidt. Description of an improved closing port for discharge in lock. *Van Nos. Eng. Mag.*, Vol. XXII., p. 151.

By General Wm. Sooy Smith. Gives details of the work at the Waugoshance Lighthouse, Omaha and Leavenworth bridges. *Trans. A. S. C. E.*, Vol. II., p. 411.

Cost of. Extract from Prof. I. O. Baker's "Treatise on Masonry Construction," giving cost of foundations of Havre de Grace Bridge; average per cu. yd., $22.69; and Blair Crossing Bridge, average per cu. yd., $20.70, and other data. *Eng. News*, Nov. 16, 1889, Vol. XXII., p. 472.

On the Frictional Resistances of. By A. Schmoll. The details given relate to cast-iron cylinders, gives tables of co-efficients of friction derived from experiments. *Van Nos. Eng. Mag.*, Vol. XX., p. 131.

Of a Tidal Basin Entrance Lock at Dieppe, France. Under over 55 feet head of water; area of caisson, about 180 feet by 115 feet. *Annales des P. & C.*, Nov., 1887.

Repair of. A Coffer Dam or Caisson without Timber or Iron in its Construction. Abstract of an interesting paper by Robert L. Harris, before the Am. Soc. C. E., describing the repairing of some bridge piers, in which a coffer dam was constructed by forcing cement grout into riprap and loose stone previously in place. *Eng. News*, March 7, 1891, pp. 220-1. Illustrations, *ibid.*, March 14, p. 249. *Eng. Record*, March 7, 1891, pp. 221-2.

Replacing Under Elevator at Providence, R. I. By A. McL. Hawks. Gives details of the method employed to put a new foundation under one side of the Columbia Elevator at Providence, R. I. *Eng. News*, Feb. 14, 1885.

Sand as a. An article compiled by W. W. Curtis, C. E., giving the results of French experiments on the action of sand; the laws of its distribution of pressure when inclosed; its reliability when inclosed, etc. Illustrated. *Eng. News*, May 15 and 29.

Foundations, *Pneumatic*, continued.

Screw Pile, on Mobile River crossing of M. & M. R. R. By Col. Wm. M. Patton. Fully illustrated, showing method of sinking. *Eng. News*, April 4, 1885.

Steam Stamp. See *Steam Stamp*.

Steel Foundations. Description of the steel foundations used in Chicago, with inset of drawings. By Congdon T. Purdy. *Eng. News*, August 8, 1891, pp. 116-17.

Sub-Aqueous.

A new system of sheathing of iron, like sheet-piling, which may be removed after the masonry is above the water-line. From Abstracts Inst. C. E., in *Van Nos. Eng. Mag.*, May, 1886; also *Am. Eng.*, April 28, 1886.

By Gabriel Jordan. Describes the process employed in constructing the piers of a bridge for the Mobile & Montgomery Railroad across the Tensas River. The piers were composed of two groups of twelve piles each, inclosed in iron cylinders, and filled in with concrete. *Trans. A. S. C. E.*, Vol. II, p. 309.

Two lectures delivered before the Royal Engineers Institute at Chatham, by W. R. Kinipple. Gives most recent practice, with examples from important works. Illustrated. Lon. *Eng.*, Oct. 10, 1890, pp. 438-40, *et seq.* Reprinted in *Eng. News*, Nov. 8, 1890, pp. 413-16, *et seq.*

Of the Northeast and Southeast Quays of the Outer Harbor of Calais, France. Translation of part of a report on the Harbor of Calais, recently made by the Chief Engineer in Charge, M. Betillart. Illustrated. *Eng. News*, Jan. 4, 1890, Vol. XXIII., p. 7.

The new Maritime Installations of the Port of Antwerp. A valuable paper by H. de Matthys, giving a full description of the caissons and method of constructing the quay-walls. 20 plates. *Annales des Travaux Publics de Belgique*, Vol. XLXVII., p. 1. Abstract in *Proc. Inst. C. E.*, Vol. CI., 1890, pp. 331-2.

Sinking a Sea Wall Foundation by a Water Jet. Hollow concrete blocks 31 ft. long successfully sunk in this way to a depth of 15 ft. at Calais, France. Illustrated description. *Eng. Record*, March 14, 1891, p. 241.

Employment of Water Under Pressure in founding the Quay-Walls of the outer Harbor of Calais. Paper by M. Bailly in *Compte Rendu des Travaux de la Societe des Ingenieurs Civils*, Oct. 1890, p. 580, describing the sinking of large blocks of masonry by means of the water jet. Abs., *Proc. Inst. C. E.*, Vol. CIV, 1891, pp. 352-3.

See *Torpedo Nausalis*.

Supporting Power of Soils. By Randall Hunt. Treats of the supporting power of soils as deduced from personal observation and the recorded examples. *Jour. Assn. Eng. Soc.*, June, 1888, Vol. VII., pp. 189-196; *Eng. News*, June 16, 1888; *Sci. Am. Sup.*, June 30, 1888; *Eng. & Build. Rec.*, June 23, 1888.

Use of Compressed Air in the Tubular Foundations, and its application at South Street Bridge, Philadelphia. D. McN. Stauffer. Illustrated. *Trans. A. S. C. E.*, Vol. VII. (1878), pp. 287-309. Discussion in Vol. VIII., pp. 186.

Underpinning Yarmouth City Hall. Description of an interesting case of underpinning. Illus. *Eng. & Build. Rec.*, Nov. 15, 1890, p. 371.

See *Aqueduct, Construction. Brick Masonry. Bridge Foundations. Coffer Dams.*

Foundry Practice. *On the Mechanical Treatment of Moulding Sand*. A paper by Walter Bagshaw, read before the the Inst. M. E., describing in a thorough manner the various processes and methods of treatment and giving best present practice. Lon. *Eng.*, February 27, 1891, pp. 299-300. *E. & M. Jour.*, February 28, 1891, p. 260.

Foundry Work.

A System of Estimating for. Translated from the German and adapted to American practice by Geo. L. Fowler, M. E. A valuable series of articles in *Am. Eng.*, beginning July 23.

Estimating Cost of. By G. L. Fowler, before the American Society of Mechanical Engineers. Gives outlines of a plan used by the writer to find the cost of castings. *Am. Eng.*, May 9, 1888.

Fourier's Law. See *Heat, etc.*

Framed Structures. Contribution to the Theory of Framed Structures. By Prof. H. Muller, Breslau. *Zeitschrift des Arch. u. Ing. V. zu Hannover*, 1885, p. 417.

Frames.

Note on the Theory of Frames. By T. Landsberg. Calculation of secondary stresses due to rigid joints. *Zeitschrift des Arch. u. Ing. V. zu Hannover*, 1886, pp. 19; 202.

The Theory of. By Prof. Mohr. *Der Civilingenieur*, 1885, Heft. 5.

See *Trusses.*

Framing. See *Joints, Holding Power of Dovetails.*

Franklin Institute *Index to Reports of the Committee on Science and the Arts.* From 1834 to 1890. Covering many reports of considerable value. *Jour. Frank. Inst.*, November 1890.

Freezing Hydrants. See *Hydrants.*

Freezing Mixtures. A list given, with resulting temperatures. *The Sanitary News*, Chicago, Sept. 19, 1885.

Freight Traffic, *Cost of.* See *Railroad.*

Freight Transfer Bridge. See *Bridges, Transfer.*

Friction.

A paper before the Institution of Naval Architects, treating of the changes of lead in the surface of the water surrounding a vessel, produced by the action of a propeller and by skin friction. Lon. *Engineer*, April 1, 1887.

A Series of Illustrated Lectures. By Prof. H. S. Hele-Shaw, London. Very valuable for students, as well as manufacturers and engineers. *Jour. Soc. Arts.* London, Vol. XXXIV., 1886. Republished in *Sci. Am. Sup.*, Dec. 18 and 25, 1886, *et seq.*

And Lubrication. A series of tests by W. O. Webber, made to determine relative value of various lubricants for railroad purposes. *Technology Quarterly*, Vol. I, No. 4, 1888, pp. 324-9.

Of Brake Shoes. See *Brake Shoes.*

Of Brakes. See *Brakes, Coefficient of Friction.*

Coefficient, with a given oil, but under varying conditions of temperature and pressure, with tables, formulas, and diagrams. A paper read before Am. Soc. of Mech. Engrs., by C. J. H. Woodbury. *Mechanics*, Jan., 1885.

Of Collar Bearing. Gives the third report of the research committee "On Friction" of the Institution of Mechanical Engineers. Report gives results of experiments on the friction of collar bearings, with description of apparatus. Lon. *Engineer*, May 4, 1888.

Due to Flow of Water in Long Pipes. See *Hydraulics, Flow of Water.*

Effect of Compression on Shaft Friction in the Triple-Expansion Engine. Methods of calculation and general results. By Geo. I. Alden. *Cassier's Mag.* Vol. 1, No. 1, Nov. 1891, pp. 8-13.

Engine. By R. A. Buel. A very valuable paper, giving two tables of friction data. *Am. Eng.*, June 8 and 15, 1887.

Of Non-condensing Steam Engines. By Prof. R. H. Thurston, Ithaca, N. Y. A paper describing experiments showing that the friction is constant at dif-

Friction, *Engine, etc.*, continued.

ferent speeds and under different loads. Read at the meeting of the American Society of Mechanical Engineers held in New York. *Jour. Frank. Inst.*, December, 1886. Reprinted in *R. R. & E. Jour.*, February, 1887. See *Engines, Steam.*

See *Engines, Steam, Friction.*

Fourth Report of the Research Committee on Friction. This report is on "Experiments on the Friction of a Pivot Bearing," read before the Inst. of Mech. Engrs. Experiments were made on a flat-ended bearing 3 in. in diameter, and were carried out for various speeds, loads, and amounts of lubrication, upon both a Manganese-bronze bearing and a white-metal bearing. Results given. *Iron*, March 27, 1891, pp. 270-1.

Of Hydraulic Capped Leather Packing. By Theodore Cooper. Gives details of tests made during the construction of the St. Louis bridge to determine the friction of the packing in the hydraulic testing machine to be used for testing the material of construction. *Trans. A. S. C. E.*, Vol. XVI., p. 30.

Investigation of One of the Laws of. By A. S. Kimball. Details experiments on sliding friction and friction of belts and journals. *Van. Nos. Eng. Mag.*, Vol. XVII., p. 86.

Journal. By W. J. M. Rankine. A paper treating of journal friction in the steam engine. *Van Nos. Eng. Mag.*, Vol. VI., p. 379.

Apparatus for testing. Illustrated. Simple and efficient. *R. R. Gaz.*, May 22, 1885.

The practical cause of hot journals, and discussion of Mr. Tower's experiments. *Lon. Engineer*, March 6, 1885.

Of Metal Coils. By Prof. Hele Shaw and Edward Shaw, before the Bath Meeting of the British Association. *Lon. Engineer*, Sept. 28, 1888.

Of Piston Packing Rings. By Prof. J. E. Denton, before the Scranton meeting of the American Society of Mechanical Engineers. Gives results of a series of experiments made with a special instrument for measuring the friction of piston packing rings in steam cylinders. *R. R. Gaz.*, Oct. 19, 1888; *Am. Eng.*, Oct. 24, 1888.

Recent Researches in. By John Hoodman, before the Students of the Institution of Civil Engineers. Gives a comparison of the results obtained by various authorities and examines the phenomena from a theoretic point of view. *Eng. News*: March 3, *et seq.*, 1888.

Recent Researches in, and the Action of Lubricants. By Prof. Hele-Shaw. A general review of the knowledge of this subject in 1885. *Trans. Liverpool Eng. Soc.*, Vol. VI., pp. 100-16.

Rolling Resistance. Well analyzed by Prof. R. A. Smith (commonly called rolling friction). *Lon. Engineer*, May 29, 1885.

Of Rotation, Theory of the sliding. By R. H. Thurston. *Van Nos. Eng. Mag.*, Dec., 1884.

Of Slide Valves. See Slide Valves.

In Tooth Gearing. By G. Lanza, before the American Society of Mechanical Engineers. Gives a mathematical investigation of friction in the involute and epicycloidal forms of gearing. *Trans. A. S. M. E.*, Vol. IX., pp. 181-228.

Of Water in Cast-Iron Pipes. Tables showing loss of head due to friction in pipes for diameters from 4 inches to 48 inches. Computed and compiled under direction of Saml. M. Gray, City Engineer, Providence, R. I. *An. Rep. City Eng. Prov., R. I.*, 1889, No. 10.

See *Air, Flowing in Pipes, Engines, Steam, Friction of. Fluid Friction. Lubrication. Sliding Friction.*

Friction and Lubrication of ***Journals.*** Abstract of paper by Prof. John Goodman, before the Manchester Assn. of Engrs., giving results of investigations and suggesting forms of bearings for journals for minimum friction. *The R. R. & Eng. Jour.*, December, 1890.

Friction Brakes. See *Brakes, Friction.*

Friction Clutch.

By Geo. Adams. A practical paper on the construction of friction clutches. Illustrated. Lon. *Eng.*, Sept. 7, 1888.

The Mechanics of. By S. M. Balch, giving the mathematical equations, practical applications and economic considerations involved. *Sch. of Mines Quar.*, Jan., 1887.

Frictional Gearing ***on a Dredge.*** By J. G. Griffith, before the Institution of Mechanical Engineers. Gives a description of the frictional gearing used on a double steam dredge in the port of Dublin. Illustrated. Lon. *Eng.*, August 24, 1888.

Frictional Resistance.

Of Engine and Shafting. Two papers before the Am. Soc. Mech. Engrs. by J. T. Henthorne and Samuel Webber. The latter shows that the belts should all be thrown from the loose pulleys when making the test. *Trans. A. S. M. E.*, Vol. VI., p. 461, and Vol. VII. The latter also in *Am. Eng.*, Dec. 3, 1885.

Of Railroad Trains. Results of experiments by C. H. Hudson for obtaining the resistance in pounds per ton on level track. *Jour. Assn. Eng. Soc.*, Vol. V., p. 44.

Of Railroad Trains in Winter and Summer. A new theory by the editor of the *R. R. Gaz.*, based on the experiments of C. J. H. Woodbury, and other known facts, that is due to the cooling of the axles in winter. *R. R. Gaz.*, April 9, 1886.

And Train Resistance, by A. M. Wellington. Abstract of results made with special apparatus designed to show the friction at various speeds and loads. Tables and diagrams given. *R. R. Gaz.*, June 5, 1885. See also *Resistance.*

Frogs and Switches. See *Railroad Track.*

Fuel.

A paper before the Iron and Steel Institute on the theoretic minimum of fuel required to produce one ton of pig iron. *Van. Nos. Eng. Mag.*, Vol. VII., p. 314.

A lecture by Dr. Siemens before the British Association at Bradford. *Van. Nos. Eng. Mag.*, Vol. IX., p. 465.

Anthracite Waste, Utilization of by Gasification in Producers. Brief paper by W. H. Blauvelt before the A. I. M. E., stating as the result of tests and experience that little can be done beyond reducing the amount of refuse. *Trans. A. I. M. E.*, 1891, pp. 3. Reprinted in *Eng. News*, Oct. 10, 1891, p. 329.

Artificial. A brief resume of the different processes followed in the production of artificial fuel. *Van Nos. Eng. Mag.*, Vol. IX., p. 81.

Coke as an Engine Fuel on the Baltimore & Ohio. Account of its introduction and present successful use. By T. H. Lymington, Inspector. *R. R. Gaz.*, May 2, 1890, pp. 299-300.

And Combustion. By R. H. Buel. Gives a general synopsis of the most important principles and data from various sources. *R. R. Gaz.*, July 13, *et seq.* 1888.

Consumed in Locomotive Engines, and corresponding work done on four American railroads, worked out under various heads and tabulated. The roads are the Michigan Central, Canada Southern, Hannibal & St. Joseph and Lake Shore. By G. C. Cunningham. *Trans. Inst. C. E.*, Vol. LXXXIII., p. 341.

By M. G. Marie, before the Inst. of Civil Engrs. Gives particulars of three-days' trials with locomotives, during which consumption 2.88 lbs. coal per indicated horse-power was obtained. *Van. Nos. Eng. Mag.*, Dec., 1884.

Fuel, continued.

Consumption. *Comparative Fuel Consumption and Designs of Compound Locomotives*, Royal Saxon State Railroads. Two articles, giving much valuable data and information concerning comparative economy of locomotives, including many indicator cards, and a tabulated statement of results of actual experiments on several locomotives. *R. R. Gaz.*, Nov. 22 and 29, 1889. Vol. XXI, p. 772.

Culm Compared with Other Fuels. A very cheap fuel according to some reliable experiments. Results given. *Eng. News*, Dec. 6, 1890, p. 514.

Economy of. By L. Lhoest. Details of experiments made at Maestricht on a boiler and 110 horse-power engine. Describes heating apparatus and engine, and examines results. *Van. Nos. Eng. Mag.*, Vol. XV., p. 128.

Energy and Utilization of Fuel, Solid, Liquid and Gaseous. Paper by Mr. W. R. Taylor read at the Washington meeting of the Am. Inst. of Mining Engineers, Feb., 1890. pp. 23. *Trans. A. I. M. E.* *Am. Mfr.*, March 7, 1890, p. 17.

Gas. See *Fuel Gas*.

Gaseous. By J. E. Dawson before the British Association for the Advancement of Science. *Am. Mfr.*, Oct., 26, 1888.

Liquid. An interesting comparison of the results obtained by some of the best methods of burning crude petroleum refuse. Illustrated. *Mechanics*, March, 1889.

An illustrated article on the burning of liquid fuel, showing arrangement of furnace, etc. Illustrated. Lon. *Eng.*, June 11, 1886.

By Percy F. Tarbutt, A. M. I. C. E. Paper read before the London Society of Engineers, Nov. 1, 1886. *Am. Eng.*, Numbers of Dec. 1 and Dec. 8.

Crude Liquid Hydrocarbons as Heating and Lighting Agents. A paper before the Liverpool Polytechnic Society, Feb. 11, 1889, by B. H. Thwaite. *Mech. World*, March 2, 1889, *et seq*.

Details of an interesting series of experiments on liquid fuel made by Mr. B. H. Thwaite. Lon. *Engineer*, Dec. 9, 1887.

Results obtained in petroleum firing on the Central Pacific R. R. ferry boats. Saving effected and ratio of oil to coal in caloric value. *Sci. Am. Sup.* April 2, 1887.

Dorsett's System. Gives result of experiments on the steamer "Reviewer." *Van Nos. Eng. Mag.*, Vol. I., p. 57. See *Fuel*.

"*Fuel Oil*" for Stationary Boilers in New York City. Valuable paper by H. F. J. Porter, M. E. Illus., pp. 20. *Sch. of Mines Quar.*, Columbia College. July, 1889, Vol. X., p. 350.

For Gas Retorts. Gives details of experiments made on burning coal tar under gas retorts with the Drury spray nozzles. Illus. Lon. *Engineer*, Sept. 7, 1888.

On Locomotives. An illustrated description of the mode of using liquid fuel on locomotives now being carried out by Mr. James Holden, the Locomotive superintendent of the Great Eastern Railway in England. Uses the liquid fuel as an auxiliary to the coal, replacing a large part of it. *R. R. & Eng. Jour.*, Jan., 1889.

On Locomotives. Gives method of employing liquid fuel on locomotives now being carried out by Mr. James Holden, locomotive superintendent of the Great Eastern Railway. Illustrated (from *Engineering*). *Am. Mfr.*, Nov. 30, 1888.

Oil Fuel in Marine Practice. Article of some value in *Eng. News*, Feb. 1, 1890, Vol. XXIII., p. 101.

Oil for Fuel. An account of experiments by the Pennsylvania Railroad Company on the use of petroleum as fuel in locomotives, with a comparison of

Fuel, *Oil for Fuel*, continued.

its cost with that of coal. *R. R. & Eng. Jour.*, August, 1887, and *R. R. Gaz.*, July 1, 1887.

Oil as Fuel. Three valuable papers, by S. S. Leonard, M. J. Francisco and C. M. Ransom, before the National Electric Light Association at Chicago, with discussion on the same. Papers give results of actual experience. *Elec. World.* March 2, 1889.

Oil in the Puddling Furnace. Description of a method of applying the use of oil fuel to puddling. Illus. *Am. Mfr.*, Aug. 16, 1889.

See *Fuel Petroleum. Oil Burners.*

Liquid and Gaseous. The Uses of. Paper by H. L. Gardner, of New Orleans, read at the Louisville Meeting American W. W. Assn., giving relative value of coal, gas and petroleum as fuel for making steam. *Eng. News.* April 27, 1889. Am. W. W. Assn., Louisville meeting, 1889, p. 16.

Peach Stones for. Used in Vacca Valley, Cal., where they bring $6 per ton and are considered equal to the best coal. *E. & M. Jour.*, May 4, 1889.

Petroleum.

Treats of the problem of burning petroleum on locomotives. Gives experience in Russia and on the Pennsylvania road in this country. *Sci. Am. Sup.*, Oct. 15, 1887.

A short article giving results obtained by the use of "fuel oil" at North Chicago Rolling Mill, where it is used under a battery of fourteen boilers and said to cost less than coal. *Am. Mfr.*, Nov. 23, 1888.

The Aerated Fuel Company's System of Burning Crude Petroleum. Gives methods and appliances in use by this company. Their method is to spray the oil by means of compressed air. Data given in article shows considerable saving in fuel bill by use of oil, as used by U. S. Cartridge Co., at Lowell, Mass. Illus. *Eng. News*, Nov. 30, 1889. Vol. XXII., p. 510.

Gives a report of the tests made at the Salem pumping station, to test the value of petroleum as fuel, when converted into and used as gaseous vapor. The oil was found more economical than coal. *Am. Mfr.*, April 6, 1883.

Note on use of Petroleum on locomotives in Penn. Weight of oil used per train mile about one-half that of coal. *Am. Mach.*, June 20, 1891.

Vs. Coal. By C. E. Ashcroft. Considers theoretically and practically the use of petroleum oil as fuel in place of coal. Gives valuable comparisons of test trips on the Russian railroads. *Am. Mfr.*, May 25, 1888.

For Locomotive Boilers. Tests made by a Board of Naval Engineers. Illus. *Eng. News*, Feb. 2, 1889.

For Locomotives. A paper by Thomas Urquhart, Russia, read before the Inst. of Mech. Engrs. Fully illustrated. *Sci. Am. Sup.*, September 27, 1890. No. 769, *et seq.* See also description of an improved petroleum motor. *Ibid.*, pp. 12282-3.

In Locomotives. A paper read before the Inst. of Mechanical Engineers, by Thos. Urquhart, Locomotive Supt. Grazi & Tsaritsin Ry., S. E. Russia. Illustrated. Coal and Petroleum refuse compared. *Lon. Eng.*, Feb 8, 1889, *et seq.*

Refuse as fuel for locomotive engines. Articles giving results of experience and tests on Indian railways. *Ind. Eng.*, July 6, 1889; July 20, 1889.

Uses of Petroleum in Prime Motors. A paper by Prof. Wm. Robinson, giving much valuable information regarding the nature of various oils and their use in oil and gas engines, and as fuel. Diagrams of vapor tensions, indicator cards, etc., given with discussion. *Jour. Soc. Arts*, May 1, 1891, pp. 492-510.

Working Locomotives with Petroleum Fuel. Paper by Mr. Thomas Urquhart, giving experience of the Grazi & Tsaritsin Railway, Russia, in the use of pe-

Fuel. *Working Locomotives, etc.*, continued.

troleum and petroleum waste as used on both locomotives and stationary engines. Several tables of cost and comparative economy of coal and petroleum are given. Illustrated. Lon. *Eng.*, Feb. 8, 1889, Vol. XLVII, p. 130.

Powdered Anthracite and Gas. Abstract of the Report of the Board of Trade of Scranton, Pa., describing the successful use of pulverized coal as fuel, resulting also in a great saving and freedom from smoke. *Eng. News*, Nov. 13, 1886.

Pressed, the Manufacture of. By E. F. Loiseau. Describes the process of manufacturing fuel from coal dust. *Van. Nos. Eng. Mag.*, Vol. XXIV., p. 41.

Racine System of Burning. Description and illustrations of tanks, piping, auxiliary boilers, etc., used by the Racine Hardware Manufacturing Company, Racine, Wis., for burning crude oil. *Elec. World*, April 6, 1889.

Relative Cost of Fluid and Solid at New York, Chicago, New Orleans, San Francisco, London, Port Natal, Sydney and Valparaiso. Fuels discussed are Anthracite and Bituminous Coal, Petroleum, Coal Gas, Generator Gas and Water Gas. By James Beatty, Jr. *Proc. Engrs. Club, Phila.*, Vol. V., p. 172, and *Am. Eng.*, Oct. 30, 1885.

And Smoke. By Oliver Lodge. Two lectures in the Royal Institution. A treatise on ventilation and combustion. *Sci. Am. Sup.*, March 19, 1887. First lecture in Engineering May 28, 1886.

Tests. See *Fuel Testing*.

See *Combustion. Gas, Coal, Wood*

Fuel Gas.

By J. M. Cutchlow, before the Ohio Gas Association. *Am. Mfr.*, Mch. 30, 1888.

By Walton Clark before the Western Gas Association. Compares the relative efficiencies of pure fuel gas and a mixture of coal, water and producer gas. *Am. Mfr.*, June 22, 1888.

A brief general article with illustrations of apparatus for burning. By R. H. Buel. *R. R. Gaz.*, Nov. 30, 1888.

And Some of Its Applications. Paper by Mr. Burdett Loomis, presented at the N. Y. Convention of the A. I. M. E. Illustrations of plant, Loomis process. *Eng. News*, Oct. 11, 1890, pp. 317-19. Abstract in *Am. Mfr.*, Oct. 10, 1890, p. 17. *Sci. Am. Sup.*, No. 774, Nov. 1, 1890, pp. 12363-5.

The Fuel of the Future. An article describing the methods of producing water and oil gas, and comparing the economy of water gas and coal as a steam-producing and forge-heating fuel. *Am. Mfr.*, June 28, 1890.

And Incandescent Gas Lighting. By Chas. M. Lungren. Gives comparison of the economy of the different methods of illumination, with figures of cost. *Sci. Am. Sup.*, March 3, 1888.

Loomis Fuel Gas Plant. Gives brief description of the Loomis fuel gas plant at Tacony, Pa., with account of the Loomis system of production. *Am. Mfr.*, Oct. 5, 1888.

The Present Outlook for Manufactured Fuel Gas. Abstract of a paper by Geo. H. Christian, Jr., read before the Ohio Assn. Gas Engrs., discussing the relative efficiencies of various fuel gases and their cost. *Eng. News*, May 30, 1891, pp. 512-13.

And the Strong Water-Gas System. By Henry Wurtz. A paper before the Am. Inst. of Min. Engrs., giving analysis of and cost of production of the Strong fuel gas. *Van. Nos. Eng. Mag.*, Vol. XXIII., p. 376.

See *Gas for Heating and Motive Power*.

Fuel Testing.

Fuel Testing. Critical Review of Efficiency Tests for Coals. Papers by William Kent containing much valuable information. *E. & M. Jour.*, Oct. 10, 17, 24, 31, 1891.

Fuel Testing, continued.

Relative Cost of Wood and Coal on the Southern Pacific Railway. Tests on a very heavy division indicated but very slight difference in cost. Table of results. *Ry. Rev.*, April 18, 1891, p. 249. *Eng. News*, April 25, 1891, p. 398.

Results of a test between bituminous coal and crude petroleum, made at Mansfield, Ohio, by Mr. J. W. Gwynn. Coal for thirty-one days, 61 1-8 tons, equals $167.02. Oil for thirty-one days, 251 barrels, equals $130.10: balance in favor of oil, $36.92, or nearly 22 per cent. Illustrated. *Mechanics*, July, 1889, Vol. VI., p. 174.

See *Calorimeters*.

Furnace Plant *at McKeesport, Pa.* Mr. F. C. Roberts, chief engineer. The arrangement of this plant seems to be very excellent. Plan, sections and full description in *Am. Mfr.*, Aug. 8, 1890.

Furnace Testing. *The Murphy Smokeless Furnace.* Results of test of its economy made at the Cambria Iron Works. The furnace is also described and illustrated. *Am. Mfr.*, July 3, 1891, pp. 22 4.

Furnaces.

Construction of, for Liquid Fuel. A valuable series of papers by Herr Dusley, in *Wochenschrift des Vereines Deutscher*, reviewing the use of liquid fuels. The methods employed are classified and a large number of various appliances of these methods are illustrated and described. Translated in Lon. *Engineer*, Feb. 10, *et seq.*, 1888: *Power-Steam*, June, *et seq.*, 1888.

Efficiency of Burning Wet Fuel. By. R. H. Thurston. Gives results of experimental investigation made upon two distinct varieties of furnaces burning spent tan-bark wet from the leaches. *Trans. A. S. C. E.*, Vol. III. (1874). pp. 290-318.

Flue Gases from Steam Boiler and Other Furnaces. Methods of analysis to determine waste. Lon. *Eng.*, Oct. 3, 1890, pp. 384, *et seq.*

Hoist for. See *Hoisting Machinery.*

Plate Thickness in. Gives a discussion of the limit of thickness of plate used in boiler construction. *R. R. Gaz.*, Nov. 11, 1887.

Safe Working Pressures for Cylindrical. By M. Longridge. Discusses the different formulae relating to the safe working pressure on cylindrical furnaces and flues, showing their discrepancies, and proposed a modified form of Fairbairn's formula. Lon. *Engineer*, Sept. 21, 1888.

Smoke-Consuming Furnace. The Weinmann's Furnace. See *Smoke-Consuming Furnace.*

Waste of Heat in. Paper by Wm. Metcalf before the Engrs. Soc. West. Pa., investigating the loss due to this cause. *Eng. News*, Nov. 5, 1881, p. 453.

See *Engines, Steam, Notes.*

Furnaces, Annealing.

An illustrated description of an annealing furnace in use at the Rhode Island Locomotive Works for annealing boiler sheets after flanging. The furnace is designed for the largest boiler sheets in use for locomotives or marine boilers. *R. R. Gaz.*, Feb. 7, 1890, Vol. XXII., p. 88.

Furnaces, Basic Open Hearth. *The Use of Magnesite in.* From a paper by Chas. Walrand, M. E., supplemented from a paper by Kurt Sorge in *Stahl und Eisen.* Illustrated article in *Am. Mfr.*, Jan. 3, 1890, p. 7.

Furnaces, Blast. See *Blast Furnaces. Compressed Air in. Coke.*

Furnaces, Gas. *Some Different Kinds of Gas Furnaces.* A descriptive paper by Bernard Dawson, read before the *Inst. Mech. Engrs.*, dealing with those kinds which are connected with iron and steel manufacture. *Iron*, February 6, 1891, pp. 120-4.

Furnaces, Regenerative. *The Development of the.* By F. Siemens. Particularly with regard to the improvements by the new Siemens method of heating. *Zeitschr. d. Oesterr. Ing. u. Arch. V.*, 1886, pp. 27-35.

Description of a continuous regenerative gas kiln for burning fire-bricks and pottery. By John Mayer. *Trans. Engineers and Shipbuilders of Scotland*, March 24, 1885.

At Munich Gasworks. By Drs. Schilling and Bunte. Gives results of the working of a bench of six ovens on the so-called "damp" system. *Van. Nos. Eng. Mag.*, Vol. XXIX., p. 311.

Gas. A new method of heating. By Fred'k Siemens. The flame is not brought in contact with the materials under treatment. Being a paper read before the Iron and Steel Institute. Lon. *Eng.*, Dec. 19, 1884. See *Combustion*.

Puddling, for Natural Gas. Illustrated description. *E. & M. Jour.*, April 20 1889.

See *Siemen's Regenerative Furnace*.

Garbage Crematories.

A short description, with cuts, of the appliances used by the U. S. Sanitary and Fertilizer Co. of Boston. *San. Engr.*, Jan. 22, 1885. An English method given in same Journal, Feb. 19, 1885.

Of Bombay. A brief description, accompanied by plans and sections, of a Beehive Refuse Destructor, for the combustion of garbage and other refuse, which is now in process of construction for the municipality of Bombay, India. *San. Eng.*, Dec. 31, 1885.

A description of Hannecken's System. The solids from streets, kitchens, and closets are dried and burned, all escaping gases being conveyed into the furnace and burned. From the Russian. *Abstr. Inst. C. E.*, Vol. LXXXIII., p. 481.

By John Zellweger. Gives drawings of a battery of two furnaces, each of seven ton capacity, in 24 hours. Estimated cost of combustion from 21 to 36 cents per tons. *Jour. Assn. Eng. Soc.*, Vol. V., p. 233; *Sci. Am. Sup.*, July 17, 1886.

Description of the "bee-hive" furnace for burning refuse, in operation at Richmond, Eng. *Eng. News*, Aug. 29, 1885. Also description of furnace used on Governor's Island, N. Y. Harbor, in *San. Eng.*, Aug. 13, 1885.

And the Destruction of Organic Matter by Fire. A valuable paper by Samuel S. Kilvington, M. D., President of Board of Health of the city of Minneapolis. *Minneapolis*, 1888.

Garbage Disposal.

Account of method used by one company at London, Eng. Lon. *Eng.*, May 15 1891, pp. 591-1.

By Thomas. Codrington. Gives a report on the different methods in use for destroying town refuse. Contains large amount of data relating to refuse. Abstracted *Eng. & Build. Rec.*, Sept. 15, 1888.

The Disposal of Refuse in American Cities. The report of Walter V. Hays, General Sanitary Officer of the Chicago Department of Health, contains a summary of the methods of disposal in the ten largest cities in the U. S. Abstract, *Eng. News*, July 18, 1891, pp. 51-2.

Disposal of Household Refuse at Chelsea, Eng. Method is by mechanical and hand sorting, part being saved and part burned. *Eng. Record*, Sept. 19, 1891, p. 250.

Economical Disposal of Ashes and Other City Refuse. A paper by Wolcott C. Foster, giving suggestions and describing machine for sorting city refuse, illustrated. *Eng. News*, March 15, 1890. Vol. XXIII., p. 243.

Gas.

By Henry Gore. Treats of the substitutes for coal in the production of illuminating gas. *Van Nos. Eng. Mag.*, Vol. IX., p. 321.

Gas, continued.

A Study of the Relative Poisonous Effects of Coal and Water Gas. By Professors Wm. T. Sedgwick and Wm. Ripley Nichols, of the Mass. Inst. of Tech., Boston. A pamphlet of 50 pp., being perhaps the most valuable, exhaustive and unbiased investigation, by acknowledged experts, ever made in America. A conclusive presentation of the case. Address the authors.

Coal Gas. A discussion of its constituents and description of manufacture. By Wm. D. Marks, before the Engrs. Club of St. Louis. *Eng. News*, Jan. 6, 13, 1877.

Companies of America. By Wm. W. Goodwin. A paper before the annual convention of the American Gas Light Association. Gives statistics of the industry; price of gas per thousand feet; amount sold, coal used, etc. *Sci. Am. Sup.*, Feb. 5, 1887.

Compressed Oil Gas and its Applications. By Arthur Ayres before the Inst. of Civ. Engrs. Discusses the application of compressed oil gas to light-houses, railroad cars, etc. Describes the Pintsch works at South Foreland. *Proc. Inst. C. E.*, Vol. XCIII., pp. 298-319; Lon. *Eng.*, April 13, 1888. Abstracted. *Sci. Am. Sup.*, May 19, 1888.

Conveyance of. A paper before the Engrs. Soc. of West. Penn. Treats especially of Natural Gas, and the means of preventing leakage. Illustrated. *Am. Eng.*, April 3, 1885.

Cost of. See *Electric Lighting, Lighting.*

Efficiency and Energy of. See *Electricity and Gas.*

First Gas Works. Brief historical note. *Eng. News*, Aug. 19, 1882, p. 188.

Flow of, Through Tubes. By S. W. Robinson. Gives the general and practical problems. *Van Nos. Eng. Mag.*, Vol. XXIV., p. 370.

For Heating and Motive Power. By J. E. Dowson. Treats of the production of water gas on a small scale and gives estimates of cost of manufacture. *Van Nos. Eng. Mag.*, Vol. XXVI., p. 319.

Illuminating Power of Gas. Instruments for testing. Not a photometer. Full illustrations. Lon. *Engineer*, Sept. 19, 1884.

Illuminations. By Dr. W. Wallace. A good article. *Van Nos. Eng. Mag.*, Vol. XX., p. 318.

Illumination and its Economy. By James Chessman. Shows the cost of making gas in various cities, and says there is no reason why gas should cost over $1 per 1,000 in large cities where the coal can be had for $5 per ton. *Van Nos. Eng. Mag.*

Pipe. See *Pipes, Gas.*

Natural. See *Gas Natural*, below.

Price of. Chart showing the net price of gas in seventy-nine cities of the United States from 1873 to 1887, inclusive. *Progressive Age*, April, 1887.

Price of Illuminating. Chart showing the net price of gas in 100 cities of the United States from 1876 to 1887, inclusive. *Progressive Age & Water Gas Journal*, March, 1887.

Substitute for Natural. Gives details of the Johnson process of manufacturing a fuel gas from crude oil, and the results obtained from burning the gas. *Am. Mfr.*, July 20, 1888.

See *Fuel Gas. Oil Gas, Compressed. Oil and Gas Fields. Petroleum.*

Gas Calorimeter. *Thwaites.* Described and illustrated. *Am. Mfr.*, Nov. 21, 1890, p. 17.

See *Calorimeters.*

Gas Coals *of the United States.* Abstract of a paper by H. C. Adams, read at a meeting of the Am. Gas Light Ass'n. *Am. Mfr.*, Dec. 19, 1890, p. 17, *et seq.*

13

Gas Engineering.

A lecture by Macpherson, giving a review of the last improvements in gas manufacturing. *Sci. Am. Sup.*, July 9, 1887.

And Modern Science. By Denny Lane before the London Gas Inst. Treats of the use of gas for motive power, light and heat; also gives a comparison between electricity and gas. *Sci. Am. Sup.*, July 3, 1886.

Gas Engines. See *Engines, Gas.*

Gas-Holders.

Description and detailed drawings of a three-lift gas-holder, having a capacity of 13,000,000 cubic feet. Lon. *Engineer*, May 28, 1886.

At Erdberg. An illustrated description of gas-holder, having a diameter over all of 216 feet. It has three lifts, each 40 feet in depth and 195, 197½, and 200 feet diameter. *Sci. Am. Sup.*, March 26, 1887.

On the failure of the columns of a large one in St. Louis. By J. B. Johnson. Giving causes of failure, with some analysis of the strength of the columns. *Jour. Assn. Eng. Soc.*, Vol. IV., p. 144.

Gasometer. The large gasometer of the Imp. Continental Gas Assoc. at Erdberg (with plates). *Zeitschr. d. Oesterr. Ing.-u. Arch. V.*, 1886, pp. 9-17.

Tanks, 240 feet in diameter, Birmingham, Eng. Plans, sections and description. Lon. *Engineer*, April 9, 1886.

Tanks. Experiences in the Construction of. Paper by Mr. G. A. Hyde, read in Cleveland, June 10, 1890. *Jour. Assn. Eng. Soc.*, July, 1890, pp. 311-3.

Three-Lift Gas-Holder at Glasgow Corporation Gas Works. An article describing this three-lift holder. Illustrated by inset plate of details. Some novel connections are used. Lon. *Eng.*, March 7, 1890, p. 296.

Without Upper Guide Frames. By T. Newbegging, before the Manchester District Institution of Gas Engineers. Describes a method of constructing gas-holders with inclined guides at the base, constructed in such a manner as to do away with a large part of the upper frames. Lon. *Engineer*, Sept. 14, 1888.

Gas Lamp. See *Siemens.*

Gas Light.

Light. The Welsbach Incandescent. Gives complete specifications, English patent (A. D. 1885, No. 15,286), for the manufacture of an illuminating appliance for gas and other burners. *Sci. Am. Sup.*, Aug. 13, 1887.

Gas Lighting. See *Lighting.*

Gas Mains.

The Detection of Leaks in. By Herr Schmidt. Description of apparatus used and result obtained in Breslau. *Journal f. Gasbel. u. Wasserv.*, 1886, pp. 714-728, 737-741.

Gas, Natural Gas.

A brief review of the history and production. *Eng. & Build. Rec.*, Dec. 6, 1890, pp. 4-5.

A good article, giving results of a ten days' visit to Pittsburg and vicinity. Gives history and shows details of the management of natural gas at that point. Fully illustrated. *Sci. Eng.*, Jan. 15, 1887.

By Charles Paine. A valuable article containing analysis of Pittsburg and Findlay, O., heating power, distribution, etc. Reprinted from *R. R. & Eng. Jour.*, Jan., 1887, in *Eng. News*, Jan. 29, 1887.

A Monograph. By Charles Paine. *R. R. & Eng. Jour.*, Jan., 1887.

Paper before the Iron and Steel Institute, Eng., by Mr. A. Carnegie. Gives methods of collection, transportation and uses. Lon. *Eng.*, July 10, 1885.

Report of a committee of the Engrs. Soc. of West. Penn., Pittsburgh, on its composition, utilization, illuminating and heating power, explosibility, etc., with discussion made May 21, 1884.

Gas, Natural Gas, continued.

Composition of. Report of Prof. Francis C. Phillips for the Geological Survey of Pennsylvania, from advance sheets of the Annual Report of the Geological Survey. *Jour. Frank. Inst.*, Oct. and Nov., 1887. Vol. CXXIV., Nos. 742 and 743.

Findlay, Ohio, gas field. Its history, development and utilization as a cheap and superior fuel; its present and future value to the manufacturing interest. A special five-page article prepared for the *Iron Trade Review* of Cleveland, April 23, 1887, by Hosea Paul, C. E. Free copies sent on application to Board of Trade, Findlay, O.

Geologic Distribution in the United States. Abstract of a paper read at the St. Louis meeting of the Am. Inst. Min. Engrs., October, 1886. By Charles A. Ashburner, Geologist Pennsylvania Survey, Philadelphia. In the *E. & M. Jour.*, of Jan. 15, 22 and 29, 1887.

Industry at Pittsburg, Pa. Gives details of the method of connecting wells to main, regulating pressure, distribution of gas, etc. *Sci. Am. Sup.* Jan. 7, 1888.

In Iron and Steel Works. By Theo. D. Morgan. A series of articles in the *Am. Mfr.*, and reprinted in the *Ry. Rev.*, May 8, *et seq.*

In Iron Making. A paper by James M. Swank, printed in "Mineral Resources of the United States," published by David T. Day, Chief of Bureau of statistics and Technology, U. S. Geological Survey. *Am. Mfr.*, Nov 30, 1888.

In Iron and Steel, Manufacture in the United States. From a paper by James M. Swank. *E. & M. Jour.*, Dec. 15, 1888.

Long Distance Transportation of Natural. Discussion on a paper read before the Eng. Soc. of West. Penn. *Am. Eng.*, May 27, June 10 and 17, 1891.

Measurement of with the Pitot Tube Gauge. A description of the use of this instrument in this connection with data as to accuracy. Illus. *Am. Mfr.*, Feb. 6, 1891, p. 120.

Measurement of Gas Wells and Other Gas Streams. By S. W. Robinson. Gives application of Pitot's tube, a shunt Bunsen effusion principle for density of shunted gas and a thermometer in an open tube. *Van Nos. Eng. Mag.*, Aug. 1886

Its Occurrence and Application at Pittsburgh. A lecture by Kurt Sorge before the North German Iron and Steel Association. *Eng. News*, Sept. 10 and 17, 1887.

And Petroleum in Ohio in 1889. Paper by Mr. Edward Orton specially prepared for *Am. Mfr.*, Jan. 24, 1890, p. 17.

Piping for. Rules issued by the Fire Marshal of Pittsburgh. *Eng. News*, May 8, 1886.

The Geology of. An attempt to explain the conditions under which it may be found. Results mostly negative. By Ch. A. Ashburner, *Trans. A. I. M. E.*, Halifax meeting, 1885. *Sci. Am. Sup.*, May 29, 1885.

Use at Pittsburgh. A description of its various uses for commercial and domestic purposes, with illustrations. *Sci. Am.*, Feb. 27, 1886.

Wells near Pittsburgh. *Am. Mfr.*, Nov. 12, 1886.

Wells. An oil or gas well rig. A good description of the ordinary apparatus for drilling for petroleum or natural gas. *Age of Steel*, April 14, 1885.

Gas Producers.

By F. J. Rowan. A highly valuable and instructive paper and discussion, of over 100 pages, on the production of gas for fuel in all applications of heat in the arts. This is what will doubtless ultimately be done. Many illustrations. *Proc. Inst. C. E.*, Vol. LXXXIV., p. 2.

A New Departure in. A new gas producer, whose most important feature is a cast-iron rammer operated by compressed air, which permits the use of a larg-

Gas Producers. *New Departure in*, continued.

er producer than could be worked by hand. Illustrated. *The Progressive Age*, March 15, 1884.

The Pittsburgh. Illustrated description of a new form of producer for making fuel gas. *Am. Mfr.*, Feb. 8, 1889.

Gas, Water Gas.

An article reprinted from *Industries*, giving analysis of the various forms of water gas. Describes the plant most generally used, chemical reactions, etc. *Am. Mfr.*, Feb. 10, 1888.

A paper by C. v. Marx, before the Eng. Soc. of Stuttgardt, treating of the yield and composition of water gas, of the loss of heat in its manufacture; also gives comparisons of the heating power of water gas, coal gas and generator gas. *Van Nos. Eng. Mag.*, Vol. XXVIII., p. 7.

By G. H. Christian, Jr., before the Ohio Gas Light Association. Gives the results of experiments to substitute Lima crude oil for naphtha in the manufacture of water gas. *Am. Mfr.*, April 20, 1893.

As Applied to Metallurgical Processes. By A. Wilson, before the Iron and Steel Institute. Describes the modification of Lowe and Strong's process developed at Essen, and its application to heating purposes and steel melting at Witkowitz, Terni, etc. *Am. Mfr.*, June 8, 1888.

Illuminating. A new method of obtaining the illuminating qualities without hydrocarbons, by inserting magnesia threads in the flame. Cost 80 cts per thousand cub. ft. *Proc. Inst. C. E.*, Vol. LXXXI., p. 385.

Its Manufacture and Utilization. Abstract of a paper by Joseph Von Lange. *Am. Mfr.*, July 4, 1890, p. 17.

For Metallurgical Purposes. By A. M. Wilson before the Iron and Steel Institute. Gives analyses of the various forms of water-gas; describes the plant most generally used for its manufacture, chemical reactions, etc. *Sci. Am. Sup.*, July 14, 1888.

Gaseous Illuminants. Cantor Lectures, by Prof. Vivian B. Lewes. *Jour. Soc. Arts*, Dec. 26, 1890, pp. 81-6, *et seq.*

Gaseous Mixtures. *On the Explosion of.* By Dugald Clerk, before the Inst. Civ. Engrs.; with discussion, occupying 53 pp. Experiments made with various proportions of two kinds of coal gas and hydrogen, separately combined with air. Important conclusions derived as to resultant pressures. *Inst. C. E.*, Vol. LXXXV., p. 1.

Gases. *Kinetic Theory of.* By H. F. Eddy. *Van. Nos Eng. Mag.*, Vol. XXVIII., p. 123.

Chimney. See *Chimney Gases.*

Flue Gases. See *Furnaces.*

Gasometer. See *Gas Holders.*

Gauge. *A Siphon Tide Gauge for Open Sea Coast.* See *Tide Gauge.*

Track. See *Railroad Track.*

Gauges. *Recording Pressure.* By Chas. A. Hague before the Minneapolis meeting of the American Water-Works Association. Discusses the uses and advantages of recording pressure gauges in water-works. *Proc. Seventh An. Meet. Am. W-W. Assn.*, pp. 24-31; *Am. Eng.*, Feb. 8 and 15, 1888.

See *Mercurial. Rain. Water.*

Gauging.

By Geo. H. Mann. Gives methods of making current observations. *Van Nos. Eng. Mag.*, Vol. VI., p. 532.

Of Floods. Rapid Approximate. By C. Ritter. Gives method based on results obtained from large number of experiments with current-meter. *Annales des P. & C.*, Vol. XII., p. 697; also *Eng. News*, Aug. 6, 1887.

Gauging, continued.

Of Streams. By Clemens Herschel. With especial reference to the units to be used and the manner of portraying the results. Plate. *Trans. Am. Soc. Civ. Engrs.*, Vol. VII., (1878), p. 236. By David M. Greene. A discussion before the eighth annual convention of the American Society of Civil Engineers. *Trans. A. S. C. E.*, Vol. V., p. 231.

Of Water. On some changes in the arrangements and methods of rating the instruments used, and in the mode of calculating the discharges. By M. Chas. Ritter. *Annales des P. & C.*, Jan., 1885, pp. 1038-1182. Quite an extensive article describing various instruments, some methods of rating them, and of calculating the discharge from measurements.

Gearing.

For Electric Street Cars. See *Electric Street Cars.*

Rapid Method of Laying Out. By S. W. Robinson. Describes a method of laying out by means of a curved ruler or templet odontograph. *Van Nos. Eng. Mag.*, Vol. XIX., p. 312.

The Walker System Explained. With many cuts. Contrasted with the Willis odontograph system. By John Walker. *Jour. Assn. Eng. Soc.*, Vol. IV., p. 41.

Gear Teeth. *A new process for Generating and Cutting the Teeth of Spur Wheels.* Series of revolving cutters move longitudinally, cutting at the same time generating the teeth. Paper by Ambrose Swasey, read before the Richmond meeting of the Am. Soc. M. E. *Ry. Rev.*, Nov. 22, 1890, pp. 700-1. *Eng. News*, Nov. 29, 1890, pp. 491-2. *Iron Age*, Nov. 27, 1890, pp. 935-6.

Pin Teeth for Bevel Gears. Description of a new form of pin teeth, for bevel gears, with suggestions as to method of forming them by machinery. By Geo. B. Grant. Illustrated. *Am. Mach.*, Dec. 5, 1889.

For Skew-Bevel Gears. An account of the geometrical construction and theory of involute teeth for skew-bevel wheels, by Geo. B. Grant. The author maintains that all other forms of teeth yet proposed for skew-bevel gears are incorrect. Illus. *Am. Mach.*, Oct. 10, 1889.

Geodesy.

Connection of the U. S. C. & G. Survey with the Lake Survey, in New York State. The checks given. *U. S. C. & G. Survey Rep.*, 1884, App. No. 9.

Determination of an Azimuth from micrometric observations of a close circumpolar star near Elongation, by means of a meridian or transit and equal altitude instrument, or by means of a Theodolite with eye-piece micrometer. By C. A. Schott, assistant. Bulletin No. 21., U. S. C. and G. Survey pp. 215-18.

Distribution of Triangulation Errors. See *Surveying.*

Literature of, for 1887. Complete list of papers and books published during the year in all the modern languages. *Zeitschrift f. Vermessungswesen*, 1888. Pages, 404-16, 425-49, 465-79, 497-510, 527-43.

Geodetic Field Work.

By Geo. Y. Wisner. A 34-page article by a practical observer on the most approved methods of determining longitude latitude and azimuth in the field, also for measurement of base lines and angles, and making the reconnaissance. *Trans. A. S. C. E.*, Vol. XII., (1883), p. 367.

A series of articles by T. T. Wright on the adjustment of condition. Observation by the method of least squares, with its application to geodetic work. *Van Nos. Eng. Mag.*, Vol. XXVIII., pp. 419 and 437.

Geodetic Night Signals. A report on an exhaustive series of experiments to determine best light to use, and advantages of the system. Report of U. S. C. & G. Survey, 1880.

Geodetic Survey *of Massachusetts.* An account and discussion of the trigonometric and other methods used in Borden's recent survey. By Prof. J. Howard Gore. *R. R. & Eng. Jour.*, April, 1887.

Geodetic Surveys *in Germany*. Lecture by Prof. Jordan, describing briefly the development of some of the German surveys. *Zeitschrift f. Vermessungswesen*, 1888, pp. 310-325.

Geographic Positions *west of the 100th Meridian*. The net results of Capt. Wheeler's surveys in this region, tabulated in a volume of 660 pp. Latitude and longitude of stations and azimuth and length of lines joining them, together with elevations above sea level, variations of the magnetic needle, etc. Issued by the Engineer Department of the U. S. Army, 1875.

Geographical Names. Bulletin No. 1, issued Dec. 31, 1890, contains decisions of the U. S. Board on Geographical names, on the spelling and pronunciation of over 800 names. Other bulletins to be issued. Published by the Smithsonian Institute.

Geography, Arctic. See *Arctic Geography.*

Geological Survey, the U. S. A paper giving a full account of its various objects, departments, work and publications, with map of U. S., by J. W. Powell, the Director. Read before the National Academy of Sciences, Oct., 1884. Published in *Am. Jour. of Sci.*, Feb., 1885.

Geology.

By Archibald Geikie. A series of articles giving a full treatment of the subject of rock formation. *Sci. Am. Sup.*, Aug. 11 *et seq.*, 1888.

Of North Carolina. By Ebenezer Emmons. Description of the geological formations used and of the mineral resources. Pp. 351. Illustrated by maps and sections. Issued by the State in 1856. (Out of print).

Of North Carolina. By Prof. W. C. Kerr. Pp. 385. Appendices issued in 1875.

Of the Pittsburgh Coal Region. By J. P. Leslie, State Geologist, before the *Am. Soc. Min. Engrs.*, at the Pittsburg meeting, Feb. 1880. Discusses the coal, oil and gas products.

Report of the American Committee on the work of the *International Congress of Geologists.* Gives the reports of the Committee on the Uniformity of the Nomenclature, and of others, with a chart of the scale of colors adopted. Dr. Persifor Fraser, Sec., Philadelphia, Pa., p. 169.

Girders.

And Beams. By A. B. W. Kennedy. Second lecture at Carpenters' Hall, London Wall. A scientific and practical view of the subject. *Sci. Am. Sup.*, Apr. 30, 1887.

And Roofs. *Weight of Framed.* By J. H. W. Buck. Discusses the various formulae for obtaining the weight of structures. *Van Nos. Eng. Mag.*, Vol. XXVII, p. 511.

Plate and Box Girders. By Louis D. Berg. This paper is Ch. X of *Safe Building*. Three original tables and diagrams showing the strength of webs, angle bars and flanges with spans from 20 ft. to 80 ft. *Am. Arch.*, Aug 9, 1889, No. 763 (cont. from No. 743), pp. 82-5.

Plate, Calculation of. By A. Munster. Gives results of investigations as to the reliability of formulas in use for calculating the flange stresses in plate girders. Presents three new formulas. Gives table showing the moments of resistance as compared by different formulas. *Jour. Assn. Eng. Soc.*, February, 1886, pp. 31-48.

Plate, Depth of. By W. R. Brown. A description of the limiting value for the depth of plate girders. *Van Nos. Eng. Mag.*, Vol. VIII., p. 316.

Plate. *Long Plate Girders and Their Economy.* An illustrated article by G. H. Thomson, Bridge Engineer New York Cen. & H.R.R.R. *R. R. Gaz.*, May 23, 1890, p. 358.

Plate-Web. A paper before the Liverpool Engineers' Society, treating of the weight and strains in plate-web girders. *Van Nos. Eng. Mag.*, Vol XXVII., p. 49.

Girders, continued.

Steel Girders, Strength of. Abstract of a report on some tests of plate girders of steel of large dimensions, made by the Dutch in 1877. Found to be weaker and more unreliable than wrought iron. Abstract covers 12 pp. in *Proc. Inst. C. E.*, Vol. LXXXIV., p. 412; also *Van Nos. Eng. Mag.*, Oct., 1886.

With Solid Buckle Plate Floor. Illustrated description of method of carrying ballast over a plate girder bridge with solid floor, as constructed by New York Central & Hudson River Railroad. W. Katte, chief engineer; George H. Thomson, bridge engineer. *Eng. News*, Nov. 23, 1889, Vol. XXII., p. 485.

Glacier *The Muir Glacier in Alaska.* Abstract of some remarks by Prof. H. F. Reid before the C. E. Club of Cleveland, concerning the motion and peculiarities of this glacier. *Jour. Assn. Eng. Soc.*, Dec., 1890, pp. 581-4.

Glass.

Some Experiments on the Transverse Breaking Strain of Plate Glass. Paper by G. W. Plympton, giving the results of ten experiments. Gives deflections and loads. *Trans. A. S. C. E.*, Vol. XXV., Aug., 1891, pp. 223-26. Reprint, *Ry. Rev.*, Nov. 7, 1891, p. 722.

Mechanical Production of Glass Bottles. History, description of various moulds, and of the application of rotary motion to glass manufacture. By C. Chassetant, *Le Genie Civil*, Vol. XVIII, 1891, pp. 357-61.

Moulding and Manufacture of Supply Pipes. By L. Appert. Various applications of glass described, also Apperts process of moulding large sizes. Illus. *Societe d' Encouragement Bulletin*, 1891, pp. 114-21.

Tempering. By Fredk. Siemens, a glass manufacturer. Read before the Soc. of Arts of London. Discusses principles and methods of a new mode of making toughened or hardened glass. *Van Nos. Eng. Mag.*, August, 1885.

Glass Beams. See *Beams*.

Gold Mining. See *Mining Gold*.

Gold and Silver.

The Parting Process used at the U. S. Mint at Philadelphia. By T. Egleston. A thirty-page article, describing the processes in great detail. *Sch. Mines Quar.*, April, 1886.

Parting of, by Means of Iron, at Lautenthal. By Thos. Egleston. A minute description of the parting works and processes in the Hartz Mountains. *Sch. Mines Quart.*, March, 1885.

Goniograph. *Double Reflecting.* A kind of double sextant whereby two angles may be simultaneously measured, the images of two objects being made to coincide with a third object viewed directly. Designed expressly to solve the Three Point Problem in Surveying and observations at sea. Illustrated. Lon. *Engineer*. Feb. 12, 1886.

Government Buildings, *The Erection of.* A review of a pamphlet entitled "History, Organization, and Functions of the Office of Supervising Architect of the Treasury Department." *San. Eng.*, April 22, 1886.

Government Engineering. *A Rational Policy of Public Works.* By L. E. Cooley. A thirteen-page article by the President of the Civil Engineers' Committee on National Public Works. The growth of the present system, its deficiencies, the necessity for reorganization, with suggestions as to what should be done by Congressional legislation, and the duties of engineers in the premises, all given with great clearness and force. *Jour. Assn. Eng. Soc.*, Vol.V., p. 69; also issued as a separate pamplet by Ex. Brd. on Pub. Works. Wm. T. Blunt, Secretary, Cleveland.

See Public Works.

Government Scientific Work. Report on the co-ordination and systematizing of the various kinds of scientific and engineering work carried on by the general Government. By J. W. Powell, Director of the U. S. Geol. Survey. This report embodies a large part of the testimony taken before the Joint Commission

Government Scientific Work, continued.

of the Senate and House during the last session of Congress. Also, about 70 pp. of descriptions of the geographical, topographical and geological surveys of other countries. Apply to Maj. Powell, Washington. On the same subject *See* an 8-page synopsis in *Science*, April 17, 1885; also review above report. *Science*, April 24, 1885.

Government Work. See *Public Works.*

Governors. See *Engines, Steam, Governors Kinemeter.*

Grade Crossings *in Germany.* See *Railroad Grade Crossings.*

Grades in Cities. See *Municipal Engineering.*

Gradient, *Ruling.* See *Railroads. Construction. Railroad Grades.*

Gramaphone.

Etching the Human Voice. A paper read before the Franklin Institute, May 16, 1888, by Emile Berliner. Sketches the history and present status of the invention. *Jour. Frank. Inst.*, June, 1888.

An apparatus for making permanent records of sounds of any kind and reproducing same as often as desired. Sounds reproduced with fidelity and so loudly that several hundred persons can hear them at the same time. Description. *Elec. Rev.*, Dec. 7, 1888.

Granite. See *Stone. Granite.*

Graphic *Representation of Strains on an Engine Shaft.* See *Engines, Steam, Shaft.*

Graphical Statics.

And its Application to Construction. By M. Maurice Levy. The translations of this valuable work is begun in *Mechanics*, January, 1881, pp. 16-17.

Applied to Retaining Walls. See *Retaining Walls.*

The graphical method is illustrated by a number of force plans, for roof trusses and framed beams. An excellent paper for students. *Mechanics*, July, 1880, pp. 161-4.

Graphics. By Wm. Bell. Contains the application of a new method of investigation to the stresses in bridges, arches, continuous beams and curved structures. *Van Nos. Eng. Mag.*, Vol. VIII., p. 193.

See *Strain Diagrams. Strains.*

Graphophone, *The.* A paper read before the Society of Arts by Henry Edmunds. Very interesting illustrated description of the Graphophone. *Jour. Soc. Arts.* Dec. 7, 1888.

Gravity. Determinations of gravity at several stations in Pennsylvania. *Appendix* 19. *Coast Survey Report for 1883.*

See *Forces.*

Greene, Benjamin H. *In Memoriam.* Interesting account of the life and works of this engineer. *Jour. Assn. Eng. Soc.*, March 1881, pp. 105-13.

Green-Houses. *Construction of*, in England, Belgium and Holland. Fully illustrated. *Zeitschrift fur Bauwesen*, 1887, Pts. I. and II.

Grinding. See *Abrasive Processes.*

Grouting. See *Aqueduct, New Croton.*

Guard Rails. See *Bridge Guards.*

Gun Cotton. *Its History, Manufacture and Use.* See *Explosives, Gun Cotton.*

Gun Factories.

In France. An Abstract of Lt. Jacques' paper in Vol. X. of *Proc. of U. S. Naval Inst.*, Illustrated. Lon. *Eng.*, Nov. 21, 1884.

Steel Gun Factories in the U. S., being the report of the Board appointed to visit the factories of Europe and report on the feasibility of the establishing of such factories in this country. Report makes Vol. X., of the U. S. Naval Inst. Written by Lieut. W. H. Jaques. Valuable.

Gun Foundry Board. Report of 1884: embodying the results of observations in England, France and Russia, with recommendations to Congress for American manufacture. Issued as House Ex. Doc. No. 97, 48th Congress, 1st Session.

Guns.

Aluminum in. See *Aluminum Bronze.*

Considered as Thermo-Dynamic Machines. By Jas. A. Longridge. A highly mathematical investigation of the reactions in a gun when fired. *Proc. Inst. C. E.*, Vol. LXXX., p. 136.

Development of the Dynamite Gun. Article by B. C. Batcheller. Illus. *Eng. Mag.*, Sept., 1891, pp. 758-71.

Machine Guns. A comparison of the Hotchkiss and Nordenfelt systems for use in repelling a torpedo attack. Lon. *Eng.*, March 20, 1885.

Magazine Guns. Development of, for Army Use. Paper by Captain A. H. Russell, U. S. A. Fully Illus. *Proc. Soc. Arts, Boston.* 1890.

The Making of Big. Popular illustrated paper by Henry L. Nelson, Sup. to *Harper's Weekly*, Aug. 23, 1890, No. 1,757, pp. 665-8.

Manufacture of Heavy, for the U. S. Navy. Illustrated description of works and manner of forging guns, at Bethlehem. *Sci. Am.*, Feb. 28, 1891, pp. 127-31.

Navy Six-inch Breech-Loading Rifled. A detailed description of the method of manufacture. By Ensign T. S. Rogers, U. S. N. This account is interesting, as it is doubtless the beginning of extensive works of this character in America. The castings and forgings made at the Midvale Steel Works at Nicetown, Pa. Illustrated. *Proc. U. S. Naval Inst.*, Vol. XII., p. 77. *See* also a letter by Wm. Metcalf on *Cast Steel Guns, in E. & M. Jour.*, May 8, 1886.

Pneumatic Dynamite Gun. An exhaustive account of the new pneumatic dynamite gun, the different guns described and illustrated, etc. Lon. *Eng.*, April 15: *Sci. Am. Sup.*, May 14, 1887.

Pneumatic Dynamite Gun. An account of a few that have been built, and of experiments with them. The projectile is a dynamite cartridge. (Illustrated.) *Am. Eng.*, April 13, 1887.

Steel Castings for the Manufacture of. See *Steel Castings.*

Steel for Guns. Abstract of a paper by Col. Eardley Maitland before the Institution of Civil Engineers. *Eng. News*, April 16, 1886.

Steel Gun Metal, Etc. By G. W. Sumner. An interesting review of the relative merits of different materials for gun making; the Whitworth system; the advisability of adhering to the use of slow burning powder discussed. *Sci. Am. Sup.*, April 16, 1887.

Steel for Heavy Guns. Proceedings of the United States Naval Institute, Vol. XIII., No. 1. A paper by Edward Bates Dorsey, C. E., and a discussion by 33 prominent steel makers and experts. The discussion contains much valuable information concerning the relative values of mild steel and hard steel for ordnance, etc.

Report upon the Practice in Europe with the heavy Armstrong, Woolwich and Krupp Rifled Guns, by the Board of U. S. A. Engineers for Fortifications, 1883 A quarto of 90 pp. and 6 plates. *Prof. Papers Corps of Engrs., U. S. A.*, No. 25.

Wire Guns. On the tension of winding wire guns. By P. R. Alger, U. S. N. *Proc. U. S. Naval Inst.*, No. 32

The Zalinski Pneumatic Torpedo Gun. Extracts from a paper by Capt. E. L. Zalinski, U. S. A., read before the U. S. Naval Institute, December, 1887, on the Naval Uses of the Pneumatic Torpedo Gun. *R. R. & Eng. Jour.*, May, 1888.

See *Artillery Experiments.*

Gunpowder. See *Explosives.*

Gyroscope. *Analysis of Rotary Motion* as applied to the Gyroscope. By Maj. J. G. Barnard. Probably the most rational and satisfactory demonstration of this problem given in the English language. The descent of the rotating body allowed for, its axis moving in cycloidal curves. Illustrated. *Van Nos. Eng. Mag.*, Dec., 1886.

Hammers.

Gas. A practical, economical, and convenient forging hammer, operated by hand, the motive power being gas explosions. One cubic foot of gas gives 300 blows of 313 foot-pounds each. Fully Illustrated. From Lon. *Engineer*, in *Sci. Am. Sup.*, Nov. 6, 1886.

Steam. By C. Chomienne, Engineer. Translated from the French. The author has had many years experience in the management of extensive iron works. *R. R. & Eng. Jour.*, June *et seq.*, [illegible].

Harbor Improvement.

Algoa Bay. By Wm. Shields. Gives details of the scheme of improvement and its successful completion. *Proc. Inst. C. E.*, Vol. LXXXVIII., p. 359; also Lon. *Engineer*, April 29, 1887.

Antwerp. By M. Strukel. Description, with illustrations. *Zeitschr. d. Oester. Ing. u. Arch. V.*, 1886, pp. 151-161.

Antwerp. Fully illustrated description of these important works. The use of a movable coffer-dam in constructing the quay wall is described. *Eng. News*, Sept. [illegible], 1883, pp. 445-8.

Bayonne. Civil Engineering at the Paris Exhibition. Illustrated article. Lon. *Eng.*, Dec. 6, 1889, Vol. XLVIII., p. 646.

Breakwaters, Concrete. Abstract from an exhaustive report by L. Y. Schermerhorn, Asst. Engr. U. S. A., describing all the principal breakwaters consisting of a concrete superstructure on a random stone sub-structure. The composition of the concrete is given, also cost. *Eng. News*, Jan. 31, 1891, pp. 1013.

Breakwater at New Haven. Eng. An account of this structure, now building. Foundations composed of concrete, laid in sacks of 100 tons each. Concrete mixer shown. Lon. *Eng.*, July 3, 1885.

Breakwater and Quay Wall. West of India Portuguese Railway and Harbor Works. Illustrated description of method of constructing breakwater and quay wall and facing same with concrete blocks of large size. *Ind. Eng.*, Dec. 7, 1889, p. 4[illegible]2.

Bordeaux. Illustrated article, one of a series on Civil Engineering at Paris Exhibition. Lon. *Eng.*, Dec. 13, 1889, Vol. XLVIII., p. 681.

Bulkhead Wall, New York City. Founded on piles. Illustrated description by C. W. Raymond. *Eng. News*, June 12, 26, 1880, pp. [illegible]-5.

Calais. Description of recent improvements, including the quay walls and the methods used in obtaining foundations. Details of the floating basin and lock entrances. Illustrated. Lon. *Eng.*, May 17, 1889. Vol. XLVII., p. 549, *et seq.*

Charleston, S. C. By Gen. Gillmore. Gives project for the permanent improvement of the entrance to the harbor by means of low jetties. *Van. Nos. Eng. Mag.*, Vol. XIX., p. 193.

And Coast Works. A lecture before the Inst. Civ. Engrs., London, by Thos. Stevenson. Discusses the heights of waves on coasts and in harbors, and some typical forms of breakwaters and coast protection works. *Van Nos. Eng. Mag.*, Feb., 1886.

Colombo, Ceylon. *Inst. C. E.*, Vol. LXXXVII., p. 76.

Colombo, Ceylon. By John Kyle. A full description of the work, including preliminary work, rock work, break-water, etc., with plates. *Ind. Eng.*, March 19, and 26, 1887.

Concrete in. A review of a number of papers before the British Inst. of C. E., on the use of concrete in harbor work. Interesting details as to cost and methods of construction. *Sci. Am. Sup.*, Jan. 1, 1887.

Harbor Improvement, continued.

Copenhagen. Described by H. C. V. Moller, in *Trans. A. S. C. E.*, Vol. XIV., p. 212.

Floating Deflectors. Equilibrium and stability of the floating current deflectors designed for harbor and river improvements, by Prof. L. M. Haupt. *Proc. Eng. Club, Phila.*, Dec., 1886, Vol. VI., pp. 69-75.

Galveston, Texas, Entrance to. A description of the works in their early stages by Chas. W. Howell. Illustrated. *Trans. A. S. C. E.*, Vol. VI., (1877), p. 223-230

Havre. By M. de Coene. Discusses the region of the estuary, and the projects for the improvement of the port. *Mem. de la Soc. des Ing. Civils*, April, 1886, pp. 343 *et seq.* Discussion in the May number, pp. 493-514, and in the June number, pp. 591-604. A further article by M. Le Brun, pp. 628-648. These articles form a very interesting discussion of the subject.

Haupt, L. M., C. E. *Jour. Frank. Inst.*, July, 1889, Vol. CXXVIII., p. 13.

Hell Gate.

An account of the Government work done and doing there, with diagrams. *Eng. News*, Sept. 12, 1885. See also *Flood Rock.*

Improvement of. An historical and technical account of the improvements, including Flood Rock, by Gen. John Newton, Chief Engineer of the work. Illus. *Pop. Sci. Monthly*, Feb., 1886.

Karachi, India. Gives memorandum of works in progress as proposed at an early date for the improvement of the harbor of Karachi, India, with map. *Ind. Eng.*, Feb. 4 *et seq.*, 1888.

Lisbon, Projected. Description of the proposed improvement of Lisbon harbor adopted by the Portuguese government, involving the construction of 37,026 ft. of quay and basins of 120 acres in area. Estimated cost, $12,000,000. Lon. *Eng.*, March 4, 1887.

Lowestoft. By A. A. Langley. Gives details of the work undertaken by the Great Eastern Railway Company for extending their harbor at Lowestoft. *Proc. Inst. C. E.*, Vol. LXXVII., p. 134.

Madras. Historical sketch of its construction, destruction and reconstruction. *Ind. Eng.*, March 7, 12, etc., 1887.

Montreal. Brief description with map of proposed improvements. These will include a guard pier 7,000 ft. long for protecting the harbor from ice, several freight piers, and a masonry wall one mile long to prevent inundation. *Eng. News*, May 2, 1891, p. 414.

Montreal, Canada, and Traffic of. Report of the Harbor Commissioners contains data as to cost of dredging, various material, amount of traffic, etc. Abstract, *Eng. News*, April 25, 1891, pp. 377-8.

Nantucket. A discussion of the work by Prof. L. M. Haupt. *Eng. News*, April 16, 1887.

Newhaven, Sussex. By A. E. Carey. Gives history of the early work at the harbor, and then gives details of the recent completed project. *Proc. Inst. C. E.*, Vol. LXXXVII., p. 92

New York. Gives a brief description of the centrifugal pumps in use on the excavator in New York harbor. *Sci. Am. Sup.*, Aug. 25, 1888.

New York, Water Front. By J. D. Van Buren, Jr. Gives characteristics of the harbor, physical features of the island, and the systems adopted for improving the water front, etc. *Trans. A. S. C. E.*, Vol. III., pp. 172-189.

Principles of, as applied at Wilmington, Cal. By Clinton B. Sears. Illustrated. *Trans. A. S. C. E.*, Vol. V. (1876), p. 358-426.

Quebec. By J. Vincent Browne. Includes outer tidal harbor, nearly inclosed and inclosed inner "wet dock." Fully illustrated by many plates. *Trans. A. S. C. E.*, Vol. IX. (1880), p. 415-460.

Harbor Improvement, continued.

Quebec. Valuable paper by St. George Boswell, M. Can. Soc. C. E. The improvements consist of a wet dock and tidal harbor at the mouth of River St. George, and a graving dock at St. Joseph on the Levis side of River St. Lawrence. *Trans. Can. S. C. E.*, Vol. I., p. 77. 1877.

Restoring the Greytown Harbor. A brief description of this work now being done. *Eng. Record*, Jan. 17, 1891, p. 107.

Rock Removal at the Entrance of Eagle Harbor, Lake Superior. Dynamite was used, holes being drilled from the surface of the water. Cost given. By L. W. Schermerhorn. *Eng. News*, March 10, 1877.

Saltburn, By-the-Sea. This article, besides describing general character of the works, gives considerable valuable data concerning the slag cement used. Chemical and physical properties are given. Lon. *Eng.*, March 14, 1890, p. 317.

South Boston Flats. By Ed. S. Philbrick. A very good description of dock wall construction and the raising of flats subject to flooding at high water. Illustrated by six large plates. *Trans. A. S. C. E.*, Vol. VII., 1878, p. 17-52.

Studies. Paper of some value by Prof. L. M. Haupt, discussing various problems relating to harbor improvements, with especial reference to the physical phenomena attending the formation of bars at the entrance to harbors in alluvial formations. *Proc. Eng. Club, Phila.*, April, 1886, Vol. V., pp. 285-327.

Wicklow. By G. W. Strype. Gives an account of the improvement of the harbor by the construction of a sea-wall of concrete in an exposed position. *Proc. Inst. C. E.*, Vol. LX.

See *Concrete in*. *Docks*.

Harbors.

Antwerp. By M. Maurice Widmer. Description of harbor, docks and cranes with a few commercial statistics. *Annales des P. & C.*, 1885, pp. 1179-1200

Artificial. By MM. Joubert and Fleury. An account of the making of a harbor in the Island of Reunion. An entirely artificial harbor on a stormy coast. Its basins and channel excavated and channel entrance protected by two jetties of concrete. Length of each jetty, 183 metres. Concrete blocks used weighed from 90 to 127 tons. The writers give many details as to composition of the blocks and machinery for handling them. Twenty-three pp. and 3 plates. *Memoires de la Soc. des Ing. Civ.* Paris, Nov., 1884.

Capacity of. Rules for determining, by Maj. C. W. Raymond. *Rep. Chf. Engrs. U. S. A.*, 1884, p. 579.

Cleveland. Facilities for Handling Coal and Ore. Paper by Augustus Mordecai, read before the Civil Engineers' Club of Cleveland. *Jour. Assn. Eng Soc.*, Feb., 1890, Vol. IX., pp. 50-58.

Cleveland, Ohio, Outside Harbor. See *Railroads, Cleveland Loop Line*.

Construction, adapted to such harbors as are found on our northern lakes. A series of illustrated articles in the *Engineering Era* (Cleveland, O.), Oct. 22, 1859, *et seq.*

The Fishing Boat, of Fraserburgh, Sandhaven and Portsay, on the Northeast Coast of Scotland. By John Willet. Gives details of the construction of the above harbors for the herring fishing fleets during the past few years. *Proc. Inst. C. E.*, Vol. LXXXVII., p. 123.

London. Description, with illustrations. *Deutsche Bauzeitung*, 1885, pp. 569-574 and 593-594.

Natural and Artificial. A paper by F. H. Cheesewright, describing many natural and artificial harbors, with the accompanying engineering works of the latter. Illustrated. *Jour. Soc. Arts*, March 20, 1891, pp. 345-60.

Physical Phenomena of Entrances to. An abstract from a lecture by Prof. L. M. Haupt before the American Philosophical Society. *Eng. News*, Feb. 25, 1888.

Harbors, continued.

Of Refuge, at Nantucket, Mass. Abstracted from report of Lt. Col. G. H. Elliot, U. S. Engrs. *Eng. News*, Jan. 8, 1887.

Of St. Petersburg and the new maritime canal. By M. Petsche. *Annales des P. & C.*, Oct., 1885, pp. 654-693.

Harbors and Docks.

Abstract of valuable papers describing these works at Antwerp, Calais, Havre and other ports, are published in *Proc. Inst. C. E.*, Vol. CI., 1890, pp. 330-41.

The German Ports of the Baltic. By Baron Suinette de Rochemont. A description of the ports and of the rivers upon which many of them are situated. 11 plates. *Annales des P. & C.*, Apr., 1891, pp. 617-734. Also by the same, *The German Ports of the North Sea*, *ibid.*, May, 1891, pp. 745-863. 12 plates.

Harbors and Waterways, *National Bureau of*. Gives text of bill recently introduced into the Senate by Senator Cullom, Ill., to establish a National Bureau of Harbor and Waterways. *Eng. & Build. Rec.*, Jan. 28, 1888.

Headlights. See *Locomotives, Headlights*.

Health.

In Cities, Condition of. A contribution to the Report of the State Board of Health of Wisconsin for 1886. Reprinted in *Building*, July 23 and 30, 1887.

Climate in Relation to. By G. V. Poore. Three lectures before the Society of Arts, London. Treats of the composition of the air, its moisture and the effect on the health; floating matter in the air, and the diseases caused thereby. *Sci. Am. Sup.*, May 7, 14, 22, 28, 1887.

Heat.

Conversion into Useful Work. A series of lectures before the Soc. of Arts, London, by Wm. Anderson, M. I. C. E. Theoretical and practical applications made to furnaces and boilers. *Sci. Am. Sup.*, July 18, 25, Aug. 1, 8, 1885.

Fourier's Law of Diffusion. By Sir Wm. Thomson before the Bath meeting of the British Association. Gives five applications of Fourier's law of diffusion, illustrated by a diagram of curves with absolute numerical values. *T. J. & Elec. Rev.*, Sept. 28, 1888.

Instruments for Measuring Radiant Heat. A course of Cantor lectures before the Society of Arts in 1889, by C. V. Boys. *Jour. Soc. Arts*, Sept. 20, 1889, Vol. XXXVII., p. 815, *et seq.*

Mechanical Applications of. Six Lectures on the. Before the Inst. Civ. Engrs., London. The titles are: The General Theory of Thermo-dynamics, by Prof. Osborne Reynolds; on the Generation of Steam and the Thermo-dynamic Problems involved, by Wm. Anderson; The Steam Engine, by E. A. Cowper; Gas and Caloric Engines, by Prof. Fleeming Jenkins; Compressed Air and other Refrigerating Machinery, by Alex. Carnegie Kirk; Heat-Action of Explosives, by Capt. Andrew Noble. These lectures reprinted in one vol. of 236 pp., 44 cuts, and 4 plates. A most valuable treatise for engineers. Published by the *Inst. C. E.*

Mechanical Equivalent of.

By E. A. Cowper and W. Anderson, before the Manchester Meeting of the Association for the Advancement of Science. Gives details and results of experiments made on a large scale. The results differ considerably from those of Prof. Joule. *Am. Eng.*, Oct. 5, 1887.

By E. A. Cowper and W. Anderson. A paper before the British Association, giving accounts of experiments on a large scale, apparatus and results attained. *Sci. Am. Sup.*, Oct. 15, 1887.

Note by Prof. R. H. Thurston, relating to Rumford's determination of the mechanical equivalent of heat. Wants more credit for Count Rumford than is usually given him. *Trans. A. S. C. E.*, Vol. II., p. 289.

Non-Conductors of. A paper by Prof. J. M. Ordway, describing the peculiar features of non-conducting materials. A table is given showing the number of

Heat, *Non-Conductors*, continued.

heat units transmitted per sq. foot per hour through various substances, with a thickness of 1 inch. *Pop. Sci. Mon.*, March, 1891, pp. 641-50.

Non-Conductors of. See *Non-Conductors.*

See *Radiation.*

Heat and Cold. *Effects of on Iron, Steel and Copper.* Experiments to determine effects of alternate heat and cold on these metals. Abstracted in *Proc. Inst. C. E.*, Vol. LXXX., p. 354. *Iron Age*, Sept. 17, 1885.

Heat and Power. *Prall System of Distribution of.* By E. D. Meier before the Engineers' Club of St. Louis. Gives details of the Prall system of distributing heat and power from a central station as carried out in Boston. *Jour. Assn. Eng. Soc.*, August, 1888, Vol. VII., pp. 301-313. *Sci. Am. Sup.*, Aug. 25, 1888.

See also *Heating Plants.*

Heat and Steam. *Notes on.* By R. H. Buel. A series of articles for practical men giving a collection of the most prominent data, with tables founded on the same. *Am. Eng.*, May 2, *et seq.*, 1888.

Heat Expenditure *in Steam Engines.* See *Steam Engines.*

Heating.

By Exhaust Steam. Article by R. C. Carpenter, Lansing, Mich., including tables of data, sizes of pipes required, etc. *Am. Eng.*, Jan. 22, 1890, p. 39.

By Flame Contact. By Thos. Fletcher, before the London Gas Institute. Describes a new boiler, by means of which flame contact can be secured, and claims an efficiency about four times that ordinarily obtained. *Lon. Engineer*, July 2, 1886, and *Sci. Am. Sup.*, July 17, 1886.

Comparative Tests of a Hot Water and Steam Heating Plant. Paper presented by R. C. Carpenter, of Lansing, Mich., giving results of a test made during winter of 1889-90, showing considerable economy in the hot water system over the steam system. *Eng. & Build. Rec.*, May 17, 1890, p. 380. *Trans. A. S. M. E.*, Vol. XI., 1890, pp. 918-39.

Hot Water Heating in the New Brooklyn Polytechnic Institute, Brooklyn, N. Y. Description of the plant, with illustrations. *Eng. Record*, March 21, 1891, pp. 262-3, *et seq.*

Hot Water Heating, and the Relation it Bears to the Master Plumbers' Trade. Lecture delivered by W. W. Mackay, M.E., Supt. Hot-Water Heating Department of Richardson & Boynton Co., N.Y., at an open meeting of N.Y. Assn. of Master Plumbers. The lecture contains many and valuable suggestions concerning details and general arrangement of hot-water heating systems. *Eng. & Build. Rec.*, Nov. 30, 1889, Vol. XX., p. 356, *et seq.*

Hot Water. Town Heating. Reprint of a paper by Arthur V. Abbott, C. E., read before Am. Inst. Mining Engineers. *Lon. Eng.*, August 30, 1889, Vol. XLVIII., p. 259.

Method of Proportioning Radiating Surfaces. A discussion of the methods and formulas employed. By R. C. Carpenter. *Eng. & Build. Rec.*, Nov. 8, 1890, pp. 417, Dec. 13, 1890, pp. 33-4.

Plant, Boston. Illustrated description of the plant of the Boston Heating Company. *Eng. News*, Nov. 12, 1887.

Plant, Boston Heating Co. By A. P. Abbott, before the Boston Society of Civil Engineers. Gives full description of the plant, the method of construction adopted in the streets, and details of fixtures. *Eng. & Build. Rec.*, May 5, *et seq.*

Steam. A paper by Wm. E. Worthen, giving a brief history of the heating of the cotton mills at Lowell. Tests of plant given. *Trans. A. S. C. E.*, Vol. XXIV., March, 1891, pp. 201-16. Discussion, pp. 216-22.

By Chas. E. Jones. Describes the plant in use at Washington University and shows the work it is doing; also gives experience with underground pipes

Heating, *Steam, etc.*, continued.

and smokeless furnaces. *Jour. Ass. Eng. Soc.*, January, 1888, pp. 14-22; *Eng. & Build. Rec.*, Feb. 18, 1888. Abstract *Proc. Inst C. E.*, Vol. XCII., pp. 461-2.

Steam, in Cities. By Chas. E. Emery. Gives a good exposition of the methods used by the New York Steam Company, with some valuable data. *Jour. Frank. Inst.*, March, 1888. *Sci. Am. Sup.*, April 7, 1888.

Surfaces of Radiators. An investigation into comparative value of various forms and arrangements of such surfaces. A preliminary report to Am. Soc. Mech. Engrs., by Wm. J. Baldwin. *San. Engr.*, Nov. 26. 1885.

Water by Flame Contact. By Thomas Fletcher. It is here shown that it is poor economy to bring flame directly in contact with boilers or boiler flues, since the temperature of these is so low. Better to conduct the heat to the water from surfaces heated to incandescence. *Van Nos. Eng. Mag.*, Aug., 1886. Also *Am. Eng.*, Aug. 4, 1886.

See *Heating and Ventilation.*

Heating and Cooling, *Effects of, on Iron, Steel and Copper.* See *Iron.*

Heating and Ventilation.

Notes by S. H. Woodbridge, including—Profit and Loss in School-house Ventilation, Ventilation of Machine-Rooms of Paper Mills, a Device for Increasing the efficiency of Large Shafts Exposed to Wind Action, and Hints on Inspection. *Tech. Quart.* Vol. I., 1888, No. 4, pp. 376-92.

Compiled from the report of Albion C. Stimers on the ventilation of the Hall of the U. S. House of Representatives. *Van. Nos. Eng. Mag.*, Vol. I., p. 313.

College of Physicians and Surgeons, New York. An article giving in considerable detail the methods employed in this building; illustrated. *Eng. & Build. Rec.*, April 5, 1890, Vol. XXI., p. 283, *et seq.*

Dwelling Houses. By F. E. Kidder. Discusses the heating and ventilation of low and medium cost houses by the different methods, and gives estimates of cost. *Building*, Feb. 5, 1887.

Factory Buildings. An article by Walter B. Snow describing methods of heating and ventilating factories. Illus. *Eng. Mag.*, July, 1891, pp. 462-74.

Homes. A popular article by Leicester Allen, M. E., giving considerable practical advice. (N. Y.) *Eng.*, April 1891, Vol. I., No. 1, pp. 57-64.

Imperial Houses of Parliament at Berlin. Prize essay, describing a method of obtaining the air, purifying it, tempering it, moistening it, etc., accompanied by a basement plan, ground plan and a section of the building. *San. Eng.* Dec. 31, 1885.

John Hopkins' Hospital, Baltimore, Md. Arrangement of systems, their operation and results, given. *Eng. Record*, July 18, 1891, pp. 111-12, *et seq.*

Large Buildings. A paper read before the Rensselaer Society of Engineers by Richard Prescott, M. E. Describes the method of heating by warm air distributed by a blowing engine. *Selected papers of the Rensselaer Soc. of Eng.*, *Troy, N. Y.*, Vol. I., No. 5.

Lenox Lyceum, New York. Interesting illustration of the use of blowers in regulating the heat and air supply in Lenox Lyceum. Illus. *Eng. & Build. Rec.* Feb. 1, 1890, Vol. XXI., p. 140.

Mass. Institute of Technology.

Gives the results of four years experience with the indirect method of steam heating and ventilating at the Massachusetts Institute of Technology. *Eng. News*, Feb. 25 *et seq.*, 1888.

Some practical results observed at the Mass. Inst. of Tech., by S. H. Woodbridge. A series of article in the *San. Eng.*, beginning Sept. 17, 1885.

A Critical Study of the Heating and Ventilating of the New Building of the Massachusetts Institute of Technology. Several special features involved.

Heating and Ventilation, *Mass. Institute*, continued.

Descriptions, both results attained. By S. H. Woodbridge, *Tech. Quart.*, Vol. II., 1889, No. 1, pp. 76-88.

New York Music Hall. Illustrated description of the system. *Eng. Record.* July 4, 1891, pp. 74-7.

Orpheum in Vienna. Description, with plans and sections. *Wochenschr. d. Oesterr. Ing. u. Arch. V.*, 1886, pp. 44-45.

State War and Naval Department Building, Washington, D. C. An illustrated article on the heating and ventilation of the, by means of hot water. The building is 500x275 feet. The method is direct-indirect radiation in the basement and first floor, and indirect radiation for the remainder of the building. *San. Eng.*, Jan. 22, 1887.

School Buildings. Paper "How shall We Warm, Ventilate and Closet our School Buildings?" By N. B. Wood, Civil Engineers' Club of Cleveland. Read June 11, 1889. With discussion. Pp. 8. *Jour. Assn. Eng. Soc.*, Aug., 1889, Vol. VIII., p. 487.

School Houses. Very valuable article composed of papers by Messrs. Wm. E. McClintock, Theo. P. Perkins and Prof. S. H. Woodbridge. *Jour. Assn. Eng. Soc.*, April, 1890, Vol. IX., pp. 147-194.

School Rooms, as practiced in France. Summary of the practice adopted in accordance with plans proposed by Society of Public Medicine and Professional Hygiene. *San. News*, Chicago, Vol. IV., p. 131.

Schools. Short account of the experiments of Prof. Reischel on the Berlin schools, with results of analysis of air and some conclusions regarding quantity of air supply, etc. *Deutsche Bauzeitung*, 1886, p. 123.

Town Heating by Hot Water. See *Heating, Hot Water.*

U. S. Senate Chamber, Washington, D. C. Illustrated description of steam plants, fans, etc. *Eng. Record*, Oct. 3, 1891, pp. 287-9.

Warehouse Building. By Henry J. Snell, before the Philadelphia meeting of the American Society of Mechanical Engineers. Describes a method practiced by the author for heating an office and warehouse building in Philadelphia. *Trans. A. S. M. E.*, Vol. IX., 1888, pp. 99-107.

Warm Air. By W. D. Snow. Discusses the uses of a forced current of warm air, and advocates the use of this method, with exhaust steam for shop warming. *Mast. Mech.*, May, 1888.

Workshops. By John Walker. Gives details of the system of hot air heating applied to some shops in Cleveland. Illus. *Jour. Assn. Eng. Soc.*, Vol. VII., pp. 1-3 (Jan., 1888.)

See *Ventilation.*

Heating of Trains. See *Car Heating.*

Heliotropes. See *Optical Telegraphy.*

Hell Gate. See *Blasting, Operation. Harbor Improvements, Hell Gate.*

Highway Engineering. See *Road.*

Highways. See *Roads, also Pavements.*

Hoisting Machinery.

Brief description with general view, plan and elevation of the hoisting machinery used in the erection of the tower of the new public buildings in Philadelphia, Pa. *Eng. News*, July 9, 1887.

Automatic Stock Hoists for Blast Furnaces. This hoist is built on an incline of 38 deg. with the horizontal, the cars being pushed up by means of pusher cars attached to a cable. Its working is very satisfactory. Illustrated description. *Eng. News*, Feb. 28, 1891, pp. 208-9.

The Calumet and Hecla Mining Co's Plant. Description of this elegant plant, with fine engravings of the 90-inch Belpaire boilers—the largest of the type ever built. *R. R. Gaz.*, March 21, 1890, pp. 190-1.

Hoisting Machinery, continued.

Hydraulic Hoisting Plant for Pier of the Brooklyn Sugar Refining Co. Paper by Louis G. Engel, read at the Richmond meeting of the Am. Soc. M. E. *Iron Age*, Nov. 20, 1890, pp. 894-9.

See *Cables*. *Contractors' Plant*. *Pneumatic Hoisting*.

Hoists, *Operated by Compressed Air*, adapted to warehouse purposes. A description of such an installation in Liverpool. *Trans. Liverpool Eng. Soc.*, Vol. VI., pp. 92-99.

See *Sheerlegs*.

Holley, Alexander Lyman.

Memorial addresses and contributions, covering 40 pp., the principal tribute being by James C. Bayles, in *Trans. A. S. M. E.*, Vol. III., p. 29. *Also* address of 40 pp. by R. W. Raymond, Vol. IV., p. 35.

Inaugural address of the Holley Memorial Statue, by James Dredge, editor of *Engineering*. Full abstract of this interesting address in *Eng. News*, Oct. 4, 1890, pp. 305-7; Oct. 11, 1890, pp. 315-16. Extract in *Eng. & Build. Rec.*, Oct. 11, 1890, p. 294. Lon. *Eng.*, Oct. 10, 1890, pp. 433-4, *et seq.*

Hospital *Ship "Castalia."* A new floating hospital for small-pox patients on the Thames. Lon. *Eng.*; *San. Eng.*, April 29, 1885.

Hospital Wards.

Circular. Abstract of the English controversy on the subject. *San. Eng.*, Nov. 12, 1885.

Hot Blast, *Theory of.* By Lowthian Bell. *Van Nos. Eng. Mag.*, Vol. VI., p. 545 and 631.

House Drainage.

By Gen. T. C. Cotton. Tries to give such a *résumé* of what has been determined in sanitary science as will be a useful and safe guide to those interested in pure air. *Van Nos. Eng. Mag.*, Vol. XVI., p. 113.

By George S. Pierson before the Michigan Engineers' Society and published in *Michigan Engineers' Annual* in 1889. *Eng. & Build. Rec.*, March 8, 1890, Vol. XXI., p. 219, *et seq.*

By T. M. Reade, before the Liverpool Arch. Soc., telling how to drain a house, and gives details of the drainage of a villa. *Van. Nos. Eng. Mag.*, Vol. XVI., p. 253.

Architect's Point of View. A paper by F. G. Corser, in the *Northwestern Architect*, giving much practical information. Reprint, *Eng. News*, Sept. 19, 1891, pp. 258-9.

Double Check System of. A paper read by Henry Masters before the Conference of the Society of Arts on Health Sewage of Towns. Describes methods employed by the author. *Van Nos. Eng. Mag.*, Vol. XXI., p. 471.

Durham System. Illustrated description by Paul Gerhard. *Eng. News*, Jan. 26, Feb. 2, 1891.

Four-inch Drains Versus Six-inch Drains. Paper by G. M. Lawford before the Congress of Hygiene and Demography. Author advocates a four-inch house drain. *Eng. News*, Sept. 26, 1891, p. 294.

Isolated Country House. Being plumbing specifications for such a house, valued at $20,000, in a populous district, supplied with water from corporation pipes, and discharging house-water into a flush-tank and sub-surface irrigation system, or into a street sewer. Illustrated. *San. Eng.*, March 26, 1886.

And Plumbing. By Wm. P. Gerhard. A series of articles in *Building*, commencing in May, 1886.

Regulations of Haverhill, Mass. Gives text of the ordinance recently drawn up by the Board of Health to regulate the character of the plumbing work. *San. Eng.*, July 30, 1887.

House Drainage, continued.

Sanitary Plumbing. By Wm. P. Gerhard. A good article, in which is clearly given the objects to be accomplished, and describes the most approved mechanical appliances. *Van Nos. Eng. Mag.*, Vol. XXVII., pp. 265, 372 and 401.

Single Trap System of House Drainage. Abstract of a paper by Latham Anderson, read before the Am. Soc. C. E., advocating the use of a single trap to be placed between the sewer and the house. *Eng. Record*, July 4, 1891, pp. 79-80.

And Street Drainage of Philadelphia. A descriptive and critical paper by Rudolph Hering before the Engrs. Club of Phila. Reprint, *Eng. News*, March 21, 1878, pp. 91-3.

And Water Supply. By Robt. Vowser. Gives brief description of the drainage and water supply of a large detached mansion. *Van Nos. Eng. Mag.*, Vol. XXII., p. 366.

See *Plumbing*.

House Drains *and Traps*. By J. Honeyman. A paper before the Sanitary Institute of Great Britain. *Eng. News*, Oct. 29, 1887.

House Foundations, Healthy,

By Glenn Brown, Arch. A series of articles elaborately treated and illustrated in *San. Eng.*, New York, Vol. X., running through many numbers.

Houses, Dwelling. *Relations of Temperature to Health in*. By Dr. D. Benjamin. Treats of the proper conditions for ventilating and prevention of injurious draughts. *Sci. Am. Sup.*, March 19, 1887.

Howe Truss. *Determination of the Length of the Strut of a Howe Truss*. An exceedingly concise analytical and graphical solution of this famous problem by Albert M. D'Armit. *Eng. News*, Nov. 16, 1889.

Brace. A new method of finding its length. By W. W. Redfield. *Jour. Am. Eng. Soc.*, Vol. IV., p. 118.

Hydrants.

Paper by George A. Stacy, Superintendent of Water-Works, Marlboro, Mass. Contains many valuable suggestions bearing mainly on difficulty experienced in avoiding freezing in winter. Discussion, pp. 9. *Jour. N. E. W.-W. Asn.*, Sept., 1889. Vol. IV., p. 32.

Fire Hydrants in Winter. Several suggestions of value concerning protecting hydrants from frost and thawing out frozen ones, by Louis Lesage, Supt. Montreal Water-Works. *Eng. & Build. Rec.*, Jan. 25, 1890, p. 123.

See *Hydraulic Machinery*.

Hydraulic *Appliances at the Forth Bridge Works*. A short description in *Proc. Inst. C. E.*, Vol. XC., p. 400.

Architecture. Describes the failure of a number of docks, and shows the want of good formulas for constructing canals, etc. *Van. Nos. Eng. Mag.*, Jan., 1884.

Buffers. By Col. H. Clerk. Gives details of experiments made with hydraulic buffers to check recoil of guns, etc. *Van Nos. Eng. Mag.*, Vol. I., p. 984.

Buffers, Webb System of. Used at the open ends of tracks, and designed to stop cars without shocks. In use on the London & Northwestern Ry. Illustrated description. *Eng. News*, Jan. 24, 1891, p. 74.

Dock. See *Dock*.

Ejector, Used at Providence, R. I. Illustrated description of this ejector, which is used for pumping water from excavations, cellars, etc. *Eng. Record*, July 18, 1891, pp. 109-110.

Elevators.

Gravel. Illustrated description of a new application of the jet principle in mining. *Eng. News*, Jan. 1, 1887.

Hydraulic, *Elevators*, continued.

Objections to their Use by Water Works, and Advantages to the Public by their Use. A brief paper by B. F. Jones, followed by an exhaustive discussion of this question. *Report 11th An. Meet. of the Am. W.-W. Assn.*, 1891, pp. 89-104.

The Otis Lifts in the Eiffel Tower. Complete description of construction and operation, including the amount of work done. Work done in raising the four lifts from bottom to top, 906 feet, is nearly 1,000 H. P. Lon. *Engineer*, July 19, 1889.

See *Hydraulic Lifts, Hydraulic Machinery, Elevators.*

Hydraulic Forging. See *Forging.*

Gear for Electric Cars. The Wenstrom system. Illustrated. *Elec. Eng.*, Oct. 12, 1892.

Hydraulic Hoisting Plant. See *Hoisting Machinery.*

Laboratory. The New Hydraulic Laboratory of the Mass. Inst. Tech. Description of the various apparatus used, with illustrations. By Prof. George F. Swain. *Tech. Quart.*, Vol. III., 1890, No. 3, pp. 203-15.

Lifts.

Canal. A two-page plate showing elevation, cross-section and details, with short description of the La Louviere hydraulic lift on the Canal du Centre, Belgium. Lift, 50.5 ft.; length, 141 ft.; weight, 1 100 tons. Lon. *Engineer*, Feb. 24, 1888.

Car. Brief illustrated description of the hydraulic car lift in the St. Lazare Station in Paris. *R. R. Gaz.*, Nov. 18, 1887.

Mersey Tunnel. Designed for lifting 102 passengers at once 100 feet in elevation. Illus. By Wm. Edmund Rich, one of the engineers. *Proc. Inst. C. E.*, Vol. LXXXVI., p. 60; also *San. Eng.*, New York, Oct. 30, 1886. Lon. *Engineer*, May 21, 1886.

Neuffosse Canal. Brief description, with sections and elevation of the hydraulic lift at Fontinettes, on the Neuffosse Canal. *Eng. & Build. Rec.*, June 23, 1888.

See *Hydraulic Elevators. Hydraulic Machinery, Elevators, Canal Lifts, Boat Railway. Dock.*

Machinery.

A series of elementary articles on the design of such machinery. *Mech. World*, March 14, 1891, p. 107, *et seq.*

Compound Presses. Four styles of rapid working presses, the high pressure used only for the final movement. Illus. *Sci. Am. Sup.*, Oct. 9, 1886.

In Engineering Establishments. Some of the advantages of the use of. By Ernest W. Naylor, M. E. *Mast. Mech.*, March, 1889, *et seq.*

For Loading and Discharging Vessels. Description and illustrations of that in use in the port of Buenos Ayres. Lon. *Engineer*, March 6, 1885.

Ram. Tables of efficiencies for various heads and elevations of discharge. *Am. Eng.*, March 18, 1886. Designed to raise water on a large scale, and acting without shock, the valve being closed gradually by a special motor. Illus. Lon. *Eng.*, April 9, 1886.

Ram. Gives proportion for hydraulic rams that have been found to work well in practice. Condensed from *Am. Artisan*, in *Van Nos. Eng. Mag.*, Vol. L., p. 48.

Riveting. An illustration showing a very compact form of hydraulic riveting plant. Lon. *Engineer*, Feb. 27, 1885.

At St. Lazare Station, Paris. Described fully in the July, 1890, number of the *Revue Generale des Chemins de Fer.* Brief Extract in *R. R. Gaz.*, Nov. 28, 1890, p. 823.

Hydraulic Hoisting Plant, *Machinery*, continued.

See Forging Press, 4,000 tons capacity.

Mill Appliances. Abstract of a paper by Chas. Hyde, before the Engrs. Soc. Western Pa., describing various appliances used in casting, rolling, etc. *Iron Age*, March 5, 1891, pp. 410-411.

Power.

Distribution of. By E. B. Ellington, before the Institution of Civil Engineers. Gives details of the distribution of hydraulic power in London. Has 27 miles of mains at a pressure of 700 lbs. per sq. in. Lon. *Eng.*, April 27, 1888; Lon. *Engineer*, May 11, 1888; *Mech. World*, May 12, 1888; *Am. Eng.*, June 27, 1888.

London and Hull. Abstract of a paper, read before the Liverpool Engineering Society, on the "Recent Progress in the Public Supply of Hydraulic Power" in Hull and London, and as about to be introduced in Liverpool. It seems to be a great commercial success. *Eng. News*, Feb. 21, 1885. Also on the recent progress in the public supply of hydraulic power, with table showing its economy in certain instances. Read before Liverpool Engineering Society, by E. B. Ellington. Lon. *Eng.*, March 13, 1885, *et seq.*

London. A paper by Ed. B. Ellington, giving full description of plant, and operation of same, with plates and discussion. *Proc. Inst. C. E.*, Vol. XCIV., pp. 1-45.

Public Supply of. By Howard Constable. Gives the results of observations on such systems in London. *Proc. Eng. Club, Phila.*, Vol. V., p. [illegible]3.

Recent Progress in the Supply of. A paper by E. B. Ellington, giving description of the plant in London, and its work and efficiency. Illustrated. *Trans. Liverpool Eng. Soc.*, Vol. VI, pp. 11-34.

Propulsion. A review of a paper and discussion on the results of experiments before the Inst. Civ. Engrs. Illustrated. Lon. *Eng.*, July 10, 1885.

Propulsion; On a new System of. By J. H. Selwyn. Gives description of a system invented by Mr. Geo. Wilson. *Van Nos. Eng. Mag.*, Vol. XXVII., p. [illegible].

Hydraulic Step, *Analysis of an.* By Prof. F. G. Hesse, of University of California. Machinery of pumping station, containing vertical shaft, supporting heavy load and rotating in gritty water at high velocity. Description of step used with experimental data. Lon. *Eng.*, Jan. 11, 1889.

Hydraulic Tables.

To facilitate computations by M. Du Buat's formula expressed in English inches, giving results slightly less than Neville's tables. *Proc. Am. W. W. Assn.*, 188[illegible]. J. H. Decker, Secy., Hannibal, Mo.

Designed to facilitate computations of flow of water through pipes, based on formulas of D'Arcy and Kutter. By P. J. Flynn. From *Trans. Tech. Soc. Pac Coast*. Reprinted in *Van Nos. Eng. Mag.*, June, 1885.

Based on Kutter's Formula. By P. J. Flynn. A valuable set of tables to facilitate the calculation of velocities, discharges, slopes and dimensions of sewers, conduits, etc. *Van Nos. Eng. Mag.*, Vol. XXVIII., p. 135; also *Sci. Series*, No. 61.

For Pipes. Molesworth's tables, based on Kutter's formula, shown to be applicable to but one diameter, and value of *n*. Also a modified form of Kutter's formulas given for pipes under one foot in diameter, which is shown to be practically correct. By P. J. Flynn, in *Trans. Tech. Soc. Pac. Coast*, Vol. III., p. 16. Also in *Van Nos. Eng. Mag.*, April, 1886.

Hydraulics.

An article by J. Schlichting, treating of bottom-velocity and velocity-scales of rivers. *Van Nos. Eng. Mag.*, Vol. XVII., p. 375.

Hydraulics, continued.

By V. Dwelshauvers-Dery.. Treats of the resistance of cylinders and spheres. Gives formulas applicable to the calculation of hydraulic presses for tubbing shaft, dams, etc. *Van. Nos. Eng. Mag.*, Vol. XIV., p. 152.

As an Exact Science. By H. Hememan. Examines the present condition of hydraulics. *Van Nos. Eng. Mag.*, Vol. VI.

Calculations of the Mean Horse-Power of a Variable Stream and the Cost of Replacing the Power Lost by a Partial Diversion of the Flow. Paper read by Mr. Wm. H. Grant. Tables and folding plates illustrating the flow of the Bronx River. *Trans. A. S. C. E.*, Vol. XXII, June, 1890. paper No. 440, pp. 378-401; discussion pp. 402-8.

Considerations on the Relation of Bed to the Variables. By J. A. Seddon. Some important fundamental principles stated, and the results of a series of experiments given, tending to show the relation of cross-section to slope and velocity for constant discharge. For a given material and discharge, the velocity shown to be nearly constant; but the cross-section taking a diminishing hydraulic mean radius with increased slope. An important beginning in the right direction to a final understanding of even the most fundamental principle of this subject. Illustrated. *Jour. Assn. Eng. Soc.*, Vol. V., p. 127.

Double Float. By H. L. Abbot. A reply to the criticism of S. W. Robinson. *Van. Nos. Eng. Mag.*, Vol. XIII., p. 330.

Experiments with Large Aperture. One of the most valuable series of experiments ever made in orifices as large as two feet square. By Theo. G. Ellis. Illustrated. Received the Norman Medal. *Trans. A. S. C. E.*, Vol. V. (1876), pp. 19-100.

Experiments at Roorkee. By Capt. Allen Cunningham. Explains in detail the instruments and method of observations used for measuring velocity. *Van Nos. Eng. Mag.*, Vol. XIV., p. 542, and Vol. XXV., p. 209.

Fluid Motion. By Samuel McElroy. The object of the paper is to discuss the general principles of hydraulic motion, as furnishing the true guide to experiments and formulas. *Jour. Assn. Eng. Soc.*, Vol. VI., p. 424.

Elements of the Mathematical Theory of Fluid Motion. By Thomas Craig. Gives mathematical investigation of some cases of motion of incompressible, frictionless fluid. *Van Nos. Eng. Mag.*, Vol. XX., p. 113.

In Open and Closed Channels and Pipes, Calculations of. By Albert Frank. Discussion of cross-sections of various shapes, with tables giving best proportions; comparison of different formulas for flow, and graphical sheets, constructed logarithmically, for use in computations for all the principal formulas. *Journal fur Gasbeleuchtung und Wasserversorgung*, 1886, pp. 258-272, 290-305, 320-334.

Flow of Water.

Comparison of different Methods of Measuring. Presented to the Society of Arts of Mass. Institute of Technology, by Prof. George F. Swain. Gives results of some interesting experiments. *Proc. Soc. Arts, Mass. Inst. of Technology*, for 1886-1887.

By James Craig, M. I. C. E. This is an interesting and suggestive paper containing several new formulas, among others a slope velocity formula which lacks demonstration. Eight tables illustrate the use of the formulas. *Ind. Eng.*, May 24, 31, 1890, pp. 411-1, 424-5. See also an editorial in the same Journal June 7, pp. 441 2.

Estimation of, from the drainage areas, by R. Iszkowski. Comparison of methods with results of measurements. Extended tables of flow of European streams. A very valuable contribution to hydrology. *Zeitschr. d. Oester. Ing.-u. Arch V.*, 1886, pp. 69-98.

The new theories of the motion of flowing water. By Dr. G. Hagen. Being a criticism on Humphreys and Abbott's Physics and Hydraulics of the Mississippi. *Van Nos. Eng. Mag.*, Vol. XVII., p. 443.

Hydraulics. *Flow of Water*, continued.

In Conduits. Formula given for the flow of water in the new Sudbury River conduit, Boston. The most reliable formula known for a conduit about ten feet in width, and an hydraulic mean radius of about 1.6 feet. *Trans. A. S. C. E.*, Vol. XII. (1883), p. 114.

In Earthern Channels. A brief paper by Ralph Sadler presenting a diagram to facilitate the design of channels of a side slope of 1·5 to 1. *Proc. Inst. C. E.*, Vol. CIII., 1891, pp. 273-7.

In Mains, as Determined by Pressure Gauges. A paper by George A. Ellis, showing how the flow in pipes may be estimated from the loss of head. *Jour. New. Eng. W-W. Assn.*, Sept., 1886.

In Open Channels. On the cause of the maximum velocity of water flowing in open channels being below the surface. By Jas. B. Francis. *Trans. A. S. C. E.*, Vol. VII. (1878), pp. 109-113 and 119.

By T. G. Ellis. A review of the Roorkee experiments on the Ganges canal, containing a description of the experiments and some of the deductions. *Eng. News*, Nov. 26, 1881, pp. 478-9.

By Theo. G. Ellis. Discusses Kutter's formula. *Trans. A. S. C. E.*, Vol. VI. (1877), pp. 250-257.

By W. J. McAlpine. Gives a comparison of the different formulæ in general use, and derives a new one. *Van. Nos. Eng. Mag.*, Vol. VIII., p. 97.

Mean Velocity of Streams Flowing in Natural Channels. By Robt. E. McMath. A discussion of value, wherein a new departure is made in the treatment of the subject. Some new functions heretofore ignored, but which always exist in natural channels, introduced. The results not wholly negative, some permanent advance being made. Paper covers 25 pp., with 10 plates. A new formula well sustained by observations. *Trans. A. S. C. E.*, Vol. XI. (1882), p. 186.

Trans. from Sonnet's "Dictionnaire de Mathematiques Appliquees." *Van Nos. Eng. Mag.*, Vol VIII., p. 1.

Recent experiments in the flow of water in rivers and canals. By K. R. Bornemann. A review of the work of Gerbenau on the Rhine, Harlacher on the Elbe, and Gardau on the Irrawaddy. *Van. Nos. Eng. Mag.*, Vol. XVI., p. 521.

By De V. Wood. A paper presented at the annual convention of the Am. Soc. of C. E., 1879. *Van. Nos. Eng. Mag.*, Vol. XXI., p. 359.

Uniform Motion in Canals and Rivers. Abstract from a review of the motion of water. By M. Ganguillet. *Van. Nos. Eng. Mag.*, Vol. II., p. 213.

In Rivers and Canals. By A. Hill. Treats of average velocities, uniform motion in canals and unequal motion in rivers. *Van. Nos. Eng. Mag.*, Vol. III., p. 118.

In Streams, Observations on the Movement. By P. Graeve. Gives some observations made on the Oder and Warthe and compares result obtained with the meter with various formulas. *Van. Nos. Eng. Mag.*, Vol. XXV., p. 124.

In Rivers. By Prof de Volson Wood. A theoretical discussion. *Trans. A. S. C. E.*, Vol. VIII. (1879), p. 173.

Why the maximum velocity is below the surface. By F. P. Stearns. He holds that it is caused by the upward flow from the sides and the surface flow toward the centre of the stream. *Trans. A. S. C. E.*, Vol. XII., p. 331.

Over Weirs. An account of some experiments made in connection with the new Boston Water Supply. The most valuable results yet obtained, considered as supplementary to Francis' Experiments, on the effects of form and width of crest, conditions of approach, end contractions and height of water on down-stream side when submerged. Fully illustrated. Tables

Hydraulics, *Flow of Water, Over Weirs*, continued.

and plotted result given. By A. Fteley and F. P. Stearns. *Trans. A. S. C. E.*, Vol. XII. (1883), p. 1.

See *Hydraulics*.

Recent Experiments. By M. Bazin. These experiments consisted in gauging a standard weir, and comparative experiments upon weirs of different heights. The weirs were all sharp crested of the full width of the channel of approach and with lateral chambers to allow the air free access below the sheet of falling water. 153 experiments were made in gauging the standard weir for heads ranging from 2 inches to 2 feet; then this weir was used to determine the discharge over others placed in the channel lower down. Seven other series of about 30 experiments each were thus made to determine the variation of the co-efficient, due to velocity of approach and for various heads. The values of the coefficient *m* in the formula $Q = m l h \sqrt{2gh}$ are tabulated, and also curves drawn for the same. These results are very reliable, as experiments were conducted with great care, and while they agree closely with those of Messrs. Fteley and Stearns, the experiments were much more numerous and on a somewhat larger scale. *Annales des Ponts et Chaussees*, Oct., 1888. Reprinted in the *Proc. Eng. Club, Phila.* Vol. VII., Jan. 1890, pp. 259-310.

Submerged Weirs. By James B. Francis. An account of some careful experiments to determine the value of the empirical constants of the Francis formula. *Trans. A. S. C. E.*, Vol. XIII., p. 303.

Submerged Weirs. Discussion of existing formulas, by Edw. Sawyer. Calls attention to some errors of M. Alfred Salles. *Van. Nos. Eng. Mag.*, March 1886.

Submerged Weirs, Problem of. By Clemens Herschel. A curve derived from previous experiments from which a table is made. No formula compared. *Trans. A. S. C. E.*, Vol. XIV., p. 189.

Through Large Gates and Over a wide Crest, Experiments on. Valuable paper by Mr. Chas. E. Haberstroh, member of the Boston Society of Civil Engineers, read May 15, 1889. Illus., p. 11. *Jour. Asm. Eng. Soc.*, Jan., 1890, Vol. IX., p. 1.

Through an Orifice Furnished with a Short Pipe. By J. P. Frizell. A theoretical study of the causes of the increased flow due to the pipe, and a mathematical development of the value of the co-efficient, which agrees very closely with the experimental values. *Jour. Frank. Inst.*, October, 1886.

Through Pipes.

By James B. Francis. Memorandum and tables exhibiting the results of some of Darcy's experiments on the flow of water through pipes. Darcy's formula and results are reduced to English measures. *Trans. A. S. C. E.*, Vol. II., p. 45.

By Hamilton Smith, Jr. The results of 88 experiments, 71 of which were made by the author. Sizes from 4 feet to ½ inch, and velocities from 20 feet to 2 inches per second. Fifteen experiments with pipes about 1 ft. in diam. and 700 ft. long; 19 on pipes from 1 to 4 ft. in diam. and 1,000 to 50,000 ft. long. The balance in small pipes. Results plotted and curves drawn. Probably the most reliable data available. *Trans. A.S.C.E.*, Vol. XII. (1883) p. 119.

At Different Temperatures. By J. G. Mair. It is found that the frictional resistance of the pipe is 15 per cent. less at 160° than at 57° F. *Proc. Inst. C. E.*, Vol. LXXXIV., p. 424.

Formulæ and tables. By W. Airy. Lon. *Eng.*, Jan 4 and 11, 1889.

Formula For, and in Open Channels. By E. C. Thrupp. Gives a modification of Hogan's formula, based on experiments applicable to pipes and open channels. Compares experiments with results obtained by calcula-

Hydraulics, *Flow of Water, Formula for*, continued.

tion. Lon. *Engineer*, Dec. 16, 1887. Abstract in *Sci. Am. Sup.*, Feb. 11, 1888.

Formula for. By E. C. Thrupp, before the Society of Engineers. Gives details of experiments with pipes and open channels, and shows method of deriving his new formula for the flow of water. *Trans. Soc. Eng.*, 1888, pp. 224-261.

Identity of Prof. Reynolds' formula with Hagen's empirical data, shown by Prof. W. C. Unwin. Lon. *Engineer*, Jan. 1, 1886.

The Results of Investigations Relative to Formulas for the. Very valuable paper by Edmund B. Weston, illustrated by 28 plates and diagrams. Discussion by several prominent members. *Trans. A. S. C. E.*, Jan., 1890, Vol. XXII., pp. 1-82.

Long Pipes. The practical calculation of the heads, diameters and discharges. By E. Sherman Gould. *Eng. News*, Feb. 16, 1889, *et seq.*

"*The Necessity of More Accurate Knowledge of the, And Some Experiments with the Water Hammer.*" By Joseph B. Rider, C. E. Some examples of flow through large pipes are also given. *Report 11th An. Meet. Am. W-W. Assn.*, 1891, pp. 106-10.

Under Pressure. By Chas. G. Darrach. An account of experiments on loss of head on mains from 20 to 36 inches in diameter, and lengths from 4,000 to 21,000 feet. *Trans. A. S. C. E.*, Vol. VII., (1878), pp. 114-130.

Unnoticed Point in Darcy's Experiments on Flow in Pipes. By Prof. W. C. Unwin. Calls attention to the difference of reading of the three piezometer on Darcy's pipes, and thinks the lower readings unreliable. Lon. *Engineer*, May 7, 1886.

Through one-inch Lead Pipe, 3,400 feet long. Results of experiments. Abst. *Inst. C. E.*, Vol. LXXXIII., p. 455.

Through a 48-inch Pipe. By F. P. Stearns; with discussion. *Trans. A. S. C. E.*, Vol. XIV., p. 1.

Through a 2½-inch Rubber Hose, with Nozzles of Various Forms and Sizes, with formulas for flow and efficiency of various forms of nozzle, all derived from many experiments. By E. B. Weston. *Trans. A. S. C. E.*, Vol. XIII., p. 326.

In Small Channels, with especial reference to sewer calculations. By Rudolph Hering. Follows Kutter's formula and gives graphical solutions. *Trans. A. S. C. E.*, Vol. VIII., (1879), p. 1; also Vol. IX., p. 326.

Through Turbines and Screw Propellers. Paper read before British Assoc. at Montreal, 1884, by Arthur Rigg, of London. Discussion based on empirical data. Lon. *Engineer*, for Sept. 12, 1884.

Under Ground. Results of gaugings on Colorado streams, showing a considerable flow from ground sources into the streams. *San. Eng.*, Sept. 30, 1886.

Flow of the West Branch of the Croton River. By J. J. R. Croes. Gives details of the rainfall and gauging of the west branch of the Croton River from 1865-74. *Trans. A. S. C. E.*, Vol. III., pp. 76-86.

Formula and Diagram for the Discharge of Service Pipes. Used at Providence, R. I., in determining size of pipes. *Eng. Record*, Nov. 7, 1891, pp. 368-9.

Formulas. A comparison of various formulas and results with Darcy's formula applied. By Nath. Hill. *Van Nos. Eng. Mag.*, Dec., 1883.

Formulas. Diagram of. Valuable folding inset compiled by J. L. Fitzgerald, M. A. S. C. E. The formulas of Kutter, Trautwine, Fanning, Freeman, Ellis, Latham and others are represented by curves from which the unknown quantity can easily be derived. Sup. to *Eng. News*, Sept. 6, 1890, p. 206.

Hydraulics, continued.

Formula, Kutter's, applied to circular Sewers. *Jour. Assn. Eng. Soc.*, Vol. VI., p. 269.

See *Hydraulic Tables*.

Height of Jets of Water. Theoretical discussion and results of experiments. *Zeitschr. des österr. Ing. u. Arch. V.* (Vienna), Vol. II., 1880, pp. 70-4.

Kutter's Formula. See *Sewers Formula*.

The Miner's Inch. Some of the various methods and standards given, all going to show that the miner's inch, like the perch in masonry, is wholly a function of one's geographical position. *E. & M. Jour.*, Feb. 6, 1886.

Researches on the Oscillations of Water and on various hydraulic appliances. Two volumes, in French, by Le Mis Anatole de Caligny, containing many original memorials contributed to various scientific societies since 1837. 968 pages and eight large plates. A valuable addition to hydraulic science. Paris, 1883.

Water. Two Manners of Motion of. By Prof. Osborne Reynolds before the Royal Institution. Divides the various phenomena of moving water into two distinct classes, according to the internal motions which become visible by means of color-bands. *Van Nos. Eng. Mag.*, July, 1884.

Waves of Translation that emanate from a submerged orifice, together with an examination of the proposed Baie Verte Canal, connecting the Bay of Fundy with the St. Lawrence. A mathematical discussion of experiments on a large scale. By Clemens Herschel. *Trans. A. S. C. E.*, Vol. IV., (1875), p 185-200.

See *Air, Flow of. Compressed Air, Fire Streams, Jets. Mill Engineering. Sewer Diagram, Weirs.*

Hydro-Dynamics. Production of phenomena analogous to magnetic attraction and repulsion by the vibration of bodies in water. Experiments made by Prof. Bjerknes. By C. A. Cooke. Lon. *Eng.*, March 27, 1885, *et seq.*

Hydrodynamic Formulæ. Shows the discrepancies obtained by using several formulæ to compute discharges on different size streams. *Van. Nos. Eng. Mag.*, Vol. IX., p. 318.,

Hydro-Dynamometer, *A New.* By M. De Perrodil. Describes a new instrument for measuring the velocity at any point of a liquid current. *Van Nos. Eng. Mag.*, Vol. XVII., p. 482.

Hydrographic Surveying. By Lawrence Bradford. Gives methods employed in his own practice, describes water gauges, floats and sounding irons, and discusses the use of the sextant. *Jour. Assn. Eng. Soc.*, Jan., 1887; also *Sci. Am. Sup.*, April 21, 1887.

Hydrographic Work *of the U. S. Navy.* A lecture before the Frank. Inst. By Lt. A. B. Wyckoff. The general scope, character and importance of this work well presented. *Jour. Frank. Inst.*, May, 1886.

Hydrology. *On the relations between systems of river regulation and the duration and rate of discharge of high water.* By Paul Klunsinger. This is essentially an attempt to solve the following problem: A *uniform* rain-fall of *h m* per second falls during a given time over a given drainage area,*which is supposed either entirely impervious or completely saturated at the beginning of the given time so that the entire rain-fall flows off at the rate of h m per second*; to find the duration and rate of discharge. Analytically treated and formulas developed; and affecting conditions discussed. In view of the assumptions made however, the formulas would seem of little value. *Zeitschr. des Oesterr. Ing. u. Arch. V.*, 1886, pp. 10-19.

See *Flood Announcements.*

Hydro-Mechanics. A series of lectures before the *Inst. C. E.*, by Evans on Physiography, Pole on water supply; Unwin on Water-Motors; Hartley on In-

Hydro-Mechanics, continued.

land Navigation in Europe; Stevenson on Tides and Coast Works; and Reed on forms of Ships, making a volume of 220 pp. and 6 plates.

Hydrophone. A description of an apparatus for detecting leaks in water mains by microphony. *Sci. Am. Sup.*, April 16, 1887.

Hydrostatic Pressures in a mass of earth, saturated with water. By L. Brennecke. Interesting discussion of the pressures to be assumed in proportioning inverts of lock chambers, beds of beton, and similar structures exposed to an upward pressure in saturated earth. The writer, who is the author of the recent excellent work on Foundations (Der Grundbau, Berlin, 1887), endeavors to determine how this pressure depends upon the character of the soil, and gives the results of his experiments on the subject. He considers it unnecessary to assume the full hydrostatic head. *Zeitschr. f. Bauwesen*, 1886, pp. 102, 118.

Hypsometry. By Prof. I. O. Baker. A valuable paper on leveling by all the practicable methods. Illus. *Van Nos. Eng. Mag.*, Vol. XXXV., pp. 413 and 457. Nov. and Dec., 1886.

Ice.

Anchor or "Frasil." A paper by Geo. H. Henshaw, with discussion, covering 23 pp. Its cause and action in stopping river channels, etc. *Trans. Can. S. C. E.*, Vol. I., (1887,) p. 1.

Crushing Strength of. Being account of the elaborate experiments by the government. Col. Wm. Ludlow. *Proc. Eng. Club. Phila.*, Vol. IV., No. 2, p. 93.

Elasticity of, carefully determined by Prof. John Trowbridge. *Am. Jour. of Sci.*, May, 1885.

Impurities in. Discussion, with analyses of several kinds. *Eng. News*, Oct. 10, 1881, pp. 430-2.

Pollution of. Supplies. Report of the Mass. State Board of Health. Tables are given of many analyses of water and ice. *Twenty-first An. Rep.*, 1889, pp. 143-223.

Removal of Anchor Ice. See *Water-Works, Chicago.*

Strength of. From German Experiments determining its tensile and compressive strength, and modulus of elasticity under various conditions. *Eng. News*, Dec. 26, 1885. *Proc. Inst. C. E.*, Vol. LXXXII., p. 391.

See *Anchor Ice.*

Ice Breaking Steamers. See *Winter Navigation.*

Ice Houses.

See *Railroad Structures.*

Ice Machines.

Binary Absorption System. The substance used in ordinary ether charged to half saturation with sulphurous dioxide. With this "binary liquid" an intense cold is produced. The system described and the plans of a two-ton machine given. By H. F. J. Porter. *Trans. A. S. M. E.*, Vol. II., p. 118.

Description of the ice plant at St. Andrew's Dock, Hull, England, capable of turning out 25 tons per 24 hours. Uses ammonia gas and brine of calcium salts. Illustrated. *Lon. Eng.*, May 28, 1886.

And Refrigerating Machinery.—By T. B. Lightfoot. A valuable paper, covering the whole ground and showing the adaptation of various methods. Illustrated. Lon. *Engineer*, May 21, 1886. *Van. Nos. Eng. Mag.*, Aug., 1886.

The Theory of. By M. Ledoux. Gives theoretical conditions of the successful working of three types—Gifford's air machine, Pictet's sulphurous acid machine and Carre's ammonia machine. Enough of the practical details are given to explain the *rationale* of the operation. *Van Nos. Eng. Mag.*, Vol. XXI., p. 89.

Ice Making.

And Machine Refrigeration. A description of the methods of ice manufacture, and the use of refrigerating machines. Illus. By F. A. Fernald. *Pop. Sci. Mon.*, May, 1891, pp. 19-30.

Ice Making, continued.

Plant, 25-Ton, at Birmingham, England. Illustrated description. Lon. *Engineer*, Jan. 25, 1889.

Plant, Linde System. The Linde works in London, said to posess the largest refrigerating apparatus in the world, are fully described and illustrated. Lon. *Engineer*. *Sci. Am. Sup.*, No. 757, July 3, 1890, pp. 12089-12.

Illumination. *Economical from Waste Oils.* By J. B. Hannay, before the Society of Arts. Describes the lucigen, an apparatus in which compressed air is used with waste oils, and its applications. *Jour. Soc. Arts*, Dec. 2, 1887. *Sci. Am. Sup.*, Jan. 14, 1888. Abstracted in Lon. *Eng.*, Dec. 9, 1887.

Impact. *Effect of.* See *Metals*.

Inclined Plane Railroads. See *Railroads, Inclined Plane*.

Indexing. See *Reading and Indexing*.

Indicators. See *Engines, Steam, Indicators*.

Ingersoll Sergeant Mining Machinery. See *Mining Machinery*.

Injectors,

And Steam Pumps, Comparative Efficiency of. Finds the relative economy and difference in amount of fuel used with a boiler fed by a pump and by an injector. *Stevens Indicator*, April 20, 1888; *Am. Eng.*, April 18, 1888.

From the French of M. Jean Pochet. Gives mathemetical investigation of the injector and its application to pumps, presses, etc. *Van. Nos. Eng. Mag.*, Vol. XVI., pp. 129 and 134; also *Van Nos. Sci. Series*, No. 29.

By P. H. Rosenkrans. Gives a brief account of the development of the Gifford injector, as well as a description of the other best known injectors. *Van. Nos. Eng. Mag.*, Vol. VIII., p. 71.

By A. Winkler. Gives the mathematical theory of the injector. *Van. Nos. Eng. Mag.*, Vol. IV., p. 352.

Injectors, Efficiency of as Compared with Steam Pumps. By D. S. Jacobus. A mathematical treatment of the subject. *Am. Eng.*, Aug. 31, 1887.

Exhaust. An explanation of the action of an exhaust injector. *Am. Mach.* October, 1887. *R. R. Gaz.*, Nov. 4, 1887.

Mechanics of the. By J. B. Webb, before the Scranton meeting of the American Society of Mechanical Engineers. *Am. Eng.*, Oct. 17, 1885.

Worked by exhaust steam as boiler feeders. *Mechanical Engineer*, April 4, 1885.

Inland Navigation.

An Enlarged Water-Way Between the Great Lakes and the Atlantic Ocean. Abstract of a paper by E. L. Corthell, before the Can. Soc. C. E., advocating a Huron-Ontario Ship Railway, and giving estimates of cost. *Ry. Rev.*, Feb. 14, 1891, pp. 99-100.

Fourteen papers on canals and inland navigation were presented before the recent Canal Conference held under the auspices of the British Society of Arts. They are mostly indexed under canals. See *Jour. Soc. of Arts*, May 25, *et seq.* 1888.

In Germany. See *Railways and Waterways*.

In Great Britain. By E. J. Lloyd, before the Society of Arts Canal Conference. Gives history of the development of inland navigation. *Jour. Soc. Arts*, May 25, 1888.

Proceedings of the 1886 International Convention for Promoting Inland Navigation, held at Vienna, Austria. *Annales des P. & C.*, June, 1888.

Suggestions for its Improvement. By M. B. Cotsworth before the Society of Arts Canal Conference. Discusses the present condition of inland navigation in the United Kingdom and gives suggestions for its improvement. *Jour. Soc. Arts*, May 25, 1888.

See *Waterways, Canals, Inland Navigation*.

Inspection.

Engineering Inspection. A lecture before the students of the Rensselaer Polytechnic Institute. By O. F. Nichols. Sets forth the work of inspector and his relation to the contractors and engineers. *Eng. Record*, March 21, 1891, pp. 264-5, *et seq.*

Of Metallic Structures. By James Sanderson. Treats especially of tension members and riveted work. *Van. Nos. Eng. Mag.*, April, 1884.

Of Materials of Construction in the United States. A paper by Geo. H. Clapp and A. E. Hunt, read at the N. Y. (1890) meeting of the Brit. Iron and Steel Inst. Gives general specifications, form of test pieces, etc. *Am. Mfr.*, Dec. 5, 1890, p. 8. Dec. 12, p. 18.

Of Materials of Construction in the United States. A valuable paper by Mr. George H. Clapp, Pittsburgh, Pa., giving specifications, methods of testing and inspecting, etc. *Am. Mfr.*, May 31, 1889.

Institution of Civil Engineers. *Its History.* A very satisfactory account of the origin, changes, progress, classes of members, distinguishing features, etc., of this famous body. The account is of especial interest to those interested in engineering society relations in America. *Proc. Inst. C. E.*, Vol. LXXXVI., p. 153.

Instruments. *Hand-Book and Illustrated Catalogue of Engineers' and Surveyors' Instruments.* Full description of the instruments with directions for their care and adjustment. Articles on the planimeter, stadia measurements, etc., p. 110. Buff & Berger, Boston.

See *Distance-Measuring Micrometers.*

Insulators, Oil. Descriptive article by Mr. F. S. Pope. Illus. *Elec. Eng.*, Oct. 28, 1891, pp. 471-2.

Integrators.

Integraph. An instrument for constructing the curve, $y = \int f(x) \, dx$. Illustrated and described. A French invention. *Sci. Am. Sup.*, Nov. 7, 1885.

Mechanical. By Prof. H. S. H. Shaw. A most exhaustive paper, giving theory, cuts and practical advantages of all existing varieties of integrators or planimeters. *Proc. Inst. C. E.*, Vol. LXXXII., p. 75; also *Van. Nos. Eng. Mag.*, Nov. 1885.

See *Planimeters.*

Interlocking and Signaling. See *Railroad Signaling.*

Interstate Commerce Commission.

And Its Work. Article discussing this subject. By Hon. Augustus Schoonmaker. *R. R. Gaz.*, Oct. 16, 1891, pp. 725-6.

For summary of third annual report and editorial comments on same. See: *R. R. Gaz.*, Jan. 10, 1890. *Ry. Rev.*, Jan. 11 and 17, 1890, *Eng. News*, Jan. 11, 1890.

Report by Prof. H. C. Adams. Abstract of tables in *R. R. Gaz.*, Aug. 22, 29, 1890, pp. 583-90, 606. *Ry. Rev.*, Aug. 23, 1890, p. 498. Ed. review in *Eng. News* Aug. 23, 1890, pp. 168-70.

Report of. Fourth annual report. Abstract of the general matter in *R. R. Gaz.*, Dec. 12, 1890, pp. 857-8. Full report, *Ry. Rev.*, Dec. 13, 1890, pp. 755-7, *et seq.*

See *Railroad Statistics.*

Inventions. See *Engineering Inventions.*

Iron.

Alloys of Iron and Silicon. A paper read at the Paris meeting of the British Iron and Steel Institute, by R. A. Hatfield, Sheffield, England. *Am. Mfr.*, Oct. 11, 1889; *et seq.*; Lon. *Eng.*, Oct. 18, 1889.

Bar, cost of production in Eastern mills, itemized. *Iron Age*, May 21, 1884.

Car Wheel, Microscopic Structure of. By F. L. Garrison. A paper, with 4 plates, before the Am. Inst. Min. Engrs., Pittsburg meeting, 1886. *Ry. Rev.*, May 8, 1886; *Jour. Frank. Inst.*, Aug., 1886.

Iron continued

Cast Iron.

For Engineering Purposes. By J. S. Brodie, before the Liverpool Engineering Society. Treats of the nature, strength and other properties of cast iron. *Van Nos. Eng. Mag.*, Vol. XXII., p. 17.

Aluminum in. See *Aluminum.*

Cast, Effects of Silica and Aluminum in. See *Silica.*

Founding of. See *Alloys.*

Fracture of. A paper on viscosity and vegetation in metals. *Van Nos. Eng. Mag.*, Vol. XXII., p. 503.

Influence of Silica in Cast Iron. Paper by W. J. Keep, Detroit, Mich. Read before the Am. Inst. Min. Eng., at New York meeting, Feb., 1889. *Trans. A. I. M. E.*, New York meeting, Feb., 1889.

Malleable Iron Castings. A description of the most improved process of making malleable iron castings. Read before the Iron and Steel Institute by T. Nordenfeldt. Lon. *Eng.*, May 15, 1885; *Am. Mach.*, June 6, 1885.

Manganese in. A paper by W. J. Keep giving the results of an extended series of experiments. Gives tables of results, diagrams and discussions. *Trans. A. I. M. E.*, 1891, pp. 26.

New method of Ornamenting Castings. Consists of lining the mold with carbunized fabric, such as lace. Abstract of remarks made by Mr. A. E. Outerbridge, Jr., before the Franklin Institute. *R. R. & Eng. Jour.*, July, 1887.

Phosphorous in Cast Iron. Paper by W. J. Keep, of Detroit, Mich. *Trans. A. I. M. E.*, Ottawa meeting, Oct., 1889.

Tests.

How to Test Cast Iron. An article by Prof. J. B. Johnson, Washington University, St. Louis, describing a new form of testing machine for cast iron, discussing methods of testing cast iron, and proposing a new rule for computing resilience; also proposed change in specifications. *R. R. Gaz.*, Sept. 13, 1889; *Eng. News*, Sept. 14, 1889; *Ry. Rev.*, Sept. 14, 1889; *Mech. World*, Oct. 19, 1889.

How to Test. See *Testing Material.*

An editorial review of Prof. J. B. Johnson's paper read before the A. S. C. E. Illus. Lon. *Eng.*, July 4, 1890, pp. 19-20.

Some Tests on the Strength of Cast Iron, made in the Laboratory of Applied Mechanics of the Mass. Inst. of Tech. By Gaetano Lanza, with various students. Includes tests for modulous of elasticity in tension and cross breaking, tests of window lintels, pulleys, and of the holding power of keys and set screws. *Tech. Quart.*, Vol. II., 1888, No. 2, pp. 145-91.

Malleable Cast. Physical Tests of. By Prof. Ricketts. Specimens tested for tension, compression, bending, with skin on and off, elastic limit and ultimate strength found. *Van Nos. Eng. Mag.*, April, 1884. See also *Malleable Castings.*

See *Castings. Car Wheels.*

Changes in Iron Produced by Thermal Treatment. Paper by E. J. Hall read before the Iron and Steel Inst., giving the results of some experiments on the relation between the temperature of maximum tenacity and the carbon in the iron. Lon. *Eng.*, May 22, 1891, pp. 630-1.

Continued Contraction of, when subjected to repeated sudden coolings. The phenomenon explained by H. M. Howe. *Trans. A. I. M. E.*, Halifax meeting, 1885.

Iron, continued.

Corrosion of.

The effects of various kinds of liquids, hot and cold, on iron, and the best means of preserving it under such conditions from corrosion. *Trans. Inst. C. E.*, Vol. LXXXV., p. 295.

A description of a process for preventing, by forming a coating of black oxide on the iron. Lecture of Prof. Barff, London. *Eng. News*, July 11, 1878, pp. 220-2. Further notes, *ibid*, Oct. 24, 1878, pp. 337-8.

By A. C. Brown, before the Edinburgh meeting of the Iron and Steel Institute. Explains the process involved in the rusting of iron. *Am. Eng.*, Sept. 26, 1888; *Mast. Mech.*, October, 1888, Lon. *Engineer*, Sept. 1, 1888.

By Smoke. See *Viaducts, Across R. R. Tracks.*

An improvement over the Bower-Barff process for small articles. The method given to the public by Col. A. R. Buffington, of the Ordnance Dept. U. S. Army. *Mechanics*, June 1885. *See* also Paints.

Protection against. By H. Haupt. Describes a new process for protecting iron from corrosion, which treats the heated metal in retorts with steam and hydrocarbon vapor. *Am. Mfr.*, July 6, 1883.

See *Iron and Steel, Bower-Barff Process.*

Corrugated. See *Corrugated Iron.*

Crushing Strength of American. By T. C. Clarke. Gives details of results attained from experiment made to find absolute crushing strength of Phœnix iron. Also discussion of the mercurial gauge. *Trans. A. S. C. E.*, Vol. II., p. 279.

Crystallization of. Causes of the. Paper by Mr. Thos. Morris, F. G. S., reprinted from *Proc. Liverpool Eng. Soc.*, *Am. Mfr.*, April 25, 1890, *et seq.*

Durability of, for Engineering Structures. By G. J. C. Dawson. A paper before the Civil and Mechanical Engineering Society, treating of the durability of cast and wrought iron structures. *Van Nos. Eng. Mag.*, Vol. V., p. 300.

Effect of Intermittent Strains upon. By B. Baker in his address before the Mechanical Section of the British Association. An historical review with some new facts from his own experiments. *Iron*, September 18, 1885. *Iron Age*, Oct. 1, 1885.

Effects of Heat and Cold on Iron, Steel and Copper. See *Heat and Cold.*

Founding, Practical. By J. E. Crane and J. R. Speck. Treats of the apparatus and tools for molding, green sand and its treatment, etc. *Am. Eng.*, April 6 and 20, 1887.

Founding, Scientific. By Thomas Turner. Paper read before the South Staffordshire Inst. of Iron and Steel Managers, Dudley, March 19, 1887. Object of the paper is to examine the material with which the ironfounder has to do and consider those chemical and mechanical characters which affect it. A good paper. Lon. *Eng.*, April 1, 1887. *Pract. Engr.*, April 8, etc., 1887.

History of Decarbonizing. Gives the abstracts of the abridgements of the specifications relating to the manufacture of iron and steel in the department of decarburizing and purifying crude iron so as to make it malleable and tenacious. *Van Nos. Eng. Mag.*, Vol. I., p. 193, 358, etc.

Internal Stresses in. By Gen. N. Kalakoutzky. A valuable series of articles giving results of original investigations. Discusses the determination of the influence of internal stresses, method of determining them, reduction, etc. Lon. *Engineer*, Dec. 9, 16, 23, 1887; *Sci. Am. Sup.*, Feb. 18, 25, 1888.

Investigations on the Influence of Heat on the Strength of Iron. Report of Professor Martens of the commission of two German societies appointed to make investigations. Experiments described and results given. *Proc. Inst. C. E.*, Vol. CIV., 1891, pp. 329-34.

Iron,' continued.

Manufacture.

Address to the Chemical Section of the British Association, by Sir Lowthian Bell, President of the Section. Lon. *Eng.* Sept. 20, 1889. Vol. XLVIII., p. 369.

Magnetic Concentrates in the Port Henry Blast-Furnaces. A paper by N. M. Langdon giving the results of his experience in the use of concentrates. *Trans. A. I. M. E.*, p. 3. Reprinted in *Eng. News*, Oct. 17, 1891, pp. 351-2.

New Direct Process of. Gives the result of experiments, extending over several months. *Van. Nos. Eng. Mag.*, Vol. XXVII., p. 191.

Probable Future of the. A paper by Sir Lowthian Bell, read before the Pittsburgh international meeting, Oct., 1890, giving also somewhat of a historical sketch. *Trans. A. I. M. E.*, 1890, pp. 17. Discussion, pp. 17-21.

Early Steps in. An historical paper by Wm. F. Durfee. Illustrated *Pop. Sci. Mon.*, Dec. 1890, pp. 145-72.

Mitis-Castings from Wrought Iron or Steel. By Peter Ostberg before the Am. Inst. Min. Engrs., at the Pittsburgh meeting, Feb., 1886. Describes methods which are now extensively used in Sweden for casting wrought iron.

Mitis Process. See *Iron and Steel Mitis Process.*

Molecular Changes Produced in, by Variations of Temperature. By Prof. R. H. Thurston. *Van Nos. Eng. Mag.*, Vol. IX., pp. 169 and 273.

Metallurgy, The Progress of. Address before British Association at Newcastle, by Sir Lowthian Bell, F. R. S. Complete reprint. *Iron* (London) Nov. 8, 1889.

Ores.

Concentration of. A paper by Mr. John Birkinbine and Thomas A. Edison, reviewing the various methods in use for concentrating iron ores. *Am. Mfr.* July 5, 1889.

Reduction of. A good paper before the Iron and Steel Institute by Sir Lowthian Bell. Treats of the reduction of the ores of iron in the blast furnace, with discussion. Lon. *Eng.*, Sept. 23, 1887.

Oxidation.

Of Iron, Some Experiments on. By F. G. Calvert. *Van Nos. Eng. Mag.*, Vol. VIII., p. 127.

Pig. A lecture before the Franklin Institute, by A. E. Outerbridge, on the production of pig iron, including the relations between its physical properties and chemical constituents. *Sci. Am. Sup.*, April 14, 1888.

Cupolas for Melting. By M. A. Gouvy, Jr. Illustrated. Translated from the French by W. F. Durfee. *E. & M. Jour.*, Feb. 16, 1889, *et seq.*

Development of the Manufacture in the United States. Lecture by Mr. John Birkinbine before Franklin Institute giving extensive tables showing production in U. S. and other countries. *Jour. Frank. Inst.*, June, 1891, pp. 337-353.

Relation between Physical Properties and Chemical Constituents of. Abstract of a lecture before the Franklin Institute, by Alex. E. Outerbridge, Jr. *Jour. Frank. Inst.*, March, 1888.

Production of Pig, of a Definite Composition. By H. Pilkington. Before the South Staffordshire Iron and Steel Institute. *Am. Mfr.*, Jan. 13, 1888.

Producing Districts in the U. S. A series of articles by John Birkinbine, Sec'y. Am. Assoc. of Charcoal Ironworkers, which promises to be of considerable value. Began in *Age of Steel*, St. Louis, March 7, 1885.

Valuation of. By A. E. Tucker, before the Society of Chemical Industry. A valuable paper. Lon. *Eng.*, July 20, 1888.

Production, Development of, in the last ten years. By F. Kupelwieser. Statistics, etc. *Zeitschr. d. Oesterr. Ing.-u. Arch. V.*, 1886, pp. 36-50.

Iron, continued.

Resistance of, to Strains. Gives brief summary of the behavior of iron under strains. *Van Nos. Eng. Mag.*, Vol. IX., p. 113.

Russian Sheet Iron, Manufacture of. Paper by F. Lynwood Garrison in *Journal of the U. S. Association of Charcoal Iron Workers.* Describes process of manufacture. *E. & M. Jour.*, Dec. 1, 1888. *Am. Mfr.*, Dec. 14, 1888.

Silicon and Sulphur in. By Thomas Turner before the Iron and Steel Institute. A record of interesting experiments on the effects of an addition of sulphur to cast iron rich in silicon. Concludes that silicon has the power of expelling sulphur from cast iron.

Spongy. Treats of the manufacture and properties of spongy irons. *Van Nos. Eng. Mag.*, Vol. XIII., p. 301.

Stay-Bolt Iron, New Test for. A special bending test of great simplicity, yet of much value. *Iron Age*, May 21, 1885.

Tensile Strength, Experiments on the, of Bar-Iron and Boiler-Plate. By C. B. Richards. Object of the paper is to give particulars and results of experiments made at the Colt's Armory testing machine on the tensile strength of American bar iron and boiler-plate; also describes a gauge for measuring elongation and contraction of specimens; also tabulates results of experiments. *Trans. A. S. C. E.*, Vol. II., p. 339.

Wrought Iron.

By J. Starkie Gardner. A lecture before the section of Applied Art of the Society of Arts, treating the use of wrought iron in ornamentation. *Jour. Soc. Arts*, Feb. 18, 1887.

Aluminum in. See *Aluminum.*

Causes of Defects in. By A. D. Elbers, in *E. & M. Jour. Ry. Rev.*, July 20, 1889.

Direct from the Ore, a process of making. By W. P. Ward. *Trans. A. I. M. E.*, Vol. XII.

Influence of Punching and Drilling on Strength of. Results of a series of experiments in Zurich. *Abstract Inst. C. E.*, Vol. LXXXV., p. 481.

Rerolling and Reheating. Gives the results of experiments on the strength of wrought iron in bars and in chains; effects of different degrees of reduction in rolling, of reheating, rerolling and hammering, comparison of chemical causes with physical results, correct form of test pieces, and miscellaneous investigations into the physical properties of rolled wrought iron. *Report of U. S. Board of Testing, etc.*, Vol. I., 1881, pp. 1-120.

Tests. Tabulated records in detail of results obtained from over 2,000 test pieces of wrought irons. *Rept. U. S. Board of Testing*, Vol. I., 1881, pp. 45-91.

Iron and Steel.

Review of the memoir of M. Considere. By M. A. Berton. (Contains a brief summary of the interesting paper of M. Considere.) *Mem. de la Soc. des Ing. Civils*, March, 1886, pp. 282-286.

Action of Tidal Streams on. See *Metals.*

Analyses. By J. J. Morgan. Gives directions for the estimation of graphite, silicon, sulphur, Manganese, phosphorus and carbon. *Sci. Am. Sup.*, July 23, 1887.

Analysis, International Standard of. A paper by Prof. John W. Langley. *Proc. Engrs. Soc. of W. Penn.*, Vol. V., 1889, pp. 20-5. Discussion, pp. 20-9.

Analysis of, as Practical in Large Industrial Works. A series of papers describing methods of Industrial Analyses. By Auguste J. Rossi. *Iron Age*, March 5, 1891, pp. 426-7, *et seq.*

Bower-Barff Process. By A. S. Bower. The object of the paper is to show, what may be done in protecting iron and steel from rust by forming upon

Iron and Steel. *Analysis, Bower-Barff Process*, continued.

their surfaces a film of magnetic oxide. *Van. Nos. Eng. Mag.*, Vol. XXIX., p. 370.

Treatment of, to Prevent Corrosion. By Prof. Barff. A lecture before the Society of Arts, describing the Barff process of protecting iron. *Van. Nos. Eng. Mag.*, Vol. XVIII., p. 380, and Vol. XX., p. 432.

Described in an illustrated paper by Geo. W. Maynard. *Trans. A. S. M. E.*, Vol. IV., p. 351.

Treated Historically and Analytically, from the proceedings of the Inst. of Civil Engrs., published in *Iron*, Nov. 25, 1884.

Burning of. A short article showing it to be due to O. reduced from the cinder in the case of iron and to the oxidation of the manganese and silicon in the case of steel. *Iron Age*, Nov. 13, 1884.

Classification of. An excellent classification and definition of the terms "iron" and "steel" and their varieties, given by William Kent, in 1883, in his expert testimony in a patent case. *R. R. & Eng. Jour.*, April, 1887.

Classification of and Specifications of Tests. Adopted by the Soc. German Iron Masters. *Eng. News*, April 8, 1882, pp. 114-16.

Conley Lancaster Process of making iron and steel direct from the ore. An illustrated article describing the process in considerable detail and giving an estimate of the cost of plant and of producing ingots. *Am. Mfr.*, Feb. 24, 1890, p. 7, April 18, 1890.

Contraction of Area of Cross-Section as an Index of Quality in Tests of Iron and Steel. A brief article by P. Kreuzpointner, asserting that contraction of area is a misleading index as to quality. *Eng. News*, June 20, 1891, pp. 79-80.

Estimating the Amount of Carbon in. A paper read before the Birmingham Phil. Soc., Eng. Various methods given. Illustrated. *Iron*, July 24, 1885.

Evolution of the American Rolling Mill. Address of R. W. Hunt, Pres. Am. Soc. M. E. Reprinted in full in *R. R. Gaz.*, Nov. 20, 1891, pp. 817-20. *E. & M. Jour.*, Nov. 21, 1891, *et seq.* *Eng. News*, Nov. 28, 1891. *Iron Age*, Nov. 19, 1891, *et seq.*

Experiments on the Effects of Pickling and Rusting on the Strength of Iron. By A. Ledebur in *Mitteilungen aus den Kor. tech. Versuch. zu Berlin*, 1890, Sup. I. Gives results of experiments on various forms in steel and iron, treated in different ways. Abs. *Proc. Inst. C. E.*, 1891, Vol. CIV., pp. 399-402. Abstract *Iron Age*, Aug. 20, 1891, pp. 282-3.

Experiments on Iron and Mild Steel. Describes experiments made at the Duhbury Bridge Works on riveted girders of mild steel and wrought iron. *Van Nos. Eng. Mag.*, Vol. XXIX., p. 379.

Hardening of. By Prof. R. Akerman. Treats of the different methods of hardening, and gives tables of strength of metals treated in different manners. *Van Nos. Eng. Mag.*, Vol. XXII., p. 485.

Increase in Elastic Limit due to high strains in tension and cross-breaking and the practical value of such increase. By R. H. Thurston. *Trans. A. S. C. E.*, Vol. IX., 173-185.

Industries of America. A paper by Sir James Kitson, containing some notes on the trip of the British Iron and Steel Inst. *Eng. Mag.*, July, 1891, pp. 454-97. Reprinted from the *Contemporary Review.*

Influence of an Increase of Temperature on the tensile strength and durability of iron and steel. A paper by E. Cornut, translated by Chief Engr. B. F. Isherwood, U. S. N. *Jour. Frank. Inst.*, April, 1885.

Low Temperatures. By J. J. Webster. Gives results of experiment at low temperatures to find the effect of cold on the strength of iron and steel. *Van Nos. Eng. Mag.*, Vol. XXIII., p. 198.

Iron and Steel, continued.

Manufacture of Continuous Sheets of Malleable Iron and Steel, Direct from Fluid Metal. Paper by Sir Henry Bessemer, before the Iron and Steel Inst., describing this process and giving an account of its development. Illus. Reprinted in *E. & M. Jour.*, Oct. 24, 1891, pp. 473-4. *Ry. Rev.*, Oct. 31, 1891, pp. 687-9. *Eng. News*, Oct. 24, 1891, pp. 381-2.

Memoirs on the Use of, in construction. By M. Considere. *Annales des P. & C.*, 1885, April, pp. 572-775.

See *Corrosion*.

Microscopic Structure of. By F. Lynwood Garrison, before the Am. Inst. Min. Engrs. Describes methods of making the examination, and gives 10 heliotypes of sections. Useful adjunct to the testing machine. *Jour. Frank. Inst.*, Oct., 1885.

Mitis Process. Presented to the U. S. Naval Institute, May 11, 1885, by W. F. Dufree, M. E. Proposes among other things, the construction of cannon with mitis iron or steel by the Rodman process of casting. *Proc. U. S. Naval Inst.* Vol. XIII., No. 3.

Mitis Process. See *Iron, Mitis Process of making castings of Malleable Iron.*

Physical Condition Iron and steel. By D. E. Hughes. A paper read before the Inst. of Mech. Engrs. *Van Nos. Eng. Mag.*, March, 1884.

Preservation of. Extracts from papers by prominent engineers concerning the use of red lead for painting iron and steel. Pamphlet, pp. 24. Published by the National Lead & Oil Co., No. 28 Burling Slip, N. Y.

Preservation of. By Woodruff Jones, A. M., mentions different methods and advocates painting with red lead. The Gresner Rust Proof Process. Illus. *R. R. & Eng. Jour.*, April, 1891. *Eng. Record*, April 15, 1891, pp. 325-6.

Paris Exhibition. Series of valuable articles containing data concerning production, composition and qualities of various irons and steels. Lon. *Engineer*. Nov. 1, 1889, Vol. LXVIII., p. 367, *et seq.*

Physical Properties of, at Higher Temperatures. A valuable series of experiments made by Mr. James C. Howard at the Watertown Arsenal, on ratio of expansion, modulus of elasticity, strength, internal strains, and also on riveted joints. *Proc. Soc. Arts*, (Boston), 1890, pp. 45-60. *R. R. Gaz.*, May 9, 1890, p. 316.

Qualities of. By Wm. Metcalf. *Trans. A. S. C. E.*, Vol. V., p. 313.

Recent Developments in. Paper by George Burlot, Member of Civil Engineers' Club of Cleveland. Read Nov. 12, 1889, p. 5. *Jour. Assn. Eng. Soc.*, Jan., 1890, Vol. IX., p. 14.

Requirements of Specifications. By Mr. James Ritchie. *Jour. Assn. Eng. Soc.*, May, 1890, pp. 238-41.

Scientific Investigations as applied to the Manufacture of Iron and Steel. Two lectures delivered by Dr. C. R. Alden Wright, F. R. S., at University College, London. Lon. *Engineer*, Dec. 27, 1890. Vol. LXVIII., p. 530; Lon. *Eng.*, Dec. 20, 1890, p. 724. *Iron*, Dec. 27, 1889, p. 546.

In Shear and Torsion. Experiments to find relation between tensile, shearing and torsional strength of wrought iron, steel and cast-iron. *Proc. Inst. C. E.*, Vol. XC., p. 382.

Specifications. Prepared at the Pittsburgh Testing Laboratory, and contributed to the Engr. Soc. of West Penn. by Alfred E. Hunt. With discussion. A very valuable paper and discussion, based on actual experience in a testing laboratory. *Am. Eng.*, May 15, 22, and 29, 1886; also *Ry. Rev.*, April 24, 1886.

Study of. By J. C. Bayles. An annual address before the Am. Inst. Min. Engrs., Vol. XIII.

Shearing and Torsional Strength of. See Shearing Strength.

See *Tests*.

Iron and Steel, continued.

Tests.

Over 700 specimens of iron and steel given in detail. Taken mostly from guns, shot and shell. Watertown Arsenal Report for 1883. *Ex. Doc.*, No. 36, 49th Cong., 1st session.

Low Temperature Tests of. Brief abstract of report of bending tests made by Profs. F. Steiner and H. Gollner, Germany. *R. R. Gaz.*, Nov. 27, 1890, pp. 837-8.

Iron and Open Hearth Steel, Effects of Punching and Reaming. Tests showing effect on elastic limit and ultimate strength, made by the Passaic Rolling Mill Co. *R. R. Gaz.*, Dec. 12, 1890, p. 866.

Testing and Analysis of Iron and Steel. A series of articles by Messrs. Leadbeater and Hodgson giving full instructions in the analysis of iron. *Mech. World*, July 4, 1891, *et seq.*

Testing by electro-magnetism. By A. Herring. Gives detail of the process of testing and of the machine employed. *Van Nos. Eng. Mag.*, Vol. XX, p. 261.

Torsional Strength of Iron and Steel. Results of experiments at University College, London. *Proc., Inst. C. E.*, Vol. XC., p. 376

Working Stress of, Some Notes on. By Benj. Baker, before the Am. Soc. of Mech. Engrs. *R. R. Gaz.*, Feb. 18, 1887; also Lon. *Engineer*, Jan. 21, 1887.

See *Metals*.

Iron and Concrete *in Building Construction.* A valuable paper by G. W. Percy, of San Francisco, on recent new uses of iron rods for the tension side of concrete when used as floors and otherwise in cross-breaking strain. Illustrated. *Tech. Soc. Pac. Coast*, Vol. V., p. 1, (June, 1888).

Iron Arches. See *Arches*.

Iron Making, *Natural Gas in.* See *Gas, Natural.*

Iron Paint. See *Paint.*

Iron Working.

With Machine Tools. This is Part IV of a series of articles on "Development of American Industries since Columbus," by Wm. F. Durfee. This number treats especially of the rolling of rails. Illus. *Pop. Sci. Mon.*, March, 1891, pp. 588-9, *et seq.*

Machinery, being the appliances for manipulating, building and drilling the great tubes of the new Forth bridge, these being as much as 12 feet in diameter; with cuts. Lon. *Engineer*, Jan. 16, 1885, and Lon. *Eng.*, Jan. 9, 1885.

Iron Works. *Description of the Works of the Bethlehem Iron Co.* By W. H. Jaques, late U. S. N. Illustrated by photo-engravings and a map of works. *Proc. U. S. Naval Inst.*, 1889, Vol. XV., No. 4, p. 531; whole No. 51.

Ironwork.

Construction in Terry's New Theatre. By Max Am Ende. Gives description, with drawings of details of the ironwork in Terry's new theatre in London. Lon. *Engineer*, Oct. 7, 1887.

Ornamentation in. A paper before the British Civ. and Mech. Engrs.' Soc. *Am. Eng.*, March 13, 1885.

Irrigation.

By E. B. Dorsey. A valuable paper. Treats of the duty of water, yield of water from water-sheds, storage, reservoirs, evaporation and seepage, yield of irrigated land, distribution of water, etc. Discussed by a large number of engineers. *Trans. A. S. C. E.*, Vol. XVI., p. 102.

For a complete outline of the methods and organization in use in State of Colorado; also remarks concerning flow of water in open ditches, etc.; also map of the water districts and discharge charts for several rivers. See *Fourth Report Biennial Report of State Engineer of Colorado*, Parts I and II., for 1887 and 1888.

Irrigation, continued.

By M. J. Mack. An interesting description of the Montezuma Valley, Colo., with illustrated details of the irrigation works now being constructed. *Eng News*, Dec. 31, 1887.

The Third Biennial Report of the *State Engineer of Colorado* for [illegible]. Contains detailed reports of the water districts; condition of irrigation system of the State, and recommendations of the State Engineer, Mr. E. S. Nettleton.

Apparatus to proportionately divide among several ditches the quantity flowing in a main ditch. *Annales des P. & C.*, August, 1855.

The Bear River Irrigation Canal, Utah. A paper by H. M. Wilson before the Am. Soc. C. E., describing this canal system. Heavy work characterizes the system. Abstract in *Eng. Record*. Sept. 19, 1891, pp. 245-6.

And Irrigation Systems in San Bernardino County, California. A somewhat detailed description of the various systems. Illus. By F. C. Finkle. *Eng. News*, July 4, 25, 1891.

Canal. Description of the Jalapa Irrigation Canal, by Mr. F. C. Finkle, C. E. *Eng. News*, May 17, 1890, Vol. XXIII., p. 463.

Canals of Colorado. From annual report of Denver Chamber of Commerce. The principal canals of the State described, and statistics and dimensions given. *Sci. Am. Sup.*, Sept. 11, 1880.

Canals. See *Canals, Irrigation.*

Ceylon. By Henry Byrne. Describes the different systems employed. *Van Nos. Eng. Mag.*, Vol. XXIII., p. 197.

By R. Abbay. Gives short history of irrigation and details of some of the work of restoring the ancient tank system in Ceylon. *Van Nos. Eng. Mag.*, Vol. XVIII., p. 13.

China. By Gen. Tcheng Ki Tong. Gives an historical description of the greatest systems in China. Translated from *Revue Scientifique*, *Pop. Sci. Mon.*, Oct. 1890 pp. 821-7.

Colorado, as see by the American Society of Civil Engineers. Illustrated. *San. Eng.*, Aug. 5, 1880.

Colorado, as discussed in the Report of the State Engineer for 1884. The various engineering problems connected with extensive systems of irrigation more or less discussed. E. S. Nettleton, State Engr., Denver, Col.

Engineering. American Irrigation Engineering. A paper by H. M. Wilson, giving a detailed description of many large irrigation canals with accessory works and storage reservoirs. Many illustrations. *Trans. A. S. C. E.*, Vol. XXV., Aug., 1891, pp. 161-226. Abstract, *Eng. Record*, Aug. 29, 1891, pp. 201-2.

Engineering. A very good paper by Mr. H. M. Wilson, reviewing the various engineering problems to be considered in an irrigation scheme. *Sch. of Mines Quar.*, Vol. XI., p. 122, Jan., 1890.

Egypt. Paper by Colin Scott-Moncrieff, describing the recent works of improvement. *Ind. Eng.*, July 11, 1891, pp. 32-4.

An abstract from a paper before the French Society of Civil Engineers. Gives illustrated description of the large centrifugal pumps being erected to carry out the programme of irrigating the Bihary district. Lon. *Eng.*, Jan. 28, 1887.

An illustrated series of articles describing works constructed in Egypt. Lon. *Engineer*, Jan. 21, etc.; also *Proc. Inst. C. E.*, Vol. XC., pp. 240-247.

Land Irrigation and Reclamation in Egypt. A letter by C. A. Siegfried, stating briefly what advances in the control of the Nile and consequent improvement of land has been made in the last few years. *The Nation*, July 18, 1889.

Irrigation, *Egypt*, continued.

And the Raiyan Reservoir. Abstract of a paper by Mr. Cope Whitehouse, read at a meeting of Western Society of Engineers, Oct. 2, 1889. *Jour. Assn. Eng. Soc.*, Nov., 1889, Vol. VIII., pp. 572-576.

Functions of the Government in a Plan for General Irrigation. Paper by George A. Brown, read before the Engrs. Club of St. Louis. *Jour. Assn. Eng. Soc.*, Vol. IX., Oct., 1890, pp. 497-505.

Godavari Delta "Completion Projects." By G. T Walsh, M. I. C. E. Gives history and results of this large irrigation and drainage project in India. *Ind. Eng.*, May 30, 1891, pp. 432-6.

"*How to Make the Desert Bloom as the Rose*." An eight-page article, being practically a brief treatise on the subject of irrigation, tracing its past history and suggesting its future possibilities, etc. San Francisco *Daily Chronicle*, Aug. 23, 1889.

Idaho. An interesting report by A. D. Foote, C. E., on the irrigation and reclaiming of certain lands in Idaho. Abstract in *Eng. News*, April 26, 1890, Vol. XXIII., p. 391.

India. By W. T. Thornton. Treats the subject with especial reference to their remunerativeness. *Van. Nos. Eng. Mag.*, Vol. XV., p. 113.

A descriptive paper by Herbert M. Wilson. *Trans. A. S. C. E.*, Vol. XXIII., 1890, Paper No. 454, pp. 217-253. Discussion, pp. 253-60.

The Ashti Tank. Illustrated description of this large storage reservoir, including an earthen dam 12,709 ft. long and 98 ft. in max. height. From *Proc. Inst. C. E.* *Eng. News*, July 19, 26, 1884.

Evils of Canal, in India. By T. H. Thornton, before the Society of Arts. Discusses the evils arising from canal irrigation in India, viz: Impoverishment of soil, waterlogging and poisoning, and malarial influences, and suggests remedies for them. *Jour. Soc. Arts*, March 23, 1888.

Injurious Effects on Health in India. By Surg. Gen. H. W. Bellew, before the Society of Arts. Gives results of personal observation of the injurious effects of canal irrigation on the health of the population of the Punjab, and their remedy. *Jour. Soc. Arts*, May 11, 1888.

In Mysore. Article by C. A. Farget, Ex. Eng. P. W. D., India, giving considerable information concerning problem to be solved and the systems adopted. *Ind. Eng.*, March 22, 1890, *et seq*.

Present Practice in the Construction of Irrigation Works in India. Abstracts from a manual by Lieut. Gen. J. Mullins. Flood discharges and formulas for, are discussed and illustrations of Indian anicuts given. *Eng. Record*, Dec. 5, 1891, *et seq*.

In Southern India. A description of the basin of the Krishna River with its irrigation works. *Van. Nos Eng. Mag.*, Vol. XVIII., p. 158.

Value and Necessity. By H. M. Wilson, M. A. S. C. E., in *Eng. News*, Nov. 21, 1891, pp. 496-7.

Literature of. A list of books, reports and articles from society publications and technical papers bearing on this subject. *Eng. News*, Jan. 23, 1893, p. 76.

Machinery on the Pacific Coast. A paper by John Richards, before the Institution of Mechanical Engineers. Describes the methods employed in irrigation, and then discusses the different kinds of machinery used. *Lon. Eng.*, Nov. 4, *et seq.*; *Sci. Am. Sup.*, Dec. 17, *et seq.*, 1887.

Measurement and Division of Water. Bulletin No. 13 of the Agricultural Experiment Station, Fort Collins, Colorado, contains much valuable information on measuring devices and customs in use in Italy and Colorado with reference to irrigation. By L. C. Carpenter, Prof. Irrigation, Eng. State Agr. Col.

Irrigation, continued.

Nebraska. Selections from a series of letters by Mr. Charles L. Stevenson, C. E., to the Salt Lake *Tribune*, containing facts and figures of some value. *Eng. News*, Oct. 5, 1889. Vol. XXII., p. [illegible].

Periyar Irrigation Works. A brief description, by Prof. A. Chatterton, of these works now in progress, involving the construction of a concrete dam 160 feet high and a cutting and tunnel [illegible] feet long to form a conduit. Lon. *Eng.*, Feb. 6, 1891, pp. 165-7.

Prevention and Reclamation of Alkaline Lands. Abstract of a paper by H. Scorgall before the Polytechnic Soc. of Utah, discussing the prevention of the deposit of alkali and its removal where it exists. *Eng. News*, Nov. 14, 1891, p. 455.

Reclamation of Arid Lands by Irrigation. Report of a committee on arid lands of the California State Board of Trade. A pamphlet of 52 pp. California State Board of Trade, No. [illegible] Market street, San Francisco.

Sluices, Automatic, for either constant level or constant delivery. Some very useful appliances for automatic control of irrigation, canals or mill races. Illustrated. From the French, in *Eng. News* March 13, 1886.

Spain. Treatise on. (Both in French and in Spanish.) By Andres Llaurado. 2nd Edition; 2 vols. Madrid, 1884. Brief review of in *Annales des P. & C.*, 1884-2-619.

Sluices, from Impounding Reservoirs. From the Italian. Chief interest is in the means used to prevent the accumulation of sedimentary deposit in reservoirs receiving mountain torrents. Scour culverts are provided for carrying off this deposit from the bottom of the reservoir. *Abst. Inst. C. E.*, Vol. LXXXIII., p. 473.

United States. A resume of the irrigation enterprises undertaken and carried out in the last few years, showing enormous increase in property values of the lands irrigated, with illustrations. *E. & M. Jour.*, Jan. 5, 1889.

See *Engineering, Agricultural. Water Meter.*

Jenkins, Prof. Fleeming. Obituary notice, giving a lengthy account of his life and works, especially in connection with *telepherage*. Lon. *Engineer*, June 19, 1885; also Lon. *Eng.*, of same date.

An obituary notice in *Proc. Inst. C. E.*, Vol. LXXXII., p. 365.

Jervis, John B., *A Biographical Sketch of*, prepared for the Am. Soc. Civ. Engrs. *Eng. News*, Nov. 28, 1885.

Jet Propellers, *Efficiency of.* By Cavaliere E. Brin. *Van Nos. Eng. Mag.*, Vol. IV., p. 641.

Jet Propulsion. Note by Prof. J. B. Webb before Providence meeting Am. Soc of Mech. Engrs., showing reaction of water jets in air and in water experimentally. Reprint, *Am. Mach.*, June 26, 1891. Discussion same, July 2, 1891.

See *Propulsion, Hydraulic.*

Jets *of Water on Curved Vanes.* By J. P. Church. Gives a good theoretical discussion of the subject. *Van Nos. Eng. Mag.*, Vol. XXII., p. 210.

Jetties.

At the Mouth of the Mississippi River. Surveys and examinations for 1883, with numerous charts and plates showing their effect on shoaling beyond their mouth, etc. *Rpt. Chf. of Engrs., U. S. A.*, 1883, Vol. I., p. 1031.

At Port of La Pallice. An interesting illustrated article describing methods used in construction, etc.; from *Le Genie Civil.* *Sci. Am. Sup.*, Jan. 15, 1890, p. 11,706.

At South Pass Miss. River.

By E. L. Corthell. With discussion. A general account of the methods used in construction and early results. Many large plates. *Trans. A. S. C. E.*, Vol. VII., (1878), p. 131-165.

Jetties, *South Pass, Miss. R.*, continued.

Notes on the consolidation and durability of the work, with a description of the concrete blocks and other constructions to secure permanence. By Max E. Schmidt. Fully illustrated. *Trans. A. S. C. E.*, Vol. VIII., p. 189.

Repairs. Illustrated article describing the nature of work to be done, and methods used in repairing these jetties in 1889. *R. R. Gaz.*, Sept. 13, 1889, Vol. XXI., p. 594.

Ten Year's Practical Teaching in River Hydraulics. Discussion of a former paper on this subject by E. L. Corthell, with direct application to the problem at Galveston Harbor. Discussion by Messrs. Merrill, Whittemore, Comstock, Boller, Post, North, McMath, Savage, Chanute, Bixby, Le Baron, Gillmore, Corthell and Eads. The whole occupies over 100 pp. in *Trans. A. S. C. E.*, April, 1886. Vol. XV., p. 203.

Full discussion of Mr. Corthell's paper before the Am. Soc. of C. E. Also includes a discussion of Galveston Harbor scheme. *Trans. A. S. C. E.*, April, 1886.

Ten Years' Practical Teachings in River and Harbor Hydraulics. By E. L. Corthell, former Engr. of the jetties. *Trans. A. S. C. E.*, Vol. XIII., p. 313.

Ocean Jetties in New South Wales. Brief description, with method of placing piles in a rock bottom. By W. A. Harper. *Proc. Inst. C. E.*, Vol. CII., 1890. Paper No. 2430, pp. 309 6.

See *River Improvement, Columbia River. Littoral Movements.*

Johnstown Disaster. Descriptions of South Fork Dam, and the causes of its failure, by Engineers who personally visited the site immediately after the failure. Illustrated. *Eng. News*, June 8, 15 and 22, 1889. *Eng. & Build. Rec.*, June 8 and 15, 1889.

Journal *and Journal Bearings.* Discussion at the meeting of the New England Railroad Club, including papers by Mr. Adams, Mr. Lauder, on above subject; Mr. Meneeley on Roller Bearings, and Mr. Shuttuck on the Trip Journal Bearing; also remarks and discussions by other members. *Mast. Mech.*, Dec., 1889, Vol. IV., p. 214.

See *Friction. Lubrication.*

Joints. *Holding Power of Dovetails.* Table of tests. *Eng. News*, Dec. 20, 1890, p. 544.

Journal Bearings.

Anti-Frictional Metal for Journal Bearings. Discussion of the subject before the Western Railway Club, Jan. 15, 1889. *Mast. Mech.*, Feb., 1889.

Discussion on the Anti-Frictional Metal for Journal Bearings and on heating cars by steam at January 15 meeting of Western Railway Club. *Mast. Mech.*, Feb., 1889.

Keels. Notes on bilge keels. By Asst. Naval Constructor F. T. Bowles, U. S. N. *Proc. U. S. Naval Inst.*, No. 32.

Kinematics.

Diagrams. By Prof. R. H. Smith. A paper before the Royal Society, 1885. Abstract of method and illustrative diagrams in Lon. *Engineer*, July 31, 1885, and April 2, 1886.

Kinematics of Machinery. See *Machine Designs.*

Kinemeter, *Hayne's.* An illustrated description of Hayne's kinemeter. It is intended to serve the purpose of a speed indicator, recorder and governor. *R. R. Gaz.*, Nov. 18, 1887.

Knots for Engineers. Twenty-nine different kinds of knots described and illustrated. *Eng. News*, Dec. 26, 1885.

Krupp Steel Works, *Germany.* By Moncure D. Conway. A popular article written after a personal inspection of the works. This privilege was not granted

Krupp Steel Works, continued.

even to the American Gun Foundry Board in 1884. A pleasing description of some technical value. Well illustrated. *Harper's Monthly Mag.*, March, 1886.

Kutter's Formula. See *Hydraulics*, *Sewers*, *Formula*

Labor *Problem*. A lecture by J. C. Bayles, Prest. Am. Inst. Min. Engrs., before Cornell University students. A clear statement of the progress of the labor movement in Europe and a most graphic description of the present situation in this country. A very readable production, which sets one to thinking. *Sci. Am. Sup.*, April 10, 1886.

See also a paper by Andrew Carnegie in *The Forum* (New York) for April, 1886. A valuable contribution to the philosophy of the subject.

Also a remarkable series of letters, articles and opinions from business and practical men, as well as from students of political economy and professional men. The best and most practical discussion of these subjects that has ever appeared in America. *Age of Steel* (St. Louis), February to May, 1886.

Premium Plan of Paying for. Paper by F. A. Halsey before Providence meeting A. S. M. E. Reprint. *Am. Mach.*, June 18, 1891. Discussion, same, July 1, 1891. *Mechanics*, Aug., 1891. *Eng. News*, July 18, 1891, pp. 61-62.

See *Railroad Labor*.

Laboratories. See *Engineering Laboratories*, *Hydraulic Laboratory*.

Lake. *The New Lake in the Desert*. Description of this lake and surrounding country. By J. W. Powell. *Scribner's Mag.*, Oct., 1891, pp. 463-8. Abstract with additional information in *Eng. News*, Oct. 3, 1891, pp. 313-14.

Lake Currents *and Projected Opening of the Breakwater*. A paper by Walter P. Rice, read before the C. E. Club of Cleveland, giving results of some experiments at Cleveland as to the relation between winds and currents, with discussion. *Jour. Assn. Eng. Soc.*, Jan., 1891, pp. 9-17.

Lake Survey. See *Survey of the*.

Lamp. See *Siemens's Gas Lamp*.

Land Division. A brief account of land grants and surveys in Ohio. By R. W. McFarland. *Report Ohio Soc. Surveyors*, 1883.

Land Locked Navigation. See *Navigation*.

Landings. See *Floating Landings*, *Ferry Landings*.

Landslide *on Boston & Maine R. R.* By Ed. S. Philbrick. Gives details of a landslide on the Boston & Maine R. R., near Dover Station. Contains details of the construction of the bridge abutments and bluffs. *Jour. Assn. Eng. Soc.*, Vol. VI., p. 453.

Caused by accumulation of water in an underground grotto. Illustrated. *Lon. Engineer*, Feb. 6, 1885.

Constitution of Earths and Slips in Clayey Soil. Constitution of earths, theory of slips and means of preventing. From *Annales des P. & C.* *Eng. News*, Oct. 16, 23, 1880.

On the Hudson River R. R. Slide covered four tracks with a depth of from 2 to 8 feet for a distance of 150 feet. View and profile in *Eng. News*, April 23, 1887.

Removal of. A paper by W. G. Curtis, read before the Am. Soc. C. E., describing the removal of 9,000 cu. yds. by means of a water jet. Illus. *Ry. Rev.*, Aug. 29, 1891, pp. 558-9. *R. R. Gaz.*, Aug. 28, 1891, p. 593.

Landslides. *Report on Various Methods of Consolidating Earth that has Slipped*. By M. Comoy, Ponts et Chaussees. Very valuable. Giving methods of treatment of several landslides. Translated and published in *Eng. News*, Sept. 22, Oct. 13, 1887.

Lantern Slides *by Contact Printing*, suitable for amateurs. *Sci. Am. Sup.*, Nov. 28, 1885.

Lathe Tools. Hints regarding the choice and care of tools for different kinds of work. *Eng. Mechanic and World of Science*, May 19, 1889, *et seq.*

Lathes. *Horse Power Required to Run Lathes.* Results of experiments with various tools and lathes. By J. J. Flather in *Am. Mach*; *Mech. World*, May 23, p. 1891.

Latimer, *Charles, Eulogy Upon the Life of.* By W. H. Searles, before the Civil Engineers' Club of Cleveland. *Jour. Assn. Eng. Soc.*, June, 1888, Vol. VII., pp. [illegible].

Latitude Determination. See *Time, Latitude and Azimuth. Determination of Latitude.*

Latitudes *and Longitudes by Horizontal Telescopes.* By Prof. J. N. Stockwell. Proposes a new instrument called the horizontal telescope, for determining the differences of latitude and longitude. *Jour. Assn. Eng. Soc.*, Vol. VI., p. 408.

Longitude and Azimuth. Formulas and factors for the computation of. The formulas derived and factors given between latitudes 23° and 65°. Appendix 7 of *Report for* 1884, *U. S. Coast & Geo. Survey.* Sent on application.

Variations of, to be investigated by fixed observatories. Methods for making such observations, together with a general discussion of the subject. By Prof. Hall. *Am. Jour. of Sci.*, March, 1885.

Launches. Drawings and description of a light-draught screw launch for use on the Nile. The propeller has guide blades. Lon. *Eng.*, April 10, 1885.

Electric. See *Electric Launches.*

Law. See *Marine International Mining Code. Mining Laws.*

Lead and Zinc. *Production of in* 1888. Resume of the industry for the year. *E. & M. Jour.*, Jan. 12, 1889.

Leaks in Gas Mains. See *Gas Mains.*

Leaks in Water Mains. *Detection of.* See *Water Mains.*

Least Squares.

On some simplifications which may be made in the application of the method of least squares. By Dr. Neil. *Zeitschr. f. Vermessungswesen*, 1887, pp 454-467.

Proof of Method. By Prof. Vogler. *Zeitschr. f. Vermessungswesen*, 1887, pp. 141-147, 183-180.

See *Observations.*

Legislative Control *of Railways.* See *Railroad Management.*

Lenses. See *Photographic Lenses.*

Levees. *Construction.* A paper before the Tech. Soc. of the Pac. Coast, by Geo. J. Specht. Gives cross-sections, methods, implements, cost, etc., also a valuable method of computing the necessary haul in R. R. earthwork. Reprinted in *Eng. News*, Aug. 22, 1885.

Davis Crevasse. By S. F. Lewis, before the American Society of Civil Engineers. Gives details of the break in Davis levee, near New Orleans, and the method adopted in its repair. *Trans. A. S. C. E.*, Oct., 1888, pp. 169-204.

On the Mississippi River. The basis of the claim in their behalf, in effectively improving the low-water channel, denied, a controversy participated in by J. B. Johnson, B. M. Harrod, and T. T. Johnston. *Am. Eng.*, May 1, May 29, June 26, July 30, Aug. 6, Aug. 27, and Oct. 1, 1885.

Of the Mississippi River. By Caleb G. Forshay. Gives outlines of history of levees on the Mississippi; discusses the forms and dimensions of what is required of a levee. A valuable paper. *Trans. A. S. C. E.*, Vol. III., (1874), pp. 267-281.

Their Relation to River Physics. By Robert E. McMath. A discussion of the subject as applied to the Mississippi River and its improvement. A valuable paper. *Jour. Assn. Eng. Soc.*, Vol. III., p. 43.

Specifications for. *Eng. News*, Sept. 1, 1877, pp. 213-4

Levees, continued.

As a System for Reclaiming Low Lands. By Geo. W. R. Bayley. A general review of their effects on European rivers, and its application to the Mississippi River. *Trans. A. S. C. E.*, Vol. V., (1876), pp. 115-146. Discussion, pp. 279-308.

Theory Tested by Facts. By Robt. E. McMath. Being a careful analysis of a great deal of data, which goes to show that confining the flood-waters by levees is not conducive to the interests of low-water navigation. *Trans. A. S. C. E.*, Vol. XIII., p. 331.

Levees. See *River Improvements.*

Level. *A New Form of Water Level.* See *Water Level.*

Level Rod,

Design for Self Reading Rod. By H. M. Baines. A good design for easy reading. Illustration, *Ind. Eng.*, Apr. 25, 1891, p. 318.

New Form of. Illustration of successful form of self-reading rod. Designed by M. Nicholson. *Eng. Record*, Sept. 26, 1891, p. 267.

Leveling.

Determining Slope of Canal. See *Canals, Erie.*

Hand instruments for leveling and measuring vertical angles. By Prof. Jordan. Description of various German instruments, and results of tests. *Zeitsch. fur Vermessungswesen*, 1887, p. 113.

Precise Swiss. Also a new method of making the bubble visible to the observer when reading the rod. Abstracted from the French. *Proc. Inst. C. E.*, Vol. LXXVIII.

Precise. Reports on field methods, with results, discrepancies, elevations and descriptions of bench-marks, from Biloxi, Miss., to Carrolton, La., and from Cairo, Ill., to Fulton, Ill. *Report Miss. River Commission*, St. Louis, 1883, p. 48.

Probable Errors of, with rules for the Treatment of Accumulated Errors. By Wilfred Airy in *Proc. Inst. C. E.* Treats of various classes of errors and methods of adjustment. *Eng. News*, March 31, April 28, 1877.

Sources of Error in Spirit Leveling. By J. B. Johnson. Gives the results of an extended experience in precise leveling on the U. S. Lake and Miss. Riv. surveys. *Jour. Assn. Eng. Soc.*, Vol. II., p. 162.

Survey of France. See *Surveys.*

Trigonometrical. A discussion of the work on the Davidson Quadrilateral in California, the elevations of the four stations, ranging from 21 to 1,173 meters above sea level. A least square reduction given. *U. S. C. & G. Survey Rep.*, 1884, App. No. 10.

Trigonometrical. The Methods and formulæ used on the *New York State Survey.* See *Report*, 1882, App. 2.

See *Barometer. Bench Marks. Hypsometry.*

Levels.

Errors in Railroad. By Howard V. Hinckley. Gives tables of checks at various places, also a table of side checks. *Trans. A. S. C. E.*, Dec., 1886.

Of the Great Lakes as affected by the proposed Lake Michigan and Mississippi waterway. See *Waterway.*

On the Missouri River. A descriptive list of bench-marks, with their elevations, from the mouth of the Missouri river to a point about 1,200 miles above; all referred to the St. Louis City Directrix and to mean Gulf level. Over seven hundred elevations in this list. *An. Rpt. Chf. Engrs., U. S. A.*, 1883, Vol. II., p. 1,328.

Level-Vials. *Irregularities in.* Shown to be due to excrescent formations insoluble in alcohol, formed on the side of the tube. *Abst. Inst. C. E.*, Vol. XC., p. 450.

Liberty, Statue of. *Plans* and elevation, with a general description, dimensions, etc., *Eng. News*, Nov. 20, 1886.

Life Boats. See *Boats*.

Light.

The Artificial Light of the Future. An interesting address by Prof. E. L. Nichols before the N. Y. Electric Club, discussing the efficiency of various sources of light. *Elec. Rev.*, (N. Y.) Nov. 29, 1890, pp. 166-7, 70-1. *Sci. Am. Sup.*, No. 780, Dec. 13, 1890, pp. 12,460-1.

On the Cheapest Form of Light. By Prof. S. P. Langley and F. W. Very. This is an interesting paper on the radiation of the fire-fly which is intended to show the possibility of producing light without heat other than that in the light itself. 3 plates. *Am. Jour. of Sci.*, Aug., 1890, No. 236, Vol. XL., pp. 97-113.

Standards of. A paper read before the Society of Arts, Dec. 21, 1888. By W. J. Dibdin. *Jour. Soc. Arts*, Dec. 21, 1888.

Velocity of. By Prof. A. A. Michelson. Treats of the velocity of light in air and refracting media. *Jour. Assn. Eng. Soc.*, Dec., 1886.

Light Tower. *Working Platform* used in the construction of Mosquito Inlet Light Tower, Florida. Gives description of a novel method of supporting the working platform. Illustrated. *Report of the Lighthouse Board* for 1887, Washington.

Lighthouse.

Construction on sand foundation in 14 feet of water off the Delaware coast. Wooden pneumatic caisson, with concrete monolithic pier encased in iron. *An. Rpt. Light-House Board*, 1886.

Eddystone. Fully illustrated description. By Wm. T. Douglass before the Inst. C. E. Reprinted in *Eng. News*, April 19, 1884, pp. 181-5.

Eddystone. Gives short history of construction of the new tower and description of the lantern. *Van Nos. Eng. Mag.*, Vol. XXVII., p. 120.

Electric, of Macquarie and of Tino. Paper read before the Inst. of C. E., London, Eng., Dec. 7, 1886, by John Hopkinson, M. A., D. Sc., F. R. S., etc. Full description, illustrated, with discussion of advantages, etc. *Electrician & Elec. Eng.*, April, 1887. *Proc. Inst. of C. E.*, Vol. LXXXVII., p. 243.

Engineering, as displayed at the Centennial Exhibition. By J. G. Barnard. Illustrates quite fully this branch of engineering as now practiced in the U. S. Twelve full-page plates. *Trans. A. S. C. E.*, Vol. VIII., pp. 55-91.

Foundations of Piles. See *Foundation*.

Illumination. Comparative experiments on gas, oil and electric light. By Dr. H. Kruss. *Jour. f. Gasbel. u. Wasserv.*, 1886, pp. 799-801.

Investigation of the stability of the proposed Diamond Shoals Lighthouse. *Eng. News*, Aug. 2, 1890, p. 100. The specifications for this lighthouse are published *ib.* Aug. 9, 1890, pp. 115-7. A folding inset shows the details of caisson and substructure as designed by the three lowest bidders.

Japan. Account of the establishing of lighthouses on the Japan coast at the time the ports were opened to foreign powers. By R. H. Brunton before Inst. C. E. Reprint, *Eng. News*, July 14, Aug. 18, 1877.

Lanterns. General and detail drawings of a 14-foot lantern designed by Sir. Jas. Douglass, Engr. Lon. *Engineer*, April 16, 1886.

Moving. See *Movement of a Lighthouse*, also *Foundations of Lighthouses*.

Proposed Diamond Shoals Lighthouse. Three of the plans which were submitted to the Lighthouse Board for a 150-ft. tower off Cape Hatteras, are described and illustrated by a folding inset. *Eng. News*, July 26, 1890, p. 76.

Stability of. See *Wave Pressure*.

Report on the European system, by Maj. Elliot. *Van Nos. Eng. Mag.*, Vol. XIV., p. 87.

Lighthouse, continued.

Report of the U. S. Lighthouse Board, 1885. Contains three appendices of value 1. Report on experiments made at the South Foreland, near Dover, Eng., to determine the comparative power of oil, gas and electricity as lighthouse illuminants, by French E. Chadwick, Commander U. S. N. Illustrated by seven plates. 2. Report on the Hell Gate electric light, Hallett's Point, N. Y., and on progress of experiments on electricity as an illuminant for lighthouses, by John Mills, Lieut. of Engineers, U. S. A. Thirteen plates. 3. Report on the construction of the Detroit River Light Station, in Lake Erie, in twenty feet of water. Five plates.

Rothersand. By O. Offergeld, before the Association of Architects and Engineers, at Hamburg. Gives details of the construction of the Rothersand lighthouse in the North Sea, with two-page plates showing plans and sections of the pneumatic caisson. Valuable. Lon. *Eng.*, Dec. 2 and 16, 1887.

Lighting.

By Compressed Oil Gas. An illustrated article reprinted from *Le Genie Civil. Sci. Am. Sup.*, No. 760, July 26, 1890, p. 12140.

Cost of Lighting by Gas and by Electricity. Address by W. H. Preece, F. R. S., giving results of practical experience. *Mech. World*, July 18, 1891, *et seq.*

Incandescent Gas Lighting. A paper by W. Mackean, F. C. S., describing several systems, and the nature of various illuminating oxides. Discusses also the difficulties encountered in the practical application. *Jour. Soc. Chem. Indus.* Reprinted in *E. & M. Jour.*, May 16, 1891, pp. 585 6.

On the Measure and Distribution of Light. By. Dr. H. Kruss. Mathematical treatment of the subject, with applications to the illumination of closed spaces. *Journal fur Gasbeleuchtung und Wasserversorgung*, 1886, pp. 66-80.

Of Streets. Tables showing the cost of lighting streets by gas, gasoline and electricity, in different cities in the U. S. *Eng. News*, June 28, 1890, p. 607.

See *Electric Lighting*.

Lightning.

And Lightning Conductors. By W. H. Preece, before the Soc. of Tel. Engrs. Treats of atmospheric electricity, its effects and methods of protection. *Van. Nos. Eng. Mag.*, Vol. VIII., p. 147.

Arresters and the Photographic Study of Self-Induction. A paper read before the American Institute of Electrical Engineers, New York, Jan. 8, 1889, by E. G. Acheson. Illustrated. Describes a series of experiments undertaken to determine the cause of the occasional "grounding" and failure of telegraph and telephone cables by electrical discharges, and some conclusions as to lightning arresters. *Elec. Eng.*, February, 1889; *Elec. World*, Jan. 19, 1889.

Arrester. See *Electric Lightning Arrester*.

Conductors. Gives some of the most important directions for the erection of lightning conductors drawn from the joint report of the English Lightning Rod Conference. *Building*, April 2, 1887.

Guards for telegraphic purposes and the protection of cables from lightning, with observations on the effect of Conducting enclosures. Paper by Dr. Oliver Lodge read at the meeting of the Institution of E. E. Illus. *Elec. Rev.*, May May 16, 23, 30, 1890, pp. 567, 592, 621.

Protection of Buildings from. By Capt. J. S. Bucknell. Treats of the present practice and advocates the use of wire ropes instead of rods, etc. *Van Nos. Eng. Mag.*, Vol. XXVII., p. 154.

Protection of Buildings from. A lecture by Prof. Oliver J. Lodge before the Society of Arts. *Jour. Soc. Arts*, June 15, 1888.

Lightship, *with Electric Lights*. Illustrated description of ship and equipment. *R. R. & Eng. Jour.*, Aug. 1891, pp. 361-4.

Lime. A *resume* of several series of experiments made by French engineers and others into the nature and treatment of lime. *Van Nos. Eng. Mag.*, Vol. XIII., p. 20.

Lime Floors. *Mexican Method of Making Hard.* By Theo. G. Ellis. Describes the methods used in Mexico to make hard lime floors. *Trans. A. S. C. E.*, Vol. II., p. 179.

Lime Sulphite Fibre. See *Fibre.*

Limes. See *Cements.*

Link Motion. *Errors of the Zeuner Diagram* as applied to the Stephenson. By W. E. Marks. *Van Nos. Eng. Mag.*, Vol. XXIII., p. 353.

See *Locomotive Tests.*

Linkages. By J. D. DeRoos. A kinematic discussion of the different forms of articulate links with especial reference to Peaucellier's cell. *Van. Nos. Eng. Mag.*, Vol. XXI., pp. 159, 239 and 318.

Linkwork. By A. B. Kempe. Treats of a general method of producing exact rectilinear motion by linkwork. *Van Nos. Eng. Mag.*, Vol. XVI., p. 103.

Liquid Fuel. See *Fuel-Liquid.*

Literature. See *Engineering and Technical Literature. Engineering Literature, Technical.*

Littoral Movement *of the New Jersey Coast*, with remarks on Beach Protection and Jetty Reaction. Paper giving results of some investigations. By Lewis M. Haupt. Illustrated by several plates. *Trans. A. S. C. E.*, Vol. XXIII., 1890. Paper No. 452, pp. 123-41. Discussion, pp. 142-54.

Lixiviation *of Argentiferous Zincblende and Galena Ore.* A valuable paper by Mr. Ottokar Hofmann. *E. & M. Jour.*, Feb. 9, 1889, *et seq.*

Lobed Wheels. *Derived from Ellipses.* Demonstration of rolling properties and graphical methods of design, by Horace B. Gale. Illus. *Jour. Frank. Inst.*, Feb., 1891.

Lock Gates. Winch to open and to shut. *Annales des P. & C.*, Nov., 1886.

Lock Masonry. *Cost of, in Detail.* See *Masonry.*

Locks *of Royal Albert Docks*, London. Fully described, and plans, sections, details given. Lon. *Engineer*, Feb. 5, 12, and 19, 1886.

Locomotive Boilers.

Auxiliary Dome and Connections—Class "O" Engines, Pennsylvania Railroad. Illustrated description of this apparatus designed to secure dry steam. *Ry. Rev.*, Aug. 15, 1891, p. 532.

Belpaire Type. Gives detailed drawings of a 52-inch straight top boiler designed for the new Mogul engines on the Chicago, Burlington & Quincy Railroad. *Mast. Mech.*, March, 1885.

Circulation in. By John Hickey, before the Western Railway Club. Discusses the proper construction of locomotive boilers and fire-boxes to obtain the most economic results. *Mast. Mech.*, Oct., 1888. Discussion on this paper in *Mast. Mech.* for November.

Circulation and Purification of Water in. A paper read before the Western Railway Club, Dec., 1888, by Mr. Herbert Hackney, giving results obtained on locomotives using his device, with short discussion. *Mast. Mech.*, Jan. 1889; *R. R. Gaz.*, Dec. 28, 1888.

Experiments to determine variations in economy due to different lengths of tubes, fire-brick arches and water arches. This is a valuable paper reporting the results of extended experiments on the Paris, Lyons & Mediterranean R. R. Table and diagrams. *R. R. Gaz.*, July 4, 1890, pp. 460-2.

For High Pressures. Description, with details, of a locomotive boiler to stand a working pressure of 180 pounds per sq. in., as constructed by Rhode Island Locomotive works. *Mast. Mech.*, March, 1890, p. 40.

Locomotive Boilers, continued.

Improvements in Locomotive Boiler Construction. Abstract of a paper by John Hickey, read before the Western Railway Club, giving present requirements of material and many recent improvements in the details of construction. Illustrated. *R. R. Gaz.*, April 3, 1891, p. 239. *Eng. News*, April 4, 1891, pp. 321-2. Full paper in *Ry. Rev.*, April 4, 1891, pp. 212-14.

Laws with regard to Locomotive Boilers. Collection and discussion of the existing laws in Germany regarding locomotive boilers. *Archiv. fur Eisenbahnwesen*, 1886, pp. 213-222.

Lentz's Stayless Locomotive Boilers. The fire-box is made of corrugated flues. Illustrated description. Lon. *Eng.*, Dec. 19, 1890, pp. 724-6. *R. R. Gaz.*, Jan. 9, 1891, pp. 22-3. *Ry. Rev.*, Jan. 24, 1891, pp. 52-3.

Multitubular Boilers. Typical specifications for *Mech. World*, July 25, 1891, p. 32.

Screw Stay Bolts for. A discussion of some value relating to size, method of inspection, materials for, and standard tests, at the annual meeting of the New Eng. R. R. Club. *Ry. Rev.*, March 21, 1891, pp. 180-1. *R. R. Gaz.*, March 20, 1891, pp. 193-4. Editorial discussion, *Eng. News*, March 21, 1891, pp. 276-8.

Strains in Locomotive Boilers. A paper read at the Nashville meeting of the American Society of Mechanical Engineers. By L. S. Randolph, Mt. Savage, Md. Showing that the failure of locomotive boilers is generally due to unequal expansion and contraction of the fire-box sheets. *Am. Eng.*, May 16, 1888.

Treatment of waters used in Locomotives to Prevent Incrustation. A paper by J. N. Barr, before the Western Ry. Club, giving some examples in which the compound used proved advantageous. Gives analyses of waters and amount of compound necessary. *Ry. Rev.*, Nov. 14, 1891, pp. 743-4. *R. R. Gaz.*, Nov. 13, 1891, pp. 803-4. Abstract *Mast. Mech.*, Dec., 1891, p. 188. Additional Notes on above paper, *R. R. Gaz.*, Nov. 27 and Dec. 4, 1891. Abstract of entire discussion. *Eng. News*, Dec. 5, 1891, pp. 531-2.

Water for, and Practice in Washing out Boilers. By G. A. Gibbs, at January meeting of Western Railway Club. Treats the subject from a practical point of view. Gives some experience gained by the Chicago, Milwaukee & St. Paul Railroad. *R. R. Gaz.*, Jan. 20, 1888.

Locomotive Brakes.

Water Brakes. Illustrated description of the Le Chatelier Water Brake in use on the Northern Pacific, the Denver and Rio Grande and the Barre Railroad in Vermont. *R. R. Gaz.*, March 15, 1889.

See *Brakes*.

Locomotive Building. See *Locomotive Construction*.

Locomotive Connecting Rods *and Crank Pins, C., B. & Q. R. R.* For class "H" Engine. Description with detail drawings. *R. R. Gaz.*, Jan. 16, 1891, p. 40.

Connecting Rod Ends. Review and discussion of various methods of constructing connecting rod ends. Illus. *Mech. World*, April 12, 1890, p. 153, *et seq.*

Coupling and Connecting Rods. A series of papers on the proper design of coupling and connecting rods for locomotives. Lon. *Engineer*, May 25, *et seq.*, 1888.

Locomotive Construction. *Controlling Expansion in Locomotives.* By E. W. Hall. Discusses the present practice in the construction of rack, reverse lever and catch, and advocates the use of double catch. *Proc. Eng. Club, Phila.*, Vol. VI., p. 116.

Cost of Rebuilding. Gives details of the cost of rebuilding a locomotive at the Ohio & Mississippi railroad shops at Vincennes, Ind. Also an editorial on the subject. *R. R. Gaz.*, Oct. 19, 1888.

English. The construction of the London, Brighton & South Coast Ry. By Wm. Stroudley. Special features to adapt to high grades and sharp curves. With discussion. Illustrated. *Proc. Inst. C. E.*, Vol. LXXXI.

Locomotive Construction, continued.

Future of. By Geo. S. Strong. An analysis of the present practice and of the future demands, with suggestions as to how these are to be met. Considers fast trains in England and America, the increase of freight traffic, train resistence, boiler power and economy of fuel; 23 pp. *Pro. Eng. Club. Phila.*, Vol. IV., No. 3.

Proportions of Locomotives, with tables of "Coefficients of the Motive Power of Locomotives" and "General Dimensions of Standard Locomotives, Penna. R. R." *R. R. Gaz.*, April 26, 1889.

Smoke-Boxes and Stacks, Improvements in, as made at Union Pacific shops, Denver, by Mast. Mech. Wertheimer. Said to be very efficient. Illus. *R. R. Gaz.*, July 10, 1885.

See *Locomotive Boilers. Steel Castings.*

Locomotive Counterbalancing.

An Account of Certain Experiments upon Several Methods of Counterbalancing the Reciprocating Parts of a Locomotive. Gaetano Lanza with certain students. Very full experiments made upon an eighth-scale model, to determine the amounts of the various motions resulting. *Tech. Quart.*, Vol. II., 1889, No. 3, pp. [illegible]-47.

By E. L. Hill. Gives history and theory of counter-balancing, and methods employed on various locomotives. *Proc. Inst. C. E.*, Vol. CIV., 1891, pp. 265-86.

By C. A. Smith. Treats the subject both theoretically and practically. *Van Nos. Eng. Mag.*, Vol. XXIII., p. 9.

By Prof. Lanza, before the Scranton meeting of the Am. Soc. of Mech. Engrs., Gives results of experiments on the effects of different methods for counterbalancing the reciprocating parts of a locomotive. *Mast. Mech.*, Nov. 1888. *R. R. Gaz.*, Oct. 19, 1888.

Counterbalancing a Main Rod. An illustrated description of a practical method of counterbalancing a main rod on locomotives. *Mast. Mech.*, Jan., 1889.

Effect of Counterbalancing on Locomotives. Experiments in relation to vertical motion, with theoretical discussion. Curves automatically drawn. By S. D. Bawden. *Technograph* (Univ. of Ill. annual), 1890-91, pp. 51-63.

Paper before the April meeting of the Western Railway Club, by Mr. W. H. Lewis, Master Mechanic C. B. & N. Illus. *R. R. Gaz.*, April 25, 1890, Vol. XXIII., p. [illegible]. *Ry. Rev.*, April 26, 1890, Vol. XXX., p. [illegible]. *Eng. News*, April 26, 1890, Vol. XXIII., p. 3[illegible]. *Mast. Mech.*, May, 1890, p. 8[illegible].

See *Locomotive Wheels.*

Locomotive Drivers. See *Locomotive Wheels.*

Locomotive Exhaust Nozzles. *Wertheimer's Variable Exhaust Nozzle.* Can be operated by the engineer. Its use for a year shows a considerable saving of fuel. Illustrated description. *R. R. Gaz.*, July 31, 1891, pp. 534-5.

Pipes and Steam Passages. Report of committee of Master Mechanics' Assoc., gives drawings of various forms of pipes with indicator cards from tests. *Ry. Rev.*, June 20, 1891, pp. 405-6. *R. R. Gaz.*, June 19, 1891, pp. 427-9. *Eng. News*, June 20, 1891, pp. 598-9.

Locomotive Fire Boxes.

Brick Arches in. Report of the committee of the Ry. Master Mechanics' Assoc. *Eng. News*, Aug. 9, 1890, p. 121.

Corrugated. Condensed from paper before German Technical Railroad Union, by Herr H. Lentz, Engineer, and published in *Glaser's Annalen.* Illustrated. *R. R. & Eng. Jour.*, April, 1890.

Firebrick. Abstract of remarks of Herr Bork, Locomotive Inspector of Hungarian Railroad, before German Technical Railroad Union, giving favorable experience with firebrick-lined fire-boxes. *R. R. & Eng. Jour.*, May, 1890.

Locomotive Fire-Boxes, continued.

Fire-Brick Lined. Article stating advantages of, as shown by experience. *R. R. & Eng. Jour.*, Oct., 1889. Vol. LXIII., p. 442.

Staying of. An article by Arthur T. Woods, giving a discussion on the proportioning of stays and stayed surfaces, and the stresses therein. *R. R. Gaz.*, April 17, 1891, pp. 266-7.

See *Locomotives, Extension Front. Locomotive Tests.*

Locomotive Fuel. See *Fuel. Fuel, Liquid. Locomotive. Burning Soft Coal.*

Headlight, Electric. Description of method of constructing and the advantages claimed for electric headlights. Signals said to have been seen easily at a distance of 1,200 feet. *R. R. Gaz.*, Jan. 17, 1890, p. 32.

Headlight, Electric. Description of a successful electric headlight. Illustrated. *Elec. World*, Sept. 27, 1890.

Locomotive Journals.

Bearing Brasses. An argument in favor of altering the form of and increasing weight of locomotives. *Am. Mach.*, March 7, 1885.

Locomotive Tests.

By H. G. Manning, C. E. A valuable article giving indicator diagrams and tabulated results of a series of tests made on passenger locomotive No. 18, on the Central Vermont Railroad. Indicator diagrams are given for four points of cut off. *R. R. Gaz.*, July 12, 1889.

By H. G. Manning. This paper discusses the proper method of making tests. *R. R. Gaz.*, July 18, 1890, p. 509.

Baldwin Ten-Wheeler on the Baltimore & Ohio Road. Full details of test and apparatus used in testing this very heavy express locomotive. *R. R. Gaz.*, Nov. 27, 1891.

Calorimeter in Testing. Paper by Geo. H. Barrus, describing its use in the tests of the Baldwin compound locomotive, May 3, 1890, with records. *R. R. Gaz.*, Oct. 17, 1890, pp. 711-12.

Comparing Radial Motion and Link Motion. By Angus Sinclair. Gives details of test trips made on the Burlington, Cedar Rapids & Northern Railroad, with engines of similar dimensions but equipped with different valve-gear. Gives 38 indicator diagrams. *Nat. Car & Loco. Builder*, April, 1888.

Details of a test made of a locomotive on the New Jersey Central Railroad, by Messrs. H. S. Wynkoop and John Wolff, with tables and indicator diagrams. *R. R. Gaz.*, Aug. 17, 1888.

Economy Tests of Various Forms of Fire-Boxes. Made on the Eastern Railway of France. Detailed account in *Revue Generale des Chemins de fer*, Feb., 1891. Abstract giving main deductions from the tests. *R. R. Gaz.*, May 22, 1891, p. 349.

Fuel Saving Devices on French Locomotives. Tests of. A paper by Messrs. Lencauchez and Durant, read before the Societe des Ingenieurs Civils giving an account of tests with the Ten Brinck fire-box, feed-water heater, steam dryer and other devices. Abstract with illustrations of this fire-box. *R. R. Gaz.*, May 8, 1891, p. 319.

Indicating a Locomotive on the Old Colony Railroad. Article describing method used and giving several cards from an 18x24-inch Old Colony Locomotive. *Mast. Mech.*, March, 1890, p. 39.

Indicating Locomotives. Paper by Mr. Lewis F. Lyne, giving practical suggestions for use of indicators on locomotives. *R. R. Gaz.*, Aug. 2, 1889.

Indicating the power of locomotives. Practical hints on. *Am. Jour. Railway Appliances*, April 1, 1885.

Indicator Diagrams from Locomotive Engineers on the Cin., N. O. & Tex. Ry. Forty sets of diagrams, with tabular statements of conditions, being for various train-loads, speeds, kinds of Engines, pressure, etc. *R. R. Gaz.*, Jan. 2, 1885.

Locomotive Tests, continued.

Method of Testing. By A. T. Woods. *R. R. Gaz.*, June 13, 1890, p. 409.

Standard Method of Testing. By F. W. Dean. Excellent paper with tabulated results. *R. R. Gaz.*, June 20 and 27, 1890, pp. 426–7, 453–4.

Of the Value of Different Proportions. By Angus Sinclair. A general account with some discussion, of experiments made on the C. B. & Q. R. R. to determine the relative values in speed and power of various working dimensions in locomotives. The experiments were careful and elaborate, and should be published in great detail. *National Car & Loco. Builder*, Dec., 1886. *Also* editorials bearing on the same subject in the *R. R. Gaz.*, Sept. 3, 1886, and Nov. 26, 1886.

See *Locomotives, Compound. Tests, Locomotive Exhaust Nozzles, Locomotive Traction Force.*

Locomotive Tires.

Breakages on German. A very valuable article condensed from the report of the Society of State Railroad Officers of the German Railroads, which contains a thorough study of the breakages of wheel tires for 1886. *R. R. Gaz.*, July 15, 1887.

Fracture of. By W. W. Beaumont. Points out what he believes to be the reasons for fractures and proposes remedy. *Van Nos. Eng. Mag.*, Vol. XVI., p. 403.

German Tire Fastenings. Cuts of the thirty different kinds of fastenings recognized by the German Railroad Union. *R. R. & Eng. Jour.*, August, 1891, p. 341.

Irregular wear of. An investigation of the forces acting on the wheel at different parts of the revolution, and discussion of their effect on tire wear. Diagrams and table. A paper by J. N. Barr, before the Western Railway Club. *Ry. Rev.*, Feb. 7, 1891, pp. 72–3. *R. R. Gaz.*, Feb. 6, 1891, pp. 92–3. *Mast. Mech.*, March, 1891, pp. 40–1.

Irregular Wear of. Diagrams showing the location of flat spots and counterbalancing of all wheels whose records have been obtained. Convenient for reference. *R. R. Gaz.*, March 20, 1891, pp. [illegible]

Steel, Influence of Chemical Composition on. By J. O. Arnold. A paper and discussion covering 52 pp. *Proc. Inst. C. E.*, Vol. XCV.

Steel. Test of. An article by E. Roussel in *Revue Generale des Chemins de Fer*, Jan., 1891, discussing methods of testing and giving some results of static and of drop tests. *R. R. Gaz.*, April 3, 1891, p. 237.

Steel Tired Wheels and the Principles to be followed in Attaching the Tire to the Center. Paper read by Mr. George W. Rhodes before the Western Railway Club. Discusses the subject of locomotive tires and gives numerous cases of failures. Illustrated by etchings. *Proc. West. R. R. Club.* Text reprinted in *Ry. Rev.*, Nov. 8, 1890, pp. 666–8. Abstracted in *Eng. News*, Nov. 8, 1890, p. 407. *R. R. Gaz.*, Nov. 7, 1890, pp. 768–9. Illustrations in *R. R. Gaz.*, Dec. 5, 1890, pp. 839–41.

Thick vs. Thin. Discussion of the subject at the meeting of the Western Railway Club Dec. 18, 1888; *Mast. Mech.*, Jan., 1889. *R. R. Gaz.*, Dec. 21, 1887; *R. R. Gaz.*, Dec. 28, 1888.

Locomotive Trucks.

Lateral Motion Trucks, Vertical Movements and Lateral Resistances. Editorial discussion. Diagrams. *R. R. Gaz.*, August 22, 29, 1890, pp. 586, 604–5.

Length of Radius Bars. An article by Arthur T. Woods, discussing the question of the proper length of radius bars and effects of shortening them. *R. R. Gaz.*, May 15, 1891, p. 336.

Valve Gear as applied to locomotive engines. A valuable paper before the R'y. Master Mechanics' Association, by Otto Gruninger. The improvements of Brown and Strong, whereby the steam chest is obviated, boiler pressure obtained in the cylinder, no back pressure, etc. Accomplished by a system of "radial valve gear." Fully illustrated. *Lon. Eng.*, Jan. 15, 1886.

Locomotive Wheels.

Driving Wheels. An investigation to determine the possibility of obtaining a perfectly balanced wheel. A valuable paper with formulæ. *Mast. Mech.*, Dec., 1888.

Imperceptible Slip of Drivers. Discussed in an editorial in *R. R. Gaz.*, Sept. 4, 1885. The conclusion reached that it does not occur with American locomotives.

Tender Truck, Standard on Union Pacific Ry. Illustrated description of standard tender truck, used under the Wootten express engines on the U. P. Ry. *R. R. Gaz.*, Dec. 7, 1883.

Truck Wheels for Fast Passenger Engines. Short article calling attention to, and suggesting remedy for, the marked tendency of truck journals to "heat" when engines are run at high speed. Wheels, in general, too small. *Mast. Mech.*, Dec., 1889, Vol. IV., p. 201.

Wheel Weights of Heavy Locomotives. Diagrams of wheel loads of nine heavy modern locomotives. *R. R. Gaz.*, May 22, 1891, pp. 347-8.

Wheels and Axles. English Tire and Axle Specifications. Approved methods of tire fastenings illustrated. *R. R. & Eng. Jour.*, Aug., 1891, pp. 369-71.

See *Car Wheels, Truing up, Locomotive Tires.*

Locomotives.

Automators and tram-cars worked by stored-up energy. Jurors report on the Antwerp trial, with formulas. Lon. *Engineer*, May 14, 1886.

American and English. Two editorials comparing the two types at considerable length. *R. R. Gaz.*, April 25 and March 9, 1892.

American Type, Evolution of the. An interesting and instructive lecture by M. N. Forney, before the Franklin Institute. Illustrated. *Jour. Frank. Inst.*, Oct., 1886.

Baldwin Ten-Wheel Express for the N. Y. L. E. & W. R. R. A full description with double inset; cross sections and dimensions. *R. R. Gaz.*, June 13, 1890. pp. 413-5.

Burning Soft Coal in. A paper read before the Western Railway Club by Mr. C. M. Higginson, of the C. B. & Q. R. R., with discussion. *R. R. Gaz.*, Nov. 23; *Mast. Mech.*, Dec.; *Am. Eng.*, Nov. 28, 1888. See also discussion in *Nat. Car Builder*, Dec., 1888; *Mast. Mech.*, Jan., 1889.

Canadian Pacific. An abstract of the paper by Mr. T. R. F. Brown, describing the engines designed by him for the Canadian Pacific R. R. *R. R. Gaz.*, June 17, 1887.

Caustic Soda. The Honigmann System. A long letter, illustrated, by W. Howard White, to the *R. R. Gaz.*, July 3, 1883. Heat is stored in solution of soda and given off to water to make steam. Is found efficient where smoke must be avoided. See also *Engine without Fire.*

C. B. & Q. A complete and well illustrated description of the new C. B. & Q. class "I" engine, designed for suburban traffic. *R. R. Gaz.*, Dec. 13, 1889, Vol. XXI., p. 816. *Ry. Rev.*, Dec. 14, 1889, Vol. XXXI., p. 717.

Compound Principle Applied to. A paper by Edgar Worthington, B. Sc., Assoc. M. Inst. C. E. *Proc. Inst. C. E.*, Vol. XCVI., p. 2.

Compound.

Paper read before the Western Railway Club by Mr. D. L. Barnes. *Mechanics*, Nov., 1889.

Paper by Angus Sinclair, and discussion of the subject before the New England Railroad Club, April 10, 1889. *R. R. Gaz.*, April 19, 1889; *Eng. News*, April 20, 1889.

Paper by Mr. A. T. Woods, read before the Engineers' Club of St. Louis. Comprehensive essay comparing different styles of compound locomotives

Locomotives, continued.

with each other and with simple locomotives. Tables of dimensions and indicator diagrams. *Jour. Am. Eng. Soc.*, August, 1890, pp. 393-407.

In American Service. A full discussion of the subject. *R. R. Gas.*, Feb. 15, 1889, *et seq.*

B. & O. Road. Description of a new Baldwin compound locomotive now in use on the B. & O. For fast passenger service. *Ry. Rev.*, Nov. 23, 1889.

Baldwin Compound Locomotives. List of the 115 locomotives of the Vauclain system ordered at these works. *R. R. Gas.*, Nov. 27, 1891, p. 836.

Correspondence by S. Kerbeds, relative to the successful working of compound freight locomotives on the Russian railroads. The proper operation of valves is discussed and several indicator diagrams given. *R. R. Gas.*, March 13, 1891, pp. 175-6

Economy of. Table showing the performance of various classes of locomotives on the Mexican Central Ry. for 12 months. Shows a gain of over 20 per cent for the compound. *R. R. Gas.*, Nov. 20, 1891, p. 812.

Express. Gives a two-page plate showing sectional elevation and plan of a compound express locomotive built for the Northeastern Railway, Eng. Lon. *Eng.*, March 30, 1888. Details of tests of these engines in Lon. *Eng.*, April 13, 1888.

Express. With single driving wheels. Northeastern Railway, Eng. Details, dimensions, indicator cards, illustrations, etc., are given. Lon. *Engineer*, March 7, 1890, p. 184.

Four-Cylinder. An illustrated description of an engine designed for heavy work on the Mexican Central R. R. The tender is provided with driver and cylinders. Lon. *Eng.*, Jan. 28, 1887.

Four-cylinder Compound Locomotive, by the Baldwin Locomotive Works. Illustrated description showing principal details and dimensions, as well as general appearance of the locomotive. *Eng. News*, May 3, 1890, Vol. XXIII., p. 424. *R. R. Gas.*, May 2, 1890, Vol. XXII., pp. 298-306. *Mast. Mech.*, May, 1890, p. 73. *Ry. Rev.*, April 26, 1890, p. 234.

Four-cylinder Compound. Description, with illustrations, of a French design used on the Northern Railroad of France, and an account of trials. *R. R. & Eng. Jour.*, March, 1888.

Freight. An illustrated description, with table of the dimensions of the first compound freight engine constructed. Lon. *Engineer*, Jan. 14, 1887.

Freight. Description of a compound freight locomotive of the Northeastern R. R., with two page plate and cuts showing details and dimensions. Lon. *Eng.*, Sept. 23, 1887.

In Germany. Article containing the opinions of several prominent German engineers and taken from *Glaser's Annalen*. *Eng. News*, May 3, 1890, Vol. XXIII., p. 424.

On the German State Railways. Paper by H. Hawkins, reprinted from Lon. *Eng.*, *Ry. Rev.*, July 6, 1890, p. 393.

History of, given in *R. R. Gas.*, Aug. 28, 1885.

Lindner Starting Gear. A description of recent improvements in the Lindner starting gear. Illus. *R. R. Gas.*, March 27, 1891, pp. 207-8.

London and N. W. Ry. Description, with engravings and double plates. Lon. *Eng.*, July 25, 1890, pp. 98-100.

Michigan Central Compound. Intercepting and reducing valves of Michigan Central compound described and illustrated in *Ry. Rev.*, January 25, 1890, p. 52.

Michigan Central Railroad. Description, details, indicator cards, etc. *Eng. News*, March 8, 1890, Vol. XXIII., p. 220.

Method of computing the mean pressures. *Mech. World*, March 10, 1888.

Locomotives, continued.

Paris Exhibition, 1889. Illustrated article describing engines in use on the Paris, Lyons and Mediterranean Ry. Lon. *Engineer*. Aug. 9. 1889.

Passenger. An illustrated description, with table of dimensions, full-page plate and indicator diagrams of a compound passenger engine for the Northeastern Railway. Lon. *Eng*., June 17, 1889.

Proportion of Cylinders for. Table giving comparative cylinder capacities of various compound locomotives. Editorial discussion. *R. R. Gaz*. Oct., 1891, p. 691.

Report of Committee to the 23rd convention of the Am. Ry. Master Mechanic's Assn. Excellent paper followed by discussion. *Ry. Rev*., June 21, 1890 pp. 351-5. *R. R. Gaz*., June 27, 1890, pp. 451-2. *Eng. News*. June 28. 1890 pp. 603-4. *Mast. Mech*., July, 1890, pp. 119-120.

In Russia. Abstract of a paper by Mr. Thomas Urquhart, read before the Inst. of Mech. Engrs. Tables, illustrations and indicator diagrams. *Eng. News*, June 14. 1890, pp. 554-6.

Schenectady. A statement of the practical workings and economy of compounds on the East Tennessee, Virginia & Georgia R. R. A saving of 31 per cent. in coal and consumption is attained. By A. T. Pitkin. *Ry. Rev*. March 14, 1891, p. 170.

Ten-Wheel Compound Locomotive; Mexican Central Railway. Drawings and table of dimensions, with results of a comparative test. *Eng. News*, Aug. 29, 1891. pp. 197-98.

Ten-Wheel Compound Locomotives—Southern Pacific Company. Table of dimensions, general views, and description, with illustrations of details, of the new starting gear and intercepting valve. *R. R. Gaz*., September 11, 1891, pp. 637-9.

Tests.

Baldwin consolidation compound on W. N. Y. & P. R. R., showing saving of 35.2 per cent. coal and 17.9 per cent. water. *R. R. & Eng. Jour*., Oct. 1891. *R. R. Gaz*., Oct. 2, 1891. *Ry. Rev*., Oct. 10, 1891.

Comparative Tests of a Johnstone Compound Locomotive. Full account of several tests on the Mexican Central Ry. Various kinds of coal were used. 30 per cent less coal was used in the compound than in the simple in doing the same work. Design of valves of compound and indicator diagrams given. *R. R. Gaz*., May 22, 1891, pp. 350-2. *Ry. Rev*., May 23, 1891, pp. 326-8.

Comparative Tests of Compound and Simple Locomotives. Data and table of results of six tests on the Mexican National Ry., gives a favorable showing for the compound. Comparative tests of Pecos and Indian Territory coal were also made. *R. R. Gaz*., Dec. 4, 1891, pp. 858-9.

East Tennessee, Virginia & Georgia Railroad. Comparative test of a single and a compound consolidation locomotive. Diagrams. *Eng. News*, Nov. 22, 1890, p. 458.

E. T. V. & G. Ry. Practical Tests of Compound Locomotives on. A paper by C. H. Hudson, before the West. Soc. Engrs., giving the results of 2 years' operation on the E. T. V. & G. Ry., comparing the working of the compounds with simple locomotives of the same type and weight. Gives also results of some short tests. *Jour. Assn. Eng. Soc*., Oct., 1891, pp. 471-9. Reprint, *Mast. Mech*., Dec., 1891, pp. 181-2. The results of the years' operation are also given in *R. R. Gaz*., Nov. 13, 1891, p. 796. *Eng. News*, Dec. 5, 1891, pp. 545-6.

High Speed Indicator Cards. A set of cards taken at speeds from 5 to 80 miles per hour, taken on a Worsdell compound. *R. R. Gaz*., March 28 1890, p. 207.

Indicator diagrams and details of a trial trip of 105 miles of the compound lo-

Locomotives, *Tests*, continued.

comotive "Dreadnaught." Also a table giving a comparison between her performance and that of other fast passenger engines. *R. R. Gaz.*, Aug. 10, 1887.

Test of a Compound. Full report by G. H. Barrus, M. E. Pamp. 8 vo. pp. 55 illustrations and tables. Pub. by Baldwin Locomotive Works. Abstract in *Eng. News*, Sept. 6, 13, 1890, pp. 216, 231-7. *Ry. Rev.*, Sept. 6, 13, 20, 1890. *R. R. Gaz.*, Sept. 12, 1890, pp. 627-8, 634-5. *Lon. Eng.*, Sept. 19, 1890, *et seq.* *Mast. Mech.*, Oct., 1890, pp. 111-3.

Union Elevated Railroad, Brooklyn, N. Y. *Eng. News*, Dec. 27, 1890, pp. 575-6, *et seq.* *Ry. Rev.*, Dec. 27, 1890, pp. 780-2.

Vauclain Compound Locomotive on the Pennsylvania and Lehigh Valley. Table of results of tests on heavy freight traffic. *R. R. Gaz.*, May 29, 1891, p. 363. Indicator cards and drawings showing details of cylinders. *ibid.* Aug. 7, 1891, pp. 541-4.

Three-cylinder Compound. Illustrated description of a new engine that has given excellent satisfaction on the Northern Ry. of France. *Mast. Mech.*, June, 1890, p. 99.

Von Borries' Starting Valve for. Illustrations and information regarding size of ports, motion of valves, with diagrams. *R. R. Gaz.*, June 12, 1891, p. 403.

Von Borries' System, used on the Hanover railroads. Eighteen brought into use since 1880, with a saving of 14 to 21 per cent. of the fuel. Fully described and illustrated in *R. R. Gaz.*, Oct. 15, 1886.

Worsdell Compound. A full illustrated description of the system of compounding locomotives, invented by Mr. T. W. Worsdell, Locomotive Supt. Northeastern Ry. (England), and Herr A. von Borries, Mech. Engr. of the State R. Rs. of Hanover, Germany. *R. R. Gaz.*, Dec. 14, 1888.

Worsdell. Drawings, description, and dimensions in detail of Worsdell's Compound locomotive for passenger traffic. *Lon. Engineer*, May 15, 1891.

Worsdell-Von Borries Compound—Grand Trunk Railway. Brief description with illustrations of valves, etc. *R. R. Gaz.*, Nov. 6, 1891, p. 776.

Worsdell-Von Borries. Plan, elevation and cross-section of a compound locomotive, Worsdell and Von Borries system, of the Bengal & Nagpur Railroad. *Lon. Engineer*, July 20, 1888.

See *Locomotives, Tank*.

Consolidation.

B. & O. R. R., class "E." 6. Description and specifications of this new and in some respects novel locomotive. Illustrated. *R. R. Gaz.*, Aug. 2, 1889.

Canadian Pacific Railway. A description of the first consolidation engines built in Canada. Illustrated with double-page plate and three cross-sections with dimensions. *R. R. Gaz.*, May 6, 1887.

Union Pacific Railroad. Baldwin Co. Description, inset plate of details, illustrations of cylinders, etc., with table of dimensions and weights. *Ry. Rev.*, May 3, 1890, Vol. XXX., p. 247.

Cost of Power in Different Countries. From a treatise on "Railway Problems," by J. S. Jeans, Secretary British Iron and Steel Association. A valuable review of the situation, occupying seven columns in the *R. R. Gaz.*, Dec. 16, 1887.

Cost of Traction, and the economy resulting from the use of various improvements in locomotives. By M. Ricour. Quite an extended discussion. The author calculates that by the introduction of certain improvements on all the locomotives of France (7,000 in number), the annual saving in cost of traction would be 16,000,000 francs. *Annales des P. & C.*, Sept., 1885, pp. 310-68.

Cylinders, Proportions of. Gives text of the report on the above subject by the

Locomotives, continued.

committee appointed by the American Railroad Master Mechanics' Association. *R. R. Gaz.*, July 1, 1887.

Decapod.

Burlington 22x28 in. Table of dimensions, with cut, of this 73-ton locomotive. *R. R. Gaz.*, July 24, 1891, p. 512.

The largest in the World. Illustrated and data given. *R. R. Gaz.*, May 1, 1885.

St. Clair Tunnel Locomotives. Illustration and table of dimensions of this new design; a "Decapod Tank Freight" with 195,000 lbs on drivers. *R. R. Gaz.*, April 3, 1891, p. 226.

Defective Parts. A review of the various defective parts in the present practice, with suggestions for improvements. By D. B. Dixon. *Age of Steel*, July 4, 1885.

Distribution of Steam in the Strong. By F. W. Dean, before the American Society of Mechanical Engineers. Abstracted. *R. R. Gaz.*, May 25, 1888.

Draft Appliances in. A paper by F. C. Smith before the April meeting of the Western Railroad Club. Gives results of experiments made on a number of engines to see what could be done to obtain a better efficiency. Discussion. *R. R. Gaz.*, April 20, 1888; *Mast. Mech.*, May, 1888; *Nat. Car & Loco. Builder*, May, 1888.

Duplex Compound Locomotive, St. Gothard Ry. Brief description with illustrations of this peculiar locomotive. *Eng. News*, Aug. 15, 1891, p. 147.

Economy, subserved by New Performance Sheets, wherein the evaporation per pound of coal shall be the basis. Suggestions as to how this may be accomplished. By C. N. Baker. *Eng. News*, Feb. 6, 1886.

English, an outline History of the. By Theo. West. A paper before the Cleveland Institution of Engineers, of much historical value. *Sci. Am. Sup.*, May 15, 1886; condensed in *Ry. Rev.*, May 15, 1886; *Am. Eng.*, April 28, 1886, *et seq.*

English and American Compared. By R. H. Burnett. The essential differences pointed out as seen from the English point of view. Lon. *Eng.*, April 2, 1886.

English and American. Comparisons of locomotives with valuable table of mileage and cost of maintenance and operation. *R. R. & Eng. Jour.*, Nov., 1891. Reprint, *R. R. Gaz.*, Nov. 27, 1891, pp. 840-1.

Efficiency of, and Resistance of Trains. By G. R. Henderson. *Proc. Eng. Club, Phila.*, Vol. VI., p. 48.

Efficiency of. Paper by W. F. Dixon, read at the Cincinnati meeting of the A. S. M. E. *Trans. A. S. M. E.*, Vol. XI., 1890, pp. 867-92. Abstracts in *Mechanics*, July, 1890, pp. 167-7. *Eng. News*, August 16, 1890.

Eight-Wheeled Locomotive, "Class O," Pennsylvania R. R. Full page plate showing side elevation. *Ry. Rev.*, July 26, 1890. Description, table of dimensions, and cross-sections, *ibid.*, Aug. 2, 1890, pp. 466-7.

Estrade's High-Speed. An illustrated description and criticism of Estrade's high-speed locomotive. It has six driving wheels 6' 3" diameter, cylinders 18½ x 27½, with a weight of 42 tons. Lon. *Engineer*, March 9, 1888; *Sci. Am. Sup.*, April 28, 1888.

Express.

Baltimore & Ohio. Gives a brief description, with two-page plate and extracts from the specifications of a new eight-wheel locomotive built for the Baltimore & Ohio Railroad. Its dimensions are: Cylinders, 19 by 20 in.; drivers, 60 in.; boiler, 53 in.; tubes, 174; weight on drivers, 70,000 lbs.; total weight, 102,000 lbs.; *Mast. Mech.*, May, 1888.

C. & N. R. R. Gives brief description, with two-page plate, giving sectional elevation, half plan and cross-section of a passenger locomotive for the Chicago & Northwestern Railroad. *R. R. Gaz.*, Dec. 23, 1887.

Locomotives. *Express*, continued.

Class "B," Chicago, Milwaukee & St. Paul Railway. An 8-wheel type with radial stay, wagon-top boiler. Description with full drawings; also extracts from new specifications for locomotives issued by this road. *R. R. Gaz.*, Feb. 20, 1891, pp. 125-7.

Description of an express locomotive for the Great Northern Railway, England, on exhibition at the Newcastle Exhibition. Double page plate, giving plan, elevation, and cross-section, with dimensions. Lon. *Eng.*, May. 20, 1887; also Lon. *Engineer*, June 3, 1887.

Lancashire & Yorkshire Railway, England. Eight-wheel type. Inset of drawings and tables of comparative dimensions of English and American locomotives of this type. *Eng. News*, March 28, 1891, pp. 222-3.

Mich. Cent. R. R. Brief description, with drawings, showing sectional elevation, half plan, cross-section of a standard eight-wheeled passenger engine for the Michigan Central Railroad. *Mast. Mech.*, August, 1887.

N. Y., L. E. & W. Gives double-page plate showing elevation and half-plan, also a smaller drawing showing sections and details with dimensions. *R. R. Gaz.*, Jan. 6, 1888.

Outlines, General Dimensions and Weight of English Express. Valuable compilation of data of 19 modern English and Scotch engines. *R. R. Gaz.*, May 9, 1890, pp. 315-323.

Ten-wheel. M. C. R. R. Gives elevation of a ten-wheel locomotive built for the Michigan Central Railroad, with specifications giving the leading dimensions. *R. R. Gaz.*, Feb. 24, 1888.

Extension Arch on C. B. & Q.'s Class H. Engines. Brief description, with detail drawings. *Mast. Mech.*, Jan., 1889.

Extension Front of. Gives sectional drawings with dimensions of the extended front on the Fremont, Elkhorn & Missouri Valley Railroad. Also, discussion of the different forms of smoke-boxes by the Western Railroad Club. *Nat. Car & Loco. Builder*, December, 1887.

Extension Fronts and Fire-Box Arches. Gives the report of the committee of the Master Car-Builders' Association on extension fronts and fire-box arches. *R. R. Gaz.*, June 22, 1888; *Nat. Car & Loco. Builder*, July, 1888; *Mast. Mech.*, July, 1888.

Fast Passenger. The problem discussed and general specifications given. By T. E. Austin. Illustrated. *Proc. Eng. Club, Phila.*, Vol. V., p. 97.

Fireless, Principal kinds of. Illustrated description. By M. Lavoinne in *Annales des P. & C.* *Eng. News*, Feb. 15, 22, 1879.

French Historical Engines. Double plate. Lon. *Eng.*, June 20, *et seq.*, 1890.

French Locomotives and their Every-Day Working. Two illustrated articles by Chas. R. King. *R. R. Gaz.*, Jan. 16, and 23, 1891.

French Passenger. Fully described. Details illustrated. Lon. *Engineer*, June 6, 13, 1890.

Freight, C. B. & N. R. R. Gives a two-page plate showing sectional elevation and half plan, other cuts showing cross-sections of a ten-wheel locomotive of the Chicago, Burlington & Northern Railroad. *Mast. Mech.*, December, 1887.

Freight, N. N. & M. V. R. R. Gives specifications for a ten-wheeled freight engine for the Newport News & Mississippi Valley Company. Drawings with dimensions showing elevation of engine, also details of coupling, connection and eccentric rods, rocker box, crank pins, links, etc. *R. R. Gaz.*, March 16, 1888.

Hauling Capacities of Freight Locomotives. Tables giving the hauling capacities on six railroads: on level tangents, on 10° curves, and on various grades. *R. R. Gaz.*, Nov. 21, 1890, p. 802.

Heavy 10-Wheel Locomotive. Description and abstract of the specifications

Locomotives, continued.

Total weight, 130,000 lbs. Double plate and other illustrations. *Mast. Mech.*, Sept., 1890, pp. 150-1.

High and Low. By Prof. A. G. Greenhill. A mathematical treatment of why a high locomotive will run with greater safety and steadiness than a low one. Lon. *Engineer*, Dec. 2, 1887.

How to Run a. By R. H. Buel. Considers the action of the stationary engine to see how nearly the same action is possible for the locomotive, and advocates a close study of the same. Gives details of some experiments made with locomotive, with observations of every change of throttle, valve and link. *R. R. & Eng. Jour.*, Feb., 1887.

Imperial Railroads, Japan. A two-paged plate showing sectional elevation and plan of a tank locomotive for the Imperial railroads of Japan. They are 3 feet 6 inches gauge, two pairs drivers 4 feet 4 inches in diameter, 14×20 inch cylinder; weight 31¼ tons, with 20 tons coupled to wheels. Lon. *Eng.*, Feb. 10, 1888.

Indian State Railroad. Gives brief description with plan and elevation of a six-coupled metre gauge locomotive for the Indian State Railroad. Lon. *Engineer*, March, 27, 1888.

Indicator Diagrams. See *Locomotive Test.*

Lancashire and Yorkshire. Gives longitudinal section and description, with dimensions of engine and tender, of a passenger engine for the Lancashire and and Yorkshire Railway Co. Lon. *Engineer*, Aug. 26, 1887.

Large and small. Comparison of a thirty-nine-ton and a three-ton locomotive. Gives their relative dimensions and gauges. *Sci. Am. Sup.*, Oct. 15, 1887.

London and Southwestern. Gives description and specifications, with a two-page plate of detailed drawings of a four-coupled, outside cylinder locomotive. Lon. *Engineer*, Nov. 4 and 11, 1887.

Lubrication. See *Lubrication.*

Mogul.

Freight. An illustrated description of an engine built for the Michigan Central Railroad, with table of dimensions. *R. R. Gaz.*, June 10, 1887.

Freight, D., L. & W. R. R. Gives short description, with elevations and cross-sections, and dimensions of a freight engine, of the Mogul type, for the Delaware, Lackawana & Western Railroad. *R. R. Gaz.*, Feb. 3, 1888.

New York & New England R. R. Description, dimensions, details, etc., of mogul locomotive as recently built. *Mast. Mech.*, Aug., 1889.

Standard, C. B. & Q. R. R. Description with principal dimensions. Several cross sections. *Eng. News*, July 5, 1890, pp. 6-7. Folding inset, with elevation and plan of the same engine, *id.*, June 28, 1890.

Standard, N. Y. C. & H. R. R. R. Description of this fine locomotive with table of dimensions and details of boiler and frame. Folding plate. *R. R. Gaz.*, July 25, 1890, pp. 519-21.

Parallel Rod, Problem of the. By Prof. J. B. Johnson. Demonstrates by analysis that the I section is the correct one, and in ordinary practice the depth is too small. *R. R. Gaz.*, Sept. 10, 1887. Discussed in *R. R. Gaz.*, Nov. 4, 1887.

Passenger.

Caledonian Railroad. Gives half plan, elevation and cross-section of a passenger locomotive for the Caledonian Railroad. It has two pairs 5 ft. 9 in. drivers, cylinders 18×26 in., and weighs 83,000 lbs. Lon. *Engineer*, April 13, 1888.

N. Y., L. E. & W. R. R. Gives elevation, half plan and three cross-sections with brief description, of a high speed passenger engine for the New York Lake Erie & Western Railroad. It has two pairs of 68-inch drivers, cylinder 19×24, and weighs 115,000 lbs., with 78,000 on the drivers. *Nat. Car & Loc. Builder*, April, 1888.

Locomotives. *Passenger*, continued.

Ten-Wheel. Gives drawing and description, with specification of ten-wheel locomotive built by the Schenectady Locomotive Works for the Colorado Midland Railroad. *R. R. Gaz.*, Nov. 25, 1887.

Philadelphia & Reading R. R. Gives outline engravings of the four classes of new locomotives to take the place of the Wootten locomotives on the Philadelphia & Reading Railroad. *Nat. Car & Loco. Builder*, April, 1888.

Pike's Peak Rack Railway. Drawings and table of dimensions. *Ry. Rev.*, Feb. 7, 1891, p. 87. *R. R. Gaz.*, Feb. 6, 1891, pp. 90-1.

Propelling Force of. See *Train Resistance.*

Qualities Essential for a Free Steaming Locomotive. A short, practical paper by A. E. Mitchell, read before the N. Y. R. R. Club. *Am. Eng.*, Jan. 31, 1891, pp. 46-8. *Ry. Rev.*, Jan. 24, 1891, pp. 51-2. *R. R. Gaz.*, Jan. 23, 1891, pp. 61-2.

Radius Bar. See *Locomotive Trucks.*

Road. See *Engines, Steam, Traction.*

Staying of. See *Locomotive Fire-Boxes.*

Shay Locomotive. Illustrated description of this locomotive driven by bevel gears, and designed for use on roads with sharp curves and steep grades as in lumbering. *Eng. News*, April 26, 1890. Vol. XXIII., p. 386.

Shay Locomotive. Illustration with brief description of a heavy Shay engine for the Montana Union Ry. *R. R. Gaz.*, May 8, 1891, pp. 315-16.

Specifications for.

Gives drawing-room specifications for express locomotives built at the Baldwin works, for the New York, New Haven and Hartford Railroad. *Mast. Mech.*, Sept., 1888.

Gives the standard specification for an 18 by 24 Mogul engine. Lon. *Eng.*, Oct. 1887.

General specifications for a heavy passenger engine, built at the Mason Works, Taunton, Mass. Full-page drawing. *R. R. Gaz.*, June 4, 1886.

Gives specifications of locomotives for the London & Southwestern R. R. Lon. *Eng.*, Sept. 2, 1887.

Statistics for 25 years, C. B. & Q. R. R. Table of statistics for 7 years with diagrams illustrating the tendency in management. *Eng. News*, May 16, 1891, p. 479.

Steam Generation in Locomotives. By Prof. A. Frank. *Zeitschr. d. Vereins deutscher Ingenieure*, 1886, pp. 573-578.

Straw-burning. Description of a locomotive exhibited at the Vienna Exhibition. *Van Nos. Eng. Mag.*, Vol. IX., p. 109.

Strong.

A brief description of the Strong locomotive. Illustrated, with reasons for the peculiar features in its design. By George S. Strong. Also a summary of the results of trials of the engine on the Lehigh Valley Railroad, made by E. D. Leavitt. *Jour. Frank. Inst.*, Feb., 1888.

Descriptive illustrated paper, by Mr. Geo. S. Strong. *Proc. Soc. Arts*, Mass. Inst. of Tech., 1887—1888.

Description of the Strong locomotive, "*A. G. Darwin*," with perspective view and sectional view of boiler. *R. R. Gaz.*, Jan. 11, 1889.

Illustrated description of the Strong locomotive *Duplex* an engine of many novel features, weighing 140,000 pounds, of which 90,000 pounds is on the drivers. *R. R. Gaz.*, Feb. 18, 1887.

Tank.

Compound. Gives a two-page plate of elevation and plan, with dimensions, and brief description of a freight locomotive, Webb's system and Joy's valve

Locomotives, *Tank*, continued.

motion for the London & Northwestern Railway. Lon. *Eng.*, Dec. 23, 1887.

Four-Wheels-Coupled Radial Tank Engine for the Bombay, Baroda, and Central India Railway. These engines are built for suburban passenger traffic. Table of dimensions with large scale drawings. *Ry. Eng.*, Feb., 1891. pp. 29-30.

For Heavy Grades, and Saloon Car; New Zealand Governmental Railways. Illustrations, inset of detailed drawings, table of dimensions, etc. *Eng. News*, Nov. 14, 1891, pp. 454-5.

North London Ry. Designed for heavy freight service. Lon. *Eng.*, Oct. 3, 1890, pp. 394-5.

Ten-Wheel Locomotive. Georgia Pacific Railway. Description, illustration, general dimensions, weight, etc. *Eng. News*, April 5, 1890. Vol. XXIII., p. 313.

Ten-Wheel Express Locomotive, Wabash. Has a peculiar arrangement of cross-head guides. Full drawings, folding inset. *R. R. Gaz.*, Oct. 17, 1890, pp. 712-13.

Tractive Force. Diagram giving the tractive force exerted by locomotives per 1 lb. mean effective pressure on piston for all standard sizes of cylinders from 14x18 inches to 22x28 inches. *Ry. Rev.*, March 1, 1890, p. 119.

Tractive Force, Tests of. By C. H. Hudson. Describes a series of experiments to determine what proportion of weight on drivers could be utilized under the most favorable circumstances. *Jour. Assn. Eng. Soc.*, Vol. V., p. 283.

Tramway.

Brief illustrated description of Burrell's tramway locomotive, that gave very good results in Birmingham. Lon. *Engineer*, Oct. 28, 1887.

Compound Condensing. Brief illustrated description of a compound condensing tramway engine at the Newcastle Exhibition. Lon. *Eng.*, September 30, 1887.

Describes a small, powerful condensing engine for street work. Illus. Lon. *Engineer*, Oct. 21, 1887.

Description, with elevations and cross-sections, of a tramway locomotive for the Wolverton and District Light Railway Co. Lon. *Eng.*, Aug. 26, 1887.

Light Tramway. Gives brief illustrated description of the locomotive used on the Cavan, Leitrim and Roscommon tramway, Ireland. Lon. *Engineer*, April 20, 1888.

Used in the Construction of the Arlberg Tunnel. Reprinted from abstracts made by Inst. C. E., London. *Am. Eng.*, Sept. 17, 1885.

Vibratory Movements of Locomotives, and on Timing Trains and Testing Railway Tracks. By John Milne and John McDonald. Gives account of some instruments for registering the oscillations and vibrations of trains, with discussion of their usefulness. *Proc. Inst. C.E.*, Vol. CIII., 1891, pp. 47-67. Correspondence and discussion illustrating several kinds of machines for this purpose, pp. 68-170. See also a paper by J. Milne in *Ry. Rev.*, March 28, 1891.

Woolf's Four-Cylinder. Illustrated description and results of experiments in France. *R. R. Gaz.*, March 8, 1889.

Wootten. Report of Franklin Institute committee on Science and Arts on the Wootten locomotive, describing various forms. *Jour. Frank. Inst.*, Sept., 1891. *R. R. & Eng. Jour.*, Oct., 1891.

Wootten, U. P. R. R. Brief description, with detailed drawings of passenger engines with the Wootten fire-box built for the Union Pacific Railroad. Cylinders, 18 by 26 inches; drivers, 63 inches; weight on drivers, 76,100 pounds; total weight, 118,500 pounds. *R. R. Gaz.*, June 15, July 6, Aug. 17, Sept. 7 and Oct. 5, 1888.

Locomotives, continued.

See *Appliances*, *Electric Locomotives*, *Electric Motors*, *Comparative Tests*, *etc.*, *Railroad Rolling Stock*.

Logarithms, *Mental*. An attempt to popularize the system of Mr. Oliver Byrne, introduced many years ago. *Van. Nos. Eng. Mag.*, Vol. XX., p. 97.

Notation. By Mansfield Merriman. Gives a brief history of logarithms, with remarks on the use of different tables. *Van. Nos. Eng. Mag.*, Vol. XXI., p. 189.

Longitude, *Determination by Telegraph*. Description of apparatus with directions for using. Many cuts. *U. S. C. & G. Survey Report*, 1880.

See *Determination of*, *Latitude*, *Time*.

Longitudes *of American Cities*. Adjusted values of all such longitudes determined by the U. S. C. & G. Survey from 1846 to 1885. *Rep. U. S. C. & G. Survey*, 1884, App. 11.

See Latitude.

Lumber.

Compound. A paper by G. A. M. Liljencranz before the Western Society of Engineers, discussing various forms of lumber made up from two or more pieces. pp. 3. *Jour. Assn. Eng. Soc.*, March, 1890, Vol. IX., p. 91.

Pine. *Comparative Value of Heart and Sap Pine*. A complete discussion of relative merits of long-leaved Southern pine and the short-leaved yellow pines, latter being commonly known as sap pine. The article includes table giving size of trees, physical properties of the timber, etc. *Eng. News*, No. 31, 1889, Vol. XXII., p. 128, *et seq.*

Lubrication.

Of Air Compressing Machinery. Condensed from a paper by John Morison read before the last meeting of the North of England Institute of Mining and Mechanical Engineers on "The Danger Attending the Use of Light Mineral Oils for Lubricating Air Compressing Machinery. *Eng. News*, Dec. 8, 1888.

Car Journals, *Cost of*. Paper read by L. S. Randolph at New York meeting of Am. Soc. M. E. Giving valuable and interesting data respecting cost of lubrication, including wear of journals and brasses. *R. R. Gaz.*, Nov. 22, 1889; *Ry. Rev.*, Nov. 23, 1889.

Of Cars and Locomotives. Abstract of Grossman's German work on the subject. *R. R. Gaz.*, Oct. 30 and Nov. 6, 1884.

Cylinders. An article giving various mechanical contrivances and table of efficiencies of various kinds of oil. *Am. Eng.*, Nov. 7, 1884.

Finance of. A paper before the Am. Soc. of Mech. Engrs., by Prof. R. H. Thurston. An exhaustive discussion of the ingredients, cost, value and use of lubricants. *Jour. Frank. Inst.*, July, 1885, *et seq.*; also *Am. Eng.*, May 19, *et seq.*; also *Van Nos. Eng. Mag.*, April, 1886.

See *Car Lubrication*, *Friction*.

Lubricants.

By J. E. Denton. A paper before the Nashville meeting of the American Society of Mechanical Engineers, discussing the mechanical significance of the determination of viscosity. *Trans. A.S. C. E.*, Vol. IX., (1888), pp. 369-376; *R. R. Gaz.*, May 11, 1888; *Am. Eng.*, June 13, 1888.

Gives tables showing proportion of various oils commonly used for lubrication. *Lon. Eng.*, April 20, 1888.

History of Attempts to Determine the Relative Value of Lubricants by Mechanical Tests. Address by James E. Denton, Vice-President section D., A. A. A. S. *Proc. A. A. A. S.*, July, 1891, pp. [illegible].

Measurement of Durability of Lubricants. Paper by Prof. J. E. Denton, read before Cincinnati convention of Am. Soc. Mech. Eng., May 13, 1890. *Ry. Rev.*, May 17, 1890, Vol. XXX., p. 274. *Trans. A. S. M. E.*, Vol. XI., 1890, pp. 1013-28.

Lubricants, continued.

Notes on the Action of. By Prof. J. E. Denton. Concise review of the subject. *Mechanics*, May, 1890, pp. 115.

Oil. Specifications for lard oil in use on the Pennsylvania Railroad, with methods of testing, etc., by C. B. Dudley, Chemist, and F. N. Pease, Asst. Chemist, of Penn. R. R. *R. R. & Eng. Jour.*, March, 1890.

Special Experiments with Lubricants. With reference to the lubrication of steam cylinders and journals subjected to heavy pressure. Description of apparatus. Abstract of a paper by J. E. Denton, read before the Richmond meeting of the Am. Soc. M. E. *Ry. Rev.*, Nov. 22, 1890, pp. 699-700.

See *Axles, Car.*

Machine Construction. *Milling Machine as a Substitute for the Planer in.* By J. J. Grant before the American Society of Mechanical Engineers. Gives data relating to the cost of work on the two machines. Shows the milling machine to be the cheaper. *Trans. A. S. M. E.*, Vol. IX., (1888), pp. 259-269; abstracted in *R. R. Gaz.*, Jan. 6, 1888. *Mech. World*, Dec. 17, 1887.

Machine Designing.

By John E. Sweet. A lecture delivered before the Franklin Institute. Illus. *Sci. Am. Sup.*, April 28, 1888; abstracted *R. R. Gaz.*, June 15, 1888.

A series of articles by Oberlin Smith in *Mechanics*, for 1885.

The Constructor, a Hand Book of Machine Designs. By F. Reuleaux. Complete translation of this great German standard work on machine designing is commenced in January number of *Mechanics*, to be continued throughout the year. *Mechanics*, Jan., 1890, p. 10, *et seq.*

Machine Tools *and Workshop Appliances* for the treatment of heavy forgings and castings. Describes some of the largest tools in England, including a forty-foot face lathe. Illustrated. *Proc. Inst. C. E.*, Vol. LXXXVI., p. 180.

Machinery.

The Kinematics of. By Prof. Alex. B. W. Kennedy. Abstract of lectures at South Kensington relating to Reuleaux methods. Treats of the principles of mechanism. *Van Nos. Eng. Mag.*, Vol. II., and XXV., p. 1 and 177.

Textile. A general review of the textile machinery at Manchester Royal Jubilee Exhibition, including all classes of cotton machinery. *Sci. Am. Sup.*, Oct. 8, 1887.

At the World's Fair in Antwerp, in 1885. A series of illustrated articles describing the steam engines and textile machinery exhibited. *Zeitsh. V. d. Deutscher Ing.*, 1886, pp. 11-15, 61-63, 63-67, 81-83, 101-106, 107-102, 149-151, 194-197, 240-244.

Cutting Speeds in Turning, Boring, Milling, etc. Table given by Mr. George Oldfield, of Johnstone, near Glasgow, of safe speeds for various kinds of work. *Mech. World*, April 2, 1887.

In U. S. See *Power and Machinery*.

Machines.

Topography of. By Oberlin Smith. A paper read before the Am. Soc. for the Adv. of Sci. A new system of phraseology advocated for describing the location of parts using the system of rectangular co-ordinates in space. *Mechanics*, Oct., 1884.

See *Mechanism*.

Magnetic Circuit. *A Theoretical Discussion, Including a Formula for Magnetism in Soft Iron.* By Edward Collins, Jr. *Tech. Quart.*, Vol. II., 1889, No. 4, pp. 385-92.

Magnetic Declination.

A report embodying the results of observations made in some 35 different parts of the State at the same time, under instructions specially sent out by the Asso-

Magnetic Declination, continued.

clation. By Benj. Thompson, W. H. Jennings and others. *Rpt. Ohio Soc. Surv.*, 1884.

Theoretical Discussion of the Variation of, based on all available information, with curves showing the same. *Report of the U. S. C. & G. Survey*, 1882.

In the U. S., being the declination for January, 188–, at *over two thousand* points in North America, mostly in the United States, all based on the most reliable and latest *observed* declinations. *Report of U. S. C. & G. Survey*, 1882.

See *Declination*.

Magnetic Needle By Thos. Job. Treats of the causes of its secular variation. *Van. Nos. Eng. Mag.*, Vol. XIX., p. 413.

Magnetic Pole. A new position assigned by Lieut. Fredk. Schwatka, with reasons. *Science*, Feb. 27, 1883.

Magnetic Reluctance. Paper by A. E. Kennelly read before Am. Inst. Elec. Eng., Oct. 27, 1891. *Elec. Eng.*, Nov. 4, 1891. Illus. pp. 308-312.

Magnetic Separator. *Edison's.* See *Separator*.

Magnetism.

Its Principles and Application to Ships and Compasses. A compilation of information on this subject from the writings of Sir W. Snow Harris, F. R. S., Rev. W. Scoresby, F. R. S., and Capt. S. T. S. Lecky, R. N. R. Published by the Bureau of Navigation, Navy Department. Naval Professional Papers, No. 13.

Of Ships and Mariner's Compass. A lecture by Wm. Bottomley, London. Illus. *Sci. Am. Sup.*, March 27, 1886.

See *Watches, Protection of*.

Magnetization *of Iron and other Magnetic Metals of very Strong Fields.* Abstract of a paper read before the Royal Society, Nov. 22, 1888. By J. A. Ewing and William Low. *Elec. Rev.*, Nov. 30, 1888.

Magnets.

On the Character of Steel used for Permanent Magnets. Paper by W. H. Preece, F. R. S., read at the Leeds meeting of the British Assoc. Adv. Sci., 1890. Tables. Lon. *Elec. Rev.*, Sept. 12, 1890, pp. 305-7.

Notes on Permanent Magnets. By S. P. Thompson. A valuable series of articles. *Electrician*, July 3, 10, 31. *Elec. World*, Aug. 1, 1891. *et seq.*

Maintenance of Way. See *Railroad Maintenance*. *Railroad Track*.

Malaria. Report in regard to, from several Connecticut towns. *Eng. News.*, May 6, 1882, pp. 144-5.

Malleable Iron Castings. See *Iron, Cast*.

Manganese Steel. See *Steel*.

Manufactures.

Cost in. By Henry Metcalf. A lecture before the students of Cornell University on the methods of obtaining the cost of work done in machine shops. *Sci. Am. Sup.*, Aug. 6, 1887.

Methods of Ascertaining the Cost of Manufactures. A paper by Prof. C. H. Benjamin, read before the C. E. Club of Cleveland, describing a system of cost accounts for manufacturing establishments. Abstract, with forms of record, etc. *Jour. Assn. Eng. Soc.*, March, 1891, pp. 114-24.

Manufactories. *A Modern Industrial Plant.* Detailed description of the establishment of Proctor & Gamble, with illustrations of buildings, details of tools, floors and various apparatus. *Eng. Record*, May 2, 1891, pp. 356-63.

Management of Men. A very good editorial on the subject. Lon. *Engineer*, Dec. 25, 1885.

Marine Propulsion. Review of a book containing three lectures by S. W. Barnaby. The object of the book is to determine the best dimensions, form and speed

Marine Propulsion, continued.

of a screw for given conditions of resistance and speed of ship. Some 510 experiments worked up. Results important. Lon. *Eng.*, April 26, 1889.

Maps.

Autographometer. A brief account of an invention intended to plat to a fixed scale the topography and levels of any given country. Illustrated. *Eng. News*, May 7, 1887; *Am. Eng.*, May 4, 1887; *Sci. Am. Sup.*, June 4, 1887.

"*What Constitutes a Map.*" Paper by Mr. Wm. G. Raymond, Mech. Tech. Soc. Pac. Coast, discussing the question concisely, yet quite fully, and proposing a method of regulating the filing of records by law. *Trans. Tech. Soc. Pac. Coast*, February, 1890, Vol. VI.

Projection of. See *Projection.*

For Railroad Surveys. Paper by Emile Low, calling attention to the great advantages of the method of drawing railway maps on comparatively small sheets as 19x24 inches. Illustrated. *Proc. Eng. Club, Phila.*, June, 1888, Vol. VI., pp. 280-292.

Topographical Maps. The Construction of Topographical Maps by Reconnaissance Methods. Paper by Arthur Winslow, Assistant State Geologist of Arkansas. pp. 8. *Trans. Ark. Soc. Eng. Archts. & Survs.* Vol. II., p. 71, 1888.

Topographical Signs on. A new method of making such signs on the original maps, an inked cylindrical rubber stamp being used, on which the various topographical signs appear, lettering and figures also put on mechanically. By J. A. Ockerson, in *Trans. A. S. C. E.*, Vol. XIV., p. 319.

See *Chart. Railroad Maps.*

Mapping. *The History of Cartography.* Address by E. G. Ravenstein before the Brit. Assn. *Eng. News*, Sept. 12, 1891, pp. 245-6.

Marble. See *Stone Quarrying.*

Marine Engineering.

Its Progress and Development. An address before the Inst. Nav. Archts., Liverpool. *Sci. Am. Sup.*, Oct. 16, 1886.

Some Details in. A paper read by Mr. Thomas Mudd, before the Inst. of Naval Architects. Several details of construction as to boilers, crankshafts, piston packing, and cylinders, are discussed. Lon. *Eng.*, Apr. 10, 1891, pp. 454-6. Lon. *Engineer*, Apr. 10, 1891, pp. 288-90.

Recent Progress in. By Wm. Parker, Chief Engineer-Surveyor Lloyd's, before the Inst. Naval Archts. Shows progress made in the use of steam. *Am. Eng.*, Aug. 25, 1886.

Review of Marine Engineering during the Past Decade. A paper by Alfred Blechynden read before the Inst. M. E. Lon. *Eng.*, Aug. 21, Sept. 18, 1891, *Mech. World*, Aug. 8, 1891, *et seq.* *Sci. Am. Sup.*, No. 820, Sept. 19, 1891.

Review of Progress During the Last Quarter of a Century. By Alfred Holt before the Inst. C. E. Reprint, *Eng. News*, May 16, Sept. 12, 1879.

Univ. Expos. of Anvers. By J. Gaudry. *Mem. de la Soc. des Ing. Civils*, November, 1885, pp. 607-668.

See *Ship-building. Sounding Machine.*

Marine International Law. Compiled by Commander Henry Glass, U. S. N., and published as No. 34 of *Proc. of the U. S. Naval Inst.*

Marking Patterns. See *Pattern Marking.*

Mars. *Facts and Speculations Regarding the Planet Mars.* A paper by Mr. N. B. Wood, member of the Civil Engineers' Club of Cleveland, read Nov. 26, 1889, p. 3. *Jour. Assn. Eng. Soc.*, Jan., 1890, Vol. IX., p. 12.

Masonry.

Abutments. By Wm. Cain. An investigation of their proper proportions and

Masonry, continued.

sizes, deduced from Rankine's general formula of earth pressure. *Van Nos. Eng. Mag.*, Vol. VII., p. 453.

Action of Air and Water upon. Lecture by D. Smith discussing this subject. *Eng. News*, May 30, 1878, pp. 172-3.

Arches. See *Arches.*

Bridge. See *Vibration.*

Cleaning of, by Sand Blast. Description of method employed in the U. S. Assay Office. Illustrated. *Eng. Record*, Oct. 3, 1891, p. 283.

Columns. See *Columns.*

Cost of, in Detail. Gives data for cost of stone culverts and bridges, lock masonry and aqueduct masonry. Extracts from an Engineer's Note Book No. 2. *Eng. News*, Dec. 8, 1888.

Cost of Construction and Masonry. A paper by Mr. E. J. Chibas in *The Polytechnic*, giving a description and cost of stone and brick masonry and concrete used in the construction of a power station at Pittsburg, Pa., *Eng. Record*, Apr. 11, 1891, pp. 313-4.

Detecting Bad Work. See *Aqueduct, New Croton.*

Masonry *of East River Bridge.* Notes on. By Francis Collingwood. Illustrated. Gives description of towers, anchorage and hoisting apparatus. *Trans. A. S. C. E.*, Vol. VI., 1877, pp. 7-27. *Van Nos. Eng. Mag.*, Vol. XVI., p. 431.

Government Specifications for. Gives the regular specifications for stone work in use in the Government architect's office. *Eng. News*, June 2, 1888.

Process for Cleaning. From the French. *Eng. News*, Sept. 27, 1884.

Proper Construction and Cost of. By T. H. McKenzie, before the Connecticut Association of Civil Engineers and Surveyors. Gives specifications, with comments, for first-class masonry. *Proc. Conn. Asso. C. E. & Surv.*, 1888, pp. 45-54.

Repairs of. By O. Chanute. An account of the applications of beton on the Erie Railway. Piers, culverts, arches and tunnel linings repaired. Illustrated. *Trans. A. S. C. E.*, Vol. X., 1881, pp. 291-307.

Repairs. See *Dock Walls.*

Settlement of. See *Cements, Compressive Strength of.*

Specifications, for stone bridge over the Mississippi River, at Minneapolis. Span of arches 125 feet. *Am. Eng.*, Oct. 29, 1883.

And Stone Cutting. By Law. Harvey. A series of articles giving instruction in the draughting of details of masonry. *Build. News*, Dec. 11, *et seq.* 1887.

Structures, for Railroads and Highways. Being thirty-five plates on a scale of 10 ft. to 1 in., of such structures as have been actually built, together with their volumes and cost. Includes several skew arches. Published by Richard B. Osborne & Sons, Civil Engrs., Philadelphia, 1884. Price $6.00.

Supports for Hanging Walls at the Tilly Foster Iron Mines. By Louis G. Engel. A lengthy article, with 17 full-plate plates. *Sch. of Mines Quar.*, May, 1885.

See *Brick Masonry. Bridges, Highway. Bridge Piers.*

Master Car Builders' Association.

Report of the twenty-fourth annual convention. Illustrated. *Ry. Rev.*, June 14, 21, 1890, pp. 336-9, 342, 344, 351-3. *R. R. Gaz.*, June 13, 20, 1890, pp. 417, 431, 434-6. *Ry. Age*, June 14, 1890, pp. 416-20. *Mast. Mech.*, July, 1890.

Proceedings of Convention at Saratoga, June, 1884. Apply to M. N. Forney, Sec'y, 73 Broadway, New York.

Master Car and Locomotive Painters' Association. Report of convention contains papers of value on "Cleaning Varnished Surfaces," "Treatment of Headlining

Master Car and Locomotive Painters' Association, continued.

Material," and the "Chemical Purity of Soap." *Ry. Rev.*, September 12, 19, 1891.

Master Mechanics' Association. Report of the twenty-third annual convention. *Ry. Rev.*, June 21, 28, 1890, pp. 352-7, 363-5, 370-1, 377-8. *R. R. Gaz.*, June 27, 1890, pp. 451-2, 461-6. *Ry. Age*, June 21, 1890, pp. 432-6. *Eng. News*, June 28, 1890.

Material, *A New View of the Resistance of.* By Prof. W. C. Unwin. Discusses Bauschinger's experiments and shows the relation between them and Wahler's experiments. Lon. *Engineer*, Jan. 7, 1887.

Materials of Construction. By R. H. Thurston. Treats of the strength and other properties of materials of construction, as deduced from the strain diagrams produced by the Thurston testing machine. A reply to criticism by Prof. Kick on a previous paper. *Trans. A. S. C. E.*, Vol. V., p. 9.

Working Loads for. See *Working Loads.*

See *Building, Materials.*

Measurements.

Discussion of the Precision of Measurements. Notes of a brief lecture course on the laws governing the precision of measurements, with application of the method of least squares. By Silas W. Holman. *Technology Quart.*, Vol. I., 1887, Nos. 2, 3, and 4.

The Refinements of Modern Measurements and Manipulations. A paper by J. A. Brashear in *Sci. Am. Sup.*, April 5, 1890, p. 11787.

See *Base-Line.*

Measures.

Recent Progress of the Metric System. Interesting paper by Dr. B. A. Gould, member of the International Committee of weights and measures. *Jour. Asn. Eng. Soc.*, June, 1890, pp. 283-9.

Standard. By F. A. Gieseler. Gives brief history of the development of standards of length, describes the present standards of the United States and the methods adapted to compare them with other standards, etc. *Jour. Frank. Inst.*, Aug., 1888; *Eng. News*, Sept. 8, 1888.

See *Meter. Tapes. Weights and Measures.*

Measuring Instruments *used in Mechanical Testing.* By Prof. W. C. Unwin. Discusses the different forms of instruments for determining the exact distance between two marks. Lon. *Engineer*, May 20, 1887.

Mechanical Engineering.

Chalk Age of. By J. T. Halloway. Fifth lecture before the students at Cornell University. Gives picture of engineering practices of days gone by and the contrast between the former and modern practice. *Sci. Am. Sup.*, June 16, 1888.

The outline of a course of instruction, with arguments by Robt. H. Thurston. *Jour. Frank. Inst.*, Vol. CXVIII., p. 188 (Sept., 1884.)

Mechanical Engineers.

Abstracts of the proceedings of the Boston Meeting. An extended account in *Mechanics*, Dec., 1885.

Advice to Young. By Prof. Perry to the students at Finsbury Technical College. A valuable paper for working engineers. *Sci. Am. Sup.*, Sept. 1, 1888.

Semi-Annual Report of the Committee of the Cleveland Engineer's Club on— By J. L. Gobeille. A review of the year's work. *Jour. Asn. Eng. Soc.*, Jan., 1887.

Special Report of the 22nd Annual Convention of the Am. Soc. M. E., held at Richmond, Va. *Mechanics*, Dec., 1890.

Report of the meeting of the Inst. of Mech. Engrs. at Sheffield, July 29, 30. The report contains valuable papers on steel rails, coal mines, the manufacture of fuel and gas and the Sheffield Water-Works. *Iron*, Aug. 1, 1890, pp. 91-96.

Mechanical Engineers, continued.

Mech. World, Aug. 1, 1890, *et seq.* Lon. *Engineer*, Aug., 1890, *et seq.* Lon. *Eng.*, Aug. 1 and 8, 1890, pp. 102-4, 136-9, 171-4.

Mechanical Equivalent of Heat. See *Heat.*

Mechanical Integrators. See *Integrators.*

Mechanical Stokers. See *Stokers.*

Mechanism. By Arthur Rigg. One of the Canton lectures under the auspices of the Society of Arts. *Van Nos. Eng. Mag.*, Vol. VII., p. 311.

Elliptic Lobed Wheels. See *Lobed Wheels.*

See *Machine Designs.*

Mercurial Gauge. A discussion of its reliability. *Trans. A. S. C. E.*, Vol. II., p. 228.

See *Pressure Gauge.*

Mercury Column. *An improved form of*, for measuring steam pressures. It is bent into vertical loops, the mercury resting in the lower ends. Thus a slight displacement is multiplied by the number of loops. Illustrated. *Trans. A. S. M. E.*, Vol. II., p. 98.

Meridian Time. An historical account of progress in its advocacy and adoption. *Van. Nos. Eng. Mag.*, March, 1886.

See *Standard Time.*

Metallic *Cross Ties.* See *Railroad Ties.*

Metalloids. See *Alloys.*

Metallurgy. See *Gold and Silver. Mining Engineering. Ore Crushing.*

Metals.

The Action of Tidal Streams on Metals. A valuable paper by Thomas Andrews, M. I. C. E., on the electrolytic disintegration during the change from fresh to salt water. Tables of the electromotive force during this process, for various kinds of iron and steel, are given. *Iron*, Jan. 23, 1891, pp. 71-5.

The Action of Tidal Streams on Metals during Diffusion of Salt and Fresh Water. Analyses of iron and steel, being the conclusion of experiments reported in *Iron*, 1891, p. 71, by Thos. Andrews. *Iron*, April 24, 1891, pp. 760-1. Abstract, *Eng. Record*, May 16, 1891, p. 389.

And Alloys. By Prof. W. C. Roberts Austin. A lecture delivered to the operative class at Birmingham. *Sci. Am. Sup.*, Feb. 5, 1885.

Corrosion of. See *Corrosion.*

Effect of Chilling on the Impact Resistance of Metals. A paper by Thomas Andrews, giving results of experiments on forgings, subject to various rates and degrees of chilling. Five sets of experiments. Results tabulated. *Proc. Inst. C. E.*, Vol. CIII., 1891, pp. 231-49. Abstracted in *Am. Mfr.*, Nov. 21, 1890, p. 10.

On the Effect of Permanent Elongation on the Cross Section of Hard-drawn Wires. Short but very interesting paper by Profs. T. Gray and C. L. Meet. The results of a few experiments are tabulated. *Phil. Mag. & Jour. of Sci.*, April, 1890, No. 179, pp. 355-8.

Expansion of, from Low Temperatures. Some interesting experiments of expansion of wrought iron, steel and cast iron between temperatures of 15° C. and 300° C. By Thomas Andrews, F. R. S. *Sci. Am. Sup.*, March 15, 1890, p. 11842.

Properties of, held in Common with Fluids. A lecture by Prof. W. C. R. Austen. Shows that the properties held in common with fluids are numerous and surprising. Illustrated. *Sci. Am. Sup.*, July 17, 1886.

Rate of Set of. By Prof. R. H. Thurston. Gives details and results of experiments made to determine the law of the rate of set of metals subjected to strain for considerable periods of time. *Trans. A. S. C. E.*, Vol. VI., p. 28.

17

Metals, continued.

Resume of the Metal Industries of the World for 1888. *E. & M. Jour.*, Jan. 19 1889.

Wear of. See *Rails*.

See *Fatigue*; also *Oxidation*.

Metallic Compounds. A long list of authorities on metallic compounds may be found in the *Rep. of U. S. Board of Testing Iron, etc.*, Vol. I., 1881, pp. 149-110.

Meteorology.

Artificial Rain Making. Paper by Prof. E. J. Houston before the Franklin Institute, discussing the possibility of producing rain artificially. Reprint, *Sci Am. Sup.*, No. 814, Oct. 17, 1891, p. 13161.

Can We Make it Rain? Papers by Gen. R. G. Dyrenforth and Prof. Simon Newcomb, discussing the recent experiments in Texas. *N. Am. Rev.*, Oct., 1891. pp. 385-404.

Conditions Causing a Cold Wave. An article by T. Russell of the U. S. Weather Bureau describing the method of making predictions. *Eng. Mag.*, Dec., 1891. pp. 351-8.

Conditions Causing a Tornado. Illustrated article by Prof. H. A. Hazen. *Eng. Mag.*, Oct., 1891, pp. 63-74.

Excessive Rainfalls Considered with Especial Reference to their Occurrence in Populous Districts. Part I. on maximum rainfall and flood-flow; Part II., on "Expediency of Making Adequate Provision in Systems of Drainage," with application to Washington, D. C. Topographical map and sections of sewers of this city given. *Trans. A. S. C. E.*, Vol. XXV., July, 1891, pp. 70-118.

Recent Advances in. Systematically arranged in form of a text-book for the Signal Service. By Wm. Ferrel. 440 pages. Discusses all questions in meteorology. A valuable treatise. *H. Rep. Ex. Doc.* 1, *Part 2, 49th Cong., 1st Session.* Being *Part 2 of Vol. IV., Rep. of Secy. of War*, 1885.

Self-Recording Instrument, invented and used by Dr. Draper at the New York Meteorological Observatory. Fully described and illustrated in Lon. *Eng.*, Dec. 4, 1885.

Work of the Weather Bureau. A review of its growth and work by Sergt. E. B. Dunn. Illus. *Eng. Mag.*, Sept. 1891, pp. 773-83.

See *Atmosphere*, *Rain*, *Temperature*, *Tornado*, *Weather Service*.

Meter, The, *as applied to Land Surveying.* By J. D. Varney. Conclusions adverse to its adoption. *Jour. Assn. Eng. Soc.*, Vol. V., p. 138.

Relation of the Yard to the Meter. See *Yard*.

Meters. See *Current Meters*, *Electric Meters*, *Water Meters*.

Metric and British Measures.

Compared for Engineering Purposes. By Arthur Hamilton-Smith, before the Inst. Civ. Engrs., with discussion. Issued also as a separate pamphlet. Discusses measures of length, capacity, weight and pecuniary values used in England. It is favorable to the metric system. Paper No. 2024. *Proc. for* 1885, *Inst. Civ. Engrs.* Abstract of Lon. *Eng.*, Jan. 23, 1885; also *Eng. News*, April 4, 1885. Also interesting particulars about the benefits of its introduction, especially in Germany. Letter from Frankfort by W. Howard White, C. E. N. Y. *Nation*, March 5, 1885; also New York *Semi-Weekly Evening Post*, March 13, 1885.

Comparative Size of, with Reference to Convenience. A plea for the metric system, with tables showing the relations between the metric units and those now in use. By Fred Brooks. Illustrated. *Jour. Assn. Eng. Soc.*, Vol. V., p. 225.

In Engineering. Its use in Railway work in Mexico, in an engineer's office in Boston, and for leveling. *Jour. Assn. Eng. Soc.*, Vol. II., pp. 321, 322 and 326.

Metric System.

The New Standards of Weight and Measure. An interesting letter from Paris

Metric System. *The New Standards, etc.*, continued.

describing the methods used in preparing the standard metre bars and standard kilogrammes for the International Bureau of Weights and Measures. *The Nation*, Oct. 24, 1889.

Weights and Measures. Chart of the system in its simplicity showing the relations of the different units to one another and the actual magnitudes of the principal units. Copies may be had with valuable tables and condensed information printed on the back from *American Metrological Society, 41 East 49th St., New York.* Price 10 cents.

See *Measures.*

Metric Tables for converting meters to feet and feet to meters; centimeters to inches and inches to centimeters; kilometers to miles and miles to kilometers. Each given for every unit from 1 to 1,000. Lon. *Engineer*, Feb. 5, 1886.

Mexico.

City of. New and Old Naming of Streets in. With diagrams showing streets and giving both the old and new names. *Eng. News*, Dec. 8, 1888.

Population Chart of, showing lines of railroads. *R. R. Gaz.*, July 10, 1885.

Railroad Building in. By Laurence Bradford. Discusses topographical peculiarities and advantages and disadvantages of the metric system in railroad work. *Jour. Assn. Eng. Soc.*, Vol. IV., p. 345.

See *Drainage. Mining Code.*

Micrometer. *The Run of.* Described and methods of determining discussed in *U. S. C. & G. Survey Rep.*, 1884, App. No. 8.

Micro-Organisms. *Some of the Conditions Affecting the Distribution of, in the Atmosphere.* By Dr. P. T. Frankland. Gives summary of the experiments of Pasteur, Tyndall, Miquel and Freudenreich. Describes methods of cultivation, etc. Gives results of experiments on the distribution of micro organisms at the different altitudes and places; also their nature. Discussed by Doctor Carpenter and others. *Jour. Soc. Arts*, March 25, *et seq.*, 1887.

Mill. See *Slag Mill.*

Mill Architecture. A lecture by C. J. Hexamer, before the Franklin Institute. Discusses floors, girders, walls, cornices, columns, roofs, fire doors, picker-houses, card rooms, transmission of power, heating, lighting, and frictional electricity. *Jour. Frank. Inst.*, July, 1885.

Mill Engineering.

Leaves from the Note Book of a Mill Engineer. Treats of water in channels and in pipes, of water motors and horse-power of water-wheels. *Sci. Am. Sup.*, Oct. 22, 1887.

A paper read by Edward Sawyer, Member of Boston Society of Civil Engineers, read September 18, 1889, p. 15. Discussion by several members, p. 24. *Jour. Assn. Eng. Soc.*, November, 1889, Vol. VIII., pp. 513-520. *Am. Arch.*, December 14, 1889. *Eng. & Build. Rec.*, March 22, 1890, p. 148.

Mills. See *Fire and Floors.*

Milling.

A two-page plate of a design for a flour mill to be built in Covinto, Chili. Water-power to be used. Lon. *Eng.*, April 24, 1885.

Modern. By Gilbert Little. An illustrated series of articles dealing with the birth and development of the Hungarian or Semolina system. Lon. *Engineer*, July 15, etc., 1887.

Roller. By Henry Simon. Gives record of tests instituted to determine the power consumed by the different machines used in a roller mill when working under different conditions. Lon. *Eng.*, July 8, 1887.

See *Flour Milling. Hydraulic Mill Appliances.*

Milling Cutters. A paper by Mr. George Addy, read before the Inst. Mech. Eng'rs. Illustrated. Lon. *Eng.*, Dec. 5, 1890, pp. 677-8.

Mineral Resources.

Of the United States for 1883 and 1884. A volume of over 1,000 pages. Issued by the United States Geol. Survey, giving statistics of production and many memoirs by technical experts on methods in use in prospecting, mining and reduction. *H. R. Misc. Doc. No. 30, 49th Congr., 1st Session, Dpt. of the Interior.*

Of the United States. By Albert Williams, Jr. Description of the areas from which the different mineral products are obtained, accompanied by tables of analysis and statistics. Also outlines of methods used in prospecting, mining, reduction and manufacture, with valuable information on the modes of occurrence. Published by U. S. Geological Survey in 1883, pp 813. Price 50 cents. Same for 1883 and 1884, pp. 1,016. Price 60 cents. *Van Nos. Eng. Mag.*, Jan., 1886.

Of the United States for 1885. H. of Rep. Misc. Doc. No. 136, 49th Congr., 2nd Sess., 1889.

Minerals.

Production for 1887, Statistics of. Good summaries of the production and prices of zinc, copper, coal, coke and pig-iron for the past year will be found in *E. & M. Jour.*, Jan. 7, 1888.

Coal. Geology of the Panther Creek Basin, with sections, plates and atlas. By C. A. Ashburner. *II. Geol. Survey Penn.* Report A. A.

Tables of the Rapid Determination of the Common Minerals by "External Signs" and by the Behavior of the Minerals before the Blowpipe. A valuable paper by A. J. Moses. *School Mines Quar.*, July, 1890, Vol. XI., No. 4, pp. 334-51.

Miner's Inch. See *Hydraulics.*

Mining.

By Maj. T. B. Brooks. An analysis of the cost and description of the methods of mining employed in the Marquette iron region, Lake Superior, Michigan. *Trans. A. S. C. E.*, Vol. II., p. 17.

Coal.

By H. M. Chance. *II. Penn. Geol. Survey.* Report A. C. A complete treatise, freely illustrated, and accompanied by maps of the mining practice of the anthracite region. Harrisburg, 1883.

Bellville Seam, No. 6, at St. Clair Co., Ill. Article by Leo Gluck, E. M., describing the details of work, especially the operation of the Harrison coal cutting machine. *Colliery Eng.*, Oct., 1891.

Methods and Appliances used in the Anthracite Coal Fields. By H. M. Chance; 574 pp., with an atlas of 25 plates, 54 page-plates, and 60 illustrations in the text. A full description of methods of prospecting, shaft and slope sinking; gangway and tunnel driving, timbering, etc.; of tools, machinery, etc., and of colliery management. *Publications Second Geol. Survey of Pa.*, AC. 1883.

Methods used in mining the thick coal of South Staffordshire, England, the vein being from 8 to 14 feet. Extracts covering 18 pp., from a paper before the British Soc. of Min. Students. Illustrated. *Sch. Mines Quar.*, April, 1886.

Nova Scotia. Paper by E. Gilpin, Jr., Deputy Commissioner and Inspector of Mines, with discussion. *Trans. Can. Soc. C. E.*, 1888, Vol. II., pp. 350-400.

Recent Improvements in the Mechanical Engineering of Coal Mines. Paper by Mr. Emerson Bainbridge, read at Sheffield before the Inst. Mech. Engrs. This paper discusses sinking of shafts, pumping, haulage, ventilation, screening, etc. *Iron*, Aug. 1, 1890, pp. 97-101.

Waste in Mining Anthracite. By Franklin Platt; with a chapter on the methods of mining, by John Price Wetherill. Numerous illustrations, 134 pp. Facts largely deduced from mine records. *Publications of Geol. Survey of Pa.*, A2, 1881.

Code, New, of Mexico. Regulates all matters relating to mining property. Ad-

Mining, *Code, New, etc.*, continued.

oped 1884. A full synopsis given in a paper before the May meeting of Inst. of Min. Engrs. By Rich E. Chism. E. M. *Iron*, July 3, 1885.

Debris in California Rivers. By A. J. Bowie. A full and elaborate treatment of the above subject. *Trans. Tech. Soc. Pac. Coast*, Vol. IV., p. 1. also briefly abstracted in *Eng. News*, July 30, 1887.

Diamond in South Africa. An article by G. D. Stonestreet, giving a brief description of the Kimberley mine and of the method of working. Illus. *Eng. Mag.*, Aug., 1891, pp. 579-81.

Drift. By T. Egleston. Describes methods used in the West to work the ancient river beds. *Sch. Mines Quar.*, Vol. VIII., pp. 204 and 289.

Engineering.

By J. L. Culley. A paper involving the general principles of mining engineering, relating particularly to coal bank work as pursued in Ohio. *Eng. News*, March 10, 1888.

Report with abstract of papers of the A. I. M. E., the Brit. Iron and Steel Inst., and the Ver. Duet. Eisenhuettenleute. Also accounts of excursions of the party to various points of engineering interest. *Eng. News*, Oct. 4, 1890, *et seq.* *R. R. Gaz.*, Oct. 3, 1890, pp. 680-1. *Ry. World*, Oct. 4, 1890, p. 941. *R. R. Gaz.*, Oct. 10, 1890, pp. 694-5. Lon. *Eng.*, Oct. 10, 1890, *et seq.* *E. & M. Jour.*, Oct. 4, 1890, pp. 392-4. Brief notice in *Am. Eng.*, Oct. 11, 1890, pp. 158-9.

Modern Progress in. An address before the Liverpool Eng. Soc., Dec., 1883. *Transactions*, Vol. IV., p. 136.

Exhaustive. A paper by W. H. Jennings before the Ohio Inst. of Min. Engrs., relating to a more effectual exhaustion of coal beds. *Ohio Mining Jour.*, Zanesville, May, 1881.

Gold. Working Placer Deposits in the United States, by T. Egleston. An historical and descriptive article of 30 pp., giving an account of the various ways in which placer mines have been worked in this country. *Sch. Mines Quar.*, Jan 1886.

Interests of the United States, Review of, for 1888. *E. & M. Jour.*, Jan. 12, 1889.

Laws of the United States and Regulations thereunder, together with State and Territorial Mining Laws and Local Mining Rules and Regulations. Quarto, 700 pp., being Vol. XIV., U. S. Census, 1880. *H. Rep. Mis. Doc. 42, Pt. 14, 47th Cong. 2d Sess.*, 1885.

Machinery.

Coal. The Ingersoll-Sergeant. Brief description, with record of work. Illustrated. *E. & M. Jour.*, Feb. 2, 1889.

Westphalia. By Messrs. Malkel, De Gournay and Suisse. Gives notes on the machinery, appliances, mode of working, etc., of the collieries of Westphalia. *Proc. Inst. C. E.*, Vol. XCII., pp. 367-376.

See *Electric Mining Machinery. Hoisting Machinery. Cable Grips.*

Winding Engines.

A description of the new winding engines and drum recently started at the Lady Windsor Colliery in Wales, with dimensions and details of construction. Illus. *Sci. Am. Sup.*, April 16, 1887.

A paper by Herbert W. Hughes, F. G. S. Reprinted from Trans. S. Staffordshire and E. Worcestershire Inst. of Min. Engrs., Vol. XII., Pt. 1. *Sch. Mines Quart.*, Columbia College, April, 1889. Vol. X., p. 258.

The Equalization of Load on Winding Engines by employment of spiral drums. By E. M. Rogers, Central City, Col. A theoretical discussion. P. 8. *Proc. A. I. M. E.*, Buffalo meeting, 1888.

Wire Rope Haulage and its Application to Mining. By F. C. Roberts. A 45-page paper, fully illustrated. Treating of hoists, inclined planes, engine

Mining *Machinery, Wire Rope Haulage, etc.*, continued.

gravity, and aerial haulage, tail-rope, counter-rope, and endless-rope systems. *Trans. A. I. M. E.*, Duluth meeting, 1887.

Mica in North Carolina. By W. B. Phillips. A series of papers describing the geology of the mining districts, formation of the veins, dressing the mica, etc. *E. & M. Jour.*, April 21 *et seq.*, 1888; *Sci. Am. Sup.*, July 21 *et seq.*, 1888.

Ore Deposits of Red Mountain, Colo. By G. E. Kedzie before the American Institute of Mining Engineers. Gives a description of the bedded ore deposits of Red Mountain mining district, Colo. *E. & M. Jour.*, Aug. 11, 1888.

Ores and their Mode of Occurrence at Aspen Mountain, Colo. By D. W. Brunton. A series of illustrated articles describing the ore deposits and the faulting of the Aspen district, Colo. *E. & M. Jour.*, July 14 *et seq.*, 1888.

Progress of the Art of Mining. An interesting and valuable lecture delivered at the Royal College of Science, London, by C. Le Neve Foster. Many references to existing literature are given. Lon. *Eng.*, February 13, 1891, pp. 201, *et seq.*

Shot-Firing in, Theory of. By M. P. F. Chalon. Gives a general theory of shot-firing with common powder or high explosives. *Eng. News*, Jan. 14, 1887.

Submarine.

A series of articles. Lt. Col. Bucknell, R. E. Gives analyses of important experiments, etc., treats of the apparatus employed for measuring the effects of submarine explosions. Lon. *Eng.*, July 1, *et seq.*, 1887.

On the Personnel for. By Lieut. J. T. Bucknell, before the Royal Inst. of Submarine Mining. Lon. *Eng.*, March 25, April 1, etc., 1887.

Surveying, in Coal Mines of Ohio. By R. S. Paul. *Ohio Mining Journal*, Nov. 1882. See also paper on the same subject in same journal by R. M. Haseltine.

Surveying. See *Surveying.*

Systems of, in Large Bodies of Soft Ore. By R. P. Rothwell, before the Boston meeting of the American Institute of Mining Engineers. Describes the system employed at the Dean River mine and proposes working the vein out from the top down instead of from the bottom up. *E. & M. Jour.*, March 10, 1888.

See *Boring, Compressed Air Applied to. Hydraulic Elevator, Gravel. Masonry Supports.*

Mines.

Accidents in.

Abstract of the Report of the British Commission to the House of Commons, from the London *Mining Journals.* *E. & M. Jour.*, April 24, 1886; *Eng. News*, Sept. 25, 1886, *et seq.*

See *Coal and Gold.*

By Sir Fredk. Aug. Abel. The first part of an able paper in which the general situation is reviewed in the light of recent researches, including the Rep. of the Commission in 1886. Statistics, causes, appliances, etc., discussed. *Proc. Inst. C. E.*, Vol. XC., pp. 160-222.

A paper by Sir Fred. Abel, being a continuation of a previous paper. With discussion, covering 150 pp. in *Proc. Inst. C. E.*, Vol. XC., p. 36.

Fire-Damp, Accidents from, and Means of Preventing. A paper reprinted from *Le Genie Civil.* By L. Parent. Discusses in detail an accident in France, whereby 28 workmen lost their lives; then considers the influence of the blasting charge; the influence of dust; the influence of currents of air, with regard to their quality and quantity, and with regard to the point of origin of accident; and the measures to be taken. *Eng. News*, Sept. 5, 1885.

Coal, Conflagration at Kidder's Slope. By Martin Coryell. Gives details of the conflagration in the mines and the methods adopted to put it out. *Trans. A. S. C. E.*, Vol. III., pp. 147-152.

Mines, continued.

Explosion, Colliery, at Dour, Belgium. Brief account. *E. & M. Jour.*, Dec. 8, 1888.

Explosions in Coal Mines caused by fire-damp forced into the mine from the surrounding earth strata as a result of a sudden fall in the barometric pressure. A rigid relation of cause and effect, thought to have been established by observations in Hungary. Abstract of report on the subject in *Eng. News*, Feb. 6, 1886.

See *Accidents*.

Coal, Spontaneous Combustion in. Causes, development and preservation. By —— Durand. From *Proc. Inst. C. E.* *Eng. News*, March 8, 1884, pp. 109-10.

Electric Haulage in. See *Electric Haulage*.

Fire Damp in. See *Fire Damp*.

Tilly Foster Iron Mine, Reopening of. Paper by F. H. McDowell, New York, describing the methods used in making the necessary excavations, viz., the use of wire cables and trolleys for the open pit. Illustrated. Pp. 9. *Proc. A. I. M. E.*, New York meeting, Feb., 1889.

Ventilation of.

Coal Mines. By Geo. G. Andre. A paper, with discussion, before the Soc. of Engrs. *Van. Nos. Eng. Mag.*, Vol. XIX., p. 369.

Colliery in Westphalia. Description, duty, efficiency, experiments, cost, etc. Abstracted in *Proc. Inst. C. E.*, Vol. XC., p. 541.

Mechanical. By Wm. Cochrane. Abstract of a paper read before the Inst. of Mech. Engrs., at Newcastle. *Van. Nos. Eng. Mag.*, Vol. I., p. 98.

See *Fire Bulkheads in*, also *Fire Damp*. *Shaft*. *Shaft Sinking*.

Miscellaneous Notes *of European Travel*. Some brief accounts of matters of engineering interest noted by members of Boston Soc. C. E., during the 1889 excursion of American engineers to Europe. pp. 9. *Jour. Assn. Eng. Soc.*, March, 1890, Vol. IX., p. 119.

Mississippi River. See *Jetties*.

Molecular Forces. See *Forces*.

Moment of Inertia,

On the determination and graphical representation of the moment of inertia of plane surfaces. By Prof. Mohr. Discussion of some very pretty geometrical methods, but of more interest from a mathematical than from an engineering point of view. *Der Civil Ingenieur*, 1887, pp. 43-67.

New and general formula. By J. W. Davis. *Van Nos. Eng. Mag.*, Vol. XXI., p. 77.

Simple Graphical Method for Determining Moment of Inertia. A paper by F. E. Turneaure, read before the Eng. Club of St. Louis, describing a method useful in cases of irregular areas. *Jour. Assn. Eng. Soc.*, April, 1890, pp. 205-11.

Monument. See *Washington*.

Moore, Henry C. A memoir of the late Henry C. Moore, read before Engineers Club of St. Louis, May 15, 1889, by E. D. Meier. Pp. 2. *Jour. Assn. Eng. Soc.*, October, 1889, Vol. VIII., p. 492.

Mooring Light-House. See *Foundation of a Light-House*.

Mortars,

By Samuel Crompton. Treats of the effects of adding saccharine matter to mortar. *Eng. News*, Jan. 1, 1887.

Chemical Ingredients and Conditions necessary to make good mortar, and what chemical changes occur. By D. P. Waters. *Jour. Assn. Eng. Soc.*, Vol. V., p. 85.

Economy of Lime-Cement Mortar. The results of some experiments indicate that any combination of lime and Portland cement is uneconomical. Article

Mortars. *Economy of Lime-Cement, etc.*, continued.

by Prof. I. O. Baker. *Am. Arch.*, Nov. 29, 1890, p. 133. The principal series of these experiments is given in a paper by D. R. Kinkead. *Selected Papers, C. E. Club, Univ. of Ill.*, 1889-90, pp. 59-60.

Effects of Freezing on. By Alfred Noble. Gives results of experience on construction of masonry during freezing weather; also gives results of experiments of mixing salt with Portland cement mortars. Discussed by a number of prominent engineers. *Trans. A. S. C. E.*, Vol. XVI., p. 79; also abstracted *Eng. News*, July 2, 1887; *Building*, July 30, 1887.

Efflorescence and Impervious. By Ira O. Baker. Gives reason of efflorescence and discusses its remedy. Also discusses the use of soap and alum to render brick and mortar impervious. *Eng. News*, April 7, 1888.

The Injurious Effects of Cement on Lime Mortar. Discussion before the C. E. Club of Cleveland. *Jour. Assn. Eng. Soc.*, March, 1890, pp. 143-7.

Making and Using. By B. F. Bowen. Contains much useful information relative to lime mortar. *Rpt. Ohio Soc. Surv. & Engrs.*, 1888, pp. 223-31.

Cement.

Behavior Under Various Contingencies of Use. By F. Collingwood. With especial reference to farther investigation on the change of volume with age and under pressure. *Trans. A. S. C. E.*, Vol. XIV., p. 491.

And Concretes, Notes on. By W. H. Grant. An article giving the results of the writer's experience in the construction of the Naval Observatory, Washington, D. C. Gives an account of many practical difficulties. *Trans. A. S. C. E.*, Vol. XXV., Sept., 1891, pp. 249-79. Discussion, pp. 279-94. Abstract, *Eng. Record*, Nov. 28, 1891, *et seq.*

Does Salt Water Increase the Strength of. A table of over 200 tests of various kinds of cement shows a marked increase. *Eng. News*, Dec. 20, 1890, p. 543.

Effect of Salt in Mixing Cement Mortar. Paper by M. I. Powers, Jr., in *The Technic*, (Univ. of Ia.) The results from a large number of tests indicate a gain in strength at first, but little or no permanent gain. Reprint, *Eng. News*, Nov. 21, 1891, p. 481. *Am. Arch.*, Dec. 5, 1891, pp. 154-5.

Economy in the Composition of. By Prof. I. O. Baker. Discusses the use of Rosendale *vs.* Portland cement, of lime and cement, strength of cement mortars, quantities of ingredients required and cost of mortars. *Eng. News*, March 10, 1888.

The Permeability of Portland Cement Mortar and its Decomposition by Sea Water. Paper by L. Durand-Claye and P. Debray in *Annales des P. & C.*, Vol. XIV., p. 326, 1887. *Proc., Inst. C. E.*, Vol. XCVII., p. 445. London, 1889.

Permeability of. Notes by Prof. L. M. Haupt, and experiment by the Board of Experts on the Washington Aqueduct Tunnel. *E. & M. Jour.*, April 27, 1889.

Porosity, Permeability, and their Decomposition by Sea Water. Very complete paper by Paul Alexandre. *Annales des P. & C.*, Sept., 1890. Abstract in *Eng. News*, Jan. 10, 1891, pp. 40-1.

Portland Cement Mortars and their Decomposition by Sea Water. Resume of a memoir on the permeability of Portland cement mortars and their decomposition under the action of sea water. By MM. Durand-Claye and Paul Debray in *Annales des P. & C.* *R. R. Gaz.*, Dec. 28, 1888.

Public Works. An abstract of a report by the Executive Board of the City of Rochester, N. Y., prepared by Emil Kuichling. *Eng. & Build. Rec.*, March 24, *et seq.*, 1888.

Strength of. By Prof. I. O. Baker. Gives tables showing the strength of cement mortar of various ages, compiled from a large number of experiments. *Eng. & Build. Rec.*, May 5, 1888.

Mortars, continued.

Composition of ancient mortars and Rosendale cements. By Arthur Beckwith. *Van Nos. Eng. Mag.*, Vol. VIII., 105.

Hydraulic. By Dr. Michaelis. Gives details of experiments made with hydraulic mortars. *Van Nos. Eng. Mag.*, Vol. XXV., p. 262.

Lime-Cement Mortar. Tests of adhesive and cohesive strength of various per cent. mixtures of lime with Portland and Rosendale cement. By various students. *Technograph* (University of Ill. annual), 1890-1, pp. 67-76.

And Mortar Joints. By A. S. Jennings. An essay on the general use of mortars. Contains many practical hints. *Building*, June 4, 1887.

Saccharine Matter in, and Building in Frost. Paper by Samuel Crompton, with results of experiments at various temperatures. Lon. *Engineer*, Jan. 18, 1889 and Feb. 8, 1889.

See *Cements*.

Moscrop Recorder. See *Speed Indicator*.

Motive Power, *small.* By H. S. H. Shaw. The working agents are classified, and the relative advantages of the agents are considered with reference to the objects to which they are applied. *Van Nos. Eng. Mag.*, Vol. XXIV., pp. 265 and 313.

Movable Sidewalk. See *Railroads, Multiple Speed, at Jackson Park, Chicago.*

Movement of a Lighthouse, weighing 440 tons, on ways, by means of jacks. Methods described. *Am. Eng.*, April 10, 1885. Inst. Civ. Engrs., Vol. LXXIX., p. 347.

Motor, *Ammonia.* See *Ammoniacal Gas.*

Motors. See *Compressed Air Motors, Engines, Water Motors, Electric Motors.*

Moulding Sand. See *Foundry Practice.*

Mountain Railways. *The Abt Rack Rail System for.* See *Railroads.*

Municipal Affairs.

Social Statistics of Cities in the United States. Compiled by Geo. E. Waring, Jr., containing a great deal of matter of interest to sanitary and municipal engineers. *Tenth Census*, Vol. XVIII. and XIX.

Statistics, for eighty-five of the largest cities in the world. Given in connection with the annual address of the President of the Inst. of Civ. Engrs., London. Among the items given are: Population; no. of houses and ratable value; length of streets and sewers; water supply, source and amount; rainfall; amount of refuse and methods of removing same; no. of police; no. public vehicles on the streets daily; length of tramways or horse railroads. no. of slaughter houses and management; method of lighting streets, and annual death rate. *Proc. Inst. C. E.*, 1884, Vol. LXXVI.

See *Registry Bureaux, Phila. for registration of lots.*

Municipal Engineering.

A very valuable series of articles prepared expressly for the *Engineering News*, begun in number for Jan. 16, 1886. They are to embrace the "duties and practice of the departments of city governments which have to do with the construction and maintenance of public works." Such information to be given concerning all the larger American cities.

Abroad. Brief notes on sewage disposal works, water supplies, and pavements of various European cities. By Rudolph Hering. *Eng. Record*, Sept. 26, 1891, pp. 261-2.

And the Management of Office. By B. Schreiner. Gives good hints relative to the management of the offices of city engineers. *Jour. Assn. Eng. Assoc.*, Jan., 1888, pp. 9-12. *Eng. & Build. Rec.*, May 12, 1888.

In Boston. Concise abstract of principal facts contained in *Annual Report of*

Municipal Engineering. *In Boston*, continued.

City Engineer, Boston for 1889. *Eng. & Build. Rec.*, April 5, 1890. Vol. XXI., p. 274.

Boston Works and Methods. The first of a series of municipal engineering articles. Illustrated. Begun in *Eng. News*, Jan. 16, 1886.

Grades and Grade Systems of Cities. By Barnabas Schreiner. Discusses the difficulties of establishing grades, and considers the points to be attained by a system of street grades. *Jour. Assn. Eng. Soc.*, Vol. VI., p. 159.

Liverpool. Paving specifications, forms of sewers and gutters, and description of artisans dwellings owned by the city. From United States Consul Sherman's Report. *Eng. News*, Oct. 4, 1890, pp. 308-9. *Am. Eng.*, Nov. 1, 1890, pp. 188-190.

London (Eng.) Municipal and Sanitary Engineering. A series of articles describing these abstracted from the municipal reports. *Eng. News*, Aug. 21, 1886.

In New York. A sample "job." detailed in connection with the repaving of Fifth avenue. *Eng. News*, Oct. 2, 1886.

The Organization of the City Engineer's Office at St. Paul, Minn. Detailed description of organization, appliances used, plan of offices, etc. *Eng. & Build. Rec.*, Dec. 6, 1890, pp. 6-7, *et seq.*

Philadelphia Works. The fourth series of municipal papers, beginning in *Eng. News*, Dec. 18, 1886. Map.

Providence, R. I. A description of the city, organization of the engineering department, and of the various works of interest to the engineer. *Eng. News*, April 17, May 22, 1880.

Shaping of Towns. Two articles discussing the best methods of laying out towns. *Eng. News*, July 7, 28, 1877.

See *Civil Engineering*.

Municipal Government.

Government of Cities in the United States. A paper by Seth Low, discussing scope of work, form of organization and methods of work of city governments. *Century Mag.*, Sept., 1891, pp. 730-36.

Our Remedy for. Paper by Pres. C. W. Eliot. Advocates permanent employment of trained experts. *Forum*, Oct., 1891, pp. 153-68.

Multiplication Tables.

Abridged from Crelle's tables, suited to numbers of two figures. 20 pp. quarto. *Bureau of Navigation, Navy Dept., U. S.*

Nails *made from Tin Scrap*. See *Tin Scrap*.

Natural Gas. See *Gas, Natural. Petroleum. Fuel, Gaseous.*

Naval Exhibition. *The Royal Naval Exhibition.* Illustrated description of the Armstrong Gallery of the Exhibition. *Lon. Eng.*, July 17, 24 and 31, 1891.

Naval Vessels. See *Ships*.

Navigation.

Early Experiments in Steam. An account of Elijah Ormsbee's first steamboat and David Grieve's first screw propeller boat (1792), compiled by Elisha Dyer and S. B. Smith. *Sci. Am. Sup.*, May 14, 1884.

Early History of Steam Navigation. Paper by G. W. Buckwell, M. I. Mar. Eng. Illus. *Mech. World*, June 21, 1890, *et seq.*

Inland, of Europe. An abstract of lecture before the Inst. Civ. Engrs. A very fair summary of over three columns in length, of the facts connected with the river and canal navigation on the Continent. *Eng. News*, Feb. 6, 1886.

In France. Description of vessels employed, details, statistics, and prices for

Navigation, *Inland, in France*, continued.

transportation, with discussion. *Mem. Soc. des Ing. Civils*, Aug., 1885, pp. 186-200.

Some Interior Ports of Germany and Bohemia. Description of the most important ports, machines for handling the various kinds of traffic, etc. 6 plates. By ——— Monet. *Annales des P. & C.*, Apr., 1891, pp. 500-616.

Study on the mode of Navigation to be applied (on the Rhone) between Lyons and Marseilles. Largely a discussion of chain towage. *Annales des P. & C.*, Aug. 1885, pp. 300-395.

Land-Locked Navigation from Long Island Sound to the Mississippi River. An investigation of the feasibility of a route near the coast as regards the portion from the Suwanee River to the Mississippi R. By Captain S. C. McCorkle. *Proc. Eng. Club, Phil.*, Vol. VIII, No. 3, July, 1891, pp. 195-218.

Steam Lanes across the Atlantic. Proposed international agreement for ships channels, by Lieut. H. Barroll, U. S. N. Illus. *R. R. & Eng. Jour.*, Oct., 1890.

Position Indicator. Paper read before the U. S. Naval Institute by Lieut. H. O. Rittenhouse, U. S. N., March 8, 1887. Describes an instrument to indicate readily on a chart the position of a vessel when navigating coasts or inland waters, where landmarks may be seen and recognized. *Proc. U. S. Naval Inst.* for 1887, Vol. XIII., No. 2.

Submarine. The Nordenfeldt Submarine Boat. Illustrated by a number of cuts, and described in *Sci. Am.*, Nov. 7, 1885; also see *Eng. News*, Oct. 24, 1885.

See *Marine Propulsion. Sailing Charts.*

Navy.

Brigade; Its Organization, Equipment and Tactics. Prize essay for 1887, presented to United States Naval Institute by Lieut. C. T. Hutchins, U. S. N. *Proc. U. S. Naval Inst.*, Vol. XIII., No. 3.

Her Majesty's Ship "Victory" Her History and Construction. Interesting account, with illustrations showing details of construction and general processes of designing. Lon. *Engineer*, May 1, 1891, pp. 333-40.

Italian. The number and class of ships, with some data of largest. Lon. *Eng.*, Feb. 13, 1885.

List of vessels authorized by Congress up to March 4, 1887. *R. R. & Eng. Jour.*, April, 1887.

The New and the Old. A number of notes from various sources in relation to the new ships and guns for the Navy, and other naval matters. *R. R. & Eng. Jour.*, Dec. 1887.

Recent Progress in the Manufacture of War Material in the United States. Full abstract of a paper by W. H. Jaques, before the Brit. Iron and Steel Inst. *Iron*, May 22, 29, 1891.

Sword-Ships for the. Proposed type of small vessels having a sword, or ram, at the bow for destroying the propellers of hostile ships. Illus. *Am. Mach.*, Dec. 4, 1890.

The U. S. and its Prospects of Rehabilitation. By Rear Admiral Edward Simpson, Pres. U. S. Naval Inst. Annual address. A 38-page article reviewing the needs and prospects of a proper equipment of the various branches of the service. *Proc. U. S. Naval Inst.*, Vol. XII., No. 1 (Dec., 1885).

United States Armored Coast Line Battle Ships. Description and drawings of one of these new ships, including plan and elevation of engines. *Iron Age*, May 7, 1891, pp. 870-3.

War Ships of the British Navy. Good general description of two new designs adopted by the British Admiralty, one for a turret ship and one for a barbette ship. Illustrated. *Mechanics*, Aug., 1882.

War Ships of the United States Navy. A paper by Horace See, giving an ac-

Navy. *War Ships of the U. S., etc.*, continued.

count of the construction, with brief description, of the navy. Illus. *Eng. Mag.*, July, 1891, pp. 423-39.

See *Engines, Marine. Signals, Naval. Torpedo Boats.*

Navy Yards.

Report of the League Island Permanent Improvement Board. Report reviews condition on League Island, nature of soil, ease of constructing foundations, size, strategic importance, etc., and recommends extensive improvements amounting in total to $9,874,155. *Eng. News*, November 23, 1889, Vol. XXII., p. 483.

Needle, *Magnetic.* See *Declination.*

New Orleans.

Drainage of. See *Drainage.*

Topography and Drainage of. By T. S. Hardee, City Surveyor. Gives description with map and profiles. *Eng. News*, May 5, 12, 1877.

Nichols, William Ripley. *A Memorial.* By Geo. F. Swain. Gives good sketch of his life and work. *Jour. Assn. Eng. Soc.*, Jan., 1887.

Nile, *Delta of.* Description of delta and artificial works. By B. Baker, in London *Times*. *Eng. News*, Aug. 26, 1882, p. 303.

Nitrate of Soda. Machinery and methods of manufacturing this fertilizer, as employed in Chili. *Proc. Inst. C. E.*, Vol. LXXXII., p. 317.

Nitrates. *Determining Nitrates in Drinking Water.* See *Water-Analysis.*

Non-Conductors.

Of Heat. Experiment to find the relative values of hair-felt, mineral wool, mineral wool and tar, sawdust, charcoal, pine wood, loam, slacked lime, coal ashes, coke, air, asbestos, and charcoal. By Chas. E. Emery. *Trans. A. S. M. E.*, Vol. II., p. 34.

Of Heat.
See *Heat. Steam Pipes.*

Non-Conducting Boiler and Pipe Covering. Abstract of a paper by W. H. Collins before the Brit. Assn. Gives results of experiments with various material. *Eng. Record*, Oct. 31, 1891, p. 346.

Nozzles. See *Water Meter.*

Obelisk.

The cause of its decay found in mechanical disintegration due to changes of temperature. *Science*, Dec. 11, 1885.

Its disintegration. By Thos. Egleston. Gives results of examinations as to causes of its decay. Due to temperature changes rather than to moisture and freezing. *Trans. A. S. C. E.*, Vol. XV., p. 79.

Preservation of the, by the application of a parafine wax compound. The method employed fully described and illustrated, with the theories on which it is based. By R. M. Coffall before the New York Academy of Sciences. *Sci. Am. Sup.*, Jan. 30, 1886.

Oblique Arches. See *Arches, Skew.*

Observation. *A Generalized Theory of the Combinations of Observations so as to obtain the Best Results.* By Simon Newcomb. This is a modified form of the Least Square method, whereby the discrepant observations are neither wholly rejected nor given full weight, but each is weighted by both *a priori* and *a posteriori* arguments, so as to obtain a final, *most probable* value, which is subject to the personal judgment of the computer. *Am. Jour. of Math's.* John Hopkins Univ., Vol. VIII., No. 4, p. 343.

Ocean Cables. See *Submarine Cables.*

Ocean Currents. *Velocity of, at Great Depths. The Measurement of.* By E. Suchier. Translated from the German. A current meter used. Illus. *Eng. News*, Nov. 14, 1885.

Ocean Jetties. See *Jetties.*

Ocean Piers. See *Piers.*

Oil. See *Fuel, Liquid.*

Oil and Gas Fields of America. By Prof. James Dewar, F. R. S. A lecture before the Society of Arts, London. A very complete account of the geographical distribution and methods of working. *Sci. Am. Sup.*, July 18, 1885.

Oil and Natural Gas,

In Illinois. By Prof. Theo. B. Comstock. Treats of the origin and relations of petroleum and natural gas; physical and chemical relations; geological distribution; range and condition. *Rep. Ill. Soc. Engrs. & Surv.*

Oil Burners.

For Steam Boilers. Abstract of a series of papers by Herr Busley, of Kiel, a marine engineer, describing the leading or typical devices which have been and are used for burning liquid fuel, under the following classification: 1. Hearth fires. 2. Gas fires. 3. Spray fires. *R. R. & Eng. Jour.*, April, May, June, 1888.

Improved System of Burning Oil. Description of Aerated Fuel Co.'s method of arranging for the use of oil fuel for boilers. *Am. Mfr.*, Aug. 9, 1889.

Oil Gas—Compressed, *and its Applications.* By Arthur Ayres. Paper and discussion covering 66 pp. in *Proc. Inst. C. E.*, Vol. XCIII.

Oil Insulators. See *Insulators.*

Oil Mixing Plant. See *Railroad Structures.*

Oil Regions *of Penn and New York.* Their product and exhaustion. By Ch. A. Ashburner. *Trans. A. I. M. E.*, Halifax meeting, 1885.

Oil Wells. See *Gas.*

Oils, *Apparatus for Testing.* Description of an ingenious registering apparatus for testing the quality of lubricating oils. Illus. *Sci. Am. Sup.*, Oct. 15, 1887.

In Heavy Seas. A 37-page Bulletin, No. 82, of the Bureau of Navigation, U. S. Hydrographic Office, describing the effects of oil in lessening the dangerous effects of heavy seas. 1886.

The Illuminating Quality of Kerosene Oils. A special report by Henry A. Kastle, State Inspector of oils, St. Paul, with special reference to the diminished illuminating power of such oils in the last few years.

Testing. For viscosity, acidity, drying, and adulteration. A description of simple methods and apparatus. *Power*, Dec., 1887.

See *Illumination from Waste Oils. Lubricants.*

Optical Glasses. *The Jena Optical Glasses.* An account of the remarkable improvements in their manufacture, with description of various apparatus used in the process. Lon. *Eng.*, Apr. 24, 1891, pp. 399-401, *et seq.*

Optical Telegraphy, *by Sun and Lamp Light.* From the French. Fully illustrated. *Sci. Am. Sup.*, Sept. 12, and 19, 1885.

Ordnance.

And War Ships. Rep. of the Select Com. of Cong. on. Many illustrations. A mine of information concerning foreign practice and home facilities for steel manufacture. Contains a list of 40 "books of reference" on this subject. 1886. 500 pp.

Firing Trial of the 110½ ton B. L. Elswick Gun. With drawings of the gun and and projectile. *Am. Mfr.*, April 1, 1887.

Materials, New. An official report by Capt. Wm. H. Bixby, not printed elsewhere. A thorough review of the present state of theory and practice. Fully illustrated. To run some three months in *Eng. News*, beginning Oct. 24, 1885.

See *Guns.*

Ore.

Paper by W. H. Hoffman, M. E., describing process, and giving practical results, including statement of cost. *Trans. A. I. M. E.*, 1891, p. 7. Reprinted in *Eng. News*, Oct. 17, 1891, pp. 353-4. *E. & M. Jour.*, Oct. 31, 1891, pp. 502-4. *Iron Age*, Oct. 15, 1891, pp. 628-9.

Crushing. Theory of. By Luther Wagener. A valuable contribution to the present meagre literature on this subject. Discusses objects, surface of the pulp, work of crushing, useful work, relative surface produced by stamps and rollers, and sliming, all on the basis of efficiency and economy. *Trans. Tech. Soc. Pac. Coast*, Vol. III., p. 45 (May, 1886).

See *Crushers*.

Dressing.

By Electricity at the Tilly Foster Mine. Paper by F. H. McDowell, presented at the N. Y. meeting of the A. I. M. E., giving conditions of operation and valuable tables of results for seven months of 1890. *Eng. News*, Oct. 4, 1890, p. 291. Description of this mine *ibid.*, April 20, 1889.

In Germany. Lon. *Engineer*, Aug. 15, 1890, *et seq.*

Of Non-Bessemer. By G. W. Maynard and W. B. Kunhardt. *E. & M. Jour.*, March 31, *et seq.*, 1887.

Deposits.

Forms of, in Limestone. By Carl Henrich. Describes the peculiar form of galena deposits in Missouri. *E. & M. Jour.*, Nov. 3, 1888.

Geology of the Aspen, Col. By L. D. Siver. *E. & M. Jour.*, March 17, *et seq.*, 1888.

Of the Precious Metals Popular Fallacies Regarding. By Albert Williams, Jr. A paper of 14 pp., of considerable value to prospectors, in *Fourth An. Report, U. S. Geol. Surv.*, 1882-3.

Sorting. By T. L. Bartlett. Gives a description of the method of ore sorting employed at Milan mine. *E. & M. Jour.*, April 14, 1888.

Ore Docks *and Machinery, The Fairport.* Full illustrated description of the docks and machinery used to handle ore and coal; 600,000 tons of ore handled in 1888. *Am. Mfr.*, March 1, 1889.

Ore Separator. See *Separator.*

Ores. *Reduction of.* See *Hot Blast. Theory of.*

See *Mining. Metallurgy.*

Organisms. Micro. See *Micro Organisms.*

Orifices, Large. See *Hydraulic Experiments.*

Overhead Wires. See *Electrical Conductors.*

Oxygen. *The Extraction of Oxygen from the Atmosphere.* Article of interest in *Eng. News*, April 12, 1890, Vol. XXIII., p. 341.

Paddle-Wheels. Notes on the history of paddle-wheel steam navigation. Read before the Inst. M. E. By Henry Sandham. Lon. *Eng.*, April 3, 1885, *et seq.*

Paine, Thomas, the original inventor of long iron bridges. This claim fairly proven in *Eng. News*, Sept. 11, 1886.

Paints.

For Exposed Metallic Surfaces. *Iron Age*, Aug. 13, 1885. See also *Iron Surfaces.*

To Prevent Iron Surfaces Rusting. By Ernest Spon. A valuable paper, giving present practice by engineers. *Sci. Am. Sup.*, Sept. 11, 1885.

See *Master Car and Locomotive Painters' Association. Varnishes.*

Pantagraph.

Its Theory and Its Use. By E. A. Gieseler. *Eng. News*, Nov. 16, 1887.

Pantagraph, continued.

Rules and tables for setting the pantagraph, as made by the American Steam Gauge Co., and by the Ashcroft Mfg. Co. *Boston Jour. Com.*, April 4, 1885.

Paper Making *by Machine*, as exemplified in the manufacture of fine writing and printing papers. *Proc. Inst. C. E.*, Vol. LXXIX., pp. 251-304.

Paper Manufacture, *Lime Sulphite Fiber Manufacture in the United States*. By Major O. E. Michaelis, before Am. Soc. C. E. *Sci. Am. Sup.*, Jan. 25, 1890, p. 11,724.

Papier Mache. A brief description of the process of manufacture. *Building*, March 26, 1887.

Parallel Motions. Articles by Prof. A. Mac Lay. Well illus. *Am. Mach.*, Sept. 17, 24, Oct. 1, 1891.

Parallel Rod. See *Locomotive, Parallel Rod.*

Parker, George Alanson. A Memoir, by the late Samuel M. Felton. Read before Boston Soc. C. E., April 17, 1889. *Jour. Assn. Eng. Soc.*, June, 1889, Vol. VIII., p. 331.

Patent Centennial. *Report of the Celebration at Washington, D. C.* Contains papers of interest on inventions, as relating to the electrical and other sciences. *Elec. World*, Apr. 18, 1891, *et seq.*

Address of O. Chanute. See *Addresses, Chanute.*

Patent Equivalents. A very valuable and pertinent paper by Emil Starck of considerable interest to patentees and corporations. *Sch. of Mines Quar.*, Vol. XI., p. 121, Jan. 1890.

Patent Systems *of the United States*. A series of articles describing the same. By Geo. O. G. Coale. In Lon. *Eng.*, Jan. 1, 1886, *et seq.*

Patents.

By A. K. Mansfield. Discusses the present practice of patent solicitors, and advocates the establishment of educational institutions, or of special courses, which shall be designed to fit men for the specialty of patent attorneys. *A. R. & Eng. Jour.*, February, 1887.

Reforms needed in the U. S. Patent Office. Report of the Legal Committee of the National Electric Light Association on the above subject at the Pittsburgh meeting. *Elec. Eng.*, March, 1888; *Elec. World*, March 3, 1888.

Pattern Marking. A system of marking patterns, tried in the shop, with some remarks upon economy as viewed in the pattern. Paper by A. J. Frith, M. Engineers' Club of St. Louis. Read April 17, 1889, pp. 4. *Jour. Assn. Eng. Soc.*, August, 1889, Vol. VIII., p. 410.

Pavements.

A paper before the Soc. of Arts on the sanitary advantages of smooth and impermeable street surfaces. *Van Nos. Eng. Mag.*, Vol. V., p. 387.

A valuable discussion in the Western Society of Engineers, Chicago, by engineers of experience. Some important and new facts concerning brick pavements. *Jour. Assn. Eng. Soc.*, Vol. VI., p. 23.

A valuable discussion before the Western Soc. of Eng., in answer to the query, "What is the best form of pavement or improved roadway for a city residence street having a limited amount of travel?" *Sci. Am. Sup.*, Jan. 20, 1887.

Abstract of a report by Capt. F. V. Greene, U. S. E., on the result of his recent examinations abroad, where he went especially to study this subject. *Eng. News*, Sept. 19, 1885.

By G. T. Deacon. Treats of the details and cost of construction of pavement of hammer-dressed stone, asphalt and wood. *Van. Nos. Eng. Mag.*, Vol. XXIV., p. 236.

Capt. F. V. Greene on the construction and care of streets. Paper read before the Commonwealth Club of New York City. *Eng. & Build. Rec.*, March 1, 1890, p. 106.

Pavements, continued.

Extracts from a paper by Francis Collingwood, read before the Academy of Sciences, at Elmyra, N.Y. This paper treats of the general features of the various kinds of paving, giving considerable data as to cost, durability and best methods of laying. *Eng. Record*, March 7, 1891, pp. 223-4.

Extracts from the report of Captain Griffin on the streets of Washington. Contains much useful information relating to pavements, especially that made from coal tar distillate. *Eng. News*, Oct. 22, 1887.

Extracts from two papers before the Inst. of C. E. *Van Nos. Eng. Mag.*, Vol. XXI., p. 35.

Gives the experience of Omaha in the use of macadam, Sioux Falls stone, Colorado sandstone and asphalt for pavements. *San. Eng.*, July 30, 1887.

And Paving Material. A popular elementary article giving considerable information. *St. Ry. Jour.*, July, 1890, pp. 331-6.

Asphalt.

A report by W. P. Rice to the Board of Improvement of Cleveland on the use of asphalt pavement. *Eng. & Build. Rec.*, May 26, 1888.

The trade publication of Warren-Scharf Asphalt Paving Co., of N. Y., gives some information on this subject. Pamph., pp. 43.

Berlin. By Leon Malo. Eighteen pp. *Mem. Soc. des Ing. Civils*, February, 1885.

Berlin. Gives amount of asphalt laid in Berlin from 1878 to 1888. Cost per square yard and cost of maintenance. *Eng. News*, Dec. 15, 1888.

Buffalo, N. Y. Specifications for. *Eng. & Build. Rec.*, Dec. 29, 1888.

Concrete, Foot. By G. R. Strachan, before the Association of Municipal and Sanitary Engineers and Surveyors, of Leicester. Gives details of experiments with asphalt walks in England, with data of durability, cost, etc. *Sci. Am. Sup.*, Dec. 31, 1887.

On Concrete Foundations in New York. Summary of specifications respecting preparation of street surface, foundations, and mixing and laying the asphalt wearing surface. *Eng. News*, Nov. 16, 1889, Vol. XXII., p. 473.

Paris. Brief description of the method of mixing and laying the concrete foundation, with a notice of the character of pavement commonly adopted in Berlin. *San. Eng.*, Jan. 7, 1886.

Details of Work at Washington, D. C. From "Specifications for laying Standard Asphalt Pavement." *Eng. Record*, Feb. 14, 1891, pp. 178-80.

See *Asphalt*.

Asphalt and Stone. *Relative Value of*. As determined by number of falls in Fire Department of Berlin for four years. From the German. *Eng. News*, March 13, 1886.

Berlin, *Granite and Asphaltum*. Abstract of methods of laying, in *Proc. Inst. C. E.*, Vol. LXXXI., p. 376

In Bremen. See *Sewer Systems*.

Brick.

By Wm. Kent. *E. & M. Jour.*, June 21, 1890, p. 702.

Article describing a method of laying brick pavements for city streets and method of constructing foundation for same. *Eng. News*, Aug. 3, 1889, Vol. XXII., p. 99.

City Streets. Paper by E. B. Shifley, describing the brick pavement of Bucyrus. *12th An. Rep. Ohio Soc. S. & C. E.*, 1891, pp. 121-4.

Columbus, O. A few notes as to the wear of brick pavements in this city. *Eng. News*, March 28, 1891, p. 294.

Durability of Brick Pavements. By Prof. I. O. Baker. Pamph., p. 46. T. A. Randall & Co., Publishers, Indianapolis, Ind. Price, 25 cents. A valuable

Pavements, *Durability of Brick, etc.*, continued.

paper. Contains a summary of results of a series of tests made at the Univ. of Ill., with a discussion of the value of different kinds of tests and the proper ones upon which to base specifications. Several examples of wear in actual use under a known traffic are given with comparisons with test on same brick. The paper contains also data on street traffic of various cities, and the results of some St. Louis experiments on pavements of various kinds.

Memphis, Tenn Specifications for, with some data as to cost. *Eng. Record*, April 4, 1891, p. 289.

Duluth, Minn. Brief note describing method of construction, abstracted from annual report for 1890. *Eng. Record*, Sept. 19, 1891, p. 245.

Methods of Laying Brick Street Pavements. See Brick street paving.

Rochester, N. Y. Description of a brick pavement with concrete foundation. *Eng. Record*, Nov. 14, 1891, p. 386.

Specifications for Nebraska City, Neb. Copy of specifications as drawn up by Mr. A. F. Nims, City Engineer of Nebraska City. *Eng. News*, March 29, 1890, Vol. XXIII., p. 293.

Specifications for Fire-Clay Brick Pavements. A copy of specifications as written and used at Cleveland, O., by Mr. John L. Culley, C. E. *Eng. News*, March 8, 1890. Vol. XXIII., p. 219.

Cedar Block.

Carriage Ways. A paper by Alan Macdougal before Canadian Society of Civil Engineers. *Trans. Can. Soc. C. E.*, 1888, Vol. II., p. 163.

"*Ceramite*," *A New Material for*. Shown at the recent Hungarian Exhibition at Buda-Pesth. Has been in use for five years. It is a kind of artificial stone burned, and harder and stronger than granite. Composition not given. From the French, in *Eng. News*, March 20, 1886.

Ceramite. By M. A. Gouvy. Method of use, and comparison with other pavements. *Mem. Soc. Ing. Civ.*, Oct., 1885, pp. 458-464.

Cleveland, Ohio.

By M. E. Rawson. Describes the methods of construction, specifications and durability of the Cleveland pavements. *Rpt. Ohio Soc. Engrs. & Surv.*, 1888, pp. 70-94.

Gives experience with many kinds of pavements. By M. E. Rawson. In *An. Rpt. Ohio Soc. Engrs. & Surv.*, 1888. C. N. Brown, Sec'y, Columbus, O.

History of those of. By M. E. Rawson. *Jour. Assn. Eng. Soc.*, Vol. III. p. 98.

Combination Iron and Oak Pavement. A paper by J. W. Cole, read before the Am. Soc. M. E., describing a form of pavement consisting of oak blocks on a cast iron plate. Abstract. *Eng. News*, Nov. 28, 1891, p. 509. *Eng. Record*, Dec. 5, 1891, p. 10.

Contract Prices for various kinds of pavements for years 1887-8. *Rpt. of Eng. Dept.*, Senate Mis. Doc. No. 15, 50th Cong. 2d Session.

Granite Block, Westminster street, Providence, R. I., 1880. Detailed description. *Eng. News*. Feb. 11, 1882, pp. 45-6.

New York. Paper by Prof. J. S. Newberry, describing the condition of New York pavements, and suggesting means of improvement. Pp. 7. *Sch. of Mines Quar.*, Columbia College, July, 1889, Vol. X., p. 289.

Notes on. By T. R. Wickenden. Object of paper is to call attention to desirable as well as objectionable qualities of various pavements in general use. *Rpt. Ohio Soc. Surv. & Engrs.*, 1888. pp. 69-75.

Past, Present and Future. By J. H. Sargent. Gives results of his experience and observations on the street pavements of Cleveland. *Jour. Assn. Eng. Soc.*, Sept., 1888.

Pavements, continued.

Kansas City. The Experience of. Description of an efficient kind of cedar block pavement on concrete base, with blocks covered with asphalt. *Sci. Am. Sup.*, Feb. 14, 1885; also *Eng. News*, Dec. 13, 1885.

Impervious Street Pavements of Liverpool. Have a cement concrete base, covered with blocks of granite or wood. *Trans. Liverpool Eng. Soc.*, Vol. III., p. 61.

Philadelphia.

Abstract of specifications for paving streets in Philadelphia with granite blocks and pitch cemented joints. *Eng. News*, Jan. 15, 1887.

Report of Board of Experts (Q. H. Gilmore, F. V. Green and E. P. North) on repaving the streets of Philadelphia, with recommendations as to kinds of pavement to be used. *Jour. Frank. Inst.*, CXVIII., p. 210 (Sept., 1884), and in other current publications.

Repairing, Cleansing and Watering. Extracts from a report of George Livingston, Surveyor, of St. George, London. Gives details of the methods employed in repairing, cleansing and watering 40 miles of macadamized streets. *Eng. & Build. Rec.*, Jan. 7, 1888.

Report on Street Pavements, by a special committee of the city council. Considerable data concerning street pavements in various cities is given, including some examples of cost and durability, and recommendations given for carrying out work at Detroit. *Eng. News*, March 19, 1892, Vol. XXIII., p. 189.

Small Cities. By George F. Wightman. Recommends first gravel, second macadam, and third brick. Gives extracts from specifications used in Peoria. *Sec. An. Rpt. Ill. Soc. Eng. & Surv.*

Specifications for.

Granite. Form of agreement and specifications used in paving at Providence, R. I. *Eng. News*, June 5, 1885.

Granite Block, Asphalt, Asphalt Block and Macadam pavements in Washington, D. C. *An. Rep. Eng. Dept. Distr. Columbia*, 1886, p. 60. *Eng. News*, Sept. 4 and 11, 1886.

Kansas City, Mo., Paving Specifications. For cedar block, asphalt and brick and stone pavement. *Eng. Record*, Aug. 29, 1891, pp. 203-4.

Material. Gives specifications for the supply and delivery of wood blocks for paving portions of the street known as King's road and Pont street, in Parish of Chelsea. *San. Eng.*, Dec. 31, 1887.

Standard Pavements and street construction material. Giving specifications for asphalt, coal for distillate, asphalt on hydraulic base, asphalt on bituminous base, asphalt block and granite block pavements. *Report of Engineer Dept.*, under Capt. Symonds. *Senate Mis. Doc.*, *No. 15*, *50th Cong.*, *2d Session.*

At Boston, Mass. Specifications for granite and asphalt pavement and for sidewalks given. *Eng. Record*, July 25, 1891, pp. 119-20; Aug. 1, 1891, pp. 136-7.

Wood. Gives specifications for wood pavement in Parish of St. George, London. Said to be one of the best recently written. *San. Eng.*, Dec. 24, 1887.

St. Louis.

A short description of the various forms of pavement in use, with statement of cost of maintenance. *Eng. Record*, Feb. 7, 1891, pp. 159-60.

By Thos. H. Mackland. Treated historically, giving benefit of experience in macadam, wooden block, Telford, granite, asphaltum, etc. *Jour. Assn. Eng. Soc.*, Vol. V., p. 213.

Stone, Wood, Macadam, Asphalt, Horse R. R. Tracks, Street Cleaning. Report of journey of inspection, by one of the engineers of the city of Paris, made to London, Liverpool, Edinburgh and Glasgow. *Annales des P. & C.* March, 1881.

Pavements, continued.

Street Pavements. Article discussing relative value of various forms. Contains extracts from paper by B. H. Cunningham at Edinburgh. Gives results of tests. *Eng. News*, Dec. 1, 1877, pp. 357-9.

Study of Street Pavements. A lecture by Prof. Lewis M. Haupt. Illustrated by examples taken from the City of Philadelphia. *Jour. Frank. Inst.*, Dec., 1889.

Sydney, N. S. W. By A. C. Mountain. Before the Institution of Civil Engineers. Gives details of the pavements, principally wooden, of the city of Sydney, New South Wales; also description of and results of tests on Australian timber. *Proc. Inst. C. E.*, Vol. XCIII., pp. [illegible].

Telford and Macadam. By A. P. Starrs. Treats of the material, cross-section, foundation, metal and maintenance. *Van. Nos. Eng. Mag.*, Vol. XIII., p. [illegible].

Telford Macadam. Article describing in detail the most approved methods of constructing this form of street pavements, together with illustrations showing cross-section of street and steam road roller for use in construction of pavement. *Eng. News*, Sept. 7, 1889, Vol. XXII., p. 221.

Treatment of Wood for. By T. J. Caldwell and T. D. Miller. Gives experience of St. Louis with the tannin and zinc-gypsum processes. *Jour. Assn. Eng. Soc.* Vol., IV., p. 132.

Washington, D. C. Specifications for granite, asphaltum and asphaltum block pavements, as laid in that city, together with a schedule of the pavements of the city, with map showing same in colors. *Report of Engr. Department, District of Columbia*, 1884. See further *Wood Pavements*, and *Street Pavements*.

Wear of and Cost of Repairs. Data gathered from many European cities, and given in Municipal Report of Rochester, N. Y. *Eng. News*, Nov. 20, 1886.

Wood.

In London. By G. H. Stayton before the Institution of Civil Engineers. Gives full particulars of the extent and construction of wood pavements in London. *San. Eng.*, Aug. 20, 27, etc., 1887.

In London. A paper by Geo. H. Stayton. There are now 53 miles of this pavement in that city. Methods, wear, cost, etc., are given. With discussion, occupies 75 pp. in *Proc. Inst. C. E.*, Vol. LXXVIII. Also *Eng. News*, Nov. 22, 1884, *et seq*.

In Paris. An abstract from a paper by M. A. Laurent in *Genie Civil*, gives details of the present practice of paving with wood in Paris. *Eng. & Build. Rec.*, April 7, 1888.

In Paris. Gives a translation of the specifications and principal instructions issued in 1886. *Eng & Build. Rec.*, June 9, 1888.

Of Chicago. Description with statements as to durability. By E. A. Fox. *Eng. News*, Jan. 10, 1875, pp. 14-15. Discussion, *ibid.*, Jan. 17, p. 22.

See *Municipal Engineering*, *Streets and Highways*.

Pavements, Sewers and Drainage *of Steep Streets*. This article is an abstract from the report of Rudolph Hering and Andrew Rosewater on contemplated improvements at Duluth, Minn. Illustrations of pavements, catch basins and forms of gutter, are given. *Eng. News*, Feb. 14, 1891, pp. 148-9.

Paving.

For Streets of High Grades. *An. Rep. Board of Public Works, Duluth, Minn., for the Year ending Feb. 28, 1890.* pp. 87-93 and plates.

Bricks. See *Bricks*.

Materials. Test of. Record of absorption and abrasion tests made on several specimens of brick and granite. By J. Herbert Shedd, City Engr., Providence, R. I. *Eng. Record*, Sept. 26, 1891, pp. 266. *Eng. News*, Sept. 26, 1891, p. 193.

Stones. Apparatus used by French Engrs. to determine the co-efficient of attri-

Paving. *Stone*, continued.

tion. By Prof. Wm. Watson. Illustrated. *Jour. Assn. Eng. Soc.*, Vol. I., p. 325.

Valuation of Road Metal and Setts for. By W. F. Stock. Discusses the salient features to be looked at in selecting road material, and gives results of examinations made with a machine for testing the abrasion resistance of road metal. *Eng. News*, Sept. 22, 1888.

Pendulum. See *Engines, Steam, Governors. Theory of.*

Pendulum, Titan. A kind of crane which handles and throws automatically large quantities of material, such as loose rock or concrete, to the distance of 100 or 150 feet Useful in constructing breakwaters. *Sci. Am. Sup.*, Nov. 14, 1885. Also, Lon. *Engineer*, Nov. 13, 1885. Also *Iron*, Nov. 13, 1885.

Permanent Way. See *Railroad Track.*

Petroleum.

By H. E. Wrigley. Gives description of the oil regions of Pennsylvania, methods of drilling, pumping and transportation, also treats of liquid fuels. *Van Nos. Eng. Mag.*, Vol. VII., pp. 225, 313, 513.

Illumination. *A Graduating Thesis* at the Stevens Institute of Technology. Discusses various burners, forms of wicks, candle power, oil consumption, etc. Illustrated. *Sci. Am. Sup.*, Nov. 9, 1886.

Pipe Lines for. See *Pipe Lines.*

And its Products. Being part of Vol. X. of U. S. Census, 1880. Quarto, 300 pp. An encyclopædia of information on this subject, giving geology and technology of gaseous and liquid fuels. *H. Rep. Misc. Doc. 42, Pt. 10, 47th Congr., 2d Sess.*, 1884.

Reception and Storage of Refined Petroleum in Bulk. A paper describing the machinery and plant for its reception from steamers, its storage and delivery. Illus. By W. T. H. Carrington. *Proc. Inst. C. E.*, Vol. CV., 1891, pp. 106-18, Discussion, pp. 119-26.

Sources of Petroleum and Natural Gas. The geological conditions under which they occur in general, and in detail for various fields is considered in paper by W. Topley, F. R. S. *Jour. Soc. Arts*, Apr. 17, 1891, pp. 421-37.

Steamers. Abstract of papers read before the Institution of Engineers and Shipbuilders at Newcastle-upon-Tyne, by Messrs. B. G. Nichol and J. Gravell, describing appliances for burning petroleum as fuel, and recent practice in the construction of petroleum-carrying steamers. *R. R. & Eng. Jour.*, Feb. 1887.

Storage Installations at Avonmouth and at Cardiff. A paper by R. Pickwell, describing these large plants. Two folding plates. *Proc. Inst. C. E.*, Vol. CIV., 1891, pp. 249-58.

Territories. By B. Redwood. An exhaustive paper on the petroleum-producing territories of the United States and Canada, giving tables of statistics and general data of value. *Sci. Am. Sup.*, Sept. 10, 1887.

Transportation of, in Russia. Gives a detailed account of the proposed 8-inch pipe line from Baku to Port of Batam. Length of line 797 miles, with 14 stations, each of which contain four 150 horse-power engines. Delivery estimated at 110 cub. ft. per minute. Lon. *Eng.*, Feb. 3, 1888.

See *Gas, Natural.*

Petroleum Engines. See *Engines, Gas. Engines, Petroleum.*

Philbrick, Edward Southwick. A memoir by Albert H. Howland, Desmond Fitz-Gerald and Walter Shepard, Committee of the Boston Society of Civil Engineers, pp. 4. *Jour. Assn. Eng. Soc.*, Aug., 1889, Vol. VIII., p. 432.

Phonograph, The. An address by Colonel Gourand before the Society of Arts, descriptive of the Edison Phonograph. *Jour. Soc. of Arts*, Nov. 30, 1888.

See *Gramophone.*

Phosphate of Lime. *Production of.* A series of articles describing various phosphate deposits, methods of working, etc. *Eng. News*, July 11, 18, 25, 1891.

Photographic Chemistry. *Cantor Lectures.* Three valuable papers by Prof. R. Meldola, F. R. S. *Jour. Soc. Arts*, Aug. 28, Sept. 4, 11, 1891.

Photographic Lenses. A very valuable paper by Thomas R. Dallmeyer discussing the theory and construction of photographic lenses. *Jour. Soc. Arts.* May 2, 1890, p. 583.

Photographic Map Reduction. By O. B. Harden. Calls attention to the method of making reductions of large maps by the use of the camera. The method is used by the Pennsylvania Geological Survey. Largest size of negative used is 21x25 inches. *Proc. Eng. Club, Phila.*, Vol. VI., p. 1.

Photography. *In Colors of Nature.* By F. E. Ives. Illustrated. Describes process, etc. *Jour. Frank. Inst.*, Jan., 1891.

Photography of Engineering. Gives outfit and practical hints. Shows its use in surveying, whereby geographical positions and relative elevations may be mapped with tolerable accuracy. Illustrated. By D. C. Humphreys. *Jour. Assn. Eng Soc.*, Vol. III., p. 110.

See *Surveying.*

Photometer, Bunsen, a new form of Dr. Hugo Kruss. By reflection prisms the opposite sides of the oiled spots are observed side by side. *Elec. Rev.*, Nov. 8, 1884.

Photometry. See *Platinum Standard of Light.*

Piece-Work, shown to be applicable to railroad shops. A series of well-written articles by J. D. Casanave in *R. R. Gaz.*, July 2, 9 and 16, 1886.

Pier Construction *of Recent Date.* Paper by G. A. M. Liljencrants before the C. E. Club of the N. W., describing the construction of several harbor piers. Illus. *Eng. News*, March 1, 1879, pp. 68-70.

Piers.

Marine Park, Boston. Short description, with full detail drawings, of the iron pier at Marine Park, Boston. It has twelve spans, 60 feet each, resting on cast-iron piers filled with concrete. *Eng. & Build. Rec.*, Jan. 28, 1888.

Ocean, at Coney Island. Its construction. Iron piles driven by water jet. Chas. Macdonald. *Trans. A. S. C. E.*, Vol. VIII. (1879), p. 227.

Ocean, Long Branch Iron Pier. Description of this pier supported on disk piles. *Eng. News*, April 19, 1879, pp. 121-2.

St. Leonards-on-Sea. Gives brief description, with drawing of details of a screw pile promenade pier 900 feet long and 25 feet wide, being built at St. Leonards-on-Sea, England. Lon. *Engineer*, May 11, 1888, *Sci. Am. Sup.*, June 23, 1888.

On Soft Ground. Design for. Paper by Joseph Ripley read before the Michigan Engineering Society. Gives general description of work including specifications under which construction was conducted. *Eng. & Build. Rec.*, March 20, 1890, Vol. XXI., p. 264.

See *Concrete Piers. Bridge Piers. Masonry Abutments.*

Pig Iron. See *Iron, Pig.*

Pile Bridges. See *Bridges, Trestle.*

Pile Driver.

An Account of the Operations of the Gunpowder. By Samuel R. Probasco. Gives experience with Shaw's patent gunpowder pile-driver. *Trans. A. S. C. E.*, Vol. II., p. 401.

Missouri Pacific Railway. Designed for effective work at a greater distance from the road bed than usual. Full detail drawings and description. *Ry. Rev.*, Oct. 25 and Nov. 8, 1890, pp. 636, 666-7.

Pendulum Pile Driver. Illustrated description of a machine in which the leaders

Pile Driver, *Pendulum*, continued.

are pivoted at the top so that all piles of a trestle bent can be driven without moving the machine. *Eng. Record*, Dec. 5, 1891, p. 11.

Revolving Hydraulic Pile Driver. A heavy truck carries a turn table on which is mounted the pile driver. With this machine piles can be driven in bents 20 ft. apart, the driver being supported on those already driven. Illustrated description. *Eng. Record*, April 4, 1891, p. 290.

Steam. Short illustrated description of a direct acting steam pile driver, whose peculiarity is a fixed piston, the cylinder forming the striking weight. *R. R. & Eng. Jour.*, Feb., 1889; *E. & M. Jour.*, Jan. 5, 1889.

Pile Driving.

By E. H. Beckler. Gives forms of records used and experience of the writer on the St. Louis River bridge at Duluth, Minn. *Jour. Assn. Eng. Soc.*, Vol. V., p. 248.

Cost of. Extract from Prof. I. O. Baker's treatise on masonry construction, giving data from actual practice of considerable value. *Eng. News*, Dec. 14, 1889, Vol. XXII., p. 515.

And Ditch Plant. Description of a new and ingenious plant designed by Mr. A. E. Buchanan which has been used on the O. & St. L. Ry. Tables show the extraordinary amount of work performed at one-fourth of the usual cost. Illustrated by an inset plate. No patents. *Eng. News*, Aug. 16, 1890, p. 140.

By Dynamite. Brief account of the successful use of this method. *Eng. News*, March 24, 1883, p. 143.

Jets. See *Dikes*.

In Sandy Soils. An elaborate report on the methods employed on the Mississippi River and elsewhere, comparisons of methods by water-jet and steam hammer, with suggestions for further improvements. By Lt. F. V. Abbot. *Rpt. Chf. of Engrs., U. S. A.*, Vol. II., 1883, p. 1249.

Service of Hammer Lines. A statement showing the service of 70 lines of various make, on the C. M. & St. P. Ry. *Eng. Record*, March 28, 1891, p. 272.

By Water-Jet and by Hammer. Two extended reports by Lt. F. V. Abbot, on this part of the Mississippi River improvement works. Methods and cost given in great detail, and comparisons drawn. *Rpts. Chf. Engrs., U. S. A.*, 1883, Pt. III., p. 1249, and 1884, Pt. II., p. 1505.

See *Water Jet*.

Pile Foundations.

And Pile-Driving Formulas. A circular of 16 pp. from the office of the Chief of Engrs., U. S. A., dated Nov. 26, 1881, embodying the combined experience of many officers of the corps, with a theoretical discussion by Lieut. Col. George H. Elliot.

In Compressible Soils, with experimental tests of Pile Driving and Formulas for Resistance deduced therefrom. Compiled by Gen. Richard Delafield. A memoir of 37 pp. Apply to Chief of Engrs., U. S. Army, Washington.

Pile Trestle.

Gives detailed drawings of the standard trestle of the Chicago, Burlington & Northern Railroad. *Eng. News*, May 5, 1888.

See *Bridges, Trestles*.

Piles.

Cutting off. Under Water. An illustrated description of a saw used for cutting off piles at the draw-bridge of Ruhrort. *Eng. News*, April 2, 1887.

Destruction of Piles by Limnoria Lignorum and Limnoria Terebrans in Boston Harbor. A special report on piles and piling made under direction of Asst. Henry Manley. Fully illustrated by heliotypes. *Rep. City Engineer*, Boston, Mass., 1878, p. 20.

Experiments on the Resistance to Horizontal Stress of Timber. By J. W. Sande-

Piles, *Experiments, etc.*, continued.

man. Gives results of a number of experiments undertaken to find the resistance to horizontal movement of piling by different strata. *Van Nos. Eng. Mag.*, Vol. XXIII., p. 223.

Formula for their Sustaining Power, Size and Disposition in Any Foundation. Compiled by Rudolph Hering. *Eng. News*, Feb. 14, March 21, 28, April 11, 18, 1878.

Notes on the Friction of, in Clay. By A. C. Hertzig. Gives formula for the extreme load to be carried by a pile after a definite amount of driving. Also the determination of the friction on piles pulled at Hull, Eng. *Van. Nos. Eng. Mag.*, Vol. XXV., p. 273.

Iron, Sunk by Water Jet. A description of the ocean pier at Coney Island. Illustrated. Charles Macdonald. *Trans. A. S. C. E.*, Vol. VIII. (1879), p. 227.

Protection of, from Limnoria and Teredo. By M. Manson, before the American Society of Civil Engineers. Gives details of the treatment of piles for the Mission street pier, San Francisco, by various methods, and the condition of piles after five years' service. An abstract in *San. Engr.*, Dec. 31, 1887; Lon. *Engineer*, Jan. 6, 1888.

Screw. Abstract from specification for the construction of a screw pile wharf at Karrachee, India. Details illustrated. Lon. *Engineer*, Jan. 21, 1887.

Sheet, Clamp for Pulling. Paper by Chas. E. Emery, M. Am. Soc. C. E. Illustrated. *Trans. A. S. C. E.*, Vol. XX., p. 118, March, 1889.

Supporting Power of.

By Prof. I. O. Baker. Discusses the formula of Mr. Trautwine, and shows its defects. Gives an empirical formula derived by Mr. Hertis from driving of over 100 piles. *R. R. Gaz.*, April 6, 1886.

An analytical investigation of the subjects to determine their resistance to loads in terms of the energy of the final blows and the distance moved; applied to the foundations of Fort Montgomery, N. Y. A pamphlet of 16 pages, issued by the Engr. Department, U. S. A.

Bearing. Formula for Safe Loads on. Correspondence from John C. Trautwine, Jr., and lengthy editorial with formulæ. *Eng. News*, Dec. 29, 1888.

Bearing Power of, Formulas for the. By Prof. I. O. Baker. Examines a number of formulas to see if the limits of the unknown and uncertain cannot be contracted a little, and proposed a new formula. *Jour. Assn. Eng. Soc.*, Vol. VI., p. 117.

Formulas. A discussion of more or less value, the gist of which is well given in editorial in Lon. *Engineer*, April 2, 1886. The exact reactions well described. Gives abstracts from a number of letters from officers of Engineer Corps to the Chief of Engineers. *Van Nos. Eng. Mag.*, Vol. XXVII., p. 21. By A. C. Harzig. Gives set of diagrams and the method by which they were obtained. *Van Nos. Eng. Mag.*, Vol. XXVII., p. 386. By Rd. Randolph. A discussion of a number of formulas. *Van Nos. Eng. Mag.*, Vol. XXVII., p. 198.

See *Jetties. River Improvements.*

Pillars.

A method of deducing formula from experiments on wrought-iron pillars. By J. D. Crehore. *Van Nos. Eng. Mag.*, Vol. XIX., p. 360.

Formula for. By John D. Crehore. An attempt to establish a formula for the strength of pillars that yield to bending, rather than to direct crushing. *Van. Nos. Eng. Mag.*, Vol. XXI., p. 501. By J. D. Crehore. Gives table of strength. *Van Nos. Eng. Mag.*, Vol. XXIII., p. 60. An article by John D. Crehore. Gives the application of a formula to pillars tested on the government machine at Watertown, Mass. *Van Nos. Eng. Mag.*, Vol. XXIX., p. 129.

Pipe.

Aqueducts Pipe, Making, Calking, Dipping and Shipping Pipe Sections, and In-

Pipe, continued.

Inspector's Pipe Testing Car. East Jersey Aqueduct. Illustrated description. *Eng. Record*, Sept. 26, 1891, pp. 264-5.

Connections.

The Moran Flexible Joint for Steam Pipe Connections. An all-metal connection made up of a combination of ball and socket joints. Illustrated description. *Eng. News*, Aug. 29, 1891, p. 191.

For Water Pipes of large diameter, by separate flanges and either rubber rings or lead packing, according to system Halberg. Dispenses with pouring joints. Brief description. *Journal fur Gasbeleuchtung und Wasserversorgung*, 1886, pp. 13-15.

Fittings, Manufacture of. By W. D. Forbes. Treats of the manufacture in the United States of pipe fittings. Describes some of the machines used and methods employed. Lon. *Eng.*, Feb. 25, 1887.

Joints. With especial reference to water pipes. Discusses spigot and socket joints, hinged joints, screwed joints and solder joints. A practical paper of value. *Trans. Liverpool Eng. Soc.*, Vol. IV., p. 103.

Laying.

Laying of Large Mains. A paper of some value by R. S. Wyld before Liverpool Engineering Soc. *Trans. Liverpool Eng. Soc.*, 1889, Vol. X., pp. 20-35.

Laying a Submerged Sewer in Paris. Brief description with illustration of the method of laying two 16 in. pipes across the Seine. *Eng. Record*, April 11, 1891, p. 313.

Laying Submerged Water-Mains. An account of the laying of a supply main, Nashville water-works. Long sections of pipe were floated into place, sunk and jointed by divers. Details of pontoons shown. *Eng. News*, Aug. 8, 1891, pp. 117-8.

Laying Submerged Water Mains. Illustrated description of a method which has been successfully used by Mr. E. C. Cook in laying submerged water mains. *Eng. & Build. Rec.*, Jan. 25, 1890, Vol. XXI., p. 122, *et seq.*

Tables for estimating the Cost of Laying Cast Iron Water Pipe. *Eng. News*, June 21, 28, 1890, pp. 580-1, 618-9.

Under the Mersey River. Account from Lon. *Engineer.* *Eng. News*, July 4, 1891, p. 5. Flexible joint used is illustrated in *Eng. Record*, July 11, 1891, p. 93.

Under Panchpara River in India. By A. Hayes, describing construction of the jointed pipe used, manner of floating and sinking, etc. *Ind. Eng.*, July 27, 1889, Vol. VI., p. 74, *et seq.*

Under Water. Brief note describing process of laying 1,000 feet of 8-inch water pipe in over 16 feet of water at Rouse's Point, N. Y. *Eng. News*, May 3, 1890, Vol. XXIII., p. 426.

"*Wallis*" *Gradient Indicator.* An exceedingly simple device by means of which, in connection with an ordinary spirit level, workmen can correctly lay pipes to any desired grade. *Mech. World*, March 1, 1890, p. 81. *Eng. News*, March 15, 1890, Vol. XXII., p. 250.

See *Water Works.*

Lines.

New Method of Constructing Large Pipe Lines. An illustrated article describing method of constructing a 36 inch steel pipe by riveting plates together similar to method of building boilers. *Am. Mfr.*, April 25, 1890.

Petroleum. A valuable paper on "The Different Methods of Transporting Crude Petroleum in the United States," was read by Mr. John H. Harris before the Brit. Asso. Adv. Sci. An extended extract giving details of the pipe lines was reprinted in *Eng. & Build. Rec.*, July 26, 1890, pp. 116-7.

Transportation of Petroleum. A general map of the main pipe-lines from the

Pipe. *Lines, Transportation, etc.*, continued.

oil regions to the sea and to Lake Erie, with a profile of the Olean-New York line, and an account of the enterprise. *Eng. News*, June 13, 1881. Reprinted in *Science*, June 19.

See *Salt Works*.

Testing Machine. Illustrated description of a machine recently constructed in England for testing water-pipe. *Eng. News*, Jan. 31, 1885.

Threads, Uniform Standard in. Abstract of final report of Committee on Standard Pipe and Pipe Threads of the Am. Soc. of Mech. Engrs. *Am. Eng.*, March 2 and 9, 1887.

Standard Pipe and Pipe Threads. Report of a committee of the Am. Soc. of Mech. Engrs., read at their New York meeting, Nov., 1886. Recommends the adoption of the Briggs standard, already in quite general use. *Trans. A. S. M. E.*, Vol. VIII., 1887.

Systems, The Efficiency of, for Furnishing Water to Fire Engines. By S. Bent Russell. This is a pretty successful attempt to show "how to furnish at any point in a city a sufficient quanity of water to supply a maximum number of fire-engines and use the least quantity (tonnage) of pipes consistent with assured safety." A theoretical investigation based on experiments. Illus. *Jour. Assn. Eng. Soc.*, Vol. V., p. 179.

See *Chemical Obstructions*; *Corrosion*; *Hydraulics*; *Hydraulic Tables*.

Pipes.

Cast Iron.

Making. Description with illustrations of molds, etc. *Iron Age*, Nov. 6, 1890, pp. 789-90.

Internal Corrosion of. Paper by B. Jamieson before the Inst. C. E., dealing with water pipes, estimating cost of removal of obstructions and loss of space and pressure due to same. Illus. *Eng. News*, Sept. 17, 1881, pp. 376-8.

Report of and Specifications proposed by a committee of the American Water Works Association. *Iron Age*, March 19, 1891, pp. 525-6.

Specifications for cast-iron pipe for Peshawar City water supply. *Ind. Eng.*, July 9, 1887.

Specifications for Cast Iron Coated Water Pipe. Paper read by Thos. W. Yardley, Feb., 1890. *Trans. A. I. M. E.* Reprinted in *Eng. & Build. Rec.*, July 26, 1890, p. 106.

Cement Concrete. Description of some tests on Cement Pipes and also Stoneware Pipes. Illus. *Ind. Eng.*, Dec. 6, 1890, pp. 452-3.

Copper. See *Engines, Steam.*

Corrosion of. See *Corrosion.*

Earthenware. Report of Tests upon, to determine fitness to replace iron piping in house sewerage, the tests being especially to determine imperviousness to gas and water after construction. By James J. Powers, Dept. of Health, Brooklyn, N. Y.

Gas Pipes for Conveying Natural Gas at Pittsburgh. Report of the Natural Gas Commission on requirements for conveying natural gas with safety. *Am. Eng.*, July 30, 1885.

Hydraulic, Remarks on. Paper by E. H. Phipps describing the manufacture and advantages of cement-lined pipes; with discussion. *Proc. Conn. Assn. C. E.*, Aug. 29, 1890, pp. 1-10.

Lead, Action of Water Upon. Extracts from report of Dr. S. White, Medical Officer of Health of Sheffield, England. *San. Eng.*, April 24, 1886.

Mannesmann Process of Making Weldless Tubes. An illustrated description of this interesting process. Tables of tests. *Eng. News*, Aug. 2, 1890, pp. 92-3.

Pipes continued.

Riveted Steel Pipes. A paper by D. J. Russell Duncan, M. I. C. E., read before the Am. W.-W. Ass'n. A valuable paper, giving the best practice in Great Britain and America, strength of pipes, and details of various kinds of joints. Illus. *Eng. News*, Apr. 25, 1891, pp. 404-5.

Service Pipes.

Discussion as to the best material for. *Jour. N. E. W.-W. Assn.*, Sept., 1891, pp. 21-7.

A paper by W. H. Richards before the N. E. W.-W. Assn., discussing the requisites of service pipes. *Eng. News*, June 28, 1884, pp. 307-8.

On the use of galvanized iron for service pipes. By Dr. H. Bunte. (With statistics and reports on its use in America, by Prof. Ripley Nichols.) *Journal f. Gasbel. u. Wasserversorgung*, 1887, pp. 61-69, 100-111, 127-133, 163-160.

For Water Supply. By W. H. Richards, before the New England Water Works Assoc. Summarizes arguments, *pro* and *con*, for the various kinds of materials. *Trans. New Eng. W.-W. Assoc.*, 1884, Albert S. Glover, Secretary, Newton, Mass.

Sewer Pipe. Description of method of manufacture of the Akron pipe. Illus. *Eng. News*, May 28, 1881, pp. 216-8.

Spirally Welded Steel Tubes. By J. C. Bayles, before the Franklin Institute, Jan. 16, 1889. A description of the method of manufacture and a cut of the machine. *Jour. Frank. Inst.*, Feb., 1889.

Steam Pipes.

Copper, Strength of a. Gives an abstract from a report for the Board of Trade Surveyors on the testing of a copper steam pipe taken from a steamship. Lon. *Eng.*, Sept. 11, 1881.

Copper, Strength of. Gives table of results of a large number of tests made at the Lancefield Engine Works to ascertain some of the mechanical properties of the copper and brazing found in ordinary high-pressure steam pipes of large size. Lon. *Eng.*, Dec. 31, 1887.

The Ferranti. A multiple pipe made of solid-drawn copper tubes, designed to withstand high pressure. Illus. description. *Eng. Record*, March 28, 1891, p. 280.

Support of Main Steam Pipes. An article illustrating the different forms of support. *Am. Eng.*, Sept. 27, Oct. 4, 1890, pp. 134, 143.

See *Boilers, a Special, etc.*

Testing Pipes and Pipe-Joints in the Open Trenches. By M. McC. Patterson. Gives method employed by the author at the Omaha Water Works. *Van Nos. Eng. Mag.*, Vol. XXI., p. 166.

Tests of Lap-Welded Steam Pipes. Hydraulic tests on two steel pipes, with tension tests on specimens cut from the broken pipes. Tables giving full details, also illustration of testing apparatus. Lon. *Eng.*, Oct. 30, 1891, pp. 509-10.

Tests. See *Sewer Pipe.*

Water Pipes.

A valuable paper by A. H. Howland, before the Engineers' Club of Philadelphia. *Proc. Eng. Club, Phila.*, Dec., 1886. Vol. VI., pp. 55-69.

Water Pipes and Pumps. Air Vessel on. Paper by Dugald Baird read at the meeting of the Mining Inst. of Scotland. Illus. *Eng. & Build. Rec.*, Sept. 1890, p. 230-2.

Specifications for coated cast-iron water pipes and other castings for the Boston Water-Works. *Eng. News*, Feb. 21, 1885.

Thickness of. A paper read before the Engineers' Club of St. Louis by S. Bent Russell. *Am. Eng.*, March 6, 1889, *et seq.*; *Jour. Assn. Eng. Soc.*, Feb., 1889.

Pipes, continued.

Water Ram in. See *Water Ram.*

Wrought-Iron Conduit. By Hamilton Smith, Jr., before the Iron and Steel Inst., London. Describes the pipes used in hydraulic mining in the West. *Eng. News*, Dec. 18, 1886.

See *Tubes. Steam Heating. Water Pipes.*

Piston Springs. A new form of piston springs in which the radial pressure is easily adjusted. Lon. *Eng.*, March 13, 1884.

Piston Rings. Illustration of a new kind. Lon. *Engineer*, March 6, 1885.

Pitot's Tube. *A New Form.* By S. W. Robinson. Treats of the previous forms of tube, gives main features of a new form, with discussion of an actual trial. *Van Nos. Eng. Mag.*, Vol. XVIII., p. 253.

Pittsburgh and Vicinity. A record of seven years' progress, by Wm. P. Shinn, before the *A. I. M. E.*, at the Pittsburgh meeting, February, 1886.

Pivot Friction. See *Friction.*

Planimeter.

Amsler's Polar. Theory of. The mathematical proof of its workings. By E. L. Ingram. *Sch. of Mines Quar.*, May, 1885.

Corade's Rolling. By Prof. F. Lieber. A description and theoretical study of Corade's rolling planimeter. *Zeitschr. f. Vermessungswesen*, 1887, pp. 377-383, 421-437.

Directions as to care, use and testing of Corade's planimeter. Translated by J. S. Elliot. *Van. Nos. Eng. Mag.*, June, 1885.

Polar. E. A. Giesler. Gives a mathematical discussion of the theory and use of the polar planimeter. *Sci. Am. Sup.*, March 17, 1888.

Polar its Use and Practice. An article by Wm. Cox, describing the practical use of this instrument. *Eng. News*, March 21, 1891, pp. 268-9.

Rolling. A new form, said to possess many advantages over Amsler's in the matters of compass and accuracy. From the German. Illustrated and formulas given. *Sci. Am. Sup.*, April 25, 1885.

Spherical. By Prof. Hele Shaw. Gives an illustrated description, with theory of action, of a new spherical planimeter. *Eng. News*, Oct. 13, 1885.

Theory of. Discussed from the Manufacturers' Stand-point. By Chas. E. Emery. *Proc. A. S. M. E.*, Vol. VI., p. 495.

Theory of the Polar. By Fred. Brooks. A new and valuable development. Illustrated. *Jour. Assn. Eng. Soc.*, Vol. III., p. 194.

Plate Girders. See *Girders. Bridges, Girders.*

Platinum.

Experiments on the Melting Platinum Standard of Light. By Charles R. Cross. Paper read at a meeting of the American Academy of Arts and Sciences, June 16, 1886. *Elec. & Elec. Eng.*, November, 1886.

Specific Heat of. From original memoirs. Of use in calorimeter measurements. *Trans. A. S. M. E.*, Vol. III., p. 174.

Plumbing.

Appliances, Improved. By J. Pickering. Treats of the development of the anti-siphon trap. Illustrated. *Building*, June 5, 1886.

In the Boston Custom House. Illustrated description. *Eng. & Build. Rec.*, May 11, 1889, *et seq.*

Evaporation of Water Traps. Paper read at a recent convention of American Institute of Architects, by Mr. Glenn Brown, giving results of a series of tests made at the Museum of Hygiene, U. S. Navy Department. *Am. Arch.*, Dec. 7, 1889, Vol. XXVI., p. 278.

Inspection in St. Louis, Mo. City ordinance relating to plumbing, with a descrip-

Plumbing. *Inspection, etc.*, continued.

tion of the system of keeping records, etc., is given in *Eng. Record*, April 4, 1891, pp. 298-9.

Inspection. See *House Drainage Regulations.*

Isolated Country House, valued at $3,000, connected with the town water supply and discharging house wastes into a subsurface irrigation system, or into a street sewer. Specifications and drawings given. *San. Eng.*, Dec. 10, 1885.

Practice at Birmingham, Ho. Drawing showing general features, with plumbing regulations. *Eng. Record*, June 13, 1891, pp. 27-8.

Providence, R. I. The text of the new ordinance on plumbing and house drainage. *San. Eng.*, Aug. 20, 1885.

Regulations Prescribed by the Health Department of Brooklyn. Earthenware soil pipes allowed under restrictions. *San. Eng.*, March 26, 1885.

Regulations adopted by the Rochester, N. Y., Board of Health. *San. Eng.*, Jan. 27, 1887.

Specifications for. Gives specifications of the Board of Health for plumbing in New York City. *Eng. & Build. Rec.*, July 21, 1888.

Tanks a Necessity for Supplying Water-Closets and Hot-Water Boilers in Houses. A paper before the New England Water-Works Association. Illustrated. *San. Eng.*, July 9, 1885.

And Ventilation Inspection in New York. An article giving in detail the requirements of the new New York law and the manner of conducting the inspection, etc. *Eng. & Build. Rec.*, April 19, 1890, Vol. XXI., p. 315.

See *Faucet, House Drainage.*

Pneumatic Dispatch. A description of the working of the pneumatic dispatch in London. *Van. Nos. Eng. Mag.*, Vol. VII., p. 180.

Pneumatic Foundations.

See *Foundations.*

Pneumatic Hoisting. Paper by H. A. Wheeler, read before the A. I. M. E., at the N. Y. meeting, describing the Blanchet system of hoisting for mines, and comparing it with other systems. *Proc. A. I. M. E.*, 1890.

Pneumatic Railway.

See *Railroads, Pneumatic.*

Pneumatic Safety Gate.

See *Railroad Crossings.*

Pneumatic Tool. *MacCoy's.* Description and details of this novel and useful invention. *Jour. Frank. Inst.*, July, 1889, Vol. CXXVIII., p. 1. *Eng. News*, July 27, 1889, Vol. XXII., p. 82.

Pneumatic Transmission.

Between London and Paris. Translated from *Le Genie Civil* and published in *Eng. News*, March 28, 1885. The proposed scheme described.

Through Tunnels and Pipes. By Robert Sabine. Extract from a paper before the British Association. Gives mathematical discussion of the subject. *Van Nos. Eng. Mag.*, Vol. III., p. 581.

Of written telegraphic messages between the central and branch offices in London. Over 30 miles of such tubes in operation. A paper by J. W. Willmot before the British Association. Illustrated. Lon. *Engineer*, November 20, 1885.

Pneumatic Tubes. Results of experiments on the movement of air tubes. Length of tube experimented with 6,700 feet, and 1.54 inch in diameter. *Van Nos. Eng. Mag.*, Vol. XIV., p. 315.

Pocket Compass, *An Improved French.* Paper by R. A. Bergier, Butte City, Mont., describing a new and convenient form of pocket compass. Illustrated, p. 6. *Proc. A. I. M. E.*, Colorado meeting, June, 1889.

Polar Exploration. Discussed by Lieut. John W. Danenhower, U. S. N., before the U. S. Naval Inst. A paper of over 80 pp., with discussions by Nares, Melville, Greely, Rink (Gov. of Greenland), Nourse, Koldeway, and Markham. The question discussed from all sides and the conclusion reached that further effort in that direction should be discouraged. A very valuable paper. *Pro. U. S. Naval Inst.* Annapolis, Md., Vol. XI., No. 4, p. 633.

Polytechnicum *in Berlin.* Description of the new building of the Polytechnic School in Berlin, with maps and plan. The building is 227.82 metres by 89.95 metres, and 4 stories high; cost about $2,000,000, with equipment. *Zeitschr. f. Bauwesen*, 1886, pp. 158-162; 321-338.

Pontoons. See *Docks.*

Population *of U. S. and Canada.* Diagram showing growth of from 1830 to 1888. See *Railroad Statistics.*

Port of Lisbon. A series of articles giving an interesting account of the harbor improvements, statistics, cost, results accomplished, etc., with maps. *Revista de Obras Publicas*, 1885; and July to October, 1886.

Portal Bracing. See *Bridge Stresses.*

Position Indicator.

See *Navigation. Electric Position Indicator.*

Powdered Metals, *Pressure Required to Solidify.* See *Pressure Required to Solidify Powdered Metals.*

Power.

Cost of. See *Engines, Steam, Compound, Non-Condensing. Steam Power.*

Importance of Economical Generation of Steam Power to the development of manufactures in St. Paul. By F. T. Hampton. A very good discussion of the means of obtaining power from coal, with special reference to the conditions of greatest economy in the use of steam. *Jour. Assn. Eng. Soc.*, Vol. V., p. 453.

Hydraulic in London. By Bagg and Ellington, before the Institution of Civil Engineers. Gives a description of plant of Hydraulic Power Company of London. Lon. *Engineer*, May 11, 1888, Lon. *Eng.*, April 27, 1888, *Am. Eng.*, June 27, 1888, *Mech. World*, May 12, 1888; *E. & M. Jour.*, Nov. 3, 1888.

Press Problems. By Oberlin Smith, before the Philadelphia meeting of the Am. Soc. of Mech. Engrs. *Trans. A. S. M. E.*, Vol. IX., (1888), pp. 161-171.

Steam vs. Water. Comparative Cost of. A paper read before Am. Soc. of M. E., by Chas. H. Manning, of Manchester, N. H. *E. & M. Jour.*, June 1, 1889.

See *Locomotives, Cost of. Motive Power. Water Power.*

Power, Transmission.

By R. F. Hartford, Columbus. *Rept. Ohio Soc. Surv.*, 1885.

By Henry Robinson. Gives results of experience respecting the transmission of power with water pressure, compressed air and wire rope; describes the methods used at various places. *Van Nos. Eng. Mag.*, Vol. XVIII., p. 210.

By M. A. Achard. Furnishes a summary of the practical results obtained by the transmission of power to a distance. *Van Nos. Eng. Mag.*, Vol. XXV., p. 339.

By Belting. By J. H. Shay. Discusses the qualities of belting for driving dynamos. *Sci. Am. Sup.*, Oct. 22, 1887.

By Belts, Cords and Wire Ropes. By George Leloutre. Gives values of coefficients of friction for various styles of belts, etc. *Van Nos. Eng. Mag.*, July 1884.

By Compressed Air.

By Prof. W. C. Unwin. States in a simple form the laws governing the transmission of power by compressed air. *Proc. Inst. C. E.*, Vol. XCIII., pp. 411-436.

Power Transmission, continued.

An account of the Birmingham (Eng.) plant, which supplies 1,000 h. p. *Trans. Liverpool Eng. Soc.*, Vol. VIII., p. 23, (1887.)

By Robert Zahner. Treats of the physical properties and laws of air, thermo-dynamic principles and formulas and their applications, etc. Also gives examples. *Van Nos. Eng. Mag.*, Vol. XIX., pp. 426 and 441.

And Electricity. Classification of air motors, and comparison with electricity. By ——— Bayet. Also a paper on the same subject by E. Gerard. *Revue des Mines*, 1891, Vol. 13, pp. 166-84 and 185-92.

Motive-Power Supply at Birmingham, Eng. A new company, with capital of $1,500,000, to supply motive power to all parts of the city; with some mathematic discussion, table showing loss of head, etc. Lon. *Engineer*, Feb. 6 *et seq.*, 1886.

Pamphlet issued by Dr. S. H. Wetsthoff, Baltimore, Md., containing considerable data on compressed air plants and their advantages. Abstract in *Sci. Am. Sup.*, Aug. 30, 1890, No. 765, pp. 12213-4.

An examination of the conditions, using as a basis the recent extensive experiments of Prof. Riedler on the compressed air system of Paris. By W. C. Unwin. *Proc. Inst. C. E.*, Vol. CV., 1891, pp. 180-214.

In Paris, Experiments upon. A paper by Alexander B. W. Kennedy, F. R. S. Lon. *Eng.*, Sept. 13, 1889. Vol. XLVIII., p. 320. Lon. *Engineer*, Sept. 20, 1889, Vol. LXVIII., p. 2;3. *Sci. Am. Sup.*, Oct. 26, 1889. *Ry. Rev.*, Sept. 1889.

Paris. Illustrated description of the new central station which will generate 24,000 horse-power. Lon. *Eng.*, March 13, 1891, pp. 297-300, *et seq.*

Popp System in Paris. Lon. *Eng.*, Feb. 15, 1889.

Distribution of, by rarefied air. A company in Paris supplies power for motors of one horse-power or less, and leases the motors. An elaborate account of the installation, with numerous plates. *Mem. Soc. des Ing. Civils.* Paris. March, 1885.

By electricity, water, air, cables, coal gas and steam discussed and compared. From the French. *Sci. Am. Sup.*, Dec. 4, 1886.

By Friction. A description of the Evans Friction Belt, a new device for transmitting power between a pair of pulleys with their surfaces in contact. Illus. *Elec. World*, Sept. 14, 1889. *Power*, Oct., 1889.

By Gearing. The results of numerous experiments made by Wm. Sellers, and presented to the Am. Soc. Mech. Engrs., to determine the efficiency of spur, bevel and worm gearings. Results plotted and mean curves drawn. Three pages of illustrations showing plotted results. Lon. *Eng.*, Dec. 25, 1885.

High Pressures, Hydraulic System of Distributing Power in Cities, with some remarks on other methods. By J. Richards, before the Tech. Soc. of the Pac. Coast. Gives English methods and experience. *Trans. Tech. Soc. Pac. Coast*, Vol. III., p. 87.

Over long distances. By Professor Reuleaux. A valuable discussion of the various methods in use. Relative economic results obtained. *Sch. Mines Quar.*, Vol. VIII., pp. 93 and 220.

By Ropes.

Rope Driving. Paper by C. W. Hunt, read before the Richmond meeting of the A. S. M. E., giving relative advantages at various speeds. Illustrated by diagrams. *Eng. News*, Nov. 15, 1890, pp. 427-8. Abstracted in *Eng. & Build. Rec.*, Nov. 15, 1890, p. 377. *Ry. Rev.*, Nov. 29, 1890, pp. 718-9. *R. R. Gaz.*, Nov. 21, 1890, pp. 802-3.

Gives illustrated description of the method of rope transmission adopted in the boiler shops of the Southern Railroad of France. *R. R. Gaz.*, Jan. 13, 1888.

Power Transmission. *By Ropes*, continued.

Full description of the apparatus and the principles involved, methods of splicing, table of power transmitted by ropes of various sizes, etc. By Loomis I. Seymour. *Power*. May and June, 1889.

Manila Ropes. Paper by John H. Gregg, M. Western Soc. C. E. Read Dec. 4, 1889. Valuable presentation of methods used and results attained. Illustrated. Pp. 6, *Jour. Assn. Eng. Soc.*, March, 1890, Vol. IX., p. 110.

By Wire Ropes. By A. W. Stahl. Gives a complete investigation, including wheels, tension and deflection of rope, limit of span and some of the practical difficulties. *Van Nos. Eng. Mag.*, Vol. XVI., pp. 68, 166 and 225.

See *Belts. Compressed Air. Crank Transmission. Electric Power Transmission. Heat and Power.*

Power Transmitter. Illustrated description of Shaw's Coil Friction Power Transmitter. Reprinted in Lon. *Engineer. Sci. Am. Sup.*, No. 752, May 31, 1890, p. 12010.

Power and Machinery *Used in Manufactures in U. S.* Compiled by Prof. W. P. Trowbridge. Contains statistics of steam and water power used in manufacture of iron and steel, machine tools and wood working machinery, wool and silk machinery, and monographs on pumps and pumping engines, manufacture of engines and boilers, marine engines and steam vessels. *Tenth Census*, Vol. XXII.

Precious Metals. Vol. XIII., U. S. Census, 1880. Statistics and Technology of the Precious Metals. Quarto, 540. *H. Rep. Misc. Doc. 42, Part 13, 47th Congr., 2d Sess.*, 1883.

Preservation of Iron.

See *Iron, Corrosion of. Iron and Steel, Bower-Barff Process.*

Pressure.

Of Flowing Liquids against an oblique plane. Discussion of different formulas. *Der Civilingenieur*, 1885, Heft 2, pp. 77-102.

Required to Solidify Powdered Metals. A brief note relating to experiments on the formation of solid metals by compressing powders of the constituent metal. Experiments made by Prof. Chandler Roberts Austin. *Mech. World*, Dec. 8, 1888.

In Water Pipes. Pressure necessary for fire purposes in German cities. By E. Grahn. Statistics regarding number of fires, amount of water required, and pressure. *Journal für Gasbeleuchtung und Wasserversorgung*, 1886, pp. 45-53, 87-90, 118-122.

Pressure Gauge.

The Mercurial Pressure Gauge on the Eiffel Tower. Description of this gauge, 984 ft. in height. Lon. *Eng.*, Apr. 17, 1891, p. 469. Reprinted by *Ry. Rev.*, May 2, 1891, pp. 277-8.

Recording. An abstract from a paper by Chas. Hague, before the American Water Works Association. *San. Eng.*, July 30, 1887.

See *Gauges.*

Prismoidal Formula. By J. W. Davis. A good discussion of the formula with some queer results. *Van. Nos. Eng. Mag.*, Vol. XX., p. 430.

Prisms. See *Angle Prisms.*

Privy Vaults. *Substitutes for.* Discussions, Editorials and Abstracts of Reports on the question. *San. Eng.*, March 12, 19, 26, *et seq.*, 1885.

Profits. *The Source of Business.* By President Francis A. Walker, of the Mass. Institute of Technology. A masterly exposition of the law determining the employer's share of the product of industry. *Proc. Soc. Arts, Mass. Inst. Tech.*, *for* 1886-7.

Projectiles.

The Electrical Determination of the Velocity of. By Prof. Ed. J. Houston. Illustrated. *Jour. Frank. Inst.*, August, 1881.

Projectiles, continued.

Velocity of. Paper by Capt. H. C. L. Holden describing methods and apparatus used in measuring velocities. Read before Brit. Iron and Steel Inst. Reprint, Lon. *Eng.*, Oct. 23, 1891, pp. 489-90.

Velocity of. H. A. Sinclair. *Elec. Eng.*, New York, April, 1884. Describes best modern methods for measuring velocity of gun shots.

Projections of Maps.

An account of various methods that may be used, with a fuller account of the method of polyconic projections used on the U. S. C. and G. Survey. See Report, 1880.

Tables for the projection of large maps by the polyconic system, using Clarke's spheroid, and computed from the equator to the pole. Values given in metres. *U. S. Coast and Geodetic Survey, Report for 1884. App. No. 6.*

Propeller Blades. *Design of.* Mathematical paper by Prof. M. F. Fitzgerald. Table and diagrams. Lon. *Eng.*, Sept. 19, 1890, pp. 325-7.

Strains on. By G. A. Calvert. A paper before the Institution of Naval Architects, discussing the forces acting upon the blade of a screw propeller. Lon. *Eng.*, April 8, 1887.

Propellers.

A paper by Thomas Drewry, read before the Inst. Marine Engrs. Describes various propellers, and discusses their special features. *Mech. World*, July 18, 1891, *et seq.*

Efficiency of Guide-Blade. By J. I. Thornycroft. M. I. N. A. Description of a series of experiments with guide blade and other propellers, with tabulated results. Also remarks as to the limitation of the use of guide-blade propellers. Naval Prof. Papers, No. 16.

On the fundamental principle of the action of a propeller. *Van Nos. Eng. Mag.*, Vol. VI., p. 618.

Experiments with Screws. By J. B. Andrews, before the Institute of Naval Architects. Gives details of experiments with four and two-bladed propellers. Lon. *Eng.*, April 13, 1888, *Sci. Am. Sup.*, May 19, 1888.

Jet. See *Jet Propellers, Efficiency of.*

Material For. See *Aluminum Bronze.*

Screw.

By A. Blechynden. A paper read before the Northeast Coast Institution of Engineers and Shipbuilders, treating of the reaction and efficiency of the screw propeller almost wholly from an experimental point of view. Lon. *Eng.*, May 13, 1887.

Effective Area of Screws. A paper by D. S. Jacobus, giving complete tests on two screws, also results from two previous tests. *Trans. A. S. M. E.*, Vol. XI., 1890, pp. 1028-37.

Efficiency of: A Description of a Method of Investigation. By R. E. Froude, M. I. N. A. A most important paper on this subject, describing the method adopted in the model screw experiments at Torquay conducted by Mr. Froude, for investigating (1) the efficiency of screws working by themselves in undisturbed water, and (2) the manner in which that efficiency is affected by the screw being brought into conjunction with the hull of a ship. With discussion by Mr. W. H. White, Sir Edward Reed, and others. Reprinted from *Trans. Inst. Naval Architects* by the Bureau of Navigation, Navy Dept. Navy Professional Papers, No. 18.

General considerations on their mode of action and on their construction. By A. Gouilly. *Mem. de la Soc. des Ing. Civils*, February, 1886, pp. 162-176.

Illustration and description of a new method of fitting with screws which is claimed to overcome many of the difficulties of twin screw propulsion. Lon. *Engineer*, April 24, 1885.

Propellers, continued.

New Method of Designing. By Christian Hoehle. Introduces some new propositions as to the slip. Illustrated. *Jour. Frank. Inst.*, Aug., 1886.

Paper by S. W. Barnaby giving a complete review of the subject and the results of recent experiments. *Proc. Inst. C. E.*, Vol. CII, 1890, paper 2401, pp. 71-92. Discussion, pp. 93-124. Correspondence, 124-152. Folding plate of diagrams.

A paper by C. Trathen on the practical geometry of the screw propeller. *Mech. World*, Feb. 4, 1888.

Physics of the Screw Propeller. Discussion of a proposed formula for computing the thrust. By John Lowe, Chief Eng., U. S. N. *Sci. Am. Sup.*, May 2, 1891, No. 800, pp. 12766-7.

Researches on the theoretical construction of screw propellers. By M. Duray de Bruignac. *Mem. de la Soc. des Ing. Civils*, 1885, July, pp. 60-118. Extended mathematical discussion.

Twin Screws. An important paper on the subject of screw propulsion. *Sci. Am. Sup.*, June 18, 1887.

Vogelsang Screw. A short illustrated description of a new form of screw propeller invented by Mr. Alex. Vogelsang, which has been in use some time with very favorable results, having increased the speed of the vessels on which it has been applied without increasing the consumption of coal. *R. R. & Eng. Jour.*, January, 1889.

See *Ferry-Boat*. *Screw Propeller*. *Marine Propulsion*. *Jet Propulsion*.

Proportional Resistance. *The Laws of Proportional Resistance*. Paper by Prof. J. F. Klein, discussing Prof. Kick's law, and giving practical applications of same to blasting, bursting and collapsing strains, forging under hammers, rolling iron and steel, etc. pp. 22. Illus. *Jour., of Eng. Soc.*, Lehigh Univ., Nov. 1889, Vol. V., p. 1.

Propulsion.

Hydraulic. A description of the latest experiments made to find the comparative efficiencies of the jet and screw. By S. W. Barnaby. *Van Nos. Eng. Mag.*, April, 1885. See also *Hydraulic Propulsion*.

Hydraulic. *On the Laws of Steamships*. Lon. *Engineer*, July 27, 1888.

Of Ships by Air Propellers. By H. Vogt before the Bath Meeting of the British Association. Gives details of experiments made with propellers working in the air instead of the water. Lon. *Engineer*, Sept. 28, 1888.

See *Jet Propulsion*.

Prospecting *in Soft Ground*. A paper by Fred. S. Ruthman. The method used was that of boring vertical holes with an artesian well-drilling machine, and proved both economical and satisfactory. *Sch. Mines Quart.*, April, 1889; *Eng. News*, July 6, 1889, Vol. XXII., p. 8.

Public Buildings. *The Proposed Law for the Erection of*. Letters from many architects upon the subject in *Building*, March 27, 1886. The text of the bill in *San. Eng.*, March 27, 1886.

Public Health Abstracts of the papers read and discussions at the meeting of the Am. P. H. Association, held at Washington, Dec. 1885. *Science*, Dec. 11 and 18, 1885.

Public Lands, *Survey of in Ohio*. By Col. Chas. Whittlesey. Eight pp., with map. Tract No. 61. *N. Ohio Hist. Society*. W. W. Williams, Printer, Cleveland, O.

Public Works.

An editorial on reorganizing the Scientific Bureau of the General Government. Presents many facts and arguments making a substantial addition to the literature on this subject. *Eng. News*, Feb. 13, 1886.

A National Board of. A report of the committee of the Technical Society of the Pacific Coast. *Eng. News*, June 12, 1886.

Public Works, continued.

Cullom-Breckenridge Bill. Arguments on the Cullom-Breckenridge Bill by a Committee of the St. Louis Engineers' Club. A pamphlet of 19 pages. Copies may be had by addressing Prof. J. B. Johnson, St. Louis, Mo.

The organization of this department in the Argentine Republic. It is purely a civil bureau and newly organized. From the German. *Eng. News*, March 13, 1886.

The Organization of the Navy Department. A full description (13 pp.) of the evils of the present organization, and an appeal for reform by Secretary Whitney in his annual Report for 1885-6, Vol., I., p. XXVII. A valuable contribution to the Public Works literature.

Proceedings of the Council of Engineering Societies on, at Cleveland, O., April, 1886. The official proceedings, a pamphlet of 16 pp., can be had by addressing John Eisenmann, Sec'y, 44 Euclid avenue, Cleveland, O.

In Prussia. Description and statistics of prominent works in process of execution in 1884, including hydraulic works, buildings, etc., etc. The complete paper extends through various numbers of the periodical *Zeitschr. f. Bauwesen* for 1886; the particularly interesting portion relating to hydraulic works on pp. 527-541 of the last number.

Report of a Committee on, with a lengthy discussion of the subject by an officer of the Engineer Corps, U. S. A. *Trans. Tech. Soc. Pac. Coast*, Vol. III., p. 65, May 1886.

Report of the Commission on Signal Service, Geol. Survey, C. & G. Survey and Hydr. Office of Navy Dept., made to Congress June 19, 1886. Does not include the testimony taken by the commission, which is published separately. *Report No. 2,340, House of Rep., 49th Congr., 1st Session.* 183 pp.

And Scientific Bureaus of the U. S. Government. Editorials on their conduct and on the question of separating them from the War and Navy Departments. *Science*, Dec. 18, 1885, and *N. Y. Evening Post*, Dec. 17, 1885.

Testimony before the Joint Commission of Congress concerning the present organization of the various scientific and technical departments of the government, 1,100 pp. *Sen. Misc. Doc. No. 82, 49th Congr., 1st Session*, 1886.

Puddle. See *Clay*.

Puddling Furnace, *Regenerative for Natural Gas.* See *Furnace*.

Pulley Blocks.

Report by Dr. R. H. Thurston on the comparative efficiency of several pulley blocks. Table of results. *Eng. News*, July 26, 1890, pp. 73-4.

Efficiency of. By R. S. Ball. Gives detailed account of experiments to determine the mechanical efficiency of different forms of pulley-blocks. *Van Nos. Eng. Mag.*, Vol. VII., p. 74.

Pulleys, *Balancing of.* Illustrated article by Joshua Rose, M. E. *Eng. News*, Jan. 3, 1880, p. 2.

See *Belts*.

Pulsometer.

Discussion of the theory of the pulsometer, showing its efficiency to be less than one per cent. of the heat of combustion of the coal used. By Prof. de Volson Wood. *Sci. Am. Sup.*, June 29, 1889.

Test of. Paper by Prof. de Volson Wood before the Am. Soc. M. E., describing and giving results of efficiency and duty tests. Full abstract in *Eng. News* Nov. 21, 1891, pp. 497-8. *Eng. Record*, Nov. 21, 1891, pp. 399-400.

Pulverizing Machines. Taken from Prof. Tischer's paper (German) and fully illustrated. The various principles in use shown and explained; a very satisfactory article. Lon. *Engineer*, Sept. 10, 1886 also *Sci. Am. Sup.*, October 30, 1886.

Pumping.

Cost of, at the Montreal Water-Works in 1889. *Eng. News*, December 6, 1890, p. [illegible].

Cost of Raising Small Quantities of Water by Steam Pump. By Alfred R. Wolff. Gives tabulated data of 34 items for eight different rates of pumping, from 370 to 13,000 gallons raised 25 feet per hour. *Eng. News*, Oct. 9, 1886.

And Drainage. Specifications. Copy of specifications for plumbing and drainage used by the division of plumbing and ventilation of New York City Health Department. *Eng. & Build. Rec.*, May 10, 1890, p. 360.

Electric in Collieries. By Frank Brain, before South Wales Institute of Engineers. Gives details of the pumping plant at the Trafalgar collieries. Gives an analysis of the work done, showing the per cent. lost in different parts of plant; also gives comparison of cost of underground haulage by electricity, cables, compressed air and hydraulics. *Mech. World*, Dec. 24, 1887, *Am. Eng.*, Jan. 11 and 18, 1888.

Electric. Brief description of an electrical pumping plant at St. John Colliery, Normanton, Eng. Efficiency, 44.4 per cent. Lon. *Engineer*, Dec. 2, 1887.

Hot Water. Some results of experiments, giving relative heights, temperatures and quantities. *Eng. News*, June 27, 1885.

Pumping Engines.

Allegheny City, Pa. Fully illustrated and described. Three cylinder compound engines of crank and fly-wheel type, made by E. P. Allis & Co., Milwaukee. Capacity, 18,000,000 gals., raised 210 feet; duty, over 100,000,000. Lon. *Eng.*, Jan. 8 and 22, 1886.

Of the Boston Main Drainage Works. Illustrated and described in *Mechanics*, Dec., 1881.

Burton-on-Trent Sewage Works. Illustrated description of four compound rotative beam engines, each, with its pump, is capable of lifting 125,000 gallons an hour 110 feet. Lon. *Engineer*, Feb. 25, 1887.

Centrifugal, for Irrigation. Capable of discharging 25,000,000 gals. per day to a height of 35 feet at 200 rev. per minute. Illustrated. *E. & M. Jour.*, Jan. 26, 1889.

Chicago. The report of Mr. F. W. Gerecke to the Commissioner of Public Works on the condition and capacity of the pumping engines of the water-works. *Am. Eng.*, March 14, *et seq.*, 1888.

Chicago, Harrison St. Brief description with inset. *Eng. News*, May 31, 1890, p. [illegible].

Comparative Merits of vertical and horizontal engines and on rotative beam-engines for pumping where large powers are required; with many illustrations. By W. E. Rich. Discussion by many members of the institution. *Proc. Inst. C. E.*, Vol. LXXVIII., p. 1.

Compound.

The Thermodynamics of Certain forms of the Worthington and other Types. A discussion of such compound engines as effect the principal parts of the expansion of the steam in the act of exhausting from the high-pressure cylinder into a receiver, from which receiver the low-pressure cylinder is supplied with steam. By Prof. S. W. Robinson, *Trans. A. S. M. E.*, Vol. III., p. 132.

A two-page plate showing the engines and pumps of the Southwark and Vauxhall Water company and other engravings, giving engine details. Engines are inverted double-cylinder compound direct-acting rotative. Lon. *Engineer*, July 22, *et seq.*, 1887.

Differential Pumping Engine, Bradley. A horizontal engine with cylinders 32 in. and 60 in., and 10 ft. stroke, and works under a 300 ft. head of water. Short description with illustrations. Lon. *Engineer*, Jan. 23, 1891, p. 72.

Pumping Engines, continued.

High-Pressure Compound Pumping. Description and detailed drawings of engines for Millwall Dock Co. Specification called for engine capable of pumping 30,000 gallons per hour at a pressure of 750 lbs. per square inch. Lon. *Eng.*, May 14, 1886.

Horizontal. For drainage of Rhymney Iron Mines, South Wales. One lift of 720 ft. Lon. *Eng.*, Sept. 19, 1890, p. 232.

Tables for Power of. By John W. Hill. *Eng. News*, April 27, 1889.

Davey's Differential Pumping. Brief description, with two-page plate and other engraving, showing plan, elevation and cross-sections of Davey's differential pumping engine for the Weston Water-Works. Lon. *Engineer*, Jan. 13, 1887.

Discussion, pp. 13. *Jour. Assn. Eng. Soc.*, July, 1889. Vol. VIII., p. 365.

Duty and Cost of. Statistics obtained by committee and tabulated. *Trans. A. S. C. E.*, Vol. IV., (1875), pp. 142-146. Discussion, pp. 223-229.

Duty and its influence in perfecting the steam engine. By John Whitelaw. *Jour. Assn. Eng. Soc.*, Vol. II., p. 76.

Duty Trials.

Abstract of Test of No. 1 Pumping Engine. At Pawtucket Water-Works, known as the Corliss High Duty Pumping Engine. The tests and report of same were made by Prof. J. E. Denton, of Stevens Institute, Hoboken, N. Y. *Tenth Annual Report of Water Comms.* Pawtucket, R. I., p. 3, 1890.

By J. W. Hill. A report on the tests for capacity and duty of a large Worthington engine at Buffalo, N. Y. *Van Nos. Eng. Mag.*, Vol. XXVIII., p. 267.

Chestnut Hill Gaskill Pumping Engines at Boston Water-Works. Complete account of tests, giving all data and records. Tests by Geo. H. Barrus. *Report City Engineer*, of Boston, Mass., 1888, p. 22.

Croydon Water-works Engine. By Prof. Kennedy. Diagrams and double plate. Lon. *Eng.*, Aug. 29, 1890, pp. 167-9.

Gaskill Compound. By John W. Hill. Report of the test of the engines erected at Evansville Water-works. *Van. Nos. Eng. Mag.*, Vol. XXIV., p. 418.

Philadelphia Water-Works. A report by J. L. Ogden, J. E. Codman and F. T. Haily on the duty and capacity of a 20,000,000 gallons Gaskill engine at the Spring Garden pumping station, Philadelphia. Gives the method of conducting test, with calculated and observed data. The engine gave a duty of 120,522,276 foot-pounds per 100 pounds of coal. *Eng. News*, Feb. 18, 1888.

Standard Method of Conducting, A Plea for a. By George H. Barrus, M. Boston Soc. C. E. A complete and valuable discussion of duty tests and the desirability of uniformity in conducting them. Pp. 20, *Jour. Assn. Eng. Soc.*, July, 1889, Vol. VIII., p. 365. *Lon. Eng.*, Feb. 8, 1889, *et seq.*

Standard Method of Conducting. Report of a committee of five to the Am. Soc. Mech. Eng. at Cincinnati convention, May 14, 1890. The proposed scheme seems to be an improvement on the ordinary method. *Eng. News*, May 17, 1890, Vol. XXIII., p. 474, *et seq.* *Eng. & Build. Rec.*, May 24, 1890, p. 390. *Trans. A. S. M. E.*, Vol. XI., 1890, pp. 645-88.

Worthington High Duty Pumping Engines at Memphis, Tenn. Abstracts of the report of the experts who conducted the tests. Gives description of the artesian water supply, engines and boilers, and results of tests. Pumping station, tunnels and wells, illustrated. *Eng. News*, Sept. 12, 1891, pp. 233-5. *Eng. Record*, Sept. 12, 1891, pp. 234-7.

Performance of a Worthington High Duty Pumping Engine, of one and one-half million gallons capacity per twenty-four hours, against a head equivalent to two thousand feet of water. Used to pump petroleum. Full details

Pumping Engines, *Performance of a Worthington High Duty, etc.*, continued.

of tests of boilers and engine. Paper by J. E. Denton before the Am. Soc. M. E. Abs., *Mechanics*, Oct., 1891, pp. 216-19.

Reports of a Trial of a Worthington High Duty Pumping Engine at the West Middlesex Water Works, Hampton, England. *Eng. & Build. Rec.*, Dec. 22 and 29, 1888. *Power*, Jan., 1889. Lon. *Eng.*, Jan. 4, 1889.

Holly-Gaskill Engines at Saratoga, N. Y. Description, illustration of details and results of tests. From Lon. *Eng.* *Eng. News*, April 26, 1884, pp. 193-99.

How to Use Steam Expansively in Direct-Acting Steam Pumps. Paper read by J. F. Holloway, of New York, at the meeting of the Am. Soc. M. E. in New York, Nov., 1889. The paper includes a good description of the new Worthington high duty attachment for pumping engines and is a valuable contribution to literature on the practical use of compound direct acting pumping engine. *Eng. & Build. Rec.*, Nov. 30, 1889. Vol. XX., p. 333 *et seq.* *Trans. A. S. M. E.*, Vol. XI., 1890. pp. 235-53.

Hydraulic.

Hydraulic. A paper by H. D. Pearsall read before the Inst. C. E., describing the pump and giving results of tests of efficiency. Reprint *Eng. Record*, Sept. 26, 1891, pp. 270-1.

Description of a direct action engine designed to be operated by high heads of water. One engine is now working in Cal., under a head of 780 feet, and shows an efficiency of 75 per cent. Illus. *Sci. Am. Sup.*, No. 757, July 5, 1890, p. 12092.

Illustrated description of a new form of pumping engine for Champaign, Ill. It is a vertical compound duplex direct, double-acting, steam pump, with outside flanges. *Eng. News*, Nov. 26.

Leicester. Specifications for the erection of pumping engines, etc., for the Beaumont Lego pumping station, Leicester, Eng. Lon. *Engineer*, August 12, 1887.

At the Lincoln Water-Works. By Henry Teague, before the Inst. of Mech. Engrs. Gives particulars respecting various mechanical details, with reasons for their adoption and experience of their practical working. Lon. *Eng.*, March 4, 1887.

Machinery for drainage purposes in the Fenlands, Eng. Various devices in use described. *Proc., Inst. C. E.*, Vol. XCIV., p. 264.

Oil Pumping and Engine Data. A paper by Prof. J. E. Denton, giving an account of the performance of a Worthington high duty oil pumping engine. Read at the Providence meeting of the Am. Soc. C. E. Illus. *Eng. Record*, July 4, 11, 18, 1891.

Performance of the Best Modern Engines, compared with that of the Cornish pumping engines of the period of 1840. A paper by Prof. J. E. Denton, Stevens Inst. Tech. *Eng. News*, Dec. 28, 1889, Vol. XXII., p. 603 *et seq.*

Saratoga, N. Y., the Gaskill. By John W. Hill. Gives details of the performance of the Gaskill compound pumping engine, which gave a duty of 113,000,000 lbs. *Van Nos. Eng. Mag.*, Vol. XXIX., p. 397.

Specifications for High Service Pumping Engine No. 5, and boilers for the St. Louis Water-Works. With two-page plate of general drawings of same. *Eng. News*, Aug. 7, 1886.

The Screw Pumping Engines at the Milwaukee Flushing Tunnel. Illustrated description of this somewhat novel and highly efficient pumping engine for flushing out large tunnels. *Eng. News*, March 8, 1890, Vol. XXIII., p. 228.

Triple-Expansion Duplex Pumping Engine. A new Worthington engine. Illustrated description. *Iron Age*, Feb. 5, 1891, pp. 234-6.

Triple Expansion, for the Mildura Irrigation Colony, Victoria. Illustrated description. Lon. *Engineer*, Jan. 18, 1889.

Pumping Engines, continued.

Umaria Colliery. Gives abstract from the specification for the construction of a colliery plant in India. Illus. Lon. *Engineer*, Aug. 19, 1887.

Underground pumping. A two-page plate of drawing showing details of a pair of pumping engines recently erected at the Denby Colliery designed to pump 1,000 gallons per minute with a lift of 332 ft. Lon. *Eng.*, March 25, 1887.

For Water Works. On the selection of steam pumps from those offered in the market. The considerations which should enable an engineer to decide for any given case. By Alfred R. Wolff, M. E. *Eng. News*, Sept. 25, 1886.

Whampoa Docks. Brief illustrated description of the large centrifugal pumping engines in use at Whampoa Docks, Hong Kong. Lon. *Engineer*, Nov. 18, 1887.

Worthington New Direct Acting, using steam expansively and giving uniform flow. An account of some elaborate and scientific tests by English engineers. Illustrated. *Proc. Inst. C. E.*, Vol. LXXXVI., p. 293.

Worthington. High-Duty. A description and sectional view of the high-duty pumping-engine at New Bedford, Mass. *Eng. News*, June, 26, 1886.

Worthington High-Duty Attachment to Pumping Engines, and How it Works. An illustrated description of this efficient device, showing indicator diagrams and giving data indicating the amount of saving effected by its use. *Eng. News*, Aug. 31, 1889, Vol. XXII., p. 193.

Worthington High-Duty Pumping Engine at West Middlesex Water Works, Hampton, England. See *Water Works*.

See *Docks, Pumping, Pumps. Sewage Pumping Plant. Water Works.*

Pumping Machinery.

Ancient and Modern. One of a series of lectures at Sibley College, Cornell Univ., by J. F. Holloway, Past Pres. Am. Soc. M. E., giving the historical review of development of pumping machinery and a description of some modern forms, notably the Worthington high-duty engine. *Sci. Am. Sup.*, July 6, 1889.

By E. D. Leavitt. Being the tenth Sibley College lecture at Cornell University. Treats of the pumping engines in use in the American cities for obtaining water supply, mine pumping machinery as well as the economy and duty of pumping machinery. *Sci. Am. Sup.*, July 31, 1886.

Raising Machines in Holland, Practical Results Obtained From. By G. Cuppari. A valuable article, giving descriptions and details of pumping machinery in use at various stations in Holland. *Van Nos. Eng. Mag.*, Oct., 1884.

Types of Hydraulic Pumping Machinery. By J. T. Fanning. A short historical and descriptive paper, with cut showing an improved form of turbine pumping machinery. *Jour. New. Eng. W.-W. Assn.*, Sept., 1886.

For Water Works, being a comparison of the relative merits of direct acting and crank and fly-wheel engines for various capacities. By Frank H. Pond. *Am. Eng.*, June 3, 1885.

See *Aqueduct, New Croton.*

Pumping Station *at Watertown, N. Y.* General description of intake wheel-pits and station. Fully illustrated. *Eng. & Build. Rec.*, Nov. 16, 1889, *et seq.*

Pumps.

Illustrations and description of new high-pressures pumping machinery. Lon. *Engineer*, March 13, 1885.

For Boiler Feeding. See *Boilers, Pumps.*

Centrifugal.

A brief account of the performance of the two 42-inch centrifugal pumps at the Mare Island Navy Yard, Cal. *Proc. Eng. Club, Phila.*, Vol. VI., p. 110.

Balancing. Abstract of a paper by J. Richards, read before the Technical Society of the Pacific Coast, discussing the methods of balancing the unequal

Pumps. *Centrifugal, Balancing*, continued.

pressures arising in various forms of pumps. *Ry. Rev.*, Jan. 17, 1891, p. 33. *Eng. News*, Jan. 31, 1891, pp. 114-15.

Compounding. Proposed method of arranging one pump to deliver into the supply pipe of the other, and thus adapt them to high heads. By W. H. Booth in *Am. Mach.* Reprinted in *Eng. News*, Sept. 12, 1891, p. 236.

Their Efficiencies. By Wm. O. Webber before the American Society of Mechanical Engineers. Gives details of experiments made upon centrifugal pumps. *Trans. A. S. M. E.*, Vol. IX., pp. 237-245.

Efficiency of Centrifugal and Reciprocating. By Wm. O. Webber. Results of a series of valuable experiments with different lifts and horse-powers. *Trans. A. S. M. E.*, Vol. VII., p. 598.

The history and theoretical laws of centrifugal pumps, as supported by experiments and their application to their design. By R. C. Parsons. Illustrated. *Van Nos. Eng. Mag.*, Vol. XVII., p. 454.

The Inventor of. An Account of the Inventor and Invention, by J. Eliot Hodgkin. Lon. *Eng.*, Dec. 5, 1890, pp. 670-1.

Will raise 318 cubic feet per second ten feet high at an expenditure of 2.5 lbs. of coal per hour per effective H. P. of water raised. The builders earned $60,-000 in premiums paid under the contract for exceeding the required duty. *Annales des P & C.*, September, 1888.

Comparative Trials of piston pumps and centrifugal pumps. By De Glehn. Abstract from *Bulletin* de la Societe Industrielle de Mulhouse. 1888, p. 661. The piston pumps gave a duty of 72 per cent. of the "engine power," while the centrifugal force gave only about 42 per cent. of the "engine power." *Proc. Inst. C. E.*, Vol. XCVI., p. 423. 1888.

Details of. The Construction of. A series of illustrated articles by Philip R. Bjorling. *Mech. World*, Jan. 3, 1891, p. 6, *et seq.*

Discharge from Reciprocating Pumps. "Table of constants which, multiplied by effective number of strokes per minute, gives the discharge in gallons per hour." *Mech. World*, April 5, 1890, p. 134.

Dow Positive Piston. An entirely new type of pump of the rotary class, claimed to realize the perfect action of a piston or plunger moving in one direction only, and to be the most economical pump yet made. *R. R. & Eng. Jour.*, Dec., 1888.

Hydraulic,

Deep mine pumps worked by water pressure from the surface. Gives cost and performance. Abstracted in *Trans. Inst. C. E.*, Vol. XC., p. 528.

A Compound-Plunger. Description and illustration of a mine pump operated by water pressure, the pressure being maintained by means of a hydraulic ram. By E. R. Woakes. *Trans. A. I. M. E.*, 1891, pp. 4.

Johnston's. Illustrated description of a novel hydraulic pump. *Eng. News*, Jan. 22, 1887.

For Indian State Railways. Two pages of illustrations showing details of pumps described in the specifications as six horse-power engines and boilers with pumps for supplying water for the use of locomotives from wells 120 feet deep. Lon. *Engineer*, March 4, 1887.

Malta Dockyard. Description, with cuts, showing the arrangement of the large centrifugal pumps, also of the supplementary pumps for the Malta Dockyard. Lon. *Eng.*, Aug. 26, 1887.

Mercurial Air. By Prof. S. P. Thompson, before the Society of Arts. A very complete paper, tracing the development of the mercurial air pump. Gives cuts of the various machines and the results attained with some of them. *Jour. Soc. Arts*, Nov. 25, 1887; *Sci. Am. Sup.* Jan. 21, *et seq.*, 1888. Abstracted Lon. *Eng.*, Nov. 25, 1887.

Pumps, continued.

And Rotary Blowers. The best dimensions of parts as the result of long experience in England. Illustrated. *Lon. Eng.*, April 16, 1886.

For Sewage and Sludge. A short description of the pump for the sewage works at Walthamstow. Illustrated. Lon. *Engineer*, June 18, 1886.

Valves for. Experiments on the motion of automatic pump valves. By Prof. C. Pach. Experiments on valves of different kinds, to determine proper weight of valve, loss of head, etc. *Zeitschr. des V. Deuts. Ing.*, 1886, pp. 421-426, 473-477.

Vacuum, The Bosco. Description of the most improved form of piston pump for exhausting the bulbs of incandescent lamps, etc. Does better work than the mercury pump. Illustrated. *Elec. Eng.*, March 4, 1891.

Vacuum, Mercury, for exhausting incandescent lamps. Illustrated. Lon. *Eng.*, Aug. 7, 1885.

Vacuum, Steam. Efficiency of pulsometer pumps as compared with piston or plunger pumps. How to determine such efficiency. By J. Foster Flagg. *Trans. A. S. C. E.*, Vol. V., 1876, p. 381.

See *Electric Mining*. *Machinery*. *Hydraulic Ejector*. *Pipes*, *Water Pipes and Pumps*. *Pulsometer*.

Punching Metals. *Improved Method*. Punch first makes hole smaller than required, and then takes a second cut to form exact size. Tests showing increase in strength of plates punched by this method. *Mechanics*, June, 1890.

See *Steel*.

Purification.

Of Sewerage and contaminated water by electrolysis. See *Sewage Disposal*.

Purification of Water.

See *Water Filtration and Purification*.

Pyrometer.

A new form by which the furnace temperature is determined by observing the change in the temperature of water passed through a short section of tube inside the furnace. *Mechanics*, Nov., 1883.

The Thermo-Electric Pyrometer of M. C. Chatelier. A complete description of this instrument, both of the form used in the laboratory and that used in industrial furnaces. By Joseph Struthers. *Sch. Mines Quart.*, January, 1891, pp. 143, 157.

Use of the Calorimeter for High Temperatures. By J. C. Hoadley. Describes a successful use and the construction of this instrument. Illustrated. *Trans. A. S. M. E.*, Vol. II., p. 43. See *Platinum* for the specific heat of this metal.

Quarrying. See *Blasting*. *Stone Quarrying*.

Radiation *at Different Temperatures*. Capt. John Ericsson gives results of experiment. Illustrated. *Van. Nos. Eng. Mag.*, Vol. VII., p. 113.

Rail Fasteners. See *Railroad Track*. *Railroad Maintenance*.

Rail Joints.

By C. Peter Sandberg, before the Inst. Civ. Engrs. Gives European experience and practice. Gives results of extended tests of different kinds of joints on the Swedish State railways; also proposes tests for rails, and points out some common and fatal errors in the design of cross-sections of rails. The author has been a rail and joint inspector for over 20 years. The article is well illustrated and valuable. Reprinted in *Eng. News*, March 27, 1886.

Practice concerning, given in tabulated form for twenty different railroad corporations. Gives style of splice and allowance for expansion. *R. R. Gaz.*, Feb. 5, 1886.

Rail Joints, continued.

Proposed Metallic. Illustrated description of tie and rail fastenings proposed for the Chinese railroad. *Eng. News*, Jan. 15, 1887.

And Splice Bars, Why They Break. A paper read before the Engrs.' Soc. West. Penn., by M. J. Becker; a careful and valuable study of the subject, with satisfactory conclusions. *Am. Eng.*, Feb. 6, 1885.

Standard 85-lb. Rail and Joint; B. & O. R. R., 1891. Drawings of rail and joint given. *Eng. News*, July 18, 1891, p. 51.

Standard Rail and Splice Bars; New York, New Haven & Hartford R. R. Description and drawings. *Eng. News*, Oct. 3, 1891, p. 305.

And Steel Rails. A paper before the Inst. of Civ. Engrs., London, by C. P. Sandberg. Gives European experience and practice bearing on the discussion in this country in 1885, in the matter of splice bars. Illustrated. *R. R. Gaz.*, Feb. 12, 19 and 26, 1886.

And Steel Rails. By C. P. Sandberg. A paper resulting from the discussion in America on rail-joints and splice-bars. Discussed from the English standpoint. A 100-lb. rail section given. Some valuable facts on the wear of steel rails. *Proc. Inst. C. E.*, Vol. LXXXIV., p. 365.

Table, abstracted from a paper by W. F. Ellis before the New Eng. Roadmaster's Assn., giving the standard joint of many railroads. *R. R. Gaz.*, Aug. 28, 1891, p. 598.

Wear of, at Joints. A series of engravings from photographs of worn joints on the Penn. Road, after nine years' use. Instructive as showing the nature of the wear in actual service. *R. R. Gaz.*, Sept. 3, 1886.

Why They Break. A careful summary of various arguments with some new theories. Illustrated. An editorial in *R. R. Gaz.*, Nov. 6, 1885.

See *Railroad Maintenance of Way.*

Rail Sections.

A set of sections designed on recent principles, by R. W. Hunt. *R. R. Gaz.*, June 7, 1889.

A Mean Set of Rail Sections. A set of designs averaging nearly the proportions of the various sets prepared by the individual members of the committee of Am. Soc. C. E. Most dimensions are either constant or uniformaly varying. Full size drawings. *Eng. News*, March 21, 1891, pp. 278-9.

Editorials in which the form and area of cross-section and weight of rail proper for various degrees of strength, stiffness and durability are well argued. *R. R. Gaz.*, March 13 and 27, 1885.

Boston & Albany 95-lb. Rail. Illustrations of section and angle splices, with specifications for chemical composition. *R. R. Gaz.*, March 13, 1891, pp. 176-7. *Eng. News*, April 4, 1891, pp. 330-1.

Comparison of different forms in use, with suggested improvements. Paper read before the American Institute of Mining Engineers, Feb., 1887, by W. F. Mattes, Scranton, Pa. *E. & M. Jour.*, April 30, May 7, and May 14, 1887; also *R. R. Gaz.*, April 8-15, 1887.

And Flange Wear. An able discussion of a paper, by M. N. Forney, Secretary Master Car-Builders' Association, taking exceptions to conclusions there reached. Illustrated. *R. R. Gaz.*, March 27, 1885.

Heads, Form of the Under Side of. Gives a number of the illustrations of worn rail and discusses the causes of the wearing. *Eng. News*, Feb. 26, 1887.

Prussian Roads, the New Rail Section for the (adopted Aug. 22, 1885), has a height of 5⅛ inches, breadth of base 4¼ inches, and thickness of web in center, .43 inch. The total width of head is 2⅝ inches, with vertical sides, rounded at the upper corners with a radius of .55 and at the lower corners with a radius of .8 inch. The slope of the under side of the head and top of foot is 1 in 4. Illustrated in *Deutsche Bauzeitung*, 1885, p. 482.

Rail Sections, continued.

Proper Design of. By Fred. A. Delano. A paper before the Am. Inst. Min. Engrs., 1889, giving essential conditions of good rolling, with proposed specifications. *Eng. News*, June 8, 1889.

Proposed. Paper by Robert W. Hunt, Chicago, giving details of proposed rail sections, attention being paid to facility of manufacture as well as to meet the demands of use. *Proc. A. I. M. E.*, New York meeting, Feb., 1889.

Proposed 100-lb. A full-sized section of the rail proposed, with a good discussion of the need of it. Also an editorial on the same subject. *R. R. Gaz.*, June 25, 1886.

Relation of the Section of Rails and Wheels. The Preliminary report of the committee of the American Society of Civil Engineers, to consider the relation to each other of the sections of railroad wheels and rails. *Trans. A. S. C. E.*, Vol. XIX., (July, 1888), pp. 1-84; *Eng. News*, Nov. 3, *et seq.*, 1888; abstracted *R. R. Gaz.*, Nov. 2, 1888.

Relation of, and Wheels. Abstract of a report of a committee of the Am. Soc. of C. E. *R. R. Gaz.*, Nov. 2, 1888.

Sandberg's Revised Goliath. Details of this new 100-lb. rail. By C. P. Sandberg. *R. R. Gaz.*, July 26, 1889. Vol. XXV., p. 491.

Standard Rail Sections. Progress Report of the Committee on Standard Rail Sections. Am. Soc. C. E. Ten different sets of designs submitted by eleven different members are illustrated, and their chief features discussed by the designer. *R. R. Gaz.*, Feb. 20, 1891, pp. 127-8. *Eng. News*, Feb. 21, 1891, pp. 177-8. *Trans. A. S. C. E.*, Vol. XXIV., 1891, pp. 1-12

Standard Rail Section and Track; South Australian Government Railways. Details of rail section, splice bars, and fastenings. *Eng. News*, Oct. 31, 1891. p. 408.

Standard 80-lb. Rail and 42-in. Angle Plate used on Mich. Cent. R. R. *R. R. Gaz.*, Dec. 7, 1883.

Standard 80 lbs. Rail Section, N. Y. L. E. & W. R. R. Illustration, full size, and description of this new form of rail section. Head of rail broad as compared to its depth, ¼ in. corner radius, vertical sides. *Eng. News*, Feb. 1, 1890, Vol. XXIII., p. 3. *R. R. Gaz.*, June 3, 1890, Vol. XXII., p. 69

Standard, 85 lb. Gives drawing with full dimensions of the 85-lb standard-rail of the Pennsylvania railroad. *R. R. Gaz.*, April 6, 1888.

Standard 90-lb. Philadelphia & Reading. Gives drawing, with full dimensions, of the 90-pound rail being placed on the Philadelphia & Reading Railroad. *R. R. Gaz.*, Aug. 24, 1888.

Standard 60-lb Rail Section. Chicago South Side Elevated Railway. Detail illustration and description. *Ry. Rev.*, April 19, 1890. p. 219.

System of Rail Sections in Series. A valuable paper presented by Mr. P. H. Dudley at the Washington meeting of the Am. Inst. of Min. Engrs., Feb., 1890. Table of tests and numerous illustrations. *Trans. A. I. M. E.* Reprinted with folding inset in *R. R. Gaz.*, Aug. 15, 1890, pp. 569-70. *Ry. Rev.*, Aug. 23, 1890, pp. 491-3.

See *Rail Joints and Sections.*

Rail Tests. See *Rails, Tests of.*

Rails.

Abrasion and Endurance of, Notes on the. A discussion of the nature of these phenomena and their relation to other practical properties. Two papers by James E. Howard. *R. R. Gaz.*, Jan. 23, 1891, pp. 51-4.

Best Methods of Fastening Rails to Wooden Ties. See *Railroad Maintenance.*

Breaking Iron. Notes by O. Chanute on the weight of rails and also on the breaking of iron rails. *Trans. A. S. C. E.*, Vol. III., pp. 111-113.

Improved Street Car. Describes the types of rails in use for street car traffic abroad. Illus. *Eng. News*, May 19, 1888.

Rails, continued.

Creeping of, on the St. Louis Bridge. By J. B. Johnson. The amount of this creeping is [illegible] feet a year in the direction of the traffic, independent of an [illegible] grade. Theory of its cause advanced. *Jour. Assn. Eng. Soc.*, Vol. IV., p. 1.

Comparison Between the Uniformity of Bessemer Steel and Iron. By J. B. Pearse. Compares the experiments made for the Central Railway of Orleans at the Terre Noire Works, and those of Peter Ashcroft, Esq., on English iron rails. *Van Nos. Eng. Mag.*, Vol. I., p. 103.

Considered Chemically and Mechanically. A valuable paper by Mr. Christer P. Sandberg, read at Sheffield before the Inst. Mech. Engrs. Synopsis of the discussion on steel rails at the International Railway Congress in Paris, 1889. Tables of tests, etc. *Iron*, Aug. 1, 1890, pp. 92-7. *R. R. Gaz.*, Aug. 15, 29, 1890, pp. 567, 600. *Ry. Rev.*, Aug. 16, 23, 1890, pp. 476, 496-8. Lon. *Engineer*, Aug. 1, 8, 1890, pp. 85, 108-9. Lon. *Eng.*, Aug. 1 and 8, 1890, pp. 141-4, 171-3. *Mech. World*, Aug. 2, 1890, *et seq.* *Am. Mfr.*, Aug. 22, 1890, pp. [illegible]. *Eng. News* Aug. 23, 1890, p. 167.

Destruction of Rails by Excessive Weights. Paper by Joseph T. Dodge, M. Am. S. C. E., with discussion. *Trans. A. S. C. E.*, April, 1889, Vol. XX., p. 181.

Economy of Light and Heavy Heads. By Ashbel Welch. Shows a great saving by increasing the depth of the head. Illustrated. *Trans. A. S. C. E.*, Vol. X. (1881), p. 251-274.

Effect of Rail Upon Wheel, and of wheel upon rail. Review of book by Boedecker. Hanover, 1887. Hahn, Publisher.

Flangers. The subject discussed before the Northwest Railroad Club. Illustrated by scale drawing of rail flanger in use on C., St. P. M. & O. Ry. *R. R. Gaz.*, Feb. 15, 1889; *Mast. Mech.*, March, 1889.

Foreign Practice as to Steel Rails and Rail Tests. Abstract of a valuable paper by M. Mussy in the *Annales des P. & C.* This essay attempts to reconcile the apparently contradictory results of experiments made by Messrs. Dudley, Gruner, Conrad and Care. A table shows the present standard specifications for rails on seven leading French railways. *Eng. News*, Aug. 23, 1890, pp. [illegible].

Hard or Soft Rails. Editorial in *R. R. Gaz.*, Oct. 18, 1890, pp. [illegible]-3.

Hard Rails, Recent Experience in making. Article by P. H. Dudley giving results of drop tests on the 95 lb. [illegible] carbon rails of the Boston & Albany R. R., and on his 75 lb. rail. Sections of these rails given. *R. R. Gaz.*, Nov. 6, 1891, pp. 778-9.

Hard Rails or Mild Rails. Kind of steel best suited for rails, chemical composition, and influence of the various impurities. By A. Stevarts. *Assoc. des Ing. de Liege*, 1891, Vol. 4, pp. 25-46.

Life of. A report by a committee on the form, weight, manufacture and life of rails. *Trans. A. S. C. E.*, Vol. III., pp. 87-110.

Manufacture and Wear of. By C. P. Sandberg. Treats of best method of manufacturing rails from common iron, and their capacity to resist wear. Also of the disposal of old rails. *Van Nos. Eng. Mag.*, Vol. VII., pp. 78, 174 and 277.

Old Utilization of. Detailed drawings of a bridge built of old rails. *Eng. News*, May 15, 1876.

Radii of Worn Surface of. Gives engraving showing the radii of sixteen pairs of rails from sections carefully taken under the direction of Mr. Thomas Rudd. *Eng. News*, Oct. 8, 1887.

And Rail Fastenings. Paper by James Churchward, C. E., before the Am. Soc. Ry. Sup's. Abstract in *Ry. Rev.*, Oct. 18, 1890, pp. [illegible]-6. Editorial discussion *ibid.* pp. 621-2.

Rolling of. See *Iron Working.*

Rails, continued.

Rolling Steel. Paper of some value by Mr. D. K. Nicholson, Steelton, Pa., read before Am. Soc. Mech. Eng. *Eng. News*, May 3, 1890, Vol. XXIII., p. 416. *Trans. A. S. M. E.*, Vol. XI., 1890, pp. 906-13.

Rolling Steel. Abstract of a paper by D. K. Nicholson before Am. Soc. M. E., giving a brief review of methods, etc. *R. R. Gaz.*, Nov. 22, 1889; *Ry. Rev.*, Nov. 23, 1889.

Specifications for Steel. By R. W. Hunt, before the American Institute of Mining Engineers. Gives specification for the manufacture of steel rails, embracing the conclusion derived from twenty years' practice in the manufacture of steel. *E. & M. Jour.*, Oct. 27, *et seq.*, 1888; *R. R. Gaz.*, Oct. 26, 1888.

Specifications for. Gives the specifications adopted by Ward & Bros., rail inspectors, Pittsburgh, for steel rails and track fastenings. *R. R. Gaz.*, Sept. 7, 1888; *Eng. News*, Sept. 1, 1888.

Specifications for their Manufacture. By Robert W. Hunt, Chicago, Ill. Illus. Pp. 16. *Proc. A. I. M. E.*, Buffalo meeting, 1888.

Statistics on the Durability of Rails. Supplement to the statistics for the years 1879-81. Published by the *Verein Deutscher Eisenbahn, Verwaltungen*, as a supplement to the "*Organ fur die Fortschritte des Eisenbahnwesens*" for 1885.

Steel.

By T. A. Delano. Gives brief discussion of some of the conditions of manufacture which may greatly influence the life of steel rails. *R. R. Gaz.*, Aug. 10, 1888.

Comparative cost of manufacturing in Pennsylvania and England. Abstract from consular report by Mr. J. Schoenhof, Oct., 1888. *Eng. News*, Feb. 9, 1889.

Discussion. By F. H. Lewis. *Proc. Eng. Club, Phila.*, April, 1886.

Endurance of. Gives the result of 19 years' experience of the London & Northwestern railway with iron and steel rails. *Van Nos. Eng. Mag.*, July, 1886.

Physical and Chemical Tests of. Thirty samples of Bessemer steel made at the Edgar Thomson Works. *Iron Age*, Oct. 1, 1885.

Tests of.

Tests Applicable to. By James E. Howard. Discusses some of the simple methods of testing rails, and the relative behavior under these tests with the more elaborate ones requiring special machinery. *R. R. Gaz.*, Sept. 7, 1888.

Russia. Perhaps the most elaborate tests in the world applied to steel rails. *Proc. Inst. C. E.*, Vol. LXXXII., p. 369.

Some Track Experiments on the Deflection of Rails. Description of some experiments to determine the deflection of a rail under heavy locomotives. With diagrams of deflection and stresses. By James E. Howard. *R. R. Gaz.*, July 3, 1891, pp. 463-4.

Wear of Rails of Different Hardness. A translation of an article by J. W. Post, in the *Revue Generale de Chemins de Fer*, giving results of experiments on hard and soft rails as actually used in the track. *Ry. Rev.*, Oct. 26, 1889. Vol. XXIX., p. 618.

Wear of Steel Rails of Different Degrees of Hardness. By J. W. Post. Tests are given covering a period of 3 years. *Organ fur die Fortschritte des Eisenbahnwesens*, 1890, p. 14. Abstract with tables of tests. *Proc. Inst. C. E.*, Vol. CII., 1890, pp. 380-1.

For Street Railways. Iron vs. Steel. By A. W. Wright. The question argued, but not sufficient evidence to reach satisfactory conclusions. *Jour. Ass. Eng. Soc.*, Vol. IV., p. 314.

Wear of, as influenced by its Chemical and Physical Properties. A valuable

Rails. *Wear of, etc.*, continued.

paper giving results of experiments on rails and actual service. By Dr. C. B. Dudley, read at the N. Y. meeting of the Brit. Iron and Steel Inst. Abstracted in *R. R. Gaz.*, Oct. 10, 1890, pp. [illegible]. *Ry. World*, Oct. 11, 1890, pp. [illegible]. *Ry. Rev.*, Oct. 25, 1890, pp. [illegible]. *Iron*, Oct. 17, 1890, pp. [illegible].

Wear of in Musconetcong Tunnel, L. V. R. R. *R. R. Gaz.*, November 28, 1890, p. [illegible].

Wear of Rails as Related to their Section. Paper read by Mr. P. H. Dudley at Ottawa meeting of Am. Inst. Mining Eng., giving illustrations and descriptions showing behavior of railway track under loads, wear of wheels and rails, and giving section of 80 pound steel rail having comparatively broad head, for New York Central and Hudson River Railroad. *Ry. Rev.*, Nov. 30, 1889, Vol. XXXI, p. [illegible]. Postscript to above paper, *Ry. Rev.*, April 12, 1890, p. [illegible]; also *R. R. Gaz.*, Feb. 28, 1890, p. 138.

Wear of in Swiss Tunnels. Results of some experiments extending over more than three years. *Eng. News*, Oct. 11, 1890, p. 313.

Weight and Form of. Report of the Com. of the Road-Masters' Assn. *Eng. News*, Nov. 7, 1885.

See *Car Wheels, Cylindrical. Rail Joints and Steel Rails. Rail Sections. Railroad Track. Railroads.*

Railroad Accidents.

In 1886. Tabulated statement of train accidents for 1886 classified; also the accidents for 14 years tabulated according to nature and cause. *R. R. Gaz.*, Jan. 28, 1887.

In 1890. Table giving annual summary with comparisons with other years. *R. R. Gaz.*, Feb. 13, 1891, p. 116.

For the year ending June 30, 1889. Table compiled by Prof. H. C. Adams, Statistician to the Interstate Commission. *R. R. Gaz.*, May 30, 1890, p. [illegible]; also *ibid.* June 13, p. [illegible]. *Ry. Age*, June 7, 1890, p. 397.

On English Railroads. A summary of special investigations concerning several noticeable accidents; contains information regarding the efficiency of signals. *R. R. Gaz.*, March 20, 1891, pp. [illegible].

In Great Britain in the Year 1890. Summary of several special investigations as to cause. *R. R. Gaz.*, May 15, 1891, pp. [illegible], May 23, p. 349.

The Quincy Accident. Massachusetts Railroad Commissioners' Report. Rules of R. R. Cos. regarding use of track jacks, also brake equipment of Mass. R. Rs. *Eng. News*, Nov. 22, 1890, p. [illegible]. *R. R. Gaz.*, Nov. 21, 1890, pp. [illegible]. Editorial discussion *ibid.*, pp. [illegible]. *Ry. Rev.*, Nov. 22, 1890, pp. [illegible]. *Ry. Age*, Nov. 22, 1890, pp. [illegible].

Regulations for the Prevention of, in Great Britain. Synopsis of laws and practice of. 12 pp. in *An. Rept. R. R. Com. of Mass.*, 1889.

Statistics of the U. S. and Great Britain. Comparative tables. *Eng. News*, June 7, 1890, pp. 535-7.

Train Accidents—Their Nature and Causes for Sixteen Years. (Table). *R. R. Gaz.*, Feb. 15, 1884.

Railroad Accounts.

Classification of Accounts, etc. By G. Mordecai, before the American Society of Civil Engineers. Gives notes on the classification of railroad accounts and the analysis of railroad rates. *Trans. A. S. C. E.*, Vol. XVIII., Feb., 1888, pp. [illegible].

And Returns. By Wm. P. Shinn. Gives suggestions derived from experience, as to the proper requirements in keeping accounts and giving returns. With discussion. *Trans. A. S. C. E.*, Vol. V., p. 315.

Railroad Axles. See *Axles.*

Railroad Brakes. See *Brakes.*

Railroad Buildings. See *Railroad Stations. Railroad Structures.*

Railroad Clearance Diagram. See *Railroad Rolling Stock.*

Railroad Coaling Stations. See *Railroad Structures.*

Railroad Construction.

A map printed in colors showing the location of railroads built in 1886, is given in *Eng. News*, July 9, 1887.

Gives a list of all railroad lines in United States, Canada and Mexico, on which track was laid during 1887, with a two-page map printed in colors, showing track laid from beginning of 1886 to June 15, 1888. *Eng. News*, June 16, 1888. See also Tabulation in *R. R. Gaz.*, June 1 and 8, 1888.

Classification of Material in. Paper by A. M. Van Auken read before Western Soc. of Engrs., July 18, 1888, with discussion. *Jour. Assn. Eng. Soc.*, January, 1889.

Covered Way, Glasgow City and District Railroad. By W. S. Wilson, before the Institution of Civil Engineers. Gives details of the construction of a covered way of which 2,602 yards were in tunnel. *Proc. Inst. C. E.*, Vol. XCII., pp. 288-91; *Eng. & Build. Rec.*, June 23, 1888.

Cross Sectioning, A method of Angular. By R. Bell. Gives method which can be used to great advantage on high cliff and rough, irregular ground. *Van Nos. Eng. Mag.*, Vol. XIV., p. 793.

Cross Sections. Graphical Method of Determining Areas of Railways. By M. Dubret. *Van. Nos. Eng. Mag.*, Vol. XXVIII., p. 1.

Earthwork.

Calculation of Overhaul. A simple method of estimating overhaul by means of a profile of quantities, is described and illustrated by T. S. Russell. *Eng. News*, March 14, 1891, pp. 254-5. Other correspondence on this subject, *ibid*, p. 355.

Computing Areas from Cross-Section Notes. Examples of the application of the rectangular co-ordinate formula, and of Wellington's diagrams in getting solidities directly. *Eng. News*, April 18, 1891, pp. 364-6. Example of overhaul computation on heavy work, by a profile of quantities. By C. W. Staniford, *Id.*, p. 377.

Diagrams and tables for simple and compound formations. By Edward Thiange and J. M. Rudiger, together with graphical determination of haul. *Proc. Eng. Club Phila.*, Vol. IV., M. 2, pp. 67-92.

Diagram of Average Heights Corresponding to the True Prismoidal Content from a Table of Level Sections. By Wm. F. Shunk. *Eng. News*, Dec. 9, 1882.

Economic. Discussion on Gordon's. *Eng. News*, July 31, 1886.

Economical. By Robert Gordon. This is a review of the American as compared with the English practice, to discover the cause of the more economical operation of American lines. Fully illustrated.

Also, *Railway's in New Countries*, by J. R. Morse and *The Construction of the Can. Pac. Ry.*, by G. C. Cunningham, followed by discussion on the entire subject, the whole covering 122 pp. in *Proc. Inst. C. E.*, Vol. LXXXV., p. 54.

Economical. By Robert Gordon, before the Inst. Civ. Engrs., London. A paper based on the American practice as seen by an Englishman. Favors our methods for new countries. *Eng. News*, April 17, 1886. *Sci. Am. Sup.*, May 1, 1886.

Elements of Construction and Maintenance. By Charles Paine. Being a series of articles appearing in the *R. R. Gaz.* Nov. 7, Drainage. Nov. 14, Real Estate Records, etc. Nov. 21, Stations. Nov. 28, Main Track and Sidings. These articles are full of valuable practical suggestions. Republished in book form by the R. R. Gaz. Co.

Embankment to be submerged. Paper by Lawson B. Bidwell, M. Boston Soc. of

Railroad Construction, *Embankment, etc.*, continued.

C. E. Read April 17, 1889; pp. 7. *Jour. Assn. Eng. Soc.*, Aug., 1889 Vol VIII., p. 447.

Far Western States. Inset map and editorial. *Eng. News*, Nov. 18, 1890, pp. 463-4, 467.

Far Western States. Progress map and statistics. *Eng. News*, Nov. 28, 1891, pp. 508-9.

From Preliminary to Track. By M. P. Paret. A paper for young engineers, giving points on the ordinary methods and routine of field and office work on railroad construction. *Eng. News*, Aug. 11, 1888.

Harlem Depression Works on the New York Central. Description of this extensive work, with illustrations of overhead station. *R. R. Gaz.*, July 3, 1891, pp. 460-1.

Method of Reducing the Cost of. By Alfred F. Sears. Advocates the building of narrow road-bed, leaving the sides of the cut untrimmed, etc. Also contains a description of the iron caisson of the pivot pier of the Hackensack draw-bridge. *Trans. A. S. C. E.*, Vol. II., p. 1.

In Mexico. See *Mexico.*

Mountain Railway Construction. Abstract of a paper by W. B. Parsons, read before the Am. Soc. C. E., describing the construction of a light lumber road giving cost, etc. *Eng. News*, May 30, 1891, p. 515. *R. R. Gaz.*, June 5, 1891, p. 384.

New Construction, nine months, table of. *R. R. Gaz.*, Nov. 9, 1888.

New York, West Shore & Buffalo Ry., and Methods Used in Its Construction. Illustrated paper by Peter Chalmers before the Inst. C. E. *Eng. News*, Sept. 27, Oct. 11, 1884.

Northwestern States. Inset map with accompanying notes and editorial. *Eng. News*, Aug. 2, 1890, pp. 95-6, 102.

Northwest. Statistics and progress map for 1890-91. *Eng. News*, Aug. 1, 1891, pp. 95-6, 102-3.

Notes on. By Theo. Low. Gives hints which may be of use to young assistant engineers on construction work. *Proc. Eng. Club, Phila.*, Vol. V., pp. 136-222 (Feb., 1886.)

Notes on the Construction and Equipment of Narrow Gauge. By A. I. Reed. Contains notes on construction and estimation of cost. *Van Nos. Eng. Mag.*, Dec., 1884.

Over the Raton Mts. By Jas. D Burr. Illustrated by maps and profiles of the Atchison, Topeka & Santa Fe Railroad. Also the performances of locomotives on the steep grades. *Trans. A. S. C. E.*, Vol. VIII., 1879, pp. 293-340.

Rapid Track-Laying in Texas. Description of organization and method of work for laying at the rate of two miles per day. By W. M. Johnson, of the Texas Pacific. *Eng. News*, Jan 21, 1882, pp. 17-19.

Regulations for the *Engineering Department* of the South Pennsylvania Railroad during construction. *Eng. News*, Jan. 15, 22, etc., 1887.

Second Track. By H. C. Thompson, before the Civil Engineers' Club of Cleveland. Discusses the question of building an additional track to a single track railroad already in operation. *Jour. Assn. Eng. Soc.*, April, 1888.

Side Leveler for Surfacing Track—Michigan Central R. R. Description and drawings of a machine built on a flat car, for leveling or shaping embankments for side tracks or other purposes. *Ry. Rev.*, May 2, 1891, pp. 278-9.

Slope Stake Setter and Universal Slope Indicator. An instrument for finding position of slope stakes, and to aid in calculating cross-sections for preliminary estimates. Description taken from *The Compass.* *Ry Rev.*, Sept. 12, 1891, p. 601. *Eng. News*, Sept. 12, 1891, p. 231.

Railroad Construction, continued.

Southern States. Progress map and statistics. *Eng. News*, June 13, 1891, pp. 569, 570-2.

Southern States. 2,810 miles of road in process of construction. Inset map. *Eng. News*, June 14, 1890, pp. 549-61, 565.

Southwestern States. Progress map and statistics. *Eng. News*, Oct. 3, 1891, pp. 306-7.

Specifications and Contract. Full specifications. *Eng. News*, Aug. 25, 1883, pp. 399-401.

States East of Chicago. Inset map with statistics of progress of construction for the year. Also editorial review. *Eng. News*, Apr. 11, 1891, pp. 343-4, 349.

Statistics.

Statistics of railway construction east of Chicago with inset map. *Eng. News*, April 12, 1890, Vol. XXIII., pp. 343-348.

A map showing the railroads constructed, surveyed and projected from Jan. 1, 1887, to date, with text giving terminal point and length of lines in the States east of Chicago. *Eng. News*, Oct. 8, 1887.

Detailed tabular statement of the number of miles of railroad in construction from Jan. 1, to Oct. 1, 1889. *R. R. Gaz.*, Oct. 11, 1889.

During the first half of 1890, in the United States, Canada and Mexico. A table of 2 pages showing lines projected and under contract as well as new track. *R. R. Gaz.*, July 4, 1890, pp. 469-478. Also *id.*, July 18, 1890, pp. 512-3.

Of 1890. With inset map and editorial. *Eng. News*, Jan. 3, 1891, pp. 13-14, 16-17. *R. R. Gaz.*, Jan. 2, 1891, pp. 6-9.

Surveys and Construction, Notes on. By H. P. Bell. Some general notes on railroad work. *Van Nos. Eng. Mag.*, Vol. XXV., p. 11.

Texas Frontier. By G. W. Rafter, C. E. Interesting account of incidents in his personal experience. Illus. *Eng. Mag.*, Oct., 1891, pp. 29-41.

See *Earthwork.*

Railroad Cranes. See *Cranes.*

Railroad Crossings.

In Chicago. Map showing a very complicated network of crossings, with notes as to proposed interlocking systems. *R. R. Gaz.*, Apr. 24, 1891, pp. 282-3.

Chicago, Vicinity of Clark and Sixteenth Streets. Large inset plan of this crossing together with a brief reading notice. *Ry. Rev.*, April 12, 1890, Vol. XXX., p. 210.

The Fontaine Crossing. This is a continuous grade crossing, made so by inserting short sections of rails at the intersections, which may be set for either track. Illustrated description. *R. R. Gaz.*, June 19, 1891, p. 428.

Grade Crossings in Germany, Law and Practice in. Translated by Prof. Geo. F. Swain, and given with many cuts in *An. Rept. R. R. Com. of Mass.*, 1889.

Pneumatic Safety Gate. An illustrated description of an improved form of gates for railroad crossing, etc. *Ry. Rev.*, Jan. 8, 1887.

Railroad Culverts.

By E. A. Hill. Gives details of the building of a culvert for the drainage of about 1,600 acres of land; shows plan, cost, etc. *Rep. Ill. Soc. Eng. & Surv.*, 1885, pp. 46-49. *R. R. Gaz.*, May 15, 1885. A continuation of the above discussion by Mr. Hill in *R. R. Gaz.*, Nov. 2, 1885.

By Ed. A. Hill. A supplemental paper relating to the Nichols Hollow Culvert. *An. Rep. Ill. Soc. Eng. & Surv.*, 1889, p. 75.

A good example of an arched stone culvert for railroad embankment, with table of cost. Illustrated. *Eng. News*, June 8, 1889.

Box, Maximum Spans of, for various loads and thicknesses of stone. By Emil Low, C. E. *Eng. News*, July 25, 1885.

Railroad Culverts, continued.

Box Culverts, Canadian Pacific Railway. With detail drawings. *Eng. News*, Dec. 28, 1888.

Description of iron culverts, with full drawings, used in Iowa on the Chicago, Burlington & Quincy Railway. *R. R. Gaz.*, Feb. 25. 1887.

Box. Standard Timber box culverts as constructed by the Chicago, Milwaukee & St. Paul R. R. Co. Detailed drawing. *R. R. Gaz.*, May 14, 1886; *Eng. News*, May 22, 1886.

Pipe Culverts on the "Plant" System. Detail drawing of standard pipe culverts, with description of methods of construction. By W. B. W. Howe, Jr. *R. R. Gaz.*, April 24, 1891, pp. 283-4.

Standard Pipe. Gives sectional drawings of pipe culverts from twelve inches to two feet in diameter, in use on the Kansas City and Omaha R. R. *Eng. News*, Oct. 22, 1887.

Water-Way for. By A. N. Talbot. Discusses the determination of water-way for bridges and culverts, proposes a new formula. *Selected Papers C. E. Club, Univ. of Ill.*, 1887-8, pp. 14-22.

See *Bridges, Railroad*.

Railroad Curves.

And Crossings. By S. W. Robinson. Gives formulæ and tables for easement curves as adapted to field practice. *Van Nos. Eng. Mag.*, Vol. XXVII., pp. 56.

Circular, for Railroads. By Wm. M. Thornton. Gives a number of approximate formulas and a small collection of tables with the object of saving time over the usual logarithmic method. *Van Nos. Eng. Mag.*, Vol. XIX., p. 10.

Compensation for. By William R. Morley, with valuable discussion, giving both theory and practice on many roads. *Trans. A. S. C. E.*, Vol. XIII., p. 188.

Easement Curves. A valuable article by Mr. Willard Beahan, with table and illustrative example. *R R. Gaz.*, May 31, 1889.

Easement. By A. P. Man, Jr. Adjusts by offsets for 100 feet each side of the P. C. Tables and cuts given, with discussion. *Trans. A. S. C. E.*, Vol. XV., p. 319.

Easement. On the Manhattan Elevated. Drawing of some of the curves on the Manhattan Elevated Railroad, showing the introduction of short reversed curves to enable the turning of right angles within the street limits. *Eng. News*, April 2, 1887.

Economy of Curvature. By Wilson Crosby. Discusses a number of formulæ for determining the economy of different curves between two given tangents, deduces a new formula and gives a table to facilitate its use. *Trans. A. S. C. E.*, Vol. II., p. 323.

Elevation of Outer Rail. See *Railroad Track Elevation*.

Holbrook Spiral as a Transition Curve. Paper by E. T. Reisler, discussing this spiral. *Jour. Eng. Soc. of Lehigh Univ.* Feb., 1890. Vol. 4, No. 21, pp 57-71.

Resistance. A Theoretical Investigation. By S. Whinery, giving results agreeing remarkably well with experiment. With discussion. *Trans. A. S. C. E.*, Vol. VII. (1878), pp. 70-108. Also Vol. VIII., p. 179.

Sharp Curves in Yards. A valuable paper by Mr. Edwin A. Hill, read at Peoria, Jan. 30, 1890. Diagrams, tables and numerous formulas. *Rep. 5th An. Meeting, Ill. Soc. Eng. & Surv.*, 50 cts., pp. 96-176.

Spiral or Transition Curves for Street Railways, operated by Mechanical Motors. Paper by E. L. Woolley, read before the Am. St. Ry. Assn. *St. Ry. Gaz.*, Oct. 1890, p. 196. *St. Ry. Jour.*, Nov. 1890. *Eng. News*, Nov. 1, 1890, p. 401.

Spirals or Helices. A description of the only bridge in existence, and of another in process of construction, with a reference to certain tunnel spirals on

Railroad Curves, *Spirals or Helices*, continued.

the St. Gothard Railroad. Maps and alternative locations given. *R. R. Gaz.*, Nov. 27, 1885.

Taper Curves. By Frank Olmsted. Gives method of locating curves with tapering ends. *Eng. News*, July 21, 1888.

Transition Curves. An excellent mathematical discussion by Prof. W. H. Echols, Rolla, Mo. Pamphlet, p. 19.

By Wm. G. Raymond. Development of working equations and corresponding tables for cubic-parabola curves. Details of methods for setting out given. Probably the most convenient tables yet published. *Trans. Tech. Soc. Pac. Coast*, Vol. III., p. 33.

Development of formulas, and tables for a general transition curve, with discussion of various other forms. By Prof. A. N. Talbot, *Technograph* (Univ. of Ill. Annual,) 1890-91, pp. 77-103.

Formulas for a spiral curve developed in a very simple and direct manner. *Eng. News*, March 11, 1882, pp. 78-9.

From the German by E. A. Giesecler. The equation of the transition curve is developed, its characteristic and important features discussed and simple methods of laying out are given. *R. R. Gaz.*, Jan. 4, 1889.

Letters by P. H. Philbrick, C. E., and Franklin Riffle, C. E. Diagrams and two tables. *R. R. & Eng. Jour.*, June, 1890, pp. 267-8.

Letter from Prof. W. H. Echols, Missouri School of Mines. [The formulæ given are simple and useful.] *Eng. News*, May 4, 1889.

A letter from Mr. G. James Morrison, giving a very simple and quite satisfactory method of setting out transition curves. *R. R. Gaz.*, Sept. 27, 1889, Vol. XXI., p. 623.

Series of articles by Charles D. Jameson, C. E., discussing the whole subject of transition or adjustment curves. *R. R. & Eng. Jour.*, May, 1889, Vol. LXIII., p. 223, *et seq.*

Turnouts.

From Combination Tracks. Discussed by E. G. Tuttle. Illustrated. *Sch. Mines Quar.*, Jan., 1884.

Exact Formulæ for their Determination, together with practical and accurate tables for direct use in the field. By W. B. Parsons. *Eng. News*, July 19, Sept. 6, 1884.

Functions of. A graphical chart for determining functions of any turnout when ordinary stub switches are used. *R. R. & Eng. Jour.*, October, 1889, LXIII., p. 459.

Good working rules for the same. *R. R. Gaz.*, May 14, 1886.

Theory of Railway. By J. R. Stevens. An illustrated article giving full solution of double and single turn-out problems. *Van. Nos. Eng. Mag.*, Vol. XXIX., p. 89.

Railroad Earthwork. See *Railroad Construction. Earthwork.*

Railroad Engineers. See *Engineering. Alpine.*

Railroad Grades.

Approximate Value of a Reduction of Ruling or Maximum. By John G. Clark. Gives a method of obtaining approximately the value of the reduction of ruling or maximum grade of a railroad especially for the use of engineers on location. The elements considered are cost of repairs, deterioration and locomotive power. *Trans. A. S. C. E.*, Vol. II., p. 379.

Economics of Momentum Grades. With especial reference to Tributary and Suburban Railways. By Mr. Frank Cooper. *Proc. Eng. Club, Phila.*, Vol. VII., No. 2, Feb., 1890, p. 113.

Railroad Grades, continued.

The Mountain of the St. Gothard Railway. Gives profiles of the mountain grades approaching the St. Gothard Tunnel. The two grades have seven spiral tunnels turned into the mountain in order to gain length of line. *Eng. News*, April 16, 1887.

Railroad Improvements.

Modern Improvements of Facilities in Railway Traveling. Paper by George Findlay, Assoc. Inst. C. E., General Manager London and Northwestern Railway. *Jour. Soc. Arts*, Feb. 14, 1890, Vol. XXXVIII., pp. 264-303.

Railroad Labor.

United States Report on. Fifth Annual Report of Commissioner of Labor. Tables of rates, annual earnings, etc., are abstracted in *R. R. Gaz.*, Dec. 5, 1890, pp. 843-4.

Railroad Location.

A Manual of Railway Field Work and Laying out of Work. A treatise by Mr. A. M. Wellington, M. Am. Soc. C. E. A series of articles preparatory to publication in book form. *Eng. News*, Oct. 12, 1889. Vol. XXII., p. 343, *et seq.*

A Piece of Difficult Railway Location on the Knoxville Southern R. R. in Polk Co., Tenn. Very heavy work avoided by an elaborate system of spirals. Contour map given. *Eng. News*, May 9, 1891, pp. 453-4.

By Samuel McElroy. Discusses railroad locations as one of the most vital questions in railroad constructions. *R. R. Gaz.*, July 6, 1888.

By J. A. L. Waddell. A paper to young engineers on exploration, preliminary surveys, locations and constructions. *Van Nos. Eng. Mag.*, Vol. XXIV. p. 281.

The Descent from Boreas Summit to Breckenridge, Col., on the Denver & South Park Railway. Plans, maps, profiles and perspective view, with description, in *R. R. Gaz.*, Dec. 24, 1886.

The American Line from Vera Cruz to the City of Mexico. By A. M. Wellington. A paper read before the annual convention of the American Society of Civil Engineers, describing the location and partial construction of the above line, with notes on the best methods of surmounting high elevations by rail. Discusses switchbacks and spirals. Accompanied by maps and profiles. *Trans. A. S. C. E.*, Nov., 1880.

Field Practice in the West. By Willard Beahan. A valuable paper on the methods of location from the standpoint of a chief of a locating party. *Jour. Assn. Eng. Soc.*, June, 1886. Vol. VII., pp. 196-201. *R. R. Gaz.*, May 11, 1888; *Sci. Am. Sup.*, June 23, 1888.

Location of the Northern Pacific Across the Rocky Mountains. By J. C. Chesbrough. A good illustration of the benefits of a good contour map. Accompanied by a large contour map of the region, with the many trial lines given. *Jour. Assn. Eng. Soc.*, Vol. III., p. 133.

Ruling Gradient. By E. Holbrook. Discusses how to determine the best gradient for a railroad. *Sci. Am. Sup.*, July 21, 1888; *R. R. Gaz.*, July 27, 1888.

Surveys. Trunk line field work on a preliminary survey for location. By Saml. McElroy. Gives field instructions to parties, and notes on the same. Contains many valuable hints. *Jour. Assn. Eng. Soc.* Vol. I., p. 157.

Western Division of the Canadian Pacific Ry. By F. P. Davis. A 15-page article in *Proc. Mich. Eng. Soc.*, 1885.

Topography in R. R. Location. By Arch. A. Schenck. A very sensible article on the use of topography in location, the writer favoring the use of slopes and distances on cross-sections in preference to the use of contours. He does not indicate how his contours were determined, however. *Eng. Era*, Jan. 14, 1886.

Railroad Maintenance of Way.

A paper by Wm. Greenhill, read before the Inst. C. E. in Ireland, giving considerable practical information relating especially to Irish Roads. *Ry. Eng.*, May, 1891, pp. 116-19.

By W. G. Curtis, Supt. Track, etc., Southern Pacific. *R. R. Gaz.*, Nov. 16, 1888.

Best method of fastening rails to wooden ties. European experience. Abstract of a paper prepared for the International Railway Congress. *Ry. Rev.*, Sept. 28, 1889; *R. R. Gaz.*, Oct. 4, 1889.

Cost of Maintenance of Prussian State Railroads. By Wm. Nordling. Shows by statistics of Austrian and Prussian railroads the relation between cost of maintenance and total traffic. *Archiv. fur Eisenbahnwesen*, 1886, pp. 45-79.

Drainage of a German Railway Embankment. A valuable and suggestive article. Illus. *Eng. News*, May 10, 1890. Vol. XXIII., p. 436.

Draining the Terraces in Cajon Pass. General view showing method of draining the slope of a 120 ft. cut by means of wooden terraces. *Ry. Rev.*, Sept. 19, 1891, p. 615.

Economics of. Prof. S. W. Robinson, Inspector of Railroads for the State of Ohio. Gives some results of observation on bridges, roadways, etc. Illus. *Trans. A. S. M. E.*, Vol. II., p. 524.

East Indies. Treats of turning and changing rails; expansion, creeping of rails on double-track roads, squaring joints, lifting road, packing sleepers, etc. *Eng. News*, Aug. 14, 1886.

Fifteen-ton derrick car on Pennsylvania R. R. See *Derrick Car*.

Land Slides, Treatment of, on the Illawarra Ry., (N. S. W.) Abstract of a paper by W. Shellshear, M. I. C. E., giving an account of a very serious case of land slides, which was effectively treated by constructing a thorough system of drainage. *Ry. Rev.*, April 11, 1891, p. 238.

Report on Use of Metal Track on Railways as a Substitute for Wooden Ties. By E. E. R. Tratman, C. E., Forestry Division, Dept. of Agr., *Bulletin No. 4*. Washington, 1889.

Some Methods of Meeting the Ordinary Requirements of Railroad Maintenance. A lecture by A. J. Swift, before the students of the Rensselaer Polytechnic Inst., discussing the efficiency of various forms of waterways, describing best methods of construction and illustrating special details. *Eng. News*, May 30, 1891, pp. 503-4, *et seq.*

See *Railroad Construction, Elements of Construction and Maintenance, Washouts.*

Railroad Management.

A "symposium" on railroad problems in the United States, is published in the *N. Y. Independent*, last week of Aug., 1890. J. P. Meany, editor of *Poor's Manual*, has a paper on history and statistics. The state control or ownership of railroads discussed by Hon. C. M. Clay, Ex-Gov. Larrabee, Prof. Richard T. Ely and others.

Discriminations from the use of Private Cars of Shippers. A paper by Hon. Aug. Schoonmaker read before the 5th annual convention of the U. S. R. R. Commissioners, giving some results of investigations, showing enormous earnings of private cars from mileage. *Ry. Rev.*, March 14, 1891, pp. 169-70. *R. R. Gaz.*, March 20, 1891, p. 197.

Free Railway Construction vs. Governmental Railways. Abstract of a paper by E. Bates Dorsey, read before the Am. Soc. C. E., discussing the question as regards rates and operating expenses. *Eng. News*, May 30, 1891, pp. 515-6. *Ry. Rev.*, May 23, 1891.

Legislative Control of. Paper by Edwin F. Woodman, Secretary C., St. P., Minn. & Omaha Ry. Read before the St. Paul Society, April 1, 1889. *Jour. Assn. Eng. Soc.*, May, 1889, Vol. VIII., p. 304.

Railroad Management, continued.

Managers and Employes, Relations of. Discussed by Dr. W. T. Barnard in *Pop. Sci. Monthly* for Sept., *et seq.*, 1885.

Operating Departments and Management of. By "Superintendent." A series of articles discussing the duties of a General Superintendent who has charge of transportation, roadway, and motive power departments. *R. R. Gaz.*, Dec. 4, 1885, *et seq.*

Our Railways and their Economical Use. A series of articles discussing the relations between railways and the industries of the country. *Eng. News*, Aug. 18, Sept. 21, 1877.

Premium to Employes. Translation of paper submitted to the International R. R. Congress at Paris by M. Bela Ambrosovics, of the Hungarian State Railroads. *R. R. & Eng. Jour.*, Nov., 1890.

See *Railroads, Gov't Control of.*

Railroad Maps.

A map showing an interesting piece of work where the *Western North Carolina* crosses the Blue Ridge Mountains, 18 miles from Asheville. *R. R. Gaz.*, March 4, 1887; also *Sci. Am. Sup.*, March 26, 1887.

Being letters from many engineers on this subject in *R. R. Gaz.*, Nov. 7, 14, and Dec. 5, 1884.

See *Maps.*

Railroad Operating Expenses.

A table showing the ratios of the items of operating expenses to the total operating expenses. Deduced from the annual report of the President of the Union Pacific Railroad. *R. R. Gaz.*, June 3, 1887.

Constant and variable operating expenses. By Lymphor. With a note by W. v. Nordling. *Archiv. f. Eisenbahnwesen*, Jan. and Feb., 1887, pp. 67–89.

Gives a table of the percentage of various items of expense to the total operating expense. The table is made from the average of two years' practice on a road operating over 3,000 miles of road. *R. R. Gaz.*, May 6, 1887.

English and American Operating Expenses Compared. A paper read before A. S. C. E. meeting Dec. 5, 1888, by Mr. Edward Bates Dorsey. Gives data showing relative cost of hauling one ton one mile on the two systems. *R. R. Gaz.*, Dec. 14, 1888; also editorial comments in same paper.

English and American Railroads Compared in Operating Expenses. By Edward Bates Dorsey, Mem. Am. Soc. C. E. *Trans. A. S. C. E.*, April, 1889, Vol. XX., p. 1.

The Attempted Division of Operating Expenses between Passenger and Freight Traffic. Address by G. L. Lansing before Assn. of Ry. Accounting Officers, at Niagara, July 11, 1889. *Ry. Rev.*, Aug. 17, 1889.

Train Dispatching. The Principles of Train Dispatching. Extracts from a paper by J. A. Anderson. *R. R. Gaz.*, Sept. 25, 1891, pp. 668–9.

Railroad Operation.

A Heavy Train. Brief editorial notice of a train of 85 cars weighing about 2,400 tons hauled by one engine. *R. R. Gaz.*, April 11, 1890, p. 250.

Modern Facilities. A paper before the Frank. Inst. By Wm. B. Le Van. With especial attention given to high-speed trains in various countries. *Jour. Frank. Inst.*, April, 1885.

Scientific. By Gen. J. H. Wilson. Reviews the present position of railroading as a science and enters a plea for a good railroad school. *R. R. Gaz.*, Dec. 30, 1887.

Speed.

An Unprecedented Run; 216.3 Miles in 229 Minutes. A time card of the Mich. Cent. R. R., Canada Division, gives the complete record of this extraordinary run. *Ry. Age*, Aug. 30, 1890, p. 615.

Railroad Operation, continued.

On Future Trunk Freight Lines. By Russell Sage. Discusses the number of engines, cars and employes that will have to be employed under various conditions of train speed. *Van Nos. Eng. Mag.*, Vol. VIII., p. 460.

An answer to some criticisms on the above article. *Van Nos. Eng. Mag.*, Vol. IX., p. 7.

On the most economical speed for freight trains, and their relations between the cost of maintenance, and the speed of trains, the grades and curves, and the amount of traffic. By Prof. Frank. *Organ fur die Fortschritte des Eisenbahnwesens* 1884, p. 165.

Stopping and Starting Trains. By E. W. Hyde. Compares the expense of stopping a moving train, and thus bringing it back to the same speed with that of maintaining the given velocity unchanged. *Van Nos. Eng. Mag.*, Vol. 16, p. 490.

Working Single Track. Abstracted from a paper by W. K. Muir before the Montreal meeting of the British Assoc. *R. R. Gaz.*, Feb. 4, 1887.

Working of *steep gradients* and sharp curves on railways, by H. W. Tyler, before the Inst. of Civ. Engrs. *Van Nos. Eng. Mag.*, Vol. XVI., p. 167.

See *Railroad Management. Railroad Maintenance. Railroad Sidings.*

Railroad Rates or Tariffs,

A Contribution to the Theory of Railway Rates. Extended abstract of an article by F. W. Taussig giving an analysis of this subject. *Ry. Rev.*, July 18, 1891, pp. 464-5.

In France. Statement of rates, with a discussion of the subject and comparison with German rates. *Archiv fur Eisenbahnwesen*, 1885, Heft 5, pp. 525-556.

And Passenger Rates, Study of, as they exist throughout the Empire of Austria-Hungary. *Annales des P. & C.*, 1884-2-549

Influence of the Rates of Toll on the Utility of Transportation Routes. An analysis of the laws concerned. From "Transports et Tarifs," by C. Colson. *R. R. Gaz.*, Sept. 4, 1891, pp. 611-13.

Passenger Tariffs. An account of an extensive and somewhat remarkable change in passenger tariffs recently put in operation on the Hungarian State Railroads. *R. R. Gaz.*, Oct. 25, 1889. Vol. XXI., p. 700.

Policy and Rates in France. By Herr Ulrich. Interesting account of discussion on this subject in the French Chamber. *Archiv fur Eisenbahnwesen*, Nov. and Dec., 1886, pp. 725-48.

Railroad Rolling Stock,

Clearance Diagram for. A letter by Archibald A. Schenck, with diagram giving clearance of rolling stock by several prominent makers. *R. R. Gaz.*, June 21, 1889.

Flexible Wheel Bases. A paper by J. D. Twinberrow, before Students' Section of Inst. C. E., discussing the relative merits of rigid and flexible wheel bases in rolling stock. pp. 13. *Proc. Inst. C. E.*, Vol. XCVIII., p. 392. 1889.

Of Italian Railroads. By S. Fadda. Gives a very complete description of engines and cars. With many cuts. *Trans. Inst. C. E.*, Vol. LXXXIII., p. 391.

Locomotives and Cars. Estrade's New Patterns. All wheels 8½ ft. diameter for both locomotive and cars. Speed 75 miles per hour. Now finished at Paris. Cuts given. *Sci. Am. Sup.*, Sept. 11, 1886.

Rolling Stock and Tramways, Guinness Brewery. By S. Geogeghan, before the Institute of Mechanical Engineers. Gives a full description, with detailed drawing, of the rolling stock and tramways at a brewery in Dublin. *Lon. Engineer.* Aug. 31, 1888.

Rolling Stock, Uniformity in. By O. Chanute. A paper of 23 pp., describing the writer's experience in introducing uniformity in the rolling stock of the Erie road so that all parts were interchangeable. Illustrated. *Trans. A. S. C. E.*, Vol. XI., (1881), p. 2.1.

See *Journals and Journal Bearings.*

Railway Safety Appliances. See *Interstate Commerce Commission.* *Railroad Signals.*

Railroad Shops.

The Cheyenne Shops. Union Pacific Railway. Interesting feature is the 40-ton electric travelling crane. Description and illustration. *Mast. Mech.*, Nov., 1890, pp. 184-8. Description of the crane in *Ry. Rev.*, Nov. 8, 1890, p. 673.

The new Switching Yards and Shops of the Union Pacific at Denver. Plan with description of buildings, etc. *R. R. Gaz.*, April 3, 1891, pp. 228-9.

The Design and Construction of Railway Shops. A paper by Mr. J. Davis Barnett, Assistant Mechanical Supt. Grand Trunk Railway. From Transactions of the Canadian Society of Civil Engineers. *Eng. News*, Oct. 19, 1889. Vol. XXII., p. 363, *et seq.*

Locomotive and Car Shops, C., St. P. & K. C. R. R. Gives general and detailed plans of the locomotive and car shops of the Chicago, St. Paul & Kansas City Railroad at St. Paul. *R. R. Gaz.*, Feb. 10, 1888.

And Machinery. The "Arrangements of Shops and Machinery for the Construction and Repair of Railroad Rolling Stock." Discussion at the meeting of the New Eng. R. R. Club, Dec. 12, 1888. Also editorial notes on same. *Mast. Mech.*, Jan., 1889; *R. R. Gaz.*, Dec. 21, 1888.

New Nashville shops of the Nashville, Chattanooga and St. Louis R. R. General description with illustrations showing general plan, section and other details of round house. *Ry. Rev.*, March 14, 1891, pp. 165-6. March 21, pp. 182-3.

Piece Work in. See *Piece Work.*

Railroad Sidings.

Edge Hill, Liverpool. A full description, with plan, of the grid-iron siding of the London & Northwestern R. R. *R. R. Gaz.*, April 15, 1887.

Lap Sidings on the Cleveland & Pittsburgh. Description of their use on single track road. By L. F. Loree. *R. R. Gaz.*, Dec. 26, 1890, pp. 898-9.

See *Railroad Shops and Switching Yards.* *Railroad Stations.*

Railroad Signals.

An abstract from the Eleventh Annual Report of the Commissioners of Mass. Gives a brief description of a number of systems of signals. *Van Nos. Eng. Mag.*, Vol. 22, p. 437.

By Wm. H. Decbaut. A practical test of operating a distant signal by a wire run through a pipe filled with air. *Proc. Eng. Club, Phila.*, April, 1886.

American Practice in Block Signaling. The first of a series of Papers giving "a brief and simple, yet full, description of the methods of signaling now in use in this country." *R. R. Gaz.*, May 2, 1890, Vol. XXII., p. 309, *et seq.*

In Austria-Hungary. Extract from report of MM. Brawe and Weiss. *Annales des P. & C.*, 1885, pp. 1038-1144.

Automatic.

Gives description of the system employed on the Boston & Albany R. R. *R. R. Gaz.*, June 24, 1887.

A paper by G. W. Blodgett, stating the principles of automatic signaling and describing several recent electric systems. *Proc. Soc. of Arts*, 1890-91, pp. 150-6.

Gives a description of the automatic signals in use on the Fitchburg Railroad, Mass. *R. R. Gaz.*, June 15, 1888.

Block Signals on the Boston and Albany. Brief description of the method of signaling that line from Boston to Springfield. *R. R. Gaz.*, May 9, 1890, p. 321.

System Aubine. Block-signal, for use at yard limits and at entrance to stations. *Annales des P. & C.*, Nov., 1882.

Railroad Signals, continued.

Block System.

On the Baltimore & Ohio. Extracts from the rules applying to single track sections. *R. R. Gaz.*, Oct. 30, 1891, p. 763.

On the Erie. Description with code of rules, of the block system in operation on a 93 mile division of the N. Y. Pa. & O., (single track road). *R. R. Gaz.*, Oct. 9, 1891, pp. 699-700.

Problem. A Solution of the. An article by Mr. H. Ward Leonard describing an ingenious electric system of signals operated by generators on the locomotives. *Elec. World*, Oct. 10, 1891, *Elect. Eng.*, Oct. 7, 1891.

Some Details in the Manipulation of. Paper by Mr. W. W. Nichols of the C. B. & Q. R. R., read at the convention of the Telegraph Superintendents Association. *Ry. Rev.*, July 5, 1890, p. 388. Abstract in *R. R. Gaz.*, July 4, 1890, pp. 469, 478.

Die Hydraulische Central-Weiche von Bianchi and Servetlag [sic]. Plate with 10 figures. *Wochenschrift des österr. Ing. u. Arch.-V.*, July 11, 1890, pp. 254-5.

Electrical.

Apparatus. Report of Board of Examiners on the exhibits at the Philadelphia Exhibition. Three systems reported on, being those of the Union Switch and Signal Co., the Hall System and the Putnam System. All these most clearly and fully described, with many cuts. Supplement to the *Jour. Frank. Inst.*, Jan., 1886.

Block Signals. Notes from a lecture by Geo. W. Blodgett on the new Hall signal system, delivered before the Mass. Inst. of Tech. *Ry. Rev.*, July 12, 1890, p. 406.

Conveniences at the Central Railroad Station, Jersey City, N. J. Description and diagram of Annunciator and Bell Circuits. *R. R. Gaz.*, Jan. 2, 1891, p. 3.

Electricity and Otherwise. Two papers by Geo. W. Blodgett and G. R. Hardy, Boston. Essential conditions discussed. *Jour. Assn. Eng. Soc.*, Vol. V., pp. 431 and 437.

Electro-Pneumatic Block Signaling. A description of several systems of signaling on railroads. *R. R. Gaz.*, June 6, 1890, pp. 393-4.

Semaphore Signal, Long's. An illustrated description. *R. R. Gaz.*, Nov. 30, and *Eng. News*, Dec. 1, 1888.

In England. By A. M. Thompson. With discussions, covering 63 pp. in *Proc. Inst. C. E.*, Vol. LXXXII., p. 166.

Evolution of Railroad Signaling. Paper by C. Herschel Koyl, before the Franklin Institute, giving a short popular review of the various forms of signals which have at various times been used. *Jour. Frank. Inst.*, Jan., 1890, Vol. XCIX., p. 36.

Fourth Avenue Tunnel, New York. Specifications and requirements for the new signaling apparatus. *R. R. Gaz.*, Aug. 21, 1891, pp. 575-6. *Eng. News*, Aug. 22, 1891, pp. 156-7.

In France. Concluding article of a long series. Plates. *Wochenschrift des österr. Ing. u. Arch. V.* (Vienna), July 25, 1890, No. 30, pp. 264-7.

Hall Block Signal on the N. Y. Central. Operation on an 8 mile section. Illustrated and described. *R. R. Gaz.*, Dec. 5, 1890, pp. 834-5.

Interlocking Systems. See *Railroad Signals, Interlocking System, Below.*

Kelsey Gong Signal. An auxiliary signal which is sounded by a train passing a signal at danger. Illustrated description. *Eng. News*, Feb. [illegible], 1891, p. 194.

Koyl Parabolic Semaphore. Illustrated description of a semaphore signal, which is claimed to present the same appearance, as to color and position, at night as during the day. *Ry. Rev.*, Oct. 5, 1889, Vol. XXIX., p. 578.

Railroad Signals, continued.

National Switch & Signaling Company's Torpedo-Placing Machine. Automatic machine used in connection with semaphores. Description with details fully illustrated. *Eng. News*, May 23, 1891, pp. 490-1.

Requirements to be Complied With in the Construction of, etc., in the State of Illinois. General rules. *Jour. Assn. Eng. Soc.*, Nov., 1890, pp. 515-6.

Switch Signal. A brief description, with plan, elevation and cross-section of a standard distant semaphore signal, being introduced on the Pennsylvania Railroad. *R. R. Gaz.*, Oct. 28, 1887.

Sykes Block Signal Apparatus. Illustrated description. *R. R. Gaz.*, Oct. 3, 1890, pp. 675-7.

Train Signal. The Mason Train Signal and Engineers' Valve. Description of a successful pneumatic signal which uses the air brake pipes for operation. Illus. *Ry. Rev.*, May 2, 1891, pp. 280-1.

Train Signals, the new system of, recommended by the General Time Convention and about to be adopted on most American railroads. *R. R. Gaz.*, Oct. 31, 1884.

Treatise on. Second edition, 1884, (first edition in 1868). By Brame & Aguillon. Standard French work. One vol., 8vo., 43 plates in atlas. Price 30 francs.

Railroad Signals, Interlocking System.

A description of the various devices in use in America. An article of some assistance in deciding upon a system for any given case. Illustrated. By E. A. Hill, Engr. H. & H. Ry., in *Trans. Conn. Assn. Eng. & Surv.*, 1885; D. C. Sanford, Secy., New Haven.

Account of a very complete and interesting system of interlocking signals at the Ash street crossing at Chicago, with rules and regulations for same. *Ry. Age*, Nov. 22, 1889.

A paper by Isham Randolph read before the Western Soc. Eng'rs, describing the Randolph Interlocking Machine. Illustrated. *Jour. Assn. Eng. Soc.*, Nov., 1890, pp. 537-44.

Paper read by Mr. Chas. Hansel at Peoria, Jan. 30, 1890. *Rep. of the 5th An. Meeting, Ill. Soc. of Eng. & Surv.*, (50 cts.) pp. 66-74.

B. & A. Ry., at Boston. With cuts, showing arrangement of track systems. By Geo. R. Hardy. *Jour. Assn. Eng. Soc.*, Vol. IV., p. 35.

And Block Signaling in Great Britain. Editorial containing comparative statistics. *Eng. News*, July 26, 1890, p. 62.

C. W. I. & Belt Ry. at Chicago. By Isham Randolph. Work done by hand-power applied to levers, etc. Illustrated. *Jour. Assn. Eng. Soc.*, Vol. IV., p. 20.

Draw-Bridge. Illustrated description of a system of interlocking in which an unusual number of connections are used, being operated from the bridge tower, New York, Pennsylvania and Ohio Railway, Cleveland, O. *R. R. Gaz.*, Jan. 10, 1890, Vol. XXII., p. 20.

Draw-Bridge Interlocking Plant at the Bridgeport Draw-Bridge in Chicago. Designed and constructed by the Union Switch and Signal Company. General description, detail illustrations, etc., are given. *Ry. Rev.*, April 5, 1890, Vol. XXX., p. 190.

Grand Central Depot, New York. Illustrated. *R. R. Gaz.*, March 13, 1885.

A description of general characteristics of this interlocking system with illustrations of signal tower, etc. *Sci. Am.*, Feb. 15, 1890, p. 97.

Hall System. Descriptive catalogue with beautiful engravings and letter press. Pamph. Folio, p. 53. Hall Signal Co., 50 Broadway, New York. Extended abstract in *Eng. News*, Sept. 13, 20, 1890. *Ry. Rev.*, Sept. 27, 1890.

Jersey City Terminals of the N. Y. L. E. & W. R. R. Plan showing signals and manner of operating from the three signal towers, with brief description. *Eng. News*, May 23, 1891, p. 486. *R. R. Gaz.*, May 22, 1891, p. 348.

Railroad Signals, *Interlocking System*, continued.

Jersey City Terminus of the Pennsylvania Railroad. Description of the plant and its operation, with diagram of the electric switch valves of the yards. By H. M. Sperry. *R. R. Gaz.*, July 31, 1891, pp. 523-5.

L. I. R. R. Gives a brief description with illustrated details of the interlocking apparatus on the Long Island Railroad. *R. R. Gaz.*, Feb. 10, 1888.

Parsons Block and Switch System and Continuous Rail Frog. Complete description of. Illustrated. *Ry. Rev.*, Aug. 17, 1889. *R. R. Gaz.*, August 16, 1889.

Progress in Interlocking Apparatus. An illustrated paper reviewing the early history of interlocking and describing a new improved machine. *Eng. News*, Aug. 23, 1890, pp. 158-9.

Semaphores and Interlocking. Abstract of a paper by C. H. Hammond before the October meeting of Am. Soc. Ry. Sup's. *Ry. Rev.*, October 18, 1890, pp. 613-4.

St. Louis Bridge and Tunnel Railroad. A paper read before the Engineers, Club of St. Louis by N. W. Eayrs, C. E. Illustrated. *R. R. Gaz.*, February 8, 1889.

Switches and Signals. By Charles R. Johnson. A series of papers showing the progress made in the use of interlocking switches and signals, and the modifications in practice. *R. R. Gaz.*, May 4, *et seq.*, 1888.

Switches and Signals. Tables and an editorial showing the extent to which they are now used on the railroads of the United States. *R. R. Gaz.*, Nov. 30, 1883.

Switch Signal Station. The Elizabethport. An illustrated and very full description of the modern switch and signal plant recently put in by the Union Switch & Signal Co. *Eng. News*, Nov. 24, 1888.

Westinghouse. System of Pneumatic Interlocking. Fully illustrated and described. *R. R. Gaz.*, Dec. 21, 1888.

See *Railroad Signals above*.

Railroad Snow Plows See *Snow Plows*.

Railroad Spikes.

Adhesion of Railroad Spikes and Lag-screws in Ties. Results of tests at the Watertown Arsenal. Detailed account of several tests upon ties of various woods, with diagrams. By James E. Howard. *R. R. Gaz.*, May 1, 1891, pp. 298-9.

And Lag Screws, Strength of. Paper by Mr. A. J. Cox in *The Transit*, giving results of tests of the holding power of various spikes and lag screws. Abstract: *Ry. Rev.*, Oct. 31, 1891, pp. 706-7.

Railroad Stations.

Boston. Revere Beach and Lynn R. R. at Boston. Design by C. A. Hammond, C. E. Description and plans of the first and second floors. *R. R. Gaz.*, Aug. 1, 1890, pp. 537-8.

Freight Station, Bishopsgate, in London. Showing arrangement of freight depot built in the interior of a large city, and connecting with several of the underground railways of London. *Zeitschr. f. Bauwesen*, 1884-311.

Grand Central Station in Chicago, Illustrated Description of the. *Ry. Rev.*, Dec. 20, 1890, pp. 763-6.

Hartford, Conn., Asylum Street Crossing. Brief description of the elevation of the tracks of two railroads, and arrangement of station facilities. By I. B. Bidwell before the Boston Soc. C. E. *Jour. Assn. Eng. Soc.*, November, 1890, pp. 530-3.

Indianapolis Union Station. Illustrated description of this new and very complete and satisfactory Union station. *R. R. Gaz.*, January 31, 1890, Vol. XXIII, p. 69.

Railroad Stations, continued.

Lighting of Stations. Abstract from a report to the International Railroad Congress, Milan Session, 1887. Discusses the use of gas and electricity. *R. R. Gaz.*, Jan. 6, 1888.

Newark, N. Y., New Passenger Station; Pennsylvania Railroad. Illustrated description of this new station and of the subway under the tracks by which passengers may pass from one side to the other. *Eng. News*, Feb. 14, 1891. pp. 146-7.

Springfield, Mass. Union Passenger Station. Account of this new station together with plan of improvements and floor plan of Lyman street building and bird's eye view. *R. R. Gaz.*, March 14, 1890, Vol. XXII., p. 174.

Standard "Overtrack" Stations. Harlem Depression, New York Central and Hudson River R. R. Novel design for stations reviewed, with details of construction and specifications for station at Melrose, soon to be constructed for. Illustrated. *Eng. News*, Feb. 8, 1890. Vol. XXIII., p. 126.

St. Lazare Station, Paris. A complete description of this station of the Western Railway of France; including details and dimensions of the elaborate hydraulic machinery in use there for moving both freight and cars. Illustrated. *Ry. Rev.*, July 27, 1889. Lon. *Eng.*, July 12, 1889, *et seq.*

St. Louis. The Union Passenger Station. Illustrated description of this proposed new passenger station. *R. R. Gaz.*, July 24, 1891, pp. 508-9.

St. Louis Union Passenger Station. General description of accepted plan, with illustrations. *Eng. News*, Oct. 3, 1891. pp. 300-1.

Railroad Stations and Yards.

Of the Chicago and South Side Rapid Transit R. R. Illustrated description giving plan, elevation and cross-section of station. *R. R. Gaz.*, April 11, 1890, Vol. XXIII., p. 244.

Design of. Folding plate and table of standards of English and Scotch railways. Paper by R. M. Parkinson. *Proc. Inst. C. E.*, Vol. CII., 1890. Paper No. 2440, pp. 249-70. Abstract, *Eng. News*, Nov. 29, 1890, pp. 476-9.

On the London and South Western Railway. By E. Andrews, M. I. C. E. Two folding plates. Lon. *Engineer*, Sept., 1890. No. 128. The general elevations and plans are reprinted in *R. R. Gaz.*, Sept. 19, 1890, pp. 646-7.

See *Railroad Terminals.*

Railroad Statistics.

Abstract from a part of the annual report of the Interstate Commerce Commission. *Eng. News*, Sept. 27, 1890, p. 275.

American Railroad Statistics. A paper read before the Am. Statistical Assn. by Prof. Arthur T. Hadley. *R. R. Gaz.*, Aug. 2, 1889, *et seq.*

Of Austria-Hungary in 1883 (statistics entirely). *Archiv. fur Eisenbahnwesen*, 1886, pp. 503-512.

Austrian-Hungarian. Gives general statistics, details of construction, operation and rolling stock. Illus. *R. R. Gaz.*, Aug. 19, 1887.

Comparisons of the mileage, population and area of the U. S. and other countries. *R. R. Gaz.*, July 18, 1890, pp. 511, 513.

German, French, Swedish and Norwegian Railroads. *Archiv. fur Eisenbahnwesen*, 1886, pp. 81-90.

For 1889 from Poor's Manual, 1890. *R. R. Gaz.*, Aug. 1, 1890, pp. 540, 542. *Ry. Rev.* Aug. 2, 1890, pp. 447-8. *Ry. World*, Aug. 2, 1890, pp. 726-7. *Eng. News*, Aug. 2, 1890, pp. 97-8. *Ry. Age*, Aug. 2, 1890, p. 543.

Operations in the United States for 1890. From "Poor's Manual." *Eng. News*, Aug. 8, 1891, pp. 119-21.

Regarding Traffic, etc., on the railroads of *England, Germany, Italy and Russia*. *Archiv. fur Eisenbahnwesen*, 1886, pp. 637-704.

Railroad Statistics, continued.

Report of the Railroad Commissioner of Rhode Island. 1890. Contains considerable matter relating to grade crossings; also annual returns, from the railroads in the state, and general R. R. statistics.

Russia. Detailed statistics from the Government report for 1885. On Jan. 1, 1884, there were 25,090 km. (about 15,680 miles) in operation. *Archiv. fur Eisenbahnwesen*, 1886, pp. 353-364.

Saxon Railroads, for 1884 of the. St. Gothard road, and other roads. *Archiv. fur Eisenbahnwesen*, 1885, pp. 272, 258.

Scandinavia. Statistics of railroads in Sweden, Norway and Denmark. *Archiv. fur Eisenbahnwesen*, 1885, Heft. 6, pp. 671-681.

Transportation on Railways in the New England States. Statistics of mileage, earnings and income, operating expenses, etc., for the decade 1880-90. *Census Bulletin No. 40*, being the first of a series of ten to be issued. By Henry C Adams. *Ry. Rev.*, April 25, 1891, pp. 260-2.

United States in 1888. Statistics of value with discussion of same, from advance sheets of Poor's Manual for 1889. *Eng. News*, August 17, 1889; *R. R. Gaz.*, August 16, 1889.

Of the World. Some statistics up to the end of 1884. The total length of operated road was 468,108 km., about 292,500 miles, an increase of 25.7 per cent. since 1881. *Archiv. fur Eisenbahnwesen*, 1886, pp. 289-297.

Of the World. Detailed statement of mileage, Dec. 31, 1888. *R. R. Gaz.*, May 30, 1890.

Of the World. Statistics reprinted from *Archiv. fur Eisenbahnwesen*. *Eng. News*, July 5, 1890, p. 9.

See *Interstate Commerce Commission.* *Railroad Construction Statistics.*

Railroad Structures.

Buildings and Structures of American Railroads. Condensed from a forthcoming book by Walter G. Berg. *Snow Sheds*, *R. R. Gaz.*, Oct. 17, 1890, pp. 717-8. *Sand Houses*, *ibid*, Nov. 28, 1890, pp. 817-8. *Signal Tower*, *ibid*, Oct. 24, 1890, pp. 734-5, *et seq.*.

Cattle Guard. Gives plan, sections and bill of material for the standard cattle guard of the Kansas City & Omaha Railroad. *Eng. News*, July 30, 1887.

Cattle Guards and Culverts. Descriptions and detail illustrations of several useful forms of these minor railway structures. *Eng. News*, Aug. 21, 1889, Vol. XXII., p. 174.

Coaling Station. Gives a brief description, with plans and cross-section, of the new coaling station of the Pennsylvania Railroad near the Hackensack River. *R. R. Gaz.*, Sept. 2, 1886.

Coaling station of the Pennsylvania Railroad near Jersey City. Description and detailed drawings. *R. R. Gaz.*, Sept. 2, 1887.

Coaling Stations

Report of the Com. of Mast. Mech. Assn. on "Coaling of Locomotives: the Various Plans in Use and Their Relative Efficiency." By J. D. Barnett. Fully illustrated. *Eng. News*, Sept. 24, 1887.

Freight House. *Standard Small Freight House—Maine Central Railroad.* These are 20 ft. by 50 ft. Full drawings in *Ry. Rev.*, Feb. 28, 1891, p. 133.

Ice House. Description of several ice houses on various railroads, with illustrations of some details. By Walter G. Berg. *R. R. Gaz.*, Jan. 9, 1891, pp. 24-6. Water Tanks, *ibid.*, Jan. 23, pp. 59-61.

Oil Mixing Plant—Lehigh Valley Railroad. Full description with illustration of the company's plant, by C. P. Coleman, chemist. *R. R. Gaz.*, April 10, 17 and 24, 1891.

Oil Storage Houses. By Walter G. Berg. *R. R. Gaz.*, December 19, 1890, pp. 874-5.

Railroad Structures, continued.

Snow Fences and Sheds. See *Snow Fences and Sheds.*

Snow Protection. A brief article by Horace A. Towne discussing the use of various kinds of snow fences. *Ry. Age*, Nov. 27, 1891, p. 917.

The Use of Wood in. A series of articles by Prof. C. D. Jameson, intended to appear afterward in book form. *R. R. & Eng. Jour.*, February, 1889, *et seq.*

See *Railroad Stations. Railroad Shops.*

Railroad Surveys. See *Railroad Location. Surveying. Surveys.*

Railroad Switches and Frogs. See *Railroad Signals Interlocking. Railroad Track.*

Railroad Tariffs. See *Railroad Rates.*

Railroad Terminals. *Brooklyn Bridge.* See *Brooklyn Bridge.*

Chicago Railway Problem II. By Max E. Schmidt. *Jour. Assn. Eng. Soc.*, July, 1890, pp. 268-72.

Chicago Railway Problem, etc. Three short papers treating of terminals, rapid transit, and the avoiding of accidents at street crossings. *Jour. Assn. Eng. Soc.*, June, 1891, pp. 273-83.

Elevated Track of Pittsburgh Junction R. R. Theodore Cooper, Engr. *R. R. Gaz.*, March 13, April 3, 1881.

Facilities of the Del. Lac. & West. R. R. at Hoboken, N. J. Fully illustrated. *R. R. Gaz.*, Oct. 16, 1885.

Of the N. Y., W. S. & B. R. R. at New York City. Illus. *R. R. Gaz.*, Nov. 6, 1885.

Of the Midland Ry., London. An area of ten acres covered by a two-story yard the lower story being 14 ft. high and lighted by electricity. Cars raised and lowered by hydraulic lifts. Upper yard wholly supported by iron columns and floor of bent plates, etc. Fully illustrated. *Lon. Engineer*, Dec. 25, 1885.

Facilities for Handling Freight at New York. Report of Com. of Am. Soc. on, O. Chanute, chairman. *Trans. A. S. C. E.*, Vol. IV., 1875, pp. 1-80. Also a series of papers by Gratz Mordecai in *R. R. Gaz.*, Jan. 23, 1885, *et seq.*

Fourth Street Extensions of the Chesapeake and Ohio Railway at Cincinnati. Illustrated description of this improvement. *Ry. Rev.*, March 22, 1890, Vol. XXX., p. 163.

Improved Railroad Terminal Facilities in Providence, R. I. A paper by Samuel L. Minot, read before the Boston Soc. C. E., giving the history of the proposed improvement, with general sketch of plans. Illustrated by maps. *Jour. Assn. Eng. Soc.*, Nov., 1890, pp. 517-30.

Jersey City Freight Terminus—Lehigh Valley Ry. Description of piers and their construction with "inset" giving general plan of terminals, and elevations of covered piers. *R. R. Gaz.*, Sept. 4, 1891, pp. 612-13.

Montreal. See *Viaduct, Approach.*

Pennsylvania Terminal Improvements in Jersey City. Description of foundation and superstructure of the Four-Track Iron Viaduct with detail drawings of the latter. *R. R. Gaz.*, Dec. 19, 1890, pp. 832-4. *Eng. News*, Dec. 27, 1890, pp. 577-3.

Pennsylvania Railroad at Jersey City. Full description of these extensive improvements. Illustrations of train shed with details and strain diagram, hydraulic elevator for baggage, etc. *Eng. News*, Sept. 26, Oct. 3, 1891.

See *Railroad Stations. Railroad Stations and Yards.*

Railroad Ties.

Average life of, for different kinds of timber. Abstract of report to Department of Agriculture. *Eng. News*, Aug. 29, 1881.

Railroad Ties, continued.

Discussion as to the proper number, length, and cross section. Editorial in *R. R. Gaz.*, April 10, 1885.

Beechwood. An account of the successful use of such ties when creosoted or treated with chloride of zinc. Special treatment is required in the seasoning. From the German. *Eng. News*, April 3, 1886.

The Best Method of Preserving Cross-Ties. A paper by H. W. Reed, read before the Roadmasters' Assn. of America. Gives data regarding life of ties, and information as to their preservation. *Eng. News*, Sept. 19, 1891, pp. [illegible].

Iron and Wood. Relative Cost. From the German. Iron ties must last at least 25 years to be as cheap as wooden ones on the Continent. *Abst. Inst. C. E.*, Vol. LXXXIII., p. 498.

Iron vs. Wooden. By J. L. Weyers. Reviews briefly the different systems proposed and gives the objections to them. *Van Nos. Eng. Mag.*, Vol. XVII., p. 523.

Metallic.

Description of the metallic cross-ties that have been used for several years on German roads. *Am. Eng.*, March 13, 1885.

Gives figures showing the cost of maintenance of track laid with metallic ties. *R. R. Gaz.*, Dec. 16, 1887.

Cast Iron. A description, with cuts, of the cast-iron sleepers now in use by the N. Y. C. & H. R. Ry., in the New York yards. Designed by J. M. Toucey. *R. R. Gaz.* Sept. 4, 1885.

"Cross ties in India." Article by an Indian engineer, describing various forms of cross ties in use and under trial on the railways of India. *R. R. Gaz.*, Aug. 23, 1889, Vol. XXI., p. 552.

Experimental Iron Cross. An editorial review of the field, and looking at the matter from a financial point of view. *R. R. Gaz.*, June 4, 1886.

Experience With, on Swiss Railroads. Abstracted from a report by M. Mayer, Chief Engineer of Western Railway of Switzerland. *R. R. & Eng. Jour.*, October, 1889, Vol. LXIII., p. 462.

Hartford Steel, for railroads. Illustrated and described. *R. R. Gaz.*, Dec. 7, 1888.

Iron, In Germany. Illustrations of several forms, and experience in use of. From *Le Genie Civil*. *Eng. News*, March 8, 1884, pp. 111-13.

Plates For. *Elastic Steel Tie Plate* in use in Sweden. Brief description. *R. R. Gaz.*, March 8, 1889.

Post's Iron. Illustrated. *R. R. Gaz.*, Sept. 9, 1887.

Standard Metal Ties. Paper by I. S. McGieham. *Proc. Engrs. Soc. of W. Penna.*, Vol. V., 1889, pp. 110-18, discussion, 118-23.

Standard Steel Ties. This tie consists of a steel channel bar, laid with the open side uppermost, and has a block of creosoted wood under each rail. Gives good satisfaction. Described and illustrated. *Eng. News*, Feb. 7, 1891, pp. 124-5.

Steel Tie, The Hartford. Illustrated description of. A test of this tie is about to be made on the Hudson River Division of the N. Y. C. & H. R. R. *R. R. Gaz.*, Dec. 7, 1888.

Steel Ties. By J. W. Post. Before the annual convention of the American Society of Civil Engineers. Gives cost of maintaining track on steel ties on the Netherland State railroads. *Eng. News*, June 30, 1883.

Types of metallic ties in use on German railroads, with their chairs and modes of fastening rails. Well illustrated. *Am. Eng.*, March 13, 1885.

See also *Iron Sleepers and Railroads.*

Types of. Illustrated description. *E. & M. Jour.*, Dec. 29, 1888.

See *Roadmasters' Association. Timber Preservation.*

Railroad Track.

Advantage of a Longitudinal Bearing System for Railway Tracks. A paper by T. C. Clarke setting forth the advantages of such a system, requisites and comparative cost. *Trans. A. S. C. E.*, Vol. XXV., Aug. 1891, pp. [illegible]-40. A description, with illustrations, of the Hohenegger system in use in Austria, is appended. *Id.*, pp. 211-3. Discussion, pp. [illegible].

Angle Bars. Why Do They Break? By J. A. Weiss, before the Club of Engrs. of Maint. of Way, South-Western System, Pennsylvania. *R. R. Gaz.*, Jan. 14, 1886.

Buffers. See *Hydraulic Buffers.*

Construction and Maintenance of Track. A paper by Julien H. Hall, giving many facts from actual experience, and discussing the best methods and forms of construction. Illustrated. *Trans. A. S. C. E.*, Vol. XXIII., 1890, paper No. [illegible], pp. [illegible].

Continuous Rail for. A brief paper by R. T. Gleaves read before the Engineer's Club of Philadelphia, describing a continuous rail three miles long laid by him, and in successful use. *Ry. Rev.*, Sept. 5, 1891, p. 578. *R. R. Gaz.*, Sept. 11, 1891, p. [illegible].

Cost of Maintenance of Trial Lengths of Line laid with Metal Sleepers on the Netherlands State Railway. Table shows the cost of maintaining six different systems. *Bulletin de la Commission Internationale du Congres des Chemins des fer.* 1889, p. 1138. Abstract in *Proc. Inst. C. E.*, Vol., XCIX, pp. 461-3.

Development of American Rail and Track. Very valuable paper, fully illustrated, by J. Elfreth Watkins. *Trans. A. S. C. E.*, April, 1890, Vol. XXII., pp. [illegible].

Elevation of Outer Rail on Curves. A good editorial in the *R. R. Gaz.*, Dec. 3, 1886.

Erie Standard. A brief description, with cross sections of the standard track of the Erie Railroad. *R. R. Gaz.*, June 3, 1887.

Fastenings in Europe. Some facts as to methods in use as brought out at the International Railway Convention in Paris. *Eng. News.* November 1, 1890, p. [illegible].

Fastening Rail to Ties. By C. C. Wenshall, C. E. This paper discusses tie plates of various kinds, including a new design by the author. Illustrated. *R. R. Gaz.*, July 18, 1890, pp. 505-6

Frogs and Switches.

A paper by W. F. Ellis read before the N. Y. R. R. Club, describing the operation and defects of various switches. *R. R. Gaz.*, May 29, 1891, pp. [illegible]. *Ry. Rev.*, May 30, 1891, p. 348.

By W. F. Ellis, before the January meeting New England Railroad Club. Discusses and advocates the use of spring-rail frogs. Discusses safety switches and advocates the split or point switch. *Mast. Mech.*, Feb. 1888.

By Geo. T. Sampson. An illustrated article which will be found helpful to inexperienced engineers charged with laying out yards or with the selection of frogs for a line of road. *Jour. Assn. Eng. Soc.*, Vol. V., p. 198.

Discussion, at the January meeting at the New England Railroad Club, of frogs and safety switches. *R. R. Gaz.*, January 20, 1888; *Mast. Mech.*, Feb., 1888.

Frog Angles. The Application of the Differential Calculus to the Determination of Frog Angles. A useful method when field measurements are difficult to obtain. By E. A. Haskell *R. R. Gaz.*, June 5, 1891, pp. [illegible].

Safety Switches and Frogs. By Geo. Richards, before the New England Railroad Club. Discusses the development of switches and frogs, and gives the qualities of a good safety switch. *Mast. Mech.*, Feb., 1888.

See *Railroad Signals. Interlocking System. Railroad Track. Switches.*

Railroad Track, continued.

Gauge.

Change of, on Mobile & Ohio R. R. Description of method employed, organization of forces, etc. *R. R. Gaz.*, May 14 and 21, 1886.

Change of Gauge of the Ohio & Mississippi Railway Co.'s Track. Paper by I. A. Smith, read before the Engineers' Club of St. Louis. Dec. [illegible], [illegible]. *Jour. Assn. Eng. Soc.*, March, 1889.

Change of, on the Southern Roads. Programme prepared by the committee for such a change on May 31, 1886. *Ry. Rev.*, Feb. 27, 1886.

Change of Gauge of Southern in 1886. By C. H. Hudson. An interesting description of the changing of the gauges of a large number of the railroads in the Southern States in 1886. Gives plans and methods in both track and machinery departments, also cost of the work. Illustrated. *Jour. Assn. Eng. Soc.*, Vol. VI., p. 3[illegible]. *R. R. Gaz.*, Nov. 11, 1887.

Changing While in Operation. An account of the work on the N. Y., P. & O. R. R. By H. C. Thompson. General and specific instructions given. Illustrated. *Jour. Assn. Eng. Soc.*, Vol. I., p. [illegible].

Early History of railroads and origin of gauge. *Van Nos. Eng. Mag.*, Vol. VII., p. [illegible].

Gauges of the World. An abstract from an article by Herr Claus in *Glaser's Annalen*, showing the history and development of the railroad gauges of the world. *R. R. Gaz.*, Sept. 14, 1888.

General Review of the Metal Track Question. Report by E. E. Russell Tratman, C. E., to the Forestry Div. of the U. S. Dept. of Agriculture. Fifty pages of Bulletin No. 4. An extended abstract of a portion of this valuable report is published in *Ry. Rev.*, August 23, 1890, *et seq.* *Ry. Age*, August 23, 30, 1890, pp. 591, 613-4. *Eng. News*, August 30, 1890, pp. 187-8. *Am. Mfr.*, August [illegible], 1890, p. 7.

Life of Ties and Rails. Gives tables of statistics of sleepers and rails on the German Railroads. *R. R. Gaz.*, Aug. 26, 1887.

Maintenance of Track. By John M. Goodwin. An attempt to show the relation existing between the cost of track maintenance and the use of steel rails. *R. R. Gaz.*, May 4, 1884.

Metallic. By M. S. Cantagrel. A very complete and interesting account, historical and descriptive, of the various systems of iron superstructure for railways, which have been proposed or used, with illustrations and statistics. *Mem. Soc. des Ing. Civils*, July, 1886, pp. 59-104.

Metallic. By E. E. Russell Tratman. A translation of a very complete and interesting article. By M. Cantogue. In *Mem. Soc. des Ing. Civils*, July, 1886. In *Eng. News*, Jan. 29, 1887.

Parson's "Track" published in *Eng. News*, Oct. 4, 1884, *et seq.*

Permanent Way.

Abstract from the report of New York State Engineer on railroads for 1881. *Van Nos. Eng. Mag.*, Vol. I., pp. 363 and 481.

Gives description of the permanent way of the London & Northwestern Railroad. *R. R. Gaz.*, July 8, 1887.

Gives a summary of returns received in reply to circular issued under a resolution pertaining to roadway adopted at a meeting of the Association of North American Railroad Superintendents, Oct. 11, 1887. *R. R. Gaz.*, April 13, 1888.

In Germany. A summarized account of rail sections, cross tie and splicing arrangements on the principal European railways. *Proc., Inst. C. E.*, Vol. XCV.

Rail Braces. Correspondence relative to advisability of using "rail braces" on sharp curves. *Eng. News*, Dec. 1, 1888.

Railroad Track, continued.

In Great Britain and Ireland, with especial reference to the use of timber, preserved and unpreserved. Compiled by John Bogart from information received from engineers in charge of those railways. Gives tabular survey of information and many illustrations. *Trans. A. S. C. E.*, Vol. VIII. (1879), p. 87.

Standards on the Baltimore & Ohio. Detail drawings and description of the warning and distance posts, etc., on the B. & O. *Eng. News*, Nov. 7, 1891, pp. 445-6.

Standards and Rules for the Government of Trackmen on the B. & O. R. R. Relates to roadbed and ballast, embankments, rails and splices, standard switches, track walking, and accidents, cross-sections of roadbed given, also table of switches. *Eng. News*, Oct. 31, 1891, pp. 416-7.

And Street Railway Track, Improvement of. By E. E. Russell Tratman, Jr., Am. Soc. C. E. Read before convention June 25, 1889. *Eng. News*, July 20, 1889. Vol. XXII., p. 52.

Switches.

Johnson Interlocking Machine. Illustrated description. *R. R. Gaz.*, Jan. 18, 1889.

German Switch Movement. Gives a translation of a lecture before the Berlin Railroad Club discussing the arrangements by which a close contact in split switch worked from a distance is obtained. *R. R. Gaz.*, Aug. 10, 1888.

Price Frogless Switch. Illustrated description. *Eng. News*, April 6, 1889.

Sliding Switch and Pivoted Frog. Illustrated description of a new device in successful use on the Brooklyn Bridge, both switch, rails and frog being shifted bodily instead of the rail being merely deflected and using a pointed frog as in common practice. *R. R. Gaz.*, Jan. 24, 1890, p. 59.

Standard Point B. & A. R. R. By C. E. Alger. Gives plan and details of the point switch in use as the standard of the Boston & Albany Railroad. *R. R. Gaz.*, March 2, 1888.

Tracklaying Machine. By D. Sweeney. A good description of the method of laying track with the Harris Machine. *Railway Service Gaz.*; also *R. R. Gaz.*, Nov. 11, 1887.

Track without Frogs. A paper by C. B. Price, before Engrs.' Soc. West Penn. The problem discussed and a number of remedies considered. *Am. Eng.*, Nov. 5 and 12, 1886.

Track Work. Essays received in competition for prize offered by the Am. Soc. of R. R. Supts. Three practical original, illustrated papers by Andrew Morrison (p. 26), H. W. Reed (p. 37) and E. A. Hill (p. 18), respectively. *Pro. of the 18th Meeting of the A. S. of R. R. Supts., held in New York, Oct. 7, 1880.* The pamphlet can be obtained through C. A. Hammond, Secy., 380 Atlantic Ave., Boston.

See *Fish Plates*. *Rail Joints*. *Rail Fasteners*. *Rails*. *Roadmasters' Association*. *Street Railway Track*.

Railroad Track Laying. See *Railroad Construction, Rapid*. *Railroad and Track*. *Track Laying Machine*. *Railroad Track, Parsons*.

Railroad Traffic and Transportation.

Cheap Transportation. By W. B. Hyde. Proposes a new method to be applied to side lines or feeders. *Van Nos. Eng. Mag.*, Vol. VIII., p. 119.

Cost of. Address by Pres. H. S. Haines before the Am. Ry. Assn. *R. R. Gaz.*, Oct. 26, 1891, pp. 781-3.

Freight: Its Classification and the Principles on which It is Based. By Commissioner E. P. Vining. An effort to found discriminations in freight rates upon economic and scientific principles. *Ry. Rev.*, Oct. 18, 1884.

Freight Traffic, the Elements of Cost of. By O. Chanute. The paper discusses

Railroad Traffic and Transportation. *Freight Traffic, etc.*, continued.

the various elements of costs and the manner in which they burden the traffic differently upon several roads; points of their numerous combination, and gives general deductions. *Trans. A. S. C. E.*, Vol. II., p. 381.

Freight Traffic, Cost of. By O. Chanute. The elements of cost analyzed. *Ry. Rev.*, May 2 and 9, 1885.

Freight Traffic on the Rhine. By Lehmann. Statistics and discussion of the amount and character of the traffic, and comparison of transportation by water and rail. *Archiv fur Eisenbahnwesen*, 1886, pp. 188-210.

Passenger, on the Prussian State Railroads. By Counciller Todt. Extended discussion of rates and receipts and comparison of the different "classes" with statistics. *Archiv fur Eisenbahnwesen*, 1886, pp. 12-64.

Production of, and Transportation of Freight and Passengers. By Martin Coryell. Sketches the developement of canals, turnpikes and railroads. *Trans. A. S. C. E.*, Vol. II., p. 240.

Relative Cost of Transporting Car-loads and less than Carload Lots. Gives the testimony of Mr. Fink before the Inter-State Commerce Commission. Submitted a statement based upon statistics. A valuable paper. *R. R. Gaz.*, Feb. 3, 1888.

Traffic receipts, locomotive and car expenditures on the principal British railways. Lon. *Engineer*, Nov. 7, 1890, p. 370. *Ry. Eng.*, November, 1890, pp. 294-5.

Transportation. How railroads can be made more efficient in this regard. Two papers by Wm. P. Shinn, with discussion by many engineers and others, contributed from Dec., 1882, to June 1883, and the whole occupying 146 pp. Of lasting value to chief engineers and managers of railroads. *Trans. A. S. C. E.* Vol. XI., p. 365, Vol. XII., pp. 126, 180, 189 and 319.

See *Railroad Rates.*

Railroad Trains. See *Train Resistance.*

Railroad Transfer Table. See *Transfer Table.*

Railroad Transportation. See *Railroad Traffic and Transportation.*

Railroad Train Service.

American and Foreign. By A. T. Hadley. Gives a comparison of the train service in the different countries. Shows the average frequency of trains and the proportion between train service and population. *R. R. Gaz.*, Nov. 25, 1887.

See *Railroad Operation.*

Railroad Turnouts. See *Railroad Curves, Turnouts.*

Railroad Turntables.

Description of the Lawthrop wrought-iron turntable, with table of dimensions and detailed drawing. *Eng. News*, April 2, 1887.

Indian Railroads. Gives drawings, in detail, with dimensions for turn-tables to be used on the Indian State Railroads. Lon. *Engineer*, Oct. 28, 1887.

Silverton Railroad in Colorado. Turn Table on the Main Track of the. Description by C. W. Gibbs. *Trans. A. S. C. E.*, Vol. XXIII., 1890, Paper No. 450, pp. 180-8.

Railroad Vision Test.

Practical. A valuable paper by R. Brudenell Carter, F. R. C. S., discussing methods of testing employes for color-blindness and other imperfections of vision. *Jour. Soc. Arts*, Jan. 24, 1890, Vol. XXXVIII., p. 100.

Railroad Water Tanks. See *Water Tanks.*

Railroad Wheels. See *Car Wheels. Locomotive Wheels. Railroad Rolling Stock Wheels.*

Railroad Wrecking Outfits, *and Handling Wrecks*. Various forms of special blocks tackles, etc., are described and illustrated. By P. W. Hymes. *R. R. Gaz.* Dec. 5, 1890, pp. 839-41, Dec. 10, pp. 871-3.

Railroad Wrecking Plant used on Great Western Railroad of Ireland. Illus. From Lon. *Engineer*. *R. R. Gaz.*, July 17, 1885.

Railroad Yards. See *Railroad Stations and Yards*. *Railroad Terminals*.

Railroads. *Abt System*. See *Railroads, Rack*.

By Jules Morandiere. Gives statistics and other information on the construction and working of narrow gauge and secondary roads. *Van Nos. Eng. Mag.* Vol. XIX., p. 168.

By S. W. Robinson. Notes, with comments, from a tour over Ohio railroads Leading topics are bridges, curves and crossings. *Van. Nos. Eng. Mag.*, Vol. XXVI., p. 385; also 59 of *Science Series*.

(British) Board of Trade and Railways. Important document giving in detail an extensive list of requirements for railways, referring particularly to the adoption by all English roads of the block system of signalling trains, limiting gradients at stations, bridge loading and design, etc. Lon. *Engineer*, Oct. 25, 1889, Vol. LXVIII., p. 356.

Algerian Railways. An account of the railways of Algiers. By Consul Grellet, of Algiers. *U. S. Consul's Reports*, April, 1889.

American and Foreign Railways. Paper by Mr. Willard Beahan, member of St. Louis Eng. Club, read March 5, 1890. The paper is a very interesting and readable review of the subject treated. pp. 12. *Jour. Assn. Eng Soc.*, March, 1890, Vol. IX., p. 98.

In Australia. Giving average cost per mile, working expenses, net earnings and mileage. *Eng. News*, March 30, 1889.

Of Austria. Abstract from a consular report, giving a brief description of the Austrian railways, and the effect of the zone passenger tariff. *Eng. News*, Nov. 14, 1891, pp. 461-2.

Of Austria-Hungary. Extract from report of MM. Brawe and Weiss. Short account of principal features. *Annales des P. & C.*, Dec., 1885, pp. 1019, 1028.

Barre Railroads. A description of an interesting application of steep grades and switchbacks in reaching several granite quarries at Barre, Vt. Illustrated. *Eng. News*, Aug. 3, 1889, Vol. XXII., p. 104.

Belgian Railways. Editorial correspondence describing and illustrating track details, and various forms of locomotives on the Belgian state railways. *Eng. News*, Jan. 18, 1891, pp. 46-8.

Belgium. By S. Sonnenschein. Organization of branch and local roads with laws respecting them. *Archiv. fur Eisenbahnwesen*, November and December, 1886, pp. 748-784; also statistics of Belgian roads for 1883-84, on pp. 785-791

Berlin Circuit Railroad. A series of articles containing a complete description of the great work. *Zeitschrift fur Bauwesen*, 1884 and 1885.

British India, and the agricultural products of the country. Descriptive account. *Arch. fur Eisenbahnwesen*, 1885, Heft 4, pp. 570-584.

Brooklyn Bridge. See *Cable Railroads*.

Canadian Pacific.

The annual Address of Mr. T. C. Keefer, President of Am. Soc. C. E., read at the annual convention of 1888. Gives details of construction of the C. P. R. Illustrated by several maps and plates. pp 33. *Trans. A. S. C. E.*, August, 1888, Vol. XIX., p. 55. Abstracted in *R. R. Gaz*, July 6, 1888. *Sci. Am. Sup.*, Aug. 4, 1888.

Article by T. Kennard Tomson giving a brief account of its history, and description of the western portion. Numerous illustrations. *Eng. Mag.*, Dec. 1891, pp. 319-23.

Railroads, continued.

An account of its location and construction, by J. C. James, the Chief Engineer. *Proc. Inst. C. E.*, 1884, Vol. LXXVI. Also a paper by F. T. Abbott, showing rate of progress made, being the most rapid, probably, on record. *Jour. Assn. Eng. Soc.*, Vol. IV., p. 150.

An account of the road from its inception to its completion, with total cost and financial standing, mileage and branch lines. *Sci. Am. Sup.*, July 24, 1886.

In 1887. Abstract of a paper read by R. E. Peary, C. E., U. S. N., at the Seabright convention of the Am. Soc. C. E. *Eng. News*, Aug. 3, 1889, Vol. XXII., p. 101, *et seq.*

Cardenas & Jucaro. Gives historical sketch of road; also treats of its operating expenses, construction contracts, cost of construction, equipments, etc. *R. R. Gaz.*, July 29 and Aug. 5, 1887.

Cause of Shock. By H. Hollerith. Discusses the shock produced in stopping trains in the light of the theory impact. *R. R. Gaz.*, April 27, 1888.

Cheapest Railroad in the World. Description of the Dublin and Wrightsville R. R., 19 miles long, costing $3,441 per mile complete. By Arthur Pew. *Trans. A. S. C. E.*, Vol. XXIII., 1890. Paper No. 469, pp. 111-19.

Chemin de Fer Glissant. Description of a novel railway, in which the trains are propelled by water pressure, at a high velocity, up steep grades. Illustrated. *Eng. News*, Aug. 31, 1889, Vol. XXII., p. 196; *Eng. News*, Oct. 26, 1889.

Chemin de Fer Glissant. A very complete illustrated description of the sliding railway exhibited at the Paris Exhibition. Paper read by Sir Douglas Galton before the Society of Arts. *Jour. Soc. Arts*, March 14, 1890, pp. 394-407. *Sci. Am. Sup.*, No. 753, June 7, 1890, pp. 12028-31; also *Mechanics*, Oct. and Nov., 1889.

Chili, Railroads in. Map and description. *R. R. Gaz.*, Nov. 23; *Eng. News* Dec. 8, 1888.

Chili. By F. W. Conn. Descriptive paper. *R. R. Gaz.*, Oct. 24, 1890, pp. 735-7.

China. Abstract of a paper on "Railways and Collieries in North China," by Mr. C. W. Kinder, M. I. C. E., giving some information on the progress of railroad construction in that country. *R. R. Gaz.*, April 10, 1891, pp. 247-8.

Cleveland Loop-Line Railway and Her Magnificent Outside Harbor. Paper by John H. Sargent, read before the Civil Engineers' Club of Cleveland. *Jour. Assn. Eng. Soc.*, Feb., 1890, Vol. IX., pp. 58-62.

And Collieries of North China. A paper by Mr. C. W. Kinder, M. I. C. E., describing briefly the condition of the various departments of construction and maintenance, traffic management, etc., with brief statement of coal deposits and workings. Reprinted from Selected Papers, *Proc. Inst. C. E.* *Iron*, April 24, 1891, pp. 355-59.

Corinth. By Armand Saint Yves. Abstract from *Annales des Ponts et Chaussées*, 6th series, Vol. XVI., 1888, p. 392. *Proc. Inst. C. E.*, Vol. XCVI., 1889, p. 379.

Darjeeling-Himalayan Railroad Rolling Stock. Brief description of the more marked peculiarities of the rolling stock on this 2-foot gauge Indian railway; also an illustration showing a train of 14 cars and engine on a curve of 70-foot radius. *Ind. Eng.*, Oct. 26, 1889.

The Denver, Colorado Canon & Pacific R. R. Project. An interesting paper on the Grand canon of the Colorado by Mr. Robert B. Stanton. Fully illustrated by maps and large inset, giving several views. *Eng. News*, Sept. 21, 1889, Vol. XXII., p. 269.

Denver, Colorado Canon and Pacific Railroad Project. Very complete discussion as to the practicability of the route. Illustrated by many engravings from photographs. By Robert B. Stanton, Chief Engineer. *Eng. News*, Oct. 18, 1890, pp. 1018-20, 351-6. Editorial, *ibid.*, pp. 349-50.

Railroads, continued.

Earnings of German Railroads per Kilometre. Tables and diagram. *Wochenschrift des österr. Ing. u. Arch. V.*, August 15, 29, 1890, pp. 283-5, 297-8.

Effect of, on the Value of Lands. By Gen. W. S. Rosecrans. Contains rules for calculating the value of land on a basis of its distance from the railroad. *Van Nos. Eng. Mag.*, Vol. VIII., p. 540.

Effect of Invention Upon. See *Addresses, Chanute.*

Electric. See *Electric Railroads, Application, etc.*

Elevated. See *Railroads, Elevated, below.*

English and American Compared.

By Ed. B. Dorsey. An article of some 80 pages, comparing methods of construction and management. *Trans. A. S. C. E.*, Vol. XV., p. 1. Abstract three columns in length in *Eng. News*, April 10, 1886.

By E. B. Dorsey. A supplementary paper. Gives tables showing the cost of transportation of freight and passengers on some of the leading English and American roads. Also cost of maintenance of way, repairs and renewal of locomotives, motive power and total operating expenses. Also discussion on the same. *Trans. A. S. C. E.*, Nov., 1886.

An additional paper by Mr. Dorsey, giving cost of maintenance of way, repairs and renewals of locomotives, motive power and total operating expenses. *Eng. News*, July 10, 1886

A discussion of Mr. Dorsey's paper. Editorial *R. R. Gaz.*, Dec. 14, 1885.

A very pleasing and profitable article, contrasting the methods and accommodations of the two countries in passenger travel. Illustrated. *Harper's Monthly Mag.*, August, 1881. Also paper by E. B. Dorsey in *Iron Age*, Aug 20, 1885.

English and Foreign Compared. Lon. *Eng.*, May 7, 1886.

English Railways. A series of articles giving many details of English railway practice. From notes by E. E. Russell Tratman. *Eng. News*, Feb. 7, 1891, pp. 126-7, *et seq.*

European, as they Appear to an American Engineer. Gives brief description of some of the most prominent features in the construction and workings of the European railroads. *Trans. A. S. C. E.*, Vol. III., p 61.

Extension of, in Prussia. Account of the project now under way. *Archiv. fur Eisenbahnwesen*, 1886, pp. 329-32.

Festiniog, as a Type. A short history of the road, with a discussion, bearing on the question of broad *vs.* narrow gauges. *Van Nos. Eng. Mag.*, Vol. VIII., p. 305.

France. (*Statistics* for 1884.) *Archiv fur Eisenbahnwesen.* 1886, pp. 319-42.

French Railways. Editorial correspondence describing the peculiar features of some French railways; with diagrams of rolling stock, etc. *Eng. News*, July 11, 1891, pp. 25-7.

Future Development. By J. W. Grover. *Van Nos. Eng. Mag.*, Vol. VIII., p. 531.

German Notes on American Railroads. Report of Herr Von Borries before the Verein fur Eisenbahnkunde, of his inspection tour in this country. From Glaser's *Annalen.* *R. R. Gaz.*, October 23, 1891, p. 740, November 6, 1891, p. 781.

German Studies of American Railroads. Interesting article giving an account of the effect of American railroads on the development of railroads in Germany. *R. R. Gaz.*, Feb. 13, 1891, pp. 111-12, *et seq.*

Germany and England in the years 1881-83. Comparison as regards statistics, finances, rolling stock, etc. *Archiv. fur Eisenbahnwesen*, 1885, pp. 355-366.

Railroads, continued.

Government Control of. By J. F. Schrober. *Arch. f. Eisenbahnwesen.* 1886, pp. 457, 478.

Greenock and Gourock Railway, Scotland. A complete description of bridges, Tunnels, stations, quays, etc., giving a very good idea of English practice in in railway construction. Lon. *Eng.*, February 22, 1889. Vol. XLVII., p. 193, *et seq.*

History of.

Early. By J. Dutton Steele. Gives short account of the early history of railways and origin of gauge in England and United States. *Trans. A. S. C. E.*, Vol. II., p. 53.

Between Cleveland and Chicago. By J. H. Sargent. Gives history of the construction of the early roads, and details some of the methods employed. *Jour. Assn. Eng. Soc.*, Sept., 1887, and *R. R. Gaz.*, Sept. 30, 1887.

L. S. & M. S. By C. P. Leland. Gives history of early roads that were consolidated to form the Lake Shore & Michigan Southern Railway. *Jour. Assn. Eng. Soc.*, Sept., 1887, and *R. R. Gaz.*, Sept. 23, 1887.

In Russia. By H. Claus: Short account of the principal roads. *Archiv fur Eisenbahnwesen*, Jan. and Feb. 1887, pp. 50-66.

Hoellenthal Railway; Details of Permanent Way. The details of the rack railway are fully illustrated. Tables of cost of tunnels and bridges are also given. Lon. *Eng.*, Jan. 30, 1891, pp. 134-40, *et seq.*

Hudson Bay Railway. Description of the road now building from Winnipeg to Hudson Bay, with maps. *Eng. News*, Dec. 18, 1886.

Hungarian. Brief historical sketch and some results of the workings of the new "zone tariff". *R. R. Gaz.*, May 9, 1890.

Inclined Plane. See *Railroads, Inclined Plane. Railroads, Rack.*

India and Burma. By H. S. Hallett. A paper before the Soc. of Arts advocating the borrowing £10,000,000 a year for 10 years for the construction of railways in India and Burma. *Jour. Soc. Arts*, March 4, 1887.

Indian Railways. The broad and narrow gauge systems contrasted. A paper by Francis J. Waring, M. Inst. C. E.; together with discussion and correspondence on same. Pp. 68. *Proc. Inst. C. E.*, Vol. XCVII., p. 106. London. 1889.

India. A long historical and descriptive paper read before the Society of Arts by Sir Theodore C. Hope, K. C. S. I. 2 tables of mileage and a map. *Jour. Soc. Arts*, June 21, 1890, pp. 717-26; discussion, pp. 726-30. *Sci. Am. Sup.*, No. 756, July 26, 1890, p. 12,076.

India, Railway and Trade in. An exhaustive paper by Sir Juland Danvers, K. C. S. I. *Jour. Soc. Arts*, April 5, 1889, Vol. XXXVII., p. 431.

Indo-European, Project. Gives map showing the location of the English route from Constantinople to India; and a description of the country. *Eng. News*, Aug. 27, 1887.

Inter-Continental. Description and map of a proposed route through Central America. *R. R. Gaz.*, April 18, 1890, p. 262.

Italy. On the tramways and narrow gauge roads of Italy. By Juttner. *Archiv fur Eisenbahnwesen*, 1886, pp. 476-502. Laws, traffic, construction and management briefly discussed.

Japan. Extract from report of Secretary of the British Legation of Tokio. Maps and descriptions. *Am. Mfr.*, Dec. 10, 1885.

Jungfrau Railway. The substance of a lecture given by Herr R. Boder before Austrian Engineers and Architects' Association. *Eng. News*, March 29, 1890, Vol., XXIII, p. 296.

Railroads, continued.

Jungfrau Railway, Projected. An illustrated article by Chas. King. It is proposed that the carriage shall form a piston moving in the tunnel as a cylinder, the latter being closed at both ends by tight doors. Lon. *Eng.*, July 4, 1890, pp. 9-10.

Lartigue System. Gives brief description of Listowel & Ballybunion Railway, Ireland. It is 10 miles long and built on the Lartigue single rail system. Gives cuts of rolling stock and details of roadbed. Lon. *Engineer*, March 2 and 9, 1888. *Sci. Am. Sup.*, April 7, 1888.

Legislation.

In England. By Ed. E. R. Tratman. Gives an acount of the surveys, maps, estimates, etc., preliminary to obtaining a charter. *Eng. News*, August 1, 1885.

Italian Railroad law of April 27, 1885, and the new traffic agreements. By Dr. Piech. A paper that will be of value to all who are interested in economic questions regarding railroad policy. *Archiv fur Eisenbahnwesen*, 1886, pp. 141-188.

Italian Railway law of April 27, is given in full in *Archiv. fur Eisenbahnwesen*, 1886, pp. 364-408.

See *Railroad Management. Railroads, Government Control of.*

Mexico. Brief history of the principal Mexican railroads. Abstracted from *Le Genie Civil.* Map and view of the Wimmer viaduct. *Sci. Am. Sup.*, No. 752, May 31, 1890, p. 12011.

Mexico. Valuable paper by William Barclay Parsons. *Trans. A. S. C. E.*, April, 1890, Vol. XXII., pp. 213-247.

The Multiple Dispatch Railway. The invention consists of three continuous platforms, of which the middle one, containing the seats, travels on the periphery of the wheels, while the outer ones are attached to the axles and thus move at half the speed of the inner one. Designed to carry large numbers of people at slow speeds, as at expositions, etc. A paper by Max E. Schmidt, before the West Soc. Engrs. *Jour. Assn. Eng. Soc.*, February, 1891, pp. 68-79.

The Multiple Speed Railway at Jackson Park, Chicago. Description with details drawings of the rolling stock of the temporary road to be built at Jackson Park. *Ry. Rev.*, Sept. 12, 1891, pp. 596-7. *R. R. Gaz.*, Sept. 11, 1891, p. 632.

Narrow Gauge. See *Railroads, Narrow Gauge* below.

Notes on Railroads and Railroad Tunnels in Wisconsin. Brief paper by Mr. Woodman before the Engrs' Clubs of St. Paul and Minneapolis, describing methods of work, progress and cost. *Jour. Assn. Eng. Soc.*, Oct., 1890, pp. 463-5.

Operating by Electricity. Paper by Mr. W. E. Hall, read before the Cincinnati meeting of the A. S. M. E. The paper showed some of the advantages of such operation. It was followed by a discussion in which Mr. Scheffler, Acting Superintendent of the Westinghouse Electric Co., estimated in details the cost of an electric railroad plant and the expense of operating it; he showed that it is impossible at present for electricity to compete with steam on long roads. *Eng. News*, July 12, 1890, pp. 27-30. *Trans. A. S. M. E.*, Vol. XI., 1890, pp. 839-67.

Pacific, and the Government. Editorials showing how the present difficulties have arisen, and examines the plans proposed for their salvation. *R. R. Gaz.*, Jan. 27, *et seq.*, 1887.

Pan-American Railroad. An article by F. W. Conn, C. E., describing a proposed route and giving illustration of many interesting features of the countries passed through. *Eng. Mag.*, Aug., 1891, pp. 627-42.

Pan-American. A proposed route with somewhat detailed description, and map. By F. W. Conn, C. E. *R. R. Gaz.*, May 29, 1891, pp. 366-7.

Railroads, continued.

Past and Prospective. A brief popular paper by J. H. Sargent, read before the C. E. Club of Cleveland *Jour. Assn. Eng. Soc.*, Dec., 1890, pp. 574-9. Discussion, pp. 576-82. Reprinted in *Ry. Rev.*, Jan. 31, 1891, pp. 67-8.

Pneumatic. An account of an ingenious method for using compressed air as a motive power. The principal point of special interest is the manner of opening and closing the slot of the air tube by a mechanical device of great originality. *Eng. News*, March 22, 1890, Vol. XXIII., p. 269.

Principles of. By Prof. C. D. Jameson. An extensive work, begun in a serial form in the November number of the *R.R. & Eng. Jour.*, 1887. Promises to be a very clear exposition of the subject, accessible to those not familiar with the higher mathematics

Progress in Transportation. See *Addresses, Chanute.*

The Projected Maduro-Mamore road in Brazil. Account of preliminary works character of the country, etc. *Wochenschrift des Oesterr. Ing. und Arch.-V.*, 1886. pp. 113-115, 141-144.

The Projected Siberian R. R. This road, 6,660 miles long, from St. Petersburgh, is described by A. Zdziarski in *R. R. & Eng. Jour.*, June, 1890. pp. 258-61.

Proposed Railway through Siberia. A paper by W. M. Cunningham giving a somewhat detailed description of this railway. *Proc. Inst. C. E.*, Vol. CVI., 1791, pp. 280-91.

Proposed Railway through Central Africa. By W. Wiseman, M. I. C. E. Discusses the importance and advantages of such a road, and gives outline of proposed route. Lon. *Engineer*, July 17, 1891, *et seq.*

Suakim-Berber Railroad, Africa. An account of a reconnoissance of the line in 1875. By H. G. Prout, of the Egyptian Army, with a plan and profile of the line, and other valuable data. *Eng. News*, March 7, 1885.

Rack. See *Railroads, Rack* Below.

Railway Stations at Nicaragua. Description, with map, of the existing proposed railways. *Eng. News*, Aug. 29, 1891, p. 180.

Ratio of Population to Mileage. By W. H. White. Gives diagram showing the ratio of population to railroad mileage, and the probable increase of mileage demanded. *R. R. Gaz.*, Oct. 5, 1888.

Relation of, to the State. By Wm. P. Shinn, C. E. A series of articles giving the latest legislative enactments and decisions. *Ry. Rev.*, March 6, 1886, *et seq.*

Reminiscences and Experiences of early engineering operations on railroads, with especial reference to steep inclines. By W. Milnor Roberts. *Trans. A. S. C. E.*, Vol. VII., (1878), p. 197-215.

Report of Railroad Commissioners for State of Rhode Island, 1890, contains considerable information and data of a somewhat local character. E. L. Freeman, Comm'r., Providence, R. I.

Report on the Applications of Electricity to Railways. By MM. E. Sartiaux and Z. Weissenbruch. This report summarizes the proceedings at the third session of the "International Railway Congress," held in 1889 at Paris. *Elec. Rev.*, November 29, 1889, p. 611.

Report of the Work of the 3rd International Railway Congress (1890). Wornes de Romilly. Account of the 28 subjects submitted with an abstract of the discussion and the text of the conclusions arrived at. Covers all branches of the departments of operation, maintenance of way, rolling stock, railroad and canal rates, application of electricity, and labor problems. *Annales des P. & C.*, Jan., 1891. pp. 1-212.

Report of the International American Conference Relative to an Inter-Continental Railroad. Washington, 1891. Large 8vo.; pp. 214. 3 maps. An appendix contains the special report of Lieut. Geo. A. Zinn, U. S. A. Brief abstract with maps in *Eng. News*, Sept. 6, 1890, p. 294.

Railroads, continued.

Right of Way for. Paper by Julien A. Hall discussing methods of securing right of way, form of records, proper officers to take charge of the matter, etc., with discussion. *Trans. A. S. C. E.*, Vol. XXV., Sept., 1891, pp. 313-40.

Schuylkill River, East Side. By H. T. Douglas. Gives details of the work at Philadelphia; shows tunnel sections, etc. *Eng. News*, March 5, 1887.

South African. Gives map showing their location and data of population, trade, etc. *R. R. Gaz.*, Nov. 18, 1887.

The State and the Railroads. A paper by L. P. Morehouse, read before the Western Soc. Eng'rs., advocating some method of State Arbitration in disputes between employers and employees. *Jour. Assn. Eng. Soc.*, Dec., 1890, pp. 566-74. *Ry. Rev.*, Jan. 31, 1891, pp. 77-8.

Railroads. *Statistics of.* See *Railroad Statistics.*

St. Gothard. Brief description of various engineering works. Illus. *Eng. News*, Jan. 6, 1883, pp. 1-3.

Street Railroads. See *Street Railroads.*

Swedish. By W. Koerner. Treats of the railroads of Europe, but more especially with the development of the railroads in Sweden. Maps and tables of revenues, etc. Lon. *Eng.*, Jan. 6, 1888.

Switchback. See *Railroads, Switchback.*

Telpher. See *Telpherage.*

Three Rail. An article giving details of tests made on the centre rail system in Brazil. *Van Nos. Eng. Mag.*, Vol. VII., p. 391.

Timber for. See *Timber.*

Transandine. An interesting account of the progress of the work. Map and views. Lon. *Eng.*, Sept. 19, 1890, pp. 329-31.

Railroad Accidents. *Derailments and their Cause.* Ed. in *Ry. Age*, June 21, 1890, p. 428.

Underground. See *Railroads, Underground.*

West Shore. Reminiscences of. By Wm. H. Searles. Many interesting particulars given. *Jour. Assn. Eng. Soc.*, Vol. IV., p. 413.

The Western Railroad Situation. Article by Mr. M. L. Scudder, Jr., discussing the economics of the present railway situation. *Belford's Mag.*, Jan., 1890, Vol. IV., p. 227.

See *Electric Motors, Comparative Tests, Electric Railroads, Interstate Commerce Commission.*

Railroads and Waterways, *Relative Advantages of, in Germany.* By M. Todt. A valuable paper giving statistics relative to the quantities of goods carried in Germany by rail and by water, ratio of tons to population, etc.; also discusses the relative advantages of the railroads and waterways. Translated from the *Bulletin du Ministere des Travaux Publics* for the *Journal Royal Statistical Society*, July, 1888; abstracted *Jour. Soc. Arts*, Aug. 31, 1888.

Railroads, Elevated.

Gives detail of iron work on the Inter-state Rapid Transit Railroad, K. C. *Eng. News*, May 19, 1888.

Berlin (elevated) City R. R. Full description of the whole line, bridges, stations and other structures, with details. One station covered with a roof, being approx. a semi-circle of 65 ft. radius, the rafters hinged at the crown and at the springing, the springing on top of wall some 20 ft. above the ground level. *Zeitschr. f. Bauwesen*, 1884 and 1885.

Berlin. Gives a complete account of the structure, with details of construction, traffic, etc. Illustrated. *Eng. & Build. Rec.*, Feb. 4, *et seq.*, 1888.

Railroads, Elevated, continued.

Berlin. The Metropolitan of. Gives details of the elevated railroad in Berlin. *R. R. Gaz.*, May 6, 1887.

Chicago and South Side Railway General description, specifications, dimensions, moments and stresses of superstructure, etc. Illustrated. *R. R. Gaz.*, April 4, 1890, Vol. XXII., p. 226.

Chicago, Lake Street. Brief illustrated article describing the general method of constructing this elevated road. J. Q. Baird, Chief Engineer. *R. R. Gaz.*, March 7, 1890, Vol. XXII., p. 154.

Liverpool Elevated Railway. Illustrated description by J. C. Robinson. *St. Ry. Rev.*, Aug. 1891, pp. 309-12.

Meigs System of. An abstract from the report of the Mass. State Board of R. R. Commissioners on "the safety and strength of the Meigs elevated railway, so-called, in Cambridge, and the rolling stock and motive power used thereon." *Eng. News*, Jan. 12, 1884.

Meigs. A description of the system as built at East Cambridge, Mass.; also the report of Gen. Stark, C. E., The same to Board of Railroad Commissioners. *Sci. Am. Sup.*, Feb. 5, 1887.

Meigs System, a description of the, as built in Cambridge, Mass. A single track on a single line of columns. A paper by F. E. Galloupe, before A. S. M. E. Illustrated. Printed in *Am. Eng.*, Nov. 19, 1885.

New York.

Gives review of the Manhattan Elevated Railroad, number of its employees, and other data and statistics. *R. R. Gaz.*, July 15, 22 and 29, 1887; *Sci. Am. Sup.*, Sept. 24, 1887.

The Construction of the Second Ave. Line, New York. By G. T. Hall. Describes details, special methods of obtaining foundations, etc. Fully illustrated. *Trans. A. S. C. E.*, Vol. X. (1881), p. 127-138; *Eng. News*, Oct. 21, 1882, pp. 363-5, 70.

Details of Foundations. By C. F. Carpenter, C. E. Illustrated description. *Eng. News*, Jan. 24, 1880, pp. 29-32.

How they are taken Care of. Article stating methods of inspection and repairing in daily use on the New York City elevated roads. Also illustrations of methods of supporting roadway while superstructure or columns are undergoing repairs. *Iron Age*, Feb. 6, 1890, p. 226.

Report of Com. of Am. Soc. on Rapid Transit and Terminal Freight Facilities at New York. O. Chanute, chairman. A valuable report. *Trans. A. S. C. E.*, Vol. IV. (1875), p. 1-80. Discussion, p. 240-263.

Two articles giving minute account of the organization and operation. *R. R. Gaz.*, July 15 and 22, 1887.

Short-Nesmith System. Now in use in Denver, Col. A series system, using conductors in a conduit, with circuit openers at intervals in the line. Description, with illustrations. *Elec. World*, April 2, 1887; also in *Electrician & Elec. Eng.*, April, 1887.

Single-Rail Elevated. The Lartigue System as now in operation in London, Eng. A very interesting new departure in this line, designed to accommodate sparsely settled regions, to serve as freight tramway, etc. Illustrated. *Iron*, Sept. 3, 1886.

Single Rail. Gives a historical review of the subject, and describes the road constructed on the Lartigue System at Westminster, with cost. *Sci. Am. Sup.*, March 12, 1887.

Vienna. Lively discussion regarding the projects that have been proposed by the Austrian Soc. of Engineers. *Wochenschr. d. Oester. Ing. u. Arch. V.*, 1886, Nos. 19, 20, 21, 22, 23.

Railroads, Inclined Plane.

To take the place of a series of canal locks on the Elbing-Oberland canal, built 1874-1881. Operated by wire rope traction. The car is run under the boats while they are afloat, the boat remains on a horizontal keel throughout, the track passing a summit and permitting the car and boat to dip into the water again. The old planes on this canal, built 1844-1860, were modeled after the old Morris & Essex Canal planes. *Zeitschr. f. Bauwesen*, 1885-63.

Gravity Plane at Moulton Hill Mine, Quebec. A plane 4,000 ft. long to transport ore from the mine to the railroad. Loaded cars attached to one end of the rope are made to furnish a portion of the power necessary to haul up the empty cars. Description by F. J. Falding. Illus. *E. & M. Jour.*, Jan 31, 1891, pp. 143-4.

Hastings Cliff Tramway. Illustrations of gear and winding machinery of this inclined railway. Lon. *Engineer*, July 24, 1891, pp. 74-5.

Lookout Mountain. Fully described and illustrated. A three-rail road, operated by Wire cable. *Trans. A. I. M. E.*, Duluth meeting, 1887.

Lookout Mountain. By W. H. Adams, before the American Society of Mechanical Engineers. Gives full description of the inclined railroad up Lookout Mountain, Tenn., with profile and plan of road, engine plant and details of cars, etc. *Eng. News*, Jan. 7, 1888; *Sci. Am. Sup.*, May 12, 1888. Lon. *Eng.*, March 30, 1888. Abstracted *Proc. Inst. C. E.*, Vol. XCII., pp. 463-4.

Madison, Ind. By M. J. Becker. An illustrated description of this road, which is 7,000 feet long, on a grade of 1 in 17. The road has been in operation some 20 years. *Trans. A. S. C. E.*, Vol. VII. (1878), pp. 68-77. Discussion, p. 216.

Pittsburgh. A description of the plant and machinery. Illustrated. *Am. Eng.*, April 16, 1884.

Proposed System for Mountain Roads. By Ed. M. Rogers. Gives a short history of inclined planes, and proposes the use of counter-weights, thus making a gravity system. *Van Nos. Eng. Mag.*, Vol. XXIV., p. 56.

Steep Inclines. *The Trincheras Steep Incline* on the Puerto Cabello & Valencia Railway, Venezuela. A brief paper by John Carruthers, M. Inst. C. E. *Proc. Inst. C. E.*, Vol. XCVI., p. 100. Discussion and correspondence, *Proc. Inst. C. E.*, Vol. XCVI., p. 149.

Their Defects. By Prof. Osborne Reynolds, before the Inst. Civ. Engrs., London. The errors from friction, inertia and stretch of cord carefully investigated, and results given. Abstract of paper in Lon. *Eng.*, June 18, 1885; also in *Am. Eng.*, July 3, 1885.

See *Railroads, Rack. Cable. Railroad Systems.*

Railroads, Narrow Gauge.

By Auguste Moreau. A minute discussion of relative economies by a confirmed narrow gauger. Much studied, but *ex parte*. 58 pages. *Mem. Soc. des Ing. Civ.*, Dec., 1884.

Construction and Equipment of. A paper by A. L. Reed before the Mich. Assn. Sur. & Engrs., describing the various branches of such work. *Eng. News*, March 15, 22, 1884.

Narrow versus Standard Gauge. By C. H. Hudson. A comparison of advantages for transportating freight, showing the great advantage of the standard gauge. *Jour. Assn. Eng. Soc.*, Vol. V., p. 6.

Railroads, Rack.

Abt Rack-Railways in the World. Brief statistics of the 16 different lines. From *Le Genie Civil*. *Eng. News*, Aug. 29, 1891, pp. 186-7.

The Abt Rack Rail System. Illustrated description of the rack-rail system for mountain railroads, devised by Herr Roman Abt, the best system of the kind yet introduced, and an account of recent progress made. *R. R. & Eng. Jour.*, Oct., 1888, and Jan., 1889.

Railroads, *Rack*, continued.

Abt System, Indian Experiments. Gives details of experiments made on a section railroad built on the Abt system over the Bolan Pass. Lon. *Engineer*, July 13, 1888.

The Abt System of, for Steep Inclines. By W. W. Evans, before the Am. Soc. of C. E. *R. R. Gaz.*, Jan. 7 and 14, 1887.

Abt Visp-Zermatt Railway. A line in Southern Switzerland which when completed will be 22 miles in length. Description in *Eng. News*, Oct. 25, 1890, pp. 367-8.

Agudio's System of Mountain Railways. In this system, a moving cable alongside the track operates the mechanism of the locomotive, which in turn is connected by a friction clutch to a cog wheel engaging in a rack placed between the rails. Illustrated description, *St. Ry. Jour.*, Jan. 1891, pp. 19-20, *et seq.*

Brakes for. See *Brakes, Friction.*

Green Mountain. Description of the rack railroad at Mount Desert, Maine. *Eng. News*, Jan. 22, 1887.

Harz Mountain Railway. The Cost of Working. A paper by Robert Wilson, M. Inst. C. E., giving detailed statement of cost of working, consumption of fuel, etc. *Proc. Inst. C. E.*, Vol. XCVI., p. 131.

Harz Mountain R. R. Combined Adhesion and Rack-Rail System of Abt. Description of the roads as constructed from Blankenburg to Tanne, in the Harz, and the freight road to Oertelsbruch, in Thuringia, with estimate of cost Illustrated. *Eng. News*, May 22, 1886.

Harz Mountain Railroad (from Blankenburg to Tanne) and the road at the quarries Oertelsbruch to Buringen, both on the combined Abt and adhesion system. *Wochenschrift des Oesterr. Ing-und Arch.-V.*, 1886, pp. 102-7.

The Harz R. R. Combined adhesion and rack rail. Description of road and locomotives. Illustrated. *Zeitschrift fur Bauwesen*, 1886, pp. 71-86.

Locomotive for. See *Locomotives, Pikes Peak.*

Mt. Pilatus, Switzerland. Brief description of this interesting mountain railway, including main dimensions of the engine and car combined which is used. The *Ry. Age*, Nov. 29, 1889.

Mount Pilatus. An illustrated description. *Mfr. & Build.*, May, 1890, p. 106.

Mount Pilatus. Gives description of the rack-railroad now being constructed up Mount Pilatus, Switzerland. Double page plate showing details of locomotive and carriage, and cuts, showing details, construction of rack-rail, etc. Lon. *Eng.*, May 13, 1887.

Mt. Pilatus. Illustrated description of the rack-way up Mt. Pilatus, near Lucerne, Switzerland. Maximum gradient is 48 per cent., and total length about 2.8 miles. *R. R. Gaz.*, Feb. 4, 1887; also *Eng. News*, Feb. 19, 1887.

Pike's Peak, Manitou Railway. Brief descriptive article on this mountain railway on the Abt System. *R. R. Gaz.*, April, 1890, Vol. XXII., p. 277; also *Eng. News*, April 19, 1890, p. 366.

Pikes Peak Rack Railway. Short review of the work with some details of the alignment, etc. Also illustration of locomotive used. *Eng. News*, Oct. 4, 1890, p. 290.

See *Railroads Inclined.*

Railroads Underground.

Gives some engineering difficulties of the London Metropolitan Railroad. *Van Nos. Eng. Mag.*, Vol. I., p. 98.

In Cities. Brief *resume* of the papers by Messrs. B. Baker and J. W. Barry before the Inst. of C. E. Illustrated. *San. Eng.*, May 20, 27, etc.

See *Cars, Rolling Stock, Snow Tunnel, Track.*

Railroads, Underground, continued.

Comparison of the Different systems in London. *Van Nos. Eng. Mag.*, Vol. I., p. 33?.

Glasgow (Scotland) Central. Illustrated description of railway soon to be built under the city of Glasgow. *Eng. & Build. Rec.*, Dec. 15, 1888.

Metropolitan, of London, Eng. History and description of the thirteen miles of subsurface railway in the heart of the city. Fifty cuts, showing profiles, cross-sections and methods of construction. Specially valuable in the matter of tunneling under large buildings. By B. Baker and J. W. Barry. With discussion *Proc. Inst. C. E.*, Vol. LXXXI.

Metropolitan, for Paris. By M. Haag. Project for a railroad, partly elevated and partly in tunnel, to cost $108,000,000, including the cutting of new streets. Sixteen pp. and folding plate. *Mem. Soc. des Ing. Civils*, Paris, Dec., 1884. Also, by Jules Garnier: Project for a system of elevated railroads, to cost 50,-000,000 francs; 17 miles. Two tracks, one above the other. *Mem. Soc. des Ing. Civ.*, Feb., 1885.

Subway, London and Southwark. By J. H. Greathead. Before the British Association. Gives description of the nature of subway, then describes some o the details of the mode of construction and the appliances devised for passing through water-bearing strata. Illustrated. *Eng. News*, Nov. 5, 1887.

See *Electrical Subways*. *Rapid Transit*.

Railways. See *Railroad*. *Railroads*. *Street Railways*.

Railroads, Switchback.

By H. S. Huson. Gives reasons for its construction, and details of location, track and locomotives, etc. *R. R. Gaz.*, Feb. 3, 1888.

Stampede Pass, N. P. R. R. A description with plans and profile of the temporary switchback of the Northern Pacific Railroad over the Cascade Mountains at Stampede Pass. *R. R. Gaz.*, Dec. 23, 1887. For details of method of running trains over this switchback see *R. R. Gaz.*, Jan. 13, 1888.

Rain. *The Cause of Rain and the Structure of the Atmosphere.* A paper by Frantz A. Velschow, discussing the relation between low pressure and rain, and putting forth a new theory as to the cause of rain. *Trans. A. S. C. E.*, Vol. XXIII., Dec., 1890, paper No. 459, pp. 303-20.

See *Meteorology*.

Rain-Fall.

The Amount Available for Water Supply. A most valuable paper before the N. E. Water-Works Association. By Desmond Fitzgerald, giving many facts from American observations as to the amount collectible, also the same ly analyzed for the various months of the year. *Eng. News*, Aug. 21, 1886. *Jour. N. E. W.-W. Assoc.*, Sept., 1886.

Analysis of that Observed at Lake Cochituate, Mass., to determine the law of variation from mean. Illustrated. *A. S. C. E.*, Vol. XIII., p. 359.

As Influenced by Forests. See *Forests*.

Distribution of during a great storm in New England in 1869. By J. B. Francis. The maximum fall was over twelve inches in two days. Map given showing distribution. *Trans. A. S. C. E.*, Vol. VII., (1878), pp. 224-235.

Effect of Wind Currents on, and on the Gauge Record. A 10-page pamphlet, with cuts, being *Signal Service Notes No. XVI.*, to be had of Sig. Serv. Bureau, Washington.

Hartford, Conn., and Vicinity. Tabular exhibit of rainfall by three observers in this section, covering a period from 1875 to 1889. *30th An. Rep. of Board of Water Comm's*, Hartford, Conn., p. 29.

The maximum that may occur in a given number of hours. The question discussed by the aid of Dr. Englemann's observations since 1831, and formula derived. By Prof. F. E. Nipher. *Am. Eng.*, May 6, 1885.

Rain-Fall, continued.

Received and Collected on the Water-Sheds of Sudbury River and Cochituate and Mystic Lakes, Mass. Records from 1863 to 1885 by months. Shows that in dry years not more than ten inches can be collected. The most complete and valuable American record. *Jour. Assn. Eng. Soc.*, Vol. V., p. 395.

See *Water Supply*.

Its Relation to the *Water Supply of a City*. By P. H. Baerman. *Van Nos. Eng. Mag.*, Vol. XXIV., p. 179.

Relation Between the Rainfall and the Discharge of Sewers in populous districts. *See Sewers*.

And River Flow. The ratio of the two as observed for three years at the Cape of Good Hope over 105 sq. miles. *Proc. Inst. C. E.*, Vol. LXXXI.

Tables and diagram showing rainfall, consumption of water, temperature of air and water and much other data of somewhat local interest, may be found in *Report of City Engineer*, of Boston, 1889, p. 40.

Water Level and, of the Great Lakes. Gives diagram showing the fluctuations of the water-surfaces, areas, etc., of the Great Lakes, with comments on the phenomena observed. *Eng. News*, Oct. 6, 1888.

Rain Gauges.

The Influence of Elevation and Wind upon Gauge Records. By Desmond Fitz-Gerald. *Jour. Assn. Eng. Soc.*, Vol. III., p. 233.

The Practical Value of Self-Recording. A paper read at a meeting of the New England Meteorological Society, April 17, 1889. *Eng. News*, May 4, 1889

Self Registering. Illustrated description of an apparatus used at Philadelphia, with a table showing maximum rainfalls for six years. By John E. Codman. *Eng. Record*, March 14, 1890, p. 246.

Self Registering, and their Use for Recording Excessive Rainfalls. An article describing several gauges, translated from the *Zeitschr. f. Bauwesen*, by Rudolph Hering. Illus. *Eng. Record*, Jan. 3, 1891, pp. 74-5.

Rain Storms *in New England*. An account of the great rainfall of Feb. 10-11, 1886, when 6.6 inches fell in 24 hours. Also the maximum fall in other great storms. By Prof. Winslow Upton, in *Science*, March 19, and *Eng. News*, April 3, 1886

Ramie *as a Textile Material*. Description of the methods of culture and manufacture of ramie, its properties and adaptation to the United States. By Jules Juvenet. *Jour. Frank. Inst.*, Nov., 1889.

Range Finding *by Electricity*. See *Electric Position Indicator*.

Rapid Transit.

A discussion of the question adapted to Philadelphia, with map showing the half hour limit for pedestrians, street cars, and elevated railroads, the respective areas being 8, 18 and 72 square miles, with supplement on Growth of Cities, by L. M. Haupt. *Proc. Eng. Club, Phila.*, Vol. IV., No. 3.

Boston. Paper by Henry Manley, Member of the Boston Soc. of Civ. Engr's. Read Feb. 20, 1889. Discussion by members. *Jour. Assn. Eng. Soc.*, May, 1889, Vol. VIII., p. 282.

Ideal Rapid Transit (N. Y.) A brief statement of the proposed systems with a description of a four tunnel scheme, and a discussion of the use of electricity for motive power. An "interview" by Frank J. Sprague, in N. Y. *Com Adv.* Reprinted in *St. Ry. Jour.*, March, 1891. *(Supplement.)* *St. Ry. Gaz.*, March, 1891.

Inter-Metropolitan. By O. T. Nichols. Gives details of the project of the road to begin at Union Square, pass through tunnel to Brooklyn, to Bushwick and Wallabout avenues, to City Halls, Brooklyn and New York, thence to Union Square. A circuit of about ten miles, three of which are over viaducts and seven miles in three different tunnels. *Trans. Soc. Engrs.*, Vol. I. p. 163.

Rapid Transit, continued.

Metropolitan Underground Ry. Co's Tunnel under the East River. Description with inset showing sections, elevators and other details connected with this proposed deep tunnel scheme. *Eng. News*, March 14, 1891, p. 249.

For New York City.

A discussion of the financial side of the question, based on the present traffic and earnings of the elevated system. By Theodore Cooper. *R. R. Gaz.*, March 20, 1891, pp. 191-2.

A well developed plan for rapid transit in New York City by means of electric cars running in deep tunnels; with map and illustrations; by Frank J. Sprague. *Elec. Eng.*, March 4, 1891.

An editorial showing the increase in the number of trips per inhabitant, etc. and points out its bearing on construction of rapid transit lines. *Eng. News*, April 23, 1887.

Borings for Rock Along Broadway. Table of results, with profile of surface, and underlying rock. *Eng. News*, July 18, 1891, pp. 62-3.

Problem in New York. A discussion and comparative estimate of the cost of various schemes. By W. Howard White. *R. R. Gaz.*, April 24, 1891, pp. 279-81.

Question in New York City. Description of plans proposed by Messrs. Worthen and Parsons, with extracts from the reports of four experts, Messrs. Bogart, Chanute, Cooper and Wilson, chosen to examine them. *R. R. Gaz.*, Oct. 9, 1891, pp. 700-3. *Eng. News*, Oct. 10, 1891, pp. 342-5, Oct. 17, pp. 372-4. Illustrations and further description of the plans of Messrs. Worthen and Parsons. *R. R. Gaz.*, Oct. 16, 1891, pp. 720-22. *Eng. News*, Oct. 24, 1891, pp. 389-91. (Inset of drawings and other illustrations, also editorial. Report of the Commission to the Council. *R. R. Gaz.*, Oct. 23, 1891, pp. 735-6. *Eng. Record*, Oct. 24, 1891, pp. 324-5.

Question. An article by T. G. Gribble, A. M. I. C. E., giving a general discussion and advocating a large subway. *St. Ry. Jour.*, March, 1891, pp. 147-9. *R. R. Gaz.*, April 3, 1891, pp. 227-8.

Reno Tunnel System for. A shallow tunnel or subway of square cross-section with four separate apartments and flat steel roof. Method of driving tunnel and construction described. Illus. By J. W. Reno. *R. R. Gaz.*, June 19, 1891, pp. 401-5. Brief account, *Elec. World*, June 20, 1891.

Suburban Rapid Transit Railroad, New York City. A comprehensive account of this elevated road, with illustrations of various structures. *R. R. Gaz.*, July 24, 1891, pp. 509-12.

Subsurface. Proposed plans for such a system on Broadway, New York. Fully illustrated. *Eng. News*, Jan. 2, 1886; also *R. R. Gaz.*, Jan. 11, 1886, and *Elec. World*, Jan. 9, 1886.

Transit in London, Rapid and Otherwise. An interesting paper by James A. Tilden, read before the Boston Soc. C. E., giving a description of the various means of transit, accommodations afforded, etc., *Jour. Assn. Eng. Soc.*, Mch., 1891, pp. 124-32. Discussion by Mr. H. D. Woods, describing briefly the Paris system, pp. 130-1. Abstract, *Ry. Rev.*, May 9, 1891, pp. 301-2.

See *Electric Railroads. Railroads Elevated. Railroads Underground.*

Reading and Indexing. Paper by Prof. Ira O. Baker. *Selected Papers C. E. Club, Univ. of Ill.*, 1889-90, pp. 11-16.

Reamers. *A Substitute for.* A new tool for sizing holes in metal by "broaching," claimed to be superior to the reamer in rapidity of working and wearing qualities. *Am. Mach.*, Jan. 21, 1889.

Reclamation of Arid Lands. See *Irrigation.*

Recording City Survey. See *Registry Bureaus, Phila.*

Refrigerating. See *Car Cooling. Ice Machines.*

Refrigerating Machinery. See *Heat, Mechanical Application of.*

Refrigerating Machine.

A paper by C. H. Peabody, describing the action of different kinds of machines giving methods of calculation, and nature and properties of various refrigerating fluids. *Tech. Quart.*, Vol. II., 1889, No. 4, pp. 313-48.

An American Ammonia. A description fully illustrated, of the Wood & Shipley machine. Lon. *Engineer*, Feb. 4, 1887.

On Board Ship. By T. B. Lightfoot. Gives early history of refrigerating machines and then describes in detail special machines for use on board of ships. *Trans. Soc. Engrs.*, 1888, pp. 105-124.

Performance of a 35-ton Refrigerating Machine of the Ammonia Absorption type. Abstract of paper by J. E. Denton, Hoboken, N. J., and read before the Am. Soc. Mech. Eng., at Erie, May, 1889. *Mechanics*, June, 1889.

Performance of a 35-ton refrigerating machine of the ammonia absorption type. Paper by Prof. J. E. Denton. *Trans. A. S. M. E.*, Vol. X., 1889.

Pontifex. A report of a board of experts appointed to test the machines erected for the Brooklyn Bridge Refrigerating and Cold Storage Company. The machines were to be equal to 40,000 lbs. of ice per 24 hours. They showed 9 per cent excess above the requirement. *Rens. Soc. Eng.*, Vol. I., p. 164.

Str. Fifeshire. Illustrated description of the most powerful refrigerating machinery ever put in a ship. It uses compressed air, and is to cool 84,000 cubic feet of space. Lon. *Engineer*, Oct. 14, 1887; *Sci. Am. Sup.*, Dec. 3, 1887.

Supplying Cold Air free from moisture. Chambers' patent, England. Described in *Iron*, July 3, 1885.

Refrigerating Plant, *Test of.* Paper read by Prof. De Volson Wood at the Cincinnati meeting of the A. S. M. E. *Trans. A. S. M. E.*, Vol. XI., 1890, pp. 830-9. Abstract in *Eng. News*, June 28, 1890, p. 617.

Refrigerating System. *Colorado Automatic Refrigerator System at Denver, Col.* A paper by A. McL. Hawks, giving a brief account of the workings of the system. *Trans. A. S. C. E.*, Vol. XXIV., May, 1891, pp. 369-92.

Refrigerators.

Tests of Refrigerating Machines. Description and tests of two German machines. Illus. Lon. *Eng.*, July 31, 1891, *et seq.*

The Theory of the Ammonia. Prof. C. M. Woodward. An exposition of the theory with application to a practical example. *Jour. Assn. Eng. Soc.*, Vol. V., p. 243.

Refuse. See *Garbage Disposal.*

Regenerative Furnace. See *Furnace, Siemens.*

Registry Bureaus. By John H. Dye. Gives details of the system used in Philadelphia for the registration of lots. A valuable article. *Eng. News*, March 18 and 19, 1889.

Re-railing Devices. See *Bridge Guards.*

Reservoir Walls.

Construction of. By M. E. Marichal, before the Philadelphia Engineers' Club. Gives diagram showing the number of cubic yards of masonry contained in dams for depth of water varying from 0 to 200 ft. *Am. Eng.*, June 8, 1887.

Failure of, at Lowell, Mass. A paving set on a puddled slope, 1½ to 1, slips when water is drawn down, after 13 years use. The puddle softened and forced out by pressure of back water coming from the natural bank. Illus. *San. Eng.*, Jan. 14, 1886.

See *Dams and Reservoirs.*

Reservoirs.

By Samuel McElroy, before the American Water-Works Association. Contains experience in construction of the Ridgewood reservoir. *Proc. Am. W.-W. Assn.*, Vol. VIII. ([illegible]), pp. 70-77.

Gives some of the most important features in the construction of large reservoirs for irrigation purposes. Abstracted from *Inst. of C. E.* in *Eng. News*, Jan. 8, 1887.

Ashland, Ky., New Reservoir at. Description, with section of the reservoir which has a capacity of 1,500,000 galls. Masonry Walls with outside earth embankments are used. *Eng. News*, April 11, 1891, p. 349.

Askti River, India, covering 4.2 sq. miles; capacity, 1,500 million cub. ft.; embankment, 1¾ miles long, with max. ht. of 98 ft., composed of earth. A full account by C. T. Burke, M. I. C. E., the Engineer. *Proc. Inst. C. E.*, 1884, Vol. LXXVI.

Athens, Ga. By C. H. Ledlie, before the Engineers' Club of St. Louis. Gives details of the construction of an earthen dam for the Athens, Ga., water-works. *Jour. Assn. Eng. Soc.*, April, [illegible]; *Eng. News*, May 1, 1885.

Bombay, India. Brief description with plan and section of the John Hay Grant reservoir of the Bombay system. The work comprises a storage basin 350 by 150 by 30, six filters of 16,000 square feet each and a clear well. *Ind. Eng.*, Sept. 15, [illegible].

Boston. A description, with plan and section, of the Fisher Hill, Boston, reservoir; also plan and sections of the gate chamber. *Eng. News*, March 24, [illegible].

Construction of. By Max Kraft. A description of the reservoirs constructed to supply power at the mines, Freiburg. *Van Nos. Eng. Mag.*, Vol. XXIX., p. 32.

Construction of. Gives details of the construction of the Dale Dyke embankment and the Druid Lake reservoir. Also points out the causes of their failure. *Van Nos. Eng. Mag.*, Vol. I., p. 226.

Covered.

A paper by Charles H. Swan, from Journal of the New England Water-Works Assn., Sept., [illegible]. *Eng. News*, Feb. 16, 1889.

Cheap. Brief Description of a small reservoir excavated in sandstone. By A. D. Foote. *Trans. A. S. C. E.*, Vol. XXV., August, 1891, pp. [illegible].

Coshocton, O. Description and illustrations of this 300,000-gallon reservoir. *Eng. Record*, Aug. 15, 1891, p. 169. *Eng. News*, Aug. 15, 1891, p. 139.

Service. A paper by W. Morris before the Inst. of Civil Engrs. Gives details of a large number of covered reservoirs in Europe. *Van Nos. Eng. Mag.*, Feb., 1884.

Service, at Nottingham, Eng. Dimensions, 180 x 141 x 16 feet. Plan, elevation, sections, and details are given. *Lon. Engineer*, July 10, 1885.

Embankments. A description of the slide of the sloped puddle walls of the Beacon St. Reservoir, Lowell water-works. Illustrated. *Jour. Assn. Eng. Soc.*, Vol. V., p. 197.

Fouling of Water in Deep Storage Reservoir. Causes and remedy. *Eng. News*, March 14, 1885.

Headwaters of the Mississippi River. A progress report, with cross-sections of dams, estimates of cost, etc. *Rep. of Chf. of Engrs.*, 1883, Vol. II., p. 1455.

Irrigation. See *Dams and Reservoirs.*

Molteno. By Chas. J. Wood. Gives details of the failure of the above reservoir at Cape Town, caused by fissures in the earth; also description of the methods used in its repairs. *Inst. of C. E.*, Vol. LXXXIX., p. [illegible].

Montmartre, Paris. These new reservoirs consist of two separate buildings, one of which contains water in two stories and the other in three. Fully de-

Reservoirs, *Montmartre, Paris*, continued.

scribed and illustrated in *Nouvelles Annales de la Construction*, 1890, p. 18. Abstract in *Proc. Inst. C. E.*, Vol. C., 1890, p. 448. See also *Eng. & Build. Rec.*, Aug. 9, 1890, p. 150-1. Lon. *Engineer*, Jan. 7, 1890, p. 40. *Sci. Am. Sup.*, March 8, 1890, p. 11819.

Naples, Galleries and Conduit in. An abstract from *Les Annales des Ponts et Chaussées* describing the reservoirs or galleries excavated in the heart of the mountains for the Naples water supply. *Eng. & Build. Rec.*, Aug. 11, 1888.

Nashville, Tenn. By H. De B. Parsons. Gives a brief description, with plan and sections of wall, of the new storage reservoir being constructed at Nashville, Tenn. *Eng. News*, June 16, 1888.

New Storage, Grand Junction Company, Ealing, Eng. Gives brief description, with plan, cross sections, elevations, etc., of a new storage reservoir, of a capacity of 51,000,000 gallons, constructed for the Grand Junction Water-Works Company, Ealing, Eng. Lon. *Engineer*, Aug. 24, 1888.

The New Prospect Storage and Distributing Reservoir, at Buffalo, N. Y. Earth embankments, capacity 135 mill. galls. Illustrated description. *Eng. News*, Jan. 10, 1891, p. 26.

Open or Closed? A paper by E. G. Beach, before the seventh annual meeting of the American Water-Works Association, discussing the question whether storage reservoirs be open or closed. Gives the experience of a number of cities. *Eng. News*, Dec. 3, 1887.

Railroads, Mills, etc. By S. F. Balcom, before the Illinois Society of Engineers and Surveyors. Gives the cost of constructing a first-class water station of the Illinois Central R. R., including reservoir. Gives directions for constructing reservoirs. *R. R. Gaz.*, Nov. 11, 1887.

Remarkable Breaks in a. By L. N. Lukens before the Philadelphia Engineers' Club. Gives details of a number of breaks and their repairs in Conshohocken Hill Reservoir. *Proc. Eng. Club, Phila.*, Vol. VI., pp. 147-50 (Dec., 1887); *Eng. News*, Aug. 15, 1888.

Reservoir "M" on the Titicus River, New York Supply. Account of this reservoir now under construction, with contoured plan of location and inset plate showing details of masonry structure. *Eng. News*, March 22, 1890, Vol. XXIII., p. 271.

Service. A most excellent paper, by Thos. Duncanson, on Service Reservoirs covered and uncovered, but mostly devoted to the former. Gives many plans and elevations. *Trans. Liverpool Eng. Soc.*, Vol. VI., pp. 118-36.

The Sodom and Bog Brook Reservoirs in the Croton Basin, N. Y. Illustrated description of the dams. *Sci. Am.*, July 12, 1890, p. 21.

Some Notes on Distributing Reservoirs. A paper by W. W. Curtis, read before the C. E. Soc. of St. Paul, setting forth requisites as to capacity, and discussing advantages of various forms, and details of construction. Illustrations of two circular reservoirs. *Jour. Assn. Eng. Soc.*, April, 1891, pp. 197-203.

Storage. A description of the building of the embankment of Ashland Basin No. 4, Boston Water Supply, by W. E. Learned, together with the methods used for the mixing and handling of concrete. Illustrated. *Jour. New Eng. W. Wks. Assn.*, Dec., 1887.

Storage Reservoirs. Removal of the Shallow Flowage from the Basin. By D. FitzGerald. Gives description of the work done on Basins Nos. II. and III. of the Boston Water-Works. *San. Eng.*, April 9 and 16, 1887.

Vyrnwy, Gauging at. See *River Gauging*.

Water Storage in the West. Article by Walter G. Bates describing briefly several of the more noteworthy Western storage reservoirs, with illustrations of noted works. *Scribner's Magazine*, Jan., 1890, Vol. VII., p. 3.

Reservoirs, continued.

For Water Supply. Capacity of Storage. Treats of Hawley's formula, the empirical method, and then gives a new graphical method for computing the capacity for storage. *Van Nos. Eng. Mag.*, Vol. XXIX., p. 67.

West Hallington Reservoir. Paper by C. G. Hensell, describing the construction of this large reservoir. Folding plate. *Proc. Inst. C. E.*, Vol. CII., 1890, paper No. 2465, pp. 271-82.

See *Irrigation, Engineering, Water-works, Weir's.*

Resilience. See *Steel.*

Resistance, *Compensated Standards of.* See *Electrical Resistance.*

Resistance Governor. See *Electric Resistance Governor.*

Resistance of Materials *to Deformation.* See *Proportional Resistances.*

Resistance *of Trains.* See *Train Resistance.*

Retaining Walls.

Algebraic determination of the lines of pressure within retaining walls of all the customary cross-sections, for quick work. *Zeitschrift f. Bauwesen*, 1885-93.

An attempt to reconcile theory with practice. By Casimir Constable. *Trans. A. S. C. E.*, Vol. III., pp. 67-75.

By William Cain. A theoretical discussion in which the earth is considered a homogeneous and incomprehensible mass, made up of little grains possessing friction but without cohesion. *Van. Nos. Eng. Mag.*, Vol. XXII., p. 265.

By Prof. C. E. Greene. A brief graphical determination of pressure at any bed joint. *The Technic*, Univer. of Michigan, 1888.

By E. S. Gould. An attempt to plan the calculation of ordinary retaining walls upon its simplest footing. *Van Nos. Eng. Mag.*, Vol. XVI., p. 11.

By E. S. Gould. A sketch of the work of Mr. Dubosque with formulæ. *Van Nos. Eng. Mag.*, Vol. XXVIII., p. 204.

By Arthur Jacob. Reviews the principles involved in determining the strength of walls to support earth-work, and gives some simple rules. *Van Nos. Eng. Mag.*, Vol. IX., p. 104.

By Samuel McElroy. Gives practical notes on the construction of retaining walls. *R. R. Gaz.*, Nov. 9, 1888.

By James S. Tate. Gives methods for finding the correct dimensions of the different forms of wall that are in general use. *Van Nos. Eng. Mag.*, Vol. IX., p. 205.

By W. M. Thornton. Gives a concise account of Lamé's theory of pressure of earth and its application to retaining wall. *Van. Nos. Eng. Mag.*, Vol. XX., p. 313.

For Earth Treated Graphically. By Prof. Chas. E. Greene, University of Michigan. *Eng. News*, Dec. 29, 1888.

Mathematical investigation of the question of the use of earth behind a retaining wall. By J. R. Allen. *Van Nos. Eng. Mag.*, Vol. XVII., p. 155.

Methods of Calculating and Designing. By C. P. Karr. A series of articles following the methods of Dr. Weyrauch and Prof. Rankine, with additional examples from the French practice. *Building*, Dec. 17, 1887, *et seq.*

Quay and Other. By J. D. Van Buren, Jr., Object of the paper is to establish practical formulæ for the dimensions of quay and other retaining walls. A mathematical discussion. *Trans. A. S. C. E.*, Vol. II., p. 193.

Southern Railroad of France. These are constructed on an entirely new pattern being shaped with triangular counterforts, saving from ⅜ to ½ of the masonry required for vertical walls. Abstracted from the French. *Proc. Inst. C. E.*, Vol. LXXXVI., p. 396.

Retaining Walls, continued.

Stability of and the Thrust of Earth upon. By A. Gobin, in *Annales des P. & C.* Vol. VI., 1883, p. 98; 62 wood cuts. An abstract given in *Proc. Inst. C. E.*, 1884, Vol. LXXXVI. The results accord with experiment, and the theory takes account of the cohesive forces of earth. P. 387.

Standard Retaining Wall Section, N. Y. C. & H. R. R. R. This section is subtially the same as was used in the approaches to New York City, and has been proven to be amply strong, when well constructed, by over fifteen years of service. Very brief illustrated article. *Eng. News*, Nov. 30, 1889, Vol. XXII., p. 513.

Study of the resistance of materials in retaining walls. By M. Heller. A lengthy article dealing with retaining walls under every conceivable condition, but apparently containing nothing new. *Annales des P. & C.*, 1885, May, pp. 793-790.

Theory and Practice. By Graham Smith. *Eng. News*, June 20, July 18, 1876.

See Masonry Abutments, Sustaining Walls.

Rheostat. See *Electric Welding.*

Rifle. *The Military, the Development of.* By Lieut. Joseph M. Califf, Third U. S. Artillery. *R. R. & Eng. Jour.*, Jan. and Feb., 1889.

River and Canal Works. A paper by L. F. Vernon Harcourt, M. A. M. Inst. C. E., giving descriptions and illustrations of a few river and canal improvements. *Proc. Inst. C. E.*, Vol. XCVI., p. 182.

Discussion and correspondence on paper by L. F. Vernon Harcourt. *Proc. Inst. C. E.*, Vol. XCVI., p. 224.

River and Harbor Works *in Northern France.* By L. Schrader. Interesting description of the works on the lower Seine, and of the harbors of Nantes, St. Nazaire, Rouen and Havre. *Zeitschr. d. Oester. Ing. u. Arch. V.*, 1887, pp. 130-134.

River and Harbor Bills *in the U. S. Congress.* By Clemens Herschel. A valuable article, showing the defects in the present system and contrasting it with that pursued in other countries. *Jour. Assn. Eng. Soc.*, Vol. IV., p. 93.

River and Harbor Improvement.

Views of Col. Craighill. An important suggestion. *Eng. News*, June 19, 1886.

On the Atlantic coast of the Southern States. Abstracts from the report of Gen. Gillmore. Gives details of the improvement at Charleston Harbor. *Van Nos. Eng. Mag.*, Vol. XXIV., p. 1.

The Port of Nantes. Detailed account of various improvements of this port, with account of progress and cost of Work. *Lon. Eng.*, July 10, 1889, *et seq.*

River Bars.

Cause of the Formation of, at the Mouths of Rivers, as Shown in the Examination of the Connecticut River. By Gen. T. G. Ellis. Gives results of the study of changes going on at the mouth of the Connecticut River. Contains information relative to the water-shed and flow of Connecticut River, and of the tidal currents of Long Island Sound. *Trans. A. S. C. E.*, Vol. II., p. 313.

At the Mouths of Tidal Estuaries. Valuable paper read by William H. Wheeler, Feb. 4, 1890. Folding plate shows plans and sections of different bars. *Proc. Inst. C. E.*, Vol C., paper No. 2,438, pp. 116-43; discussion, pp. 144-78, correspondence, pp. 179-216.

See River Hydraulics.

River Basins.

Basin and Regimen of the Mississippi. By C. M. Woodward. Gives much information relating to the river and its action. *Van Nos. Eng. Mag.*, Vol. XXVII., p. 18.

The Illinois River Basin and Its Relation to Sanitary Engineering. A paper by L. E. Cooley, C. E., discussing the conditions existing at present in the Illinois River basin. *Prelim. Rep. Ill. State Board of Health*, 1889, pp. 49-81.

River Discharge.

By David Stevenson. Remarks on the methods employed to determine velocity, direction of currents, sediment, etc., of rivers. *Van Nos. Eng. Mag.*, Vol. VII., p. 580.

Computation of. Paper by Dr. Geo. H Johnston, C. E., discussing the best method of computing the discharge when several velocities are observed. Original formulas. *Eng. News*, Sept. 6, 1890, p. 215.

Obstruction to, by Bridge Piers. By Gen. Q. A. Gillmore. Gives a short history of the case of the Chemung River, N. Y., and gives some interesting comparisons between the Chemung and Mississippi rivers. *Van Nos. Eng. Mag.*, Vol. XXVI., p. 441.

Reports on methods and results of eight parties on the Mississippi River, each for one year's observations, 1881 and 1882; together with computed daily discharges of the river at many other points for the year 1882. *Rep. Miss. Com.*, St. Louis. 1883, p. 176.

See *Flood. Floods. River Hydraulics. River Improvements.*

River Gauging.

Account of Recent, in Holland. These measurements show the inaccuracy of loaded tubes in ordinary rivers, the results they give being always too large, sometimes by 16 per cent. Various other interesting results regarding accuracy of methods of gauging. *Wochenschrift d. Oester. Ing. u. Arch. V.*, 1886, pp. 233-239, 243-247.

And the Double Float. By S. W. Robinson. A brief history of the double float is given, after which the author treats of the discrepancies between the values obtained by floats and meter and their causes. *Van Nos. Eng. Mag.*, Vol. XIII., p. 99.

On the Elbe in 1886-7. By A. Riogel. [Made with current meters, essentially like Harlacher's.] *Der Civilingenieur*, 1888, pp. 505-528.

At Vyrnwy Reservoir. By J. H. Parkin, before the Students' Institution of Civil Engineers. Gives details of the gauging to determine the daily discharge of the Vyrnwy River. *Proc. Inst. C. E.*, Vol. XCII., pp. 353-367.

See *River Hydraulics.*

River Hydraulics.

Determination of the Flood Discharge of Rivers and the Back Water caused by Contraction. By Wm. R. Hutton; with discussion by Theo. R. Ellis and Robt. E. McMath. This is the *Elmira Crossing case*, between the Lackawanna and Erie roads. The discussion mostly valuable in showing the worthlessness of all formulas for such uses. *Trans. A. S. C. E.*, Vol. XI. (1882), p. 211. Same case and formulas discussed by Gen. Q. A. Gillmore in *Van Nos. Eng. Mag.*, Vol. XXVI., June, 1882, p. 441.

See *Hydraulics.*

Mississippi River. By James B. Eads. A review of the report of Humphreys and Abbot on the Mississippi River. *Van Nos. Eng. Mag.*, Vol XIX., p. 311.

Of the Mississippi River. A discussion of the relation between velocity, suspended earthy matter, and of the bed of the Mississippi River. By H. L. Abbot, in reply to J. B. Eads. *Van Nos. Eng. Mag.*, Vol., XX., p. 1.

A reply to the above paper. By J. B. Eads. *Van Nos. Eng. Mag.*, Vol. XX., p. 154.

The Mississippi as a Silt-Bearer. By Robert C. McMath. *Van Nos. Eng. Mag.*, Vol. XX., p. 219.

Physics and Hydraulics of the Mississippi. A reply to criticism made by Dr. Hagen, by Gen. Humphreys and Gen. Abbot. *Van Nos. Eng. Mag.*, Vol. XVIII., p. 1.

River Hydraulics, continued.

Practical Consequences of the Variation of Wet Section of Rivers. A very valuable study of rivers of variable stage. By Robert E. McMath. *Trans. A. S. C. E.*, Vol. IV., (1880), pp. 377-380.

Profile and Cross-section of. By Opel. The writer thinks that the normal profile of every stream not flowing in a rocky bed is parabolic. *Deutsche Bauzeitung*, 1885, pp. 134-138, 147-151.

Silt Movement in the Mississippi River. By R. E. McMath. Gives results of observations at St. Louis. *Jour. Assn. Eng. Soc.*, Vol. I., p. 266.

Three Problems in River Physics. A paper by J. B. Johnson before the Am. Assoc. Adv. Science. Discusses (1) The Transportation of Sediment and the Formation and Removal of Sand Bars; (2) The Flow of Water in Natural Channels; (3) The Relation of Levees to Great Floods and to the Low-Water Navigation of Rivers. Reprinted in *Eng. News*, Aug. 1, 1885.

See *Bridge Piers. Hydraulics. Floods. River Discharge. River Gauging. River Improvements. Rivers.*

River Improvement.

An account, with plates, of a very successful improvement of the mouth of a tidal river. *Proc. Inst. C. E.*, Vol. XC., p. 344.

Article by W. H. Wheeler, M. I. C. E., describing the "Eroder," a new machine for breaking up bars and deposits and mixing the material with the water. Illus. Lon. *Engineer*, July 18, 1890, pp. 42-3.

Discussion on, by Messrs. J. H. Streidinger, M. Am. Soc. C. E.; William E. Worthen, Past Pres. Am. Soc. C. E., and George H. Henshaw. *Trans. A. S. C. E.*, May, 1889, Vol. XX., p. 229.

And Waterways, The Regulation of, with a View to the Prevention of Floods. By Gustav Ritter von Wex. A valuable article. *Van Nos. Eng. Mag.*, Vol. XXVIII., p. 1, 8.

Adaptation of Movable Dams to the Improvement of the Low-Water Navigation in the Ohio and Kanawha Rivers. By Prof. L. M. Haupt. Data taken from the reports of Chief of Engrs. of U. S. Army, and brought together in good shape. Thirty-three pp., 5 plates. *Proc. Eng. Club, Phila.*, Vol. IV., No. 1.

Application of the Transporting Power of Water to the Deepening of Rivers. Paper by W. H. Wheeler, M. Inst. C. E., showing that the transporting power of water can be economically applied to deepening and improving rivers. Lon. *Engineer*, Oct. 25, 1889, Vol. LXVIII., p. 343, *et seq.*

On the Atlantic Coast. By W. P. Craighill, before Annual Convention, 1888. Gives description of the treatment of several tidal rivers on the Atlantic Coast. Illustrated, pp. 20. Discussion, p. 19. *Trans. A. S. C. E.*, Nov. 1888, Vol. XIX., p. 233. *Eng. & Build. Rec.*, Aug. 11, 1888.

Bank Protection.

By W. S. Chaplin. Gives illustrated descriptions of some of the methods employed in Japan. *Van Nos. Eng. Mag.*, Vol. XIX., p. 119.

Methods pursued on the Missouri River, with many cuts showing mattresses in process of construction, mattress boats, etc. *Rep. Chf. of Engrs.*, 1883, Vol. II., p. 1197.

See also *Shore.*

On the Mississippi River. Illustrated article in *Eng. News*, Dec. 14, 1889, Vol. XXII., p. 558.

Use of Fascines in the Public Works of Holland. Detailed account. By T. C. Watson before the Inst. C. E. Reprint, *Eng. News*, June 23, 30, 1887.

Bavaria. An illustrated article condensed from the *Journal of the Austrian Society of Engineers and Architects*, No. 43, 1887. Shows method of constructing embankment, shore protection and dikes. Gives rule for the proper proportion of depth and width of channel at lowest water line. *Eng. News*, March 17, 1888.

River Improvement, continued.

Blasting of Iron Gates (River Danube.) Paper describing this work and the plant now established for doing the work. *Verhandlungen des Vereins zur Beforderung des Gewerbfleisses*, Feb. 1891, p. 119. Abs., *Proc. Inst. C. E.*, Vol. CIV., 1891, pp. 323-4. Abstract in *Eng. News*, July 11, 1891, pp. 30-1. *Sci. Am. Sup.*, No. 810, July 11, 1891.

Brazos, At the Mouth of the. Pamphlet issued by "The Brazos River Channel and Dock Company," including reports on the work by Mr. E. L. Corthell, Chief Engineer, and Mr. Geo. Y. Wisner, Resident Engineer. *The Brazos River Channel and Dock Company*, Equitable Building, Boston.

Canalization of the Mosel, from Metz to Koblenz. Fall of stream is 309 feet in the distance of 188 miles. Thirty-two dams are projected, from 7.25 to 9.80 feet high, with a total height of 280 feet. the navigable depth to be at least 6.5 feet The cost, with locks for single boats only, is estimated at about $2,600,000. *Deutsche Bauzeitung*, 1886, pp. 178-179.

The Canalization of Rivers. Abstract of a valuable paper by M. A. Route., Ing. en Chef des Ponts et Chaussees. Third Congress Frankfurt-on-Main, 1888. Lon. *Engineer*, Jan. 18, 1889, *et seq.*

Cassel. Short extract from *Centralblatt der Bauverwaltung*, giving illustrated description of methods employed, including a bridge and "needle dam." *Eng. News*, May 3, 1890, Vol. XXIII., p. 416.

Of Channels in Sedimentary Rivers. Paper by George H. Henshaw, with discussion. Illustrated. *Trans. A. S. C. E.*, March, 1889, Vol. XX., p. 109.

Ravine du Sud in the Island of Hayti, Characteristics of the, and Plan for Averting its Overflow. A paper by J. Foster Crowell, describing plans adopted for this torrential stream. Methods were by excavation for a channel, and bank protection. Discussion on discharge of such streams and on their control. *Trans. A. S. C. E.*, Vol. XXIV., June, 1891, pp. 470-92.

Clyde. A paper by C. A. Stevenson before the Institution of Mechanical Engineers, describing the dredging of the lower Clyde estuary. Gives full description of the hopper dredge used there. Lon. *Eng.*, Sept. 9, 1887.

Columbia River Jetty. This is a pile and mattress jetty about 4½ miles long. Method of construction described, and illustration of the revolving pile driver used, is given. *Eng. News*, April 15, 1891, pp. 368-9.

Converting the River Main, between Frankfort and Mayence, into a slack-water navigation, by means of five movable dams, locks, channels for rafts and other works. To be finished Oct. 1, 1886, at a cost of about $1,375,000. *Zeitschrift. d. V. D. Ing.*, Oct. 25, 1884.

Coosa. Design, with accompanying map, for improving the river by pumping into it a large volume of water. *Ind. Eng.*, May 16, 1891, pp. 392-3.

Danube. Improvement of at Vienna. A lecture before the Society of Austrian Engineers and architects. Trans. by Gen. G. Weitzel. *Van Nos. Eng. Mag.*, Vol. XXVI., p. 193.

Danube; two lectures by Gustav Wex, Chief Director of the work. Translated and issued by the Engr. Dept., U. S. A., as two separate pamphlets of 25 pp. and 60 pp. respectively.

Danube, Mouths of the, and Improvement of the Mouths of Rivers in Non-Tidal Seas. Paper by Major Stokes describing various methods that have been employed. *Eng. News*, May 26, June 9, 1877.

Danube and its Trade. A valuable paper by Lieut. Gen. Sir John Stokes, touching briefly on improvements, as well as giving a discussion of the commercial features of the subject. *Jour. Soc. Arts*, May 2, 1890, p. 550.

In France. By Wm. Watson. Gives an account of an excursion on the Main River, and describes Desfontaine's Drum Weirs. *Van Nos. Eng. Mag.*, Vol. XVII., p. 253 By Wm. Watson. Gives an illustrated description of Poirees

River Improvement, *In France*, continued.

system of movable dams. *Van Nos. Eng. Mag.*, Vol. XVIII., p. 339. By Wm. Watson. An illustrated description of the Chanoine system of falling gates *Van Nos. Eng. Mag.*, Vol. XVIII., p. 458. By Wm Watson. Illustrated description of the Chanoine movable dam at Paris. *Van Nos. Eng. Mag.*, Vol. XVIII., p. 481.

At Harlem, Mo., and Kansas City, Mo. An account of recent improvements, cost, methods employed, including experience sinking piles by water jet, etc. Fully illustrated. Details of nozzle used in pile-sinking given. *Report of Missouri River Commission, Chief of Eng. Rep.*, 1889. Appendix WW, p. 2785.

Irrawaddy Embankments. A series of illustrated papers giving a brief description of these embankments, with notes as to their efficiency, etc. *Ind. Eng.*, Dec. 27, 1890, *et seq.*

Kanawha River, Lock and Dam No. 6. Described by the Res. Engr. *Eng. News*, Nov. 20, 1886.

Methods employed and results obtained on the River Tees. Illustrated. Channel obtained over bar at mouth. *Proc. Inst. C.E.*, Vol. XC., pp. 344-54.

Mississippi.

A discussion of what has been done, of the results obtained, and of what may be expected. Estimates $100,000,000 necessary to obtain low water depth of 12 ft. to Vicksburg, 10 ft. to Cairo, 8 ft. to St. Louis, 5 ft. to Keokuk and 3 ft. to St. Paul. Undertakes to show why greater results are unattainable by the methods employed, also that these methods are the only ones available for the Mississippi. *R. R. Gaz.*, April 22, 29, etc., 1887.

A paper prepared for the University of the City of New York, by William Starling, M. Am. Soc. C. E. Illustrated *Trans. A. S. C. E.*, March, 1889. Vol. XX., p. 85.

General review of methods employed with illustrations of dikes and revetments. Supplement to *Harper's Weekly*, Oct. 4, 1890.

Cut-offs on. Their effect on the Channel above and below. By C. G. Forshey. *Trans. A. S. C. E.*, Vol. V., (1876), p. 317-22.

Cut-offs. By C. G. Forshey. Gives particulars of Shreve's or Red River, Racourt, Terrapin Neck, and Palmyra cut-offs. *Trans. A. S. C. E.*, Vol. V., p. 317. See *Mississippi River.*

Effect of Levees on Stage and Discharge. A strong paper by Gen. C. B. Comstock, Pres. Miss. Riv. Commission, combating the theory that levees cause an increased discharge at a lower stage on the Mississippi River. *An. Report Chief Engrs., U. S. A.*, 1888. Part IV., p. 2270.

Improvement of Upper Mississippi. Extracts from annual report of Capt. Mackenzie. *Eng. News*, Aug. 21.

Keeping the Mississippi Within Her Banks By William Starling, C. E. A popular article describing the methods of bank protection and levee building. (N. Y.) *Eng.*, April, 1891, Vol. I., No. 1, pp. 111.

Methods employed on the Miss. River below Cairo. *Rep. Miss. Riv. Com., St. Louis*, 1883, pp. 349-475.

Its Mouth. A paper by J. G. Barnard. *Trans. Am. Soc. C. E.*, Vol. IV., (1875), p. 104-21. Discussion, p. 180-209; also paper by W. Milnor Roberts, pp. 311-33; and discussion, Vol. V., p. 275-97.

Protection of the Lower Mississippi Valley from Overflow. Condensed reprint of a paper by Prof. J. B. Johnson, read in 1884. *Eng. News*, April 19, 1890, Vol. XXIII., p. 364.

Protection of the Lowlands below Cairo from Overflow. By J. B. Johnson. The present conditions stated and some remedies suggested. *Jour. Am. Eng. Soc.*, Vol. III., p. 169.

River Improvement, continued.

Report of the Mis. Riv. Com. for 1884, including reports from the Secretary and other officers in charge of works and surveys; also results of high water observations on the floods of 1882, 1883 and 1884; precise levels from Fulton, Ill. to Chicago, Ill.; also description and elevation of permanent bench marks from Greenville, Miss., to Carrollton, La., results of experimental work with current meters at various discharge stations. Apply to Sec. Miss. Riv. Com., St. Louis, Mo.

Wing Dams on the Upper Mississippi. By Ed. P. North. Gives methods, cost and results. *Trans. A. S. C. E.*, Vol. VI., (1877), p. 258-276.

Missouri River Commission, Report of the year ending June 30, 1890. Contains report on borings in the Missouri river Valley, with 27 plates, description and elevations of bench marks and reports on triangulation and various improvement works.

New Channel from Rotterdam to the North Sea. From the German. A fine account of this important work, with maps, plans, sections of jetties, etc. Consists of the cutting of a channel three miles long through low lands and sand dunes; the closing of the old river; the construction of two parallel jetties far into the sea; the regulation of the river above the new cut; the construction of a large dyke ten miles long, and the construction of a canal with locks. All described in *Eng. News.*, Aug. 28, 1886.

A paper by C. DeGraffenried, giving a short sketch of the regulation of the waters of the Jura river. *Van Nos. Eng. Mag.*, Vol. XXIV., p. 58.

An article on the improvement of rivers having considerable fall, and with beds liable to scour. Proposes to use low training banks and movable weirs. *Van Nos. Eng. Mag.*, Vol. XXVII., p. 102.

Newer, on the Isar. By A. Wolf. Protection of banks, training walls, etc. With illustrations. *Zeitsch. f. Bauwesen*, 1884, pp. 516-527.

Nile. By Benj. Baker. Paper supplementary to the article in Beardmore's "Hydrology." The data being from Egyptian Gov. document. *Van Nos. Eng. Mag.*, Vol. XXIII., p. 405.

Nile. See *Nile.*

Non-Tidal. Illustrated by the River Tiber, by Wm. Shelford. Contrasts tidal and non-tidal rivers, and discusses remedies against floods. Favors levees but admits their deteriorating influence on the channel, with discussions. *Proc. Inst. C. E.*, Vol. LXXXII., p. 2.

Non-Tidal Rivers, being translations from the Russian, German and French, and fairly covering the whole ground of river improvements as now practiced in Europe. Issued by the Engr. Dept., U. S. A., as a folio pamphlet of 500 pp. with many plates. See also *Low Water Navigation.*

At Omaha, Sioux City, Nebraska City, St. Joseph and Atchison. The latter being given in considerable detail, with illustrations and methods of conducting the work, plans, etc. *Report of Missouri River Commission*, 1889, or *Appendix WW of Rept. Chf. of Eng.*, 1889.

Ottawa River Dam and New Canal at Carillon, Que. Detailed description of wooden dam and canal. Illus. By Andrew Bell. *Eng. News*, March 17, 1883, pp. 122-4.

Permeable System of Works. Theory and Application of, for the improvement of Silt Bearing Rivers. By R. E. McMath. Presents arguments, relative to. *Eng. News*, Nov. 1, 1879, pp. 353-5.

Potomac River Flats in front of Washington. A complete history of the river's changes for 100 years; an account of all former attempts at reclamation; a history of the long bridge and its effects; together with a full account of the methods now in operation for reclaiming them. *Rpt. Chf. of Engrs., U. S. A.*, 1882, Vol. I., p. 770.

River Improvement, continued.

Project for the Vienna River at Vienna. Proposed to carry the stream through the city in a subterranean covered channel. Maximum discharge about 10,000 cubic feet per second. *Wochenschrift d. Oesterr. Ing. u. Arch. V.* 1880, pp. 87-90.

Rhone, Mouth of. The problem of improving, discussed at length by Adolphe Guerard. Translated from the French. *Proc. Inst. C. E.*, Vol. LXXXII., p. 305.

Sand Bar Removal.

By Harrows. Bikoff's system. *Ingenieur*, St. Petersburg, Vol. IV., 1887, p. 379. An abstract of the method in *Proc. Inst. C. E.*, 1886, Vol. LXXXVI., p. 393.

By Propeller Sluicing. An account of sand-bar removal on the Columbia River, Oregon, by means of a screw propeller, commercially successful. *Trans. Inst. C. E.* Vol. LXXXIII., p. 386.

By Sluicing, and mechanical appliances, at Liverpool. Illustrated. *Proc. Inst. C. E.*, Vol. XC., pp. 308-318.

Seine. Improvement of. By L. F. Vernon-Harcourt. A paper read and discussed before the Inst. of C. E. Gives a full account of the methods employed and the work done. *Proc. Inst. C. E.*, *Eng. News*, July 3-31, 1886.

Seine. By Vernon-Harcourt. A paper which considers. I. The Hydrology of the Seine. II. Inland Navigation Improvement Works. III. Estuary works. This river furnishes one of the most interesting studies to be found in the department, river physics. Floods and their prediction, movable dams and training works, and the jetties and other works at its mouth are among the problems here discussed. The whole occupies over 150 pages in *Proc. Inst. C. E.*, Vol. LXXXIV., p. 210.

St. John's, (Fla.) Abstract of a report to the Municipal Board of Jacksonville, Fla. By J. B. Eads. *Van Nos. Eng. Mag.*, Vol. XVIII., p. 419.

Sulla sistemazione dei fiumi e torrenti della Carinzia. 9 tables and a handsome folding plates showing the details of this work. *Giornale del Genio Civile*, Anno XXVIII., April, 1890, pp. 169-99.

Thames. A description of the river and various improvement works. By J. B. Redman before the Inst. C. E. Reprint, *Eng. News*, Sept. 8, Oct. 6, 1877.

Tiber, Discharge and Regulation of the. By T. Montanari. An examination of the measurements made at different times, with a view to the construction of a new scale for the river; also discusses mean velocities, rainfalls, etc. *Van Nos. Eng. Mag.*, Vol. XXIX., p. 502.

Tidal Works on the Seine and other Rivers. A summary of 16 pp. *Trans. Liverpool Eng. Soc.*, p. 83 (1886).

Tidal River, the Successful Work of Improvement of the Mouth of a. The Welham in England. Illustrated. *Proc. Inst. C. E.*, Vol. XCV.

Timber Structures for Controlling Water. Proposed system, by H. C. Herton. Description with examples of application to breakwaters and wing dams. *Eng. News*, April 10, 1880, pp. 172-3.

Weaver, Eng. By J. A. Sauer, before the Society of Arts Canal Conference. Gives a short description of the improvement of the River Weaver, England. *Jour. Soc. Arts*, June 1, 1888.

See *Appliances. Blasting. Dikes. Harbor and River. Jetties. Levee Construction. Sounding Apparatus.*

River Physics. See *River Hydraulics. River Improvements. Rivers.*

River Pollution.

By J. P. Kirkwood. An abstract from the report of the Massachusetts Board of Health. A valuable article. *Van Nos. Eng. Mag.*, Vol. XVI., p. 146.

River Pollution, continued.

By Manufacturing Wastes. Report of Samuel M. Gray, City Engineer, Providence, R. I., giving valuable data and facts, chemical analyses etc., concerning properties and disposal of manufacturing wastes. *Rep. City Eng., Prov., R. I.*, 1884. No. 19.

In the United States. The valuable paper by Chas. C. Brown, read before the Eng. Club of St. Louis and published in *Jour. Assn. Eng. Soc.*, Oct., 1890, is reprinted in *Eng. News*, March 21, 1891, pp. 283-4, *et seq.*

See *Water Filtration and Purification. Water Supply Abstract, etc.*

Rivers.

By Edward Easton. Contains review of ancient water-works and treats of the English rivers. Advocates a local government board for dealing with every watershed in England. *Van Nos. Eng. Mag.*, Vol. XIX., p. 145.

By W. H. Wheeler. Treats of the rivers in the Eastern Midland district of England, the Withan, Wiland, Ouse, etc.; also of floods of their remedy. *Van Nos. Eng. Mag.*, Vol. XXVII., p. 181.

Backwater in. See *Backwater*.

Canalization of. See *Canals. River Improvement.*

And Canals. A short description of the more recent methods of transport on rivers and canals, and details of their employment in Germany. *Van Nos. Eng. Mag.*, Vol. XXVIII., p. 205.

Of China. Gives discharge, sediment, and other data for the Yang-Ste, Hoang-ho and Pei-ho rivers. *Van Nos. Eng. Mag.*, Vol. XXIV., p. 63.

Control and Management of. By J. C. Hawkshaw, before the Dublin meeting of the British Assn. Shows some of the difficulties encountered in river work. *Van Nos. Eng. Mag.*, Vol. XX., p. 491.

Conservancy. By J. C. Hawkshaw. Illustrated by drainage administration in Holland. *Van Nos. Eng. Mag.*, Vol. XXIII., p. 250.

Debris in. See *Mining. Debris*

Mississippi; See *Rivers, The Mississippi River*, below.

Tidal. The Horizontal Range of Flow of a given particle of water during the ebb and flow of a single tide. The discussion bears on the discharge of sewage with such streams. Illustrated. *Proc. Inst. C. E.*, Vol. LXXXVI., p. 253.

Tidal. The relative value of tidal and upland waters in maintaining rivers, estuaries and harbors, by Walter R. Browne, before the Inst. Civ. Engrs. Endeavors to show when embankments or jetties limiting the tidal basin may be advantageous and when not. *Van Nos. Eng. Mag.*, Sept., 1885.

Rivers, The Mississippi River.

Dangers Threatening its Navigation. By B. M. Harrod. Favors bank protection and levees on the theory that confining the flood waters will not increase flood heights. *Trans. A. S. C. E.*, Vol. VII. [1878], p. 213.

Flood Heights in Mississippi River. A paper by Mr. William Starling, M. Am. Soc. C. E. Pp. 33; plates, 6. *Trans. A. S. C. E.*, May, 1889, Vol. XX., p. 195.

Floods of Earlier Times. By J. A. Ockerson. With an elegant lithographic map of 130 miles of the river from Vicksburg to Natchez. Article shows that the floods of former times were not materially different in height from those of to-day. With discussion by Robt. E. McMath. *Jour. Assn. Eng. Soc.*, Vol. IV., p. 177.

Geology of, Bottoms. Being a report on numerous borings made throughout the lower river bottoms, with important conclusions. *Rep. Miss. Riv. Com., St Louis*, 1883, p. 479.

Map of, from Cairo to the Passes, in Thirty-two Sheets, besides three Index Sheets. Scale, one inch to the mile. Prepared under the direction of the *Mississippi*

Rivers. **The Mississippi River,** *Map of, etc.*, continued.

River Commission, from the most careful and accurate topographical surveys ever made in this country over so large an area. The sheets were first drawn to a scale of 1:10,000, and on all these 3 foot or 5 foot contours have been drawn. These contours have been carefully located from one to one and a half miles back from either bank. These are not shown on the reduced map, however, but the topography is drawn, and all landings, position of channel at time survey was made, and distances from Cairo was given. Address Secretary Miss. Riv. Com., 1828 Washington avenue, St. Louis, Mo.

Silt Movement by the Mississippi, its volume, cause and condition. By R. E. McMath. The results of a series of observations made at St. Louis in 1879. *Van Nos. Eng. Mag.*, Vol. XXVIII., p. 20.

True Source of. As determined by Captain Willard Glazier, July, 1881, with map, being in Lake Glazier 3 feet above Lake Itasca. *Proc. Royal Geog. Soc.*, Jan 1885.

See *River Improvement, Mississippi River, Hydraulics.*

Rivers and Harbors.

Annual Report of the Chief of Engrs., U. S. Army, 1884. Four volumes, with table of contents and alphabetical index. Gives engineering and financial information concerning all public works under this department, including the report of the Miss. Riv. Com. for 1883. Apply to Chief of Engrs. U. S. Army, Washington.

Improvements of. See *Public Works.*

Rivet Hole Punching. See *Punching, Improved Method.*

Rivet Holes *in steel plates.* Results of experiments carried out under the direction of the English Board of trade on the relative effect of punching and drilling steel plates of different thicknesses. *Van Nos. Eng. Mag.*, Vol. XXV., p. 155.

Riveted Joints.

An account of 180 experiments summarized in a paper before the Inst. of Mech. Engrs., England, by Prof. A. B. W. Kennedy. The conclusions relate to "joints made in soft steel plates with steel rivets, the holes all drilled and the plates unannealed." Tabular proportions and formulas given. Lon. *Eng.*, July 3, and 10, 1885. Also *Van Nos. Eng. Mag.*, July, 1885.

For Boiler Work; account of some tests of. By C. H. Moberly. Full description and results of experiments made to find the best form of joints for steel boiler plates. *Van. Nos. Eng. Mag.*, Jan., 1884.

In Boiler Shells. A paper by W. B. Le Van read before the Frank. Inst. The author gives the result of his experiments in designing joints. Numerous illustrations. Reprinted in *Sci. Am. Sup.*, Nos. 809-10, July 4, 11, 1891.

Details of tests of 96 specimens of O. H. steel plates, 11 specimens of rivet metal, and of 194 riveted joints made at the *Watertown Arsenal* in 1885. Report for that year. *Ex. Doc. No. 36, 49th Cong., 1st Session.*

Kensington Nat'l Agricultural Hall. Design for a truss of 160 ft. span. *Sci. Am Sup.*, June 5, 1886.

Some Recent Tests on Iron and Steel Riveted Joints. Paper by A. W. Brightmore before Liverpool Eng. Soc. *Trans. Liverpool Eng. Soc.*, 1888. Vol IX, pp. 1-14.

Tests of. As to the rivets filling the holes when the number of the plates is large. Illustrated by sections through the line of rivets. *R. R. Gaz.*, Oct. 10, 1890, 695.

Tests of Riveted Joints for Boiler Work. Paper by C. H. Moberly before the Inst. C. E., giving the results of many tests with discussion. *Eng. News*, May 26, June 2, 1883.

Their Proportions and Strength. A lecture at Cornell University by J. M. Allen, of Hartford, Conn. *Sci. Am. Sup.*, No. 811, July 18, 1891, pp. 12958-9.

See *Copper Joints.*

Riveting.

Ch. 1. Sec. III. of *The Constructor*, Illustrated *Mechanics*, July, 1890, *et seq.*

Abstracted from M. Considere's article in *Annales des P. & C.* giving the results of a series of experiments made to ascertain the additional strength obtained in riveted connection by the frictional resistance caused by the shrinking of the rivets. *R. R. Gaz.*, Feb. 11, 1887.

A 200-Ton Hydraulic Riveting Plant. Description, with detailed illustrations of a 200-ton hydraulic riveter capable of successfully upsetting rivets 3 inches in diameter. Lon. *Engineer*, Feb. 14, 1890, p. 125.

Boilers. See *Boiler Construction.*

Net Section in Riveted Work. Short paper by Theodore Cooper. *R. R. Gaz.*, Aug. 22, 1890, p. 583.

Riveting Machine. A description of the tube riveting machine used on the Forth Bridge. *San. Eng.*, Oct. 21, 1887. Lon. *Eng.*, Sept. 9, 1887.

See *Hydraulic Riveting Machine.*

Riveting Tests. To determine effect of length of rivet or thickness of metal on hand and machine driven rivets. Made by Charles W. Bucholtz. *Eng. News*, Dec. 6, 1889, p. 500.

Rivets, *Pitch of.* By Theodore Cooper. Gives methods of computing pitch of rivets and thickness of the web in riveted plate girders. *Van. Nos. Eng. Mag.*, Vol. XVII., p. 209.

Steel for Boilers. See *Boiler Construction, Steel Rivets for.*

Rivets and Riveting. By Martin Baicke. *Van Nos. Eng. Mag.* Vol. I., p. 593.

Road.

And Drainage Construction in Boston Parks. Gives brief description of the methods of road construction and drainage adopted in the Boston park system. Illus. *Eng. News*, Sept. 15, 1888.

Road Construction.

Abstract of a paper by Prof. J. V. Hazen, Hanover, N. H., giving a concise statement of the general principles of road making, and including considerable valuable data concerning actual cost, etc., of certain recently constructed roads. *Eng. & Build. Rec.*, April 26, 1890, Vol. XXI., p. 326, *et seq.*

A paper by C. Frank Allen, read before the Boston Soc. C. E., giving a general discussion of the subject. *Jour. Assn. Eng. Soc.*, May 1891, pp. 203-33. Discussion, pp. 233-48.

Abstract of an exhaustive paper on this subject, by Capt. Francis V. Green, in *Harper's Weekly*, Aug. 10, 1889. *Eng. & Build. Rec.*, Aug. 10, 1889, Vol. XX., p. 145, *et seq.*

Prize essay for *Eng. & Build. Rec.*, by I. V. Pope of Austin, Texas. *Eng. & Build. Rec.*, April 12, 1890, Vol. XXI., p. 294, *et seq.*

Maintenance of. Paper by E. P. North before the Am. Soc. C. E. Treats also of pavements. For illustrated description of road rollers see *ibid.*, Aug. 9, 1879. Reprint, *Eng. News*, July 12, Aug. 2, 1879.

And Maintenance. Prize essay for *Eng. & Build. Rec.*, by Mr. S. C. Thompson of New York City. This is a concise and valuable treatment of the subject. *Eng. & Build. Rec.*, March 29, 1890, Vol. XXI., p. 266, *et seq.*

And Maintenance of. By E. B. Ellice-Clark. Gives a good general review of the subject. *Van Nos. Eng. Mag.*, Vol. XV., p. 515.

And Maintenance. By Ed. P. North. Gives American and European practice in the case of Earth, Macadam, Stone, Wood and Asphalt pavements; also specifications, cost, cuts of tools, rollers, &c. *Trans. A. S. C. E.*, Vol. VIII., (1879), pp 95-147. Discussion, pp. 133-360.

Road Construction, continued.

In Richmond County, N. Y. Description and specifications of the roads there being built. *Eng. News*, Dec 13, 1890. pp. 526-7.

Science of. A prize essay by Clemens Herschel. A valuable paper. Republished with additional notes in *Eng. News*, June 9, Sept. 1, 1877.

Turnpiking and Underdraining Common Roads. By R. C. Carpenter. Illus. *Eng. News*, Oct. 27, Nov. 3, 1877.

Road Engines. See *Engines, Steam, Traction.*

Road Improvement.

By C. G. Elliott, before the Illinois Association of Engineers. The building of country roads considered as an engineering problem. *Eng. News*, March 13, 1886.

As Applied to Country Roads. Eng. News. Nov. 14, 1885.

Suggestions for. As derived from experience in Western Ohio. By S. A. Buchanan. *Report Ohio Soc. Surv.*, 1883.

See *Macadam*, also *Gravel Roads.*

Road Machine Contest.

Mexico, Mo. Results of a comparative trial of six machines, as to mechanical construction and general operation, and ability to perform special work. *Eng. News*, Oct. 24, 1891. pp. 394-5.

Railroad Maintenance.

See *Ditch Apportionment. Road Construction and Maintenance.*

Road Material.

Of Ohio. By Ed. Orton. Describes the stone in Ohio available for road-making material. *Rpt. Ohio Soc. Surv. & Engrs.*, 1888, pp. 60-68.

Road Metal.

Metal and Paving Setts, Valuation of. By W. F. K. Stack. Discusses the proper method of testing road-making material, and gives details of tests made on duration by means of a machine. Illustrated. Lon. *Engineer*, Aug. 31, 1888.

Road Roller.

New Harrisburg Steam Road Roller. Illustration and description of a new and improved form of steam road roller. *Eng. & Build. Rec.*, Feb. 15, 1890. Vol. XXI., p. 170.

Roadmasters' Association of America.

Minneapolis Convention. Report contains President's address on "Preservation of Cross Ties," and reports of committee on Track Work, Track Jacks, Interlocking, and Construction of Frogs. *R. R. Gaz.*, Sept. 18, 1891. pp. 647-8. *Ry. Rev.*, Sept. 19, 1891. pp. 591-5.

Roadmasters' Assn. of America. Report of the 8th annual convention, Sept. 9, 10, 1890. *Eng. News*, Sept. 13, 1890, pp. 232-3. *R.R. Gaz.*, Sept. 19, 1890. pp. 647 8.

Roads.

The Common Roads of Europe and America. By Isaac B. Potter, C. E. Compares the common roads of Europe and America and discusses advantages of good roads. Illus. *Eng. Mag.*, Aug., 1891, pp. 613-26.

Common, in France. Gives notes on the administration of the public roads in France. *Eng. & Build. Rec.*, Aug. 18, 1888.

Building of. See *Addresses, Baker. Road Construction.*

Cost of Bad Roads. Paper by Prof. I. O. Baker giving a brief analysis of the economy of good roads. *Rept. 6th An. Meeting Ill. Soc. Eng. & Surv.*, 1891, pp. 59-64.

Roads, continued.

County Highways in Penn. Report on proposed laws. Excellent paper read by Thos. H. Johnson. *Proc. Eng. Soc. of W. Penn.* Vol. V., 1889, pp. 36-55; discussion, pp. 55-8.

Gravel.

A paper by Chas. C. Brown, before the Indiana Assn. of Surveyors and Engineers. Discusses advantages, roadway, line, road-bed, side slopes, side ditches, crossings, shrinkage, stone culverts, pipe culverts, bridges, gravel, and repairs. *Proc. of the Assn.* for 1881; L. S. Alter, Rensselaer, Ind., Sec'y. *Eng. News*, Apr. 24, 1886.

Cost of. Paper by A. E. Harvey, giving the cost of some Illinois roads. *Selected papers of C. E. Club, Univ. of Ill.*, 1889-90, pp. 23-26.

In England, Repair and Maintenance of. Pamphlet by W. H. Wheeler, M. I. C. E. Extracts in *Eng. News*, Sept. 13, 1890, pp. 219-20.

Highway Engineering. A Course of Instruction in. A paper by Prof. C. F. Allen discussing the importance of good roads, and the various branches connected with a course of instruction in this subject. *Proc. Soc. Arts*, 1890-91, pp. 79-97.

Highways and National Prosperity. An article by E. P. North treating of transportation facilities in general, and showing the necessity for better common roads and water transportation. (N. Y.) *Eng.* April, 1891, Vol., I. No. 1. pp. 47-56.

Macadam.

At Bridgeport, Conn. Article by Edward P. North, describing process in use at Bridgeport under direction of Mr. B. D. Pierce. Macadam only four inches thick has been used with marked success and economy, costing less than 40 cents per square yard. *Eng. & Build. Rec.*, December 28, 1889, Vol. XXI., p. 53.

Maintenance and Rolling of. Gives the cost of rolling by steam at about half that horse-power. *Van Nos. Eng. Mag.*, April, 1884.

Quantity of macadam required per annum to maintain road surfaces in proper order. This is shown to vary largely with the quality of the stone used, but is found to follow a simple law for every kind of stone, depending further on the amount of travel; and the formulas proposed are borne out by experience on highways whose daily traffic ranged from only 30 up to 1,000 draught animals. *Zeitschr. f. Bauwesen*, 1854-447.

In Union County, N. Y. Brief article of some value. *Eng. & Build. Rec.* Jan. 18, 1890, Vol. XXI., p. 101.

Proposed Tennessee Highway Law. This law proposes to place the construction and repairs of roads into the hands of County Engineers, and also to create a State Board of three Engineers. Prepared by the Highway Reform Committee, Nashville, Tenn., 1891.

Repair and Maintenance of. By W. H. Wheeler. Contains many good suggestions on the maintenance of roads. *San. Eng.*, June 25, 1887.

Report of Committee of the Engineers' Society of Western Pennsylvania. Valuable report giving a proposed law providing for a division of roads, etc., and the appointment of a county road commission and a county engineer, who shall have direct charge of all public roads. *Trans. Eng. Soc. W. Penn.* Meeting of March 19, 1889, p. 31, *et seq.*

And Road Drainage. Brief article by F. S. McClannahan, describing a proposed method of road drainage. *Rept. 6th An. Meeting Ill. Soc. E. & Surv.*, 1891, pp. 44-50.

And Road Maintenance. Construction of Roads. Paper by John P. Prichard, of Quincy, Mass., for *Eng. & Build. Rec.*, prize series. *Eng. & Build. Rec.*, April 19, 1890, Vol. XXI., p. 310, *et seq.*

See *Pavements*, *Paving*.

Rock Blasting *under Water.* See *Blasting.*

Rollers. See *Bridge Rollers.*

Rolling Mills.

Drawings and descriptions of a new geared rolling mill with balanced top roll in use at the Ebbw Vale Steel Works. Lon. *Eng.*, April 24, 1885.

Of the Roanoke Rolling Mill Company, Virginia. Illustrated description of roll trains. *Am. Mfr.*, Feb. 22, 1889.

Test of Power in. By Henry Simon. Gives records of tests as to the power consumed by various machines used in a roller mill. Lon. *Engineer*, July 8, 1887.

Universal. Gives description of Sacks, improved universal rolling mill, adapted to rolling double angle, star, H. T., and similar sections. *Sci. Am. Sup.*, Jan. 7, 1888.

Universal Rolling Mills for the Rolling of Girders and Cruciform Section. By Hugo Sach, Duisberg-on-the-Rhine. Gives reasons why I-beams can only be rolled with difficulty when the section becomes great, and shows that by use of a universal four-roll mill this difficulty is overcome. Illustrated. *Sci. Am. Sup.*, July 20, 1889.

See *Iron and Steel. Steel.*

Roof Rafters.

Problem of. By De Volson Wood. *Van Nos. Eng. Mag.*, Vol. VI., p. 233.

Strains in Rafters. By S. H. Shreve. Derives formulæ, and makes practical application of the same. *Van Nos. Eng. Mag.*, Vol. III., p. 473.

Roof Trusses.

A brief description with full detailed drawing of main trusses of the station of the South Brooklyn Railroad and Terminal Company. The trusses have a span of 147 feet, rise of 30 feet, and 27 feet effective depth. *Eng. & Build. Rec.*, March 17, 1888.

Armory, Buffalo. A description, with plans, elevation and details, of the 10 roof trusses, 180 feet span, at the 74th Regiment Armory, Buffalo, N. Y.; also details of the rolling scaffold used in erecting them. *San. Eng.*, Oct. 29, 1887.

Bandora Station. Brief description with detailed drawing of a roof truss for Bandora station, India. Lon. *Engineer*, Dec. 8, 1887.

Cantilever. Gives details of a cantilever roof erected by the Berlin Bridge Co., over their girder shop. *Eng. & Build. Rec.*, Aug. 4, 1888.

The Cleveland Arcade Roof. These roof trusses are 3-hinged arches with about a 50-ft. span. Details illustrated. *Eng. Record*, March 21, 1891, pp. 256-7, *et seq.*

Dead Weight of iron roof-trusses. By Prof. Landsberg. *Zeitschr. f. Bauwesen*, 1885, pp. 103 and 243.

Depot of the Central Railroad of New Jersey. Gives plan and elevation of the new depot of Jersey Central Railroad at Communipaw, etc., also half section showing roof truss, with dimensions. Its span is 142 feet. *Eng. News*, Oct. 6, 1888.

The French. By P. H. Philbrick. Gives an analysis of the truss. *Van Nos. Eng. Mag.*, Vol. XXIII., p. 89.

Of the Fink pattern, with the rod, analyized graphically by using the equilibrium polygon in place of the Maxwell diagrams. By De Volson Wood, in *Van Nos. Eng. Mag.*, Vol. XXXI., p. 177 (September, 1884).

Liverpool International Exposition, description and detailed drawings of the truss at. Lon. *Engineer*, June 11, 1886.

Madison Square Garden. General plan, elevation and details. *Eng. Record*, Jan. 17, 1891, pp. 110-11, *et seq.*

Roof Trusses, continued.

Main Drill Room of the Twenty-Second Regiment Armory, New York City. Illustrated description of arched-steel roof of 176-ft. span and 60-ft. clear rise with strain diagram. *Eng. News*, April 13, 1889.

New Market Hall, La Plata. Three parallel roofs cover an area of 9½ acres. The principal members are iron arches. Described and illustrated with plates of details, by Max am Ende. Lon. *Engineer*, Nov. 28, 1890. pp. 419-30.

Most Economical Spacing of. Formulas for weights of trusses are derived and a table of weights given for various spans and spacing. By W. W. Robertson. *Ind. Eng.*, Dec. 20, 1890, pp. 490-93.

Paris Exhibition. Description, with elevation and details, of the roof truss over the fine Arts Court at the Paris Exhibition, 1889. Lon. *Engineer*, Sept. 2, 1887; also Sept. 16.

Paris Exhibition. A two-page plate showing details of roof truss and other iron work of the galleries of miscellaneous exhibits at the Paris Exhibition of 1889, Lon. *Eng.*, Dec. 30, 1887.

Phœnix Bridge Co. Gives short description, with general plans, elevation diagrams and details of the roof trusses of the new girder shop of the Phœnix Bridge Works. *Eng. & Build. Rec.*, April 7, 1888.

Renewal of, at King's Cross Terminus, G. N. R. By R. M. Bancroft before the Soc. of Engrs. Gives details of the renewal of the roof truss over the departure platform of King's Cross terminus of the Great Northern Railroad. Nine plates. *Trans. Soc. Eng.*, 1888. pp. 123-145.

Semicircular Timber designed by Captain F. Fowke, R. E. Described and illustrated in *Proc. Inst. C. E.*, Vol. LXXXII., p. 301.

Twelfth Regiment Armory, New York. Brief description, with illustrations of the riveted arch roof trusses of the armory of the Twelfth Regiment in New York City. *Eng. & Build. Rec.*, Jan. 7, 1888.

See *Howe Truss*, *Railroad Terminals*, *Trusses*.

Roof and Bridge Trusses. *Diagrams, Formulas and Tables* for the Use of Engineers, Draughtsmen and Architects. A collection of many useful formulas, standards, etc., with tables and diagrams to aid in their application. *Mechanics*, Jan., 1891, pp. 20-31, *et seq.*

Roofs.

Conical. By Thos. Doane. Description of such roofs over tanks 90 feet in diameter, without interior support. Illustrated. *Jour. Assn. Eng. Soc.*, Vol. II., p. 73.

Description of the collapsed roof of the Norway Iron-Works, South Boston. *San. Eng.*, March 19, 1887.

Iron. Illustrates various forms of iron roofs and gives tables of dimensions of iron proper for each member on different lengths of span. *Mech. World*, Dec. 8, 1888.

On the Fall of the Huddersfield Station Roof, England, resulting in the death of four men and the wounding of several others. A good example of how not to do it. Illustrated. Lon. *Engineer*, Nov. 13, 1885; also Lon. *Eng.*, Oct. 16, 1885.

On the design and proportions of painted roofs. *Van Nos. Eng. Mag.*, Vol. VIII., p. 62.

Removal of Roof Water from Buildings. An article by Dwight Porter, Ph. B., describing some of the methods in use on large buildings for overcoming special difficulties—notably that arising from ice formation. *Tech. Quar.*, Vol. II., 1889, No. 3, pp. 220-28.

Some Celebrated Timber Roofs. A lecture by Prof. T. Rodger Smith. London Illustrated. *Am. Arch.*, June 6 and 13, 1885.

See *Roof Trusses.*

Roorkee Experiments. See *Hydraulics*.

Rope Driving.

A paper by Louis J. Seymour. Illustrated. *Power*, May, 1889, *et seq*.

See *Crane*, *High Speed*.

Rope Gear.

For Electric Light Work. A paper by W. H. Booth and Frank B. Lea discussing advantages of its use. *Elec. Rev.*, June 26, 1891, pp. 810–4.

Rope Railways.

Illustrated description of construction and operation of the Bluchert Wire Rope Tramway. *Science*, Feb. 14, 1890.

Muller's. Gives a short description of the above system, which is adapted for factory use. *Van Nos. Eng. Mag.*, Vol. IX., p. 118.

Otto Wire Ropeway. A profusely illustrated article descriptive of this novel wire rope railway as used at Gottesegen Colliery, Antenienhutte, Westphalia. *Sci. Am. Sup.*, June 22, 1889. Lon. *Engineer*, Feb. 8, 1889.

Suspended Wire Rope Tramway. Chandler's system operated by an electric motor, is fully described and illustrated. *Eng. News*, July 12, 1890, p. 26.

In Transylvania. From the French. The line is about twenty miles long and carries iron ore and charcoal in tubs, each tub carrying about 600 lbs. Lately put in operation. Grades heavy. Cost about 30 cents per ton-mile. *Proc. Inst. C. E.*, Vol. LXXXVI., p. 415.

Two Rope Haulage Systems. Description of an overhead rope tramway built for hauling mine cars at Wilkesbarre, Pa. Many special details, described and illustrated. Abstract of a paper by R. Van A. Norris, read before the Am. Soc. M. E. *Eng. Record*, June 27, 1891, pp. 54–5, *et seq*.

Way between Vajda-Hunyad and Vadudobri. By Messrs. Bochart and Leberton. Abstracted from a paper read before the Inst. of C. E. Gives details of a wire rope-way about 19 miles long used to transport ironstone and charcoal. *Eng. News*, Jan. 1, 1887.

Wire Rope Transport. By W. T. H. Carrington. Describes three systems of wire rope transport, and refers to examples in detail. *Van Nos. Eng. Mag.*, Vol. XXII., p. 177.

See *Cable Railways*.

Ropeways.

Aerial Wire Ropeways. By J. Pohlig, Germany. Description of the important features of the Otto system and of many details of several lines in actual operation. Illustrated. *Trans. A. I. M. E.*, 1891, p. 31.

A Few Facts about Wire Ropeways, with Notes on the Plomosas Line. A paper by B. McIntire, before the Tech. Soc. Pac. Coast, describing many features of construction and difficulties encountered. Condensed and printed in *Eng. News*, March 21, 1891, pp. 264–72. *Ry. Rev.*, Jan. 17, 1891, pp. 40–1.

Cable Plant for Construction of a Dam. Illustrated description of the plant of machinery used by the contractors in the construction of the "Sodom Dam," New York Water Supply. *Sci. Am. Sup.*, Jan. 5, 1884. See *Sodom Dam*.

The Garrucha Aerial Ropeway. Is of the Otto System. General description with illustrations of details. From *Industries*. *Eng. News*, July 25, 1891, pp. 71–2.

Improved Rope Railway.—Aerial Carrier at the Landain Sugar Works. Special features are the oscillating tower and smooth cable. General description and illustrations. From *Le Genie Civil*. *Sci. Am. Sup.*, No. 806, June 13, 1891, pp. 12874–5.

Rock Excavation at Tilly Foster Mine, Brewster Station, N. Y. Illustrates wire rope haulage plant for disposing of the spoil; 300,000 cubic yards of rock to be removed. *Eng. News*, April 20, 1889.

Ropes.

Manila Rope. Properties, splicing and knotting. Abstract from a pamphlet issued by the C. W. Hunt Co., New York. *Ry. Rev.*, Nov. 14, 1891, pp. 712-3.

Strength of Manila and Hemp Ropes, and Rope Power-Transmission. By T. Spencer Miller, discussing various formulas and proposing a curve of co-efficients; also a discussion of Mr. Hunt's paper on Rope Driving. *Eng. News*, Dec. 6, 1890, pp. 500-1.

Used for Pile Driving. See *Pile Driving.*

Rumford's Determination of the Mechanical Equivalent of Heat. See *Heat.*

Safe Deposit Vault, *The Modern Construction of.* Described and illustrated in *San. Eng.*, Sept. 23, 1882.

Safety Appliances *on Railways, Federal Regulation of.* See *Interstate Commerce Commission.*

Sahara, *Flooding the.* By Geo. W. Plympton. Gives a brief *resume* of the project, showing that the total area of the inland sea is about 3,100 square miles, less than one-half the size of Lake Ontario. *Van Nos. Eng. Mag.*, August, 1886.

Sailing Chart. For readily obtaining the great circle joining any two points, with instructions for using. The course is a circle on the chart. By Richard A. Proctor. Illustrated. *Sci. Am. Sup.*, Aug. 8, 1885.

Salt Manufacture in England. Methods used to dissolve rock salt and pump the brine from a depth of 1,300 feet. Illustrated. *Proc. Inst. C. E.*, Vol. XC., pp. 131-158; abstracted, *Sci. Am. Sup.*, October 1, 1887; also *San. Eng.*, Oct. 29, 1887.

Salt Works. *One Way of Obtaining Brine.* Description of reservoir and pipe line at the Syracuse Salt Works. Sections and profile. By Chas. B. Brush. *Trans. A. S. C. E.*, Vol. XXIII., 1890, paper No. 447, pp. 95-100.

Sand. *A Study of the Movement of.* Experiments on the movement of sand under various conditions. Layers of different colored sand were used and by this means the relative movement of the various layers was observed. Illus. *Eng. News*, March 3, 1883.

See *Cement Tests, Wet and Dry Sand.*

Sand-Bag Embankment *on the Jersey Coast.* A temporary device for securing a deposit of sand to close an inlet. Details of the work, with cuts. By P. T. Osborne. *Proc. Eng. Club. Phila.* Vol. V., p. 117.

Sand Banks *and Sand Hills.* By H. Keller. Treats of the formation of sand banks and hills; also of the construction and maintenance of harbors on sandy coasts. *Van Nos. Eng. Mag.*, Vol. XXVII., p. 71.

Sand Pump *used on the B. & O. R. R. Bridge Foundations* at Havre de Grace, Md. Sectional view and method of operating, together with a full description of the sinking of the caissons. *Eng. News*, April 18, 1885.

San Francisco *Water Front, Structures on.* See *Water Front.*

Sanitary.

A review of the sanitary condition in 1886, and of the progress made, being the annual address of Captain Douglas Galton, before the Society of Arts. *Jour. Soc. Arts*, Nov. 19, 1886.

Administration in Paris. Notes by Mr. Till on the water supply, gas, paving and sewerage of Paris. *Van Nos. Eng. Mag.*, Vol. XXIV., p. 296.

Advantages of Smooth and Impermeable Street Surfaces. By Edwin Chadwick, before the Soc. of Arts. Advocates the use of asphalt paving. *Van Nos. Eng. Mag.*, Vol. V., p. 356.

Appliances in use in Denver, Colo. A description of the sewerage and water systems, but especially a novel and excellent Flushing Apparatus and Water Trap. Illustrated. *San. Eng.*, July 29, 1886.

Sanitary, continued.

Arrangements at Montreal during the small-pox epidemic, 1885. Describes how the Exposition buildings were fitted up for hospital purposes. Many interesting problems in sanitary engineering successfully solved. Illustrated. By J. W. Hughes, Engineer. *San. Eng.*, Nov. 6, 1886.

Arrangements of Pullman, Ill. Extracts from a full report of the State Board of Health for 1885. *San. News*, Nov. 14, 1885, *et seq.*

Condition of Coney Island, and need of improvement. A special investigation by Wm. Paul Gerhard. An exhaustive investigation by a competent engineer, forming a series of articles in *Eng. News*, beginning Sept. 19, 1885.

Condition of Soil under Cities. By Dr. H. Maris Davy. Discusses the contamination and purification of the ground water of cities. *Eng. News*, April 9, 1887.

Construction and Arrangement of Dwelling Houses. By W. H. Corfield. A valuable contribution to the subject. *Van Nos. Eng. Mag.*, Vol. XXII., pp. 177 and 281.

Defects in Houses. Abstracted from a lecture by Dr. Chas. Kelly, at the Parkes Museum of Hygeine. *Van Nos. Eng. Mag.*, Feb., 1884.

Disposition of the Dead. Address by Dr. C. A. Harvey, of New York, before the Franklin Institute. The importance and bearing of this question on the public health is evident. *Jour. Frank. Inst.*, Nov., 1889. Abstract in *Am. Arch.*, Dec. 7, 1889, Vol. XXVI., p. 264.

Engineering. See *Sanitary Engineering* below.

Entombment. Dry Air System. Description of the system of the New Mausoleum Company, consisting essentially of passing a current of dry air through the tomb, and afterward through the flame of a furnace. *Jour. Frank. Inst.*, April, 1890.

Fallacies. An address delivered at Croydon by Prof. W. H. Corfield. *Van Nos. Eng. Mag.*, Vol. XXII., p. 28.

Inspection.

Health Inspector's Guide. Instructions in the various details of inspection, by E. M. Hunt, Sec'y. *Report N. Y. State Board of Health*, 1890, pp. 33-97.

Service in Chicago. Abstract of report on this service for 1881, showing a large amount of efficient service rendered. *San. News*, Oct. 10, 1885.

Of Houses and premises. By Dr. Henry Mitchell, Health Officer of Asbury Park, N. J. Designed as a guide to health officers. *San. News*, Dec. 5, 1885.

Revelations of. By E. C. Robins. *Van Nos. Eng. Mag.*, Vol. XXV., p. 508.

Lessons Reviewed and New Lessons Considered. Extract from a paper by Mr Robert Rawlinson, read before the Exeter Congress of the Sanitary Inst. *Van Nos. Eng. Mag.*, Vol. XXIII., p. 526.

Laws.

The Public Health Act of 1885 of New York State. Review of the law and its action by Arthur Hollick in *Sch. of Mines Quart.*, Vol. XI., p. 114, Jan., 1890.

In Eighteen States, given in App. C. of *An. Rep. National Board of Health*, 1883.

Matters, including Gaugings of Dry Weather Flow in Sewers; Water Supply of New Orleans and Mobile; Water Analysis for sanitary purposes; Sanitary Surveys of Memphis and Baltimore, etc., found in *An. Rep. Nat'l Board of Health*, 1880.

Matters in Isolated Country Houses. By E. W. Bowditch. Some practical considerations of value. *Jour. Assn. Eng. Soc.*, Vol. IV., p. 131.

Ordinances. A model schedule of provisions proposed by Wm. Paul Gerhard. *Eng. News*, Sept. 26, 1885.

Sanitary, continued.

Ordinances of Asbury Park, N. J. The full text of these ordinances which are actually in force in this model village so far as sanitary arrangements are concerned. Also an account of the growth of the village itself. *Sup. to San. News*, Oct. 17, 1881.

Registration of Buildings. The text of a bill now before the British Parliament, providing for the sanitary registration of houses in London. *San. Eng.*, May 7, 1887.

Science. A City of Health. By B. W. Richardson. Gives sketch of an ideal city built with all the improvements of modern sanitary science. *Van Nos. Eng. Mag.*, Vol. XIV., p. 31.

Science of July 15, 1883, is called a "sanitary number." It has [illegible] short papers on topics connected with sanitary studies.

Supervision of Dwellings. By Lewis Angell. A paper read before the Assn. of Municipal and Sanitary Engrs. and Surv., Birmingham, June, 1881. *Van Nos. Eng. Mag.*, Vol. XXV., p. [illegible].

Survey, of St. Louis, Mo. A series of valuable papers by city officials and local sanitarians. Reprinted from *Trans. Am. Pub. Health Assn.*, Vol. X for the St. Louis Local Committee. Address John D. Stevenson, Health Commissioner, St. Louis.

Sanitary Engineering.

By Prof. William Cain, C. E. Pp. 90. Pamphlet published by the N. C. Board of Health, 1885. Brief discussion of the subjects of drainage, ventilation, water supply and sewerage, with particular reference to North Carolina. Illustrated.

By Capt. D. Galton, before the Sanitary Congress at Croydon. *Van Nos. Eng. Mag.*, Vol. XXII., p. 128.

Baldwin Latham's work on Sanitary Engineering published as a supplement to *Eng. News*, 1877.

A paper by Theo. Rosenberg on the drainage and plumbing of dwellings. *Jour. Assn. Eng. Soc.*, Dec. 1876.

Essay on importance of, with examples. By J. E. Strawn. *Rep. Ohio Soc. Surv.*, 1883. A continuation of same, with remarks on heating and ventilation. Report of 1884.

Papers on sewerage, drainage and water pollution are contained in the tenth annual report of the N. Y. State Board of Health, Albany, 1890, pp. [illegible].

Report of the fourth Kansas State Sanitary Convention, including interesting papers on water pollution and other subjects. *Fifth Annual Report of the Kansas State Board of Health*. J. W. Redden, M. D., Secretary, Topeka.

The Illinois River Basin and Its Relation to Sanitary Engineering. See *River Basins*.

Influence of Ground Water Upon Health. By Baldwin Latham before the Congress of Hygiene and Demography. *Eng. News*, Sept. 19, 1891, p. 270.

Sanitation of Bombay. An investigation of the climatic conditions and sanitary requirements of Bombay. By Baldwin Latham. *Ind. Eng.*, July 4, 1891, *et seq.*

Sanitation.

Household. An address before the Rochester, N. Y., Academy of Sciences, by W. S. Hoyt, C. E. *Building*, May 15 and 22, 1886.

In Schools. Full abstract, with illustrations, of the report of committee on sanitation in the public schools at Lynn, Mass. *Eng. Record*, April 25, May 2, 1891.

Of Towns. By J. Gordon. A presidential address before Society of Municipal and Sanitary Engineers and Surveyors of England. *Sci. Am. Sup.*, Nov. 19, 1887.

Sanitation, continued.

In the United States. A review by A. R. Leeds. *Van Nos. Eng. Mag.*, Vol. XX., p. 6.

See *Air. Flushing. Plumbing.*

Scale *of variable proportions.* A rubber band with graduations marked on it. By extending the same the whole scale is extended in equal proportional amounts in all its parts. *Annales des P. & C.*, 1884-3-684.

Scale. See *Weighbridge, Twenty Ton.*

School. *Drainage and Sewerage of the Lawrenceville School.* By F. S. Odell. Gives the details of the system of water supply, drainage and sewerage put in at a School at Lawrenceville, N. J. *Trans. A. S. C. E.*, Vol. XVI., p. 66.

School Buildings. See *Ventilation of Habitations.*

School Buildings and Grounds *of the Lawrenceville School for Boys, N. J.* A recent reconstruction of buildings, grounds and sanitary arrangements under best architects and engineers. Illustrated. *San. Eng.*, Dec. 4, 1886.

School House *Construction.* A series of articles on the planning and construction of school houses of all grades. By W. R. Briggs. *Building*, Feb. 6, 1886, *et seq.*

Primary. By the City Architect of Boston. Three floor plans and two elevation of a new building, with description of plumbing, heating, ventilation, etc. Heated by steam, indirect system, closets in basement. *San. Eng.*, May 6, 1886.

See *Heating, etc.*

Scientific Departments *of the Government*, being extracts from the Report of the Committee of the National Academy on the subject, with the testimony of Maj. Powell, Director of the U. S. Geol. Surv. *Science*, Jan. 16, 1885. See also *Government Scientific Work.*

Scow.

Automatic Rock Dumping Scow. Used for automatically dumping rock or other material in dyke construction in the harbor works at Newburyport, Mass. Illustrated description. *Eng. News*, Nov. 24, 1888.

For Dumping. Designed to dump rock, by allowing the upper platform to run down ways far enough to dump the load. Cuts shown. *An. Rep. Chief of Engrs., U. S. A.*, 1883, p. 498.

Self Dumping. Described and illustrated in *San. Eng.*, Jan. 1, 1886.

Screw. *Ball-Bearing for.* See *Ball-Bearing.*

Screw Forging Machine. An illustrated article describing a machine for forging the threads on large screws and thus saving material and labor. *Lon. Engineer*, in *Am. Eng.*, April 15, 1886.

Screw Threads, *New System of.* Recites objections to systems in use, and proposes a new system giving an increase of strength of 17 per cent. over the Sellers thread. By John L. Gill, Jr., Phila. *Jour. Frank. Inst.*, March, 1883.

Screw Piles. See *Foundations. Piers. Piles.*

Screw Propeller. See *Ferry-Boat; Propeller Blades, Design of. Propellers.*

Screw Threads. See *Bolts.*

Screws.

A Practical Solution of the Perfect Screw Problem. By Wm. A. Rogers. Shows how a perfect screw may be made. *Trans. A. S. M. E.*, Vol. V., p. 216.

Systems of Screw Threads. Chapter IV. of "The Constructor." Formulas and tables. *Mechanics*, Sept. and Oct., 1890.

Secondary Batteries. See *Electric Batteries.*

Sediment-Bearing Mountain Streams, *Treatment of.* By Geo. J. Specht, before Tech. Soc. Pac. Coast. Discusses how the formation of detritus may be prevented and how it may be stored near its source. *Eng. News*, Nov. 7, 1885.

Sedimentation. See *Falling Velocities.*

Seismography. *Studying Earthquakes by Electricity.* See *Earthquakes.*

Self-Purification *of Peaty Rivers.* By W. N. Hartley. A good treatment of the subject, from a chemical point of view. *Van Nos. Eng. Mag.*, Vol. XXIX., p. 17.

Self-Recording *Rain Gauges.* Value of. See *Rain Gauges.*

Sea Walls.

Scarborough. A short description of the sea wall recently commenced at Scarborough. Length, 1,200 yards; width of base, 17 feet; at 5 feet face to be curved on radius of 17 feet and constructed of concrete block, 2 feet x 12 in. x 18 in. The outer wall to be made of Portland cement. Lon. *Engineer*, Jan. 17, 1887.

The Use of Asphaltum in Building Sea Walls. A paper by W. C. Ambrose, describing the use of asphaltum as mortar, with discussion. *Trans. A. S. C. E.*, Vol. XXIV., March, 1891, pp. 227-9 Reprinted in *Eng. Record*, May 16, 1891, pp. 395-6

Sea Water. Its effect on iron piles as found at the Brandywine Lighthouse, Delaware Bay, after a period of thirty-five years. *Proc. Eng. Club, Phila.*, Vol. IV., No. 3.

Separator.

Edison's Magnetic. For ores. Illustrated description, with results of experiments on various ores. *E. & M. Jour.* Dec. 8, 1888.

Venstrom Magnetic Separator. The illustrated description of the Venstrom magnetic separator in operation at the metallurgical works of Mr E. N. Riotte, New York. *E. & M. Jour.*, Nov. 24, 1888.

Settling Basins. See *Water-Works.*

Sewage.

An abstract from the report on the *experiments made in Paris* on the application and purification of sewage. *Van Nos. Eng. Mag.*, Vol. II., p. 153.

As a Fertilizer of land and land as a purifier of sewage. A paper read before the Society of Arts, by J. Bailey Denton. A valuable article. *Van Nos. Eng. Mag.*, Vol. VI., p. 184.

A general article on *old and new systems*, requirements, capacity, ventilation of sewers and purification of sewage. Compiled from various sources. *Van Nos. Eng. Mag.*, Vol. I., p. 225.

A paper read by Mr. R. W. P. Birch before the Society of Municipal Engineers on the examination of some recent experiments on sewage treatment made by Mr. W. J. Dibdin. *Eng. News*, Sept. 17, 1887.

And Irrigation in Germany. Gives brief accounts of the works at Danzig and Berlin. *Van Nos. Eng. Mag.*, Vol. XX., p. 263.

And What Shall be Done With it. Illustrated paper by G. W. Hosmer, M. D., discussing in a popular way both the sanitary and engineering details of sewage conveyance and disposal. Sup. to *Harper's Weekly*, July 19, 1890, pp. 561-8.

Berlin. Analysis of the affluent from the Berlin sewage irrigation works. *Van Nos. Eng. Mag.*, July, 1886.

Contributions to Our Knowledge of. By Prof. Wm. Ripley Nichols and C. R. Allen. An analysis of 17 samples of Boston sewage. *Jour. Frank. Inst.*, Aug. 1875.

Crops. A brief review of the claims of sewage crops. A few details given of the farming of the Romford farm. *Van Nos. Eng. Mag.*, Vol. IV., p. 381.

Composition of. Gives statistics relating to London sewage and its composition. *Van Nos. Eng. Mag.*, Vol. I., p. 793.

Discharge of, into Tidal Rivers. By H. Law. Gives statistics relating to London sewage and the Thames River. *Van Nos. Eng. Mag.*, Vol. XIX., p. 524.

Sewage, continued.

Flow of Chiswick, Eng. Automatic registration of flow for several years. *Proc. Inst. C. E.*, Vol. XCIV., p. [illegible]

Precipitates, a new method of dealing with. By H. Y. D. Scott. Advocates the use of lime and clay as a precipitant, and the conversion of the slush into cement. *Van Nos. Eng. Mag.*, Vol. VII., p. [illegible]

Question. By C. M. Bazalgette, before the Inst. of C. E. The object of the paper is to limit and define the proper application of the various systems of sewage and to call attention to subordinate questions arising upon the practical working of these systems. *Van Nos. Eng. Mag.*, Vol. XVII., pp. [illegible] and [illegible]. The above article abstracted in *Van Nos. Eng. Mag.*, Vol. XVI., p. [illegible]. Comments on the above paper. *Van Nos. Eng. Mag.*, Vol. XVII., p. [illegible]

Treatment of.

And Utilization of. Abstracted from a report made to the British Association by a commission composed of prominent engineers. *Van Nos. Eng. Mag.*, Vol. I., p. [illegible]

Discussion of a paper at the Society of Arts. Contains particulars relating to the new scheme for the Thames purification. *Jour. Soc. Arts*, May 7, 1886.

By Capt. L. Flower. Gives notes relating to the sewage of town on the River Lee, Eng. *Van Nos. Eng. Mag.*, Vol. XV., p. 76.

By Dr. C. M. Tidy. A paper before the Society of Arts. A general article contains much information. *Van Nos. Eng. Mag.*, July, 1886.

By Dr. C. Meymott Tidy. This is a paper of over 60 double-column pages treating the subject in a scientific, economical and practical way. It is a mine of valuable information not readily accessible elsewhere. *Jour. Soc. Arts*, London, Oct. 8., 1886.

A brief description of the Astrop dried sewage process. *Eng. News*, April 9th, 1887.

Chemical Treatment of Mystic. By W. T. Learned. Before the Boston Soc. C. E. Gives results of a study of the chemical treatment of Mystic sewage. Precipitant used was crude sulphate of alumina. *Jour. Assn. Eng. Soc.*, July, 1888, Vol. VII., pp. 244-248.

Deodorization of London. A report by Sir Henry Roscoe to the Metropolitan Board of Works on the deodorization of London sewage. *Eng. & Build Rec.*, Sept. 4, *et sec.*, 1888.

Electrical Treatment of, a paper by W. Webster, read before the London Section of the Society of Chemical Industry. Abstract in *Electrician*, Dec. [illegible] 1890, p. 146; Lon. *Engineer*, Dec. 5, 1890, p. 452.

Electrical Treatment of. An account of a practical test of Mr. William Webster, of London, on a scale of 12,000 gals. per hour. *Eng. News*, April 13, 1889.

By the Lime Process. Gives results of its application; also cites a number of reports, wheremore details may be found. *Van Nos. Eng. Mag.*, Vol. IV., p. 595.

Utilization of Town. By Dr. A. Carpenter. A reply to Dr. Tidy's paper. Gives much valuable information regarding sewage farming, of which he is a strong advocate. Also contains reports from a number of farms. *Jour. Soc. Arts*, Feb. 4, 1887.

Willbread's Process. A brief account of the modification of the Phosphate Sewage Company's scheme, essentially a lime process. *Van Nos. Eng. Mag.*, Vol. IX., p. 313.

See *Water Supply, and Sewerage. Garbage.*

Sewage Disposal.

A paper by Rudolph Hering written for the State Board of Health of California. This paper takes up briefly the various methods of disposal explaining the conditions under which purification goes on, and gives the general results, regarding filtration, of the experiments made by the Mass. State Board of Health. *Eng. Record*, March 7 and 14, 1891.

By C. A. Allen, before the annual convention of the American Society of Civil Engineers. Gives a brief review of the history of sewage purification in England, and a review of the different methods employed at present. *Trans. A. S. C. E.*, Vol. XVIII., Jan., 1888, pp. 9-23, and a discussion, pp. 24-41.

Discussion of conditions and recommendations of State Board of Health for Brockton, Ware, Haverhill, Winthrop, Pittsfield, Framingham, Westfield, Quincy, Northampton and Lenox, Mass. Pp. 16, *20th An. Rep. State Board of Health*, Mass., 1889, p. 10.

Discussion on the paper presented to the Soc. of Arts by Dr. Tidy. *Jour. Soc. Arts*, Dec. 3 and 17, 1886.

The A B C Process Vindicated. Antagonizes the report of the Royal Rivers Pollution Commission by the investigations of Profs. Dewar and Tidy. *Jour. of Science*, London, Aug., 1885.

A brief history of *the A B C Process*, with comments on its application. The Leeds authorities adopt the method. *Van Nos. Eng. Mag.*, Vol. VIII., pp. 156 and 276.

The A B C Process at Aylesbury, Eng. An account of some thorough experiments on the purification of very foul sewage and the production of salable manure. *Eng. News*, July 25, and *Am. Arch.*, July 11, 1885.

At Acton, England. Precipitation and filtration is the method employed. Brief illustrated description. *Eng. Record*, June 20, 1891, p. 43.

At Almshouse and Insane Asylum, near Providence, R. I. Brief description of plan and methods of disposal by irrigation. Illustrated. *San. Eng.*, March 4, 1886.

American Sewage Disposal Company's methods of settlement and filtration are described and illustrated by several folding plates, in an 8 page pamphlet issued by the Company. Milwaukee, Wis., 1890.

Of Berlin. A short description, from the French. The entire rainfall and sewage proper disposed of by irrigation, except the overflow in times of heavy rains. Eminently successful. *Eng. News*, Oct. 7, 1885.

Carbonized Refuse System. This process is used by the Rivers Co., Leeds. Dry refuse is carbonized by slow combustion, and the product is used for making filters which thoroughly clarify the sewage. The plant is illustrated by a general plan and cross sections. Reprinted from *Industries*. *Sci. Am. Sup.*, No. 756, June 28, 1890, p. 12,081.

The Chemical Precipitation of Sewage. A paper by Allen Hazen, giving the results of an important series of experiments made at the Lawrence Experimental Station to determine the proper relation between the composition of the sewage and the amount of the precipitant, and the relative value of various precipitants. *Jour. Assn. Eng. Soc.*, Aug., 1891, pp. 385-99.

At Chicago. *The Problem.* The questions discussed by a committee of the Citizens' Association. Remedies suggested. Abstract in *Am. Eng.*, Sept. 3, 1885. Also printed in pamphlet form as a supplement to *San. News*.

At Cleveland, O. Recommendations of the City Eng., involving the intercepting of the sewage now flowing into the river and carrying it by gravity flow through a conduit to the lake. *Rep. of the City Civil Eng.*, 1884, Cleveland, O.

For a Country House. Description of a system adapted to a sharp hill slope. Details illustrated. *Eng. Record*, Aug. 22, 1891, pp. 189-90.

Sewage Disposal, continued.

East Orange, N. J., Disposal Works. Their efficiency as compared with other methods. By Carroll Phillips Bassett, M. Am. Soc. C. E., Engineer of the Works. *Eng. News*, Feb. 15, 1890, Vol. XXIII, p. 160.

In Europe. By W. Howard White, before Am. Soc. Civ. Engrs. A short but satisfactory resume of the present European methods of sewage and garbage disposal. *Trans. A. S. C. E.*, Vol. XV.; also *Eng. News*, March 27, 1886.

European Practice. By C. H. Swan, before the Boston Society of Civil Engineers. Gives interesting statistics relative to the amount of sewage that may be applied to given areas of land, as shown by experiments at Paris, Berlin, Croydon, etc. *Jour. Assn. Eng. Soc.*, July, 1888, Vol. VII., pp. 245-258.

And Garbage Removal, European. By W. Howard White. Gives the methods in vogue in a number of European cities, with cost of operating the same. Also discusses the disposal of garbage and describes Fryer Patent Destructor. Illustrated. *Trans. A. S. C. E.*, Dec., 1886.

Farms.

Farm at Saltley, Birmingham, Eng. This is a farm of over 1,200 acres, receiving the drainage from over half a million people. The sewage first treated by the lime process. Probably the best example of sewage farming in England. A paper read before the Br. Assn. Adv. Sc. and reprinted in *San. Eng.*, Oct. 7, 1886, *et seq.*

Their sanitary and agricultural importance. By Dr. Holdefleiss and R. Klopsch, from investigations at Breslau. *Jour. f. Gasbeleuchtung u. Wasserversorgung*, 1885, pp. 132-134.

Luton. Brief description f the Luton, Eng., sewage farm. Lon. *Engineer*, Nov. 18, 1887.

Pullman, Ill. By Benezette Williams. Gives a careful description of this system, which is the first example of sewage farming in America. Illus. *Jour. Assn. Eng. Soc.*, Vol. I., p. 311.

Inland Sewage Disposal, with special Reference to the East Orange, N. J. Works. Paper by C. Ph. Bassett, reviewing briefly the various methods of sewage purification, and giving a complete description of the East Orange works where the method used is that of chemical treatment with filtration. Fully illustrated. *Trans. A. S. C. E.*, Vol. XXV., Aug., 1891, pp. 125-160.

Of London. Discussion of the Report of the Royal Commission on the disposal of. By Capt. Douglas Galton, C. B., F. R. S. Gives an historical review of the treatment of the question and reviews the late Report of the Commission on the fouling of the Thames River. *Van Nos. Eng. Mag.*, April, 1885. Also, Resume of Report in *San. Eng.*, March 26, *et seq.*, 1885.

Review of final report of the royal commission, giving a careful abstract of the arguments and conclusions. The document itself of great value. *Eng. News*, April 18, 1885.

Suggestions for dealing with the sewage. By Gen. Scott. *Van Nos. Eng Mag.*, Vol. XX., p. 49.

To be treated with lime, and the sludge pressed and transported by ships of special design and deposited in the German Ocean. Lon. *Engineer*, April 16, 1886, editorial.

The London Sewage Question. Paper by Mr. Crawford Barlow, M. I. C. E., read at the Liverpool meeting of the Asso. of Municipal and Sanitary Engrs. June 26, 1890. *Iron*, Aug. 15, 1890, *et seq.*

For the Lower Thames Valley. A review of the questions of sewage diposal submitted to the Thames Valley Drainage Board, appointed eight years ago and now dissolved. By Henry Robinson, M. I. C. E. Lon. *Engineer*, July 17, 1885.

Sewage Disposal, continued.

In Massachusetts. By F. P. Stearns, before the Annual Convention of the American Society of Civil Engineers. Gives a statement of the present status of the question in Massachusetts, with a brief reference to the action of the State in the past. *Trans. A. S. C. E.*, Vol. XVIII., January, 1888, pp. 1-7.

At the Massachusetts Reformatory. From report of William Wheeler, Engineer. Involves pumping and disposal on land. Many good features. Illustrated. *San. Eng.*, Dec. 18, 1886.

At Medfield, Mass. By Fred Brooks, before the Boston Society of Civil Engineers. Describes the intermittent, downward filtration sewerage system at Medfield, Mass., with map and plans of basins, etc. *19th An. Rep. Mass. Board of Health; Jour. Assn. Eng. Soc.*, July, 1888, Vol. VII., pp. 239-244; *Eng. & Build. Rec.*, July 16, 1888.

At the Minnesota Hospital for the Insane. By sedimentation and filtration in summer, and precipitation in winter. Illustration of precipitation tank and brief description. *Eng. Record*, Jan. 3, 1891, pp. 72-3.

Of Paris. A note in regard to the irrigation, or sewage farming, now in use. Gives some definite and successful results. From *Am. Arch.*, in *Eng. News*, June 20, 1885.

Of Paris. By Alfred Durand-Claye. Description of the present method of sewage disposal at Paris by irrigation, and plan for disposing of the remainder of the sewage of the city in the same manner. *Annales des P. & C.*, September, 1885, pp. 477-510.

The Petrie System. Report of Dr. C. Dischoff on the efficiency of the system which is in use in Berlin. It consists in filtrations through turf, dust, and gravel, and then precipitation by lime. 5 pp. in *An. Rep. Nat'l Board of Health*, 1883.

Plant, The Hardie. Illustrated description. *Eng. News*, Feb. 2, 1889.

Pneumatic. A system of pumping sewage mud by exhaust from tank on shore into vessels with 600 ton tanks for conveyance out to sea. Illustrated. *Sci. Am. Sup.*, May 22, 1886.

Problem, Solution of. A review of the methods of filtration, irrigation and destruction, which are condemned. Advocates the A B C process of precipitation. *Van Nos. Eng. Mag.*, Vol. VIII., p. 229.

Proposal for the abolition of water-carriage in the removal of effete organic matter from towns. By Dr. Thos. Hawksley, London. Describes a new kind of earth closet to be used in dwellings on any floor. *Jour. Soc. Arts*, April 17, 1885.

Proposed Plan for Providence, R. I. A very complete and valuable report by Samuel M. Gray, City Engineer, Providence, R. I. *Rep. City Eng.*, Providence, R. I., 1884, No. 25.

Of Providence, R. I. A full abstract of the report of the committee appointed to examine and report upon the plans of Mr. S. M. Gray for the disposal of city sewage. *San. Eng.*, Jan. 8, 1887. Also abstracted in *Eng. News*, Jan. 1, 1887.

Report of the Commission of Engineers on the collection and Final Disposal of the Sewage and on the Water Supply of the City of Milwaukee, Wis. Includes an appendix on chemical and mechanical treatment of sewage. *Report of the Com.* Milwaukee, 1889.

Richmond Main Drainage Works, England. Description of pumping plant, precipitation tanks and filter beds. Folding inset of details, from Lon. *Eng. News*, Oct. 24, 1891, pp. 400-3. Sewer connections and sludge chambers illustrated. *Eng. Record*, Oct. 24, 1891.

Sludge and its Disposal. By W. J. Dibdin. Abstracted from a paper before the Inst. C. E. Lon. *Engineer*, Feb. 25, 1887.

Sewage Disposal, continued.

Stone and Ault System at Rangoon Town, India. By H. F. White. A report by order of Chief Engineer of British Burma on the proposed Stone and Ault system of sewage disposal for Rangoon Town. The report is favorable, answering each objection seriatim. *Ind. Eng.*, Nov. 5, 1887.

Southampton, Eng. An abstract of a paper by W. B. G. Bennett, before the Institution of Civil Engineers. Gives a description and cost of operation of the sewage clarification and house-refuse disposal works at Southampton, Eng., with plans and sections. *Eng. & Build. Rec.*, April 28, 1888.

By the Sub-surface Irrigation System. A paper before the New Jersey Sanitary Association. By Dr. J. W. Parkham; giving testimony of 37 engineers on the subject. *Am. Arch. & Build. News*, April 25, 1885.

Upright Precipitating Tanks. Illustrated description of several systems now in successful use in Germany. Abstract from Baumeister's "Street Cleaning and Sewage," shortly to be published. *Eng. News*, Feb. 7, 1891, pp. 137-8.

The Sewerage and Sewage Disposal Works of the Borough of Dudley, Eng. The sewage of this mining town is carried several miles in a 13-inch main following the irregularities of the ground. A sewage farm is employed. Described by E. D. Marten. *Proc. Inst. C. E.*, Vol. CIV., 1891, pp. 238-48.

Works of South Framingham, Mass. Brief description of the construction and operation of this system of sewage disposal. The system is by irrigation and drainage. Sewage farm of 70 acres, divided into fields about 500 ft. by 700 ft. *Eng. News*, Nov. 23, 1889, Vol. XXII., p. 487.

Works at Worcester, Mass. Interesting description of these works and their operation. Illustrated. *Eng. News*, Nov. 15, 1890, pp. 432-4.

At Worcester, Mass. An illustrated description of these works, including the construction of a large outfall sewer. Chemical precipitation is used. *Eng. Record*, March 21, 1891, pp. 258-60.

Of Worcester, Mass. Description of the construction of a large outfall sewer as well as details of the works, with account of the process used which is by chemical treatment. Principal details illustrated. *Mfr. & Build.*, April, 1891, pp. 84-8.

See *Drainage*.

Sewage Filtration.

Paper by J. H. T. Turner before the Liverpool Engineering Society. Brief discussion of the subject is given, with "conclusions." Discussion. *Trans. Liverpool Eng. Soc.*, 1888, Vol. IX., pp. 28-40.

Of Town. Contains standard of impurity beyond which the Royal Commission think the waste should not be admitted into water-courses. *Van Nos. Eng. Mag.*, Vol. III., p. 123.

Filtering Experiments of Massachusetts State Board of Health conducted at Lawrence, Mass. Abstract. *Eng. & Build. Rec.*, Nov. 23, 1889.

Filter Presses for the Treatment of Sewage Sludge. By W. S. Crimp. Abstracted from a paper before the Institute of Civil Engineers. Deals particularly with filter presses as now adapted to the treatment of sludge. At Wimbledon, 250 tons are reduced by means of two presses to 50 tons containing 50 per cent. of moisture. *Lon. Engineer*, Feb. 25, 1887. *Sci. Am. Sup.*, April 9, 1887.

Sewage Pumping.

A brief illustrated description of pumps for sewage and sludge at Walthamstow, Eng. *San. Eng.*, Jan. 8, 1887.

Pumping Engines for Buenos Ayres. Compound condensing engines, containing many novel features. Illustrated. Lon. *Engineer*, April 9, 1886.

Plant at Aberdeen, S. Dak. Two motors are driven by the natural pressure from an artesian well. Described and illustrated. *Eng. News*, Nov. 29, 1890, p. 476.

Sewage Pumping, continued.

Plant at Boston. By E. D. Leavitt. A paper giving a description of the high-duty engines, the low-duty storm engines, the boilers and their appendages, and the foundations and pump wells. Fully illustrated. *Jour. Assn. Eng. Soc.*, Vol. III., p. 37.

Sewage Purification.

By Aeration and Oxidation. Shown to be due more to the action of living organisms than to chemical action. *Jour. of Science*, London, Sept., 1885.

By Electricity. A description of the methods used and the experimental works of Mr. Webster in England. Estimated cost of treatment, 12s. per million gallons. *Elec. World*, April 27, 1889.

Electrical Purification of. A brief paper by Mr. Frank M. Gilley, describing the method employed in some works near London, and also a test at the same place. *Proc. Soc. Arts*, 1890, pp. 76-81. *Eng. Record*, Jan. 17, 1891, p. 114.

Electrical Purification of Sewage and Contaminated Water. A paper by Wm. Webster, read before the Chemical Society, London, giving some information as to the details and cost of his system of treatment, also giving the results of trials as to efficiency, advantages claimed, etc., *Sci. Am. Sup.*, No. 799, April 25, 1891, pp. 12771-3.

And Contaminated Water by Electrolysis. A valuable paper treating recent researches in this line of engineering. Read by William Webster, F. C. S., before Section G., British Association, at Newcastle-on-Tyne, Sept., 1889. Illustrated. *Elec. Rev.*, Oct. 11, 1889; abstracted in *Eng. News*, Oct. 26, 1889, Vol. XXII., page 387. *Sci. Am. Sup.*, Nov. 23, 1889.

At Frankfort-on-the-Main, as proposed by the engineer, W. H. Lindley. Translated from the German by Wm. Paul Gerhard. Illustrated by plan and section of settling basins, etc. *Eng. News*, Feb. 21, 1885.

By the waters of a natural chalybeate springs at Buxton, England. A paper before the Soc. of Arts, London, with discussion. Reprinted in *Van Nos. Eng. Mag.*, July, 1885.

And Intermittent Filtration of Water. Experimental investigations of the State Board of Health of Massachusetts, being part II., of report of Water Supply and Sewerage. 8vo, pp. 910, Boston, 1890. Contains report of experiments on filtration of sewage and water and chemical precipitation on sewage made at the Lawrence experiment station; report on the chemical and biological work of the station, and investigations upon nitrification and the nitrifying organism, the whole forming a very valuable work. Abstracts of above work concerning filtration and chemical precipitation of sewage. *Eng. News*, July 11, 1891; *Eng. Record*, July 18, Aug. 15, 1891.

Gives results of the process at a number of places and condemns it. *Van Nos. Eng. Mag.*, Vol. VIII., p. 490.

Natural. Paper by Hiram F. Mills giving some important deductions from experiments by Mass. State Board of Health, on intermittent filtration through sand, *Land a Home*, Sept., 1890, pp. 598-606.

See *Water Filtration and Purification.*

Sewage Works.

An illustrated article giving the leading features of the extensive and remarkable works about to be constructed at the northern outfall of the London main drainage, for the purpose of purifying that part of London sewage. Lon. *Engineer*, Feb. 4 and 11, 1887.

Acton. Brief description of the Acton sewerage works, with ground plan. Lon. *Engineer*, Sept. 9, 1887.

The Birmingham and Edmonton. By Thos. Cole. Gives results of observations made at these two works. *Van Nos. Eng. Mag.*, Vol. XVII., p. 42.

Sewage Works, continued.

Of Clapham, Eng. Sir Jos. Bazalgette, Engr. Fully described by a hundred or more cuts in Lon. *Engineer*, March 12, 19 and April [illegible], [illegible].

By Baldwin Latham. A short description of the *Croydon Works*. *Van Nos. Eng. Mag.*, Vol. XXI., p. [illegible].

Hendon. An illustrated description of the new sewerage works at Welsh Harp, near Hendon, recently constructed. The plan of treatment is, first; straining to remove coarse matter; second addition of lime; third, removal by settlement of the suspended matter; fourth, decantation of clear water from tanks and its purification by filtration. Lon. *Engineer*, May 6, [illegible].

At Kingston on Thames. Illustrated description in *Eng. & Build. Rec.*, Aug. 23, 1890, p. 181.

And Pumping Engines at Hull, Eng. Illustrated descriptions of these pumping works with considerable other data concerning cost of building and maintenance. Lon. *Engineer*, Dec. 27, [illegible], Vol. LXVIII., p. [illegible]. *Eng. & Build. Rec.*, July 26, 1890, pp. 118-20.

For the Mystic Valley, Mass. Described and illustrated. Sewage pumped and treated with precipitant. *Report of Boston Water Board*, 1888.

A short description of the precipitation works at Sheffield, Eng. *San. Eng.*, July 1, 1886.

Wednesbury. Gives brief description of the works at Wednesbury. The sewage is treated with sulphate of alumina and lime. Lon. *Engineer*, Dec. 16, [illegible].

See *Water Works, Frankfort-on-the-Main*.

Sewers.

Appurtenances and apparatus for cleaning sewers, used at Boston. *Eng. News*, Feb. 27 and March 6, 1886.

Appurtenances. Notes of discussion of Conn. Assoc. of Civ. Engrs. and Surveyors, Jan., 1885. Contains much practical information, mostly of an elementary character. *Eng. News*, April 11, 1885.

Arches. Brick Arches for Large Sewers. By Rudolph Hering. Gives cases of failure of such arches and improved designs. Illustrated. *Trans. A. S. C. E.*, Vol. VII., (1878), p. 252.

Disinfection. A paper by Dr. G. W. Wight, giving an account of the method used in disinfecting the Detroit sewers by means of burning sulphur, and also some account of the gross defects of the Detroit sewer system. *San. News*, Dec. 12, 1884.

Discharge of Circular and Egg-Shaped. By W. T. Olive, before the Inst. of C. E. Gives diagrams, based on Beardman's formulas for finding the discharge of circular and egg-shaped sewers. *Proc. Inst. C. E.*, Vol. XCIII., pp. [illegible]-[illegible].

Diagrams for facilitating the Calculation of Velocity and Discharge of. By Wm. P. Gerhard. *Jour. Assn. Eng. Soc.*, Vol. I., p. 147.

Diagram. Diagram for Approximate Sewer Calculations. A paper by Edgar S. Dorr, read before the Boston Soc. of C. E. The diagram gives the flow from any given area, and also the capacity of various sized sewers at various inclinations. *Jour. Assn. Eng. Soc.*, July, 1891, pp. 357-[illegible].

Green Avenue Relief Sewer, Brooklyn, N. Y. Brief description of this large sewer now under construction. 12,[illegible] ft. long and drains an area of 1,[illegible] acres. *Eng. & Build. Rec.*, Nov. 29, 1890, pp. 407-8.

Intercepting.

Description of the intercepting sewers, nearly completed, at New London, Conn. Illus. *San. Eng.*, March 19, 1887.

Sewers, *Intercepting*, continued.

The Manayunk Intercepting Sewer of Philadelphia. Four feet in diameter and seven miles long, designed to intercept sewage discharge into the Schuylkill River above the dam. Fully illustrated. *Eng. Era*, Nov. 19, 1885.

Mill Creek, of St. Louis. A full account of this great work, with profile, plan and cross-sections. This is probably the largest sewer in the world, when cross-section and length are considered. By William Wise. *Jour. Assn. Eng. Soc.*, Vol. IV., p. 263.

Sewer Pipe. The methods of its manufacture described, as practiced at Milwaukee, Wis. Illus. *San. News*, Oct. 10, 1885.

Pipe. Some Experiments to Determine the Strength of American Vitrified Sewer Pipe. A paper by Malverd A. Howe, read before the Eng. Club of St. Louis. The experiments consisted in hydrostatic, drop, concentrated and uniform load, and cement joint tests on pipe from fifteen different manufacturers. Methods and apparatus described and illustrated, and results tabulated. *Jour. Assn. Eng. Soc.*, June, 1891, pp. 284-303.

Sewer Gas.

Gas, Its Effect on Iron Pipe shown, by cut, in *San. News*, Chicago, Jan. 1, 1885. The pipe in use eight years and entirely eaten away; did not come in contact with liquids.

Abstracts from papers by Dr. Soyka and Dr. Rozsahegyi on sewer gas as a factor in the spread of epidemic diseases. *Van Nos. Eng. Mag.*, Vol. XXVII., p. 403.

On the Exclusion of. By Richard Weaver. A good article. *Van Nos. Eng. Mag.*, Vol. XVII., p. 411.

Penetration of into Dwellings. By Dr. Lissauer. Gives results of ten experiments made to ascertain the conditions under which sewer gas may gain admission to dwellings. *Van Nos. Eng. Mag.*, Vol. XXVI., p. 166.

Sewerage.

And Drainage of the Mystic and Charles River Valleys. Report of the Massachusetts State Board of Health, 1889, with special reports by Jos. P. Davis, H. A. Carson, F. P. Stearns, Chas. H. Swan and Fred Brooks. A valuable document of 140 pp. Maps and index. Treats of classification and filtration and gives estimates of cost of plants.

Resume of the work of a commission appointed in 1882, and resolutions passed by said commission, who report in favor of prohibiting cess-pools, carrying all house drainage to the sewers and disposal by irrigation. *Annales des P. & C.*, Sept., 1883, pp. 458-476.

Of American Cities and Towns. A review of a report by C. H. Latrobe on a plan for sewerage of Baltimore. Treats mainly of the comparative merits of the combined and separate systems. *Eng. News*, Dec. 3, 17, 1881.

Aqueducts of Ancient Rome. An abstract from a letter by Prof. Rodolpho Lanciani, giving a brief description of the works and sanitary regulations of the city. *San. Eng.*, April 2, 1887.

Assessments.

Gives the decision in the case of the New London sewer assessments. The assessments hold good. *Eng. & Build. Rec.*, March 30, 1888.

Costs of. By T. W. Whitlock, before the Connecticut Association of Civil Engineers and Surveyors. Discusses the proper method of assessing property for sewerage improvements. *Proc. Conn. Assn. C. E. and Surv.*, 1888, pp. 17-60.

In Fall River, Mass. Discusses the best method of levying sewer taxes and gives the ordinance adopted at Fall River. *Eng. News*, Jan. 8, 1887.

Chiswick. By Joseph Hetherington. Gives a detailed description of the works, with diagrams showing the effects of rain, tide and leakage on sewage flow, and

Sewerage, *Chiswick*, continued.

particulars of quantities of sewage, sludge, etc. *San. Eng.*, August 20, etc., 1887.

And House Drainage. By E. S. Philbrick, C. E. A short article, illustrated, upon methods of draining suburban residences. Arch. and Builders' edition of *Sci. Am.*, April, 1886.

Influence of Sewerage and Water Supply on the Death-Rate in Cities. By Erwin F. Smith. Pp. 90. In proceedings and addresses at the Sanitary Convention of Ypsilanti, Mich., June, 1884. *Sup. to An. Rep., Mich. State Board of Health*, 1885.

Kurrachi City. Gives scheme of J. Strachan for the sewering of Karachi City and its suburbs, India. Recommends irrigation. *Ind. Eng.*, May 7, *et seq.*, 1887.

Plans for the city of Brooklyn, White Plains and other villages. *Tenth An. Rep. New York State Board of Health*, Albany, 1890, pp. 47-101, 183-58.

Rules and regulations adopted by the City of Providence. *Eng. News.*, Nov. 1., 1886.

Mechanical Appliances in. By Geo. D. Waring. A lecture before the Franklin Inst., wherein the merits of the separate system are duly set forth. *Jour. Frank. Inst.*, April, 1886.

Sewerage Systems.

A short description of the systems used by a number of towns in England. *Van Nos. Eng. Mag.*, Vol. VII., 483.

By Rudolph Hering. A valuable article, giving results of a study of the systems in use in Europe, and a general review of the different methods available. *Trans. A. S. C. E.*, Vol. X. (1881), pp. 361-386.

Atlanta. Report of Rudolph Hering on a system of sewerage, containing valuable information upon the requirements, methods of assessments for its cost, mode of construction, etc. *Eng. Record*, April 4, 1891, pp. 234-6, *et seq.*, Abstract. *Eng. News*, April 4, 1891, pp. 328-9.

Boston Drainage System. Very well described and illustrated in Lon. *Eng.*, Dec. 11, 1885. Reprinted in *Sci. Am. Sup.*, Jan. 16, 1886.

Boston. Concrete Tank Sewer, illustrated and described. *Eng. News*, Feb. 10, 1883, pp. 64-5.

Boston. Main Drainage. Paper by Eliot C. Clarke, read before the Boston Soc. C. E., describing this great work. Many illustrations. *Eng. News*, May 8, 1880, pp. 159-65.

In Bremen. Pavements and Tramways. Description, with illustrations. *Deutsche Bauzeitung*, 1885, pp. 529-31 and 553-54.

Of Cincinnati. Paper read by A. S. Hobby before the Cincinnati Medical Society, describing the physical and sanitary condition of the sewers. *Eng. News*, Nov. 16, 1878, pp. 377-8.

Cost and Construction of Thirty-Five Sewerage Systems. Kind of system, capacity, method of construction and various items of cost, tabulated. From report of Public Improvement Com., Troy, N. Y. *Eng. Record*, Oct. 24, 1891, p. 331. *Eng. News*, Oct. 24, 1891, p. 397.

District of Columbia. Report of the Board of Sanitary Engineers. House Rep. Ex. Doc., No. 445, 51st Con., 1st Sess. Abstracted in *Eng. & Build. Rec.*, Nov. 22, 1890, p. 393.

Duluth, Minn. By Messrs. Rudolph Hering and Andrew Rosewater, Consulting Engineers. *An. Rep. Board of Public Works, Duluth, for the year ending Feb. 28, 1890.* Pp. 67-92.

Inset of detailed drawings of *Approved Plans for Storm Sewers at Duluth, Minn.* W. B. Fuller, City Eng. *Eng. News*, Oct. 25, 1890.

Sewerage Systems, continued.

East Orange, N. J., with illustrated description of sewage disposal plant in use there. *Eng. News*, Jan. 18, 1889.

European System of. Abstract of a voluminous report by Rudolph Hering to the National Board of Health. The abstract gives the general matter on various systems, requirements and cost. *Eng. News*, Jan. 28, 1882.

Foreign Cities. Abstract of a report to Senate Committee of Dist. Columbia on the sewerage of several foreign cities. *Eng. News*, July 26, 1879, pp. 237-8.

Fort of Mysore, India. By Standish Lee. Gives description, with detailed drawing, of sewerage system at the Fort of Mysore, India. *Ind. Eng.*, June 30, *et seq.*, 1883.

Frankfort-on-the-Main. Gives details of the sewerage scheme being carried out at Frankfort-on-the-Main, with drawing showing details. *Eng. & Build. Rec.*, Aug. 11, *et seq.*, 1888.

Town of Framingham, Mass. Reports of committee on Drainage and Sewerage and construction of Sewerage System, etc. Compiled by W. A. Brown, clerk to Selectmen. *Report, System of Sewage Disposal*, 1889, Union Pub. Co., So. Framingham, Mass.

Henley-on-Thames. Gives brief description of the Shone system of sewerage at Henley-on-Thames. Lon. *Engineer*, Oct. 11, 1889.

Henley-on-Thames. A short description of the Shone system as applied at Henley-on-Thames, England, to an area of 245 acres, with 4,000 population, and a flow of 1,800 galls. per diem. Cost, $86,400. *Eng. & Build. Rec.*, Feb. 18, 1888.

Of House Drainage. By Osborne Reynolds. Illustrated. *Van Nos. Eng. Mag.*, Vol. VII., p. 100.

Kansas City. Separate vs. Combined System. Address by O. Chanute before the Academy Sci., Kansas City. *Eng. News*, Feb. 16, 1884, pp. 82-4. Review of above paper by Robert Moore. *Eng. News*, April 26, 1884. Reply by O. Chanute. *id.*, May 3, 10, 1884.

A second paper by Robert Moore, *id.*, May 17, 1884.

Kansas City. By Robert Moore. Being a review of a paper on this subject by O. Chanute. Together with a reply by Mr. Chanute. Subject hinges mainly on the relative advantages of the combined and separate systems. *Jour. Assn. Eng. Soc.*, Vol. III., p. 67. The discussion continued on pp. 157 and 183.

The Leeds (England) Works. A brief account of the experiment and its results, with cost of a year's work. *Eng. News*, March 7, 1885; from London *San. World.*

Of Leicester. A review of the projects submitted to the authorities since 1854, and details and plan now about to be carried out. The scheme involves pumping the sewage six feet through 1½ miles of pipe. Lon. *Engineer*, July 9, 1886.

London Sewerage and Sewage. Paper of considerable value by Sir Robert Rawlinson, Vice-Pres. Inst. C. E., with discussion. *Jour. Soc. Arts*, Dec. 20, 1889, Vol. XXXVIII., pp. 60-82.

London Sewerage and Sewage. Appendix and discussion to a paper by Sir Robert Rawlinson. *Jour. Soc. Arts*, Jan. 17, 1890, p. 182. *Eng. & Build. Rec.*, Feb. 8, 1890, Vol. XXI., p. 143.

Memphis.

By Rudolph Hering. Gives conclusions arrived at from a recent inspection of the Memphis sewerage system. *San. Eng.*, Nov. 14, 1887.

By F. S. Odell. The separate system first used here on a large scale in America. *Trans. A. S. C. E.* Vol. X., 1881, pp. 21-32.

A description of the system together with the sewer reports for 1889 and 1890. Pamph., pp. 31. By James H. Elliott, Asst. Eng., under the supervision of Niles Meriwether, City Engr. Abstract *Eng. Record*, March 28, 1891, pp. 277-8.

Sewerage Systems, continued.

Memphis, Norfolk and Charleston. A special report of the City Surveyor of Savannah, giving details as to dimensions, cost and operation of sewers in above named cities. By Mr. W. J. Wilson. *Eng. & Build. Rec.*, Aug. 12, 1889, Vol. XX., p. 143.

At Newark, N. J. By F. Collingwood. Gives description of the improved system of sewerage at Newark, N. J., with illustrations showing details of construction. *San. Eng.*, May 7, and 15, 1887.

Of New London, Conn. First An. Rep. of the Comm'rs., including the report of the engineer, W. H. Richards; of the consulting engineer, Rudolph Hering; method of assessments; rules for connections, and the city ordinance relating to sewers. Distinguishing features of the system, which includes intercepting sewers for system of glazed tile drains, carrying house drainage only. Discharges only in low tide. Apply to the Engr., New London, Conn.

Milwaukee, Wis. See *Flushing Tunnel.*

North Minneapolis Sewage. Tunnel and Intercepting Sewers. Details and cross-sections illustrating this important work, with brief description. *Eng. Record*, May 9, 1891, pp. 375-6.

New York. Twenty-Third Ward Sewerage. Illustrated description. By J. J. R. Croes. *Eng. News*, Jan. 25, 1879, pp. 26-8.

Norfolk, Va. Description of this system with account of laying of the pipes. *Eng. News*, May 5, 1883, pp. 212-14.

Omaha. By G. W. Tillson. Contains description of the Waring system and its working at Omaha. *Proc. Neb. Assn. Eng. & Surv.*, Vol. I.

Of Paris.

Abstract of report of the Drainage Commission, and a paper on the purification of the Seine, describing the irrigation systems. *Abstr. Inst. C. E.*, Vol. LXXXIII., pp. 476 and 478.

A short description of the system; also shows its extent. *Van Nos. Eng. Mag.*, Vol. XIX., p. 124.

Described, and methods of construction and cleansing given. *Eng. News*, Nov. 6, 1886.

At Pasadena, Cal. Illustrated description of the details. *Eng. & Build. Rec.*, May 31, 1890, p. 406.

Plainfield Sewerage Franchise. Agreement as to rates, penalties, outline of system, and specifications for the construction of the system given. Full text of this proposed agreement with editorial criticism. *Eng. Record*, June 27, 1891.

Pawtucket, R. I. Report of A. R. Sweet, City Engineer, on the drainage of the town.

Potsdam, Germany, Proposed Scheme for. Separate systems for house drainage and excreta, and purification schemes by downward filtration and oxidation. *San. Eng.*, April 2, 1885.

Prague, and the plans proposed. *Wochenschrift des Oesterr. Ing-und Arch.-V.*, 1886, pp. 137-141, 147-150, 158-161.

Providence, R. I. Proposed for. Being a volume of 116 pp. and many plates and cuts. By City Engineer Samuel M. Gray. Printed by the city. Includes the results of the author's investigation in Europe on the subject of sewage disposal, where he was sent by the City of Providence in 1884.

Pullman, Ill. Sewage is disposed of upon a sewage farm. Description of system by Benezette Williams before the Western Soc. Engrs. *Eng. News*, June 17, 1882, pp. 203-4.

Of Rangoon, British Burmah. Article describing these interesting sewerage works on the Shone system, the sewage being ejected from receiving tanks to a sewer at a higher level by use of compressed air. Illus. *Eng. & Build. Rec.*, March 15, 1890, Vol. XXI., p. 232.

Sewerage Systems, continued.

Of Rome. *Il collettore basso delle fogne di Roma a sinistra del Tevere.* Five handsome folding plates showing many cross-sections and details. *Giornale del Genio Civile.* Anno XXVIII., Feb.-March, 1890, pp. 69-105.

Report on a Proposed System of Sewerage for St. John's, Newfoundland, by Rudolph Hering and C. J. Harvey, Town Engineer, including plumbing and house draining regulations. *Report on Proposed Sewerage.* St. John's, Newfoundland.

Of Schenectady, N. Y. Brief general description, with one illustration. *San. Eng.*, April 22, 1886.

Separate System. Size and Inclination of Sewers. By A. E. White before the Inst. C. E. *Eng. News*, Nov. 15, 1884, pp. 236-7.

See *House Drainage. Drainage Water Works, Frankfort.*

Shreveport, La. Extract from the specifications for the sewer system. *Eng. News*, June 19, 1886.

Shone Hydro-Pneumatic System. By Edwin Ault, before the Society of Engineers. Gives full description of the Shone hydro-pneumatic system of sewerage, with details of the works at Eastborne, House of Parliament and Henley-on-Thames. *Trans. Soc. Eng.* 1888, pp. 63-103.

Of a Small City. Abstract of a report on the sewerage of Marlboro, Mass. By Ed. S. Philbrick, of Boston. Recommends sewers for house drainage only and disposal by downward filtration. *San. Eng.*, Feb. 26, 1885.

In St. Louis, and House Drainage. By Robert Moore. An historical and statistical article, giving also the present status of the city. *Jour. Assn. Eng. Soc.*, Vol. IV., p. 139.

Of Stratford-on-Avon, Eng. The irrigation and intermittent downward filtration systems combined. Fully illustrated. Lon. *Engineer*, Jan. 1, and 15, 1886.

Of Toulon. Note on the project of M. Dyrion. By A. Hanet. Pop., 70,000; mortality, 33.8 per 1,000. Project is to exclude rain-water and pump the sewage to the sea, with possible use for irrigation to some extent. Est. cost, 3,000,000 fr. *Mem. de la Soc. des Ing. Civils*, Dec., 1885, pp. 766-76.

Trenton, N. J. Brief description of the conditions met with, and the system as built, with map. By Rudolph Hering. *Fourteenth Annual Report of the Board of Health of New Jersey.* Trenton, N. J., 1890, pp. 99-104.

In the U. S. and Canada. Abstract of statistics from the *Manual of Am. Water-Works. Eng. News*, Aug. 23, 1890, p. 170.

Of Vancouver, B. C. Paper by E. Mohun, with discussion by several other engineers. *Trans. Can. Soc. C. E.*, 1888, Vol. II., p. 243-67.

Of the Wandle Valley, Surrey, Eng. Embracing 10,000 acres, 10 miles of sewers, separate system. Sewage pumped and disposed of by natural filtration over an area of 28 acres. A full account of the construction, which has been recently completed, by the Resident Engineer, W. S. Crimp, A. M. I. C. E., *Proc. Inst. C. E.*, 1884, Vol. LXXXVI.

Of West Bay City, Mich. Brief description. Illus. *Eng. News*, Dec. 27, 1890, pp. 564-5.

Of West Troy, N. Y. Brief descriptions with illustrations of details. *Eng. News*, Jan. 3, 1891, pp. 1-4.

Sewers.

A resume of a memoir of M. C. H. Schneider upon the normal condition controlling the quantity of water required in sewers. *Eng. News*, Sept. 3, 1887.

Arches for. See *Arch Curves.*

Gives details of the receiving and catch-basins at Waterbury, Mass. *San. Eng.* July 9, 1887.

Sewers, continued.

Capacity for Storm Water. Some observations made in Washington, D. C., on the rapidity with which storm water finds its way into the sewers. A self-recording gauge of the height of water in the sewer used in the experiments. A capacity of 2 inches rainfall an hour found to be none too large for well built up and paved sections. *An. Rep. Eng. Dept. Distr. Columbia*, 1883, p. 26. Maj. G. J. Lydecker, Eng. Comr.

Care of Sewers, Notes on the. Paper by F. Floyd Weld. This short paper contains many valuable suggestions concerning both construction and caring for sewers. *Can. Assn. C. E. & Surv.*, Feb., 1890. Proc. meeting of Sept. 11, 1889, pp. 43-49.

Chenoweth System of Jointless Sewers, Drains, Electrical Conduits, etc. A concrete conduit, built around a removable wooden core. Description and results of trials. *Eng. News*, Oct. 17, 1891, p. 369.

Cleaning of. Tools used in. Illustrated description. *Eng. News*, May 22, 1880, pp. 176-7.

Concrete Sewer at Mt. Vernon. A paper by Wm. E. Worthen. The sewer consists of a concrete channel, constructed in a trench excavated through rock and has a stone arch covering. With discussion as to the use of concrete. Section of sewer shown. *Trans. A. S. C. E.*, Vol. XXIV., May, 1891, pp. 363-6.

Construction of.

In Running Sand. By S. M. Reade. Describes the methods used in some portions of England; also proposes a system with drain for subsoil water in the invert of the sewer. *Van Nos. Eng. Mag.*, Vol. IV., p. 86.

In Madras, India. By Hormusji Nowrosji. Records some of the construction details of sewers of Black Town, Madras, India. The three sewers have a total length of 3½ miles, varying in diameter from 12 to 30 in. It is on the separate system. *Ind. Eng.*, March 31, 1888.

Notes on the. By Chas. A. Judson. Gives methods and standards used at Sandusky, O. Illustrated. *12th Annual Report Ohio Soc. S. and C. E.*, 1891, pp. 92-7.

"*Rust on the.*" Paper by C. H. Rust, M. Can. Soc. C. E., with an extended discussion, the whole forming a paper of considerable value. *Trans. Can. Soc. C. E.*, 1888, Vol. II., p. 302-349.

And Sewage Disposal. A series of articles by Ed. S. Philbrick, begun in *San. Eng.*, Dec. 4., 1886.

Cost of Cleaning Small Separate Sewers at Memphis, Tenn. Gives cost of cleaning of Waring System, also describes an accumulation of a peculiar fungus and method of its removal. From a letter by J. F. Elliott. *Eng. Record.* Nov. 14, 1891, pp. 387-8.

Design and Construction of. By Graham Smith. A paper before Liverpool Eng. Soc. *Van Nos. Eng. Mag.*, Vol. XX., p. 390.

Discharge of.

Multipliers and Curves for ascertaining the discharge, etc., at various depths in the same sewer. By Robert M. Gloyne. *Eng. News*, April 6, 1889.

Relation Between the Rainfall and the Discharge of Sewers in populous Districts. A very valuable paper by Emil Kuichling, M. Am. Soc. C. E., pp. 56. Gives a great amount of data in tabular form. Illustrated. *Trans. A. S. C. E.*, Jan. 1889, Vol. XX., p. 1.

Egg-Shaped. Design and construction table, giving radii and areas, by C. G. Force, Jr. Illustrated. *Trans. A. S. C. E.*, Vol. IX. (1880), p. 201-203.

Flushing. Parenty's Apparatus. Illustrated. From the French. *Sci. Am. Sup.* March 27, 1886.

Sewers, continued.

Formula.

Kutter's Formula adapted to Circular Conduits. By Robert Moore and Julius Baier. This is a valuable study, both of the laws of the maximum flow in a circular closed conduit, and of the application of Kutter's formula to such cases. Two very valuable tables accompany the paper, one giving numerical values of the co-efficient in Kutter's formula, and the other diameters for various inclinations and maximum rates of discharge. Illustrated. *Jour. Assn. Eng. Soc.*, Vol. V., p. 349.

Some New Formulas. By R. F. Hartford. *Jour. Assn. Eng. Soc.*, Vol. IV., p. [illegible].

Mean Velocity of Flow in, Formula for. By Rudolph Hering. *Eng. News*, Dec. 15, 1888.

For Mean Velocity of Flow in Sewers. By Rudolph Hering, being an adaptation of Kutter's formula for easy computation. *Proc. Eng. Club., Phila.*, Aug. 1888, Vol. VII., pp. 16-18.

Flow of Air in. Gives details of experiments by W. E. McClintock on the flow of air in pipe sewers, and its effect on traps at the foot of soil-pipe. *Eng. & Build. Rec.*, Feb. 11, 1888.

Gaugings of. Report by Geo. E. Waring, Jr., on the gaugings of sewers, carrying house drainage alone in eight cities in the U. S. App. F. of the *An. Rep. National Board of Health, 1880: H. Rep. Ex. Doc. No. 8, 46th Congr., 3d Session.*

Gauging of the "Dry Weather Flow" in Compton Ave. and Ohio Ave. Sewers, St. Louis, Mo. Account of observations and results. By Julius Moulton before the Engrs. Club of St. Louis. *Eng. News*, Sept. 4, 1880, pp. [illegible].

The reconstruction of the main drainage system of the House of Parliament, London. Abstract from the report of the engineer, Mr. Isaac Shane. *San. Eng.*, April 2, 1889.

Of the Massachusetts School for the Feeble-Minded. Description of this system of disposal by intermittent downward filtration on cultivated land. *Eng. & Build. Rec.*, April 12, 1890, Vol. XXI., p. 300.

Proportion of Rainfall Reaching the Sewers. Paper by Emile Kuichling. *Eng. News*, March 16, 1889, *et seq.*

Repairing Mill Creek Sewer, St. Louis, Mo. Proposal and specifications for replacing a portion of the old sewer with a new one. An important work. *Eng. News*, Oct. 16, 1880, pp. 348 *q.*

Separate System, a general description of in a paper by G. S. Pierson, before the Mich. Eng. Soc. *Am. Eng.*, Aug. 13, 1881.

Separate System of Sewers as Proposed for the City of Dayton, Ohio. A paper by F. J. Cellarius, describing in detail this system. Illustrated. *12th An. Rpt. Ohio Soc. S. and C. E.*, 1891, pp. 81-91.

Separate vs. the Combined System, as exemplified in the drainage of Hyde Park, Ill. By Benezette Williams. A description of the system proposed for Hyde Park with map (pumping with downward filtration), together with a review of the arguments *pro* and *con* for the Separate and Combined Systems. A valuable paper. *Jour. Assn. Eng Soc.*, Vol. IV., p. 175.

Siphons of St. Louis Island, Paris. Illustrated description of the large siphons constructed to convey the sewage water under the river Seine. *Sci. Am. Sup.*, No. 770, Nov. 15, 1890, pp. [illegible]. Taken from *Le Genie Civil.*

Siphon. A brief description of the syphon outlet for a low sewer district, Norfolk, Va., *San. Eng.*, April 9, 1889.

Size and Inclination of. By A. E. White. Contains tables to facilitate calculation of sewers. *Van Nos. Eng. Mag.*, Dec. 1874.

Sewers, continued.

Specifications for.

Being specifications for a vitrified stone ware pipe sewer, prepared by E. Kuichling, Rochester, N. Y. They are unusually *specific* as to defects in pipes and as to methods of laying. They evince a thorough acquaintance with the subject. *San. Eng.*, July 30, 1881.

Boston, Mass. For main and outfall sewers. Abstract. *Eng. News*, Aug. 29, 1891, pp. 193-4.

Washington City. *An. Rpt. Eng. Dept. Distr. Columbia*, 1885, p. 76.

Washington, D. C. Extracts from, showing main requirements as to trenches, repaving brick work, mortar and concrete. *Eng. Record*, Jan. 31, 1891, p 138.

Washington, D. C. Specifications for various sizes now under construction in Washington, including one 20 feet in diameter, with cuts showing cross-sections of all sizes. *Rpt. of Eng. Dept.*, District of Columbia, 1884.

Size of, Determination of the. By R. E. McMath. Undertakes to determine the size of a storm water sewer, under the conditions that it shall carry off the water of the great storms of the locality, and that no excess of size and cost be allowed. Contains a large amount of data, diagrams showing capacity of sewers, etc. *Trans. A. S. C. E.*, Vol. XVI., p. 179.

On Steep Streets. See *Pavements, Sewers, and Drainage.*

Tables for determining the proper size. By Robert Moore. Based on Weisbach's formulas. Gives diameters for area in acres, and slope given for rainfall discharge of 1 inch an hour. *Jour. Assn. Eng. Soc.*, Vol. I., p. 61.

Trap. By D. J. Erbetts. A brief review of the experiments of Mr. Glenn Brown. *Building*, June 18, 1887.

Trenches for, Sheeting and Bracing. Abstract of a paper by Wm. Whittaker before the Boston Soc. of C. E., giving considerable practical information. *Eng. News*, Oct. 20, 1851, pp. 427-8.

Trenching and Pipe Laying. Article by J. L. Fitzgerald, M. Am. Soc. C. E., giving results of his own experience. *Eng. & Build. Rec.*, June 15, 1889.

Tunnelling in sandy soil under buildings in Brooklyn. A new method used by Mr. John Anderson somewhat on the plan of the Hudson River tunnel. Illus. *Eng. News.*, May 1, 1886.

Ventilation of.

Abstracts from a report prepared by Sir J. Bazalgette on the sewerage of Brighton, with discussion. *Van Nos. Eng. Mag.*, Vol. XXVII, p. 409.

Ventilation of. A paper by W. S. Crisp, Assoc. M. Inst. C. E., giving results and discussion of experiments on the movement of air in the sewers at Wimbledon, 1888. A valuable study of this subject. Pp. 8, *Proc. Inst. C. E.*, Vol. XCVII., p. 383, London, 1889.

New method of ventilating sewers by means of a small fresh air pipe laid longitudinally through the top of the sewers. Illustrated. Reprinted from Lon. *Sanitary Record*, *Sci. Am. Sup.*, No. 762, Aug. 9, 1890, p. 12178, *Eng. News*, Aug. 2, 1890, p. 94.

By Means of Cowls. Some very satisfactory results in London. Cowls placed on tops of adjoining buildings. *Proc. Inst. C. E.*, Vol. LXXXIV., p. 322.

By J. G. Winton. Advocates the use of the suction fan. *Van Nos. Eng. Mag.* Vol. XXI., p. 466.

Description of a method successfully applied in England. *Eng. News*, Jan. 30, 1886.

Sewer and Drain Ventilation. A paper read by R. Read, Assoc. M. I. C. E., read before the Congress of Hygiene and Demography. Describes various methods. *Eng. News*, Oct. 3, 1891, pp. 318-19.

See *Municipal Engineering.*

Sextant. See *Goniograph.*

Shaft. *The 3,000 Feet Vertical Shaft on the Comstock Lode*, described and compared with other deep shafts. *Eng. News*, Aug. 29, 1885.

Shaft-Sinking.

At Guerstenau Colliery. An abstract from an account furnished by Mr. H. Tomson, director of the colliery, contained in a memoir by Mr. H. Suri, in *Zeitschrift fur das Berg-Hutten und Salinen-Wesen*, Vol. XXXV. Gives full description of the methods employed and cost of operations. *Eng. News*, Nov. 5, 1887.

Experiments on the Poetsch Method. Describes the apparatus employed and gives results of experiments directed toward testing the formation of ice around the refrigerating tubes and the resistance of frozen ground. *Annales des Mines*, Vol. XI., p. 96; *Eng. News*, Oct. 15, 1887.

Shaft Sinking.

The Poetsch System, by congelation. A description of some interesting experiments with. Abstract in *Proc. Inst. C. E.*, Vol. XC., p. 531.

Poetsch Freezing Process. Description of process with temperature observations on rate of freezing. *Eng. News*, June 7, 1884, pp. 261-3. Further illustration of the process, *ibid.*, July 5, 1884.

Through Quicksand or Soft Soils. Poetsch's method with freezing mixtures in pipes. Illustrated. From *Annales des Mines* in *Sci. Am. Sup.*, April 17, 1886.

Through Loose Materials. By A. McC. Parker. A method employed successfully at "Tilly Foster" mine. Illustrated. *Sch. Mines Quar.*, Oct., 1887, p. 116.

Through Loose Water-bearing Materials. A new method. By J. E. Mills, Quincy, Cal. A paper read before the Am. Inst. Mining Engrs., and republished in Lon. *Engineer*, Feb. 6, 1885; many cuts; also in *Van Nos. Eng. Mag.*, March, 1885.

Through Water-bearing Strata, near Dortmund. Abstracted from the German. *Proc. Inst. C. E.*, Vol. XC., p. 330-336.

Through Watery, Running Ground. Describes the method invented by Herr Poetsch, which consists of freezing the ground before the excavation of the material. *Van Nos. Eng. Mag.*, Vol. XXIX., p. 491.

Through Wet Gravel and Quicksand Near Norway, Mich. Account of the sinking of a shaft by means of a caisson or drop-shaft. Much difficulty was experienced with water. Illus. By William Kelly. In *Trans. A. I. M. E.* *Eng. News*, Aug. 8, 1891, pp. 117-118.

Under Difficulties at Dorchester Bay Tunnel, of the Boston Main Drainage. By D. McN. Stauffer. Shaft composed of iron at top and timber at bottom, sunk in the bay to a depth of 150 feet. Illustrated. *Trans. A. S. C. E.*, Vol. X., (1881), p. 343-360.

Shafting.

And its Bearing. The difficulties of the problem and the common failings with suggestions. Lon. *Engineer*, Oct. 6, 1885.

Flexible. By J. T. Hall. An interesting and valuable paper on the use of flexible crank and propeller shafting. *Sci. Am. Sup.*, Oct. 22, 1887.

For Stern Wheel Boats. See *Steamboat Shafts.*

Shafts.

Cracking of Steel. Gives a discussion of a meeting of Engineers' Club of Western Pennsylvania, which embraces interesting and valuable suggestions. *Eng. News*, Aug. 13, 1887.

Engine, Graphic Representations of Strains in. See *Engines, Steam, Shafts.*

Propeller, for Marine Engines. Treats of the material from which they are manufactured, investigates the strains and proper proportions. *Mech. World*, Dec. 17, 1887.

Shafts, continued.

For Screw Steamers. Causes of breaking of shafts and suggestions of method by which liability to break may be lessened. Lon. *Engineer*, March 6, 1885.

Steamboat, Improved. By the insertion of a universal joint to allow for erroneous alignment. Lon. *Engineer*, Oct. 2 and Nov. 6, 1885, and *Sci. Am. Sup.*, Nov. 21, 1885.

Steamboat for Stern-Wheel Steamers. A discussion of the method of supporting the forces acting upon, and the materials used in the construction of shafts. Also a table giving data regarding the shafts of twenty-three different steamers. By H. W. Baker. *Jour. Assn. Eng. Soc.*, May, 1885.

Strength of Crank Shafting. A comparison of iron and steel shafts as regards liability to fracture after a crack has started. *Am. Eng.*, March 20, 1885.

Shearing Strength *of Iron and Steel*. See *Iron and Steel*.

Shearlegs, *120-Ton*, at Clydebank ship building-yard. Brief description, illustrated, with details of construction. *E. & M. Jour.*, Feb. 9, 1889.

Sheet Piling. See *Piles*.

Shells. See *Projectiles*.

Ship Building. *Modern Practice of*. By Robert Duncan, before the Inst. of Engrs. and Ship Build. of Scotland. *Van Nos. Eng. Mag.*, Vol. VIII., p. 195.

On the Pacific Coast. Paper by Lieut.-Commander F. P. Gilmore, U. S. N. Illustrated, pp. 6. *U. S. Naval Inst.*, Vol. XV., p. 443, 1889. No. 50.

Sheathed or Unsheathed Ships? By Naval Constructor Philip Hichborn, U. S. N. An argument for sheathing iron and steel vessels with copper. *Proc. U. S. Naval Inst.*, Vol. XV., No. 1, 1889.

Steel and Iron. A paper by D. A. Quiggin on certain vital features in ship designing and construction heretofore neglected. A good paper. Illustrated. *Trans. Liverpool Eng. Soc.*, Vol. VI., pp. 137-49.

Stresses in Steel Plating Due to Water Pressure. Paper by J. A. Yates, read before the Inst. Nav. Arch., giving a method of calculating these heretofore neglected stresses, with application. Lon. *Eng.*, May 22, 1891, pp. 629-30.

See *Steel Castings*. *Steel, Mild*. *Navy*.

Ship Canal,

To connect Chesapeake and Delaware Bays. Surveys, borings, estimates, comparison of routes, specifications, etc. *Rep. Chf. of Engrs. U. S. A.*, 1885. Vol. I., p. 725.

Report of the Pennsylvania Ship Canal Commission. Abstract giving a statement of the feasibility of various routes, with estimate of cost for a canal from Conneaut Harbor, to Rochester on the Ohio river. Profile of this route given. *Am. Mfr.*, Feb. 27, 1891, pp. 175-6.

See *Canals, Ship*.

Ship Railway,

Chignecto. A paper by John F. O'Rourke, describing briefly the progress of this work. *Trans. A. S. C. E.*, Feb., 1891, Vol. XXIV. Paper No. 462, pp. 13-30.

Chignecto. Maps showing location and a general description of this proposed ship railway from Baie Verte to Bay of Fundy; also abstract of report by Mr. Benjamin Baker. *Eng. News*, Sept. 7, 1889, Vol XXII., p. 218.

Tehuantepec.

A pamphlet of ten pages, reprinted from London *Times* of Aug. 20, 1884, being a full description of the scheme. Apply to E. L. Corthell, care American Soc. C. E., 127 E. 23rd. st., New York.

See *Ship Railway*.

Tehuantepec. Considered commercially, politically, and constructively. By E. L. Corthell, Chf. Engr. A more complete presentation of the case than

Ship Railway. *Tehuantepec*, continued.

any heretofore published. A pamphlet of 80 pp. and seven plates. Address the author, care Secy. Am. Soc. C. E., 127 E. 23d St., New York.

Tehuantepec of Capt. Eads, fully described and illustrated by cuts of models. Lon. *Eng.*, Jan. 9 and 23, 1885.

By E. L. Corthell. An address before the Franklin Inst., Dec., 1884. Issued as a separate pamphlet of 32 pp. by the author, as printed in the *Jour. Frank. Inst.*, for June, 1885. The various engineering, economical and commercial problems carefully considered. Many illustrations, showing working of plant. Also, further, an address by E. L. Corthell before the Am. Assn. for Adv. Sci. Issued in pamphlet form, with six large plates. A valuable and authoritative presentation of the problem. Address the author, care Am. Soc. C. E., 127 E. 23d St., New York.

Venetian. By E. L. Corthell, before the Philadelphia Engineers' Club. Gives an interesting sketch of ship railroad project carried out in Venice in the 15th century. *Proc. Eng. Club, Phila.*, Dec., 1887, Vol. VI., pp. 153-64.

See *Boat Railway. Inland Navigation.*

Ship Transfer.

Present Aspect of the Problem of American Inter-Oceanic Ship Transfer. Read before the Engineers' Club of St. Louis, March 2, 1887, by Robert Moore. A complete and interesting exposition of the subject. *Jour. Assn. Eng. Soc.*, Feb., 1888.

By E. L. Corthell. A review of the above paper, and a reply to the review by Robert Moore. *Jour. Assn. Eng. Soc.*, May, 1888.

Sinking Funds. *Formulæ Derived. Jour. New Eng. W.-W'ks. Assn.*, June, 1887.

Ships.

Discussion of the question of butt fastenings of iron vessels. By Stavely Taylor. *Trans. Inst. Engrs. & Shipbuilders of Scotland.* March 24, 1885.

The Effect of Loading on the Form of Vessels. A valuable paper by Thomas Phillips, read before the Inst. of Naval Architects giving the results of a series of experiments made on four different vessels. Lon. *Eng.*, April 10, 1891, pp. 466-9.

Hospital. See *Hospital Ship for Small-Pox Patients.*

Method of Calculating the Stability of Ships. By D. W. Taylor, U. S. N. Gives examples of its application, also tables to aid in the work. *Proc. U. S. N. Inst.*, Vol. XVII., 1891, No. 1, pp. 157-238.

Modern War Ships. Report of a lecture by W. H. White, Director of Naval Construction in the British Navy, delivered at London, in January, 1887. Gives valuable data regarding the dimensions, speed, power, etc., of modern war vessels, with tables. *R.R. & Eng. Jour.*, March and April Nos., 1887.

Protection of the Hulls of Vessels by Laquer. By Lieut. J. B. Murdock, U. S. N. *Proc. U. S. N. Inst.*, 1890, Vol. XVI., No. 4, pp. 457-72.

Protection of Iron and Steel Ships Against Foundering from Injury to their Shells, Including the Use of Armor. A paper by Sir Nathaniel Barnaby read at the Pittsburgh International meeting, Oct., 1890. The question of thickness of shells, and construction of bulkheads is discussed. *Trans. A. I. M. E.*, 1890, pp. 13. Discussion, pp. 13-18.

Steering Gear on Vessels for Lake Service. A paper by Walter Miller before the C. E. Club of Cleveland describing Hand and Steam steering gear. Illus. *Jour. Assn. Eng. Soc.*, Oct., 1891, pp. 477-85.

Strength of. Paper of some value and interest; by Prof. P. Jenkins, read before the Inst. of Naval Arch'ts. Lon. *Eng.*, April 11, 1890, p. 445.

Tonnage of. A good review of the use of the terms ton and tonnage as employed in maritime commercial transactions. Lon. *Engineer*, December 30, 1887.

Ships, continued.

Volumes of, Different Rules for Computing. Application made of Simpson's Rules and Integrators, with geometrical illustrations. *Proc. Liverpool Eng. Soc.*, Vol. VI., pp. 1-10.

War. Special Trials of Recent. By W. H. White, before the Inst. Naval Architects. Tables and curves of speed, horse-power, revolutions, etc. Lon. *Eng.*, April 16, 1886.

War, American. By W. John, before the Institution of Naval Architects. Gives description of a competitive design which was accepted by the U. S. Naval Dept. Lon. *Eng.*, March 30, 1888.

Wrecked. The Raising of the Loretty in New York Harbor in 84 feet of water. An account of the methods used. *Mech. Eng.*, Nov. 28, 1885.

See *Keels. Lightship. Navy Speed Trials. Steam Ships. Steering Gear. Vessels.*

Shipyard. *The New Shipyard at Newport News, Va.* Description of this large shipyard with illustrations of the machine shop, dry dock, and the quadruple expansion engines of the tug boat "Dorothy" recently completed. These latter are described at length. *Eng. News*, May 16, 1891, pp. 465-8.

Shop Order System of Accounts. Three papers and a discussion, in all 60 pages concerning the keeping of shop accounts of various kinds. Very valuable to superintendents. *Trans. A. S. M. E.*, Vol. VII., p. 468.

Shops. *Work Shops. Their Design and Construction.* Paper by Mr. J. Davis Barnett, with discussion. *Trans. Can. Soc. C. E.*, 1889, Vol. III., p. 151-201.

See *Railway Shops.*

Shore Protection *on Exposed Sandy Coasts.* Being notes on the protection of the Pointe de Grave, France: with many colored engravings, showing nature of methods used. A pamphlet of 10 pp., issued by the Engr. Dept., U. S. A.

Shoring of Buildings. A lecture by Th. Blashill, London. Illustrated. *Sci. Am. Sup.*, April 25, 1885.

Shoring Trenches. See *Sewers.*

Siemens Regenerative Gas Furnace, *Recent Improvements in.* By E. F. Bamber, C. E., London, Eng. An excellent description of the modern types of furnace and gas producer, with a clear explanation of the main principles involved. Illustrated. *Am. Mfr.*, April 1, 1887.

Siemens Gas Lamp, on the principle of the Siemens furnace, as exhibited at the Electrical Exhibition, with cuts. Lamps burn from 12 to 100 ft. per hour, and produce from 135 to 1,200 candle power. *Am. Gas Light Jour.*, November 3, 1884.

Signals. *Naval Coast.* By Lieut. Richard Wainwright, U. S. N. Describes and compares a number of different methods for signaling between the shore and a fleet to be used in a system of coast defense; including semaphores, electric flash lights, the heliograph, carrier pigeons, etc. *Proc. U. S. Naval Inst.*, Vol. XV., No. 1, 1889.

Railroad. See *Railroad Signals.*

Silica *and Aluminum, Effects of in Working Iron.* Gives result of tests made by Prof. J. B. Johnson, showing gain in tensile strength of five to forty-eight per cent. and a gain in resilience of ninety per cent., due to the use of "Silica Process." A comparison is made between the above results and those obtained by the use of aluminum, described in a paper, "Influence of Aluminum on Cast Iron," by W. J. Keep, before A. A. A. S., August meeting. The results show marked similarity. *R. R. Gaz.*, Dec. 7, 1888.

Silicon *in Pig Iron Production.* A brief paragraph noting a "remarkable discovery" by Mr. Charles Wood, Prest. Cleveland Institute of Engineers, of the fact that silicon has the power to convert combined carbon into uncombined carbon. *Am. Mfr.*, Dec. 14, 1888.

In Iron Founding. See *Alloys.*

Siphon.

Automatic Intermittent. Gives a description of a self-priming siphon for sanitary or manufacturing purposes where a periodical flush is required. *Sci. Am. Sup.*, March 31, 1883.

Fritsoe. Gives description of a large siphon at the Fritsoe Works, Laurvik, Norway. The siphons have a diameter of 3 ft. 2 in., and a length of 82 feet. Suction height, 3 ft. 11 in., and height of fall of 4 ft. 8 in.; is made of ⅜ in. steel plate, with cast-iron mouth pieces. Lon. *Eng.*, July 8, 1887. *Sci. Am. Sup.*, Sept. 3, 1887.

A Large Siphon. Illustrations showing details of a large siphon for passing water over the Mississippi levee. *Eng. News*, Nov. 28, 1891, p. 519.

See *Aqueduct, New Croton.*

Slag. *The Utilization of Slag.* Paper by Gilbert Redgrave. Several possible uses of ordinary iron-furnace slag are given; among them use as paving blocks and road metal, "slag wool" or "silicate cotton," as a fire-proof non-conductor of heat or sound, as hydraulic cement, with a table of strength developed; also the possible use of the slag from the basic process in steel manufacture as a phosphate. *Jour. Soc. Arts*, Jan. 31, 1890, Vol. XXXVIII., pp. 221-234.

Slag Brick. See *Bricks.*

Slag Cement.

Blast-Furnace Slag and Slag Cement, and its value as compared with Portland cement. By Dr. Lehmann. A short account and discussion of some experiments on so-called slag cement, which show that Portland cement is not only better in every respect, but also cheaper. *Deutsche Bauzeitung*, 1886, pp. 14-19.

The Manufacture of. Abstract of an article in Annales Industrielles, July 21, 1889, p. 89. *Proc. Inst. C. E.*, Vol. XCVIII., 1889, p. 419.

The Manufacture of Slag Cement. Short article of considerable value by J. Groselaude, in *Annales Industrielles*, July 21, 1889, as tabulated in *Proc. Inst. C. E.* *Eng. News*, Dec. 7, 1889, Vol. XXIL., p. 531.

See *Cements, Slag.*

Slag Mill. *A Dustless Basic Slag Grinding Mill.* A new type of mill, designed to get rid of the very injurious dust produced in grinding. Illus. Lon. *Engineer*, April 24, 1891, pp. 315-16.

Slags. *Uses of Blast Furnace.* By T. Egleston. Describes the different methods adopted in Europe to utilize slags. *Van Nos. Eng. Mag.*, Vol. VII., p. 35.

Slide Rule.

By E. A. Gieseler. Gives description and rules for using this valuable instrument. *R. R. Gaz.*, March 9, 1883.

Computations. Examples of the use of the Mannheim Slide Rule, by William Cox. *Eng. News*, Dec. 20, 1890, pp. 543-4. Also a list of *Equivalents or Useful Numbers for Simplifying Calculations, and for Slide Rule Practice*, by the same. *Eng. News*, Jan. 3, 1891, pp. 5-6. Correspondence, *ibid.*, pp. 15-16.

Thatcher's. A wonderfully convenient and expeditious calculating instrument. Described and illustrated in *Eng. News*, Dec. 18, 1886.

By H. Vischer. A discussion of the principles of construction, use, and accuracy of the logarithmic slide rule. Illustrated. *Trans. Tech. Soc. Pac. Coast*, Vol. II., p. 271. (Dec., 1885.)

A plea for their general use and a description of several forms. By C. V. Boys. *Van Nos. Eng. Mag.*, Dec., 1885.

See *Calculating Machines.*

Slide Valves.

Friction of. An account of some very accurate and valuable experiments made by J. A. F. Aspinwall. Illustrated, with discussion covering 28 pp. in *Proc. Inst. C. E.*, Vol. XCV.

Slide Valves, continued.

On the Design of. By Prof. R. H. Smith. Rules obtained from diagram. From Lon. *Engineer*, in *Sci. Am. Sup.*, Dec. 25, 1886.

Sliding Contact. Extracts from a lecture delivered before the Society of Arts on the substitution of statical for sliding contact. *Am. Eng.*, Jan. 10, 1887.

Sliding Friction. *Theory of.* By R. H. Thurston. A paper read before the Am. Soc. of Mech. Engrs. A valuable paper on this subject, giving the intensity of pressure on journals, work lost in friction, etc. *Van Nos. Eng. Mag.*, Dec., 1880, and Lon. *Engineer*, Nov. 21, 1884.

Slow Burning. See *Fire Proof Buildings. Building Construction.*

Sluices. See *Irrigation. Sluices.*

Smoke Consumer. *Hutchinson.* Steam with air injected over the fire. Described and illustrated in *Am. Eng.*, Jan. 3, 1891, pp. 1-2.

Smoke-Consuming Furnace. *The Wilmsmann's Furnace*, a great success in Westphalia. An apron wall causes all gases to pass through coal bed on way to flues. *Journal fur Gasbeleuchtung und Wasserversorgung*, 1888, p. 135. Abstract in *Proc. Inst. C. E.*, Vol. XCV., p. 464.

Smoke Preventer. *The Ackroyd-Willoughby Smoke Preventer and Fuel Economizer.* Description and illustration, with results of a thorough trial, showing a good efficiency. *Am. Mfr.*, June 5, 1891, pp. 457-0. Lon. *Engineer*, May 15, 1891, pp. 395-6.

Smoke Prevention.

A paper by Robert Moore, read before the Engineers' Club of St. Louis, Nov. 31, 1888. *Jour. Assn. Eng. Soc.*, April, 1889.

Steam Boiler Furnaces for. By J. W. Hill. Description of the furnaces and detail of test made at the Cincinnati Industrial Exhibition by the Committee on Smoke Prevention. *Van Nos. Eng. Mag.*, Vol. XXII., p. 60.

Report on the adaptation of the Williams Furnace to one of the boilers of the heating plant of Washington University, St. Louis. Of value as showing the *efficiency* is maintained, but that there is a serious loss in the extreme *capacity*, resulting from the introduction of either hot or cold air above the fire. *Jour. Assn. Eng. Soc.*, Vol. V., p. 301.

The Steam Jet a Means of Preventing Smoke Boiler Furnaces. A discussion of the merits of several devices, with accompanying description. Illus. *Eng. News*, April 4, 1891, pp. 314-5.

See *Heating, Steam.*

Snow Fence. *An Automatic.* Illustrated. *Ry. Rev.*, Jan. 8, 1887.

Snow Plows.

Paper by Mr. F. E. Siekels, member Eng. Club of Kansas City, pp. 3. *Jour. Assn. Eng. Soc.*, March, 1890, Vol. IX.

Discussion of the subject by the Northwest Railroad Club. *R. R. Gaz.*, Jan. 11, 1889.

And Flangers. Discussion of the subject by the Northwest Railroad Club. *Mast. Mech.*, Feb., 1889.

Competitive Trials of the Rotary and Other Machine Snow Plows. Illustrated review of this series of tests. *R. R. Gaz.*, May 23, 1890, p. 353. *Ry. Rev.*, May 24, 1890, p. 301.

The Jull Centrifugal for railroads. Illustrated description. *R. R. Gaz.*, March 29; *Eng. News*, March 30, 1889.

The Russell Snow Plow. Description with detailed illustration of this large plow. *R. R. Gaz.*, Oct. 23, 1891, pp. 742-3.

Snow Sheds.

Canadian Pacific Railroad. Gives description with drawings of the different forms of snow-sheds in use on the Canadian Pacific Railroad. *Eng. News*, Jan. 21, 1883.

Canadian Pacific Railroad. Gives plan and sections of the standard snow-shed in use on the Canadian Pacific Railroad. *Eng. News*, Oct. 8, 1887.

Of the Canadian Pacific Railway. Illustrated description. *E. & M. Jour.*, March 2, 1889.

See *Railroad Structures.*

Rotary Steam, for clearing railways, with cut showing it in use. *R. R. Gaz.*, Apr. 24, 1885, and *Ry. Rev.*, Apr. 25, 1885.

Snow Tunnel *for Railroads.* A proposed form, being a full circle arch of timbers, laid longitudinally, supported by outer timber cribwork. A snow slide will increase its stability. Illustrated. *Eng. News*, Sept. 4, 1886.

Social Questions ably discussed, in their various phases, especially with reference to relations between manufacturers and their employees, by many writers in *Age of Steel*, Jan. 1, 1887.

Softening of Water. By Baldwin Latham. Various methods now employed, including Clark's process. *Van Nos. Eng. Mag.*, Vol. XXXI., p. 311, Oct., 1884.

See also *Water.*

Solar Azimuth *by Transit Attachment.* Results of two days' observations with a Fauth attachment on a Buff & Berger Transit. By J. B. Johnson. Includes tables for correcting observations for erroneous setting of either the latitude or declination, and also for correcting the declination or refraction. *Jour. Assn. Eng. Soc.* Vol. V. p. 35.

Sounding Apparatus.

An Apparatus Automatically Registering the Profile of River Bottoms. Gives satisfactory results. Illustrated description. *Annales des P. et C.*, April, 1891. Reprinted from *Eng. News* July 25, 1891, p. 76.

Sounding Machine *of the U. S. S. "Ranger"* The original machine of Sir William Thompson, as improved by Ensign Harry Phelps, U. S. N., so as to produce a serviceable and reliable machine that is cheap, and not liable to get out of order. Description, with illustration, method of using, etc. *Proc. U. S. Naval Inst.*, Vol. XV., No. 1, 1889.

Soundings in Pipes *by Metallic Potassium.* The depth determined by the explosion of the potassium at the end of the tape line when it touched the water. An accurate method used at Boston. *Eng. News*, Sept. 12, 1885. Also *Abst. Inst. C. E.*, Vol. LXXXIII., p. 454.

Speed *on Railroads.* An account of the fastest run ever made, being on the New York & West Shore road, from Buffalo to Weehawken (New York), July 9, 1885. Described in *R. R. Gaz.*, July 17 and Aug. 28, 1885. The run fully analyzed in latter number, with speed plotted and tabulated.

See *Railroad Operation. Steamship Speed.*

Speed Indicator.

For showing and recording the speed of the shaft of a screw propeller. Awarded a gold medal at the Inventions Exhibition, London. Illustrated. Lon. *Engineer*, Nov. 20, 1885.

Moscrop Recorder. Description of an instrument designed to indicate by a continuous record the speed of steam engines and the amount of any variations and the time they occur. Illustrated. *Power*, Aug., 1889.

See *Kinemeter.*

Speed Trials *of British War Ships.* A detailed account of the recent trials of H. M. S. S. "Camperdown" and "Anson." *Proc. Inst. C. E.*, Vol. XCV.

Progressive. On Mr. Mansel's and Mr. Froude's methods of analysing the results of progressive speed trials. *Trans. Inst. Engrs. & Shipbuilders, Scotland*, Feb. 24, 1885.

"Speeding Up." By J. T. Hobart. An ingenious and simple method of calculating the speed of complicated systems of gearing. *Sci. Am. Sup.*, Oct. 22, 1887.

Spherical Projection. Special solutions in steographic projections. By L. H. Bernard. *Proc. Eng. Club, Phila.*, Vol. V., p. 108.

Spikes. See *Drift Bolts.*

Spontaneous Combustion.

In Collieries, giving the Causes of such Action as found in the Collieries of France. *Van Nos. Eng. Mag.*, Nov., 1884.

In Factories and Shops. By C. W. Vincent. *Van Nos. Eng. Mag.*, Vol. XVII., p. 515.

See *Mines, Coal.*

Springs.

By A. De Bonneville, Jr. Treats of the materials, forms, sizes, uses and manufacture of springs in general; the helical spring in particular and in detail, in connection with rubber springs. Also gives gives methods and results of tests. *Van Nos. Eng. Mag.*, Vol. XVIII., pp. 391 and 319.

The Deflection of Spiral Springs. Theoretical paper by A. E. Young, containing the results of experiments on a volute buffer spring. Diagram and tables. *Proc. Inst. C. E.*, Vol. CI., 1890, students' paper No. 258, pp. 161-75; appendix containing tables and summary of formulas, pp. 176-83.

Formulas for. A table giving the formulas used by the Pennsylvania Railroad, and calculated by F. Reuleux. *R. R. Gaz.*, Nov. 15, 1889, Vol. XXI., p. 447.

Helical. An investigation into the principles of their design. By J. W. Cloud. *Trans. A. S. M. E.*, Vol. V., p. 173.

Spiral. Formulas for, in compression or tension. By Oberlin Smith. *Trans. A. S. M. E.*, Vol. IV., p. 335.

See *Steel, Resilience of. Piston Springs.*

Sprinklers. *Automatic.* An account of some elaborate experiments upon the efficiency of such fire prevention devices, by C. J. H. Woodbury, Boston. *Lon. Eng.*, Jan. 9 and 16, 1885.

Spur Wheels, *Strength of.* By Francis Compin. *Van Nos. Eng. Mag.*, Vol. VI., p. 208.

Stability. *Determining Curves of.* By John H. Heck. A paper read before the Institute of Naval Architects giving a new method of using paper sections for determining the cross curves of stability. *Lon. Eng.*, April 6, 1887.

Stable Construction. Discusses the arrangements of stables from a sanitary standpoint, by A. W. Wright, Chicago. *Jour. Assn. Eng. Soc.*, Vol. III., p. 309.

Stables, *The Ventilation of.* By A. W. Wright. An attempt to determine the amount of fresh air to be supplied to a horse. *Jour. Assn. Eng. Soc.*, Vol. IV., p. 193.

Stairways. Illustrated description of many famous stairways with a discussion of architectural features. From *Encyclopédie de l'Architecture et de la Construction.* *Am. Arch.*, No. 819, July 4, 1891, *et seq.*

Standard of Length, *Wave Length of Sodium as a.* By Prof. A. A. Michelson and E. W. Morly, before the Civil Engineers' Club of Cleveland. Gives a method for making the wave length of sodium light the actual and practical standard of length. *Jour. Assn. Eng. Soc.*, May, 1888.

Standard Time. A paper recommending its adoption. By Sandford Fleming. *Trans. A. S. C. E.*, Vol. X. (1881), p. 387.

See *Meridian Time.*

Standards.

Of Length and Their Subdivisions. By Geo. M. Bond. Lon. *Engineer*, Feb. 20, 1885, *et seq.*

Of Measure. Short paper. By T. C. Mendenhall. *Rep. Ohio Sci. Surv.*, 1884.

Stand-Pipes.

By J. T. Ward. Gives method of computing thickness of metal, etc.; also some general hints on the construction. *Eng. News*, July 24, 1886.

Paper by A. H. Howland, discussing stability of stand pipes, and giving three valuable tables for use in designing stand pipes. *Proc. Eng. Club., Phila.*, March, 1887, Vol. VI., pp. 110-115.

By A. H. Howland. Gives tables showing capacity, minimum thickness of metal, etc.; also details for a stand-pipe 102 feet in hight and 25 feet in diameter. *Proc. Eng. Club, Phila.*, Vol. VI, p. 110.

A paper by Wynkoop Kiersted, C. E. Discusses strength and designing of lap joints, thickness of plates necessary to resist water pressure and wind stresses, etc. Illustrated by diagram. *Selected Papers, Rensselaer Soc. of Eng.*, Vol. II, No. 2, p. 49, June, 1889.

A paper by Wynkoop Kiersted, read before the Rensselaer Soc. Engrs. giving some brief practical formulas for designing standing-pipes. Full abstract, *Eng. Record*, April 25, May 2, 1891.

By B. T. Stevens, before the American Water-Works Association. A valuable paper, giving details of a large number of stand pipes. It also gives practical suggestions on their construction and details of a number of failures. *Proc. Am. Water-Works Assn.*, 1888, pp. 102-114; *Eng. & Build. Rec.*, May 12, 1888; *Eng. News*, Oct. 6, 1888.

Destruction of the Defiance Stand-Pipe. Account of the bursting of a stand-pipe 140 feet high by 22 ft. in diameter, cracks previously developed from excessive pressure due to the blowing out of a heavy coating of ice seems to have been one cause of final failure. Illustrated by photographs. *Eng. Record*, April 11, 1891, pp. 309-10.

Failing. A brief account of the failure of three stand-pipes. *Building*, March 5, 1887.

Foundation for. Illustrations of the extended base and connecting brackets of the Jefferson City stand-pipe. *Eng. Record*, April 11, 1891, p. 308.

Seneca Falls. Gives description of the failure of the steel stand-pipe at Seneca Falls, N. Y. *San. Eng.*, Nov. 5 to 12, 1887.

Thickness of Plates for. Formula for. By Kenneth Allen. *Eng. News*, Dec. 15, 1888; *Eng. & Build. Rec.*, Jan. 5, 1889.

Relative Economy of Stand-Pipe and Trestle Tower for Water-Works. Paper by H. S. Crocker. Reprinted from *The Technic.* *Eng. & Build. Rec.*, July 19, 1890, p. 116.

See *Water Tanks.* *Water Towers.*

Station. See *Electric Light Station. Electric Power Station. Railroad Station.*

Statue. See *Bartholdi Statue.*

Bronze. See *Bronze Casting.*

Stay-Bolts. See *Iron. Stay-Bolt Iron. Tests for.*

Steam.

Condensation in Pipes. Results of some 2,000 experiments in France. Mean results plotted for different steam pressures. *Mech. Eng.*, Nov. 28, 1885.

Steam, continued.

Condensation from Steam Pipes and Cylinders. Utilization of the Water of. Paper by Mr. Peter Fyfe, read before the Institution of Engineers and Shipbuilders in Scotland. *Mech. World*, Jan. 11, 1890, p. 17.

Dry. On the Identification of Dry Steam. Paper by Prof. J. E. Denton, describing tests of quality of steam by observing appearance of jet. *Trans. A. S. M. E.*, Vol. X., 1889.

Economic Use of. By Prof. Chas. A. Smith read before the A. A. A. S. A discussion of the various points of the problem. *Eng. News*, Nov. 21, 28, 1878.

Effect of Circulation on. By G. H. Barrus, before the American Society of Mechanical Engineers. Gives experience in the effect of circulation in steam boilers on the quality of the steam. *Am. Eng.*, May 9, 1889.

Efficiency of High Pressure. By W. W. Beaumont, before the British Association. Object of paper is to show that Carnot's theorem is limited in its application to the steam engine, and that high pressure steam must theoretically, as well as practically, be more efficient than low pressure. *R. R. Gaz.*, Nov. 6, 1888.

Efficiency of. How to Measure. *Sci. Am. Sup.*, March 14, 1891.

Expansion.

The Proper Method and Regulation of the Engine. By Prof. Thurston. Advocates governing the engine by throttling the exhaust instead of the steam supply or changing the point of cut-off. *Trans. A. S. M. E.*, Vol. II., p. 358.

Ratio of for Maximum Efficiency. A paper by Prof. Thurston. *Trans. A. S. M. E.*, Vol. II., p. 128.

Experiments on the Condensation and Re-evaporation of Steam in Jacketed Cylinder. Paper read before the Institution of Mechanical Engineers. Oct., 1889, by Major Thomas English, R. E., Supt. Royal Carriage Dept., Woolwich, Eng. Giving results and discussion of an extensive set of experiments. Lon. *Engineer*, Nov. 8, 1889, Vol. LXVIII., p. 358. Lon. *Eng.*, Nov. 8, 1889, Vol XLVIII., p. 555, *et seq.*

Experiments on the Discharge of Steam Through Orifices. Many experiments to determine discharge through various forms of orifices. Tables and diagrams. By Strickland L. Kneass. *Proc. Eng. Club, Phila.*, Vol. VIII., No. 3, July, 1891, pp. 170-87. Abs., *Eng. News*, Nov. 21, 1891, p. 484.

Exhaust, Use of. By C. H. Manning. A suggestive treatment of the subject of interest to all engineers and manufacturers. *Sci. Am. Sup.*, Oct. 29, 1887.

Exhaust, Use of. See *Heating.*

Flow of. Through Orifices. A brief paper by C. H. Peabody and L. H. Kunhardt, giving the results of some experiments on pressure and flow of steam. *Trans. A. S. M. E.*, Vol. XI., 1890, pp. 187-92.

By Friction, for obtaining fresh water in shipwrecked boats at sea. Apparatus described and illustrated. *Sci. Am. Sup.*, Sept. 11, 1886.

Generation of. By G. H. Babcock. A Sibley College lecture. Treats of the generation of steam in tubular boilers. Illustrates and describes the different boilers with furnaces for burning coal, wood, gas, bagasse, etc. *Sci. Am. Sup.*, Dec. 17 and 24, 1887.

Generation of, and the Thermodynamic Problems Involved. See *Heat.*

High Pressure.

Economy of. By W. W. Beaumont before the British Association. Abstract of and editorial on. *R. R. Gaz.*, Nov. 9, 1888.

Economy in the use of high-pressure steam. By Wm. B. Le Van. *Jour. Frank. Inst.*, May, 1885.

Outflow of. By Prof. Rankine. *Van Nos. Eng. Mag.*, Vol. II., p. 116.

Steam, continued.

The Quality of. By John W. Hill. A paper read before the Am. Railway M. M. Assoc., Providence, June, 1881. Treats of entrained water. *Van Nos. Eng. Mag.*, Vol. XXV., p. 129.

Sale of. By Chas. E. Emery. The third Sibley College lecture. Treats of the meter system and the use of the polar planimeter in calculating the consumption of steam by customers; also discusses the quantity of steam required to heat buildings. *Sci. Am. Sup.*, Jan. 15, 1887.

Saturation Point of. By G. A. Hirn. A mathematical discussion of the determination of the point of saturation of steam. *Van Nos. Eng. Mag.*, Vol. II., p. 314.

Substitutes for. By G. H. Babcock. Gives a brief history of the working of engines, with their advantages and disadvantages, using other fluids than steam. *Ry. Rev.*, June 5, 1886.

Temperature and Elasticity of. By Alex. Morton. Gives table of comparison of Regnault's "graphic curve" with Rankine's formula. *Van. Nos. Eng. Mag.*, Vol. VI., p. 43.

Transmission of. A lecture by Chas. E. Emery before the students at Cornell University. *Sci. Am. Sup.*, May 29, 1886.

Steam and Hot Water. The comparative value of, for Transmitting Heat and Power. By Charles E. Emery. Read before the Am. Soc. of Mech. Engrs. Gives an elaborate discussion. *Sci. Am. Sup.*, Oct. 15, 1887. *San. Eng.*, June 11, 1887.

Steam Boilers. See *Boilers.*

Steam Calorimeters. *An Experimental Study of the Errors of Different Types of Calorimeters.* A paper by C. H. Peabody with A. L. Williston, giving the results of some experiments on four forms of calorimeters. *Trans. A. S. M. E.*, Vol. XI., 1890, pp. 193–209.

See *Calorimeters.*

Steam Condensers.

Surface Condensers. By J. M. Whitham, Fayetteville, Ark. A paper giving history of surface condensers, their advantages and disadvantages, experimental data, etc. *Mech. World*, Nov. 17, 1888.

Steam Distribution.

Cities. By W. P. Shinn, in *Trans. A. I. M. E.*, Vol. XII.

Direct Steam Supply, or Heating Buildings by Steam from a Central Source. By J. H. Bartlett, being a pamphlet containing a paper read at British Assn. at Montreal. Gives methods used at various places, cost, causes of failure, etc. Address the author at 17 Hamilton Chambers, Montreal, Can.

The District Distribution of Steam in the United States. A paper by Chas. E. Emery, Ph. D.; also discussion and correspondence on the paper. Pp. 84. *Proc. Inst. C. E.*, Vol. XCVII., p. 96. London, 1889.

See *Heat and Power. Steam Heating.*

Steam Engine. See *Engine, Steam. Friction of Non-Condensing.*

Steam Fitting. See *Steam Heating and Fitting.*

Steam-Hammer. *The Creusot 100-ton Hammer*, with other cuts from the Creusot Steel Works. *Sci. Am. Sup.*, May 9, 1885.

See *Hammers.*

Steam Heating.

A discussion of value in *Trans. A. S. M. E.*, Vol. VI., p. 837.

A report of the trial of the steam heating boilers at Lehigh University. *Jour. Eng. Soc., Lehigh Univ.*, Vol. II., p. 120.

Steam Heating, continued.

By Robert Briggs. A practical treatise on the American practice of warming buildings by steam, with the mechanical details illustrated. *Van Nos. Eng. Mag.*, Vol. XXVIII., p. 378, etc.; also *Van Nos. Science Series*, No. 68.

By E. D. Meier. Describes the plant designed and used for the heating of the St. Louis Armory Building. *Jour. Assn. Eng. Soc.*, Vol. II., p. 128.

By Prof. C. A. Smith. Discusses the Theory; Various Systems in use in the U. S.; Experiments made by the author; and a project of heating a cotton mill by steam. *Jour. Assn. Eng. Soc.*, Vol. I., p. 379.

And Fitting. A series of articles by "Thermus" in *San. Eng.*, Vols. XII., and XIII., 1885.

Buildings from a Central Source. A paper read before the British Association, 1884. By Jas. H. Bartlett, and printed separately by J. Lovell & Son, Montreal. Gives much practical information, with tabulated list of American systems in operation, with particulars as to the same; 21 pp. Illustrated.

Improvements in. Abstract of a paper by Frederick Tudor before the Society of Arts, Boston. Discusses hammering in pipes and control of temperature. *San. Eng.*, Dec. 18, 1886.

See *Heating.*

Method employed in Manhattan Co.'s Bank Building, New York, being the "graduated" system, with cut. *San. Eng.*, Feb. 12 and 19, 1885.

Pipes, Sizes of. By Alfred R. Wolff. Discusses the Towne-Briggs and the Babcock rules, and takes the mean of the two. *Building*, March 13, 1886.

Private Dwellings. The indirect system recommended. By Alfred R. Wolff, M. E., who describes in general what may now be done in this direction. *Building*, April 10, 1886.

Towns and Villages. By D. Galton. A description of the combination system. *Van Nos. Eng. Mag.*, Vol. XXVI., p. 298.

See *Heating. Steam Distribution.*

Steam Horse Power. *Cost of.* Article by Thomas Pray, Jr., of Boston, gives methods and data for calculation in different cases. *Jour. Frank. Inst.*, July, 1890.

Steam Jackets. See *Engines, Steam, Cylinders.*

Steam Loop. A lecture by W. C. Kerr before the Frank. Inst., describing its action, efficiency and limitations. Reprint *Eng. Record*, Oct. 24, 1891, pp. 329-30, 343-3.

Steam Pipes.

Non-Conducting Covering for. The methods and results of a very elaborate investigation into the relative and absolute values of such coverings. Fully illustrated. By Prof. J. M. Ordway. *Trans. A. S. M. E.*, Vol. V., p. 73, 211 and Vol. VI., p. 168.

Non-Conductors for. Abstract from a report by Prof. John M. Ordway, giving tables of relative conductivity of various materials. *Eng. Rec.*, Jan 3, 1891, pp. 71-2.

Covering for.
See *Non-Conductors.*

Support of. See *Pipes.*

"*Water-hammer*" *in.* By Prof Thurston. The cause and effects explained. *Trans. A. S. M. E.*, Vol. IV., p. 202.

Steam Plants.

District Steam Systems. A paper by Charles E. Emery giving a description and general statement of condition of the plant of the N. Y. Steam Company and results of eight years' experience in its operation. Illus. *Trans. A. S. C. E.*, Vol. XXIV., March 1891, pp. 118-205. Discussion, pp. 216-22.

Steam Plants, continued.

For Electric Service. Paper by Wm. H. Bryan, read before St. Louis Engineers Club, April 3, 1889, pp. 8. *Jour. Assn. Eng. Soc.*, August, 1889, Vol. VIII., p. 419.

And Smoke Stacks at the Edison Central Station, Brooklyn, N. Y. Illustrated description of this new steam plant, showing an arrangement which can be used on a limited area. *Am. Eng.*, Jan. 22, 1890, p. 34.

Station J, New York Steam Co. Gives a full description of the plant of the New York Steam Company at Station J. Shows plans of building, pipe arrangements, etc. *Eng. & Build. Rec.*, April 7, 1888.

Station J, of the New York Steam Company. Good description of this plant. Illus. *Eng. & Build. Rec.*, May 10, 1890, p. 362.

For Textile Manufactories. A review of the subject, including the consideration of high and low pressure, types of boilers, engines, system of working, personnel, etc. *Sci. Am. Sup.*, April 23, 1887.

Of the Washington Mills, Lawrence, Mass. See *Engine, Steam.*

Steam Power.

The Cost of. Lecture by Thos. Pray, Jr., C. E. *Jour. Frank. Inst.*, June, 1890, No. 774, *et seq.*, pp. 478-483.

The Cost of, in America. By Chas. E. Emery. A schedule given showing cost of such power on certain assumptions true for New England in 1874. Very valuable in making estimates. *Trans. A. S. C. E.*, Vol. XII. (1883), p. 425.

Cost of vs. Water Power. See *Water Power.*

Economy of. By John W. Hill. A resume of tests made at the National Rubber Works, Bristol, R. I., with a view to improving the steam plant. *Van Nos. Eng. Mag.*, Feb. 1884.

And Water Power, Cost of. The paper by Chas. T. Main before the Am. Soc. M. E., in 1889, is republished in *Eng. News*, Jan. 24, 1891, pp. 75-7.

And Water Power, Comparative Cost of. Abstract of paper by Charles H. Manning, Manchester, N. H. Valuable comparison of cost of steam and water power for Lawrence and Lowell, Mass., and Manchester, N. H. *Mechanics*, June, 1889.

Versus Water Power. A comparison of the cost, based upon mill practice for various prices of coal, and different amounts of steam used for heating. Paper by Charles T. Main before Am. Soc. Mech. Engineers at New York meeting. *Mechanics*, December, 1889. *Trans. A. S. M. E.*, Vol. XI., 1890, pp. 108-26.

Versus Water Power. Paper by Charles H. Manning, Manchester, N. H., on the comparative cost of steam and water power, presented at the Erie meeting Am. Society of Mechanical Engineers. *Am. Mach.*, May 16, 1889.

Steam Separators.

Account of two new forms of devices for separating entrained water from steam, before admitting the steam to the cylinder. Illus. *Eng. News*, Jan. 18, 1890, Vol. XXIII., p. 53.

Test of. Report of tests of six separators made at Sibley College, Cornell University. *Power*, July, 1891, p. 9. *Eng. News*, Sept. 12, 1891, p. 233.

Steam Stamps. *Solid Foundations vs. Spring Timbers for.* Brief editorial giving two cases in recent practice. *E. & M. Jour.*, Dec. 1, 1888.

Steam Trap. Description of a new invention. *Mfr. & Build.*, March 13, 1885.

Steam Turbines.

Compound. Gives a description and discussion of a motor composed of a series of 15 turbines acted upon by a current of steam. *Sci. Am. Sup.*, Feb. 18, 1888.

Steam Turbines, continued.

Description of Dow's Steam Turbine Motor by the inventor. Illustrated. The 25-horse power motor is 9 x 5 inches in size, makes 10,000 revolutions per minute, and uses 41 lbs. of steam per horse-power at 100 lbs pressure. *Elec. World*, Feb. 28, 1891.

Notes on a New Compound Steam Turbine. Description and illustrations of a novel steam turbine which has been successfully used to drive elevators, dynamos, etc. *R. R. Gaz.*, Jan. 24, 1890. p. 60.

Parsons Compound. For running dynamos. Makes 10,000 to 18,000 revolutions per minute. *E. & M. Jour.*, Jan. 19, 1889.

Parson's Steam Turbine. Gives description with the results of the practical working of Parson's steam turbine. The best results so far attained are a consumption of 32 pounds of steam an hour for each electric horse-power with steam at 90 pounds pressure above the atmosphere. Lon. *Eng.*, Jan. 13, 1888; *Sci. Am. Sup.*, Feb. 11, 1887.

Steamboats.

Anchor Gear For. See *Anchor Gear.*

Portable. Description of two portable steam launches built for M. de Brazza for use on the Congo. Illustrated. *Mechanics*, Aug., 1889.

The "Whaleback" Cargo Steamers. Illustrated description of this new type of steamers in use on the great lakes. Lon. *Eng.*, July 31, 1891. pp. 123-4.

Stern-Wheel. A description of a stern-wheel steamer, 150 ft. long by 28 ft. 6 in. beam, recently constructed for the navigation of Maddalena River, South America. Plan and elevation. Lon. *Engineer*, Sept. 30, 1887; *Sci. Am. Sup.*, Nov. 12, 1887.

See *Ferry-boat. Paddle Wheels.*

Steamships.

Advantages of Increased Proportions of Beam to Length in. By J. H. Biles, M. I. N. A. A paper giving some of the advantages actually obtained by adopting increased proportions of beam to length in some steamers built by the firm of J. & G. Thompson, of Glasgow; with discussion. Reprinted from *Trans. Inst. Naval Architects*, by the Bureau of Navigation, Navy Department. *Naval Professional Papers*, No. 16.

A mechanical method of measuring a vessel's stability. Read before the Inst. N. A., by John H. Heck. Lon. *Eng.*, April 10, 1885.

America. A description of the distinguishing features of this fast sailing ship. By Robert Gordon. Longitudinal and transverse sections shown. *Trans. A. S. C. E.*, Vol. XV., p. 352.

Atlantic. Valuable data concerning sixteen standard steamers in tabulated form. Also a descriptive paper and discussion abstracted in *Sci. Am. Sup.*, Oct. 2, 1886.

The Designing of. By H. C. Plassons. A very good paper. Illus. *Proc. Mech. Eng. Soc.*, 1884.

Efficiency. Gives details concerning the performance of two vessels with patent feathering paddle wheels plying between New Haven and Dieppe. Lon. *Engineer*, May 27, 1884.

Modern Merchant. By James Dunn, M. I. N. A. A paper giving an account of the recent improvements in merchant steamers, read before the Inst. of Naval Architects. Reprinted by the Bureau of Navigation, Navy Dept. *Naval Professional Papers*, No. 19.

Nulli Secundus. An elaborate account of the largest freight steamer afloat, having a capacity of 7,000 allowing a speed of 11 knots on a coal consumption of 30 tons a day, triple expansion engines, etc. Illustrated. *Sci. Am. Sup.*, Jan. 1, 1887.

Steamships, continued.

Oroya. Double page plate showing plans of decks and sections of a new steamer 460 feet long, 49 feet wide, and 35 feet deep, molded and having a gross register tonnage of about 6,000 tons. Lon. *Engineer*, March 25, 1887.

Propulsion.

On the Laws of Steamship propulsion. A paper by Robert Mansel, of Glasgow, of a somewhat theoretical nature. Lon. *Engineer*, Aug. 2, 1889.

Laws of. An article by Robert Mansel, discussing the applicability of the usual formulas and showing to what extent they have erred in some cases. Lon. *Eng.*, March 6, 1891, pp. 266-7.

The Laws of. Discussion of the subject, with experimental data. By Robert Mansel. Lon. *Eng.*, Jan. 25, 1889.

On the Determination of the Coefficients of the Relation Between the Power and Speed in the Propulsion of Steam Vessels. A theoretical discussion with examples from actual trials. By Robert Mansel. Lon. *Engineer*, April 10, 1891, p. 274.

Raising of the "Peer of the Realm." A steamship of 1,813 tons register, 300 feet long, laden with 2,800 tons of coal successfully raised. Lon. *Eng.*, May 21, 1884.

Speed.

"*The Six-Day Time Passed.*" Article giving considerable information of general interest concerning the steamships *City of Paris* and *City of New York*; with a table giving principal dimensions of 17 Atlantic Steamers. Indicated H. P., stroke, heating surface, fire-grate area, working steam pressure, are included. *The Locomotive*, June, 1889, Vol. X., p. 85.

In Ocean Steamers. A paper by A. E. Seaton discussing conditions for greatest speed, and giving an account of the development of the present fast steamers. Table of speeds of Atlantic steamships given. *Scribner's Mag.*, July, 1891, pp. 3-22.

Table in *R. R. & Eng. Jour.*, for April, 1887.

Trials of the Royal Italian Ironclad Lepanto. Complete account of interesting tests made of this vessel, by Major Nabor Soliani, Royal Italian Navy, read before Institution of Naval Architects, July 26, 1888. *Proc. U. S. Naval Inst.*, Vol. XV., No. 4, p. 563, (whole No. 51.)

White Star Steamship Teutonic. Illustrated description of the interior of the elegant steamer. Lon. *Engineer*, Dec. 19, 1890, pp. 489-93. Machinery described and illustrated, *ibid*, pp. 498-500.

See *Electric Lighting for.*

Steel.

By H. M. Howe. This work, which is probably by far the most complete and valuable treatise on steel yet published, is begun in *E. & M. Jour.*, of March 5, 1887, to extend through a year, at the rate of about five pages each week.

Effects of alternate heating and cooling. See *Iron*.

From the Ingot to the Finished Tool. By A. Y. Jacobs. Gives English practice of steel metallurgy; defects in ingots; treatment of final product; economy of processes, etc. *Sci. Am. Sup.*, Jan. 29, 1887.

Gives brief review of an extensive series of experiments made by British officers on basic steel. *Am. Eng.*, Aug. 31, 1887.

By J. Riley. A paper before the Naval and Marine Exhibition at Glasgow. Describes the different processes of manufacturing steel. *Van Nos. Eng. Mag.*, Vol. XXV., p. 10.

Summary of the regulations relating to the use of steel recently issued by the Russian Ministry of Roads. *Building*, June 18, 1887.

Steel, continued.

Final report on experiments bearing on the question as to the state in which carbon exists in steel. Read before Inst. Mech. Engrs. Lon. *Eng.*, Feb. 20, 1881.

Adams Direct Steel Process. Full description of this process which compares very favorably with other methods. Illustrated. *Am. Mfr.*, Oct. 3, 1890. pp. 7-8. *Iron Age*, Nov. 13, 1890, pp. 835-7.

Aluminum Steel. A valuable paper on the properties of this alloy, both cast and forged. Presented by Mr. R. A. Hadfield, of Sheffield, Eng., to the New York convention of the A. I. M. E. Full abstract with diagrams in *Eng. News*, October 11, 1890, pp. 321-7. Full paper in *Iron*, November 14, 1890, pp. 420-30.

Aluminum Alloys in Making. A discussion by the Engineers' Society of Western Pennsylvania on the use of aluminum alloys in steel making. *R. R. Gaz.*, May 11, 1888.

Behavior of, Under Mechanical Stress. By C. A. Carus Wilson. Article says that automatic disturbances increase with stress, increasing in permanency with the increase of the strain. Variation of magnetic qualities discussed. Illus. *Proc. Physical Soc.*, Vol. X., pp. 331-41. 3 plates.

Basic Bessemer.

Process. By Prof. T. Egleston. A paper of 50 pp., illustrated, describing the plants now in use in England and on the continent. *Trans. A. S. M. E.*, Vol. VII., p. 34.

Thomas-Gilchrist Process as practiced in Germany, France, and Belgium, using Spanish ores. A careful description of ingredients, apparatus and methods. From the German. *Van Nos. Eng. Mag.*, Dec., 1881.

Fully illustrated description of the Pottstown works, method of manufacture, and results of a large series of tests on the steel produced. *Iron Age*, May 28, 1891, pp. 1012-5.

Manufacture and Cost of. By J. B. Nau, E. M. 8 tables. *Iron Age*, Sept. 4-18, 1890, pp. 385-6, 440-3.

Basic Open Hearth.

By J. W. Wailes, before the Iron and Steel Institution. Illustrated. Lon. *Eng.*, Oct. 21, 1887.

Article of considerable value from *Industries*, giving tests, analyses, etc., of several samples of steel. Tabular exhibit of qualities of 37 samples. *Am. Mfr.*, April 25, 1890.

Manufacture of. A paper by Mr. James Davis, giving full chemical and mechanical results of a few charges of basic steel. Lon. *Engineer.* Dec. 19, 1890, pp. 516-7. *Iron Age*, Jan. 1, 1891, pp. 10-11.

Basic Siemens Process. A paper by F. W. Harbord, before the Iron and Steel Works Managers' Institute, giving a description of the Basic-Siemens process of making steel. *Mech. World*, Dec. 31, 1887; Lon. *Eng.*, Dec. 16, 1887; *Am. Mfr.*, Jan. 6, 1888. A description of the Batho furnace to employ this process will be found in the same number of *Am. Mfr.*

Basic, for Ship Building. By B. Martell. Treats of the present position occupied by basic steel as a material for ship building. Lon. *Eng.*, July 29, 1887.

Bessemer.

Sir Henry Bessemer's Story of the Invention of the Bessemer Process. A very interesting letter from the distinguished engineer, presented at the Pittsburgh meeting of the Iron and Steel Inst. *Eng. News*, Oct. 18, 1890, pp. 337-8. *E. & M. Jour.*, Oct. 11, 1890, pp. 427-30.

Converting House without a casting pit. By L. G. Laureau. *Iron Age*, March 19, 1885, and *E. & M. Jour.*, April 4, 1885.

Steel, *Bessemer*, continued.

Early Days of. An account of the introduction of the process. Also paper by Mr. Bessemer read in 1856 describing his invention. Lon. *Eng.*, *Eng. News*, Sept. 10, 1881, pp. 364-7.

Manufacture of. A lecture before the Franklin Institute, Jan. 21, 1889, by Robert W. Hunt. Gives history of development of manufacture, chemical composition and physical manipulation. *Jour. Frank. Inst.*, May, 1889.

Manufacture of. A paper read before the Franklin Institute, Jan. 21, 1889, by Robert W. Hunt. *Am. Mfr.*, May 24, 1889, *et seq.*

See *Steel.*

Modifications of the Process. Abstract of a lecture before the Franklin Inst., Jan., 3, 1887, by C. Hanford Henderson. Treats especially of the Clapp-Griffiths and the Thomas and Gilchrist methods. *Jour. Frank. Inst.*, June, 1887, Vol. CXIII., No. 738.

Notes on the Bessemer Process as carried out in Robert Vessels. Diagrams and tables. *E. & M. Jour.*, Aug. 30, 1890, pp. 242-4.

On the Generation of Heat During the. A good paper by Mr. A. Akerman, before the Institution of Civil Engineers in Sweden. *Van Nos. Eng. Mag.*, Vol. VIII., p. 148.

Variation of the Bessemer process that is used in Sweden. *Iron Age*, April 2, 1885.

Bridge, Discussion on. Gives a discussion on bridge steel that took place at an Edinburgh meeting of the Iron and Steel Institute. *R. R. Gaz.*, September 14, 1888.

Converter. Later Practice and Commercial Results. By J. P. Witherow. A paper before the *Am. Inst. M. E.*, Pittsburgh meeting, 1886. Illustrated.

Clapp-Griffiths.

Cuts showing converters in action, and also sectional views, accompanied by detailed description. *Sci. Am.*, March 27, 1886.

Process. A paper before the Am. Inst. Min. Engrs., Feb., 1885; describes a process that is already supplanting the Bessemer methods, and which is likely to revolutionize the iron and steel industries. *Trans. A. I. M. E.*, Vol. XIII.; *Am. Eng.*, April 10, 1885; *Iron Age*, March 12, 1885.

Description of the new Clapp-Griffiths Steel plant in Pittsburgh. *Iron Age*, March 12, 1885.

Colors Produced in Tempering. Influence of time of heating etc. Paper read before the Birmingham Philosophical Society, by Thomas Turner. *Am. Mach.*, Feb. 13, 1890. *Am. Mfr.*, April 11, 1890, p. 10, *et seq.*

Condition in which Carbon Exists in. Results of two series of experiments made for the Committee on Steel of the Inst. of Mech. Engrs. *Van Nos. Eng. Mag.*, March, 1884.

In Construction, Russian Rules for Use of, adopted in 1885. *Abstract Inst. C. E.*, Vol. LXXXV., p. 120.

Converter, Gordon. Illustrations and description of. *Am. Eng.*, May 22, 1885.

Cracks and Annealing of. By A. C. Kirk, M. I. N. A. A paper in which the writer holds that cracks in steel plates are due to lines of weakness possessed by the steel from the ingot state, not to be cured by annealing; with account of experiments and a valuable discussion. Reprinted from Trans. Inst. Naval Architects, by the Bureau of Navigation, Navy Dept. *Naval Professional Papers, No. 14.*

Darby Process of Recarburization of Steel. The method is by adding carbon in the form of powdered coke. Condensed from a paper by A. Thielen, read at the Pittsburgh International Meeting of Metallurgical Engineers. *Eng. News*, Oct. 18, 1890, pp. 338-9. Editorial, *ibid.* pp. 348-9. *Sci. Am. Sup.*, No. 777, Nov. 22, 1890, pp. 12,410-12.

Steel, continued.

Direct from the Ore. By F. L. Garrison, before the Boston meeting of Mining Engineers. Gives the results of investigation of the attempts to produce steel direct from the ores. Describes in detail the development of Hasgafoel's improved high bloomary (a modification of the old Stuckofen process) for the production of iron and steel direct from ores. Illustrated. *Am. Mfr.*, April 6, [illegible].

Domestic Steel for Naval Purposes. A paper by Lieut.-Com. J. G. Eaton, U. S. N., giving brief descriptions of the bessemer, open hearth and basic open hearth process; together with government specifications for hull plates and rivets, steel castings, boiler plates, stays and rivets, engines and shafting, anchors, steel ropes and steels for guns. *Proc. U. S. Naval Inst.*, Annapolis, Md., Vol. XV., No. 2. Whole No. 49, 1889. *Proc. Soc. Arts*, 1890, pp. [illegible].

Effects of Heat and Cold on Iron, Steel and Copper. See *Heat and Cold.*

Hardening and Tempering of Steel. A lecture before the British Association at Newcastle, by Prof. W. C. Roberts Austen, F. R. S. *Sci. Am. Sup.*, Oct. 26, 1889. Lon. *Eng.*, Sept. 27, 1889.

Hardening of. Paper before the West Soc. Pa., by Wm. Metcalf explaining the principles of the hardening of steel. *Eng. News*, Feb. 28, 1880, pp. 78-9.

Harvey Process of Hardening Steel. Detailed description of this process used in hardening armor plates and steel guns. From patent specifications. *Eng. News*, Oct. 24, 1891, pp. 377-8.

Heaton Process. A report by M. L. Gruner on the above process. *Van Nos. Eng. Mag.*, Vol. II., pp. 84, 129, etc.

Influence of Copper on Tensile Strength. A paper prepared by E. J. Ball and A. Wingham for the Iron and Steel Institute. Gives results of experiments made to ascertain the effect of copper on tensile strength of steel and iron. It appears to render them extremely hard. *Am. Mfr.*, Sept. 28, [illegible].

Influence of High Temperature on the Mechanical Properties of Mild Steel. A valuable series of tests, reported in the *Mittheilungen aus dem Koeniglichen Technischen Versuchsanstalten*, VIII. 1890. 4. Abstract in *Iron Age*, Dec. 4, 1890, pp. 990-3.

Manganese.

A paper by Henry M. Howe, read before the Am. Soc. M. E., describing its various properties, giving results of tests, and durability under different circumstances. Illustrations of armor plate tests given. *Eng. News*, June 20, 1891, pp. 573-4.

By R. A. Hadfield, before the Inst. of C. E. Two papers, "Manganese in its Application to Metallurgy," and "Some Newly Discovered Properties of Iron and Manganese." Iron with from 2.75 to 7 per cent. of manganese is very brittle; between 7 and 20 per cent. of manganese gives a very strong and tough material, specimens of which broke with a tensile strain of 63 tons, and showed an elongation of 50 per cent. *Proc. Inst. C. E.*, Vol. XCIII., pp. 1-108: abstract Lon. *Eng.*, March 9, 1888; *R. R. Gaz.*, March 30 and Sept. 7, 1888; *Sci. Am. Sup.*, April 7, 1888; *Mech. World*, March 10, 1888; Lon. *Engineer*, March 16, 1888; *Am. Mfr.*, April 6, 1888; *T. J. & Elec. Rev.*, Aug. 10, 1888; *Mast. Mech.*, Oct., 1888.

Metallurgy of. A lecture by Pedro G. Salone. Describes the various processes of manufacture, cost, nature of product, etc. Illus. *Jour. Frank. Inst.*, Sept., 1884.

Mild.

Applied to Naval and Military Purposes. A valuable paper, by Major G. Mackinlay, R. A., before the Royal Service Institution. Fully illustrated. Reprinted in *Sci. Am. Sup.*, Sept. 5, 1885.

Steel. *Mild*, continued.

Mild. Effect of Hot Rolling and of Cold Wire Drawing. Experiments showing complete physical effects at the several stages of hot rolling and cold wire drawing. Results tabulated and plotted. *Proc. Inst. C. E.*, Vol. XCIV., p. 235.

Endurance of, and iron when exposed to corrosive influence. By David Phillips. Gives results of a number of experiments with different kinds of steel and iron, nearly all of which gave a result in favor of iron. *Van Nos. Eng. Mag.*, Feb. and March, 1884.

Made by the Thomas Process, and its Employment in Building. Brief statement of general method of manufacture and general qualities attained. *Eng. News*, April 20, 1890, Vol. XXIII., p. 387.

Report of the Naval Advisory Board on Mild Steel used in the construction of four steel vessels. A valuable and exhaustive record of the studies and tests made by the Board. 200 pp. Bound separately. Also with Vol. I. *Messages and Documents, Navy Dept.*, 1885-6.

For Ship Building. A brief review of the progress of mild steel, and the result of eight years' experience in its use for shipbuilding purposes, before the Inst. Naval Architects. Lon. *Eng.*, April 23, 1886.

Momentary Depression of the Elastic Limit of Steel at Two Critical Temperatures. Tests showing some of the laws governing this phenomenon. By Henry M. Howe. *Tech. Quart.*, Vol. II., 1888, No. 2, pp. 136-42.

Nickel Steel. See *Alloys*.

Open Hearth.

For Use in Bridge Building. Brief notes selected from a paper, by N. W. Shed, M. E., read at Colorado meeting of the Am. Inst. of Mining Engineers. *Eng. News*, Oct. 5, 1889, Vol. XXII., p. 322.

For Boilermaking. By H. Goodall, before the Inst. of Civil Engineers. The paper gives the experience of the author in the use of open-hearth steel for boilermaking since 1875, and describes numerous experiments to ascertain the cause of difficulties met with in working the plates. *Proc. Inst. C. E.*, Vol. XCII., pp. 2-72; abstract in Lon. *Eng.*, Jan. 13, 1888; *Mech. World*, Jan. 24, 1888.

Experiments with the Imperatori Open-Hearth Process of Steel Making. Abstracts from a paper by Mr. J. B. Nau before the Cleveland meeting of the A. I. M. E., describing this method and giving results obtained. Reprinted in *Eng. News*, Oct. 17, 1891, pp. 354-6.

Fifteen Years of Open-Hearth Experience. By W. E. Kock before the Engrs. Soc. of West. Penn. Traces the development of the open-hearth process, and gives much interesting information concerning the use and working of steel. Abstracted *Eng. & Build. Rec.*, May 19, 1888, and *Eng. News*, June 9, 1888; *Am. Eng.*, June 27, 1888; *R. R. Gaz.*, Aug. 17, 1888.

Improvements in Open-Hearth Practice. By A. E. Hunt, before the Boston meeting of the Am. Inst. of Min. Engrs. Gives a good description of the process of making wrought iron direct from the ore employed by the Carbon Iron Company at Pittsburgh. *Eng. News*, March 24, 1888.

Notes on the Manufacture of Open-Hearth Bridge Steel. By N. W. Shed, Phoenixville, Pa., pp. 3. *Proc. A. I. M. E.*, Colorado meeting, June, 1889.

Statistics of the Open-Hearth Steel Industry of Great Britain for 1885, with table showing growth of the United States and England during the last ten years. *Am. Mfr.*, April 1, 1887.

See *Steel, Basic, Open Hearth*.

Peculiar Phenomena in the Heating of Open-Hearth and Bessemer. A paper read before the Pittsburgh meeting of the American Institute of Mining Engineers. *E. & M. Jour.*, May 15, 1886.

Steel, continued.

Physical Properties of Steel at Very Low Temperatures. By Capt. Bernardon in *Revue d'Artillerie*, Sept., 1890, p. 481. Abstract giving summary of results. *Proc. Inst. C. E.*, Vol. CIII. 1891. Foreign Abstracts, pp. 478-80.

Pressed Steel in Car Construction. A paper by Chas. T. Schoen, read before the N. Y. R. R. Club, giving an account of what is being done in this line and results attained. *R. R. Gaz.*, March 27, 1891, pp. 211-18. *Eng. News*, March 28, 1891, pp. 297-3. *Ry. Rev.*, March 28, 1891, p. 201.

Pressed Steel in Car Construction. Report of Committee of M. C. B. Assoc. Illustrations of fittings, and tests of malleable iron given. *Ry. Rev.*, June 13, 1891, pp. 383-4. *R. R. Gaz.*, June 12, 1891, p. 406. *Eng. News*, June 13, 1891, p. 563.

Pressed. See *Car Truck. Cars.*

Production of the United States. Census Bulletin No. 12. Abstract in *Am. Mfr.*, Nov. 7, 1890, p. 18. *Eng. News*, Nov. 8, 1890, p. 416. *Ry. Rev.*, Nov. 15, 1890, p. 691.

Properties of, at Low Temperature. Some tests made by the French Government on bars cooled to 75° to 100° below zero, showed an increase of elastic limit and ultimate strength but also a considerable increase in brittleness. Abstracted from Lon. *Eng.*, *Eng. Record*, Jan. 10, 1891, pp. 91-2.

Properties of. Its Use in Structures and Heavy Guns. By Wm. Metcalf, before the American Society of Civil Engineers. A valuable contribution on steel, and a plea for the Rodman gun. Discussion covering 90 pages. *Trans. A. S. C. E.*, Vol. XVI., pp. 283-389. June, 1887. Abstracted in *R. R. Gaz.*, March 18, 1887; *R. R. & Eng. Jour.*, April, 1887; *Am. Eng.*, April 6 and 13, 1887.

Punching Holes in Soft Steel, Influence of. Translated from the Russian for Theodore Cooper and printed in *Eng. News*, Dec. 13, 1884. The experiments were numerous and valuable. The article illustrated by many cuts.

Recent Improvements in its Manufacture. By P. G. Salom, before the Franklin Institute, with especial reference to the Clapp-Griffiths, the Davy, the Gordon and the Avesta processes. All these being modifications of the Bessemer process. Fully illustrated, showing typical plants. *Jour. Frank. Inst.*, November, 1885.

Resilience of. By Wilfred Lewis. Some experiments towards determining the value of steel springs for storing power. *Proc. Eng. Club. Phila.*, Vol. IV., No. 4.

Ship Steel. The Manipulation of Ship Steel. Article calling attention to the danger of making an otherwise satisfactory material brittle by hammering and bending too cold. *Am. Mfr.*, Aug. 23, 1889. Lon. *Engineer*, Aug. 2, 1889.

Strength of, Increased by Cold. Six experiments in Russia, showing an increase in strength and elongation from +67° to −22° F. of about six and ten per cent, respectively. *Abstr. Inst. C. E.*, Vol. LXXXIII., p. 413.

Structural.

Being a summary of results of an elaborate investigation into the relative merits of different kinds of steel for structural purposes. By Edw. B. Dorsey, with discussion. *Trans. A. S. C. E.*, Vol. XIII., p. 41.

By Ewing Matheson. A paper before the Institution of Civil Engineers, advocating the use of steel for structural purposes. *Van Nos. Eng. Mag.*, Vol. XXVIII., p. 307.

By Edward B. Dorsey. A supplementary paper, giving results of two recent trips to Europe. Contains the steel specifications for the Forth Bridge, notes on steel castings, etc. *Trans. A. S. C. E.*, Vol. XIV., p. 197.

Effect of Temperature upon Structural Iron and Steel. By Jos. Ramsey before the Engineers' Society of Western Pennsylvania. Gives the results of some tests and investigations. *Eng. News*, Dec. 3, 1887.

Steel. *Structural*, continued.

Russian Rules for the use of. *Eng. News*, Dec. 18, 1886.

Some Constants of Structural, by P. C. Ricketts. Gives results of a large number of experiments made with steel bars, varying in carbon from 0.100 to .40 100. Co-efficients of elasticity in tension and compression were obtained from adjacent pieces of the same bar. *Trans. A. S. C. E.*, Vol. XVI., p. 138.

Strength and Elasticity of. By James Christy. An account of some elaborate and careful experiments at the Pencoyd Iron Works, with tabulated and plotted results on beams and long columns. *Trans. A. S. C. E.*, Vol. XIII., p. 253. A discussion of the above in same volume, p. 367.

Tempering of.

And Annealing of, investigated by means of its varying electrical and magnetic properties. A memoir of 230 pp., giving an account of original investigations in the Laboratory of the U. S. Geological Survey. Rational theories of tempering, annealing derived. By Carl Barus and Vincent Strouhal. *Bulletin No. 14, U. S. Geol. Survey*, 1885.

By Compression. *Abstracts of Papers, Inst. C. E.*, 1886.

Three papers containing practical information on this subject. *Mech. World*, Sept. 25, 1891, *et seq.*

See *Steel Hardening.*

Testing of.

By A. L. Holley. Points out why mechanical tests of steel, as ordinarily made, are not, alone, of any special value to engineers. *Van Nos. Eng. Mag.*, Vol. XIII., p. 289.

Compressive Strength of Iron and Steel. By C. A. Marshall. Gives results of a large number of tests made to discover the relation which compressive strength bears to tensile strength. *Trans. A. S. C. E.*, Vol. XVII., pp. 53-110 (August, 1887).

Effects of Rate of Speed in Testing Steel. Interesting series of experiments made to ascertain the effect of variation in speed in making tests of steel. By Bertram P. Flint. *Eng. News*, April 19, 1890, Vol. XXIII., p. 363.

How Shall We Test It? By F. Collingwood. The requirements to use in specifications, and methods of testing structural steel. A short article giving the most recent practice. *Eng. News*, Jan. 16, 1886.

Results from Tests made shortly after Rolling. By F. C. Felton, before the Philadelphia meeting of the American Society of Mechanical Engineers. Gives notes on the results of a large number of tests made to determine the effects of tests of steel shortly after the rolling. *Trans. A. S. M. E.*, Vol. IX., (1888), pp. 38-50.

Tests of the New Direct Process Open Hearth. Gives results of tests made by G. H. Thomson of the new direct process open-hearth steel, of the Carbon Iron Co., made at the shops of the Union Bridge Co. It stood severe tests, and bids fair to become an important production. *Eng. News*, Jan. 21, 188[illegible].

On Tests for Steel Used in the Manufacture of Artillery. Paper by W. Anderson before the Iron and Steel Inst. Lon. *Eng.*, May 15, 1891, pp. 599-00. Lon. *Engineer*, May 15, 1891, pp. 388. *Iron*, May 8, 1891, pp. 402-5.

See *Tests.*

Tool Steels. Report on the quality of. By the U. S. Testing Board. Abridged in the *Am. Eng.* of Jan. 2 and 9, 1885. This abstract is very full, and gives tables and diagrams. The report is valuable, as it shows what are the characteristics of steel best adapted to various purposes.

Steel, continued.

Treated by Hydraulic Pressure. A paper by W. H. Greenwood showing effect of hydraulic pressure on steel ingots. Test pieces cut from ingot longitudinally showed an increase of elastic limit of 3 per cent and strength of 1.2 per cent. over impressed ingots. Pieces cut diametrically from pressed ingot showed increase of elastic limit of 8.4 per cent. and strength of 5 per cent. Pp. 62. *Proc. Inst. C. E.*, Vol. XCVIII., p. 85, 1889. *Eng. News*, July 6, 1889, Vol. XXII., p. 9. *Sci. Am. Sup.*, Aug. 10, 1889. *Discussion. Proc. Inst. C. E.*, Vol. XCVIII., p. 145, 1889.

Use of, for Bridges. By Theodore Cooper. With discussion. A general review of the situation in 1879. *Trans. A. S. C. E.*, Vol. VIII., (1879), p. 263-294.

Use of, in Construction. By A. F. Hill before the Engrs. Soc. West Pa., giving results of experiments on eye-bars, plates, riveting, etc. *Eng. News*, May 15, 1880, pp. 169-73.

Use of Steel in the Construction of Bridges. Mild steel and hard steel. By Dechamps. *Ann. des Ingenieurs de Liege*, Ser. V., Vol. III., pp. 259-98.

Warming of, with especial reference to the injurious effects of a blue heat upon. Paper by C. E. Stromeyer, and discussion covering 100 pp. in *Proc. Inst. C. E.*, Vol. LXXXIV., p. 114.

Working Machinery. A complete study of all classes of stone-working machinery, saws, grinders, crushers, etc. Abridged from *Annales des P. & C.* in *Sci. Am. Sup.*, Jan. 15, 1887.

See *Electrical Resistance. Iron and Steel. Strength of Steel.*

Steel Castings.

Manufacture of Crucible. By Henry Seebohm. A good article, giving details of the various kinds of steel. *Van Nos. Eng. Mag.*, Dec., 1884.

Crucible. A description of the methods of manufacture as followed at Sheffield, England. The paper was read before the Iron and Steel Institute at its own request, and has considerable value as giving the various stages of the processes with considerable minuteness. *Van Nos. Eng. Mag.*, Dec., 1884.

A paper read at the Chattanooga Meeting of the Mining Engineers, by Wm. Hainsworth. Gives the difficulties and how they may be overcome in practice. The author speaks from personal knowledge. Illustrated. *Iron Age*, Oct. 1, 1885.

A paper read before the South Staffordshire Institute of Iron and Steel Works' Managers. Gives details of the present English practice. *Am. Eng.*, July 14, 1886, *et seq.*

A paper by H. L. Gantt, read before the Am. Soc. M. E., describing the many difficulties encountered in their manufacture. With discussion. *Eng. News*, June 1, 1891, pp. 506-8.

Paper by H. L. Gantt before Providence Meeting A. S. M. E. Reprint with discussion. *Am. Mach.*, Aug. 6, 1891.

Castings. By E. C. Warren. On the use of steel castings in lieu of iron forgings and brass casting in building and fitting ships. *Sci. Am. Sup.*, July 10, 1886.

For the Manufacture of Guns. By Major L. Cubillo, Spanish Artillery. From Trans. Iron and Steel Inst. *Am. Mfr.*, Dec. 7, 1888.

In Locomotive Construction. Illustrated. *Proc. Inst. C. E.*, Vol. XC., pp. 358-365.

Manufacture of Martin-Siemens Steel Castings. By P. Mohler. Deals with annealing, moulding—furnaces and open hearth and Bessemer castings. Illus. *Le Genie Civil*, Vol. XVIII., pp. 187-90.

Present Value of. The results of numerous experiments on products from all the steel-casting manufacturers in America. By Arthur V. Abbott. *Trans. A. I. M. E.*, 1885.

Steel Castings, continued.

Use of in Building and Fitting Ships. By E. C. Warren, before the Institute of Naval Architects. Gives the results of a large number of tests made on the smaller fittings of a vessel. Lon. *Engineer*, May 14 and 21, 1886.

Whitworth Process. Casting under pressure by. By T. Egleston. To produce a superior quality by using a pressure of six tons to the square inch on the fluid ingot. *Sch. Mines Quart.*, No. 3, Vol. VI.; also *Iron Age*, May 7, 1885.

See *Car Wheels. Aluminum.*

Steel Flooring See *Flooring.*

Steel Guns.

See *Guns.*

Steel Industries. See *Iron and Steel.*

Steel Ingots.

Bessemer. The Segregation of Impurities in, on cooling. By B. N. Cheever. *Trans. A. I. M. E.*, Vol. XII., and *Jour. Iron & Steel Inst.*, 1881.

Bamlton Apparatus for Compressing. Illustrated description. *E. & M. Jour.*, May 18, 1889.

Defects in, due to physical and chemical causes, and precautions to be observed to avoid defects resulting from the heating of steel. From the German. Illus. *Van Nos. Eng. Mag.*, Nov. 1884.

Structure of. By D. K. Tchernoff. Illustrated. *Van. Nos. Eng Mag.*, Vol. XXIII., p. 123.

Steel Manufacture.

By Wm. F. Durfee. An historical and descriptive paper, being No. VIII., of a series of articles on "Development of American Industries Since Columbus." *Pop. Sci. Mon.*, Oct., 1891, pp. 729-49, *et seq.*

New Methods of Refining Steel. A valuable paper by Mons. U. Le Verrier in *Revue Generale des Sciences*, Sept. 30, 1891. Discusses the principles involved especially of the Darby and aluminium processes. Reprint. *Iron.* Oct. 23.

Of Plates. A paper by James Riley, before the Iron and Steel Institute. Gives some investigations as to the effects of different methods of treatment of mild steel in the manufacture of plate. Lon. *Eng.*, July 1, 1887.

Properties and Uses. Paper read before the Greenock Philosophical Society by James Riley, General Manager of the Steel Company of Scotland, etc. This paper traces the development of the steel industry. *Am. Mfr.*, Feb. 28, 1890, p. 18, *et seq.*

Recent Developments in the Manufacture of. The Future Sites of These Industries. By Horace A. Keefer before the Engrs. Club, Kansas City. *Jour. Assn. Eng. Soc.*, Sept. 1891. pp. 437-45.

Methods and Manufacture. By Thomas W. Fitch. Describes the Bessemer and Basic Processes, as employed at various American works. *Jour. Assn. Eng. Soc.*, Vol. I., p. 406, and Vol. II., p. 14.

See *Iron and Steel.*

Steel Rails. See *Rails.*

Steel Tapes. See *Tapes.*

Steel Works.

The Homestead Steel Works. A description of these works, owned by Carnegie, Phipps & Co., at Pittsburgh, by W. Richards and J. A. Potter. Illustrated. *U. S. Naval Inst.*, Vol., XV., p. 431, 1889; whole No. 50.

South Chicago. By E. C. Potter. A full description of the above works read before the Iron and Steel Institute of Great Britain. Lon. *Eng.*, June 17, 1887.

Steel Works, continued.

Terni, Machinery for the. By H. Savage, before the Institute of Civil Engineers. Gives description of the plant at the new Terni, Italy, steelworks. *Proc. Inst. C. E.*, Vol. XCIII., pp. 390-404.

The Terni. By Sir B. Samuelson. A paper before the Iron and Steel Institute descriptive of works for the manufacture of ordnance, of armor plates and general steel work at Terni, Italy. Lon. *Eng.*, June 10, 1887.

See *Krupp.*

Steering Gear, *a New System of*, and the rudder strains recorded by it. A paper before the Inst. of Naval Archts. Illustrated. Lon. *Eng.*, April 30, 1886.

Steps. *Length of*, from 256 experiments in Germany. Diagram of frequency given. *Eng. News*, July 25, 1885.

Stone.

Artificial. A paper before the British Assoc. By Fred. Ransome. *Van Nos. Eng. Mag.*, Vol. VII., p. 967.

Ferroid. A New Artificial Stone. Interesting paper by Herman Poole, read before the C. E. Club of Cleveland. The properties of this valuable compound are fully described. Table of the results of tests. *Jour. Assn. Eng. Soc.*, Aug. 1890, pp. 395-92.

Dressing.

Cost of. Paper by R. J. Cooke, giving some data on this subject. *Selected Papers C. E. Club of Univ. of Ill.*, 1889-90, pp. 8-10.

Granite Industry of the United States. Census Bulletin containing statistics of production, methods of gauging, cost, etc. Abstract, *Eng. News*, May 2, 1891, pp. 417-18. *E. & M. Jour.*, April 25, 1891, pp. 496-7.

See also *Paving Stones.*

Stone Arches. See *Arches.*

Stone Breaker. See *Crushers.*

Stone Bridges, *Cost of, in detail.* See *Masonry.*

Stone Car, *Hurst & Treasor's.* An ingeniously designed railroad car for loading and transporting large blocks of stone. *R. R. & Eng. Jour.*, February, 1889.

Stone Cutting. Tools, and details of work described by Prof. W. P. Trowbridge in *Sch. Mines Quart.* *Eng. News*, June 23, 1883, pp. 294-9

Stone-Lewis. Two automatic acting devices used on the Amsterdam canal locks. *Eng. News*, Jan. 25, 1886.

Stone Quarrying.

Liverpool Quarry. Gives method of quarrying, blasting, flagging and dressing stone, with total cost, for the Liverpool water works. Lon. *Eng.*, March 30, 1888.

Marble Quarrying in the United States. Description of methods, by E. R. Morse. *Eng. Mag.*, Oct., 1891, pp. 42-7.

Set Quarrying in Wales. A lengthy description of the methods used. *Trans. Liverpool Eng. Soc.*, Vol. VI., pp. 64-74.

And Sawing by Steel Wires. Illustrated description of method in use at the Traigneaux marble quarries in Belgium. *E. & M. Jour.*, April 6, 1889, May 25, 1889.

Stone, Sawing, by Helicoidal Wire Rope. An ingenious method of using wire rope in sawing stones instead of the usual method. *Jour. Soc. Arts*, Aug. 9, 1889, Vol. XXXVII., p. 746; *Am. Mfr.*, Aug. 23, 1889.

Stones, Building.

A paper by H. A. Cutting, on a series of experiments made to determine the weight, specific gravity, rates of absorption and capability of standing heat of building stone. *Van Nos. Eng. Mag.*, Vol. XXIV., p. 441.

Stones, Building, continued.

And Ornamental, in the United States. By George P. Merrill, in the *Pop. Sci. Monthly* for Aug., 1885.

By Geo. P. Merrill. A series of articles promising to be the most complete work on American building stones. Quarrying and dressing are treated. In *Building*, March 26, 1887.

Chicago Building Stones. Description of the various kinds including discussion of requisites. By W. L. B. Jenney before the Chicago Academy of Sciences. *Eng. News*, Jan. 5, 1884, pp. 1-3.

Decay in Building. By C. P. Townsley. Gives results of large number of experiments with building stone to find the effects of frost. *Van Nos. Eng. Mag.*, Vol. XIV., p. 331.

Experimental Tests of. By Geo. Hatfield. Gives tables of breaking weights of a variety of building stones and bricks. Also shows the effect of the sulphate of soda tests. *Trans. A. S. C. E.*, Vol. II., p. 145.

Of Minnesota. By W. A. Truesdell. Describes nine kinds of building stones, the sources of supply, qualities as determined by actual use, etc. *Jour. Assn. Eng. Soc.*, Vol. IV., p. 322.

Nomenclature of Building Stone and Stone Masonry. By J. J. R. Croes, Wm. E. Merrill and Edgar B. Van Winkle. Gives American Practice, with full illustrations. *Trans. A. S. C. E.*, Vol. VI., (1877), pp. 297-304. Discussion, Vol. VII., p. 184.

Report on the compressive strength, specific gravity, and ratio of absorption of the building stones in most general use in the United States. A pamphlet of about 40 pp., with tables and illustrations. By Gen. Q. A. Gilmore. Issued by the Eng. Dept., U. S. A.

Resistance of, to Crushing. By C. B. Richards. Gives details of the methods used and results obtained from a series of experiments made at the Colt Company's armory to ascertain the relative resistance of American building stone to crushing. *Trans. A. S. C. E.*, Vol. II., p. 187.

The Resistance of building materials to frost. A very efficient, simple, and cheap manner of testing by means of artificial freezing and thawing when saturated with water. Results of German experiments. *Van Nos. Eng. Mag.*, Jan., 1886.

Strength of some Building Stone in the vicinity of Cleveland, O. By Prof. J. Eisenmann. Crushing and cross breaking strength, and chemical analysis of stones from Newburgh, Berea, Medina, Twinsburgh and Amherst. Tabulated results given. *Jour. Assn. Eng. Soc.*, Vol. V., p. 176.

Testing of. By George P. Merrill. *Mfr. & Build.*, April, 1890, *et seq.*, pp. 78, 104, 128. See also, in the same numbers, *The Physical and Chemical Properties of Rocks*.

Tests and Observations on Building Stones. Paper by J. A. L. Waddell, giving and discussing the results of a series of tests on Western building stone. *Jour. Assn. Eng. Soc.*, Feb., 1890, Vol. IX., pp. 33-43.

Of the U. S. Report on in Vol. X., U. S. Census 1880. Quarto, 400 pp., and 58 full page colored plates. Gives composition, distribution, physical properties and quarry statistics. *H. Rep. Misc. Doc. 42, pt. 10, 47th Congress, 2d Sess.*, 1884.

Use of. By James Gowans before the Edinburgh Arch. Assn. A practical paper treating of the various branches of stone building construction. *Eng. News*, March 14, 1883, pp. 133-5.

Used in New York City. Paper by Dr. Alexis A. Julien, describing the varieties of stone, their localities and the edifices constructed of each. *Mfr. & Build.*, July, 1890, *et seq.*

Stones, Building, continued.

Used for Structural Purposes in Allegheny Co., Pa. Paper by A. E. Hunt. Table of qualities. *Proc. Engrs. Soc. of W. Penn.*, Vol. V., 1889, pp. 149-64; discussion, pp. 194-9.

Weight. Specific Gravity, Rates of Absorption and Capabilities of Standing Heat of Various Building Stones. Table of results of many experiments made by H. A. Cutting, State Geologist, Vt. *Eng. News*, Nov. 27, 1880, pp. 404-5.

Stokers, Mechanical.

A paper by Wm. R. Roney, read before the Am. Soc. M. E., giving a brief description of the various types of stokers, and discussing the requirements to be met. *St. Ry. Jour.*, July, 1891, pp. 353-6.

The Roney Mechanical Stoker and Smokeless Furnace. An illustrated article describing the extensive plant of Claus Spreckels' Sugar Refinery, Philadelphia. *R. R. Gaz.*, Aug. 29, 1890, pp. 597-9. *Am. Eng.*, Sept. 3, 1890, pp. 155-7. *Am. Mfr.*, Sept. 5, 1890, p. 18. *Sci. Am. Sup.*, Sept. 27, 1890, No. 769, pp. 12279-81.

By J. F. Spencer. Gives brief history of the subject, and describes fully a successfully working machine of each type. Two plates. *Proc. Inst. C. E.*, Vol. CIV., 1891, Pt. II., pp. 54-67. Discussion: pp. 67-86.

Stop Valve. *A Standard 48 inch.* Brief illustrated description of the standard 48-in. stop-valve used in the public works of New York City. *Eng. News*, Dec. 15, 1888.

Storage Batteries. See *Electric Batteries.*

Stoves. *Design and Construction of Cook's Stoves.* By J. L. Goblelle. A description of some of the theoretical and mechanical problems involved. Illustrated. *Jour. Assn. Eng. Soc.*, Vol. V., p. 204.

Strain Diagrams. The proper nomenclature to use in describing their various portions. A discussion. Lon. *Engineer*, May 22 and 29, 1885.

Strain Indicator *for use at Sea.* An instrument for determining the strains in various parts of a ship or other structure. By C. E. Stromeyer. Before the Inst. Nav. Archts. Illustrated. Lon. *Eng.*, April 30, 1886.

Strains.

In a Cast-Iron Disk. By G. Leverich before the American Society of Civil Engineers. Gives details and results of an investigation to determine the strains in a cast-iron hollow disk cut from the sinking head of a casting of a Rodman gun. *Trans. A. S. C. E.*, Vol. XVIII., Feb., 1888, pp. 43-50.

In Curved Dams. Correspondence curved vs. straight dams. See *Dams.*

On Engine Shaft. Graphical representation of. See *Engine Shaft.*

Graphical Determination of Wind Strains in a Tower. Diagrams with a brief explanation, by Chas. Steiner. *Eng. News*, March 21, 1891, p. 267.

In Iron and Steel, Permissible. *Annales des P. & C.*, Dec., 1887.

In Materials. See *Proportional Resistances.*

Measurement of. A paper read before the Brit. Assn., by A. Mallock, describing a very delicate instrument for measuring strains in mechanical structures. It can be read to 1/100,000 part of an inch and gives very satisfactory results. *Ry. Rev.*, Jan. 10, 1891, p. 20.

In Metals. *On the Permanent Effects of Strain in Metals; on their Self-Registration, and Mutual Interactions.* A paper by Prof. R. H. Thurston describing some tests made on steel specimens previously strained in various ways, and discussing effects of these strains. *Trans. A. S. C. E.*, Vol. XXIV., March, 159-74. Discussion, pp. 174-97. Reprinted in *Ry. Rev.*, May 23, 1891, pp. 325-7.

See *Stresses.*

Street Cleaning.

And Disposal of Refuse. Collection of data from a number of cities. *Eng. & Build. Rec.*, Nov. 29, 1890, pp. 410-11.

Street Cleaning, continued.

Glasgow. Description of methods in use. *Eng. News*, August 30, 1884, p. 99.

London Street Sweeping. An account of the system. *Eng. News*, June 9, 1877, pp. 151-2.

In New York. A description of the work, which is under a special commissioner, costing over a million a year. *San. Eng.*, May 28, 1885.

Paris. A descriptive paper by L. Soulerin being a translation from the French. *Eng. News*, July 4, 1878, p. 213.

In Paris. By H. Vivarrez. Examination of the methods adopted by the municipal government of Paris for the removal of mud, etc., from the streets. *Lon. Eng.*, Jan. 7, 14, etc., 1887.

In Paris. Brief article from N. Y. *Herald*, describing methods employed. *Eng. News*, Aug. 22, 1891, pp. 173-4. *Sci. Am. Sup.*, No. 816, Sept. 5, 1891.

Street Sweeping Machine which elevates the sweepings into a cart. Described and illustrated. Not patented. *Lon. Eng.*, Oct. 2, 1885.

Street Sweeping Machine Used in Providence, R. I. Described and illustrated in *Eng. News*, July 31, 1886.

See *Pavements, Stone, Wood, etc.*

Street Construction.

Laws Regulating Taxation for Street Construction and Maintenance in American Cities. A digest of the laws of several cities, relating to paving, sewers, water pipe, etc. *Eng. News*, March 21, 1891, pp. 281-2.

Street Improvement in Omaha. By A. J. Grover. Gives a short history and cost of the improvements at Omaha. *Proc. Neb. Assn. Eng. and Sur.*, Vol. I.

Street Lighting.

By T. A. Shelton. Treats of the use of lenses and reflectors. *Van Nos. Eng. Mag.*, Vol. IX., p. 225.

See *Lighting.*

Street Pavements.

See *Pavements. Paving.*

Street Railway Accidents *in St Louis.* List of accidents from Jan. to Aug., 1891. *Eng. News*, Sept. 26, 1891, pp. 281-2.

Street Railway Cars.

Car Fenders. Abstract from report of the Mass. Railroad Commission, as to their efficiency, and adaptability to four wheel cars. Illustrations of proposed forms are given. *Eng. News*, Feb. 21, 1891, pp. 170-2.

Funeral Cars. Several papers describing funeral cars of various lines. Illus. *St. Ry. Rev.*, Aug., 1891, pp. 314-17, *et seq.*

Rowan Steam Carriage, Full Description of the, which was awarded a gold medal at the Antwerp tramway trial. Various forms of carriage and motor, their capacity and fuel consumption. *Sci. Am. Sup.*, May 15, 1886.

Street Railway Curves.

And Switches. Measurements of Curves and Switches for Street Railways. A valuable article by Wm. Rydler, C. E., giving several practical suggestions concerning this sometimes troublesome task. *St Ry. Jour.*, March, 1890, Vol. VI., p. 108.

Measurements of Curves and Switches for Street Railways. A brief note of some interest explaining methods of measuring existing curves on street railways with tape alone. *St. Ry. Jour.*, Oct., 1889.

Resistance on St. Railways. Wright. *Jour. Assn. Eng. Soc.*, Vol. III, p. 152.

Street Railway Rails.

See *Rails.*

Street Railway Track. A paper by T. G. Gribble, describing several systems of construction used in England, discussing various elements of stability and giving "General Conclusions" as to requirements. A method of laying out spirals and turnouts is also given with accompanying tables. *Trans. A. S. C. E.*, Vol. XXIV., Feb., 1891. Paper No. 461, pp. 80-128. Discussion, pp. 129-158.

Extracts from a paper by T. G. Gribble, read before the Am. Soc. C. E. discussing the stability of various types of construction. *St. Ry. Jour.*, Feb. 1891, pp. 83-6.

Improvement of. Paper by E. E. Russell Tratman. This is a valuable and timely discussion of this subject. The paper is illustrated by 6 plates. *Trans. A. S. C. E.*, March. 1890, Vol. XXII., pp. 135-208.

Joints. By A. W. Wright. Illustrated. *Jour. Assn. Eng. Soc.*, Vol. III. p. 264.

New York Approved. Cross section of track and substructure (from Park Engineer M. A. Kellogg's report of Nov. 5, 1888), which has been accepted, and its specifications adopted for street railroads in New York City. Gives rail section and cost of laying. *Eng. & Build. Rec.*, Jan. 5, 1889.

Providence, R. I., Girder Rails, which rest on a cement foundation. Illustrated. *Eng. News*, July 31, 1886.

See *Electric Railroads.*

Street Railways.

An illustrated description of the Mekarski compressed air car, several of which are now working on a London road. Lon. *Eng.*, March 13, 1888.

By W. B. Adams before the Soc. of Arts. Treats of the construction of tramways, vehicles, haulage and uses. *Van Nos. Eng. Mag.*, Vol. VI., p. 177.

By D Galton. Before Soc. of Arts. Discusses the cost of operating by horses, and then considers the various methods which have been proposed for furnishing motive power. *Van Nos. Eng. Mag.*, Vol. XVI., p. 301.

By Robinson Souttar before the Inst. C. E. Part I, Construction; Part II, Mechanical Motive Power. *Eng. News*, October—Dec. 8, 1877.

Birmingham Steam. A comprehensive account of the working of thirty-three miles of steam tramways in Birmingham, England. Lon. *Engineer*, March 18, 1887.

Car Propulsion, Horse-power used in. By Aug. W. Wright. An attempt to determine the power actually consumed on street car propulsion, both for starting and for maintaining uniform motion. Experimental data on Chicago car lines used in the discussion. An important contribution on a subject on which exact information is very meagre. *Jour. Assn. Eng. Soc.*, Vol. V., p. 8.

Cars, Spring Propulsion for. By E. H. Leveaux. Gives details of a scheme for the use of springs for motive power. *Van Nos. Eng. Mag.*, Vol. XV., p. 11.

Compressed Air. A review of the Vincennes-Ville Evrard compressed air tramway, which has been in operation for nine years. Illustrated. *Sci. Am. Sup.*, March 17, 1888.

Convention. Held at Minneapolis, October, 1889. Complete reports of proceedings, covering much matter of interest, in *St. Ry. Jour.*, November, 1889. *St. Ry. Gaz.*, October, 1889.

Construction of. By W. Wharton, Jr. A consideration of the details of construction, with plates of rail sections. *Proc. Eng. Club. Phila.*, May, 1883.

For Country Roads. By Ed. E. R. Tratman. Description, with estimate of cost per mile. To be operated by steam. Illustrated. *Eng. News*, April 3, 1886.

Electric, Cable, Horse and Other. How to build, maintain and operate. A series of articles by C. B. Fairchild of considerable value and interest. (Written in a popular style.) Fully Illustrated. *St. Ry. Jour.*, March, 1890. Vol. VI., p. 113, *et seq.*

Street Railways, continued.

Electricity as Applied to Street Railways. Paper by Frank J. Sprague before Kansas City Electric Light Convention. Reprint in *Elec. Rev.*, March 21, 1890, p. 130.

In England. Translation of a report upon by the Chief Engineer of Public Ways, Paris, with especial reference to styles of rails and supports. Illustrated. *Eng. News*, Jan. 16, 1886.

European Street Railway Practice. An elaborate report by Osborne Howes, Jr. to the Boston Transit Commission. Subjects investigated are: Street-Cars, and Conditions controlling their operation, underground roads, elevated roads, railroad terminals, rates of fare and street widenings. The first part is given in *Eng. News*, Nov. 21, 1891, pp. 428-3. Full report in *Eng. Rec.*, Nov. 21, 28, and Dec. 5, 1891. The latter portion of the report dealing with the underground railways of London and the Berlin and Liverpool elevated railways, is given in *Eng. News*, Dec. 5, pp. 532-4, 549-50. Abstract. *St. Ry. Jour.*, Dec., 1891, pp. 678-82.

Of Great Britain. A letter from Mr. Rufus Martin describing English practice in street railways and street paving. *St. Ry. Jour.*, Nov., 1889.

Kansas City and Omaha. *St. Ry. Jour.*, March, 1890, Vol. VI., p. 113.

Of Liverpool. Description of the construction and operation of the city tramways. Abstracted from the report of the City Engineer. Lon. *Engineer*, May 21, 1886.

Mechanical Traction on. By J. L. Hadden. A paper before the Soc. of Arts, giving conditions that should be imposed upon the machine and systems generally. *Van Nos. Eng. Mag.*, Vol. XVIII., p. 439.

Motive Power for Street Railways. Comparison of methods by R. J. McCarty, at the October Convention of the Am. St. Ry. Assn. *St. Ry. Jour.*, Nov., 1890. Also report of Committee on *Electric Motive Power Technically Considered*, pp. 521-4. Abstracted in *Eng. News*, Nov. 1, 1890, pp. 394-5.

Motor for Street Railways. Results of experiments by Jury on Railway Appliances at the Antwerp Exhibition. Steam, electric, and compressed air motors compared. *Van Nos. Eng. Mag.*, March, 1886; also *Sci. Am. Sup.*, Feb. 27, 1886.

Motors other than Animal, Cable and Electric. A report on steam, gas, compressed air and chemical motors, as applied to street car propulsion, by Mr. H. H. Windsor, Secretary Chicago City Railway Company, read at Convention of Am. St. Ry. Assn. at Minneapolis, 1889. *St. Ry. Gaz.*, Oct. 1889.

Operating Expenses.

And Receipts of the Chicago City Railway. Annual Report for 1890, of President C. B. Holmes of the Chicago City Railway. A few extracts in *Eng. News*, Jan. 31, 1891, p. 105.

Cost of Traction on Street Railways in England. Receipts and operating expenses of the steam, horse, electric and cable cars of Birmingham, Eng. From Lon. *Eng.* *Eng. News*, Oct. 24, 1891, pp. 395-6.

In New York and Brooklyn. Table giving an itemized statement of cost of operating the twenty-six lines. *St. Ry. Jour.*, July, 1891, p. 347. Compiled from Report of N. Y. R. R. Commissioners.

Of the Principal Street Railways in N. Y. State. Year ending June 30, 1890. Table compiled from reports of railroad commissioners. *St. Ry. Jour.*, Aug., 1891, p. 405.

See *Electric Railroads, Street.*

Permanent Way. A new paper by James More, discussing the merits of various systems, and the best methods of construction. Illus. *Proc. Inst. C. E.*, Vol. CIII., 1891, pp. 201-30.

Street Railways, continued.

Of Pittsburgh and Allegheny, Pa. Map and descriptions, illustrations, etc. *St. Ry. Jour.*, May, 1890, p. 234. Oct., 1891. *Souvenir No.* pp. 1-60.

The Relative Economy of Cable, Electric and Animal Motive Power for Street Railways. Tables giving description, cost and operating expenses, of 10 cable lines, 10 electric lines and 30 lines operated with animal power, with comparisons in various items. From Census Bulletin No. 55. By C. H. Cooley. Abstract, *Ry. Rev.* May 9, 1891, pp. 293-2. *Eng. News*, May 23, 1891, pp. 487-9.

Relative Economy of Electric, Cable and Horse Railways. A paper by J. S. Badger, read at the St. Ry. Convention, Pittsburg. A comparison of the best available data shows electricity to be the most economical power. Tables of data given. Abstract. *Eng. News*, Oct. 24, 1891, p. 394.

Relative Value of Steam and Street Railways. A paper by Geo. W. Mansfield, read before the Boston Soc. Arts, giving some facts regarding the passenger traffic of steam and street railroads. Seven tables. *St. Ry. Jour.*, May, 1891, pp. 261-4.

Report of the ninth annual convention of the Am. St. Ry. Assoc., at Buffalo. October, 1890. *St. Ry. Jour.*, Nov., 1890. *St. Ry. Gaz.*, Oct., 1890. *Convention Number*.

Robinson Radial Car System. Adapted especially to short curves such as occur on street railways. Illustrated. *Elec. Eng.*, Oct. 1, 1891.

Steam. Two papers with long discussion, giving requisite conditions and describing those at Sydney, Australia, in *Proc. Inst. C. E.*, Vol. LXXIX., pp. 68-215.

Of St. Louis, Mo. A review of principal street car lines in St. Louis, with map of city showing location of various systems. *St. Ry. Jour.*, Feb. 1890, p. 56.

The Tramway Spiral. Paper by Mr. T. G. Gribble. Location of curves by ordinates. *St. Ry. Jour.*, Oct., 1890, pp. 497-4.

Street Traffic.

Results of Some Observations on, as to tonnage and falls, for granite, wood and asphalt pavements. Observations made in ten cities in the United States, under the direction of Capt. F. V. Greene. *Trans. A. S. C. E.*, Vol. XV. p. 123.

Growth of City. Gives five diagrams showing the internal passenger movement in New York, Brooklyn and London, with comment on the same. *Eng. News*, Oct. 15 and 22, 1887.

Streets.

Grade Intersections. By F. A. Calkins. Gives a number of examples of actual practice in the city of New York. *Eng. News*, Feb. 26, 1887.

See *Municipal Engineering. Mexico. Trees.*

Streets and Highways. *European Streets and Highways.* Full review of a valuable, consular report of over 600 pages, concerning streets, pavements and roads. *Eng. Record*, Dec. 5, 1891, p. 3.

Strength, Elasticity, Ductility and Resilience *of Materials of Machine Construction.* By Prof. R. H. Thurston. Gives particulars of the testing of a large number of woods and metals on an automatic registering machine. Describes the machine, method of operation and interpretation of diagrams. *Trans. A. S. C. E.*, Vol. II., p. 349.

Strength of Alloys *at Different Temperatures.* Abstract of a paper by Prof. W. C. Unwin, and giving strength, elongation and reduction of area, etc., for various alloys at different temperatures. *Mechanics*, Nov., 1889.

Strength of Beams. See *Beams*.

Strength of Materials.

By S. W. Robinson. A discussion of beams of uniform resistance, the beam forming part of the loading. *Van Nos. Eng. Mag.*, Vol. XVI., p. 199.

By S. W. Robinson. Treats of the general theory of the beam and strength of wrought-iron bridge members, with practical formulas for beams, struts, columns, etc. Gives extended comparisons of formulæ with experiments. *Van Nos. Eng. Mag.*, Vol. XXVI., pp. 409 and 487. Also p. 60 of *Science Series*.

A paper from the German of Ludwig Spangberg, giving details of experiments made to ascertain the behavior of metals under repeated strains. Describes the apparatus and gives tables of tests. *Van Nos. Eng. Mag.*, Vol. XV., p. 449 to 548.

By Wm. Kent. Gives in a very clear manner the present state of our knowledge on the strength of engineering material. Urges the importance of better tests *Van Nos. Eng. Mag.*, Vol. XX., pp. 39 and 189. Also *Science Series* No. 41.

Atomic Volume and Tensile Strength. A discussion of their apparent relation. By R. A. Fessenden. *Elec. World*, Aug. 22, 1891, pp. 123-5.

Calculations of Dimensions as Depending on the Ultimate Working. By Dr. Weyrauch. Treats of Wohler's law, Launhardt's and Weyrauch's formulæ, etc. *Van Nos. Eng. Mag.*, Vol. XXIV., p. 321.

And Elasticity of Material. By Wm. J. Millar. Describing some experiments upon the strength and elasticity of iron. *Van Nos. Eng. Mag.*, Vol. XXII., p. 203.

Endurance Tests of Metals. By C. A. Marshall. Gives discussion of various experiments on the effect of repeated stresses. Illustrated, pp. 17. *The Technic*, Univ. of Mich., No. 2, new series, p. 7, 1887.

Gives the tests that material supplied for locomotives on the Missouri Pacific Railroad must stand. *Nat. Car & Loco. Build.*, Dec., 1887.

Normal Working Loads for Building Materials and Structures adopted by the Austrian Association of Engineers and Architects, a summary of a committee appointed by above Association, giving in tabular form the principal conclusions. Abstracted from *Wochenschrift des Osterreichischen Ingenieur und Architekten Vereins*, 1889, *Proc. Inst. C. E.*, Vol. XCVII., p. 428, London, 1889.

Observations on the Stresses Developed in Metallic Bars by Applied Forces. By Theodore Cooper. Gives an account of the lines of shearing stress as shown by a change in the external appearance of the specimens. Illustrated. *Trans. A. S. C. E.*, Vol. VII., (1878), pp. 174-81.

Rate of Set of Metals, subjected to stress for considerable periods of time. Illustrated. *Trans. A. S. C. E.*, Vol. VI., p. 28 36.

Repeated Loads. Some new and important facts covered by Prof. Bauschinger, reviewed by Prof. Unwin in Lon. *Engineer*, Dec. 10, 1886.

Tenacity of Metals and their Resistance to Torsion, by R. H. Thurston. *Trans. A. S. C. E.*, Vol. VII., (1878), pp. 109-73.

Some Tests on, at the laboratory of the Univ. of Minn. Gives tension, compression, and transverse tests of white pine and of brick and stone in compression and cross breaking. By Prof. Wm. A. Pike. *Jour. Assn. Eng. Soc.*, Vol. IV., p. 161.

Shearing Strength of Iron and Steel. Results of experiments at University College, London. *Proc. Inst. C. E.*, Vol. XC., p. 382.

Under Repeated Stress. The derivation of new formulas from the experiments of Wohler and Spangenburg, by Prof. Merriman. *Van Nos. Eng. Mag.*, Feb., 1883.

Weyrauch Formulæ. A discussion by a number of engineers of the formulæ for the strength of material. *Van. Nos. Eng. Mag.*, Vol. XXVII., p. 113.

See *Iron, Crushing Strength of Metals.*

Strength of Slates.

Tests of Rutland and Washington County Slates. By J. T. Williams. Contains results of experiments made on testing-machine at Rensselaer Polytechnic Inst. *Van Nos. Eng. Mag.*, Aug., 1886.

Strength of Steel.

An account of a few severe drifting, punching and bending tests on a soft steel low in phosphorus. A tension test gave: elastic limit, 33,740 per sq. in.; ultimate strength, 49,944 per sq. inch; elongation in 8 degrees, 31½ per cent.; reduction in area, 67.37 per cent. *Eng. News*, Sept. 28, 1889, Vol. XXII., p. 294.

Drifting Tests of Structural Steel. Extract from letter of Mr. A. C. Cunningham, C. E. Illustrated. *Eng. News*, May 18, 1889.

Of Structural Steel. A letter by Theodore Cooper showing no advantage in reaming punched holes in soft steel, and recommending the use of drifting test. *Eng. News*, May 25, 1889.

See *Steel, Tests.*

Strength *of Welds in Wire Ropes and Wire.* Tests. See *Wire Ropes.*

Stresses.

In Bridges for Concentrated Loads. See *Graphical Method.*

Deflection of Framed Structures and Distribution of Stresses over Redundant Members. By Prof. J. B. Johnson. *Jour. Assn. Eng. Soc.*, May, 1890, pp. 249-53.

Determination of the Stresses in Elastic Systems by the Method of Least Work. By Wm. Cain. Demonstration and application to arches, beams, and trusses with redundant members. *Trans. A. S. C. E.*, Vol. XXIV., April, 1891, pp. 265-308.

Elevation of the Limit of. A paper describing a series of experiments to determine facts in regard to the operation of the law called the elevation of the limit of stress, with miscellaneous experiments to determine physical phenomena accompanying rapid alternation of strain and rest. *Report of the U. S. Board of Testing, etc.*, Vol. I., 1881, pp. 107-21.

In Single Intersection Trusses. Position of live load, giving maximum values, analytically determined, and tables computed. By Wm. Cain. *Van Nos. Eng. Mag.*, March, 1885.

In Statically Indeterminate Structures. Articles giving demonstrations and applications of various methods of determining. *Ind. Eng.*, Aug. 29, Sept. 12, & Oct. 17, 24, 1891, *et seq.*

See *Bridge Stresses. Strains.*

Structural Designing, *Some Conditions Affecting.* By Wm. H. Burr. From Rens. Soc. Papers. Discusses many of the present unsolved difficulties in the way of rational designing of engineering structures. *Van Nos. Eng. Mag.*, May, 1885.

Structural Steel, *Tests of.* See *Strength of Materials.*

Structures *on the Water Front of San Francisco, Cal.* See *Water Front.*

Working Loads for. See *Working Loads.*

See *Inspection.*

Struts.

Stiffness of. Gives results of experiments made at Mason College to investigate the influences of variation of load, of length, and of eccentricity of thrust from centre of end section. Lon. *Engineer*, Jan. 6, 1888.

Working Strength and Stiffness. By Prof. W. H. Smith. Lon. *Engineer*, Oct. 14 and 28, 1887.

Subaqueous Foundations. See *Foundations.*

Subaqueous Underpinning. By A. G. Menocal. Describes the underpinning of a granite quay wall where bearing piles had been destroyed by teredo. Excavated in six-foot sections and concrete used in sacks. *Trans. A. S. C. E.*, Vol. XI., (1882), p. 181.

Submarine Boats.

See *Navigation, Submarine.*

Submarine Cables. A paper before the Royal Inst. by Fleeming Jenkins on the construction of the French Atlantic cables. *Van Nos. Eng. Mag.* Vol. I., p. 683.

Sub-Scales, *Including Verniers.* By H. H. Ludlow. The theory of all Vernier measurements are concisely stated and the instruments described and illustrated. *Van. Nos. Eng. Mag.*, Vol. XXVII., p. 323.

Subways. See *Conduit. Cable Conduits. Electric Subways. Railroads Underground.*

Superintendence. See *Management of Men.*

Surveying. Paper by J. D. Varney. *Rep. Ohio Soc. Surv.*, 1884.

On the Increasing Accuracy of a System of Magnetic Bearings. By O. B. Landreth. Gives method of assigning magnetic bearing to a survey in which the horizontal angles have been measured with transit. *Trans. A. S. C. E.*, May, 1886.

Adjustment of a traverse by the method of least squares. *Zeitsch. f. Vermessungswesen*, 1887, pp. 149-171, 287-297.

"*Approximate Times of Culminations and Elongations*, and of the Azimuths at Elongation of Polaris for the years between 1889 and 1910." Prepared by Chas. A. Schott, Assistant U. S. Coast and Geodetic Survey. *U. S. C. & G. Surv., Bulletin No. 14.*

Boundary Lines. End Lines and Side Lines in the U. S. Mining Law Paper. By R. W. Raymond, New York, at the New York meeting, February, 1889, pp. 22. *Proc. A. I. M. E.*, New York meeting, 1889.

Chaining, Errors in. By Prof. I. O. Baker. Discusses the various sources of error, and gives examples of accuracy attained. *Eng. News*, Sept. 19, 1885.

Chaining, Errors in. A Report on, by Prof. Mansfield Merriman, giving probable errors for different distances. *Land Office Report of Secretary of Internal Affairs of Pennsylvania*, 1884.

Chaining, Errors in, Due to Lengthening of Chain, by straightening of the links. *Eng. News*, Oct. 3, 1885.

Chaining, Precision of. An investigation of the probable error of chaining, due to erroneous manipulation and not from erroneous length of standard. By Mansfield Merrimann, *An. Rep. of Sec'y of Int. Affairs for Penn.*, 1884.

City. Laying Out and Monumenting the 23rd and 24th Wards and Part of the 12th Ward of New York City. Detailed account. *Eng. News*, Feb. 12, Apr. 9, 1881.

Contour Lines. By B. Feind. *Jour. Assn. Eng. Soc.*, Vol. VII., pp. 89-91. (March, 1888).

Co-ordinate. By Henry F. Walling. Points out a method by which the coast survey work may be made available in the ordinary operations of land surveyors and engineers within the districts of the triangulation. *Trans. A. S. C. E.*, Vol. VI., p. 88.

Co-ordinate System. By Wm. E. McClintock. Contains much practical matter, and useful tables. *Jour. Assn. Eng. Soc.*, Vol. I., p. 290.

The Curve Ranger. Illustrated description of an instrument involving similar principles to the sextant, and by means of which a single observer can trace out curves whenever the points of tangency are visible from several points on curve. *Sci. Am. Sup.*, Jan. 11, 1890 p. 11,694. *Eng. News*, Jan. 18, 1890, p. 54.

Surveying, continued.

Descriptions. A paper read before the Association of Provincial Land Surveyors, by Otto J. Klotz. Printed by C. B. Robinson, No. 5 Jordan street, Toronto, Ont., 1889.

Elimination of the Effects of Local Attractions in Surveying with the Magnetic Compass. Papers by Messrs. D. H. Davidson and I. O. Baker, read at Peoria, Jan. 30, 1890. These papers have particular reference to mine surveying. *Rep. of the 5th Annual Meeting of the Ill. Soc. of Engrs. and Surveyors.* (50 cts.) Pp. 37-40, 83-90. *How to Establish a True Meridian. Ibid.*, pp. 91-3.

Elimination of Errors in Field Work. By Wm. Bouton. Contains valuable facts and tables relating to accurate work in cities, with transit and steel tape in establishing property lines. *Jour. Assn. Eng. Soc.*, Vol. V., p. 151.

Establishment of Old Lot Lines and Adjusting Discrepancies. By S. Willet Hoag, Jr. Illustrated. *Eng. News*, May 9 and 16, 1885.

Exploratory. See *Canon.*

Exterior Boundary of Townships. By Z. A. Enos. Discusses whether the exterior boundary line of a township, as first run or modified by subdivisional lines and corners, is to be regarded the true boundary line. *Rep. Ill. Soc. Eng. & Surv.*, 1885, pp. 90-101.

Field Notes. Their record and preservation. By R. S. Weitzell. *Report Ohio Soc. Surveyors*, 1884.

Field Party. The Size of a. By A. M. Van Auken. *Jour. Assn. Eng. Soc.*, Vol. VI., p. 307.

And Geodesy. Literature of, for the year 1886, by R. Gerke. A complete list of books and papers on these subjects for 1886. Valuable for those desiring to keep posted on this subject. *Zeitschr. f. Vermessungswesen*, 1887, pp. 473 and 514-519.

Geodetic. See *Base-Line.*

Hydrographic. New Sounding Apparatus for Sub-Marine Researches. Illustrated description of a simple apparatus for this work. By E. Belloc, in *Le Genie Civil.* Reprinted in *Sci. Am. Sup.*, No. 826, Oct. 31, 1891, pp. 13195-6.

Hydrographic. Practical Notes on. By A. C. Both. Describes best methods of making such a survey. *Eng. News*, July 29, 1882, pp. 257-9.

Instruments, Their Construction and Use. By Prof. I. O. Baker. A series of articles in *Eng. News*, Aug. 7, 1886, *et seq.*

Instruments for Measuring Distance. Discusses chains, tapes, etc., and telemetrical instruments including the Porro telescope. From *Jackson's Survey Practice.* *Eng. News*, April 24, May 22, June 1, 1880.

Instruments. See *Goniograph (double sextant). Tapes.*

Levelling. See *Hypsometry.*

Locating Piers. See *Bridges.*

Logarithmic Traverse Tables, four places, from 0 to 360°, to be used in reducing or coordinating topographical surveys where the azimuth is read up to 360°. By Benj. Smith Lyman, in *E. & M. Jour.*, Dec. 26, 1885.

Micrometer Measurement of distances. A constant base is used, some 12 or 14 feet long, held either vertically or horizontally, and the angle subtended at the telescope noted on a micrometer. Is a modified form of stadia surveying. Fully described and experiments given showing accuracy, by Wm. Ogilvie. *Proc. Provincial Land Surveyors*, Canada. Willis Chipman, Sec., Brockville, Can.

Notes on Engineering Field Work. By Wm. Cain. On levelling, barometric levelling and transit work. Gives a table of elevations, comparing barometric with trigonometric work. *Eng. News*, April 2, 1881, pp. 133-4.

Surveying, continued.

Photographic Surveying. Article read before society by H. K. Landis, treating this subject in a general way in its principal features, pp. 8. Illus. *Jour. Eng. Soc.*, Lehigh Univ., Nov., 1889, Vol. V., p. 26.

Photography Applied to Surveying. Paper by G. W. Pearson, read before the Engrs. Club of Kansas City. *Jour. Asso. Eng. Soc.*, Vol. IX., Oct., 1890, pp. 489-93.

Photography Applied to Surveying. An abstract of the methods suggested in a treatise on this subject by Dr. C. Koppe, Brunschweig. Gives a simple and concise exposition of principles involved. *Mechanics*, July, 1889, *et seq.*, Vol. XI., p. 168.

The Rectangular Coordinate Method. By H. F. Walling. Recommends the general adoption of rectangular co-ordinates referred to points fixed by geodetic triangulation. *Trans. A. S. C. E.*, Vol. VI., (1877), pp. 98-116.

Triangulation Errors, Notes on Short Method of Distributing. Paper by A. E. Wackrill, on the approximate adjustment of angles in geodetic surveying. *Proc. Inst. C. E.*, Vol. CI., 1890, Paper No. 2,471, pp. 249-56; appendix containing 4 tables, pp. 257-60.

See *Civil Engineering, Engineering, Leveling, Mining Surveying, Railroad Location, Solar Azimuth, Village Plats.*

Surveying Land.

Land Report of Committee on, with questions, answers, and problems, etc. *Rept. Ohio Soc. Surv.*, 1884.

New Rules for Calculating the Contents of. By J. W. Davis. *Van Nos. Eng. Mag.*, Vol. XX., p. 169.

About Corners. By F. Hodgman. A paper describing means of identifying and restoring lost land-marks. *Rept. Mich. Assoc. Surv.*, 1881. See also *Surveying.*

In the Dominion. A series of articles giving the personal experience of a surveyor. Gives full information on this subject. *Eng. News*, July 7, Sept. 29, 1877.

Manual of Instructions for the Survey of the Public Lands of the United States. This work contains several tables for latitude and longitude, and magnetic declination, specimen field notes and eight folding plates. Cloth. pp. 160. Government Printing office. Washington, 1890.

Method of eliminating local attraction in a land survey with needle compass; and the limits of precision of such work, by I. O. Baker. *Eng. News*, Oct. 31, 1885.

Surveys of the Public Lands. How to tell Township and Section Corners. Description of method of surveying with details concerning monuments. *Eng. News*, Jan. 27, Feb. 10, 1877.

Surveying Mines.

Methods of Working and Surveying the Mines of the Longdale Iron Company, Virginia. By Guy R. Johnson. Description of methods used, especially the use of a hanging compass and clinometer in surveying some narrow chutes. *Trans. A. I. M. E.*, 1891, pp. 11. *Am. Mfr.*, June 26, 1891, pp. 515-16. *Eng. News*, June 13, 1891, pp. 559-60.

By H. B. Sturtevant. Gives description of the Lake Superior iron region, and then treats of the methods employed in subterraneous surveys. *Proc. Mich. Eng. Soc.*, for 1886-87.

Shaft-Surveying at the Przibram Mines. Abstract of a paper in *Oester. Zeitschrift fur Berg- und Huettenwesen*, 1890, p. 501, giving a fully illustrated account of some extremely difficult mine surveying. *Proc. Inst. C. E.*, Vol. CIV., 1891, pp. 387-9.

Surveying Plane Table.

An exhaustive memoir on the subject, as practiced on the U. S. C. and G. Survey. See *Report*, 1880.

By Josiah Pierce, before the Institution of Civil Engineers. Discusses the economic use of the plane table in topographical surveying, with discussion. *Proc. Inst. C. E.*, Vol. XCII., pp. 187-236.

Methods Used by the United States Geological Survey in Western Massachusetts in 1886. A paper by Louis F. Cutter, read before the Boston Soc. C. E., describing methods used, and giving data as to accuracy and cost. Includes discussion of barometric work. *Jour. Assn. Eng. Soc.*, July, 1891, pp. 358-69.

Surveying Railroads.

Railroad Leveling, by means of two rodmen and a hand-car, whereby over 30 miles of levels are run in a day. Th. S. Hardee. *Trans. A. S. C. E.*, Vol. VI. (1877), pp. 158-162.

Outfit for Railroad. By Benj. Thompson. Gives description of the outfit of a railroad surveying party, and tells how the work was done. *Rep. Ohio Soc. Surv. & Eng.*, 1885, pp. 137-144.

Party, Organization of, for Preliminary R. R. Survey. By A. W. Wright. Describes the organization of a party which surveyed 22 miles of the U. P. R. R. in one day. *Jour. Assn. Eng. Soc.*, Vol. V., p. 382.

Topography for. Description of method of 3 ft. rod and hand level. By Theodore Low. *Eng. News*, Oct. 11, 1884, pp. 171-3.

Topography. Its Use and Abuse on Railroad Surveys. Being the summary of an extended discussion on the subject. *R. R. Gaz.*, Jan. 9, 1885.

Transit Work in R. R. Surveying. By T. Appleton. Gives many practical suggestions for field location surveys. *Jour. Assn. Eng. Soc.*, Vol. II., p. 147.

Surveying, Stadia.

By A. Winslow. Theory with reduction tables. *II. Geol. Survey Penna., Report A. A.* Also *Van Nos. Eng. Mag.*, April, 1884.

Description of a method in which the vertical arc of the instrument is graduated according to the tangent. *Eng. News*, Dec. 10, 1881, pp. 498-500.

By Means of the Stadia or telemeter, or tacheometry. A very good paper, describing the various ways in which it is applied, in *Proc. Inst. C. E.*, Vol. XC., pp. 281-319.

Diagram for reducing distances and elevations in stadia measurements. The result obtained by one reading of the diagram. *Sch. Mines Quart.*, July, 1886.

The Theory of. By Arthur Winslow. Treats of the theory and practice of stadia surveying; gives tables of horizontal distances and differences of level for the reduction of field observations. The formulas and tables are such as are used on the geological survey of Pennsylvania. *Van Nos. Eng. Mag.*, Vol. XXX., p. 313; also, *Van Nos. Science Ser.*, No. 77.

Reduction Tables. By Arthur Winslow. Corrections for horizontal distances and elevations, for vertical angles increasing by 1' up to 30°. *Proc. Eng. Club, Phila.*, Vol. IV., No. 4. Also *Van Nos. Eng. Mag.*, April, 1884.

Stadia Rods.

Or Level Rods. Designs for. The numerals are so constructed as to themselves form appropriate diagrams from which tenths (or hundredths) can be read. Very ingenious, and doubtless satisfactory on short sights. By Benj. S. Lyman, in *E. & M. Jour.*, January, 30, and *R. R. Gaz.*, February 12, 1886.

By G. A. M. Liljencrantz. *Jour. Assn. Eng. Soc.*, Vol. V., p. 360.

By I. O. Baker. Gives brief history of the stadia, discusses the principle and formulas, gives designs for boards and diagrams for reduction of work. *Eng. News*, March 12 and 19, 1887.

Surveying, Stadia, continued.

By Geo. J. Specht. Description of a good pattern for stadia rods for long distances. *Eng. News*, Jan. 1, 1887.

The Porro Telescope for Stadia Work. In this by means of an auxiliary lens placed back of the objective the interval measured by the wires is made proportional to the distance of the rod from the *center* of the instrument. By J. B. Johnson, Washington University. *Eng. News*, Nov. 8, 1890, pp. 412-13.

Railroad. By J. B. Johnson. A series of articles showing how the stadia rods may be used with advantage in preliminary railroad surveys. *R. R. Gaz.*, Feb. 2 *et seq.*, 1883.

Surveying Topographical.

By George J. Specht. Gives the stadia method of doing topographical work. *Van Nos. Eng. Mag.*, Vol. XXII., p. 131.

By the Plane-table and Stadia, with methods of its use for surveys of railways, roads, etc., as practiced in Switzerland and France. Very simple presentation of the theory, with cost of such surveys, fully illustrated. By M. J. Meyer, Chief Engineer of Railways of Western Switzerland. *Proc. Soc. C. E., Paris*, Oct., 1884.

And Records for Drainage Purposes. By C. G. Elliott. Gives experience and practice of drainage surveying. A good paper. Report Second Annual Meeting *Ill. Soc. of Eng. & Surv.*, p. 43.

By the Transit and Stadia. By William Bell Dawson. Describes such a survey of 180 square miles in Nova Scotia, at a cost of $16.75 per square mile. The map made on a scale of 500 feet to the inch. *Trans. A. S. C. E.*, Vol. XI. (1882), p. 397.

Surveying, Tunnel.

Tunnel Surveying on Division No. 6, New Croton Aqueduct. Instructive paper giving methods employed and accuracy attained. Illus. By F. W. Watkins, M. Am. Soc. C. E. *Trans. A. S. C. E.*, Vol. XXIII., 1890, pp. 17-31. Discussion, including the description, with illustration, of an instrument used for taking diagrams on the Sudbury River Conduit, *ibid*, pp. 31-8.

Alignment. Paper by F. P. Stearns giving an account of some problems on the Dorchester Bay Tunnel. *Eng. News*, Aug. 16, 1884, pp. 71-6.

Surveyors.

By H. D. Bartholomew. A short account of their legal status in Michigan and other States. *Rep. Mich. Assn. Surv.*, 1881.

County. History of the office of County Surveyor in Ohio. By R. S. Paul. A full account of legislation relating to the office. *Rep. Ohio Soc. Surv.*, 1881.

And His Field Notes. A short but interesting paper describing modern practice in valuable land. By John L. Culley. *Rep. Ohio Soc. Surv.*, 1883.

Judicial Functions of. By Judge T. M. Cooley. A valuable and interesting paper relating to the legal principles to be followed in establishing and identifying land lines. *Rep. Mich. Assn. Surv.*, 1881. Also reprinted in *Report*, 1883. *Eng. News*, April 16, 1881, pp. 156-8.

His Legal Status in Michigan. By L. S. Montague, in *Proc. Mich. Eng. Soc.*, 1885.

Licensed. Propositions to require examination and license of persons desiring to practice land surveying. See Reports various State societies, Ohio, Michigan, etc.

The License System for. By Samuel S. Greeley. Gives a brief history of legislation relating to land surveying in Illinois, and advocates the system of licensing surveyors. Also gives text of a bill, which was introduced into the General Assembly of Illinois, Feb. 15, 1887. *Second Annual Report Ill. Soc. Eng. & Surv.*, p. 26.

Surveys,

Of the City of Berlin. With description of the means used for preserving the triangulation points, etc. *Zeitschrift f. Vermessungswesen*, p. 193-202.

Names of, in Ohio. See *Rep. Ohio Soc. Surv.*, 1882 and 1883.

Bottom Main Drainage. A very fully illustrated article, reprinted from Lon. *Engineer*, in *Sci. Am. Sup.*, Jan. 16, 1886.

British Ordnance Survey. History of in *Blackwood's Magazine* for October, 1886. Abstract given in *Eng. News*, Nov. 6, 1886.

Coast Survey Matters. An article on Political Scientists, with replies to the same, one of which is by Prof. Agassiz. *The Nation*, Aug. 20, Sept. 3, Sept. 17, and Sept. 24, 1885.

Of the Coast. By Lt. Geo. L. Dyer, U. S. N. This is a 42-page review of the history of the survey of the present contest between the Treasury and Navy departments for its control, as viewed by a naval officer of Hydrographic Office. Very full extracts are given from the testimony before the Congressional committee on the subject. *Proc. U. S. Naval Inst.*, Vol. XII, (1886), p. 190.

Geodetic, of the State of New York. The final report of this work, which covers about two-thirds of the State, includes valuable appendices on reconnaissance and signal building, station marks, field work and party organization, repeating instruments and methods, trigonometrical leveling, geodetic co-ordinates, tables for map projection, etc. Illustrated and accompanied with fine maps. 215 pp. quarto. *Saml. B. Ward, Pres. Brd. of Commissioners, Albany, N. Y.*

Geographical. By T. DeV. Carpenter. Presents a scheme for the organization, development and prosecution of a geographical survey in connection with its geographical commission. *Van Nos. Eng. Mag.*, Vol. XIX., pp. 52 and 173.

Government of Canada. A series of interesting articles, giving in the main the personal reminiscences of a surveyor, but including a careful description of the methods used. Ludicrously illustrated. *Eng. News*, July 10, 1886, *et seq.*

Of the Great Lakes. A description of the methods and work of the U. S. Lake Survey. By Prof. John Eisenmann. *Jour. Assn. Eng. Soc.*, Vol. I., p. 360.

Hydrographic. *Boat Survey of Peace Creek and Caloosahatchie River, Fla.* Description of survey. *Eng. News*, Sept. 9, 1882, p. 317.

Hydrographic and Topographic. See *River Improvements.*

Accuracy of the Hamburg City Survey. Probable error computed. *Zeitschrift f. Vermessungswesen*, 1888, pp. 353-74.

Of India. An historical and descriptive account by Gen. J. T. Walker. *Van Nos. Eng. Mag.*, Jan., 1886. By F. C. Danvers. Gives history of the trigonometrical survey, with short account of its methods. *Van Nos. Eng. Mag.*, Vol. XVII., p. 547. By Gen. J. T. Walker, before the British Assn. for Adv. Sci., 1885. A very minute and satisfactory account of this work. *Eng. News*, Sept. 4, 1886, *et seq.*

Lake Survey. *Final Report upon the Primary Triangulation of the Great Lakes.* By Gen. C. B. Comstock, Superintendent of the Survey. This is the most valuable and important report of American geodetic operations yet published. It is a quarto of over 900 pp., and XXX. plates, giving a history of the survey, the field methods, reductions and results of base-line measurements, triangulation, precise leveling, etc. Also an account of the elaborate intercomparison of yard and meter standards, being one of the most complete and accurate that has ever been made. A standard work of great excellence and value. *Prof. Papers, Corps of Engrs. U. S. A.* No. 24.

Land Surveys. *Origin of the U. S. System.* Two articles of value. First by Col. H. C. Moore, of St. Louis; second by Col. Chas. Whittlesey, of Cleveland. *Jour. Assn. Eng. Soc.*, Vol. II., p. 282 and Vol. III., p. 275.

Surveys, continued.

Of the City of Leipzig. Description of methods used, complete tables and calculations of co-ordinates. Method of least squares applied. By Prof. A. Nagle. *Der Civil Ingenieur*, 1887, pp. 1-42.

The Nature and Public Utility of Trigonometrical, General and Cadastral Surveys. By G. H. Knibbs. University of Sidney, N. S. W.; pamph. pp. 45, published by the Institute of Surveyors, Sidney. This essay treats the subject very fully and with considerable detail.

New Jersey State. An extract from the Report of the State Survey of New Jersey, describing the work done and the manner of preparing the results for publication. *Eng. News*, April 14, 1888.

The New Levelling Survey of France. Paper by C. Jimels, giving details of this very important work. *Le Genie Civil.* Vol. XVII, p. 5. Abstract in *Proc. Inst. C. E.*, Vol. CI., 1890, p. 313. See also *Wochenschrift des Osterr. Ing. u. Arch. V.*, Aug. 22, 1890, No. 34, pp. 290-1.

New Method of Making Running Surveys, by using a plane table on shipboard, running at a uniform rate, thus getting distances on a base line. By J. H. Fillmore. U. S. Navy. *Proc. U. S. Naval Inst.*, Annapolis, Md., Vol. XI.

Railway Survey and Resurveys. By Hosea Paul, Chief Engr., T. & W. Ry. Discusses value and use of old surveys. *Eng. News*, Sept. 12, 1891.

Report of the Regent's Boundary Commission upon the New York and Pennsylvania boundary. Gives full report of the field work, and description of the location of each mile-stone. Illustrated with many maps and sketches.

Report of Standing Committee on Permanent Markings of Surveys. Somewhat brief report by A. L. Brabazon, D. L. S., suggesting a method of marking surveys permanent by a novel method. Refer all important points in a survey to an origin, by co-ordinates; when if *any one point* can afterward be determined or identified, the location of all others may be found from recorded co-ordinates. *Proc. Assn. D. L. S.*, Ottawa, Feb. 1889, p. 39.

St. Louis. See *Sanitary Survey of St. Louis.*

State. By Charles C. Brown. Discusses the need of state surveys and gives the cost of such work in different states. *Jour Assn. Eng. Soc.*, May, 1888.

Topographical.

On the necessity of co-operation between the National and State Governments for this purpose; the arguments well stated, and facts of foreign topographical surveys are given. By H. F. Walling. A paper read before the Am. Soc. of Civ. Engrs. at the recent Buffalo meeting. *Van Nos. Eng. Mag.*, Vol. XXXI., p. 331, (Oct., 1884).

And Cadastral Survey of Illinois. Report of a Committee of the Western Society of Engineers presenting in detail the advantages from such work to Agricultural, Mining, Transportation, Municipal affairs, Science and Education, giving estimates of cost and brief descriptions of similar work in other countries. *Jour. Assn. Eng. Soc.*, Oct., 1891, pp. 465-71.

And Related Surveys as Prosecuted by the Government. By C. W. Clark. A very good summary and detailed description of the Coast and Geodetic Lake, and the Mississippi River Surveys. *Rep. Ill. Soc. Eng. & Surv.*, 1886.

For States. By H. F. Walling. A 13-page article in which are intelligently discussed the methods of making State topographical surveys, with contours of "nohupes," as the author calls them. *Jour. Assn. Eng. Soc.*, Vol. V., p. 163.

Sketch of the Topographical Survey of Massachusetts by the United States Geological Survey, 1884 to 1890. A brief paper by E. W. F. Natter, read before the Boston Soc. C. E. *Jour. Assn. Eng. Soc.*, July, 1891, pp. 356-9.

Surveys, continued.

Of St. Louis, Mo. Brief description of methods, with statement of cost to date. By O. W. Connet in the *Technograph*. Abstract, *Eng. Record*, Oct. 3, 1891, pp. [illegible].

Of the U. S. Lake Survey. Method by Transit and Stadia. See Primary Triangulation of the U. S. Lake Survey. *Prof. Papers Corps of Engr., U. S. A.*, No. 24.

Town Boundary Survey of Massachusetts. A paper by C. H. Van Orden, read before the Boston Soc. C. E., describing the triangulation used in fixing town monuments. *Jour. Assn. Eng. Soc.*, July, 1892, pp. [illegible].

See *Land*.

Sun, *Physical Constitution of.* A lecture by Prof. C. A. Young, giving the results of the latest investigations. *Jour. Frank. Inst.*, March, 1883.

Sustaining Walls. An elaborate study of the French practice of building sustaining walls with arches and arch buttresses. Fully illustrated. *Sci. Am. Sup.*, Feb. 26, 1887.

Switches. See *Railroad Track. Railroad Switches.*

Tabulating System. See *Electric Tabulating System.*

Tacheometry. See *Surveying.*

Tanning. *Electricity Applied to.* See *Electricity in the Art of Tanning.*

Tapes, Steel.

By J. L. Culley. Gives description of the tapes, with prices, manufactured by different firms, and makes suggestions relating to their use. *Jour. Assn. Eng. Soc.*, August, 1887; *Eng. News*, Sept. 3, 1887. See *Base Measurement*.

Accurate Measurements by the Steel Tape and Steel and Brass Wires. "Geodatische Langenmessung mit Stahlbandern und Metalldrahten von Edv. Jaderin, Stockholm, 1885." This is a fifty-seven page pamphlet reprint in German, from the proceedings of the Royal Swedish Academy for 1885, by Mr. Jaderin, the author of a new system of accurate measurements by the steel tape and metallic wires. It gives a careful account of the various tests, determinations of constants, and results of measurements of primary bases previously measured by primary base apparatus. No preliminary work required on the line, the wires being stretched between tripods, which are moved along with the work. Elevations of tripods taken by a spirit leveling instrument a 25 meter tape used and stretched by a spring balance. Many reduction tables given. When the steel and brass wires are used in conjunction, an accuracy of one in one million is attainable even in clear weather. It is probable that these results, together with some equally accurate obtained recently on the Missouri River Survey, will lead to new modifications of geodetic methods. Cuts of apparatus given.

Also "*Exposé Elementaire de la Nouvelle Methode de M. Edouard Jaderin pour la mesure des Droites Geodesiques au moyen de Bandes d'Acier et de Fils Metalliques.* Par E. P. E. Bergstrand, Ingenieur au Bureau Central d'Arpentage a Stockholm, 1885." This a forty-eight page pamphlet in French, describing the methods of Mr. Jaderin, with additional tables for reduction. Mr. Bergstrand assisted in Mr. Jaderin's observations. Address the author.

Base Line Measurements by means of a 300-ft. steel tape. Methods by which such measurements may be made to an accuracy of 1 in 300,000 or more. By J. B. Johnson. *Jour. Assoc. Eng. Soc.*, Vol. V., p. 40.

Base Line Measurement. See *Base-Line Measurement*

Table of Standard Steel Tapes. This table gives the length of eleven steel tapes as compared or compiled by Missouri River Commission. *Rept. of Missouri River Comm.*, 1889, p. 2759, or *Appendix WW of Rept. Chief of Engineers*, 1889.

Tapes, Steel, continued.

Use of, in Surveying. By J. B. Johnson, before the Ohio Society of Surveyors and Civil Engineers. Discusses the different errors and their effects: the limits of error; standardizing tapes, etc. *Eng. News*, March 3 and 10, 1888.

Tar. See *Coal Tar.*

Tar and Ammonia.

From Blast Furnaces. Dempster's apparatus for recovering these products described and illustrated. *Iron*, Sept. 18, 1885.

From Blast Furnaces. The present position and prospects of processes for their recovery. A paper by Wm. Jones, before Iron and Steel Inst., Glasgow. Lon. *Eng.*, Oct. 16, 1885, *et seq.*

Technical Education.

Paper by Alex. L. Holley, before the American Institute of Mining Engineers, and Discussion together with a Joint Discussion, before this body and the American Society of Civil Engineers, June, 1876. The whole covering 146 pp., and containing the views of the leading engineers and technical educators in America. Mostly given to methods of combining theory and practice. Accompanying Vol. V., (1876), *Trans. A. S. C. E.*, and corresponding Vol. of *Inst. Min. Engrs.*, See *Education.*

With Reference to the Apprenticeship System. A paper by Henry Cunynghame before the Society of Arts, London, with discussion. A very valuable review of the present anomalous condition of things, whereby the middle ground between pure theory and pure practice is mostly neglected in England and America. *Iron*, July 17, 1885; also *Van Nos. Eng. Mag.*, Aug., 1885.

See *Education.*

Technical Literature, *Handy List of.* Electricity and Magnetism, Telegraph, Gas, etc. Reference catalogue of books printed in English from 1880 to Oct., 1880, Published by H. E. Haferkorn, Milwaukee, Wis. Cloth, $1.00.

Telegraph.

The Inventor of. Some new discoveries showing that the electric telegraph was invented about 1752 by one Charles Morrison. Reprinted from the *Celtic Magazine* for September by *Elec. Rev.*, Nov. 6, 1886.

Submarine. The latest methods in this kind of work. Th. W. Rae. Illus. *Trans. A. S. M. E.*, Vol. II., p. 349.

Telegraph Lines. *New Method for Improving the Working of, at Great Distances.* By M. Fernand Godfrey. *Elec. Rev.*, Nov. 30, 1888.

Telegraph Wires. See *Electrical Conductors.*

Telegraphing from Railway Trains. *Elec. Rev.*, N. Y.; *Elec. Rev.*, London, March 7, 1885.

Telegraphy.

Cable. A new method of increasing the speed of transmission of cable messages, by which it is claimed that thirty words per minute may be sent, using the ordinary alphabet. By P. B. Delaney, before the Franklin Institute, March 11, 1889. Illustrated. *Jour. Frank. Inst.*, July, 1889.

Fac-simile. A paper by Prof. Ed. J. Houston on Delaney apparatus. Illus. Shows how writing, sketches, maps, etc., produced at one end of a telegraphic line, are automatically reproduced at the other. Illustrated. *Jour. Frank. Inst.*, Dec., 1885.

Fifty Years' Progress in. By W. H. Preece. Reviews progress made in all branches of telegraphy. Gives tables showing increase of offices, miles of wire, number of messages, etc. *Jour. Soc. Arts*, May 20, 1887.

On the New System of, to and from Moving Train. By P. H. Van der Weyde. A lecture before the Frank. Inst. *Jour. Frank. Inst.*, Sept. 1885.

Optical. See *Optical Telegraphy.*

Telegraphy, continued.

Synchronous Multiplex. Being an account of the operation of a line between Boston and Providence, capable of simultaneous use of *seventy-two* distinct circuits over one wire; or into six slow or twelve fast Morse circuits; or into thirty-six or seventy-two printing circuits. Illustrated. By Prof. Edwin J. Houston. *Jour. Frank. Inst.*, Vol. CXVIII., p. 161 (Sept. 1884).

Synchronous Multiplex. A New System. Paper read before the American Institute of Electrical Engineers by Lieut. F. J. Patten, describing his invention, whose salient feature is the employment of synchronizing devices which are in themselves not isochronous, and which therefore depend entirely upon one another. *Elec. World*, Feb. 23, 1889. Discussion in number of March 9, 1889.

Telemeters.

Gautier's Pocket Telemeter. Description of an instrument designed for rapidly measuring distances by sights taken at two ends of a base. Errors discussed. By A. G. Robbins and James H. Stanwood. *Tech. Quart.*, Vol. II., 1889. No. 4, pp. 351-66.

Reflection. By Prof. Jordan. Interesting historical and descriptive notices. *Zeitschr. f. Vermessungswesen*, 1887. pp. 317-326.

The Vernier Telemeter. Described and illustrated by Mr. Neville B. Craig. *Proc. Eng. Club. Phila.*, Dec. 1889, Vol. VII., pp. 225-32.

Telemetrograph: a combination of telescope and camera lucida, enabling the observer to draw to scale any distant view as seen through the telescope. Instrument illustrated and views taken in siege of Paris given. *Sci. Am. Sup.*, Aug. 1, 1885.

Telephone.

Invented by Reis. An article by Prof. Houston, with opinions of twenty-one eminent physicists and electricians, all going to show that Reis was the original inventor of the speaking telephone. *Jour. Frank. Inst.*, Jan., 1886.

Investigations. By Prof. Silvanus P. Thompson, D. Sc. An extremely able paper, read before the Society of Telegraph Engineers and Electricians, June 27, 1887. A valuable contribution to the development of the true telephonic theory and the perfect telephone. *Elec. World*, March 19, 26, and April 2 and 9, 1887.

Researches of Dolbear. By E. J. Houston. *Electrician* (N. Y.), Feb., 1885.

Without Wires. A description of the device by Prof. A. E. Dolbear for electric communication without wires. *Electrician & Elec. Eng.*, Nov. 1886.

Telephone Exchange *of the M. T. and T. Co., New York.* Illustrated description. *Elec. Rev.* (N. Y.), Sept. 13, 1890, p. 25.

Telephone Service *in New York City and Hartford, Conn.* Illustrated articles. *Elec. Engr.*, Sept. 10, 1890. pp. 272-5, 278-80.

Telephone Systems. By Prof. A. E. Dolbear. A lecture before the Franklin Inst. Describes the various systems, especially those of Reis, Dolbear, Edison and Bell. Illustrated. *Jour. Frank. Inst.*, Jan. 1886.

Telephony.

On the Telephoning of Great Cities. A paper by A. R. Bennett, before the Brit. Assn., discussing the merits of various plans. *Lon. Eng.*, Sept. 11, 1891, pp. 309-11.

Some Recent Advances in. By Thos. D. Lockwood, before Am. Inst. of Elec. Engrs. *Elec. Engr.*, June, 1890.

Telescopes

By Howard Grub. Treats of the testing of optical glasses, calculation and measurement of their curves of flexure and polishing. *Sci. Am. Sup.*, July 3, 1886.

Telescopes, continued.

Great, of the World. By J. C. Rees, before the New York Academy of Science. Gives a popular account of the great telescopes of the world, and discusses their construction, powers and future prospects. *Sci. Am. Sup.*, March 3, 18[illegible].

The principles Involved in the Construction of. By Thos. Nolan. Treats the optical principles involved in the construction of the various eye-pieces in a clear manner. Also considers the various problems encountered in dealing with secondary spectrums of large refractory telescopes. *Van Nos. Eng. Mag.*, Vol. XXV., p. 88 to 103 and 178; also *Van Nos. Sci. Series*, No. 51.

Progress Made in Those for Astronomical Use. By W. R. Warner. Historical and statistical, together with a discussion of the natural limitations imposed and the present state of the science. Illustrated. *Jour. Assn. Eng. Soc.*, Vol. IV., p. 183.

For Stellar Photography. By A. Grubb, before the Society of Arts. Describes the telescopes to be used in the proposed international survey of the heavens. *Jour. Soc. Arts*, April 20, 18[illegible].

Telpherage.

An illustrated account of Prof. Fleeming Jenkin's telpherage system, or transportation on suspended lines, by means of electricity. *Mechanics*, New York, Sept. 18[illegible].

A lecture by Prof. Fleeming Jenkin, the author of the system, with an especial account of the Glynde Line. Illustrated. *Elec. Rev.*, London, October 24, 1885.

A short description of the lines now in operation in England. Illustrated. *R. R. Gaz.*, Nov. 20, 1885.

The Glynde telpher line in operation. Illustrated. *Sci. Am. Sup.*, Jan. 9, 1886.

In Practical Use. By F. A. Fernald. The application of electricity described in this paper is a development of the "wire-rope haulage" system. The inventions of Prof. Jenkin and Perry are described in detail and fully illustrated. *Pop. Sci. Mon.*, July, 1890, pp. [illegible]-91.

Roman. By Prof. Fleeming Jenkin, London. An illustrated paper describing the method of transportation of goods by cars moved by electricity, the cars being suspended from wires. *Van Nos. Eng. Mag.*, May, 1885.

The telpher railway, which was the most interesting feature of the Edinburg exhibition, was well described and illustrated in Lon. *Engineer*. Reprinted in *Sci. Am. Sup.*, Aug. 30, 1890, No. 765, pp. [illegible]-4.

Temperature.

Of Our Atlantic Sea Board. A proposition to change it by cutting the polar current through the Straits of Belle Isle at the northern end of Newfoundland by a dam 10 miles long and 150 feet deep. J. C. Goodridge, Jr., in *Sci. Am.*, Oct. 31, 1885. Full page chart given.

Fluctuations of, in Different Years. Table giving mean monthly and yearly temperatures in Philadelphia for a period of 64 years. *Eng. News*, Dec. 8, 18[illegible].

Of Water at Various Depths, in Reservoirs, Lakes and Oceans. By Hamilton Smith, Jr., with discussion. *Trans. A. S. C. E.*, Vol. XIII., p. 73.

In the Wheeling Well. Table of temperatures at various depths. From a paper by Prof. Wm. Hallock. *Eng. News*, Sept. 12, 1891, p. 246.

Temperature Regulator. See *Car Heating, Temperatures, etc.*

Tempering Steel. See *Steel Hardening. Steel Tempering.*

Tenders. See *Locomotives.*

Terrestrial Magnetism. By J. S. Perry, before the Royal Soc. *Van Nos. Eng. Mag.*, Vol. VIII., p. 234.

Test Specimens.

By J. B. Johnson. Discusses the effect of size and shape of specimens on the ultimate elongation and reduction of area. *R. R. Gaz.*, June 17, 1887.

Forms and Proportions of. A paper showing by experiment the correct form and proportions of test pieces to procure correctly the tenacity, elastic limit, etc., of various metals. *Report of U. S. Board on Testing, etc.*, 1881, Vol. I., pp. 91-103.

Incorrectly Proportioned. A paper showing the errors in estimating results of tension tests due to the use of incorrectly proportioned test pieces. *Rep. U. S. Board of Testing*, Vol. I., pp. 91-104.

The Proper Form of. By William Hackney, A. M. I. C. E.: with discussion. A very exhaustive treatment of the subject, covering 91 pp., including a discussion by the best engineers and experimenters of England. *Inst. C. E.*, Vol. LXXVI., p. 72.

Testing.

Flow of Metals. See *Metals.*

Strength of Engineering Material. By Prof. J. B. Johnson. Gives summary of the present state of knowledge relating to certain materials and indicates how tests may be made useful in designing. *Jour. Assn. Eng. Soc.*, Vol VII., pp. 91-101, March, 1888.

And Testing Machines. By Theodore Cooper. Discusses effect of size of specimen, and various kinds of tests. *Eng. News*, Aug. 11, 1880, pp. 181-2.

Wohler's Experiments on. Gives results of tests made on ten steel tires. *R. R. Gaz.*, Sept. 2, 1887.

See *Measuring Instruments used in Mechanical Testing.*

See *Measurement of Strains.*

Testing Laboratory.

For Railways. Report of Committee of Master Mechanics Assn. Methods of investigation, plan, equipment, cost, etc., are discussed. *Ry. Rev.*, June 20, 1891, pp. 406-8. *Eng. News*, June 27, 1891, pp. 621-2.

Testing Machines.

An elementary paper by A. E. Hunt. *Proc. Eng. Soc. W. Penn.*, Vol. V., 1889, pp. 64-71: discussion, pp. 71-8.

Apparatus used in, for determining changes in length with accuracy. A paper before the Am. Soc of Mech. Engrs., by Gus. C. Henning. Illustrated. *Am. Eng.*, June 12, 19 and 26, 1885.

Autographic Diagrams. By Prof. W. C. Unwin. A description of six kinds of apparatus for making such diagrams. Illustrated. *Jour. Soc. Arts*, London, Feb. 26, 1886. Also Lon. *Eng.*, Feb. 19, 1886.

For Cement. Description of a machine with full detailed drawings that can be made by any blacksmith at a cost of $10 or $15. By P. T. Brendlinger. *Proc. Eng. Club. Phila.*, Vol. V. *Eng. News*, May 15, 1886.

Cement. Drawings of the cement testing machine in use at Poughkeepsie bridge. *Eng. & Build. Rec.*, Jan. 21, 1888.

Description of Dr. Bohme's automatic machine for the making of uniformly compressed test cubes of cement, same as mortar. *Eng. News*, March 26, 1887.

Description of a Testing Machine, with results of some tests. By W. Hansen. Tests of iron and steel. No results of novelty. *Zeitschrift des Vereins deutscher Ingenieure*, 1886, pp. 121-132

Discussion on the construction of. *Trans. A. S. C. E.*, Vol. II., p. 223.

Their History, Construction, and Use. By A. V. Abbott. Gives history of the development of testing machines; describes the prominent machines and gives notes on their use Illustrated. *Van Nos. Eng. Mag.*, Vol. XXX., pp. 204, 325 and 382. Also *Van Nos. Science Series*, No. 74.

Testing Machines, continued.

In Prof. Kennedy's Laboratory. Illustrated. Lon. *Engineer*, July 18, 1890, p. 54.

The Olsen. Report of Committee of Franklin Institute describing machine, which produces a graphic record of the test, and has many ingenious and valuable features. Illus. *Jour. Frank. Inst.*, Feb., 1891.

200-Ton. An illustrated description of a machine designed for use in iron and steel works, constructed by Adamson & Co., Dunkinfield, Eng. Lon. *Eng.*, June 17, 1887.

Pennsylvania Railroad Company's Drop Testing Machine for Cast Iron Car Wheels. Illustrated description with specifications of test. *Age of Steel*, July 4, 1891, p. 3.

Six-Hundred Ton. By Chas. MacDonald. Gives description of a testing machine capable of exerting a tensile strain of 1,200,000 pounds recently constructed at the works of the Union Bridge Company, at Athens, Pa. Also gives results of the testing of a full-sized eye-bar of the Hawksbury bridge. Illustrated. *Trans. A. S. C. E.*, Vol. XVI., p. 1. Also *R. R. & Eng. Jour.*, Feb., 1887.

Riehle Latest Improved Power Torsional Testing Machine. The moment of torsion is indicated directly on a scale beam. Description and cut of Machine. *Mfr. & Build.*, January, 1891, p. 6.

The 1,200-Ton Testing Machine of the Phœnix Iron Company, at Phœnixville, Pa. Fully Illustrated. Description. Folding Inset. *Eng. News*, Jan. 10, 1891, pp. 42-3. *Ry. Rev.*, Jan. 31, 1891, p. 69.

At Watertown Arsenal, Mass. The work of the United States testing machine. By J. E. Howard. *Jour. Assn. Eng. Soc.*, Vol. V., p. 261.

See *Belts, Testing Machine for*. *Friction*.

Testing of Materials.

Report of Committee on Standard Tests and Methods of Testing Materials. This report includes the testing of wrought iron and steel, cast iron, copper, bronze and other metals, and woods, with general recommendations on methods of testing and testing machines. *Trans. A. S. M. E.*, Vol. XI., 1890, pp. [illegible]. The report also contains a valuable appendix giving the Resolutions of the Conferences held at Munich and Dresden, 1884-6, relative to Uniform Methods of testing. Methods of testing are recommended on the above materials, and also on materials used in ship-building, stone and hydraulic cements. *Trans. A. S. M. E.*, Vol. XI., 1890, pp. 527-72.

Annual Report of tests at Watertown Arsenal, 1884, 500 pp. Includes 38 tests on full-sized wrought-iron bridge columns (40 were given in 1883 report); 31 brick piers; a great many tests on crushing strength of neat cement cubes; mortars of different composition; terra cotta; cast iron, steel and bronze for gun construction; steel and wrought-iron straps, bars and bolts; adhesion of nails and screws; strength of cordage and rubber springs. *Sen. Ex. Doc. No. 35, 49th Congr., 1st Sess.*

Concerning the Kirkaldy System of Testing Materials. A review of Kirkaldy's work on "Strength and Properties of Materials, etc.," by A. E. Hunt. Compares his recommendations with American practice. *Eng. News*, Sept. 19, 1891, pp. 255-7.

The Flow of Metals and its Relation to Testing. By Paul Kreuzpointner. *Iron Age*. July 3, 1890, pp. 2-3. Oct. 23, 1890, pp. 700-1.

How to Test Cast Iron. Article by Prof. J. B. Johnson, Washington University, St Louis. See *Tests*.

Impact on Iron. A paper describing a series of impact tests upon various irons, with illustrated description of the hammer and method of use; tabulated details of tests and physical phenomena observed during the work. *Rept. of Board of U. S. Testing*, etc., 1881, Vol. I, pp. 123-146.

Testing of Materials, continued.

On the Unification of Methods of Testing Materials of Construction. By E. L. Caudlot. Article treats of limes and cements. *Societe des Ing. Civils*, Feb., 1891, pp. 112-28. Article by L. Durant on iron and steel, *ibid.*, March, 1891, pp. 219-38.

Standard Tests and Methods of Testing Material. Abstract of report of third conference of German, Austrian and Prussian Engineers. Contains recommendations as to testing copper, bronze, etc., and standards for porosity of bricks and tiles. Also a valuable paper by Prof. Belelubsky on the "Effect of Shape and Size of the Test-Piece on its Elongation." From *Trans. A. S. M. E.*, *Eng. News*, Aug. 20, 1891, pp. 158-61.

Testing Works *Of the Belgian Government for the Departments of Railways, Ports, Telegraphs and Marine.* A short account of them in *Proc. Inst. C. E.*, Vol. XC., p. 392.

Tests,

Cement. See *Cement, Mortars.*

To Determine Economy of Non-Condensing, Simple, Compound Engines. See *Engine Trials.*

Of Iron and Steel for use in boiler making, as recommended by a committee of American Boiler Maker's Association. See *Boilers.*

See *Iron, Cast. Iron, Stay-bolts. Iron and Steel. Steel. Strength of Materials.*

Tests of Iron, Steel and Alloys. *Report of the Government Testing Board.* 1881. In two volumes. Vol. I. includes Rep. of Com. on Chain Cables, Malleable Iron, Reheating and Rerolling Wrought Iron. Results of Tests, etc.; comparing chemical composition with physical results. Also Report of Preliminary Investigation of Copper-Tin Alloys. 575 pp. and 100 plates.

Vol. II. contains report on Copper-Zinc Alloys, Copper-Tin-Zinc Alloys, Iron Girders and columns, Tension and Torsion Tests of Steel and Chemical Analyses of same, Tests of Quality of Steels for Tools, and a memoir by Tresca on the Planing of Metals. 700 pp. and about 200 plates.

These two volumes contain all the results of the work of the U. S. Testing Board from 1875 to 1876, when the Board was left without further appropriations by Congress. *Report of the Board on Testing Iron, Steel and Other Metals*, 2 vols.

Thermodynamics,

A lecture by Prof. De Volson Wood on Rankine's treatment, in which he undertakes to clear up some of the dark phases of that discussion. Illus. It will be found helpful. *Van Nos. Eng. Mag.*, May, 1886.

A popular paper on the general theory of thermodynamics. Read before Inst. of C. E. by Prof. Osborne Reynolds. *Am. Eng.*, April, 17, 1885, *et seq.*

Abstract of paper by J. Macfarlane Gray. From considerations connected with the specific heats of liquids and gases, the writer concludes that the second law of thermodynamics is not true. Lon. *Eng.*, March 6, 1885.

By F. T. Eddy. A compact and clear statement of the laws and their signification. *Van Nos. Eng. Mag.*, Vol. XX., p. 193. Also *Van Nos. Science Series*, No. 45.

By Prof. Wood. Some new propositions, and a further discussion of the Second Law. *Van Nos. Eng. Mag.*, Dec., 1886.

The General Theory of. By Prof. Osborne Reynolds. A lecture before the Inst. of Civ. Engrs., London. Gives a very lucid exposition of the subject in its relations to the steam engine. Reprinted in *Van Nos. Eng. Mag.*, for June, 1885.

The Second Law of, by Prof. J. Burkitt Webb, before the Am. Assoc. for Adv. of Science. Defends and explains Rankine's definitions. *Van. Nos. Eng. Mag.*, Oct., 1885.

Thermodynamics, continued.

Second Law of. By Prof. J. B. Webb. Address before the Am. Assoc. Adv. of Sci., 1885. Shows that Rankine's statement of this law is correct and intelligible. Issued as a separate pamphlet by the author, Hoboken, N. J.

See *Heat, Mechanical Applications of.*

Thermometers.

An engraving showing a good method of applying the thermometer for measuring the temperature of water and steam. *Boston Jour. Commerce*, April 4, 1885.

Calibration of. By C. C. Brown. Gives Hansen's method and its application to a practical example. *Van Nos. Eng. Mag.*, Vol. XXIX., p. 1.

See *Heat, Instruments for Measuring Radiant.*

Thomson's *Electric Welding Apparatus.* See *Electric Welding.*

Tidal Drains. *The Present and Proposed Tidal Drains of Charleston, S. C.* A brief account of the working and efficiency of the old system, and modifications recommended. *Eng. News*, Oct. 25, 1890, pp. 364-5.

Tidal Wave *and the "Mascaret."* An article containing formulas relating to wave motion. *Van Nos. Eng. Mag.*, Sept., 1884.

Tide Gauge.

A Syphon Tide Gauge for Open Sea Coast. Illustrated description of a successful form of Syphon Tide Gauge. *U. S. C. & G. Survey*, Bulletin No. 12, March 30, 1889.

Automatic, invented by Mr. Roberts, of the "Nautical Almanack" Office, London. It is self-registering and involves several new features. Illustrated. *Sci. Am. Sup.*, Dec. 12, 1885.

Tide Marshes. *The Reclamation of.* A memoir of 250 pp. to the Dept. of Agr. by D. M. Nesbit. Gives areas, methods and cost of reclamation, etc. *Dept. of Agr. Misc. Special Rep. No. 7*, 1885.

Tides.

The Range of Tide in Rivers and Estuaries. By E. A. Gieseler. An investigation, from data from 11 different rivers, of the laws governing the change in range of tide. *Jour. Frank. Inst.*, Aug., 1891, pp. 101-11. Abstract, *Eng. News*, Aug. 29, 1891, pp. 181-6.

Theory of and Prediction of Heights. A mathematical treatment, founded on the investigation of Newton, Wheewell, Lentz and Lubbock. A very satisfactory paper by E. A. Gieseler, Sup. 4th Lighthouse Dist. *Jour. Frank. Inst.*, March and Oct., 1885.

And Tidal Scour. By Joseph Boult before the Liverpool Eng. Soc. Gives results obtained at various points by tidal scour. *Van Nos. Eng. Mag.*, Vol. XXVIII., p. 148.

Ties, *Railroad.* See *Railroad Ties.*

Timber.

Adaptability of the Various Kinds of Timber in Arkansas for Railway Purposes. A paper by C. E. Buchanan. Gives description and main characteristics of a great variety of Arkansas timber, and suggests the uses to which each variety is adapted; pp. 17. *Trans. Arkansas Soc. Eng., Arch't and Surv's*, Vol. II., p. 43, 1888.

Beechwood, the Use of, in Building. *Deutsche Bauzeitung*, 1885. Pp. 390, 397, 406, 416, 420, 430.

Consumption of Wood by Railroads and Practicable Economy in Its Use. A valuable work by B. E. Fernow, Chief of Forestry Div. Pub. Doc., U. S. Dept. of Agriculture; Bulletin No. 4. Large 8vo., pp. 350, 30 plates. Abstract of tables, etc., in *Ry. Age*, August 23, 30, 1890, pp. 593, 603-4. *Ry. Rev.*, August 23, 1890, *et seq.* *Eng. News*, August 30, 1890, pp. 187-8. *Am. Mfr.*, August 29, 1890, p. 7.

Timber, continued.

The Conversion of, in the U. S., by circular and band saws. A very good description. Illustrated. *Proc. Inst. C. E.*, Vol. XC., pp. 105-130.

Creosoting. See *Timber Preservation.*

Its Decay Induced by Fungi. A lecture before the New York Academy of Sciences. By P. H. Dudley. The laws of this growth treated and methods of prevention. *Sci. Am. Sup.*, April 24, 1885.

Its Diseases. A paper by Marshall Ward giving a full discussion of the growth of timber and its diseases, with an investigation of the causes of decay. *Sci. Am. Sup.*, March 17, *et seq.*, 1888.

Dry Rot in.

Treats of the cause and remedies for dry rot. *Van Nos. Eng. Mag.*, Vol. XIII., p. 137.

Recent Investigations Concerning the Dry Rot Fungus. Abstracted from the German. Some very important conclusions as to the laws of its growth, which will be found of value to those interested in the decay of timber. *Proc. Inst. C. E.*, Vol. LXXXVI., p. 381.

By W. A. Bidlake. A report presented to the Science Standing Committee of the Royal Institute of British Architects. It is based upon 17 cases of dry rot. *R. I. B. A. Jour.*, June 14, 1888; *Eng. & Build. Rec.*, July 14, 1888.

Economy in the Consumption of Timber for Railway Purposes. A paper by E. E. R. Tratman, presented at the American Forestry Congress, Philadelphia, Pa., Oct. 17, 1889. *Eng. News*, Nov. 2, 1889, Vol. XXII., p. 410; *Ry. Rev.*, Nov. 9, 1889, Vol. XXIX., p. 420.

Expansion of. See *Wood.*

In Marine Works. Discussion as to the use of, durability, protection from the Teredo, etc. A cut of this worm is given with specimen of its work. By C. Graham Smith in Lon. *Engineer.* *Eng. News*, April 5, 19, 1879.

Pine. By C. G. Smith. Gives brief description of the various pines used in construction and gives results of a number of tests for strength. *Van Nos. Eng. Mag.*, Vol. XIII., p. 443.

On Rendering it Non-Inflammable. A synopsis of the literature on the subject, with recommendations by the authors. From the *Building News* in *Van Nos. Eng. Mag.*, for July, 1885; also *Am. Arch.*, May 30, *et seq.*

Practical Notes on the Seasoning of Building. *Van Nos. Eng. Mag.*, Vol. XXVIII., p. 180.

Preservation of. See *Timber Preservation.*

Report of Experiments in Wood Seasoning. Made by G. H. Ellis for C., B. & Q. R. R. Co., giving "the fluctuations of moisture in wood during seasoning," both by tables and diagrams. *Preliminary report*, Forestry Division, Dept. of Agriculture, Bulletin No. 3, Washington, 1889.

Strength of Round. Results of elaborate experiments made by the department of postal telegraphy, England, on full-sized poles. The conditions of actual practice duplicated in the experiments. Formulæ and tables given. Probably the first experiments ever made on round natural timbers of this size. Lon. *Engineer*, Oct. 31, 1885.

Strength and Stiffness of Large Spruce Beams. An account of experiments at the Mass. Inst. of Tech., by Prof. Lanza. *Trans. A. S. M. E.*, Vol. IV., p. 119.

Tests of. See *Timber Tests.*

Transverse Strength of Southern and White Pine. By T. E. Kidder. Gives results of experiments on the strength of these timbers. *Van Nos. Eng. Mag.*, Vol. XXII., p. 166.

See *Lumber. Wood. Woods.*

Timber Preservation.

An abstract of a paper before the Philosophical Society of Glasgow. Explains the methods used for the preservation of timber from decay and destruction by worms, etc. *Van Nos. Eng. Mag.*, Vol. I., p. 406.

A lecture by Octave Chanute, before the students of the Rensselaer Polytechnic Institute. Nearly full report in *R. R. Gaz.*, Dec. 12, 1890, pp. 853-4. Abstract, *Eng. News*, Dec. 13, 1890, p. 528.

A short review of the subject. *Eng. & Build. Rec.*, Dec. 20, 1890, pp. 45-6.

By Herman Haupt. Gives results of his investigation into a number of processes. *Van Nos. Eng. Mag.*, Vol. VI., p. 181.

Gives details of experiments made to find the value of tar as a preservative. *Van Nos. Eng. Mag.*, Vol. VIII., p. 305.

Some interesting data concerning the evaporation from creosoted timber. *Eng. News*, Feb. 15, 1890, Vol. XXIII., p. 159. *Elec. World*, March 8, 1890.

Antiseptic Treatment. By Samuel B. Boulton, Assoc. Inst. C. E. A valuable paper by one who has had 34 years' experience in the business. *Trans. Inst. C. E.*, Vol. LXXVIII., 115 pages. Reprinted in *Van Nos. Eng. Mag.*, July, 1885; also discussion of above. *Van Nos. Eng. Mag.*, August, 1884.

As Related to Their Destructive Fungi. By Prof. P. H. Dudley, in *Pop. Sci. Mon.*, August, 1886.

Creosoting.

A paper by an English Engineer. *R. R. Gaz.*, Dec. 26, 1890, pp. 893-8, 9.

Apparatus for Treating R. R. Ties, designed to be portable. Used in France. Fully illustrated by 12 cuts. *Sci. Am. Sup.*, Nov. 21, 1885.

A valuable short paper, being an abstract of a report by Dr. C. M. Tidy to the Directors of the Gas Light and Coke Co., London. Its preserving qualities analyzed. *Van Nos. Eng. Mag.*, Dec., 1881.

The practice and results of the Carolina Oil and Creosote Co., of Wilmington, N. C. *Eng. News*, Jan. 16, 1886.

Creosoting. Works at Fernandina, Fla. *R. R. Gaz.*, Nov. 16, 1888.

In New Zealand. By Wm. Sharp, before the Institute of Civil Engineers. Gives description of the first plant erected in Australasian colonies for creosoting timber; also gives specimens of 50 creosoted sleepers. *Proc. Inst. C. E.*, Vol. XCIII., pp. 408-420.

Report on Experiments with the Seely, Bethell and Hayford processes of creosoting, with descriptions of the methods, by Gen. Q. A. Gilmore. Issued as a separate pamphlet of 20 pp. by the *Engr. Dept., U. S. A.*

Effect of Preservative Agents on Mine Timber. Results of experiments on various timbers with various agents. From *Comptes Rendus Mensuels de la Société de l'Industrie Minérale*, Nov., 1890. *E. & M. Jour.*, May 30, 1891, p. 633.

Report of the Am. Soc. C. E., with discussion. This is the work of several years on the part of the committee, Mr. O. Chanute being chairman. It is the most important contribution to this subject ever made in this country. It fills three numbers of the *Transactions*. There are twenty appendices by other engineers; the whole making a discussion of great value. *Trans. A. S. C. E.*, Vol. XIV., pp. 247-398. Abstract in *Eng. News*, July 11, 1885.

Result of tests on wooden discs subjected to various treatments. From *Comptes Rendus*. Also results of the application of the Zinc-Tannin process to railroad ties, by O. Chanute. *Eng. Record*, Nov. 21, 1891, p. 400.

With Salts of Copper. By M. Rottier. Gives details of experiments with strips of wood treated with copper salts. *Van Nos. Eng. Mag.*, Vol. XVI., p. 35.

See *Creosote*.

The Zinc Tannin Process of Tie Preservation. An examination of ties after being in service for 9 or 10 years, shows very favorable results. *Eng. News*, Aug. 1, 1891, p. 93.

Timber Tests. *Experiments on Tension Joints in Timber Work.* By Professor B. Fletcher. Describes tests on different forms of joints giving results. Illus. *Eng. News*, Jan. 18, Feb. 1, 8, 1879.

The Government Timber Tests. A description of the various tests being made, in the exhaustive series recently begun at the Washington University Testing Laboratory, under the auspices of the Divison of Forestry of the U. S. Dept. of Agr. By Prof. J. B. Johnson. *Eng. News*, Aug. 8, 1891. pp. 115-6.

The Government Timber Tests. Detailed description of the tests, by Prof. J. B. Johnson, with reprint of a circular of information issued by the Dept. of Agr. Photographs of the broken specimens are shown. *St. Louis Lumberman*, Oct. 1891, pp. 32-4. Above circular reprinted in *Eng. News*, October 10, 1891. pp. 326-7.

Government Timber Tests. Brief illustrated description of cross breaking machines used, and photographs of broken specimens of the various kinds of tests. *Eng. News*, Nov. 28, 1891, pp. 506-7. Same as above with description and discussion of the work by Prof. J. B. Johnson. *R. R. Gaz.*, December 5, 1891, pp 855-7.

Of Wood Treatments. Gives details of tests to ascertain the relative life and value of wood treated with various antiseptics and untreated timber in resisting the ravages of the teredo. *Eng. News*, Sept. 1, 1888.

See *Bridge Members*.

Time. See *Determination of. Meridian. Standard Time.*

Time, Latitude and Azimuth. *Determination of.* By L. Wagoner. Gives method of determining the above elements by means of equal altitude and horizontal angle between two stars. *Van Nos. Eng. Mag.*, Vol. XVIII., p. 243.

Time Reckoning. *A Plea for a Universal or World Time.* By W. H. M. Christie Astronomer Royal, Eng. Proposes to use Greenwich time as the local time for the whole world. From *Nature*, in *Van Nos. Eng. Mag.*, May, 1886.

Time System. *Western Union.* See *Western Union.*

Tin Scrap, *Making Nails from.* A paper by Oberlin Smith, before the American Inst. of Mining Engineers, describing a new process for the utilization of tin scrap by forming it into nails by pressure. *Am. Mach.*, June 6, 1889.

Tires. See *Car Wheels, Tires. Locomotive Tires. Wheels.*

Tools. See *Lathe Tools, Milling Cutters.*

Tools, Cast Iron *for Cutting Metals.* A paper by Oberlin Smith, describing their use, which seems to possess considerable advantage over the use of forged tools except for furnishing cuts. *Trans. A. I. M. E.*, 1890, pp. 5. Abstract, *Ry. Rev.*, April 11, 1891, p. 235.

Topographical *Models, Their Construction and Use.* By A. E. Lehman before the Halifax meeting of the *Am. Inst. of Min. Engrs.*, 1885. Describes many ways of making reliefs models from contour maps and the uses to which they may be put.

Topography.

See *New Orleans R. R. Location. Surveying.*

Teredo Navalis, *or Ship Worm.* By G. W. R. Bayley. Gives the experience with the teredo navalis on the bridge piling and foundation of a railroad from New Orleans to Mobile. *Trans. A. S. C. E.*, Vol. III., pp. 155-171.

Tornado Study. A lecture by Lieut. J. P. Finley, before the Frank. Inst. An elementary analysis and classification of wind-storms as to locality, cause, duration, etc. *Jour. Frank. Inst.* April, 1886.

Torpedo Boats.

"*Ariette.*" A description of the Spanish boat "Ariette," which recently made 26 knots per hour on the trial trips. It is 147 feet 6 inches long and 14 feet 6 inches beam, of steel, with 12 compartments. A two-page plate shows plan, elevation and sections. Lon. *Eng.*, July 15, 1887.

Torpedo Boats, continued.

Details of the trial trip of a torpedo boat 140 feet long 97 tons displacement, fitted with twin screws driven by compound engines The mean of 6 trial trips gave a speed of 26 miles per hour, the greatest speed being at the rate of 27.4 miles per hour. *Sci. Am. Sup.*, May 28, 1887.

"*Fatum*." A brief description of the station torpedo boat "Fatum," with two-page plate showing plans, elevations and cross-sections. Lon. *Eng.*, Jan. 20, 1888.

Griswold's. Description of the electrical steering apparatus, and an account of its performance at Providence, R. I., recently. *Electrician & Elec. Eng.*, Jan., 1887.

Griswold's Torpedo-boat. An account of the remarkable performance of a model boat, with drawings of boat. The vessel is steered electrically without any connection with the shore. *Electrician & Elec. Eng.*, January, 1887. See *Torpedo Boat*.

Some Experiments to Test the Resistance of. By A. F. Yarrow, M. I. N. A. A paper giving results of experiments with a torpedo-boat of 40 tons displacement, at speeds ranging from 8 to 22 knots, showing indicated powers at the various speeds, and resistances at the various speeds, when propelled by the engines and when towed. With discussion by Mr. R. E. Froude, Sir Edw. Reed, Mr. W. H. White and others. Reprinted from *Trans. Inst. Naval Architects* by the Bureau of Navigation Navy Dept. *Naval Professional Papers*, No. 16.

Structural Strength of. By Lieut. F. J. Drake, U. S. N An interesting discussion of the stresses in the various members of a torpedo boat subjected to still-water pressure, also when among waves. Illustrated by several stress diagrams. *Sci. Am. Sup.*, July 6, 1889.

Torpedoes.

A series of popular articles on the Whitehead torpedo. Lon. *Eng.*, Feb. 6, 1885, *et seq.*

A brief description of the various kinds used in naval defenses and elsewhere, with a full account of the *Whitehead torpedo*. Lon. *Engineer*, Feb. 6, 1885, *et seq.*

Fish. A full account of the Fish or Whitehead torpedo, with description of its mechanism. *Sci. Am. Sup.*, May 7, 1887.

Howell's. Gives full comparison of the Howell with the Whitehead and Brennan torpedo. Shows the Howell to be the best. Lon. *Eng.*, Jan. 20, 1888.

Howell Automobile Torpedo. A complete description. Folding plates. *Proc. U. S. N. I.*, Vol. XVI., No. 3, 1890, pp. 353-60.

The Use of in War. A lecture by Commander E. P. Gallwey, before the Royal United Service Inst. Gives the progress of this science to date. Fully illustrated. *Sci. Am. Sup.*, April 10, 1886, *et seq.*

Victoria Torpedo. Description. Details illustrated. Lon. *Engineer*, Aug. 29, 1890, pp. 246-7.

The Whitehead Torpedo. Detailed description with full illustrations. Lon. *Engineer*, Aug. 7, 1891, pp. 111-12, *et seq.*

Torsional Strength *of Iron and Steel*. See *Iron and Steel*.

Tow-Boat *Operated by the Current to Tow Vessels up Stream*. *Annales des P. & C.*, Nov., 1887.

Towers.

Proposed Tower for the World's Columbian Exposition. Diagrams with description of a proposed tower, 1120 feet high. Designed by George S. Morison. *Eng. News*, Dec. 5, 1891, pp. 528-9. *R. R. Gaz.*, Dec. 5, 1891, pp. 881-3.

Towers, continued.

Of Braced Iron. An analytical method for determining the strains in a great octagonal iron tower with dome. Diagrams. *Eng. News*, June 21, 1890, pp. 583-4.

In London. A few of the designs offered for the prize are described and represented in twenty-one figures. Reprinted from *Industries*. *Sci. Am. Sup.*, No. 753, June 7, 1890, pp. 12,036-8. For a criticism of the designs see *Am. Arch.*, June 14, 1890, pp. 161-3.

Design for the London Tower. Illustrated description and estimate for a tower 1200 feet high, designed by Mr. Edward S. Shaw of Boston. *Eng. & Build. Rec.*, Aug. 23, 1890, p. 183.

See *Electric Lighting*, *Tower System*. *Viaduct*.

Tower, Eiffel.

Three hundred metres high, of iron. Projected for the Paris exposition of 1889, for electric lighting, etc. Discussion of materials, cost, etc., with plates of details. *Mem. de la Soc. des Ing. Civils.* Paris, March, 1885.

A very complete article, showing details of foundations and superstructure, together with methods of erection, etc. Fully illustrated. Lon. *Eng.*, May 3, 1889, Vol. XLVII., p. 415, *et seq.*

Illustrated description. Details of Construction. Lon. *Engineer*, Jan. 4, 1889; Jan. 11, 1889; Jan. 18, 1889.

Illustrated description of the 300 meter tower now being erected at Paris. *Eng. & Build. Rec.*, Dec. 15, 1888; (Wind Pressures), Dec. 22, 1888; (Elevators and Cost), Dec. 29, 1888.

From Foundation to Lantern. An interesting descriptive paper read by Mr. Ambrose Swasey before the C. E. Club of Cleveland. Illustrated. *Jour. Assn. Eng. Soc.*, Aug., 1890, pp. 411-26.

The Lifts in. Illustrated descriptions of the hydraulic lifts in the Eiffel tower. Lon. *Engineer*, July 19, 1889. Lon. *Eng.*, July 5 and 12, 1889.

Towing. See *Cable Towing*, *Canal Traction*.

Track Work. See *Railroad Track*.

Traction.

Fell System of Traction on Rimutaka Incline, New Zealand. Brief paper by J. P. Maxwell, M. Inst. C. E. *Proc. Inst. C. E.*, Vol. XCVI., p. 170.

Resistance to Traction on Road. Paper read before Phila. Engrs. Club, by Rudolph Hering, March 18, 1882. Table of actual resistances to traction for numerous road surfaces and velocities is given with authority for figures. *Eng. & Build. Rec.*, Jan. 25, 1890, p. 131.

Traction Engines. See *Engines*, *Steam Traction*.

Tractive Force. See *Locomotives*, *Tractive Force*.

Trade Winds, *Cause of*. By Franz A. Velschow. *Trans. A. S. C. E.*, Vol. XXIII., 1890. Paper No. 446, pp. 102-10.

Train Lighting. See *Car Lighting*.

Train Resistance.

Abstract of a paper in *Annales des Mines* by —— Desdouits. Gives results of a long series of experimental investigations of the resistance of locomotives and trains on railways. *R. R. & Eng. Jour.*, February, 1887.

A four-page abstract of some valuable experiments made in France. The resistance determined and analyzed into its component parts. *Proc. Inst. C. E.*, Vol. LXXXVI., p. 382.

By M. Desdouits. Gives the results of a long series of experimental investigation of the resistance of locomotives and trains and railways. Abstracted from *Annales des Mines*, Vol. VIII., p. 481, in *R. R. & Eng. Jour.*, Feb., 1887.

Train Resistance, continued.

By Wm. P. Shinn. Gives data obtained while experimenting with the dynagraph on the L. S. & M. S. R. R. *Trans. A. S. C. E.*, Vol. V., p. 141.

In Cold Weather. Correspondence relative to increased frictional resistance to trains at high speed, as the temperature decreases. Accounting for the marked decrease in hauling capacity at high speeds in cold weather. *Eng. News.* Dec. 1, 1888.

Experiments on. By A. M. Wellington, under the direction of Chas. Paine. Fully illustrated. A valuable series of experiments. *Trans. A. S. C. E.*, Vol. VIII., (1879), p. 21-54.

Experiments on. By —— Desdouits. *Revue Generale des Chemins de Fer*, May, 1890, p. 271. Abstract in *Proc. Inst. C. E.*, Vol. CII., 1890, pp. 380-2.

Experiments of the State Railroad of Saxony on the resistance of cars on normal gauge tracks. By F. Hoffmann. *Organ fur die Fortschritte des Eisenbahnwesens*, 1885, pp. 174-178, 202-208.

And Journal Friction. An address before the March meeting of the Western Railway Club by G. W. Rhodes. *Mast. Mech.*, April, 1888.

And Locomotives. An abstract of a paper by Prof. A. Frank; translated from the the German for the Inst. of Civ. Engrs. The most important formulas, tables and conclusions given. Theoretical treatment based on a great many carefully made experiments. Probably the most reliable data of the kind yet published. *Ry. Rev.*, Chicago, March 7, 1885; also *Van. Nos. Eng. Mag.*, March, 1885.

See also *Friction.*

And the Propelling Force of Locomotives. Two useful diagrams from Godwin's *Railroad Engineers Field Book.* A peculiar arrangement gives as an area the total distance traveled while accelerating the speed from one velocity to another. *R. R. Gaz.*, Oct. 24, 1890, pp. 731-2.

Summary of a series of tests recently made on the C., B. & Q. Ry. to determine the resistance of trains at high speeds. *Eng. News*, Nov. 24, 1888.

See *Car Wheels, Conical. Friction and Train Resistance. Frictional Resistance. Locomotives, Efficiency of.*

Tramways. See *Street Railways. Electric Railroads.*

Transfer Steamer "*Solano.*" See *Ferry Boat.*

Transfer Tables,

A cheap and satisfactory one described by Wm P. Shinn. Illustrated. *Trans. A. S. C. E.*, Vol. V., (1876), p. 377.

Electric. Illustrated description of the Sprague electric transfer table at Altoona, Pa., for Pennsylvania Railroad shops. *Mast. Mech.*, March, 1889.

Electric Transfer Table at Fitchburg, Mass. Description of transfer table, 70 ft. by 10 ft., designed to accomodate some 48 different trucks at the Fitchburg Railroad Company's shop. Illustrated. *R. R. Gaz.*, Dec. 6, 1889, Vol. XXI., p. 809.

On the Pennsylvania R. R. Illustrated description of two of the most complete and ingenious transfer tables in the country, one driven by wire rope and the other by electricity. *R. R. Gaz.*, March 22, 1889.

Electric Transfer Table. See *Electric Transfer Table.*

Transformers, *Electric.* See *Electric Current, Distribution.*

Transit Instrument. A discussion of the sources of error in its use, and the equations showing relations of errors of adjustment to the resulting errors in the work. By John Eisenmann. *Jour. Assn. Eng. Soc.*, Vol. IV., p. 196.

Transition Curves. See *Railroad Curves.*

Transmission *of Power.* See *Electric Power Transmission. Power Transmission.*

Transportation.

Cheap vs. Rapid Transit and Delivery. By Martin Coryell. A comparison of water with rail transportation. *Trans. A. S. C. E.*, Vol. IV., (1880), pp. 401-6. See *Railroad.*

Its Early History in Mass. By George L. Vose. Being an account of the evolution of the carrying trade, from the stage coach of the last century to the present time, including the construction and operation of canals. *Jour. Assn. Eng. Soc.*, Vol. IV., p. 53.

Transportation Facilities. Some historical notes going to show that the earliest canals, turnpikes, railroads and locomotives in this country were constructed in Pennsylvania. *Jour. Frank. Inst.*, Aug., 1885.

Transportation in Northern Ohio. An historical account of its development, including the building of canals and railroads. By J. H. Sargeant. *Jour. Assn. Eng. Soc.*, Vol. V., p. 138.

Trap-Syphonage. An extended series of experiments at the Museum of Hygiene of the U. S. Navy. Reported to the American Institute of Architects. Of great value in determining the conditions under which syphoning takes place. Illustrated. *San. Eng.*, Dec. 11, 1886.

Trees. *Transplanting those of Large Size.* Report of such work successfully performed at Washington, on trees as large as one foot in diameter. *Am. Arch. & Build. News*, April 11, 1885.

Trestles. See *Bridges, Trestles.*

Trials *of Speed.* See *Speed Trials.*

Triangulation.

Field Work. A very elaborate article giving an account of all the different parts of the field work of a general scheme of triangulation, including the measurement of bases. Cuts of observation stations U. S. C. and G. Survey Report, 1882. See also article by G. Y. Wisner in *Proc. A. S. C. E.*, Vol. XII., p. 267, July, 1883. See also *Lake Survey, Geodetic Signals. Observing Tripods.*

The field operations of the primary triangulation of the Prussian survey. By Erfurth. General description, outfit, signals. *Zeitschrift f. Vermessungswesen*, 1887, pp. 377-83, 421-37.

Stations, Method of Marking on the New York State Survey. Glazed pots used for the underground marks. *New York State Survey Report*, 1880 and 1881.

Observing Tripods and Scaffolds, for Triangulation Work. A practical description, with many cuts, of the exact methods employed on the *U. S. C. and G. Survey. Report*, 1882.

Triangulation of the Forth Bridge. A detailed description of the work by R. E. Middleton. Lon. *Engineer*; *Eng. News*, Nov. 20, 1886; also *R. R. Gaz.*, Dec. 17, 1886.

Trip Hammers. See *Hammers.*

Trucks. See *Locomotive Trucks.*

Trusses.

Apparatus for Illustrating the Action of Loads upon. By R. Fletcher. Illustrated *Van Nos. Eng. Mag.*, Vol. XX., p. 161.

Componential Trusses for Traveling Crane. This is a "through" bridge carrying the traveler on the lower chord, and to give sufficient clearance the main truss is inclined outward. This necessitates a horizontal truss system; and also a vertical system to carry dead load. Each truss has thus three chords and three web systems. Description, table of stresses and illustrations, by Henry B. Seaman. *Trans. A. S. C. E.*, Vol. XXIII., December, 1890, paper No. 456, pp. 277-84.

Formulas for Strains in, and their Practical Application. By S. H. Shreve. Derives practical formulas for different forms of trusses and makes a practical application of each to a truss. *Van Nos. Eng. Mag.*, Vol. III., p. 193.

Trusses, continued.

Wooden and Iron. Chap. XII., of Safe Building, by Louis DeCoppet Berg. *Am. Arch.*, Dec. 6, 1890, pp. 152–4.

With Superfluous Members. By Wm. Cain. Gives M. Levy's theorem concerning superfluous pieces in a simple and elementary manner, with examples in detail. *Van. Nos. Eng. Mag.*, Vol. XXVII., p. 314.

See *Framed Structures. Howe Truss.*

Tubbing. By Ralph Moore. Gives a method of tubbing for stopping back feeders of water met in sinking shafts and describes method of putting it in. *Van Nos. Eng. Mag.*, Vol. XVIII., p. 348.

Tubes,

Mannesmann Process. See *Pipes.*

Methods of Welding. See *Welding Metals.*

Rolling Seamless Tubes from Solid Ingots. By Frederick Siemens before the Bath Meeting of the Brit. Assoc. for the Adv. of Sci. Describes the Mannesmann process of rolling seamless steel tubes from a solid ingot. *Am. Eng.*, Oct. 10, 1888.

Seamless and Brazed Brass and Copper Tubes. A description of the works of Randolph and Clowes and of the method of manufacture. Illus. *Iron Age*, April 2, 1891, pp. 630–4.

Seamless, from Solid Blanks. An account, with illustrations, of a recent method used in Germany. *R. R. Gaz.*, Oct. 14, 1881.

Seamless Tubes Made from Solid Blanks. A novel method of making a tube from a solid ingot by passing between rolls described and illustrated. *Am. Mach.*, Oct. 13, 1887; *Sci. Am. Sup.*, March 24, 1888; *R. R. Gaz.*, Oct. 17, 1887, and Lon. *Engineer*, Nov. 11, 1887.

Spirally Welded. Abstract of remarks by J. C. Bayles before the Franklin Institute, Phila., with cut of machine for making the tubes. *R. R. & Eng. Jour.*, March 1889; *Mechanics*, Feb., 1889.

Spirally Welded Steel Tubes. Paper by James C. Bayles, read before the Iron and Steel Inst., N. Y. *Sci. Am. Sup.*, No. 773, Nov. 8, 1890, pp. 12352–4. *Mech. World*, Nov. 15, 1890, p. 198, *et seq.*

Spirally-Welded. See *Pipe.*

See *Pipes.*

Tube Wells. A description of some of the so-called Abyssinian tube wells. *Van Nos. Eng. Mag.*, Vol. XXIII., p. 281.

Tunnel,

Boston Water-Works. See *Water-Works, Boston.*

Cable Railway Tunnel Under the Chicago River. Article describing this tunnel, giving general view, sections and detail illustrations of method of lining, track laying, etc., etc. *Eng. News*, Feb. 22, 1890, Vol. XXIII., p. 170; *Sci. Am. Sup.*, March 22, 1890, p. 11,834.

New Chicago Lake. Gives description of the new tunnel for the Chicago waterworks. *Eng. News*, Oct. 29, 1887; *San. Eng.*, Oct. 29, 1887.

New Chicago River Tunnel. Description with illustrations showing section, plan, profile, etc. *St. Ry. Rev.*, Nov., 1892, pp. 511–13.

Chicago, Alterations in the Washington Street Tunnel. A short paper describing this difficult work. Illustrated. Read by Mr. S. C. Colton at Peoria, Jan. 31, 1890. *Rep. of the 5th An. Meet. of the Ill. Soc. Eng. and Surv.* (50 cents). Pp. 75–8. *Eng. & Build. Rec.*, January 10, 1891, p. 32.

At Cochem, on the Mosel R. R. By Herr Lengeling. Complete description of this tunnel, 4204.75 meters long, built 1874–77, with plates and statistics of cost. *Zeitschr. fur Bauwesen*, 1880, pp. 26–35.

Tunnel, continued.

Construction of a Tunnel for Foot Passengers through a ridge in the most populous part of Stockholm, Sweden. Lightfoot's dry air refrigerator process employed. *Sci. Am. Sup.*, May 22, 1886.

Cassa Mountain. Gives brief description, with section, of the tunnel on the Columbus & Western R. R., near Birmingham, Ala. Length 2,134 ft.; width, 10 ft. with centre height of 21½ ft. *R. R. Gaz.*, Aug. 3, 1888.

Detroit River. By E. S. Chesbrough. Gives his experience with sand and water irruptions in the trial tunnels between Detroit and Windsor. *Trans. A. S. C. E.*, Vol. II., p. 233. *Van Nos. Eng. Mag.*, Vol. VII., p. 362.

Drainage Tunnel for City of Mexico. See *Drainage Tunnel.*

The Duluth, Minn., Tunnel. Abstract of report of Wm. Sooy Smith on this proposed tunnel under the canal across Minnesota Point, as to the materials to be excavated, best method to adopt, and utility of the tunnel. *Eng. Record*, Feb. 21, 1891, pp. 193-4.

Electric Light in. See *Electric Light.* See *Rapid Transit.*

English Channel. A paper by J. C. Hawkshaw before the British Assoc., giving a history and description of this project. Map and profiles. *Eng. News*, Oct. 7, 1882, pp. 347-9.

For Foot Passengers at Stockholm. Driven by freezing the ground by cold air blast at night, this discontinued through the day. Illustrated. Lon. *Engineer*, April 9, 1886; also *San. Eng.*, May 6, 1886.

Glasgow, Under the River Clyde at. Illustrated account of this proposed tunnel, giving the principal dimensions and general outline of proposed methods of construction. Lon. *Eng.*, Jan. 17, 1890, p. 70. Brief description, *Eng. & Build. Rec.*, March 29, 1890, p. 264.

Gould's, on the Pittsburg, Cincinnati and St. Louis Ry. Specifications for the tunnel, 3,300 ft. long. *Eng. News*, Nov. 13, 1880, pp. 383-7.

Greek Tunnel of the Sixth Century B. C. By A. C. Merriam. The tunnel at Samos, described by Herodotus, surveyed in 1882 by Ernst Fabricius. The tunnel is 1,000 metres long and generally 1.75 metres square, approached by conduits aggregating [illegible] metres in length; partly in tunnel, partly in open cutting. *Sch. of Mines Quar.*, March, 1885.

Hudson River.

By Wm. Sooy Smith. A description of the methods employed, compressed air being used. Illustrated by three large plates. *Trans. A. S. C. E.*, Vol. XI., (1882,) p. 314.

Brief account of the methods that have been used here by small advance pilot and later by Greathead shield. By S. D. V. Burr. Illus. *R. R. Gaz.*, March 14, 1890, Vol. XXII., p. 174.

Brief description of methods used in reopening this work in 1889, and the subsequent progress made. *R. R. Gaz.*, August 31, 1889, Vol. XXI., p. 569.

Accident in. Testimony and verdict of the coroners jury. *Eng. News*, Oct. 9, 1880, pp. 335-7. Statement of Chas. B. Brush, *ibid.*, Oct. 16, pp. 346-7.

Caisson sunk after the accident described and its strength discussed. *Eng. News*, Oct. 23, 1880, pp. 351-2. Also *ibid.*, Nov. 13, 1880, pp. 381-2.

Description of tunnel, air lock and condition of work at date. Paper by A. Spielmann and Chas. B. Brush before the Am. Soc. C. E. Reprint, *Eng. News*, Aug. 7, 1880, pp. 265-7.

Fire in the Hudson River. Account of a fire in the east caisson. *R. R. Gaz.*, April 11, 1890, p. 246.

The methods used in the early stages of this work described and illustrated by Messrs. Spielmann & Brush contractors. *Trans. A. S. C. E.*, Vol. IV., 1880, pp. 259-277.

Tunnel, continued.

New York Works. Illustrated description of caisson, bulkhead, etc. *Eng. News*, Sept. 2, 1882.

Shields. Illustrated description of shield as designed by Sir Benjamin Baker, C. E., and in use in the tunnel. *Eng. News*, March 15, 1890, Vol. XXIII., p. 257.

Shield. Brief description with three cuts showing details. Reprinted from *Industries. Eng. News*, July 19, 1890, p. 54.

Iron Centering for. Some illustrations of types of tunnel centres made of old rails. *R. R. Gaz.*, Oct. 14, 1887.

The Location of those on the Chimbote Railway, Peru. The road and tunnels lie in a mountain canon. Maps and profiles. O. F. Nichols. *Trans. A. S. C. E.*, Vol. IX., (1880), pp. 363-376.

The Marot. Abstract of a paper in *Ann. des Trav. Publiques.* Illustrated by about 20 cuts in *Sci. Am. Sup.*, Feb. 14, 1885.

The Mersey. A full account of this, the largest tunnel under water. Describes the drainage and pumping machinery, explosives, covered ways, and retaining walls, stations, hydraulic lifts, ventilation, engines, and fans, etc. Illustrated. By Francis Fox, one of the engineers. *Proc. Inst. C. E.*, Vol. LXXXVI., p. 40; also *San. Eng.*, Oct. 7, 1886, *et seq.*

Under the Mersey at Liverpool. Sections and details shown and described *Sci. Am. Sup.*, April 10, 1886.

Mont Cenis.

By Francis Kossuth. Treats of the condition of location, geology of the tunnel, workshops and water supply. *Van Nos. Eng. Mag.*, Vol. V., pp. 113, 115, etc.

Gives results of experiments with compressed air made by order of the Italian Government. *Van Nos. Eng. Mag.*, Vol. VI., p. 61.

By W. Pole. A brief description of the ventilation of the tunnel. *Van Nos. Eng. Mag.*, Vol. XIX., p. 309.

New Croton Aqueduct. Gives details of the meeting of the headings of the tunnel. *San. Eng.*, Nov. 26, 1887.

New York and Long Island. Profile and sketches of a proposed tunnel under the East River. *R. R. Gaz.*, May 9, 1890, p. 318. *Sci. Am. Sup.*, June 21, 1890, No. 751, p. 12,001.

On the S. Penn. R. R. Method of working, taken from official reports. A valuable series of articles, fully illustrated. *Eng. News*, Nov. 14, 1885, *et seq.*

Proposed Detroit River. Description of method proposed by H. H. Hall for constructing the tunnel. *Eng. News*, March 26, 1887.

The Proposed Irish Channel Tunnel. A paper by Sir Roger Lethbridge, discussing the feasibility, and setting forth the advantages of such a tunnel. *Jour. Soc. Arts*, Feb. 13, 1891, pp. 235-43.

Projected Simplon Tunnel Temperatures discussed with reference to the effect of high peaks. Also *Temperature Observations in the St. Gothard Tunnel. Eng. News*, Nov. 1, 1890, pp. 389-390.

Proposed Simplon. A review of the various projects for crossing the Alps, with special reference to the Simplon tunnel. Abstracted from *Le Genie Civil*, May 28, 1887, in *Eng. News*, July 2 and 9, 1887.

Proposed Simplon. Gives brief review of the history and merits of the proposed Simplon Tunnel between Switzerland and Italy. Illustrated with profiles and plan. *R. R. Gaz.*, Aug. 17, 1888.

Reconstruction and Enlargement of the Cork Run Tunnel on the Pittsburgh, Cin. & St. Louis Ry., by Max J. Becker. Illustrated. *Trans. A. S. C. E.* Vol. VI. (1877), p. 1.

Tunnel, continued.

Sections on the Canadian Pacific Railway. Illustrated with sections showing masonry and timbering, with dimensions. *Eng. News*, Nov. 24, 1888.

The Severn.

By T. A. Walker, before the British Association. Gives an interesting account of the driving of the tunnel and the difficulties met with. Lon. *Eng.*, Sept. 9, 1887.

Construction of the. Gives abstracts from an address of Mr. Charles, originator and engineer of the Severn Tunnel, at Bristol, England, upon the history of the work. *Eng. News*, Aug. 13, 1887.

Profile and sections, with an account of the work. *Eng. News*, April 17, 1886. also *R. R. Gaz.*, April 16, 1886.

Sewer. See *Sewers.*

The Stampede or Cascade Tunnel on the Northern Pacific R. R. Description with map and profile of tunnel, and account of its construction. Description also of a switch-back used as a temporary line. By C. W. Hobart. *Eng. News*, Oct. 3, 10, 1891. Progress profile, estimate sheets for November, and other notes, *ibid*, Oct. 10, 17, 1891.

The St. Clair River.

A brief account of the tunnel proposed under the St. Clair River at Port Huron. Total length, 5,280 feet, of which 2,310 feet will be under the river. Clear internal diameter, 20 feet. *R. R. Gaz.*, Feb. 25, 1887.

Gives method employed in the construction of the St. Clair tunnel on the Pennsylvania Schuylkill Valley Railroad, one and one-half miles west of Pottsville, Pa. Shows plan, numerous cross-sections, progress profile, etc. *Eng. News*, Aug. 20, 1887.

Series of papers giving complete details, and methods and machinery employed. *Eng. News*, Oct. 4, 1890, pp. 291-3, *et seq.*

Very interesting description of this work, profiles and cross-sections. The Beach hydraulic shield is fully described and illustrated. *Sci. Am.*, Aug. 2, 1890, pp. 79, 87 8. *Sci. Am. Sup.*, Aug. 23, 1890, No. 764, pp. 12199 201. *E. & M. Jour.*, Aug. 16, 1890, p. 188.

Very interesting description of this work. Details fully illustrated. *R. R. Gaz.*, Sept. 26, 1890, pp. 659-62.

St. Gothard.

By Prof. Colladon. Describes the source of motive power, air compressors, and engines put in motion by the compressed air. *Van Nos. Eng. Mag.*, Vol. XVII., p. 37.

The paper contains a record of the temperature, air movements and smoke in the tunnel during its construction and since its completion. *Schweizerische Bauzeitung* Vol. XV., 1890, p. 41. Abstract in *Proc. Inst. C. E.*, Vol. C., 1890, pp. 436-40.

Brief illustrated description of some of the engineering works on the line. *Eng. News*, Aug. 3, 1889, pp. 270-1.

Circular and Other Tunnels of. Gives description of the spiral tunnels and method of working. *Eng. News*, Nov. 10, 1883, pp. 543-4.

Under East River. See *Rapid Transit.*

Under Thames River. Description of methods and details of construction of the new iron tunnel now being built in London. *Sci. Am. Sup.*, June 4, 1887.

Vosburg, Construction of the. A pamphlet of 56 pages, with plates showing sections of the tunnel at various stages, systems of timbering, drilling, firing, etc., and letter-press giving details of construction, cost, etc. Address the author, J. Von Rosenburgh, 33 Broadway, New York. Abstract in Lon. *Eng.*, April 6, 1888.

Tunnel, continued.

The Vosburg on the Penn. & N. Y. Ry. This has been recently completed, is 1900 feet long, for double track. Rock drills driven by compressed air used. Maps, sections, etc., shown. *R. R. Gaz.*, Oct. 1 and 8, 1886.

The Washington Aqueduct Tunnel. Its Mismanagement. Poor Supervision and Failure under United States Army Engineers. Illustrated. *Am. Arch.*, Oct. 10, 1889.

Weehawken, of the West Shore Road. Illustrated description. *Eng. News*, June 27, 1885, pp. 197-8.

The West Point. By Wm. H. Searles. Read before Civil Engineers' Club of Cleveland, Aug. 14, 1888. *Jour. Assn. Eng. Soc.*, Feb., 1889; abstract in *R. R. Gaz.*, April 19, 1889.

Wickes Tunnel, on the Montana Central Railway, Cost of. A brief paragraph. *Eng. News*, Nov. 24, 1888.

The Wickes. Gives details of the proposed Wickes Tunnel on the Montana Central R. R. *Jour. Assn. Eng. Soc.*, Aug., 1887; *R. R. Gaz.*, Aug. 19, 1887, and *Eng. News*, Aug. 27, 1887.

Of the Worcester (Mass.) Sewerage System. Details of cost and methods given. By Wm. Whittaker. *Jour. Assn. Eng. Soc.*, Vol. V., p. 147.

Zumpango Tunnel Works. Description of this part of the great work for the drainage of the City of Mexico. Section showing heading, arrangement of car track, etc. *Eng. News*, Nov. 1, 1890, pp. 380-7.

Zig-Zag Tunnel; New York, Ontario & Western Railway. 1,617 ft. long. Brief description with illustrations of sections and portals. *Eng. News*, Jan. 10, 1891, pp. 26-7.

See Cable Railroads. Flushing Tunnel, Milwaukee.

Tunnel Alignment. By F. P. Stearns. Being an account of the methods used on the Dorchester Bay Tunnel, Boston. Illustrated. The results were very accurate. Shaft [illegible] feet deep, and tunnel 6,000 feet long. *Jour Assn. Eng. Soc.*, Vol. III., p. 190.

Nepean, N. S. W. By T. W. Keele before the Institution of Civil Engineers. Gives details of the alignment of the Nepean Tunnel for the Sidney water supply, New South Wales. Length of tunnel, 23,507 ft., 7½ ft. high, and 9½ ft. wide. Error in alignment for [illegible] ft. was ⅝ in., and in levels, ⅜ in. *Proc. Inst. C. E.*, Vol. XCII., pp. 359-67.

Tunnel Surveying. See *Surveying*.

Tunnel Ventilation.

A paper by N. W. Eayrs, describing the natural ventilation of the Hoosac Tunnel and the ventilation of the St. Louis Tunnel by mechanical means. Illustrated. *Trans. A. S. C. E.*, Vol. XXIII., 1890, paper No. 458, pp. 288-300. Abstracted in *Ry. Rev.*, Feb. 21, 1891, pp. 114-5. *Eng. Record*, February 28, 1891, pp. 209-11.

And Lighting of the Croton Aqueduct Tunnels. Abstracted from the report of R. W. Raymond and W. H. McQuail to the aqueduct commissioners. *San. Eng.*, Jan. 8, 1887.

Railroads, Tunnel. Editorial discussion of the efficiency of various methods. *R. R. Gaz.*, Oct. 30, 1891, pp. 767-8.

Results of observations on the Pfaffensprung spiral tunnel as to temperatures, air, etc. From Swiss journal *Eisenbahn*. *Eng. News*, Sept. 17, 1881, pp. 371-2.

Tunneling.

A description, with cuts showing details of an instrument for measuring cross sections in tunnels, in use at the Croton acqueduct. *San. Eng.*, June 11, 1887.

By B. H. Hall. Sketches progress in tunneling in United States, with details of a number of tunnels. *Report Conn. Assn. Civ. Engrs.*, 1887, p. 19.

Tunnelling, continued.

Bottom or Top Heading in. By C. L. Kalmbach. Discusses the use of top heading and advocates the use of bottom heading. Illustrated. *Eng. News*, March 14, 1888.

Caisson for Subaqueous Work. A new method of laying water pipe or building a subaqueous tunnel through loose material. Has been used in Australia. Illustrated. *R. R. Gaz.*, April 17, 1885, and *Am. Eng.* of same date.

In Colorado. A report of the Golden River Mining Co., giving detailed cost of a tunnel 3,000 feet long and 7x8 section. *Eng. News*, April 2, 1887.

On the Croton Aqueduct. Gives records of progress made in tunneling, also the arrangement of drill-hole. *R. R. Gaz.*, July 29; *Eng. News*, Aug. 6, 1887.

Drills of the Arlberg Tunnel. Methods of drilling and efficiency. Illustrated. A valuable paper by G. J. Specht. *Eng. News*, Jan. 19, 26, 1884.

Excavating of N. Y. Aqueduct. Abstract of a paper by J. P. Carson before the A. I. M. E. Describes the tunneling through some bad ground. *Eng. & Build. Rec.* Oct. 4, 1890, p. 280.

Haase's System of Sinking Shafts through Quicksand. Used in the Guenin Colliery, Germany. A series of wrought iron tubes were driven side by side to form a cribbing for the shaft. Illustrated. *E. & M. Jour.*, June 21, 1890, pp. 702-3. The King-Chaudron and other methods of shaft sinking described: *ibid.* pp. 717-32.

Hoisting Cage. Detailed drawings. *Eng. News*, March 31, 1883, p. 149.

Influence of Temperature in, through High Mountains. By Dr. T. M. Stapff. Treats of the highest temperatures at which men can work and at what depth this temperature is likely to be reached. Also gives general remarks on tunnel work from experience in the St. Gothard tunnel. *Van Nos. Eng. Mag.*, Vol. XXIV., p. 31.

Paper by A. E. Baldwin before the Inst. C. E. General notes, form of section, tunneling in London clay. *Eng. News*, Oct. 3, 27, 1877.

Under Heavy Pressures. An article by Archibald A. Schenck, discussing the stability of arches in soft ground and causes of failures, and advocating a stronger construction and better methods of packing. *R. R. Gaz.*, April 10, 1891, pp. 252-3.

And Modern Methods of Tunneling. By E. W. Moir, A. M. I. C. E. Examples given from numerous works. Well illustrated. *Eng. Mag.*, September, 1891, pp. 818-30.

The Best Method of Constructing Long. By G. Bredel. Contains notes on the construction of a number of European tunnels. Author believes base heading is preferable to crown heading. *Van Nos. Eng. Mag.*, Feb., 1884.

Under Rivers. A proposed method for constructing the new Thames River Tunnel. To be constructed in sections and lowered to place on a caisson, the excavations being made in the compressed air chamber. Illustrated. Lon. *Eng.*, Aug. 14, 1885.

See *Aqueduct, New Croton. Washington.*

Tunneling Plant.

Transandine Railway Tunneling Plant. Detailed description of motors, dynamos, drills, etc. Illus. Lon. *Eng.*, May 22, 1891, pp. 603-4.

The Transandine Railway Tunneling Plant. Maps and profile showing the location of the several tunnels and the arrangement of the installations. Power is furnished by turbines and transmitted electrically. Description of plant given. Lon. *Eng.*, Apr. 24, 1891, pp. 481-2. Abstract, *R. R. Gaz.*, May 8, 1891, p. 320.

Turbines. See *Steam Turbines. Water Motors.*

Turrets. See *Armored Turrets.*

Vacuum Pumps. See *Pumps, Vacuum.*

Valley of Mexico. *Drainage of.* See *Drainage.*

Valve Gear.

Comparative Merits of Valve Gear. Paper by John Essen, read before the Aberdeen Mechanical Society. *Mech. World,* Nov. 1, 1890, p. 184, *et seq.* Reprinted in *Am. Eng.*, Nov. 15, 1890, pp. 215-16, *et seq.*

Slide Valve Gears. A paper by Prof. A. Mackay, M. I. M. E., read before the Inst. of Engrs. and Shipbuilders. Illustrated. *Mech. World,* Jan. 10, 1891, pp. 17-18, *et seq.*

See *Locomotives, Valve Gear.*

Valves. *The Slide Valve.* Illustrated description and discussion of the various forms of slide valves. By Robert Grimshaw. *Mfr. & Builder,* Feb., 1891, pp. 30-1, *et seq.*

See *Slide Valves.*

Vanes. See *Jets of Water on Curved Vanes.*

Van Nostrand, D. An account of his life and works, in *Van Nos. Eng. Mag.*, Vol. XXXV., p. 441.

Varnishes, *Methods of Testing.* A paper read before the Master Car-Painters' Convention. *Ry. Rev.*, Nov. 15, 1884.

Ventilating Fans. See *Heating and Ventilation, U. S. Senate Chamber.*

Ventilation.

Abstract from a report on the heating and ventilation of the Chamber of Deputies, Paris, made for the information of a committee of House of Commons. *San. Eng.*, July 22, 1886.

An article treating the heat of the human body as an element to be regarded in arranging the ventilation and warming of buildings. *Van Nos. Eng. Mag.*, Vol. XXI., p. 374.

A paper by Capt. D. Galton on the maintenance of pure air in dwellings. *Van Nos. Eng. Mag.*, Vol. XXV., p. 344.

By W. N. Hanley. Treats of the pollution of air and ventilation of buildings. *Van Nos. Eng. Mag.* Vol. XIII., p. 109.

Of Audience Halls. By Robt. Briggs. An authoritative article on heating and ventilating large halls, covering 54 pp., with special index to topics. Illus. *Trans. A. S. C. E.*, Vol. X., p. 53-106.

Of Buildings. By W. F. Butler. A valuable paper. Gives principles which are applicable in nearly all cases. *Van Nos. Eng. Mag.*, Vol. IX., p. 315.

Of Buildings. By J. Neville Porter. Paper read before the Society of Archts., London, with discussion. The writer shows an acquaintance with the means employed in all parts of England. A first rate article on this subject. *Van Nos. Eng. Mag.*, Jan., 1885. See also *Heating.*

Of Buildings. Valuable pamphlet discussing this subject by Alfred R. Wolff, M. E., Consulting Engineer, New York City. Potter Building.

Of Buildings, The Quantity of Air Required in the Ventilation. A formula given based on the experiments of Parkes, Pettenkofer, De Chaumont and Angus Smith. *Jour. Assn. Eng. Soc.*, Vol. IV., p. 418.

By Chimneys and by Fans. Their relative economy. By Prof. W. P. Trowbridge, before the Am. Soc. Mech. Engrs. The fan shown to be much more economical in fuel for low chimneys or shafts. *Sch. Mines Quar.*, July, 1885.

Of Dye Houses. A method of preventing vapor in dye houses by properly arranged currents of hot air. *The Locomotive,* Dec., 1888.

Of Habitations, Particularly Schools and Public Buildings. Abstract of a lecture delivered by Wm. J. Baldwin, M. Am. Soc. of C. E., before Engineering Dept. of Brooklyn Institute. *Eng. & Build. Rec.*, August 31, 1889. Vol. XX., p. 190, *et seq.*

Ventilation, continued.

Hints on. By C. T. Wingate. A brief discussion of the ventilation of dwellings. *Building*, June 5, 1886.

How Much? A valuable article by J. L. Greenleaf, on the amount of fresh air required under various conditions. (*Van Nos. Eng. Mag.*, Nov., 1881.)

The Mechanics of. By George Rafter. A thoroughly practical essay treating of the laws of air in motion, and the application of those laws to the removal of foul air and the supply of fresh air to building. *Van Nos. Eng. Mag.*, Vol. XVII., pp. 354 and 401.

By Means of Flues. The air being artificially heated before entering the flue. A theoretical study of this problem, when the air which is to be passed out of the flue is heated by steam coils as it enters the flue. The object of the investigation is to find the area of heating surface required to pass a given quantity of air under given conditions. By Wm. P. Trowbridge. The system shown to be more economical than fans. *Trans. A. S. M. E.*, Vol. III., p. 67.

Relative Economy of ventilation by heated chimneys and by fans. A paper before the Am. Soc. of Mech. Eng. at Chicago. *San. Eng.*, July 29, 1886.

The Science of Ventilation as applied to the interior of buildings. By D. G. Hoey, with appendix by Dr. William Wallace and discussion by members. *Jour. Soc. Arts*, May 31, 1889, Vol. XXXVII., p. 607.

Of Sewers. See *Sewers.*

Of Stables. See *Stables.*

Of Theatres. By J. P. Seddons. A lecture delivered at the Parkes Museum of Hygiene. *Van Nos. Eng. Mag.*, April, 1886.

The Theory of. By Sur.-Gen. F. de Chaumont. An attempt to establish a positive basis for the calculation of the amount of fresh air required for an inhabited air space. *Van Nos. Eng. Mag.*, Vol. XIII., p. 332.

See *Damp*, *Heating and Ventilation*. *Plumbing and Ventilation Inspection in N. Y. Tunnel.*

Ventilators.

Centrifugal Ventilators. A paper by R. Van A. Norris, describing experiments on 25 ventilating fans. Gives diagrams of the fans and a few general deductions. *Trans. A. I. M. E.*, 1891, pp. 34.

Verniers. See *Sub-Scales.*

Vessels.

Resistance of. Report of the Committee of the British Association, treating of the resistance which ships offer to propulsion and their behavior in respect to rolling. *Van Nos. Eng. Mag.*, Vol. II., p. 276.

Track of. *Determination of the Actual.* Translation of a paper by Lieut. L. C. de Benko. Diagrams and formulas. *Proc. U. S. N. I.*, Vol. XVI., No. 3, 1890, pp. 431-6.

Viaduct.

Description of the Brick R. R. Viaduct over the river Esk, England. Greatest height, 120 feet; length, 915 feet. Foundations sunk to rock by open brick walls on drum curbs. Large tree trunks removed by divers. There are 13 circular arches of about 60 feet span. Fully illustrated. *Proc. Inst. C. E.*, Vol. LXXXVI, 303.

Cleveland. Illustrated description of the Superior street viaduct, consisting of 8 stone arches with plate girders and truss spans. *Eng. News*, Jan. 11, 1879, pp. 10-11.

Cruzine Viaduct, Southern Railway of France. Six arches of 82.5 feet span. Description and detail drawings. Lon. *Engineer*, June 5, 1891, pp. 441-4. From *Le Genie Civil.*

Viaduct, continued.

Erection of the Buffalo Face Viaduct, near Frankfort, Ky., on the Kentucky Midland Railroad. This viaduct consist of 30 feet towers, with 60 ft. plate girder spans, and in erection a traveler was used having a 94 ft. projecting arm trussed by rods over a central tower, the anchor arm being 51 ft. long. *Eng. & Build. Rec.*, Dec. 14, 1889, Vol. XXI, p. 22.

Erection of the Superstructure of the Cervena Viaduct. Six folding plates. *Zeitschrift des, Oester. Ing. u. Arch. V.*, Vienna. Vol. II., 1890, pp. 180-90.

Garabit. Illustrated description. An iron arch of 541 feet span. *Eng. News* Aug. 9, 1884, pp. 64-5. Further details, *ibid*, Aug. 30, 1884.

The Garabit. This long viaduct includes a two-hinged arch with a span of 540 feet and rise 166 feet. Details of this arch illustrated. *Eng. Record*, March 7, 1891, pp. 224-5.

High, History of a. Paper by S. D. Mason, read before the Civil Engineers Society of St. Paul. *Jour. Assn. Eng. Soc.*, April, 1889; *R. R. Gaz.*, April 19, 1889.

Iron Viaduct. See *Railroad Terminals*.

Kinzua Viaduct. Short paper by Thos. C. Clarke, C. E., read at the Cincinnati meeting of the A. S. M. E. *Trans. A. S. M. E.*, Vol. XI., 1890, pp. 961-74. *Eng. News*, July 5, 1890, p. 7.

Malleco Viaduct. Collipulli, Chili. This steel viaduct is about 1,100 ft. long and 300 ft. above the bed of the stream. Illustrated and dimensions given in *Sci. Am.*, Jan. 10, 1891, pp. 17-21.

Marble Creek Viaduct. A 770 ft. iron trestle 210 ft. high at the center. General plan, elevations, and several details show. *Eng. Record*, April 18, 1891, pp. 306-7.

Marent Gulch Viaduct. A paper by George S. Morison describing this structure, a viaduct 800 feet long and 200 feet in extreme height. Eleven folding plates giving full details of masonry, piers and iron work. *Trans. A. S. C. E.*, Vol. XXV., Sept. 1, 1891, pp. 305-12. Expansion joint and connections illustrated. *Eng. Record*, Dec. 5, 1891, p. 12.

Moldau Viaduct. Very complete and well illustrated account of the Moldau Viaduct at Cervena, in Bohemia. *Eng. News*, April 19, 1890, Vol. XXIII., p. 362.

The New Tay. Abstract from a paper by Mr. S. S. Kelsey before the Inst. of Mechanical Engineers, giving a description of the work of construction. *R. R. Gaz.*, Sept. 2, 1887.

New Tay. By S. S. Kelsey, before the Inst. of Mech. Engrs. Gives full description of the new Tay viaduct and some of the methods employed in its erection. Lon. *Engineer*, Sept. 2, 1887.

New Portage. A viaduct 820 ft. long and 234 ft. high, built of wrought-iron, in 1875. Geo. S. Morison. Illustrated. *Trans. A. S. C. E.*, Vol. V., (1876), p. 1-8

Proposed Viaur Viaduct, a railway line between Carmaux and Rodez, France, with inset plate. The structure is made up of two half girders, hinged at middle and at abutments with cantilever projections back of abutments, carrying independent shore spans. Total length of iron work 1,344.8 feet: central span, 820.0 feet, side spans, each, 262.4 feet; height of rail above bottom of valley, 380.1 feet. *Eng. News*, Nov. 30, 1889, Vol. XXII., p. 507.

Over the River Retiro, Brasil. Plan, elevation and details of a viaduct 347 feet long, composed of two stone arches 20 feet in diameter, and five intermediate iron arches 49 feet in diameter. Lon. *Eng.*, Feb. 18, 1887; also *Sci. Am. Sup.*, Feb. 26, 1887.

Viaduct, continued.

Sixteenth Street, Omaha. Specifications for a structure having a length of 1,500 feet, composed of four Howe trusses, ten trapezoidal trusses and pile bents. *Eng. News*, May 22, 1886.

Soulevre Viaduct. 1,200 ft. long, 200-ft spans. Detailed description of structure and method of erection. Illus. Condensed from *Le Genie Civil*. *R. R. & Eng. Jour.*, March, 1890; *Eng. Record*, Jan. 10 and 17, 1891; *Sci. Am.*, April 5, 1890, p. 11882. See also *Bridge Piers*.

Soulevre. See *Bridge Piers*.

Starrucca, N. Y., L. E. & W. R. R. Gives details of the Starrucca viaduct on the New York, Lake Erie & Western Railroad taken from an old letter book of its designer, Mr. J. W. Adams. *Eng. News*, Sept. 1, 1888.

St. Giustina Viaduct over the Noce Schluct, in Southern Tyrol. Description of this arch bridge over a chasm 460 ft. deep and only 200 ft. wide. The arch was erected as a pair of cantilever arms and anchored to heavy masses of masonry at ends. Illus. *Eng. News*, Feb. 1, 1890, Vol. XXIII., p. 102.

Subaqueous Viaduct. Illustrated description, with tables of weights of material of a proposed form of submerged viaduct or tubular bridge. *Eng. News*, Feb. 28, 1891, pp. 198-9.

Tapti Viaduct, India. This viaduct consists of thirty-four stone arches and five 140-feet girders. Folding plate giving elevation and details of arches. *Ind. Eng.*, Feb. 14, 1891, p. 130.

Tavistock and Shillamill Masonry Viaducts. Described and illustrated. Lon. *Engineer*, May 31, June 6, 1890, pp. 444-56.

Timber Viaduct at Manawatu, New Zealand. Illustrated general description of this wooden structure. Details of numerous joints are also given. *Eng. & Build. Rec.*, Jan. 4, 1890, Vol. XXI., p. 72.

Val Saint Leger, Paris Belt Railway. Brief description of this iron viaduct. It is 1000 ft. long and supported on masonry piers. Illus. *Eng. News*, Nov. 18, 1882, pp. 586-7.

The Verrugas, Erection of. By L. L. Buck. A viaduct 250 ft. high erected without false works, by means of a suspension cable. Illus. *Trans. A. S. C. E.*, Vol. V., (1876), p. 103-106. Discussion on p. 240.

Verrugas. Illustrations showing dimensions, and brief description of this new cantilever bridge. *Eng. News*, May 9, 1891, p. 441.

Viaducts.

The American Railway Viaduct.—Its Origin and Evolution. Abstract of paper by J. E. Greiner, read before the Am. Soc. C. E., describing and illustrating the successive forms leading up to the present standard form. *Eng. News*, June 6, 1891, pp. 536-8. *R. R. Gaz.*, June 5, 1891, pp. 388-9.

Approach and Terminus, Montreal. A brief illustrated description of the masonry approach and terminus of the Canadian Pacific Railroad at Montreal. Also shows the cost of maintaining an iron and masonry viaduct. *Eng. News*, March 3, 1888.

Antofagasta Railway, Bolivia, viaduct over the river Loa. A valuable and complete description of the viaduct, showing details of construction and results of compression tests of two of the main pillars when in a horizontal position in machine. Fully illustrated. Lon. *Engineer*, April 19, 1889, *et seq.*

Of Blaauw-Krantz, Cape Colony. A unique construction, being in reality a trussed arch of 280 feet span, the moment of inertia varying nearly as the bending moment coming upon the various sections of the structure. *Eng. News*, Jan. 24, 1885, and *Am. Eng.*, Jan. 23, 1885.

The Cleveland Central. Illustrated description of the new viaduct recently completed for the city of Cleveland. *Eng. News*, Dec. 22, 1888.

Viaducts, continued.

Across the Railroad Tracks. A paper by Carl Gaylor, read before the Engineers' Club of St. Louis, giving the results of experience, chiefly in regard to pier foundations, effect of smoke on iron work, and general stability of the structures. *Jour. Assn. Eng. Soc.*, July, 1881, pp. 341-4.

Resistance of, to Sudden Gusts of Wind. By Jules Gaudard. Gives a good discussion of the subject, citing numerous examples. *Van Nos. Eng. Mag.*, Vol. XXVII., p. 213.

See *Bridges. Cable Railroad. Viaduct.*

Vibration, *or the effect of Passing Trains on Bridge Masonry*. By J. L. Randolph, Theo. Cooper and Chas. E. Emery. *Trans. A. S. C. E.*, Vol. XII., (1883), p. 444.

Village Plats. *Legal Requirements in Michigan.* *Proc. Mich. Eng. Soc.*, 1885.

Warehouses. See *Building Construction. Docks.*

Warfare. See *Electric Position Indicator.*

Washington Monument.

Official Report by Col. Thos. L. Casey, giving history, description, foundations, weight, settlement, cost, plan, sections, etc. H. Rep. Misc. Doc. No. 8, 48th Cong., 2d Session, 1884. Also, discussion of its stability. *Eng. News*, March 14, 1885.

By C. E. Greene. The re-enforcing of the foundation and completion of the obelisk. A brief paper with sketches. *Science*, Feb. 20, 1885. Also a discussion of the stability of the foundation, by J. C. Goodridge. *Eng. News*, March 14, 1885.

Reinforcing Foundations of. Illustrated account of this work. *Eng. News*, Feb. 7, 1880, pp. 49-50.

Washouts.

Of R. R. Bridges and Culverts. By Chas. W. Folsom. A valuable study of the number and cause of such washouts and also of the proper size of water ways. *Jour. Assn. Eng. Soc.*, Vol. V., p. 324.

Their Prevention and Treatment. By W. B. Parsons. Shows what can be done to prevent washouts, what to save damage during their occurrence, and especially what is to be done after they have taken place. *R. R. Gaz.*, April 20, 1883.

Watches.

Magnetism in. By C. K. Giles, before the Alexandria Bay meeting of the American Railway Master Mechanics' Association. Gives the results of four years investigation of the effect of magnetism in watches. *Mast. Mech.*, Sept., 1888.

Paillard's Non-Magnetic Compensating Balances and Hair-Springs for. By Prof. Edwin J. Houston. Describes Paillard's non-magnetic watches, gives composition of alloys used, and results of careful tests of the watches. A distinct advance in the construction of accurate timepieces. *Jour. Frank. Inst.*, March, 1885.

Protection of Watches against Magnetism. Also, a convenient method of demagnetizing. Paper by Dr. P. Lange, read before the National Electric Light Association, Pittsburgh. *Elec. Eng.*, March, 1888; *Elec. World*, March 3, 1888.

Rating and Trials at the new observatory, with results attained in *uniformity* of rate in many hundred cases. Lon. *Engineer*, Jan. 1, 1886.

Water.

Aeration of. See *Water Supply*.

The absorbing and Transporting Power of. By T. Login. Treats of the mechanical properties and friction: also draws practical conclusions. *Van Nos. Eng. Mag.*, Vol. III., p. 163.

Action of Boston, on Service Pipes. By Wm. R. Nichols and L. K. Russell. Gives details of experiments made to discover the amount of zinc taken up from galvanized iron water pipes. Results show that the zinc coating is slowly but continuously dissolved. *Jour. Assn. Eng. Soc.*, Jan., 1888, pp. 11-14. *Eng. News*, Feb., 1888; *Sci. Am. Sup.*, March 3, 1888; *Eng. & Build. Rec.*, March 31, 1888. Abstracted *Proc. Inst. C. E.*, Vol. XCIII., p. 480.

Analysis. See *Water Analysis* below.

Charges for. See *Water Rates*.

Of Cities. A lecture before the Inst. Civ. Engrs., London. By Wm. Pole. The whole ground gone over in an article of 25 pp. Treated, of course, from an English point of view. *Van Nos. Eng. Mag.*, Feb., 1886.

Coefficient of Contraction. Note on the determination of the co-efficient of contraction of the fluid vein. By M. Ed. Collignon. Some considerations regarding the theoretical determination of the coefficient. *Annales des P. & C.*, 1885, March, pp. 295-302.

Collection, Purification and Examination of. Lectures delivered at the Royal Engineers' Institute by Col. Francis Bolton and Dr. P. F. Franklin. *Eng. News*, June 12, 1886.

Companies, Private. By H. F. Dunham, before the American Water-Works Association. Treats of the relation of private water companies to municipal authorities and the public. *San. Eng.*, July 30, 1887.

Commissioners' Report of New London, Conn., 1887. Contains reports on prevention of waste and the improvement of the character of the water as to taste and smell. 30 pp. Address *W. H. Richards, Engr.*

For Domestic Purposes. The Creque System of Defecating, Storing, Circulating and employing Hot and Cold Water in Houses. A paper the inventor read before the Society of Arts, Boston. Illustrated. *San. Eng.*, November 27, 1886.

Expansion of. By Alex. Morton. Derives a simple formula which gives results agreeing very nearly with the mean determination of a great many observers. *Van Nos. Eng. Mag.*, Vol. VII., p. 437.

Flow of. See *Hydraulics, Flow of Water*.

In "Great Ponds." To Whom Does it Belong? The question argued in favor of the public, as against the riparian owners below. *Jour. New England W.-W. Assn.*, Dec., 1886.

Hardness of. An abstract from an article by H. J. Hardy, giving method for determining the total hardness of water. *Eng. News*, June 11, 1887.

Hardness of Water. Paper by Prof. A. R. Leeds, presented and read before Am. Water Works Assn. This is a valuable paper, giving several complete analyses of water from various parts of United States. *Am. W. W. Assn.*, Louisville meeting, 1889, p. 120.

Impure Water and the Public Health. A popular article by Dr. Floyd Davis discussing this subject. *Eng. Mag.*, Dec. 1891, pp. 359-66.

Impurities in. By G. S. Johnson. Treats of the so-called impurities in a chemical sense, and shows the useful effect of some of them. *Van Nos. Eng. Mag.*, Vol. XXV., p. 494.

Impurities of. By A. E. Hunt and Geo. H. Clapp, Pittsburgh, Pa. Pp. 18. *Proc. A. I. M. E.*, Buffalo meeting, 1888.

Water, continued.

In Japan. A Description of Japanese Water Supply Systems. By Prof. W. S. Chaplin. *Jour. New England W.-W. Assn.*, Dec., 1886.

Lifted by a New Method. The water is lifted through pipes by forcing air into them at bottom. *Proc. Inst. C. E.*, Vol. LXXXI., p. 400.

Odor and Color of Surface Waters. A most excellent paper and valuable discussion before the New England Water Works Association. By Prof. Thos. M. Drown. Gives methods of determination and results of the official examination of the potable waters of Massachusetts, the whole covering 27 pages in the *Jour. New Eng. W. W. Assn.*, March, 1888. *Technology Quart.*, Vol. I., No. 3.

Palatable Water. Paper of some value by Prof. Floyd Davis, Chemist of Iowa State Board of Health. *Fifth Biennial Rep., Board of Health, State of Iowa*, 1889, pp. 215-251.

Purity of. Abstracts of several valuable papers on this subject are published in *Proc. Inst. C. E.*, Vol. CI., 1890, pp. 349-57.

Potable. By F. P. Venables. Gives data as to danger and presence of zinc in using galvanized pipe and tanks. *Iron Age*, April 30, 1885.

Pressure in Soil. Its diminution due to the capillary attraction, which is shown to be very great. *Abstracts Inst. C. E.*, Vol. LXXXV., p. 423.

Surface Waters. Article by J. C. Thomson, of Albany, N. Y., giving a condensed statement of laws and decisions bearing on use, appropriation and contamination of surface waters, etc. Gives 109 references of cases and decisions. *Sci. Am. Sup.*, March 1, 1890, p. 11,801, *et seq.*

Tests of. See *Water Analysis* below.

Utilization of Surface Water for Drinking Purposes. Address by Prof. Wm. T. Sedgwick. Gives results of experience with sand filters in Europe taken from important paper by W. H. Lindley of Frankfort, published in *Deutsche Vierteljahr fur Oeffentliche Gesundheitspflage.* Bd. XXII., 1890. *Jour. New. Eng. W.-W. Assn.*, Sept. 1890, pp. 33-44. Abstract in *Eng. News*, Nov. 1, 1890, p. 400.

Waste of. See *Water Waste.*

From Wells. Abstracts from four papers before the Institute of Civil Engineers giving results obtained from wells in various parts of England. Lon. *Eng.* April 22, and Lon. *Engineer*, May 20, 1887.

Water Analysis.

Biological. Paper by Prof. W. T. Sedgwick, revised from paper published in the Journal N. E. Water-Works Association, Sept., 1889. *Proc. Soc. Arts*, Boston, 1890.

Biological Examination of a. By W. T. Sedgwick, before the New England Water-Works Association. Discusses the methods, results and their interpreting of biological examination of water. Discussion *N. Eng. W. Works Assn.*, June, 1888, Vol. II., pp. 7-26.

Biological Examination of. By F. Hueppe. Extended discussion of the subject. *Jour. f. Gasbel u. Wasserversorgung*, 1887, pp. 321-32, 354-65, 421-29, 463-69.

Biological and Chemical Basis. By Percy F. Frankland. A general review of the several processes in use in England and the results that may be expected from them. From *Inst. C. E.*, in *Van Nos. Eng. Mag.*, Oct. 1886.

Chemical Examination of Drinking Water. A valuable paper by Prof. T. M. Drown. *Proc. Soc. Arts*, Mass. Inst. of Technology, 1887-1888.

Chemical Test of. Enumerates the conditions which the Commission of Vienna thinks water ought to fulfill if it be used in important towns. *Van Nos. Eng. Mag.*, Vol. XIV., p. 138.

Water Analysis, continued.

Determining Nitrates in Potable Waters. A rapid colorimetric method. By Samuel C. Hooker, Ph. D. Read before Chemical Section of Franklin Institute, March 19, 1889. *Jour. Frank. Inst.*, June, 1889.

Examination by the State Board of Health of the Water Supplies and Inland Waters of Massachusetts, 1886-90. Part I., of Report on Water Supply and Sewerage, 8vo, p. 837. Boston, 1890. This report contains besides the chemical and biological examination of the water supplies of the state, many other valuable papers by members of the board. The subjects of these are: the chemical examination of waters including methods of analysis and interpretation, organisms in water, their distribution, relation to odor and means of prevention, water supply and rainfall, flow of streams, effects of storage, deep ponds, natural filtration and the pollution and self-purification of streams.

Examination of in Massachusetts. Abstract of the report of F. P. Stearns, Chief Engineer to the State Board of Health of Massachusetts, for 1887. Discusses filter galleries and covered reservoirs. *Eng. & Build. Rec.*, April 14, 1888.

Examination of, for Micro-Organisms. By Dr. G. Bischof. A detailed description given. *Eng. News*, March 6, 1886.

Examination of Water to determine its fitness for domestic purposes. A paper by Prof. V. C. Vaughan in *The Technic*, 1890. (Univ. of Mich. Annual, 50 cents.) Extended abstract in *Eng. News*, Aug. 16, 1890, p. 150.

Gives the tests which have been regularly carried on at the Berlin Water-Works; also the conclusions drawn. *Van Nos. Eng. Mag.*, July, 1886.

For Living Organisms. A new method by Dr. Dupré, by observing amount of oxygen consumed. Likely to lead to new methods of water analysis. Editorial in Lon. *Engineer*, Nov. 20, 1885.

Mississippi River. Abstract from the report of Dr. Charles Smart giving the result of a series of analyses made for the State Board of Health of Minnesota of the waters of the Mississippi River and some of its tributaries. *Eng. & Build. Rec.*

Of Potable. By C. W. Folkard. An abstract from a paper before the Inst. of C. E. *Van Nos. Eng. Mag.*, Vol. XXVI., p. 331.

Of Potable Water. By Chas. W. Folkard, read before the Inst. of C. E. Gives methods of water analysis with special reference to the determination of the previous sewage contamination. *Van Nos. Eng. Mag.*, Vol. XXVII., pp. 143 and 228. Also *Van Nos. Science Series*, No. 61.

Processes for Determining the Organic Purity of Potable. By Dr. Tidy, before the Chemical Society. Comments upon the various methods employed. *Van Nos. Eng. Mag.*, Vol. XX., p. 172.

For Railroads. A paper by George Gibbs before January meeting Western Railroad Club. Discusses the subject from a practical standpoint and gives the experience of the C., M. & St. P. R. R. Gives method of gathering data, results of analysis, etc. *Mast. Mech.*, Feb., 1889.

Recent Progress in Biological Water Analysis. A valuable paper by Prof. Wm. T. Sedgwick, Mass. Inst. of Tech., with a discussion by Prof. Drown and others. Illustrated by diagram. Pp. 16. *Jour. N. E. W.-W. Assn.*, Sept. [illegible], Vol. IV., p. 80.

For Sanitary Purposes. An exhaustive memoir of 70 pp., by Dr. Chas. Smart, Assist. Surg. U. S. A. Includes [illegible] analyses. *An. Rep. Nat. Board of Health*, 1880, pp. [illegible]-311. By A. A. Breneman. Discusses the necessity of standard conditions in the application of Wanklyn's method of analysis and proposes a modification which may eliminate many of the uncertainties of the process. *Eng. News*, June 4, 1881.

Water and Sickness.

Cholera. A review of the late Spanish epidemic in its relation to the water supply, showing a vital (or fatal) relation between the two. By Geo. Higgin. *Van Nos. Eng. Mag.*, Aug., 1886.

Relation of Ground Water to the Health of the Community. A paper by Col. Geo. E. Waring, Jr., discussing the extent of its pollution and its effect in communicating disease. *Fourteenth An. Rep. Board of Health, N. J.*, Trenton, 1890, pp. 131-4.

Sickness caused by contaminated Water. A clear case reported to New Hampshire Sanitary Board of Health. Contamination from sink and privy. Cut given. *San. News*, Oct. 10, 1885.

Typhoid Fever in its Relation to Water Supplies. Statistics of typhoid mortality and account of several special cases. By Hiram F. Mills. *Jour. New Eng. W.-W. Assn.*, June, 1891, pp. 149-56. Discussion, pp. 156 to.

Typhoid Fever and Water as related to each other at Louisville, Ky. Report by Dr. J. N. McCormack. *San. Eng.*, May 21, 1885. See further *Epidemic.*

Typhoid Fever. By Dr. S. Benjamin. A paper before the New Jersey Sanitary Assn. on the relation between drinking water and typhoid fever. *Sci. Am. Sup.*, June 11, 1887.

Zurich. Extended abstract of report of a commission to investigate its relation to the epidemic of typhoid fever in 1884. The outbreak was traced to a contamination of the water supply. Report has been separately published. *Jour. f. Gasbeleuchtung und Wasserversorgung*, 1886, pp. 80-83, 112-117.

See *Water Analysis.* *Water Contamination.*

Water-Brake. See *Locomotives, Water-Brake on.*

Water Closets. See *Privy Vaults.* *Plumbing.*

Water Conduits.

Farm Pond Conduit, Boston W. W. By Hy. H. Carter. Read before Boston Soc. C. E., Oct. 17, 1888. *Jour. Assn. Eng. Soc.*, Feb., 1889.

Farm Pond. A description of the Farm Pond conduit for the Boston water supply and the raising of a 48-inch main will be found in the *City Engr's. Rep.*, Boston, Mass.

Under Pressure. By J. T. Fanning. A wooden tube 6 ft. in diameter, used at Manchester, N. H., for conveying 65,000,000 gallons a day a distance of 6 0ft. under a head of 9 ft. Made of staves with iron hoops. Illustrated. *Trans. A. S. C. E.*, Vol. VI., (1877), p. 69.

Stability of Brick Conduits. By A. Fteley, including method of measuring their distortion. Illustrated. *Jour. Assn. Eng. Soc.*, Vol. II., p. 103.

See *Water Mains.* *Water Pipes.* *Water Supply.*

Water Consumption.

And Capacity in thirty cities and towns for the year 1886. *Jour. New Eng. W.-W. Assn.*, for March, 1888.

By Cities and Towns. By H. W. Ayres, before the American Water-Works Association. Gives table showing the consumption of water in thirty large cities, extending over a series of years. *Proc. Am. W.-W. Assn.*, Vol. VIII., (1888), pp. 60-69.

The German Rules for Estimating, wherein the consumption is analyzed with minuteness. *Eng. News*, Oct. 31, 1885.

And Waste. By J. H. Harlow. Data given for many American cities. *Trans. A. S. C. E.*, Vol. VI., (1877,) pp. 107-126.

Water Contamination and Pollution.

Animal and Vegetable Growth Affecting Water Supplies. Report of the Committee on. Gives results of experience with various water supplies, and the methods of dealing with them. *Rep. 11th An. Meeting of Am. W.-W. Assn.*, 1891, pp. 27-45.

Water Contamination and Pollution, continued.

In Certain Storage Reservoirs on the Pacific Coast and the Palliatives Resorted to. A paper by Mr. L. J. Le Conte, read May 23, 1890. *Rep. 10th An. Meeting Am. W.-W. Assn.*, pp. 56-65. Fully abstracted in *Eng. News*, Aug. 23, 1890, pp. 160-2.

From Sewage. An interesting report of the suit of the Newark Aqueduct Board *vs.* the City of Passaic. Abstract of the expert evidence and the Chancellor's decision. *Eng. & Build. Rec.*, Aug. 16, 1890.

Of New York, Present and Prospective. Report of a committee of the Medical Society of the County of New York. Gives results of investigations of the sources of pollution, with observations and discussion on the advisability of building the Quaker Bridge Dam. *Sanitarian*, April, 1886, pp. 32.

Polluted Water. Paper by Prof. F. H. Snow and *A Spectroscopic Method of Detecting Sources of Well-Water Pollution*, a paper by Prof. L. J. Blake. *Fifth An. Rep. Kansas State Board of Health*, J. W. Redden, M. D., Secretary, Topeka.

Pollution of Air and. By A. E. Fletcher, before the Society of Arts on the present state of the law concerning the pollution of air and water. Shows what the law has done, and what is still hoped for from its further action. *Jour. Soc. Arts*, April 13, 1888.

In Storage Reservoir. By A. H. Tyson and C. G. Darrach. *Proc. Eng. Club, Phila.*, May, 1885.

Storage Water. A paper presented by Mr. L. J. Le Conte at the meeting of the Am. Water Works Assn. Extended extracts in *Eng. & Build. Rec.*, June 14, 1890, p. 24.

Trouble from Volvox in Middletown, Conn., Water. Report of Prof. H. W. Conn, which says that the fishy odor and taste observable in the water for several days was due to the decomposition in the mains of immense numbers of the Volvox plant. *Eng. Record*, Jan. 31, 1891, pp. 143-4.

See *River Pollution, Ice Supplies, Sanitary Engineering.*

Water Distribution Systems.

Graphical Method of Studying Efficiency of, and the performance of fire engines, as illustrated in a great fire at St. Louis, Mo. By M. L. Holman. *Jour. Assn. Eng. Soc.*, Vol. I., p. 111.

By Geo. C. Francis. A good summary of the pros. and cons of the five different systems common in America. *Eng. News*, March 13, 1886.

The Greathead System of Fire Protection, described by Howard Constable. High-pressure mains connect with fire-plugs on the ordinary service mains. The water under high pressure draws from the service mains on the principle of the steam injector or sand pump. *Proc. Eng. Club, Phila.*, Vol. V., p. 232.

Of Boston, Mass. By Dexter Brackett. A very complete description of all the interesting engineering details, with five large plates. *Jour. Assn. Eng. Soc.*, Vol. V., p. 107.

Paris. A short description of the water distribution of Paris, France. Also showing the Chameroy intermittent-cock and a leakage gauge. *Eng. News*, April 9, 1887.

Water Examination. See *Water Analysis.*

Water Filtration and Purification.

By Gustav Bischof. An article before the Soc. of Arts relating to the purification of water for sanitary purposes. Discusses spongy iron as a filtering material. *Van Nos. Eng. Mag.*, Vol. XIX., p. 28.

Paper by Mr. C. N. Hunt before the Northwest Railroad Club discussing methods of purifying boiler feed water mainly. *Ry. Rev.*, Dec. 14, 1889, Vol. XXXI., p. 724.

Water Filtration and Purification, continued.

Papers by Mr. J. O. Pattee and Mr. C. N. Hunt, with discussions. December meeting of Northwest Railroad Club. *Mast. Mech.*, Jan., 1890, p. 7.

Paper by Mr. J. O. Pattee, before the Northwest Railroad Club. *Ry. Rev.*, Dec. 14, 1889, Vol. XXXI., p. 723.

By Col. Wm. Tweeddale, C. E. Brief description of different methods: Sedimentation, Filtration, Chemical Process and Distillation. *First An. Rep., Kansas State Board of Health.* 1885.

By Aeration. Results of experiments by the Hackensack Water Company, New York. Air pumped into the mains under pressure of 125 pounds. Proved efficient. *Trans. A. S. C. E.*, Vol. XV., p. 139.

Aeration and Filtration. Paper by Charles B. Brush, with discussion. Many new and pertinent facts presented, especially in the discussion by Dr. Leeds. *Jour. New Eng. W. W. Assn.*, Sept., 1887.

By Agitation with Iron and Sand Filtration. By W. Anderson. Gives account of Prof. Bischof's system applied to purifying the water supply of Antwerp, Belgium. Analytical results of examinations of water before and after filtration. *Sci. Am. Sup.*, Feb. 12, 1887. *Jour. Soc. Arts*, Nov. 20, 1886. Abstracted in *Eng. News*, Jan. 11, 1887. For continuation of above process, see *Proc. Inst. C. E.*, Vol. LXXII., p. 24. Vol. LXXXI., p. 279 and p. 287. *Jour. Royal Agricultural Soc.*, Vol. XX., part II., (1884), p. 681. Also see "Rapport sur la qualite de l'eau de la ville d'Anvers pendant 1885."

By Alum. By Prof. P. T. Austen. Treats of its particular application to localities where purification of the water is only necessary at certain seasons of the year, and where the expense of large pumps and filter would be a serious tax. *Sci. Am. Sup.*, July 17, 1886.

By Artificial Means. A paper read before the Liverpool Eng. Soc. Gives common practice in England. *Eng. News*, April 18, 1885.

Its Biological and Chemical Basis. By Percy F. Frankland before the Inst. Civ. Engrs. A general review of all methods found efficient and valuable, viewed from the scientific standpoint. With discussion. A valuable paper. *Inst. C. E.*, Vol. LXXXV., p. 197.

For Domestic and Manufacturing Purposes. By T. S. Crane. Gives description of the Hyatt filter. Illustrated. *San. Eng.*, June 24.

Experiments on with some remarks on the composition of the water of the River Plate. By G. Higgin before the Inst. C. E. Water was very difficult to clear by filtration. Extended experiments described. *Eng. News*, Oct. 4, 1879.

Filter Beds. Short description of a new filter at the water-works in Brieg. The two filters have an area of 110 square metres each, but the large velocity of filtration (20 inches per hour) rendered their action unsatisfactory. The new apparatus was designed to increase their efficiency, and is said to have succeeded. *Centralblatt der Bauverwaltung*, 1886, p. 42.

Filter Gallery Washing, at Eureka, Cal. At this place the water supply is taken from an infiltration gallery extending under a river bed. The system is briefly described together with an ingenious method of back flushing for cleansing the filter bed. Illus. *Eng. News*, April 11, 1891, p. 339.

Filtered Water Supply from Rivers. A system proposed by M. Lefort of France, containing several special features, among which are the thorough oxidation at the surface of the filter, also the obtaining of the water from the upper stratum of the river. From *Le Genie Civil. Eng. News*, July 18, 1891, pp. 52-3.

Filtering Plants of American Water Works. Editorial discussion of data from *Manual. Eng. News*, July 12, 1890, pp. 38-40, July 26, p. 83.

By Filtration. By Th. S. Crane. Describes various types of filters and materials. *Trans. A. S. M. E.*, Vol. VII., p. 617.

Water Filtration and Purification, continued.

Filtration. By Geo. Higgins. Details some experiments made at Buenos Ayres on the filtration of water from the river Plate. *Van Nos. Eng. Mag.*, Vol. XXII., p. 72.

Filtration. A paper before the Liverpool Engineering Society. By H. T. Turner. The most approved principles and methods of constructing artificial filter beds discussed. *Eng. News*, April 18, 1885.

Filtration of Public Water Supplies in America. Paper by J. J. R. Croes before the Am. W.-W. Assn., giving details of construction, cost and efficiency of various plants. *Eng. News*, June 16, 1883, pp. 277-8.

By Iron. Description of process employed in purifying the water supply of Antwerp, Belgium. *E. & M. Jour.*, Jan. 5, 1889.

Iron on a Large Scale. By Wm. Anderson, M. I. C. E. Revolving tanks substituted for the iron filter beds, the iron filings being placed in the tank and removed from the beds, these being used as sand filters only. Antwerp Works. Illustrated. With discussion by Geo. Henry Ogston, showing chemical action on the water. *Eng. News*, Sept. 5, 1885. Lon. *Engineer*, June 19, 1885.

By Means of Metallic Iron. A paper read by Mr. Easton Devonshire, Assoc. M. Inst. C. E. The method in use at Antwerp, Belgium, is described. Three folding plates. *Jour. Frank. Inst.*, June, 1890, No. 774, pp. 449-61. Abstract *Eng. Record*, Feb. 14, 1891, pp. 177-8.

By Metallic Iron. Brief article by Dr. Henry Leffman, describing the Anderson process and giving its advantages. *Jour. Anal. and App. Chem. Sci. Am. Sup.*, No. 830, Nov. 28, 1891, p. 13270.

National Filter Plant at Terre Haute, Ind. This plant consists of twelve filters, with a total capacity of 3,000,000 galls. per day. Illustrated description. *Eng. News*, Feb. 7, 1891, pp. 127-8.

The Present State of Our Knowledge Concerning the Self-Purification of Rivers. A paper by Percy F. Frankland read before the International Congress of Hygiene and Demography. Dr. Frankland maintains that sedimentation is the main cause of any self-purification of the water. *Eng. News*, Sept. 5, 1891, pp. 218-19.

Proposed for Washington. Water to be forced through about 8 feet of a mixture of sand and powdered coke, after being intimately mixed with about one-fifth its bulk of air. *San. Eng.*, May 13, 1886.

Self-Purification of Flowing Water and the Influence of Polluted Water in the Causation of Disease. (A biological study.) A valuable paper by Dr. Chas. G. Currier. The purity of ground water is discussed as well as that of several rivers and other water supplies polluted at places by sewage. An appendix contains a paper by the same author read before the N. Y. Academy of Medicine, on "The Efficacy of Filters and of Other Means Employed to Purify Drinking Water," being a bacteriological study. *Trans. A. S. C. E.*, Vol. XXIV., Feb., 1891. Paper No. 463, pp. 21-49. Discussion, pp. 50-79.

By Settlement, Observation and Theory. A paper read before the St. Louis Engineers' Club, May 15, 1889, by Mr. Jas. A. Seddon. Illustrated by several plates. Pp. 15. *Jour. Assn. Eng. Soc.*, October, 1889, Vol. VIII., p. 477.

Clearing Water by Settlement, Observations and Theories. An abstract made by author of a "Report on the treatment of the St. Louis Water Supply," by James A. Seddon, C. E. Illus. *Eng. News*, Dec. 28, 1889, Vol. XXII., p. 607, *et seq.*

For the Supply of Towns. By P. Lauriol. Paper I—Filtration. Lindley's system; Natural and Artificial Filtration. Illustrated. Other papers to follow. *La Nature*, 1891, March 7, pp. 209-11.

Of Water and Sewage by the Magnetic Spongy Carbon Process. Paper by Charles H. Beloe, M. Inst. C. E., before Liverpool Engineering Society. Illustrated by plates. *Trans. Liverpool Eng. Soc.* 1889. Vol. X., pp. 118-42.

Water Filtration and Purification, continued.

See *Filters. Sediment Bearing Streams. Self Purification of Peaty Rivers. Water Supply, Bethel, Conn. Sewage Purification.*

Water Front.

Of San Francisco, Cal., Structures on. Abstract of a paper by Marsden Manson, C. E., Engineer of the State Harbor Commissioners, California. Illustrated by maps showing water front of San Francisco, California. *Eng. News*, Dec. 29, 1888.

Water Gas. See *Gas, Water. Fuel.*

Water Gauges.

Telemetrie. Apparatus to record the readings of gauges at distant points from a central station. A Belgian invention. *Annales des P. & C.*, Oct., 1883.

Water Hammer.

A Discussion, giving the experience of water-works superintendents. *Jour. New Eng. W.-W. Assn.*, Sept., 1886.

Some Experience with. Brief paper by E. E. Farnham describing method of remedying a case of water hammer. The discussion cites another case. *Jour. N. E. W.-W. Assn.*, Sept., 1891, pp. 16-9.

Water Jet. *In Engineering Construction.* Being an historical sketch of its application to the sinking of piles and caissons, and the removal of sand bars, and alluvial deposits, with a detailed account of the method of using it in driving sheet piling. By L. Y. Schermerhorn. Issued as a separate pamphlet by Eng. Dept., U. S. A.

Water Level. *A New Water Level.* By P. Kahle. Abstract from *Zeitschrift fur Vermessungswesen*, 1888, p. 183. Excellent results are shown, and the time required for work was much less than with spirit level. *Proc. Inst. C. E.*, Vol. XCVI., 1889, p. 355.

Water Mains.

Note on the lowering of a large main several feet while in service, without any injurious effects. By S. Bent Russell. *Jour. Assn. Eng. Soc.*, Vol. IV., p. 166.

Care of Water Mains in Relation to the Quality of the Supply. A paper by Geo. F. Chace, giving some practical experience on this subject. *Jour. New Eng. W.-W. Assn.*, March, 1891, pp. 131-4.

Cleansing. By J. H. H. Swiney, before the Institution of Civil Engineers. Gives details of the cleansing of the water mains at Omagh. The pipe was coated with about one inch of peat. Scrapers were sent through the pipe. Lon. *Eng.*, April 13, 1888.

Description of arrangement for controlling the pressure in mains, and demonstration how initial high pressure can be simultaneously applied. By W. Key. *Royal Scottish Society of Arts; Transactions.* Vol. XII., 1891, pp. 223-30.

Detection of Leaks in. By Joseph Francis. Lon. *Engineer*, July 13, 1888.

Drainage and Filling. By S. B. Russell, before the Engineers' Club of St. Louis. Discusses the precautions that should be taken in draining and filling water-mains. *Jour. Assn. Eng. Soc.*, Vol. VI., 1887, pp. 298-305; *San. Eng.*, Dec. 24, 1887.

Laid Under Streams. Describing methods, with cuts. *Proc. Am. W.-W. Assn.*, 1886, J. H. Decker, Sec'y., Hannibal, Mo.

Laying Across River. By W. R. Coats. Gives description of the method adopted for laying a 16-inch water main across Grand River at Grand Rapids, Mich. *Proc. Mich. Eng. Soc.*, for 1887, p. 84.

Leaks in and their Detection. Methods used by four London companies. *Proc. Inst. C. E.*, Vol. XC., p. 414. *Eng. News*, Oct. 22, 1887.

Water Mains, continued.

Pressure in. Table showing pressure in the mains of various cities. *Eng. News*, May 29, 1886.

Pressure in the Mains of Various American Cities. Tabular values of Difference of Level in Town; Day Pressure; Night Pressure; and Fire Pressure. Compiled by J. J. R. Croes. *Eng. News*, Jan. 16, 1886, *et seq*.

Raised Eighteen Feet. The main was forty inches in diameter, and was raised empty by means of screws working through caps across pile bents. Illustrated. *Eng. News*, Jan. 23, 1886.

Removal of Incrustations from. By E. H. Keating. An account of some successful experiments in removing hard incrustations by mechanical scrapers propelled by water pressure. Illustrated. *Trans. A. S. C. E.*, Vol. XI., (1882), p. 187.

Repairs to, under water. By E. Ruoff. Pipe crossing a stream, 14 feet under water, was repaired by sinking a caisson by the pneumatic process over the leak. *Jour. f. Gasbeleuchtung und Wasserversorgung*, 1886, pp. 193-207.

Wooden Water Mains. An article by A. McL. Hawks, discussing their use and describing methods of construction and special details. Illustrated. *Eng. News*, June 13, 1891, pp. 596-9.

See *Discharge of Mains*. *Pipe Laying*. *Water Conduits*. *Water Pipes*.

Water Mains and Meters. By Chas. B. Brush, M. Am. Soc. C. E. Gives "facts in relation to friction, waste and loss of water in mains;" and also discusses the theory of friction of water flowing in pipes very briefly. Pp. 16. *Trans. A. S. C. E.*, Sept., 1888, Vol. XIX., p. 89.

Discussion on Brush on water mains and meters. pp. 20. *Trans. A. S. C. E.*, Sept., 1888, Vol. XIX., p. 106.

Water Meters.

A paper by J. Nelson Tubbs, discussing the advisability of the use of meters, and its effects on the receipts of the owners of the works. *Rep. 11th An. Meet. Am. W.-W. Assn.*, 1891, pp. 58-60.

Paper by John Coleman describing the water meter system and requisites for a successful meter. Abstract, *Eng. News*, Sept. 2, 1890, pp. 311-2.

By John Thomson. Gives his personal experience, statistics, description and illustrations of various forms of meters, and discussion. *Trans. A. S. C. E.*, Vol. XXV., July, 1891, pp. 4-69.

By T. E. Bodkin, before Soc. of Arts. *Van Nos. Eng. Mag.*, Vol. VI., p. 390.

By C. H. C. Gives results of the use of the meter in different cities, and advocates their general use. *Van Nos. Eng. Mag.*, Vol. VII., p. 417.

A history of their development as a mechanical device, with discussion of the essential features of a successful meter. By J. A. Tilden. *Jour. Assn. Eng. Soc.*, Vol. VI., p. 17.

A discussion covering 11 pages in the *Jour. New Eng. W.-W. Assn.*, for Sept., 1886.

Capacity of 1,500,000 gals. per day. Consists of two buckets which tip alternately. Used for measuring feed water to the Liverpool Canal. *Mechanics*, November, 1883.

Comparative tests of accuracy, capacity, wear, etc. By Ross E. Browne. Five types discussed at length, with conclusions. All styles illustrated and tabular results and diagrams given. *Trans. Tech. Soc. of Pac. Coast*. Reprinted in *Van Nos. Eng. Mag.*, July, 1885.

Distinctive features and comparative tests of accuracy, delivery, etc., of different meters. Ross E. Browne. *Trans. Tech. Soc. Pac. Coast*, April, 1885.

Experiments with at Weisbaden. Author points out the practical results from the trials of a large number of meters. *Van Nos. Eng. Mag.*, Vol. XVI., p. 189.

Water Meters, continued.

For registering small flows. By J. J. Tylor. Gives a short account of chief water meter used in England, and shows how far the design is effective in measuring small quantities. *Van Nos. Eng. Mag.*, Vol. XXV., p. 394.

Frost's. Gives illustrated descriptions. Frost's positive water meter. *Lon. Eng.* Sept. 16, 1887.

History of their Invention, description of the various classes and statement of underlying principles. By Phineas Ball. *Eng. News.* Oct. 18, 25, 1879. Also article by L. H. Nash, *ibid.*, Nov. 8 and 29, 1879. Also article by S. J. Burr, *ibid.*, Nov. 15, 1879, and the Harton & West meter, *ibid.*, Nov. 22, 1879.

For Irrigation. A very simple and effective appliance for measuring miners inches. Illustrated. No patent. Designed by A. D. Foote. *Trans. A. S. C. E.*; also *San. Eng.*, Nov. 13, 1886, and *Eng. News*, Nov. 13, 1886.

For Irrigation. By A. D. Foote. Gives description of a cheap and satisfactory water meter for measuring miners' inches in irrigation works. *Trans A. S. C. E.*, Vol. XVI., p. 134.

The Nozzle as an Accurate Water Meter. A valuable paper by John R. Freeman, describing some careful experiments on nozzles up to 2½ inches in diameter. The results indicate a remarkable accuracy. Paper illus. *Trans. A. S. C. E.*, Vol. XXIV., June, 1891, pp. 492-513. Discussion, pp. 513-27.

Proportional Water Meter to Inferentially Measure the Total Discharge of Nozzles. Paper by John Thomson describing this apparatus. The flow from a nozzle is determined from the recorded flow through a meter, of a small stream diverted from the main pipe. *Trans. A. S. C. E.*, Vol. XXIV., June, 1891, pp. [illegible]. Discussion, pp. [illegible].

Recent Tests at Boston. By L. F. Rice, before the Boston Society of Civil Engineers. Gives details of the methods and apparatus used in "making a full test and report" upon the merits of water-meters for the Boston Water Board. *Jour. Assn. Eng. Soc.*, Aug., 1888, Vol. VII., pp. 285-297.

System of, Providence, R. I. By E. B. Weston, before the Boston Society of Civil Engineers. Gives details of the management of the water meter system at Providence, R. I., where 58 per cent. of the consumers have meters. *Jour. Assn. Eng. Soc.*, August, 1888, Vol. VII., pp. 297-314; *Eng. News*, August 11, 1888.

Testing—for purposes of experiment. Illustrated. By J. C. Hoadley. *Trans. A. S. M. E.*, Vol. V., p. 63.

Value of. A plea for their adoption in cities. By Chas. W. Baker. *Eng. News*, April 17, 1886.

Venturi. By Clemens Herschel, before the American Society of Civil Engineers, giving details of experiments made with a meter, embodying the property of Venturi tubes, applied to pipes from one to nine feet in diameter. Eight plates. *Trans. A. S. C. E.*, Vol. XVII. (November, 1887), pp. 228-258; abstracted *Proc. Inst. C. E.*, Vol. XCIII., pp. 515-516; abstracted in *San. Eng.*, Dec. 24, 1887.

And Water-Meter System. By John Coleman. Gives requisites for a water meter, and advocates the use of them to stop waste. *Van Nos. Eng. Mag.*, Vol. XXVII., p. 415.

See *Gauging*.

Water Motors, Turbines, and Wheels.

A lecture by Prof. W. C. Unwin before the Inst. Civ. Engrs. A general description, modes of action, and efficiencies, of various kinds of water wheels and engines. Illus. *Van Nos. Eng. Mag.*, Jan. 1886.

A High Service Motor. By F. H. Parker, before the Sixth Annual Convention of the New England Water-Works Association. Gives a description of a hydraulic motor at Burlington, Vt. The motor is worked by the flow of water in the main of the low service works. *Eng. News*, Oct. 1, 1887.

Water Motors, Turbines, and Wheels, continued.

Improved Oscillating. A description, with sections, of the motor of Schattenbrand & Moller, for household use. *Sci. Am. Sup.*, Sept. 17, 1887.

Motor, which works automatically in a supply pipe to the reservoir, with 5 pounds pressure, whether towards or from the reservoir, pumping a percentage of the water passed to a level 60 feet higher. Has been in successful operation at Burlington, Vt., for six years. Described by F. H. Parker, Superintendent of Water-Works. Illustrated. *Jour. New England W. W. Assn.*, September, 1887.

Motors, Cost of Running, and their Economy in Connection with Water-Works Systems. Full Abstract of a paper by Mr. S. E. Babcock read before the Am. W.-W. Assn. The various motors on the market are briefly described, and a detailed statement given of the amount of water used by machines operated by water motors, at Little Falls, N. Y. *Eng. News*, April 25, 1891. pp. 393-4.

Noiseless Rover. Gives description of a new motor for utilizing the waste power of streams. *Sci. Am. Sup.*, July 2, 1887.

The Pearsall Hydraulic Engine. Illustrated description of this novel form of water motor, its action depending on the same principle as the common hydraulic ram. *Eng. News*, Sept. 28, 1889, Vol. XXII., p. 292.

Turbine Wheels.

Abstract of a lecture before the Inst. C. E. on the design of turbines. By W. C. Unwin. *Lon. Eng.*, April 10, 1885.

An account of some elaborate experiments upon, by James B. Francis, giving tabulated results, cuts, etc. *Trans. A. S. C. E.*, Vol. XIII., p. 295.

A graduating thesis by J. L. Woodbridge, with an introduction by Prof. de Volson Wood. The discussion based on the *pressures* of elementary volumes and their movements. Probably the most general and mathematically satisfactory solution yet made. *Jour. Frank. Inst.*, Nov., 1886, *et seq.*

By Prof. W. P. Trowbridge. On the inapplicability of the investigation of the turbine wheel, as given by Rankine, Weisbach, etc., to the modern construction introduced by Bayden and Francis. *Van Nos. Eng. Mag.*, Vol. XX., p. 145; also *Van Nos. Sci. Ser.*, No. 44.

A review of the above article, by Prof. Wm. H. Burr. *Van Nos. Eng. Mag.*, Vol. XX., p. 346.

A reply to the above article, by Professor Trowbridge. *Van Nos. Eng. Mag.*, Vol. XX., p. 372.

Reply, by Prof. Burr. *Van Nos. Eng. Mag.*, Vol. XX., p. 460.

Determination of the Horse-Power of. A series of articles by Prof. R. H. Smith, of great practical value. He takes account of the unsteady flow through the wheel. *Lon. Engineer*, Sept. 4 and 25.

Their efficiency as affected by the form of the gate. By Samuel Webber. Tabulated and plotted results of experiments showing evil effects of gates with sharp edge. *Trans. A. S. M. E.*, Vol. III., p. 98.

Propeller Turbine. By Andrew Murray. *Van Nos. Eng. Mag.*, Vol. VI., p. 454.

Setting at Niagara Falls. Three Victor Turbines working a 75-foot head from one iron penstock, power being transmitted separately. Illustrated. *E. & M. Jour.*, July 18, 1885.

Terni Steel Works. Gives illustrated description of turbines at the steel works of Terni, Italy. Lon. *Engineer*, Sept. 23, 1887.

Theory of the Turbine. By Gustaf Atterberg. A new theory derived by assuming the normal velocity of the water at the inlet of the wheel as the first requisite for further calculation. *Van Nos. Eng. Mag.*, Vol. XXVI., pp. 138 and 23·.

Water Motors, Turbines, and Wheels, continued.

In Willimantic, Horizontal. By John Graham. Gives brief description of turbines at the Willimantic Water-Works. *Proc. Eng. Club, Phila.*, Vol. VI., p. 35.

Wheels.

A new kind of water-wheel working small pumps used on London, Chatham & Dover Ry. Wheel works under head of four feet and forces water 1,500 feet and to a height of 50 feet. With cut. *Iron*, Nov. 21, 1884.

The Sagebien. Description of the wheel. Illustrated, with results of experiments made to determine efficiency. *Van Nos. Eng. Mag.*, Vol. XIV., p. 83.

Water Motor. See *Water Works, New London.*

Water Pipes.

Abstract from a paper before the Philadelphia Engineers' Club. Treats of the different kinds of pipes, their qualities and availabilities. *Sci. Am. Sup.*, Feb. 12, 1887.

By A. H. Howland. Treats of the different kinds of pipes, their qualities and availabilities. *Proc. Eng. Club, Phila.*, Vol. VI., p. 55; *Am. Eng.*, April 20, 1887; *Sci. Am. Sup.*, Feb. 12 and Nov. 5, 1887.

Cast-Iron. Gives a table of dimensions of cast-iron water pipes as derived from the practice of an experienced engineer. *Mech. World*, Sept. 22, 1888.

Chemical Obstructions in Iron Water Pipe, describing samples in Philadelphia, with *fac-simile* lithographic plate. By Col. Wm. Ludlow. *Proc. Eng. Club, Phila.*, Vol. IV., No. 2, p. 61.

Dimensions of. By J. E. Codman, before the Philadelphia Engineers' Club. Discusses the diameter of pipe-flanges, diameter of bolt circle, size and number of bolts and thickness of cast-iron pipe. Gives diagram to show graphically the above points. *Proc. Eng. Club, Phila.*, Dec., 1887, Vol. VI., pp. 150-152.

Drawing of the standard Philadelphia water-pipe, with table giving dimensions of the parts for various sizes. *Eng. News*, Feb. 5, 1887.

Galvanized Iron for Artesian Wells and for the Conveyance of Drinking Water. Chemical analyses and discussion. Paper by Reuben Haines. *Jour. Frank. Inst.*, November, 1890, pp. 393-401. *Eng. & Build. Rec.*, Dec. 13, 1890, pp. 27-9.

Scraper. A very efficient device which has been successfully used in England. Illustrated. *Eng. News*, April 10, 1886.

Subaqueous Work. Repairing Submerged Water Pipe. By J. J. de Kinder. A brief description, with drawings, of coffer-dam, etc., used in repairing a main across the Schuylkill. *Proc. Eng. Club, Phila.*, Vol. IV., No. 4.

Use of Vitrified Pipe. By S. E. Babcock, before the American Water-Works Association. Gives experience in the use of salt glazed vitrified pipe for conduit. Describes the conduit at Amsterdam and Little Falls, N. Y. Discussion contains description of a conduit of redwood. *Proc. Am. W.-W. Assn.*, 1888, pp. 19-26; *Eng. News*, April 28, 1888; *Eng. & Build. Rec.*, May 26, 1888.

See *Pipes. Pipe Laying. Water Conduits. Water Mains.*

Water Pollution. See *Water Contamination and Pollution. Epidemics. Storage Reservoirs.*

Water Power.

Chart by Mr. F. R. Hart, showing available energy of water powers efficiency of .75, falls from 5 to 40 feet, and volumes up to 10,000 cu. ft. per min. *Elec. Eng.*, Nov. 18, 1891.

Comparative Cost of Steam and Water Power. Paper by Charles H. Manning, giving results of experience in New England factories. *Trans. A. S. M. E.*, Vol. X., 1889.

Water Power, continued.

Cost of. See *Steam Power.*

On the Economy of Power Motors for. By A. R. Wolff. Treats of the economy of steam pipes, hot air, gas engines and wind-mills. *Building*, June 5, 1886.

For Electric Central Stations. Central Station Operated by Water Power. Paper by Geo. A. Redman before Nat. Elec. Light Assn., describing plants in Rochester, N. Y. *Elec. World*, Sept. 19, 1891, pp. 101-3.

With High Pressure and Wrought-Iron Pipes. Gives methods of utilizing small flow under great heads, as found in the mining regions, with plates and discussion. By Hamilton Smith, Jr. *Trans. A. S. C. E.*, Vol. XIII., p. 15.

At Niagara.

An account of the recent appraisement of the portion of this power now utilized and its purchase by the State of New York. By Samuel McElroy. *Jour. Assn. Eng. Soc.*, Vol. IV., p. 393.

A Scheme for the Electrical Utilization of. A description of the plan of Mr. S. H. Hamilton, of Washington, D. C., for driving dynamos by means of turbines located in chambers behind the great cataract, catching a part of the water in flumes. *Elec. World*, Feb. 9, 1884.

Niagara River Power Co. Article giving considerable information concerning this scheme for utilizing the water power at Niagara. *Eng. News*, May 17, 1890, Vol. XXIII., p. 462.

The Niagara Falls Power Co. Drawings showing details adopted, of cross-section, wheel pits, etc. *Eng. News*, Nov. 8, 1890, p. 418.

In New England. By Samuel Webber. An account of its development. Illus. *Eng. Mag.*, July, 1891, pp. 421-32.

Of the Falls of the Ohio River. By M. S. Belnap. Gives four projects for utilizing the power at the Louisville Falls and estimates the cost of the power obtained. *Trans. A. S. C. E.*, Vol. II., p. 261.

To a Stamp Mill. By W. A. Biddle. Descriptive notes showing conditions under which formulæ were applied, with notes on Rutter's formulæ. *Van Nos. Eng. Mag.*, Vol. XIX., p. 392.

Of the United States.

Being Vol. XVI. of the U. S. Census for 1880. 870 pp. quarto. Valuable to hydraulic engineers, in that it gives details of many of the present water-power plants. *H. Rep. Misc. Doc.*, 42, *Pt.* 16, 47*th Cong.*, 2*d Sess.*, 1883.

Abstract of this 900 page volume of the tenth U. S. Census Report, which is by Prof. W. P. Trowbridge, assisted by Prof. G. F. Swain, Dwight Porter and J. L. Greenleaf. Some of the more interesting facts given in abstract. *Eng. News*, Sept. 4, 1886. Also *San. Eng.*, July 15, 1886.

Compiled under the direction of Professor Trowbridge. Containing much valuable matter concerning the development and use of water power, which is of service to engineers. *Tenth Census*, Vols. XVI. and XVII.

Used in Operating Air Compressor. See *Air Compressor.*

Utilized by Means of an Inverted Siphon Tunnel, at Marseilles, Ill. Project described and illustrated by E. J. Ward, in *Jour. Assn. Eng. Soc.*, Vol. V., p. 131.

Value of Water Power. A paper by Chas. P. Main before the Am. Soc. M. E., stating the various elements upon which it depends, and discussing the method of estimating. Abstract. *Eng. News*, Dec. 5, 1891, pp. 529-30.

See *Steam Power vs. Water Power. Water Meters.*

Water Proof

Coverings. By F. Collingwood. Tests of coal tar and felt; soapstone, petroleum and resin; Trinidad bitumen; asphaltic mastics; Trinidad bitumen with alternate layers of felt or paper; Trinidad bitumen refined and mixed with residuum oil. *Trans. A. S. C. E.*, Vol. IX., pp. 338-42.

Water Purification. See *Water Filtration and Purification.*

Water-Ram in Pipes.

An account of some experiments made in Providence. Tables and diagrams given. *Trans. A. S. C. E.*, Vol. XIV., p. 1[illegible].

A theoretical discussion by Prof. Irving P. Church, Ithaca, N. Y. *Jour. Frank. Inst.*, April and May, 1890.

Momentum and Force of Impact of Water Moving in Pipes. By J. W. Nystrom, *Iron Age*, Aug. 7, 1884.

See *Hydraulics.*

Water Rates.

Discusses the charges for water by measure and assessment. *Eng. News*, July 28, 1883.

Family and Meter Water Rates in the U. S. and Canada. Table reprinted from the *Manual of Am. W.-W. Eng. News*, August 9, and 16, 1890. pp. 117, 147.

Hydrant Service. A discussion on the proper charge for hydrant service in *Jour. New Eng. W.-W. Assn.*, Sept., 1886.

Hydrant Service, Rental Value of. By C. E. Chandler, before the Connecticut Association of Civil Engineers and Surveyors. Discusses the question of the proper price to pay for hydrant service. *Proc. Conn. Assn. of C. E. & Surv.*, 1888, pp. 10-14. *Eng. News*, Nov. 19, 1887.

Hydrant Service, Value of. By J. M. Tubbs, before the Cleveland Convention of the American Water-Works Association. A valuable paper on the methods for an approximate determination of the yearly rental value of fire hydrants as connected with any system of water-works. *Proc. Am. W.-W. Assn.*, 1888, pp. 137-48; *Eng. News*, May 5, 1888; *Eng. & Build. Rec.*, April 28, 1888.

Uniform, Classification of. Report of a special committee and discussion of same *Jour. New Eng. W.-W. Assn.*, Sept., 1887.

Water Service. See *Pipes.*

Water Softening Apparatus.

Applicable to large or small quantities, replaces Clark's process, and removes the permanent hardness as well as the temporary. *Lon. Engineer*, October 24, 1884.

A paper by Baldwin Latham, giving historical review and the various processes now used. *Eng. News*, Jan. 31 and Feb. 7, 1885. Also *Van. Nos. Eng. Mag.*, Vol. XXXI., p. 344.

By Baldwin Latham. Contains descriptions of the various methods employed to soften water. *Van Nos. Eng. Mag.*, Oct., 1884.

Clark's Process of Softening. An editorial giving a summary of a long correspondence on the softening of water by the Clark process. Lon. *Engineer*, Oct. 21, 1887.

And Filtering Apparatus for Locomotive purposes at the Penarth Dock Station of Taff Vale Railway Company. A paper of some interest by Wm. W. F. Pullen, Stud. Inst. C. E. Gives details of arrangements and cost of operation. Pp. 14. *Proc. Inst. C. E.*, Vol. XCVII., p. 351, London, 1889.

Two articles, showing how to accurately determine the degree of hardness of water, and by what means it has been proposed to artificially soften it for domestic use. Discusses the Porter-Clark, Stanhope Company's and the Anti-Calcaire processes. Lon. *Engineer*, Feb. 3 and 10, 1883.

See *Softening.*

Water Storage.

Table taken from a paper read by F. P. Stearns before the Boston Soc. C. E., giving amount of storage required for different daily volumes used, corrected for evaporation. Table based on records of Sudbury River. *Eng. Record*, Oct. 17, 1891, pp. 315-6.

Water Storage, continued.

Croton Storage.

Correspondence from Samuel McElroy, C. E., with editorial criticisms. *Eng News*, Dec. 22, 1888.

A paper read in 1889 before the Western Society of Engineers, by Mr. Samuel McElroy. Contains much valuable data respecting the storage of water in this important basin. Pp. 11. *Jour. Assn. Eng. Soc.*, Oct., 1889, Vol. VIII. p. 494.

Deterioration of Water in Reservoirs and Conduits, Its Causes and Modes of Prevention. Two papers published in the *Fourteenth Annual Report of the Board of Health of N. J., 1890.* Trenton, N. J.: I., by Charles B. Brush, pp. 107-10, giving a brief discussion; II., by George W. Rafter, citing numerous cases of deterioration and proposing remedies, pp. 111-22.

The Effect of Storage Upon the Quality of Water. A valuable paper by F. P. Stearns, Engr., Mass. State Board of Health, giving results of investigation of the effects of different classes of storage upon various waters. Analyses given. *Jour. New Eng. W.-W. Assn.*, March, 1891, pp. 115-25.

Water Supply.

A comparison of the cost of supply per million gallons in the cities of Chicago, New York, San Francisco and some English cities, together with other pertinent facts. From the *Builder*, in *Van Nos. Eng. Mag.*, April, 1886.

A paper by A. H. Denman, attorney, before the American Water-Works Association at Minneapolis, on the legal relations existing between water companies and consumers. A valuable digest of a large number of decisions. *Rep. Proc. 7th An. Meet. Am. W.-W. Assn.*, pp. 80-92. *San. Eng.*, Dec. 31, 1887.

A paper before the Soc. of Arts. By G. F. Deacon. On the constant supply system and the waste of water. *Van Nos. Eng. Mag.*, Vol. XXVII., p. 115.

Abstracts of papers on *Filtration, River Pollution and Water Supply. Proc Inst. C. E.*, Vol. CII., 1890, pp. 363-71.

An abstract from the report of a committee for investigating the circulation of underground water, and the relative value of water of different districts. *Van Nos. Eng. Mag.*, Vol. XVII., p. 500.

By Kenneth Allen. Treats the subject with especial reference to work preliminary to the introduction of a water supply. *Trans. Soc. of Engrs.*, Vol. I. p. 113.

Papers read before the Conference of the Soc. of Arts on the water supply held at the International Health Exhibition, July, 1884. *Van Nos. Eng. Mag.*, Nov. and Dec., 1884.

Aeration of. By S. E. Babcock, before the New England Water-Works Association. Describes the method of aeration adopted at Little Falls, N. Y. It consists of a series of mains constructed in an open paved channel. *Eng. & Build. Rec.*, Jan. 30, 1888; *Sci. Am. Sup.*, July 14, 1888.

Aeration of a Gravity Water Supply. Air was forced through the mains from power furnished by a small turbine. Account by Chas. B. Brush. *Rep. 11th An. Meet. Am. W.-W. Assn.*, 1891, pp. 73-5. *Eng. News*, April 8, 1891 p. 379.

Amsterdam Water Supply from the Vecht. Article abstracted from *der Zeitschr. des Vereins Deutscher Ing. Eng. News*, April 19, 1890, Vol. XXIII., p. 319.

Ancient Roman City. From an address by Prof. W. H. Corfield. Eng. Describes the ancient aqueducts leading into Lyons, France. *Am. Eng.*, Sept. 17. 1885.

Australia. A paper by W. E. Cox describing in considerable detail the various sources of water supply. The location and drilling of wells is the chief feature. Lon. *Engineer*, Sept. 18, 1891, pp. 225-6, *et seq.* Abstract of the portion relating to the drilling of wells by the pole system, in *Eng. Record*, Oct. 31, 1891, pp. 347-8.

Water Supply, continued.

Available Rainfall. See *Rainfall.*

Baltimore. History and description. *Eng. News*, Oct. 22, 1881.

Bethel, Conn., New Dam and Filter Bed of. The new dam is 18½ feet below the old one, the filter bed being between them. Estimated daily consumption, 120,000 gallons. Illustrated description. *Eng. News*, March 7, 1891, p. 218.

Bombay, India. Brief general description of this extensive work, including 25 miles of conduits, syphons of cast iron pipe, bridges, tunnels, etc. Data as to quantities and total cost given. *Eng. News*, May 23, 1891, pp. 502-3.

Of Bombay, The Tansa. By K. Hedges. Gives details of the work in progress to give the city of Bombay a supply of 60,000,000 gallons per day. Gives profile of dam to be 8,500 feet long and 118 feet in height. Lon. *Engineer*, July 15, 1887. *Eng. & Build. Rec.*, Dec. 31, 1887.

Boston. Complete description of the works with illustrations of conduits, etc. *Eng. News*, Jan. 3, May 1, 1878. Tables of rainfall, etc., *ibid*, May 23.

Of Brockton, Mass., and various other Massachusetts towns, with discussion of methods of purifying supply, together with numerous recommendations of State Board of Health. *Twentieth An. Rep. State Board of Health, Mass.*, p. 6, etc.

Capacity of Drainage Ground in Time of Drought. A study of the capacity of the Sudbury and Lake Cochituate watersheds in time of drought. By Desmond Fitzgerald. Reprinted in *Jour. N. E. W.-W. Assn.*, December, 1887.

In Cape Colony. Gives results of 11 years' experience in town and irrigation works. *Proc. Inst. C. E.*, Vol. XC., pp. 265-290.

Chicago.

And Drainage. An extract from a lecture by L. E. Cooley before the Chicago Academy of Sciences, treating of the water supply and drainage of Chicago. *Am. Eng.*, March 9 and 16, 1886.

Drainage and, of Chicago. By Rudolph Hering. A statement of the problem and some of the observed facts bearing on it. *Proc. Eng. Club, Phila.*, Vol. II., p. 222.

City of Charlottetown, P. E. I. Paper by S. R. Lea, *Stud. Can. Soc. C. E. Trans. Can. Soc. C. E.*, Vol. II., p. 274, 1888.

City of Worms. Supply from the River Rhine direct, then filtered, the filter and reservoir for filtered water being arched over. Water is then pumped into a water tower with a capacity of 1,000 cu. meters; pop. 20,000; quantity allowed 150 litres (40 gallons) per head per day; estim. cost, $158,000. *Civil Ingenieur*, 1885, p. 387-413.

Of Cities, Purification of the. By A. R. Leeds. A lecture before the Franklin Institute. A good review, with special reference to the process of purification by air. *Sci. Am. Sup.*, March 5, 1887.

Of Cleveland, in its Sanitary and Chemical Aspects. By N. B. Wood. Accompanied by tables of analyses. *Jour. Assn. Eng. Soc.*, Vol. IV., p. 284.

Cleveland. Full description. *Eng. News*, Apr. 26, June 14, 1879.

The Constant and Intermittent System and the Prevention of Waste. By G. F. Deacon, before the Inst. of Civ. Engrs. Treats of the sources of waste and effects on the intermittent service. Objection to the intermittent system. Advocates the use of meters. *Van Nos. Eng. Mag.*, Vol. XVI., p. 307.

Of Constantinople. By Henry A. Homes. Gives a description of the ancient works as well as of those now in use. *Van Nos. Eng. Mag.*, Vol. IX., p. 407.

Water Supply, continued.

Co-Operative. Description of a co-operative system established by a German district. The towns are divided into groups each with its supply system. By J. J. R. Croes. *Eng. News*, March 29, 1879, pp. 98-100.

Covington, Ky. A valuable article giving a description of these new water works designed by Mr. G. Bouscaren, C. E. Illustrated. *Eng. News*, May 17, 1890. Vol. XXIII., p. 458. *Eng. & Build. Rec.*, February 8, 1890, pp. 131, *et seq.*

And Its Development for Small Cities in the West. A paper by Wynkoop Kiersted, read before the Eng's Club of Kansas City, treating the subject in a popular way. *Jour. Assn. Eng. Soc.*, Dec., 1890, pp. 357-61. Abstract in *Eng. Record*, Jan. 24, 1891, pp. 116-7. *Eng. News*, Dec. 24, 1890.

Driven Wells as a Source of Water Supply for Cities. By Albert F. Noyes. A valuable contribution to the subject, giving many important facts and generalisations. *Jour. N. E. W.-W. Assn.*, June, 1887.

English Towns. Discusses the bearing of limited water-sheds upon the future supply of English towns, and suggests the prevention of waste of water. Gives experience in the use of waste meters and house-to-house inspection. Lon. *Engineer*, June 15, *et seq.*, 1888.

Fishkill and Matteawan, N. Y. Additional Water Supply for. Brief description. Illus. *Eng. & Build. Rec.*, May 10, 1890, p. 360.

Genoa, Italy. Brief description, with an account of the failure of a dam 100 ft. high. *Eng. Record*, Jan. 24, 1891, p. 123.

Golden, Colo. Increased Water Supply for. The increased supply is obtained from a gallery pipe built in a gravelly formation. Plan and profile given. *Eng. News*, June 27, 1891, p. 610.

Head Required to Produce Velocities in Pipes. By Chas. B. Brush, before the Annual Convention of the American Society of Civil Engineers for 1888. Gives results obtained in forcing water through a 20-inch main, 75,000 feet long. *Eng. & Build. Rec.*, July 14, 1888, abstracted *Eng. News*, July 7, 1888.

History. Water Supply of Cities in Ancient Times. An account of some of the ancient works. By W. Atlee. *Eng. News*, Oct. 27, 1883, pp. 507-9.

Improvement and Protection of. A paper by J. B. Strawn giving a discussion on the necessity and methods of improvement. *12th An. Rep. Ohio Soc. S. and C. E.*, 1891, pp. 110-20.

Lincoln, Neb. By J. P. Walton. Description of the well in Lincoln Park for the water supply of the city. Estimated flow, 75,000 gallons per hour. *Proc. Neb. Soc. Assn. Eng. & Surv.*, Vol. I.

Liverpool. A series of illustrated articles giving details of the construction of Vyrnwy Lake and masonry dam and the Liverpool water supply. Lon. *Engineer*, Jan. 24, 1887.

Long Island. Notes on the Quantity of Water Available for a Water Supply for New Lots, at the Western End of Long Island. The supply is from ground water. Report of investigating committee, with map. *Fire and Water*, No. 18, 1891, pp. 228-30.

Long Island, Boston Harbor. An interesting description of the methods used in laying pipes to this island by means of flexible joints, etc. Illustrated. *Eng. & Build. Rec.*, Nov. 2, 1889.

London.

A short history of the works, with tables of general statistics of the various establishments for 1867. *Van Nos. Eng. Mag.*, Vol. II., p. 83.

Gives statistics, principally financial, relating to the London water supply. *Van Nos. Eng. Mag.*, Vol. XVII., p. 63.

Water Supply, continued.

And Amsterdam. By M. Ed. Couche. *Annales des P. & C.*, July, 1883, pp. 153-220. Treating in considerable detail of sources, composition, purification and distribution of the water supply of London, with a very brief note on that of Amsterdam.

From the Chalk Formation about London, Eng. Wells bored six feet in diameter. Four papers with discussion, covering 102 pp. in *Proc. Inst. C. E.*, Vol. XC.

From Ground About London. Several papers and discussion. Illustrated. Giving methods and results in obtaining such a supply—the whole covering about 100 pp. The water obtained from chalk formation. *Proc. Inst. C. E.*, Vol. XC., p. 1.

On the Subterranean Water in the Chalk Formation of the Upper Thames, and its Relation to the Supply of London. A paper by J. T. Harrison. Gives considerable information regarding underground water supplies. Discussion contains tables of rainfall and percolation through various depths of soil, for twenty years. *Proc. Inst. C. E.*, Vol. CV., 1891, pp. 2-99.

Montpelier, Vt. Description of a private system supplied entirely by springs; also of the corporation system. *Eng. News*, May 22, 1886.

Newark, N. J. A somewhat detailed description of the work and its progress, especially of the large steel pipe line. A profile of the latter is given. *Eng. News*, Aug. 1, 1891, pp. 96-7, *et seq.* *Eng. Record*, Aug. 1, 1891, pp. 138-40, *et seq.*

Newark, N. J. Maps of Passaic watershed, and general description of the work. *Eng. News*, Oct. 1890, pp. 312-13.

Newark, N. J. Report of engineers giving results of surveys and examinations regarding the several schemes of water supply. By J. J. R. Croes and G. W. Howell. *Eng. News*, March 15, 1879, pp. 83-4.

New Orleans. Report of City Surveyor, T. S. Hardee describing conditions and giving estimates. *Eng. News*, June 2, 1877, pp. 143-4.

New York City.

A report on the sanitary condition of the Croton drainage basin, by Prof. Chas. C. Brown. A pamphlet of 160 pages and 72 plates. Extract from ninth (1888) *An. Rep. of State Board of Health.* *Jour. Assn. Eng. Soc.*, July, 1889, Vol. VIII., p. 398.

Article by R. D. A. Parrott, discussing the conditions of Croton water-shed and noting the effect of storage basins on the population of valley. *Sci. Am. Sup.*, May 11, 1889.

The Bartlett Project. Description of project for furnishing 50,000,000 gals. per day of pure water from the Passaic (N. J.) water-shed, 40 miles from the city. *Eng. News*, Dec. 12, 1885.

The Bartlett Project. To furnish 50,000,000 gallons daily to lower New York. Brief statement of the project. *Eng. News*, Dec. 8, 1888.

History of, and Recent Explorations for New Reservoirs. Valuable paper by E. P. Roberts, Asst. Engr., Aqueduct Commission. Folding plate giving maps and profiles. *Eng. News*, Nov. 22, 1890, pp. 455-6, and Nov. 29, 1890, pp. 480-2.

Cemeteries in Their Relation to the Palatable Water Supply of the City of New York. Article by Robert Grimshaw, Ph. D., C. E., in *Sanitarian*, February, 1897, pp. 148-153.

With map of the available water-sheds east and west of the Hudson. A careful analysis of the sources of supply, aqueducts, history of present works, and description of works of extension. A series of articles, begun in *San. Eng.*, March 26, 1885.

Water Supply, continued.

Observations upon the Cape Fear River as a Source of. By Thomas F. Wood, M. D., Wilmington, N. C. A study into the character of Southern river water, particularly of the seaboard region; and of the characteristics of water contaminated with tarry matters and products of vegetable decomposition. Pp. 10. *Sanitarian*, March, 1882.

Paris. General description of supply and works. By Wm. R. Hutton. *Eng. News*, Jan. 1, 8, 1881.

Peshawar City Water Supply Project. An article by Rai Bahadur Ganga Ram, M. I. C. E., giving calculations for filter beds, cost, system of distribution, etc. *Ind. Eng.*, March 7, 1891, pp. 190-4.

Of Philadelphia.

Abstract of Mr. Hering's recent report in *San. Eng.*, Nov. 27, 1886.

A discussion as to the future supply. By J. J. R. Croes. *Eng. News*, March 20, 1880, pp. 103-4.

The annual report of the Chief Engineer (Col. Wm. Ludlow) for 1884. A bound book of about 400 pages, containing appendices on pumping tests, surveys for new supply, chemical tests, sanitary survey of Schuylkill Valley, etc. A valuable document.

Remarks on the future supply by Col. Wm. Ludlow, Chief Engineer of the Phila. Water Dept., before the Franklin Institute. *Jour. Frank. Inst.* July, 1885.

Studies for the Future Supply of Philadelphia, given in the annual report of the Philadelphia Water Department. The report covers 400 pp. and is accompanied by numerous maps, profiles, graphical presentations of consumption, tables, etc. It contains numerous reports by specialists, as well as that of Col. Ludlow, the Chief Engr. of the Water Dept.; also Rudolph Hering's report on the surveys and investigations for a new supply and proposed areas, aqueducts, etc. A most valuable report.

Piping for. An analysis and description of the piping usually required for water-works pumping stations. By Charles H. Fitch, D. E., *Mechanics*, April.

Proposed Works of the Pueblo Gravity Water Supply Co. The supply is to be from the underground flow through the gravel below the bed of a stream, and the conduits to be of California redwood. System of collection is illustrated and described. *Eng. News*, Jan. 17, 1891, pp. 53-4.

Railroad Carlstadt-Fiume. By J. R. V. Finetti. This road traverses a very dry region, and the obtaining of a supply of water was a not uninteresting problem. *Zeitschr. d. Oester. Ing. u. Arch. V.*, 1886, pp. 99-106.

Report of Stephen E. Babcock, on the advisability of taking the waters of Skaneateles Lake as a supply for the city of Syracuse, N. Y. Complete discussion of question. *Pamphlet*, pp. 72, Little Falls, N. Y., 1889.

Rochester, N. Y. Report of J. T. Fanning and Alphonse Fteley, Consulting Engrs. Valuable paper including 17 tables and 8 maps. *An. Rep. of the Executive Board, Rochester, 1890.* Pp. 30-62. Review and criticism of the report by J. N. Tubbs, Chief Engr. of the water-works, *ibid*, pp. 63-6. An appendix contains a report on the pollution of water by typhoid bacilli; illustrated by photomicrographs, *ibid*, pp. 107-27.

Rogers, Arkansas. Review of water supply for Rogers, Arkansas, read before the Ark. Soc. of Engineers, Architects and Surveyors, at second annual meeting, by Jay M. Whitham, p. 23. *Trans. Ark. Soc. Eng., Archts. and Surv.* Vol. II., p. 93, Nov., 1888.

Rural. By C. L. Hett, Assoc. M. I. C. E. Describes various systems for small works. *Eng. News*, March 15, April 5, May 10, 17, 1884.

Sandusky, O. Described in a paper by C. A. Judson. *Proc. Ohio. Soc. Surv. & Civ. Eng.*, 1887.

Water Supply, continued.

San Francisco from Lake Merced. By P. J. Flynn. A brief report on the quantity and quality of the water. *Van Nos. Eng. Mag.*, Vol. XXIX., p. 410.

Sanitary Protection of. A paper by J. M. Tubbs, before the American Water-Works Association. Describes the method adopted for the sanitary protection of the water-shed of Hemlock Lake, supplying water to Rochester, N. Y. *Proc. Am. W.-W. Assn.*, [illegible], pp. 18-23; *Eng. News*, April [illegible], [illegible].

Selection of Sources of Water Supply. A paper by F. P. Stearns before the Boston Soc. C. E., giving statistics of yield of ground water sources based on observations in Mass., amount of storage required, quantity and quality of ground and surface waters. *Jour. Assn. Eng. Soc.* Oct., 1891, pp. 485-510. Reprint of the above matter as contained in the *Report Mass. State Board of Health. Eng. News*, Nov. 7, 1891, pp. 432-4.

And Sewerage. Conditions of several Massachusetts cities and towns and recommendations of the State Board of Health. *Twenty-first Annual Rep.*, 1892, pp. 1-69.

Sewerage and Lighting of the cities of Salzburg, Munich, Stuttgart, Frankfort and Hamburg. *Zeitschrift des Arch.- und Ing.-Vereins zu Hannover*, [illegible], pp. 174-85.

And Sewerage Report of Mass. State Board of Health, giving advice to cities and towns, analysis of water from various rivers and the Lawrence experiments on purification of sewage. *Rep. Mass. State Board of Health*, Jan., [illegible].

And Sewage Disposal for Milwaukee, Wis. See *Sewage Disposal.*

Of Springfield, Ill. A well supplemented by a filter gallery lined with timber. Illustrated. *An. Rep. Ill. Soc. Eng. & Surv.*, 1889.

Sub-Surface. By Prof. W. P. Trowbridge. Treats of obtaining water supply from driven wells. Gives details of the extension of the plant for the Brooklyn supply, etc. *Sch. Mines Quar.*, April, 1887; *Eng. News*, May 21, 1887; *San. Eng.*, May 26.

Sudbury River.

Report on the Capacity of the Sudbury River and Lake Cochituate Water-Sheds in Time of Drought. By Desmond FitzGerald. A very good example of what such an investigation should be. A pamphlet of 61 pp. Apply to the author, *City Eng. Dept., Boston.*

The flow of the Sudbury River, Mass., for the years 1875-1879. By Alphonse Fteley. With discussion. Illustrated. Treats of rainfall and flow of streams. *Trans. A. S. C. E.*, Vol. X., 1881, pp. 225-50.

Yield of the Sudbury River Water-shed in the Freshet of Feb. 10-13, 1886. Statistics of rain fall and flow of water from the water-shed, with diagrams. By Desmond FitzGerald. *Trans. A. S. C. E.*, Vol. XXV., Sept., 1891, pp. 253-8.

Syracuse, N. Y. A very complete and valuable report by J. J. R. Croes, C. E. *Report of Commissioners on Sources of Water Supply for the City of Syracuse, N. Y.*, Feb. 1, 1889.

Tokio, Japan. By Yeiji Nakahama. Gives a good history and description of the water supply of Tokio, Japan, with map and illustrations of wood pipes. *Proc. Am. W.-W. Assn.*, 1888, pp. 50-55; *Eng. & Build. Rec.*, March 17, [illegible].

Treatment and Sources. Address of President J. T. Fanning at the eighth annual meeting of of the American Water-Works Association. Treats of artificial and natural clarification, deep well supplies, and the protection of sources of supply. *Proc. Am. W.-W. Assn.*, [illegible], pp. 8-17; *Eng. News*, April [illegible], [illegible]. *Eng. & Build. Rec.*, May 5, [illegible].

Water Supply, continued.

Versailles Water Supply. A brief description of these interesting works, including the pumping works, and the famous fountains, by J. T. Fanning, submitted to the Philadelphia meeting of the Am. W.-W. Assn. *Eng. News*, May 2, 1891, p. [illegible].

Village. By S. H. Terry, before Bolton Congress of Sanitary Institute of Great Britain. Gives examples of villages with the different systems of supply, and shows the cost of the works. Lon. *Eng.*, Dec. 2, 1887; Lon. *Engineer*, Jan. 20, 1888.

Ware, Mass. A well and reservoir system for town of 6,000. A paper by F. L. Fuller, giving a full description and cost. *Jour. N. E. W.-W. Assn.*, Sept., 1887.

Water Supply. See *Artesian Wells. Dams and Reservoirs. Reservoirs. Wells.*

Water Tanks,

Determination of the Efficiency of Large Wrought Iron. By John D. Crehore. Gives a number of failures of iron tanks, and then gives method to be employed in computing the strength of tanks. *Van Nos. Eng. Mag.*, Vol. XXV., p. [illegible].

Iron. A paper by P. Kieran, and discussion on same. Describes 1,000,000 gallon tank at Fall River, Mass. *Jour. N. E. W.-W. Assn.*, September, 1887.

Iron. Various forms of existing and proposed reservoirs, tanks and towers in Europe. Many illustrations, with a theoretical study of stresses. *Sci. Am. Sup.*, Aug. 28, 1886.

New Design of the U. P. R. R. Plan, elevation, section and details. *R. R. Gaz.*, Sept. 5, 1879, pp. 614-5.

Proper Design for a Hoop Joint of a. By Prof. J. B. Johnson. Gives the results of tests made on the old forms, and proposes a new design. *R. R. Gaz.*, Jan. 6, 1883.

Sluice, with Equilibrium Valve. Illustrated description of a simple balanced valve for drawing water from a reservoir into a sluice. From *Ind. Eng.*, Nov. 22, 1890. *Eng. News*, Feb. 14, 1891, p. 163.

See *Stand Pipes*.

Water Towers,

And Pumping Station. Successful designs in the recent competition. Illus. *Eng. & Build. Rec.*, March 15, 1890, Vol. XXI., p. 225.

Bursting of, at Gravesend, L. I. A very satisfactory account of this accident with many cuts showing manner of erection and conditions of the wreck. *San. Eng.*, Oct. 23, 1886.

Collapse of, at Victoria, Texas. Top bulged inwards by wind. Cut given. *San. Eng.*, Nov. 27, 1886.

Failure of, at Kankakee, Ill. Causes given, computation made, and photographic cut. *San. Eng.*, Oct. 30, 1886 and Jan. 1, 1887.

Franklin, Mass. Illustrated description of the water tower that failed on Oct. 27, 1887. *San. Eng.*, Nov. 5, 1887.

Gives detail drawings of a tower 100 feet high, built of gas pipe, for the purpose of elevating a 6,000-gallon water tank at San Antonio, Texas. *Eng. & Build. Rec.*, March 24, 1888.

A New Form of. An elevated tank with hemispherical bottom. Illustrated description. *Eng. News*, Aug. 15, 1891, p. 135.

The Norton Tower of the Vyrnwy Aqueduct, Liverpool. Consisting of a tank supported on masonry walls. Description with illustration of details. *Eng. Record*, Oct. 3, 1891, pp. 275-80.

Water Towers, continued.

Of Weehawken, N. J. Elevations, sections, plan and perspective view. At once a residence, tank, support and ornament to the landscape. Well worthy of imitation. *Eng. News*, Nov. 6, 1886.

Yonkers, N. Y. Brick tower 98 ft. high surrounding an iron tank. Illus. *Eng. News*, Oct. 18, 1884, pp. [illegible] 184-5.

See *Stand Pipes. Water Tanks.*

Water Turbines. See *Water Motors.*

Water Waste.

A paper by Dexter Brackett. Causes, methods of prevention and results accomplished. *Jour. N. E. W. W. Assn.*, Dec., 1886.

Mr. J. Parry, in a report to the Water Committee of Liverpool, shows the great difference in waste between a district with good fittings and one with bad fittings. Editorial in *San. Eng.*, Sept. 24, 1885.

By Peter Milne, Jr., Water Purveyor of Brooklyn. A paper before the Am. Water Works Assn. Discusses various phases of the subject. *San. Eng.*, May 14, 1885.

Of Allegheny. Paper by T. P. Roberts, with discussion, read before and published by the *Engr's. Soc. West. Penn.*, Pittsburgh.

And its Detection. By W. Kummel. Result of experiments in Altona, Germany, with an acoustic apparatus which is said to be very delicate. *Jour. fur Gasbel. u. Wasserversorg.*, 1886, pp. 685-692. The apparatus used, so-called "Hydrophon," is described on pp. 865-6, 926-8 of the same journal.

Experiments made with the Deacon waste-water meter system at Boston. Illus. *Jour. Assn. Eng. Soc.*, Vol. I., p. 253.

At Frankfort-on-Main. From the German. Gives account of the amount of the waste and remedy by means of the Deacon district metre. *Eng. News*, May 1, 1885.

House to House Inspection, to Prevent. By M. L. Holman. Describes the methods so effectively used in St. Louis. No expensive apparatus required. *Jour. Assn. Eng. Soc.*, Vol. IV., p. 38.

The Prevention of. By Thomas Stewart. Awarded the Miller prize by the Inst. of C. E. Treats of the sources of waste and the methods of detecting and preventing the same. *Van Nos. Eng. Mag.*, Vol. XXVI., p. 119.

Prevention of Waste of. Paper by Thomas Stewart before the Inst. of C. E., describing methods of preventing waste by defective pipes and fittings. *Eng. News*, Dec. 17, 1880, pp. 507-8.

In Towns. Abstract from the report of Mr. A. R. Binnie to the Bradford Corporation upon the amount of water saved during thirteen months of inspection. Cost of inspection, including cost of all necessary apparatus, less than one-third of value of water saved. *R. R. & Eng. Jour.*, Feb., 1887.

See *Water Consumption.*

Water Wheels. See *Water Motors.*

Water-Works.

Auburn, N. Y., Suction and Siphon Pipe of. A 24-in. suction pipe 9,960 feet long, accompanied by an air pipe to exhaust the air at crests. Profile and description. *Eng. News*, Nov. 1, 1885, p. 287.

Barrie, Ont. Report on the proposed system of water-works and sewerage for town of Barrie, Ont. By Willis Chipman, C. E., Toronto, Ont. *Report*, 8vo. Toronto, Ont.

Boston.

Beacon Street Tunnel. Boston Water-Works. Short report on, accompanied by several plates and illustrations, giving an idea of manner of constructing, details of cross-section, arch centering, concrete arches and backing, etc. *Rep. City Engineer*, of Boston, 1881, p. [illegible] 38.

Water Works, *Boston*, continued.

Method Used to determine the Best Capacity to Give to Basin No. 5. Boston Water-Works. Includes consideration of area, topography and cost. By Desmond FitzGerald before the Boston Soc. C. E. *Jour. Assn. Eng. Soc.*, Sept. 1891, pp. 431-3. Abstract, *Eng. Record*, Oct. 24, 1891, pp. 331-2.

The New High Service Works of the Boston Water Supply. Description of Fisher Hill Reservoir, with detailed record of tests of Gaskill Pumping Engines and two boilers, at Chestnut Hill pumping station; also statement of cost of reservoir. Illustrated, *Eng. & Build. Record*, June 1, 1889.

Report of City Engineer of Boston for 1890, contains results of boiler trial of pumping station, duty of engines, progress of construction on a reservoir dam, and the usual tables of rainfall, consumption, yield of water-sheds, etc. pp. 17-56.

Bremen. Gives brief description of the works and machinery. *Van Nos. Eng. Mag.*, Vol., XVII., p. 399.

Bridgeport, Conn. A description of the water-works being built at Bridgeport by a new water company. Drawings, showing section of dam, plan of gate-chamber and profile of tunnel. *Eng. News*, April 9, 1887.

Brooklyn, N. Y., Extension.

Baldwin Storage Reservoir. Description, and detail drawings showing cross section through embankment and inlet chambers. *Eng. News*, July 25, 1891, p. 74.

Description of supply ponds, conduit waste weirs, etc. Details of dams and waste weirs on large folding inset. *Eng. News*, May 9.

A general review of the water works system followed by a detailed description of the works of the extension. An inset of details of the conduits and culverts, also general map is given. *Eng. News*, March 7, 1891, pp. 225-6, *et seq.*

Calais, Me. Gives short description, with drawings, of the water works at Calais, Me., costing $78,000. *San. Eng.*, Oct. 8, 1887.

Care of Hydrants. See *Hydrants.*

Charlottesville, Va. A gravity system, with reservoir 5¼ miles from the town, of 5,500 inhabitants. Has waste weir across the dam; well culvert for pipes, etc. Illustrated. *Jour. Assn. Eng. Soc.*, Vol. V., p. 82.

Charlottesville, Va. Description of the above work, with section through gate-house and dam. *San. Eng.*, May 21, 1887.

Chicago, Ills.

By Bernhard Feind, First Asst. City Engr. Interesting and valuable paper, describing the new lake and land tunnels, cribs, shafts, pumping stations and other engineering details of this important work. Illustrated with engravings and folding inset. *Eng. News*, July 5, 1890, pp. 3-5. See also a letter on the the same subject by Mr. Rudolph Hering, *id.* July 26, p. 83.

A Contrivance for Removing Anchor Ice at Chicago, Ill. Compressed air is forced through a small pipe at the mouth of the intake. Illustrated description. *Eng News*, June 27, 1891, p. 622.

The Four Mile Crib of the Chicago Water-Works. Illustrated description. By Simeon C. Colton, *Technograph* (Univ. of Ill. Annual), 1890-91, pp. 17-41.

In China and Japan. Papers by John W. Hart, James Orange and John H. T. Turner, describing the Water-Works of Shanghai, Hong-Kong and Yokohama. 5 folding plates and tables of dimensions, quantities, cost, tests, etc. *Proc Inst. C. E.*, Vol. C., 1890, pp. 217-89. Discussion, pp. 290-307. Correspondence, pp. 307-14.

Water Works, continued.

Circleville, Ohio. A brief description of the Circleville, Ohio, water-works. *Rept. Ohio Soc. Surv. & Eng.*, 1888, pp. 181-183.

Cleveland, O. By J. Whitelaw, before the American Water-Works Association. Gives a short description and history of the Cleveland water-works. *Proc. Amer. Water-Works Assn.*, 1888, pp. 114-126; *Eng. & Build. Rec.*, April 28, 1888.

Cohasset, Mass. Brief description of the Cohasset, Mass., water-works with plan showing arrangement of wells, collecting chamber, etc. *San. Eng.*, Dec. 3, 1887.

Of Colmar in Alsace. Description of the preliminary works, with drawings. *Nouvelles Annales de la Construction*, 1885, pp. 1-8. Also description of the machinery in *Bulletin de la Societe Industrielle de Mulhouse*, 1885, pp. 130-143.

Columbus, O., Annual Report of. Water-Works, 1889, besides the usual statistics of pumping, cost, etc., contains the reports of chief engineer and State geologist, on a new water supply proposed to be taken from a subterranean channel in the valley of Alum Creek.

Constantinople. By Fred. Brittault. Gives brief description of the old works and describes the works erected in 1883 by a private corporation. *Proc. Inst. C. E.*, Vol. LXXXVII., p. 331.

Cornell University Water-Works. Illustrated description. *Eng. & Build. Rec.*, Nov. 15, 1890, pp. 376-7.

Covered Reservoirs. See *Reservoirs.*

Croton.

By B. S. Church. Gives notes and suggestions on the Croton water-works and supply for the future, with discussion. *Trans. A. S. C. E.*, Vol. V., p. 167.

Head House at Shaft 17 on the New Croton Aqueduct. Illustrated description. *Eng. & Build. Rec.*, Nov. 17, 1888

Denver. By S. Fortier. Gives a full description of the water-works of Denver, with drawings of many of the details of construction. *Eng. News*, Sept. 22-29, 1888.

Dover, N. H. Report of Mr. P. M. Blake to the Mayor of Dover on the sources of water supply available for the city. Gives detailed estimate of cost. *Eng. News*, Nov. 26, 1887.

Early American Pumping and Distributing Plant. An illustrated description of the plant erected in Philadelphia in 1801. *Eng. News*, April 16, 1887.

East Orange. Description of work of the East Orange and Bloomfield Water Company, with details of the building over one of the wells. *San. Eng.*, June 23, 1887.

Edinburgh. Gives description of the Moorfoot works of the Edinburgh system, with cuts showing section of the embankments. Lon. *Eng.*, July 29, 1887.

Egyptian, Great. A description of the Katatbeh water-works: the original installation of ten Archimedes' screws, their failure and the present centrifugal apparatus. *Sci. Am. Sup.*, April 30, 1887.

Erfurt (Germany); and some new water-towers, oil-tanks and gas-holders. By Professor Intze. Description of the works, the water being obtained from the ground water by means of a well, and previous collecting pipes. Water is pumped into a tank holding over 145,000 gallons in the top of a tower, about 115 feet high. Cost of works about $90,000. Also description of a number of similar towers, gas-holders, etc. Illustrated. *Zeitschrift des Vereins deutscher Ingenieure*, 1886, pp. 15-31.

Water Works. continued.

Fall River, Mass. Detailed description. *Eng. News*, June 6. August 1, 1878.

Frankfort-on-the-Main. Description of the water-works and sewage systems, and the sewage disposal works, 3 folding plates. By ——— Monet. *Annales des P. & C.*, April, 1891, pp. 488-519.

Geneseo, N. Y., Water-Works. The water supply is pumped from a lake through cast iron mains into a reservoir three miles distant. Illustrations of reservoir and intake. *Eng. Record*, April 4, 1891, pp. 294-5.

Geneva, N. Y. Illustrations showing details of the 600 ft. intake pipe, pump-house, etc. *Eng. Record*, March 14, 1891, pp. 244-5.

Gouverneur, N. Y. Total pumpage is 300,000 gallons per day. A standpipe system with water taken from a river. The laying of mains across a river is described and general data given. *Eng. News*, Jan. 24, 1891, pp. 89-90.

Grand Rapids, Mich. An abstract from a paper by Mr. W. R. Coates, describing the method employed to carry the water mains for Grand Rapids water supply across Grand River. *San. Eng.*, June 18, 1887.

History and Statistics of American Water-Works. A valuable series of articles by J. J. R. Croes, describing in detail the water-works of the 19 cities of over 100,000 inhabitants with statistics, followed by briefer descriptions of those of smaller towns. *Eng. News*, March 5, 1879 and running through 1880, '81, '82, '83 and '84.

Hyde Park, Mass. Short description, accompanied by a few general and detailed illustrations. Supply taken from ground water by means of 64 pipe wells, 2 inches in diameter. Blake pumping engine of 1,500,000 gallons capacity per 24 hours. 14-inch cast-iron main. Reservoir with capacity of 1,800,000 gallons, 210 feet average head. Cost, $150,000. Population about 8,000. *San. Eng.*, Jan. 7, 1886.

Hyde Park, Ill., *W. W. Tunnel and Inlet Crib.* Illustrated description, the work by C. McLennan, C. E., Engineer in charge of the work. *Eng. News*, Jan. 19, 1889.

Irvington, N. Y. Specifications for, which includes a dam, reservoir and pipe system. *Eng. News*, Oct. 14, 1882, pp. 359-60.

Isle of Jersey. A description of the works on the Isle of Jersey for the supply of the towns of St. Heliers and St. Aubins, with 48 figures showing general plans of the works, various sections of the filter beds, pure water tank and high service reservoir. *Lon. Eng.*, July 8 and 22, 1887.

Janesville, Wis. Description giving details of artesian well, analysis of water, details of storage reservoir, pump house, screen house, valve well and distribution system. A clear and satisfactory description. *Eng. News*, July 20, 1889, Vol. XXII., p. 51.

Kansas City.

By G. W. Pearsons. A series of papers describing the construction of new works at Kansas City. *Eng. News*, Nov. 5, *et seq.*, 1887.

Siphon of the. By G. W. Pearsons, before the Annual Convention of the Am. Soc. of C. E. Gives description of the siphon constructed at the first water-works in Kansas City. *Trans. A. S. C. E.*, Vol. XVIII., May, 1888, pp. 137-152.

Some Details of Valves and Other Apparatus in use by the National Water-works Company at Kansas City, Mo. A paper by Frederick E. Sickels presenting drawings of a special pipe tapping machine, pressure regulator, air valves, etc. 7 plates. *Trans. A. S. C. E.*, Vol. XXIV., May, 1891, pp. 351-8.

Report of Col. Henry Flad and T. J. Whitman on their condition upon completion. Gives items of cost. *Eng. News*, April 21, 1877, pp. 91-3.

Water Works, *Kansas City*, continued.

Report of commission of experts on condition of. *Eng. News*, Aug. 6, 1881, pp. 316-19.

Kingston, Jamaica. A paper by Felix Target, detailing experiments on the height, etc., of jets from hydrants of the Kingston Water-works, Jamaica. *Van Nos. Eng. Mag.*, Vol. XIX., p. 193.

Lawrence, Mass. Full description. *Eng. News*, Sept. 5 and 19, 1878.

Lawrence, Mass. A paper by H. W. Rogers. Illustrated. *Jour. N. E. W-W. Assn.*, Dec., 1886.

Leyden, Holland. Illustrated description of engine house and water tower. *Eng. News*, Dec. 8, 1883.

Liberty, Va. Brief description, with illustrations of dam. Supply taken from a small stream. Natural head. Pipe line, 35,000 feet long; 6 inch pipe. Cost of construction, $10,728; cost of whole system, about $42,000. Population, 2,500. *San. Eng.*, Jan. 21, 1886.

Little Falls, N. Y. Complete description and report on construction of these water works by Stephen E. Babcock, the chief engineer. *First An. Rep. Water Commissioners*, Little Falls, N. Y., 1889.

Liverpool (Vyrnwy) Water Works. Description, giving history, principal dimensions and statement of cost of entire system. Illustrated. *Lon. Eng.*, Feb. 1, 1889, *et seq.*

Malden, Mass. Brief description of these water works, using 2½ inch wells as a source of supply. Illustrated. *Eng. & Build. Rec.*, Oct. 26, 1889.

Manchester, Mass. History and general description, with details of concrete dam, map of impounding area, etc., *Lon. Eng.*, Oct. 16, 1891, pp. 435-8. Details of outlet and straining well, valves and hydraulic apparatus for operating. *id.*, Oct. 30, pp. 495-8. Abstract showing sections of dam. *Eng. Record*, Nov. 14, 1891, p. 387.

Memphis, Tenn. Description of subterranean supply, geology of the region, method of boring artesian wells, details of tubing, etc. Illustrated. *Eng. Record*, Dec. 5, 1891, pp. 6-7.

Mercer, Pa. Gives brief description of the water-works at Mercer, Pa., with plans of engine house, well and filter. *Eng. & Build. Rec.*, June 9, 1888.

Of the Middle States. Table showing present status of. From advance sheet of Eng. News W. W. Manual. *Eng. News*, Dec. 15, 1888.

Of the Middle States, with table showing relative reservoir capacity and consumption of 66 works. List of W. W. Tunnels. *Eng. News*, December 22, 1888.

Middletown, Conn., Effluent Pipe. By E. P. Augur, before the Connecticut Association of Civil Engineers and Surveyors. Describes in detail the adjustable effluent pipe placed in the bottom of reservoir of Middletown water-works. *Proc. Conn. Assn. C. E. & Surv.*, 1888, pp. 24-35.

Milwaukee. The New Michigan Intake of the. 3050 feet of tunnel, and 800 feet of 60 inch pipe laid in a trench on the lake bottom, reaching out to 40 feet of water. Plan and profile. *Eng. News*, Oct. 25, 1890, pp. 366-7.

Minneapolis Pumps. Gives brief illustrated description of the pumps in use at the Minneapolis water works. *Eng. & Build. Rec.*, September 15, 1888.

Montevideo Water Works. Paper descriptive of this system of Water Works in the Republic of Uruguay. By William Galway, M. Inst. C. E. *Proc. Inst. C. E.*, Vol. XCVI., p. 297.

Nagpur, India. Gives a general description of the works for supplying Nagpur, India, with 15 gallons per capita per diem by a gravity system; also contains much information in regard to rainfall, evaporation, discharge and consump-

30

Water Works. *Nagpur, India*, continued.

tion. *Ind. Eng.*, March 10 *et seq.*, 1888. Abstracted *Proc. Inst. C. E.*, Vol. XCIII, p. 532.

Naples, Italy.

The new works, completed in 1885, convey the waters of the Serino springs yielding over 170,000 cubic meters in 24 hours to the city. An aqueduct 80 km. long, of which 14.5 km. is in tunnel, 1.8 km. bridges, and 1.168 km. siphons, leads to Cancello. Section 1.60 by 2.00 meters; grade 1 : 3000. Vertical or nearly vertical falls, aggregating 300 meters, are arranged so as to be utilized for power. Manholes and ventilators every 300 m. A single siphon 19 km. long, consisting of 3 pipes, extends from Cancello to Naples, leading to two reservoirs, one for the high and one for the low service. These reservoirs are excavated in the rock 30 meters below the surface, to preserve a low temperature. *Wochenschrift des Oesterr. Ing. und Arch.-V.*, 1886, pp. 4-6.

A two-page abstract. Collecting galleries 2,000 feet in length, masonry conduit 37 miles long, three iron pipes under pressure for 14 miles, and reservoirs at different levels are some of the distinguishing features. *Proc. Inst. C. E.*, Vol. LXXXIV., p. 468.

Condensed description taken from *Le Genie Civil*, of the collecting galleries, conduits, aqueducts, siphons, gate houses, etc. Illus. *Eng. & Build Rec.*, Dec. 6, 13, 1890.

New London, Conn.

Water Works, High Service. Description of the motor which is driven from the gravity main and operates the pumps for the high service system. *Eng. & Build. Rec.*, Oct. 11, 1890, pp. 298-9.

Report for 1890, of the Board of Water Commissioners of New London, Conn. The report contains also a brief description of the mains, tower and tank, motor and pump, of the new works for additional supply, just completed.

Newark, N. J. Full description of the works and supply. *Eng. News*, Aug. 9, Oct. 18, 1879.

Newton, Mass. Extension at. Illustrated description of the various parts of this work. Filter gallery and covered reservoir described. *Eng. Record*, Nov. 28, 1891, pp. 418-19.

Ottawa, Canada. Full description. *Eng. News*, Feb. 14, March 7, 1878.

Paris. The Water-Works Constructed from 1880 to 1880 to Increase the Water Supply of Paris. A description of the pumping works and reservoirs. Illustrated by 6 folding plates. By —— Bechmann. *Annales des P & C.* Feb., 1891, pp. 133-300.

Pawtucket, R. I. Description of dam and artificial filter gallery, with detailed account of cost. *Eng. News*, June 1, 1889.

Peoria Water Works. Detailed description of construction of well, pumping house, and reservoir. By J. A. Harman. *Report of the 6th Annual Meeting of the Ill. Soc. Eng. and Surveyors*, 1891, pp. 38-44.

Pressburg (Austria). City has a population of 48,000; water obtained from a well 8 feet 4 inches in diameter on the bank of the Danube. Pumped out of a covered reservoir. *Wochenschr. d. Oester. Ing. u. Arch. Vereins*, 1886, pp. 219-222.

Prague. Description of the Prague water-works machinery with two-page plate and cuts showing details. Lon. *Eng.*, Sept. 9, *et seq.*, 1887.

Putnam, Conn. Brief description of the system, capacity 1,000,000 galls. in 24 hours, 2½ miles of 10-inch pipe leading to a wrought-iron stand-pipe. *San. Eng.*, April 9, 1887.

Quebec. Illustrated description. *Eng. News*, May 9, 30, 1878.

Water Works, continued.

Racine. By G. A. Ellis, before the Boston Society of Civil Engineers. Gives a very full description of the water-works at Racine, Wis., and describes method used in the construction of the same. *Jour. Assn. Eng. Soc.*, April, [illegible]; *Eng. News*, May 12, [illegible]

Of Red Bank, N. J. Described and illustrated. Consists of a well fifteen feet in diameter and [illegible] feet deep, pumps, curbing, etc. *Eng. News*, August [illegible], [illegible].

Remscheid. Supply taken from underground perforated pipes. The drainage ground limited by retaining walls to solid rock, water drawn off as desired. Lift of over [illegible] ft. Population, 35 to 50 thousand. Translated from the German. Lon. *Engineer*, April 10, [illegible].

Richmond, Va. Part I: History. General Description. Vertical Water Wheels, Pumps, and Adjustable Journals. Illus. *Eng. Record*, May 23, 1891, pp. [illegible].

Richmond, Va. A paper by Charles E. Bolling, superintendent. A description of the water-works from 1880 to date. Discussion, pp. 5. *Jour. N. E. Water Works Assn.*, Sept., [illegible], Vol. IV., p. 39.

Rochester, N. Y. Construction Details of. Eng. & Build. Rec., Nov. [illegible], [illegible], pp. 412-13, *et seq.*

Rochester, N. Y. Descriptive article. *Eng. & Build. Rec.*, Aug. 15, [illegible], p. [illegible].

Rosario. Two compound beam engines with surface condensers, two direct acting bucket and plunger pumps, two sets deep well continuous action pumps. Illustrated. Lon. *Engineer*, May 14, [illegible].

San Mateo. Spring Valley Water-Works, San Francisco. Full illustrated description. *Eng. News*, March 2, [illegible].

Seattle, Wash. Report of a system of works by Benezette Williams. Includes a dam, reservoir, aqueduct [illegible] miles long, with estimate of cost. Pamphlet of [illegible] pages; Feb., [illegible].

Shreveport, La. Extracts from the Specifications for the Proposed Works at. Eng. News, June [illegible], [illegible].

Small Cities. By G. W. Pearsons. A brief popular essay on the subject. *Report Mich. Assn. Surveyors*, [illegible].

Of the Southern States. Eng. News, Feb. 9, [illegible].

Southington, Conn. By Theo. H. McKenzie. Description of the works constructed by a private corporation, consisting of a dam [illegible] ft. long, [illegible] ft. wide, and 36 ft. high, of earth, with masonry and puddle heart, forming storage reservoir of 23 acres or [illegible] galls. capacity; distributing reservoir, with masonry dam, 170 ft. long, [illegible] ft. high, and 5 ft. wide at top; 11½ miles of pipe. Illustrated with detailed drawings. *Trans. A. S. C. E.*, Dec., [illegible].

Southwark and Vauxhall. Full description, with details of pumping engines, etc. Lon. *Engineer*, July 1, etc., 1887.

Stand Pipes. See *Stand Pipes.*

St. Denis (France). Descriptive account of the works for supplying this town, of about 48,000 inhabitants, with water. *Memoires de la Societe des Ingenieurs Civils*, May, [illegible], pp. [illegible].

St. Louis.

New Inlet Tunnel and Tower of. Description, and many of detail drawings. The coffer dam is also illustrated and described. *Eng. News*, July 4, [illegible], pp. 4-5.

Settling Basins for the Low Service Extension of the St. Louis, Mo., Water-Works. The settling basins are six in number, each [illegible] ft. by [illegible] ft. Brief description, with general plan of the works, and inset showing many details, of the basins. *Eng. News*, April 18, 1891, p. [illegible]

Water Works, continued.

Stop Valve. See *Stop Valves.*

Stratford-on-Avon. Full detail drawings and description. Gravitation system, with reservoir. Town of 8,000 inhabitants. Lon. *Engineer*, November 27, 1885.

Tables for Power of Compound Pumping Engines. See *Pumping Engines.*

Water-Works, Toronto. Annual Report of the superintendent of the Water Works. Contains Maintenance Accounts, Record of pumping and coal consumed, with comparative statement for various years, number of hydrants, etc. By Wm. Hamilton, Supt. Year ending Dec., 1889.

In the U. S. and Canada. Summary of data in the *Manual of American Water Works*. Editorial, 7 tables and statistics. *Eng. News*, June 28 and July 5, 1890, pp. 623-5, 13.

Of the United States and Canada. A study of the statistics in the forthcoming Manual of American Water-Works, being in substance the introduction to that work. *Eng. News*, Nov. 28, 1891, *et seq.*

Vancouver, B. C. By H. B. Smith. This paper contains a full description of the dam and distributing mains; five folding plates. *Trans. Can. Soc. C. E.*, Vol. III., Pt. II., 1890, pp. 313-57.

Ware, Mass., with plans and details of cost. *San. Eng.*, July 16, 1887.

Watertown, N. Y. General description of Pumping Station. See *Pumping Station.*

Wellesley, Mass. By F. L. Fuller, Engr. Interesting details of value in designing new works for small towns. Illustrated. *Jour. Assn. Eng. Soc.*, Vol. IV., p. 401.

Wem, England. A well designed system of works for small towns; includes well and covered reservoir. Fully illustrated. Total cost about $67,000. Lon. *Engineer*, Sept. 18, 1885.

Widnes. Illustrated description of the new pumps and method of setting, for the Widnes Water-Works. Lon. *Engineer*, Aug. 26, 1887.

Ypsilanti, Mich. A condensed article from a paper by W. R. Coates, read before Mich. Eng. Society, describing and discussing these works; stating cost, capacity, etc. Illustrated. *Eng. News*, May 10, 1890, Vol. XXIII., p. 439.

See *Hydrants. Pumping Engines. Water Distribution Systems. Water Supply.*

Water-Works Association.

American. Report of the tenth annual convention. *Eng. News*, May 31 and June 7, 1890.

Report of the Philadelphia meeting of the Am. Water-Works Assn. Contains brief extracts from various papers of interest. *Eng. Record*, April 25, 1891, pp. 340-3.

Trans. of the N. E. Assn., 1883. Contains valuable discussions on weights of cast-iron pipes, flushing street-mains and house-tanks. A graphical exhibit is given of the weights of pipe used in over 250 cities; also the weights given by various formulas. Albert S. Glover, Sec'y., Newton, Mass.

Water-Works Construction.

Details of. By William R. Billings. A series of articles on the practical details of constructing a water-works plant. *San. Eng.*, Sept. 24, 1887.

Water-Works Head Gate. *The Roller-Curtain Pattern.* Illustrated. *Eng. News*, Dec. 11, 1886.

Water-Works Management.

Practical Details of the Management of. A valuable paper by Edwin Darling before the N. E. W.-W. Assn. *Jour. New Eng. W.-W. Assn.*, September, 1887.

Water-Works Pressure Records.

Abstract of a paper by Charles A. Hague read before the Am. W.-W. Assn., noting the many advantages of such a record. *Eng. Record*, April 25, 1891, pp. 345-6.

Water-Works Records. Paper by Mr. J. M. Diven treating this important subject from a practical standpoint. *Am. W.-W. Assn.*, Louisville meeting, 1889, p. 18.

Water-Works Safety Valves.

Safety Valves. By S. E. Babcock, before the New Eng. W.-W. Assn. Discusses the relief and safety valves in water-works distribution system. Abstracted *Eng. & Build. Rec.*, Sept. 1, 1888.

Water-Works Securities. An analysis of this subject, by William Reinecke. *Report of the 11th annual meeting of the Am. W.-W. Assn.*, 1891, pp. 67-73.

Water-Works Statistics.

For many New England cities, for the year 1886, compiled in *Jour. New Eng. W.-W. Assn.*, June, 1887.

U. S. Compiled under the direction of Prof. Trowbridge; 260 pp. of valuable matter, giving details of all the more important works in the country in 1880, with illustrations. *Tenth Census Report*, Vol. XVII.

Water-Works and Population of United States and Canada. Diagrams showing growth of, from 1850 to 1888. *Eng. News*, April 27, 1889.

Comparison of Water Supply Systems from a Financial Point of View. By J. Leland Fitz Gerald. Gives statistics from many cities with reference to the kind of system, with comparison and discussion of efficiency. *Trans. A. S. C. E.*, Vol. XXIV., April, 1891, pp. 247-59. Discussion, pp. 259-64.

Water-Works, Test of.

By the use of Pressure Gauges. Paper by Robt. J. Johnson. Read before the St. Paul Society of Civil Engineers, Feb. 4, 1889. *Jour. Assn. Eng. Soc.*, June, 1889, Vol. VIII., p. 351.

Waterways.

Internal Improvement. A brief account of the recent convention held at St. Paul to consider the improvement of the Northwestern waterways. Eight states and two territories, represented by 926 delegates. Very important, as indicating the probable course of public policy concerning the river and harbor works. Editorial in *Bradstreet's*, Sept. 12, 1885.

Lake Michigan and Mississippi Proposed Water-way. Levels of the Lakes as affected by.

See *Water-Way.*

Waterways and Railroads. A paper by U. A. Forbes, before the Soc. of Arts' Canal Conference. Gives the history of the use and progress of waterways and railroads in England and Wales, and their mutual influence on each other. *Jour. Soc. Arts*, May 21, 1888.

Wave Impact *and the Stability of the Superstructure of Breakwaters.* Brief article by L. d'Auria, deducing a formula for the force of impact. *Jour. Frank. Inst.*, Nov., 1890, pp. 373-6.

Wave Motor. An ingenious device for compressing air by the lifting action of waves on a floating "wave motor," which is then transmitted to shore by means of a flexible tube. A pamphlet of 13 pp., describing same, sent by the inventor, T. Duffy, 228 Geary street, San Francisco.

Wave Pressure *on Exposed Structures.* Editorial suggested by the plans for the Diamond Shoals Lighthouse. *Eng. News*, September 6, 1890, pp. 211-4.

See also discussion in *Eng. News*, Nov. 15, 1890.

Wave Reaction. Remarks upon by W. P. Rice. Considered in its relations to the forms of breakwaters. *Jour. Assn. Eng. Soc.*, Vol. IV., p. 319.

Waves.

A resume of our present knowledge of wave motion. *Sci. Am. Sup.*, Nov. 19, 1887.

Memoir on the Experimental Study of. By M. L. E. Bertin. *Van Nos. Eng. Mag.*, Vol. VIII., p. 491.

Ship. By Sir William Thomson. An important lecture delivered before the Institution of Mechanical Engineers, at Edinburgh. Treats of the forms and motions of waves produced by ships. *Sci. Am. Sup.*, Oct. 15, 1887.

Transporting Power of. A good summary of facts is given in an editorial in Lon. *Eng.*, July 15, 1887. *Sci. Am. Sup.*, Sept. 10, 1883.

Weather Service. Two papers on, describing the U. S. Signal Service, and another on Forcasting Weather. *Proc. Ohio Soc. Eng. & Surv.*, 1887.

See *Meteorology.*

Weighbridge, Twenty-Ton *at the Paris Exposition*, 1889. Description of an interesting automatic weighing machine. Lon. *Eng.*, September 6, 1889. Vol. XLVIII., p. 289.

Weighing Machines, *Automatic.* For weighing grain. *Trans. Liverpool Eng. Soc.*, Vol. VIII., p. 96, (1887).

See *Electric Weighing Machines.*

Weights and Measures.

Labors of the International Bureau of Weights and Measures. Valuable and interesting article from *La Nature.* Illustrated. *Sci. Am. Sup.*, March 8, 1890. p. 11,841.

The Necessity of a Definite and Determinate System of. By Chas C. Breed. Read before the Western Society of Engineers, Nov. 14, 1888. *Jour. Asso. Eng. Soc.*, March, 1889.

Report of the Committee of the Boston Society on. Comprising a canvass of the society regarding metric reform, with opinions of members and a notice of the recent act of Congress. *Jour. Asso. Eng. Soc.*, July, 1888, Vol. VII., pp. 264-271.

Weirs.

Automatic Waste. By A. D. Foote, before the Am. Soc. of C. E. Gives description with detailed drawing of an automatic waste weir. *Trans. A. S. C. E.*, Vol. XVIII., Sept., 1888, pp. 59-73.

Effect on the Discharge of the Height of Crest above Bottom of Canal. Based on a new series of experiments by Bazin. *Annales des P. & C.*, October, 1888.

New Experiments on the Flow over Weirs. By H. Bazin. Abstract from *Annales des P. & C.*, Vol. XVI., 1888, p. 393. *Proc. Inst. C. E.*, Vol. XCVI., 1888, p. 372.

Bazin's Experiments with. Very complete experiments on a large scale. *Annales des P. & C.*, Oct., 1888. Translated for the *Proc. Eng. Club.*, *Phila.* Condensed and printed with several tables and diagrams in *Eng. News*, Dec. 27, 1890, pp. 577-8.

Formulas for. A compilation of the plotted curves of the co-efficients to be applied to the theoretical to obtain the actual discharge, as obtained by fifteen different experimenters, including the best American experiments. *Trans. Inst. C. E.*, Vol. LXXXIII., p. 377.

Three Recent Designs for Movable Weirs. By Herr Nakong in *Centralblatt der Bauverwaltung.* Abstract with drawings in *Eng. News*, October 18, 1890, pp. 315-6.

Submerged. Co-efficient of discharge found in large rivers in India. The discharges were computed from cross-sections, and slopes found from high-water marks, and calculated from Humphrey and Abbot's formula for flood stages,

Weirs, *Submerged*, continued.

the dams then becoming submerged weirs. The result of questionable value from the uncertainties in computing the discharges. *Trans. Inst. C. E.*, Vol. LXXXV., p. 307.

Waste Weirs. A Method of Determining the Capacity of the Waste Weir of a storage reservoir of a torrential stream. A paper by Wm. J. McAlpine, Hon. M. Am. Soc. C. E. showing that, when properly constructed, earth or "dirt dams," are perfectly safe structures. *Eng. News*, June 29, 1889.

Welding Metals. Description of Barington's method of welding metals and spinning and shaping tubes. The rod or tube is forced through a revolving die the heat for welding being generated by friction of die on tube. *Jour. Frank. Inst.*, Nov. 1891, pp. 321-328.

By Electricity. See *Electric Welding.*

Well Boring.

A description of methods used in boring two wells, 6 feet diameter and 70 feet deep, in chalk formation, at Southampton Water Works. These the largest wells ever bored at one operation. *Proc. Inst. C. E.*, Vol. XC., p. 33.

By Steam With a Spring-Pole. By Benj. Smith Lyman. An unpatented device whereby this useful attachment may be utilized when steam is used. *E. & M. Jour.*, Feb. 20, 1886.

See *Artesian Wells.*

Well-Sinking. *The Wagner Process of Sinking Wells*, or underground reservoirs for water supplies. The process is by use of steam under pressure, and a very interesting account of recent well-sinking is given in a report by Geo. E. Pond, Assistant Quartermaster, U. S. A. *Am. Eng.*, March 5, 1890, p. 100.

Wells.

Capping Flowing. Gives a simple and effective way of shutting off flowing wells. *Sci. Am. Sup.*, Oct. 22, 1887.

For Cities. A discussion by A. A. Buneman on the use of driven wells. *Eng. News*, May 19, 1888.

Driven. A paper by J. C. Hoadley, being a theoretical discussion of its capacity for water supply. *Proc. Soc. Arts*, Boston, for 1882-3, p. 113.

Driven Wells as a Source of Water-Supply. A paper by F. F. Forbes describing briefly some special details in the piping, intended to secure durability, with discussion of considerable value. *Jour. New Eng. W.-W. Assn.*, March 1891, pp. 141-4.

Driven Wells as a Source of. By Albert F. Noyes. A paper read before the American Water-Works Association on the driven well system as a source of or a means of obtaining a water supply. *San. Eng.*, Aug. 6, 1887.

Temperatures in. See *Temperatures.*

Tubular, for Domestic Water Supply. By J. C. Hoadley, C. E. A paper devoted chiefly to a consideration of the nature and conditions of flow of subsurface or ground water. Accompanied by tables of flow and results of pumpings at different wells, with suggestions upon the sinking of tubular wells. Illustrated by a number of diagrams and sections. Pp. 90. Supplement to V. *Annual Report State Board of Health, etc.*, of Massachusetts, 1884.

Yield of Wells Sunk in Permeable Soils. A theoretical discussion based on certain assumptions and experimental data. By Fossa Mancini in *Annales des P. & C.*, June, 1890. Abstract. *Eng. News*, March 21, 1891, p. 269.

Wells. See *Artesian Wells. Tube Wells. Water-Works.*

Welsbach, *Incandescent Gas Light.* Description of. Gives results of tests, gas consumption and candle power. *Mech. World*, Dec. 8, 1888.

See *Gas Light.*

Western Union Time System. By F. R. Hardett, C. E., and R. P. Miller, C. E. A valuable description of the Western Union Time Service. *Sch. Mines Quart.* Columbia College, April, 1889, Vol. X., p. 218.

Wheel Bases. *Length of Rigid Wheel Bases.* New England Railroad Club. *R. R. Gaz.*, May 16, 1890, p. 337.

Wheel Tires. *Wheel Tires and Rails, Wear of.* By Richard Helmholz. Discusses the wear on wheel-tires and rails on curves, and rolling stock appliances for reducing the same. *Zeitschr. d. V. deutscher Ingenieur*, 1885, pp. 330-353; abstracted *Proc. Inst. C. E.*, pp. 549-554.

Wheels.

Vauclain's Wrought Iron Wheel Centers. Illustrated description of the method of their manufacture; being the report of a committee of the Franklin Institute. *R. R. Gaz.*, July 10, 1891, pp. 477-8.

See *Locomotive Wheels, Car Wheels.*

Wheels and Axles. See *Axles.*

Wheels and Rails.

Cylindrical Wheels and Flat-Topped Rails. Abstract of a paper read by Mr. Don. J. Whittemore, Civ. Engr., C., M. & St. P. Ry., before Am. Soc. Civ. Engrs. In the nature of a discussion of the recent report of the "Committee on the Proper Relation to Each Other of the Sections of Railway Wheels and Rails." *Eng. News*, Feb. 9, 1889, *et seq.*

Cylindrical Wheels and Flat-Topped Rails. Paper by D. J. Whittemore, Chf. Engr. C., M. & St. P. Ry., at Jan. 16 meeting, A. S. C. E. *R. R. Gaz.*, Jan. 25, 1889.

Discussion on cylindrical wheels and Flat-Topped Rails is given in which many prominent engineers took part. Also the final report of the committee of *Am. Soc. C. E.*, on the Proper Relation to each other of the Section of Wheels and Rails; together with appendices *A* and *B* including summary of questions and answers, and correspondence on the same subject are joined in. *Trans. A. S. C. E.*, Oct., 1889, Vol. XXI., p. 153-300.

On Southern Pacific. Letter from W. G. Curtis, to the committee of Am. Soc. C. E., giving valuable data and information concerning wheel records on Southern Pacific Ry. *Ry. Rev.*, Feb., 1890, p. 68.

Windmill Sails. *Best Angles for Various Speeds*, By A. R. Wolff. The table given in *Am. Eng.*, June 5, 1885.

Wind-Mills *for Water Supply.* By A. R. Wolff. Gives tables of capacity and economy of different sized wheels for a 16-mile wind, when this power may be developed on the average 8 hours a day. *Eng. News*, Dec. 5, 1885.

Wind Pressure.

A table giving the relation between the velocity and the pressure at different temperatures. *Am. Eng.*, May 1, 1885.

An article giving formulæ and tables for wind pressure. *Van Nos. Eng. Mag.*, Vol. XXVI., p. 89.

By Wm. Ferrel. Gives formulæ for wind pressure. *Van Nos. Eng. Mag.*, Vol. XXVII., p. 140.

Brief article calling attention to actual wind pressure of from 40.0 to 65.8 lbs per sq. ft., and observed velocities of from 57 to 78 miles per hour. Lon. *Eng.*, March 14, 1890, p. 333.

Short article on maximum observed pressures. Reprinted from Lon. *Eng.*; *Am. Arch.*, July 19, 1890, p. 45.

Data concerning actual wind pressure at Forth Bridge at various dates during construction from 1883 to 1890. *R. R. Gaz.*, April 18, 1890, p. 65.

An Examination into the Method of Determining. By F. Collingwood. Discusses a number of the formulæ in common use and points out some of their errors; also gives tables of pressures computed by different formulæ for comparison. *Van Nos. Eng. Mag.*, Vol. XXV., p. 218.

Extract from evidence of Wm. Pole, before the Tay Bridge Commission. *Van Nos. Eng. Mag.*, Vol. XXIII., p. 163.

Wind Pressure, continued.

On Railway Structures. Report of the committee of London Board of Trade to consider the question of wind pressure on railway structures. *Van Nos. Eng. Mag.*, Vol. XXV., p. 467.

And the Measurement of Wind Velocities. An article by Asst. Prof. C. F. Marvin, U. S. Sig. Service, giving some experiments on Wind Pressure made on Mt. Washington, and discussing reduction formulas for anemometers. *Eng. News*, Dec. 13, 1890, pp. 525-1.

General Discussion of, upon Spherical, Cylindrical and Conical Surfaces. By P. H. Philbrick. *Van Nos. Eng. Mag.*, December, 1884.

Velocity and Pressure. A table showing the average monthly movement of, in miles per hour at a number of cities; also a table showing the relation between velocity and pressure. *Building*, April 2, 1887.

See *Atmosphere. Air Resistance. Chimneys. Lighthouses.*

Wind Pressure on Bridges.

By C. Shaler-Smith, with discussion. The paper and discussion is the standard authority on this subject in America. *Trans. A. S. C. E.*, Vol. X., (1881), p. 139. Also paper by F. Collingwood, p. 172.

Also paper by Ashbel Welch, with discussion in Vol. IX., p. 391-400.

Winding Engines. See *Mining Machinery.*

Winter Navigation *and Ice Breaking*, steamers designed for. Paper by Robert Runeberg, Assoc. M. Inst. C. E. Gives theoretic discussion of H. P. required to break ice of various thicknesses, and gives a few details of construction. Pp. 24. *Proc. Inst. C. E.*, Vol. XCVII., pp. 277, London, 1889.

Wire.

A formula giving the relation between the diameter in inches and the gauge, number (American standard) and the use of this formula in electrical designing. By Carl Hering. *Elec. Eng.*, (N. Y.,) April, 1886.

Steel Wire of High Tenacity. By Dr. Percy, before the Iron and Steel Institute Gives results of tests and chemical composition of specimens of steel wire having a breaking strength of about 150,000 lbs. Lon. *Eng.*, July 23, 1886.

Permanent Elongation in. See *Metals.*

Wire Gauges.

Chart showing properties of all wire gauges in use. Compiled by S. S. Wheeler. The most complete exposition of the subject yet made. *Elec. World*, Nov. 12, 1887.

Report of the Committee of the National Electric Light Association at Philadelphia meeting, February, 1887. Containing probably the best recommendations as to a standard gauge that have yet been made. *Elec. World*, Feb. 26, 1887; also in *Electrician & Elec. Eng.* for March.

Table of all the principal wire gauges, giving diameters and cross-sections. By Carl Hering. Very useful in electrical designing. *Elec. Eng.*, (N. Y.), May, 1886.

Wire Rods.

The manufacture in the United States. A paper translated from the *Revue Universelle des Mines*, being the result of the observations of two Belgian engineers in this country. Fully illustrated. *Iron Age*, Sept. 3, 1885.

Wire Rope.

A paper by Wm. B. Palmer, giving actual practice in use of wire rope hoisting and guys for derricks; deducing the stress on the ropes in several cases and comparing the strength of rope with stress, etc., etc. Pp. 10. *Conn. Assn. C. E. & Surveyor's*, 1889.

Gives description of the process of manufacturing, from drawing the wire to twisting the rope. *Sci. Am. Sup.*, July 2, 1887.

Wire Rope, continued.

The Locked Coil. A new style of wire rope, being smooth on the outside like a round rod, and the wires made of special shapes so as to interlock with each other, thus preventing any wire from moving from its place. Illustrated. *Iron*, Sept. 18, 1885.

Treats of the use of wires of special sections producing cables of true cylindrical form. *Sci. Am. Sup.*, June 11, 1887.

And Wire, Strength of Welds in. A description of some tests recently made by Herr A. Martens, under the direction of the German Minister of Public Works. *R. R. Gaz.*, Dec. 7, 1888.

See *Cables*.

Wiring Chart. Diagram for use for calculating the length of conductor necessary to give a certain drop in voltage. By Thos. J. Fay. *Elec. World*, Aug. 1, 1891, p. 70. *Eng. News*, Sept. 19, 1891, p. 242. *Elec. Eng.*, Oct. 8, 1890.

Wire Rope Haulage. See *Mining Machinery*.

Wire Ropeway. See *Cements of the Gate, etc.*

Wires. See *Electrical Conductors*.

Wood.

Expansion of, by Absorption of Moisture in the direction of the fibres. Paper by Prof. DeVolson Wood, giving results of experiments. *Trans. A. S. M. E.*, Vol. X., 1889.

Fluctuations of Moisture in Wood during Seasoning. Details of experiments made to observe the variation in moisture in different kinds of wood, and learn the months during which the greatest amount of seasoning takes place. *Eng. News*, April 2 and 16.

Preservation of. See *Timber*.

Preserved Under Water. Recent discovery of remains of old Roman bridge piers in river Trent, the oaken timbers being in good preservation. Lon. *Engineer*, Nov. 7, 1884.

See *Timber*.

Wood Pavements. See *Pavements*.

Wood Pulp.

Some of its Peculiarities. By M. L. Deering. Cleveland. Describes methods used in his own factory and the difficulty to be overcome. With discussion. *Jour. Assn. Eng. Soc.*, Vol. V., p. 414.

Vulcanisation. By M. L. Deering. Describes a new method of treating fibrous material. *Jour. Assn. Eng. Soc.*, Feb., 1888, pp. 59-55.

Wood-Working Machinery on exhibition at World's Fair, New Orleans. *Mfr. & Build.*, March, 1885.

A description of the recent improvements in wood-working machinery in England, and their application to cheapening the cost of pattern-making. Mr. Geo. Richards. Lon. *Eng.* Feb. 13, 1881.

Wooden Beams.

Transverse Strength of, being a new formula by R. F. Hartford, C. E., which harmonizes with latest experiments, including those of Lanza, Rodman, Haskell, Graham, Smith and others. *Am. Eng.*, Dec. 12, 1884.

Transverse Strength of. By Prof. Lanza. Gives results of experiments on large beams. *Jour. Assn. Eng. Soc.*, p. 133, Vol. II.

See *Beams*.

Wooden Columns.

Full Size. Strength of, as tested by Prof. Lanza on the Watertown Arsenal Machine, for the Boston Manufacturers' Mutual Fire Insurance Co., in 1882. Probably the most extensive and satisfactory set of tests of wooden columns

Wooden Columns, *Full size*, continued.

ever made. A supplement of 51 pp. and 56 cuts, with details of tests, appended to *Trans. Soc. Arts*, Boston, 1881-2.

vs. Cast Iron Posts. Brief article describing behavior of pitch-pine columns during a severe fire in a large warehouse in London. Columns burned on outside, but supported their load. *Mechanics*, August, 1889, Vol. XI., p. 193.

Woodite. By Sir Edward Reed. Discusses the use of a new structural material, the base of which is rubber. It appears to be coming into general use in many ways. *Sci. Am. Sup.*, March 31, 1888.

Woods.

American. Modulus of Elasticity as Determined by Vibrations. By M. C. Ihlseng. Gives details of method employed and table of values of the modulus of elasticity. *Van Nos. Eng. Mag.*, Vol. XIX., p. 8.

Cuban. By Esteban D. Estroda. An investigation of the strength and other properties of Cuban woods used in engineering construction. *Van. Nos. Eng Mag.*, Vol. XXIX., pp. 417 and 443.

Michigan. The Strength of. A paper read before the Michigan Engineering Society by Prof. R. C. Carpenter, January, 1889. *Am. Eng.*, March 6, 1889.

Of Nicaragua. An Investigation of the Strength and other properties of some of the Nicaragua woods used in engineering construction and in the arts, conducted at Cornell University. By Rufus Flint. Thirty-three kinds examined. A valuable study. *Sch. Mines Quart.*, Oct., 1887, p. 633.

See *Timber*.

World's Columbian Exposition. See *Exposition*.

Work *Developed in Rail and Girder Rolling.* Experiments made in St. Petersburg. *Proc. Inst. C. E.*, Vol. LXXXI., p. 414.

Working Loads. Natural working loads for building materials and structures, adopted by the Austrian Association of Engineers and Architects. *Eng. & Build. Rec.* Dec. 14, 1889, Vol. XXI., p. 25.

Yard. *On the Relation of the Yard to the Metre.* A report to the U. S. Coast and Geodetic Survey. By O. H. Tittmann, assistant in charge of weights and measures. *U. S. C. and G. Survey Bulletin No. 9*, June 15, 1889.

Yacht. *Grace Darling.* Gives a brief description, with two page plate, showing longitudinal section, deck plan and cabin plan of the steam yacht "Grace Darling." Length over all, 157 ft., breadth, 19½ ft., depth, 11 ft.; draught 8 ft.; tonnage, 239 tons, engines, quadruple expansion; cylinders, 10 in., 14 in., 20 in. and 28 in. diameter, with 20 in. stroke; 160 revolutions per minute; 300 horse-power, with boiler pressure of 180 lbs. *Lon. Engineer*, March 16, 1888.

Yachts. *Racing and Cruising.* Remarks on the length, beam and sail area of racing and cruising yachts, with suggestions for defining cruisers and for regulating races. Gives tables showing leading dimensions and antics of British and American yachts. *Lon. Eng.*, Nov. 25, *et seq.*, 1887.

Z-Bars. See *Columns*.

Zinc. Its unfitness for standards of length from its slow response to changes in temperature. In other words, its length is not truly indicated by its temperature when the temperature changes rapidly, as in field observations. Prof Papers Corps of Engrs. U. S. A., No. 24. Triangulation of U. S. Lake Survey, p. 180.

in Boilers. See *Boilers, Marine*.

Zinc and Lead. *Production of in 1888.* See *Lead*.

www.ingramcontent.com/pod-product-compliance
Lightning Source LLC
Chambersburg PA
CBHW020900210326
41598CB00018B/1725
* 9 7 8 3 7 4 2 8 0 0 1 6 9 *